Benchmark Papers in Electrical Engineering and Computer Science

Editor: John B. Thomas—Princeton University

A selection from the published volumes in this series

SYSTEM SENSITIVITY ANALYSIS / *José B. Cruz, Jr.*
RANDOM PROCESSES: Multiplicity Theory and Canonical Decompositions / *Anthony Ephremides and John B. Thomas*
ALGEBRAIC CODING THEORY: History and Development / *Ian F. Blake*
COMPUTER-AIDED CIRCUIT DESIGN: Simulation and Optimization / *S. W. Director*
ENVIRONMENTAL MODELING: Analysis and Management / *Douglas Daetz and Richard H. Pantell*
CIRCUIT THEORY: Foundations and Classical Contributions / *M. E. Van Valkenburg*
TWO-DIMENSIONAL DIGITAL SIGNAL PROCESSING / *Sanjit K. Mitra and Michael P. Ekstrom*
COMPUTER ARITHMETIC / *Earl E. Swartzlander, Jr.*
ARRAY PROCESSING: Applications to Radar / *Simon Haykin*
NONPARAMETRIC DETECTION: Theory and Applications / *Saleem A. Kassam and John B. Thomas*
DECONVOLUTION OF SEISMIC DATA / *V. K. Arya and J. K. Aggarwal*
REPRODUCING KERNEL HILBERT SPACES: Applications in Statistical Signal Processing / *Howard L. Weinert*
DISTRIBUTED PARAMETER SYSTEMS THEORY, PART I: Control / *Peter Stavroulakis*
DISTRIBUTED PARAMETER SYSTEMS THEORY, PART II: Estimation / *Peter Stavroulakis*
DISCRETE TRANSFORMS AND THEIR APPLICATIONS / *K. R. Rao*
QUANTIZATION / *Peter F. Swaszek*
TELECONFERENCING / *K. R. Rao and Ram Srinivasan*

WALSH FUNCTIONS IN SIGNAL AND SYSTEMS ANALYSIS AND DESIGN

Edited by
SPYROS G. TZAFESTAS
National Technical University, Greece

A Hutchinson Ross Publication

VNR VAN NOSTRAND REINHOLD COMPANY
New York

Benchmark Papers in Electrical Engineering and Computer Science, Volume 31
Library of Congress Catalog Card Number: 85-3236
ISBN: 0-442-28298-2

Manufactured in the United States of America.

Published by Van Nostrand Reinhold Company Inc.
135 West 50th Street
New York, New York 10020

Van Nostrand Reinhold Company Limited
Molly Millars Lane
Wokingham, Berkshire RG11 2PY, England

Van Nostrand Reinhold
480 Latrobe Street
Melbourne, Victoria 3000, Australia

Macmillan of Canada
Division of Gage Publishing Limited
164 Commander Boulevard
Agincourt, Ontario MIS 3C7, Canada

15 14 13 12 11 10 9 8 7 6 5 4 3 2 1

Library of Congress Cataloging in Publication Data
Main entry under title:
Walsh functions in signal and systems analysis and design.
"A Hutchinson Ross Benchmark book."
Includes indexes.
1. Electronics—Mathematics—Addresses, essays, lectures. 2. Electronic control—Mathematics—Addresses, essays, lectures. 3. Walsh functions—Addresses, essays, lectures. I. Tzafestas, S. G., 1939-
TK7835.W297 1985 620.7'2'0151555 85-3236
ISBN 0-442-28298-2

CONTENTS

SERIES EDITOR'S FOREWORD

This Benchmark Series in Electrical Engineering and Computer Science is aimed at sifting, organizing, and making readily accessible to the reader the vast literature that has accumulated. Although the series is not intended as a complete substitute for a study of this literature, it will serve at least three major critical purposes. In the first place, it provides a practical point of entry into a given area of research. Each volume offers an expert's selection of the critical papers on a given topic as well as his views on its structure, development, and present status. In the second place, the series provides a convenient and time-saving means for study in areas related to but not contiguous with one's principal interests. Last, but by no means least, the series allows the collection, in a particularly compact and convenient form, of the major works on which present research activities and interests are based.

Each volume in the series has been collected, organized, and edited by an authority in the area to which it pertains. In order to present a unified view of the area, the volume editor has prepared an introduction to the subject, has included his comments on each article, and has provided a subject index to facilitate access to the papers.

We believe that this series will provide a manageable working library of the most important technical articles in electrical engineering and computer science. We hope that it will be equally valuable to students, teachers, and researchers.

Walsh Functions in Signal and Systems Analysis and Design is edited by S. G. Tzafestas of the National Technical University in Athens, Greece. It consists of thirty-eight papers and editor's comments on Walsh functions—their definition, generation, computation and application to system analysis, identification, and design.

JOHN B. THOMAS

PREFACE

Walsh functions constitute a set of orthogonal functions that have received increasing attention in recent years in a variety of engineering areas such as communication, signal processing, system analysis, and control. Since Walsh functions and transforms are naturally more suited for digital computation, an effort has been made to gradually replace the Fourier transform by Walsh-type transforms. Actually, the Walsh function field has experienced a significant development, and a large amount of theoretical and applied results are presently available.

This volume is the outcome of the editor's feeling that a cohesive book is needed to bring together the major efforts and results of the technical literature on Walsh functions. Thirty-eight papers have been reprinted and additional works are mentioned in the editor's comments.

The book is organized into eight parts. Each part contains original works that have appeared in various technical journals. Existing textbooks on Walsh functions cover only specialized aspects and present the theory from particular angles of attack, while *Walsh Functions in Signal and Systems Analysis and Design* involves a large diversity of results and explains how these results have been achieved. My effort here has been to compose a volume containing a well balanced spectrum of works carried out since the appearance of Harmuth's pioneering paper in 1968 (Paper 1). It is hoped that the book will provide a useful source to researchers and practitioners in the field of Walsh and block pulse functions and their applications.

SPYROS G. TZAFESTAS

CONTENTS BY AUTHOR

WALSH FUNCTIONS
IN SIGNAL AND SYSTEMS
ANALYSIS AND DESIGN

INTRODUCTION

Walsh functions belong to the class of piece-wise constant basis functions (PCBF) that have been developed in the twentieth century and have played an important role in scientific and engineering applications. The mathematical techniques of studying functions, signals, and systems through series expansions in orthogonal-complete sets of basis functions are now a standard tool in all branches of science and engineering. Actually, the signals involved in Morse telegraphy are PCBFs, but no mathematical study of these signals was made prior to the beginning of the twentieth century.

The origin of the mathematical study of PCBFs is due to Alfred Haar (1910; 1912), who used a set of functions bearing his name. These functions have not found much use in comparison to the Walsh and block pulse functions considered in this book (Harmuth, 1969; Ahmed and Rao, 1975; Beauchamp, 1975; Prasada Rao, 1983; Tzafestas, 1983; Prasada Rao and Tzafestas, 1985). The development and utilization of Walsh functions has been strongly influenced by the parallel developments in digital electronics and computer science and engineering. Efforts to replace Fourier transforms by Walsh-type transforms have been made in communication, signal processing, image processing, pattern recognition, and so forth. The entrance of Walsh functions into the systems and control field was only about a decade ago, the developments since then occurring rapidly.

The basic definitions of Walsh functions and the ways of generating them by hardware devices are given in Part I. Algorithms for Walsh transform computation, as well as hardware implementations of these algorithms, are discussed in Part II, which includes the works on the sampling principle expressed in terms of the frequency in the Walsh domain (sequency).

The analysis of dynamic systems via Walsh function expansions is presented in Part III. Lumped parameter, distributed parameter, linear, nonlinear, multivariable, time delay, and integral equation-type systems are considered. The work on system parameter estimation via Walsh functions is presented in Part IV where the types of systems considered in Part III are studied.

The problems of designing optimal controllers, observers, and filters via Walsh functions are discussed in Part V. A diversity of results are included, covering most cases studied thus far. Lumped parameter, distributed parameter, and time delay systems are studied.

Part VI discusses the application of block pulse functions to systems analysis, identification, and control. Block pulse functions give results similar to those obtained through Walsh functions and offer considerable computational advantage, Block functions can be used for systems discretization in the same way that the bilinear transformation and the state transition matrix are used. A comparison of their relative characteristics is provided by Sinha and Zhou (1983). The one-shot operational matrices of Prasada Rao and Palanisamy (Paper 15) play an important role in reducing the errors when using Walsh and block pulse functions in high-order system analysis and design. Also noteworthy are the multidimensional block pulse functions that are useful in the study of distributed parameter systems described by partial differential equations (Nath and Lee, 1983; Prasada Rao and Srinivasan, 1980).

The properties of Walsh transforms and the process of Walsh-to-Fourier transform conversion are studied in Part VII. Aspects such as Walsh transform expression in terms of function derivatives, relation of cyclic autocorrelation with Walsh-Hadamard transform, and relation between arithmetic and logical autocorrelation functions are considered.

A representative set of Walsh transform applications is provided in Part VIII. These applications include biomedical signal (such as the EEG signal) processing, image coding, speech coding, statistical analysis, Boolean function classification, neutron transport theory, nuclear reactor test input design, and system-response time measurement.

REFERENCES

Ahmed, N., and K. R. Rao, 1975, *Orthogonal Transforms for Digital Signal Processing,* Springer Verlag, New York.

Beauchamp, K. G., 1975, *Walsh Functions and Their Applications,* Academic Press, London.

Haar, A., 1910, Zur theorie der orthogonalen Funktionensysteme, *Math. Ann.* **69:**331-371.

Haar, A., 1912, Zur theorie der orthogonalen Funktionensysteme, *Math. Ann.* **71:**38-53.

Harmuth, H. F., 1969, *Transmission of Information by Orthogonal Functions,* Springer Verlag, New York.

Nath, A. K., and T. T. Lee, 1983, On the Multidimensional Extension of Block-Pulse Functions and Their Applications, *Int. J. Syst. Sci.* **14:**201-208.

Prasada Rao, G., 1983, *Piecewise Constant Orthogonal Functions and Their Application to Systems and Control,* Lecture Notes in Control and Information Sciences Series, vol. 55, Springer Verlag, Berlin, 260p.

Prasada Rao, G., and T. Srinivasan, 1980, Multidimensional Block-Pulse Functions and Their Use in the Study of Distributed-Parameter Systems, *Int. J. Syst. Sci.* **11:**689-708.

Prasada Rao, G., and S. G. Tzafestas, 1985, A Decade of Piecewise Constant Orthogonal Functions in Systems and Control, in *Orthogonal Expansions in Computation, Simulation and Control,* S. G. Tzafestas, ed., Mathematics and Computers in Simulation Special Issue, vol. 27, North Holland, Amsterdam, The Netherlands, 240p.

Sinha, N. K., and Q. Zhou, 1983, Discrete-Time Approximation of Multivariable Continuous-Time Systems, *Inst. Electr. Eng. Proc.* **130:**103-110.

Tzafestas, S. G., 1983, Walsh Transform Theory and Its Application to Systems Analysis and Control: An Overview, *Math. Computers Simulation* **25:**214-225.

Part I

WALSH SIGNAL DEFINITION AND GENERATION

Editor's Comments on Papers 1, 2, and 3

1 **HARMUTH**
A Generalized Concept of Frequency and Some Applications

2 **AHMED, SCHREIBER, and LOPRESTI**
On Notation and Definition of Terms Related to a Class of Complete Orthogonal Functions

3 **GAUBATZ and KITAI**
A Programmable Walsh Function Generator for Orthogonal Sequency Pairs

The foundations of the Walsh functions field were made by Rademacher (1922), Walsh, (1923), Fine (1949, 1950), Paley (1952), and Kaczmarz and Steinhaus (1951). The engineering approach to the study and utilization of these functions was originated by Harmuth (1969*a*, 1969*b*), who introduced the concept of sequency to represent the associated, generalized frequency defined as one-half the mean rate of zero crossings. The variety of Walsh function definitions is due to the existence of different orderings. In the sequency ordering (or Walsh ordering), which is popular in communication engineering, Walsh functions are ordered according to the zero crossings (or sign changes). This sequency ordering implies that the *i*th Walsh function *wal* (*i, t*) has *i* zero crossings in the interval $t\epsilon[0,1]$, and, obviously, is directly related to the sequency concept. The Paley ordering (Paley, 1952) is characterized by the fact that in this form Walsh functions are represented by products of Rademacher functions, which lead to useful, recursive, Walsh-signal generation algorithms. A third ordering was proposed by Henderson (1964, 1970) and is merely Paley's ordering in reversed binary. Henderson's ordering is computationally attractive and occurs when one computes fast Walsh transforms (FWT) without sorting. Yuen (1972) called the index *i*, in *wal* (*i, t*), the "zequency" of *wal* (*i, t*) to show that it represents both the generalized frequency and the number of zeros of *wal* (*i, t*).

Paper 1 provides the background of Walsh transform theory, defines and uses the sequency concept, and introduces the *sal* and *cal* functions, which are analogous to the *sine* and *cosine* functions. Paper 2 presents a set of terms and definitions that standardize the analysis of Walsh functions. These definitions are nonambiguous and can assist workers and authors of papers to present their results in a unifying way. Paper 3 provides the logic design and implementation of a useful, programmable Walsh function generator.

Other definitions of Walsh functions have been given by Lackey and Meltzer (1971) through the Rademacher function products, and by Butin (1972) using polynomial expansions. The Butin definition leads to an easy way of deriving certain Walsh function properties, for example, their connection with the shifting theorem or the Gray code. Proof of the Lackey and Meltzer technique for selecting the combination of Rademacher functions generating a desired Walsh function is given in a paper by Davies (1972). In the same paper a digital Walsh-function generation scheme is also developed. Durrani and Nightingale (1971) provide sequential circuits for the generation of three sets of discrete Walsh functions without reference to any alternative sequency representation. Kitai and Siemens (1972) developed a sequency-ordered, Walsh-function generation scheme, which employs a Gray code counter and is free from hazards. Tzafestas, Frangakis, and Pimenidis (1976) gave a unified presentation of some Walsh function definitions, and on the basis of these definitions they designed global, digital, Walsh-function generating circuits.

REFERENCES

Butin, H., 1972, A Compact Definition of Walsh Functions, *IEEE Trans. Comput.* **C-21:**590–592.

Davies, A. C., 1972, On the Definition and Generation of Walsh Functions, *IEEE Trans. Comput.* **C-21:**187–189.

Durrani, T. S., and J. M. Nightingale, 1971, Sequential Generation of Binary Orthogonal Functions, *Electron. Lett.* **7:**377–380.

Fine, N. J., 1949, On the Walsh Functions, *Am. Math. Soc. Trans.* **65:**372–414.

Fine, N. J., 1950, The Generalized Walsh Functions, *Am. Math. Soc. Trans.* **69:**66–77.

Harmuth, H. F., 1969*a, Transmission of Information by Orthogonal Functions,* Springer, New York.

Harmuth, H. F., 1969*b,* Applications of Walsh Functions in Communications, *IEEE Spectrum,* **6:**82–91.

Henderson, K. W., 1964, Some Notes on the Walsh Functions, *IEEE Trans. Comput.* **EC-13:**50–52.

Henderson, K. W., 1970, Comment on Computation of the Fast Walsh-Fourier Transform, *IEEE Trans. Comput.* **C-19:**850–851.

Kaczmarz, S., and H. Steinhaus, 1951, *Theorie der Orthogonalreihen,* Chelsea, New York.

Kitai, R., and K. H. Siemens, 1972, A Hazard-Free Walsh-Function Generator, *IEEE Trans. Instrum. Meas.* **IM-22:**80–83.

Lackey, R., and D. Meltzer, 1971, A Simplified Definition of Walsh Functions, *IEEE Trans. Comput.* **C-20:**211–213.

Paley, R.E.A.C., 1952, A Remarkable Series of Orthogonal Functions, *London Math. Soc. Proc.* **2:**241–279.

Rademacher, H., 1922, Einige Sätze von Allgemeinen Orthogonal funktionen, *Math. Ann.* **87:**122–138.

Tzafestas, S., G. Frangakis, and T. Pimenidis, 1976, Global Walsh Function Generators, *Electron. Eng.* **48:**585.

Walsh, J. L., 1923, A Closed Set of Normal Orthogonal Functions, *Am. J. Math.* **45:**5–24.

Yuen, C. K., 1972, Remarks on the Ordering of Walsh Functions, *IEEE Trans. Comput.* **C-21:**1452.

1

Reprinted with permission from *IEEE Trans. Inf. Theory* **IT-14:**375-382 (1968)

A Generalized Concept of Frequency and Some Applications

HENNING F. HARMUTH, MEMBER, IEEE

***Abstract*—The system of sine and cosine functions has been distinguished historically in communications. Whenever the term frequency is used, reference is made implicitly to these functions; hence the generally used theory of communication is based on the system of sine and cosine functions. In recent years other complete systems of orthogonal functions have been used for theoretical investigations as well as for equipment design. Analogs to Fourier series, Fourier transform, frequency, power spectra, and amplitude, phase, and frequency modulation exist for many systems of orthogonal functions. This implies that theories of communication can be worked out on the basis of these systems. Most of these theories are of academic interest only. However, for the complete system of the orthogonal Walsh functions, the implementation of circuits by modern semiconductor techniques appears to be competitive in a number of applications with the implementation of circuits for the system of sine and cosine functions.**

Introduction

THE SYSTEM of sine and cosine functions plays a distinguished role in communications. There are a number of historical and practical reasons for this. From the theoretical point of view, one of the major reasons is that Fourier series and Fourier transform permit the representation of a large class of functions by a superposition of sine and cosine functions. This representation makes it possible to apply the concept of frequency, which was originally defined for sine and cosine only, to other functions.

In recent years more general classes of complete systems of orthogonal functions have been used for theoretical investigations as well as equipment design.[1]-[7],[27] Furthermore, semiconductor devices have made it practical to use linear time-variable circuits instead of linear time-invariant ones. While sine and cosine have indisputable advantages for linear time-invariant circuits they often lead to unnecessary complications if used to analyze time-variable circuits.

The purpose of this paper is to show that analogs exist to Fourier series, Fourier transform, frequency, power spectra, and amplitude, phase, and frequency modulation for many systems of orthogonal functions. For one such system, that of the Walsh functions, the experimental level has been reached for the analogs to frequency multiplex telephony and other applications.[1] Hence, we will be mainly concerned with these Walsh functions.[8]-[16]

Generalized Fourier Transform

It is well known that the analog to the Fourier series exists for many systems of orthogonal functions. Examples are series expansions in Bessel functions, spherical functions, orthogonal polynomials, etc. It is also known that analogs to the Fourier transform exist for many systems of functions.[17],[18] However, most of the generally used complete systems of orthogonal functions are defined by linear differential equations of second order, and it can be shown that the generalized Fourier transform is in this case essentially the same as the ordinary Fourier transform. Consider, e.g., the differential equation of the Legendre polynomials:

$$(1 - x^2)y'' - 2xy' + j(j + 1) = 0. \qquad (1)$$

The generalized Fourier transform of a function that vanishes outside a finite interval $x_1 < x < x_2$ consists of a superposition of Legendre polynomials with large value of j and small values of x. The differential equation (1) reduces in this case to that of the sine and cosine functions, and the Legendre–Fourier transform to the ordinary Fourier transform except for scale factors.[19]

Walsh functions can be defined by a difference equation rather than a differential equation, and the generalized Fourier transform or Walsh–Fourier transform is quite different from the ordinary Fourier transform.[15] Such a transform has been known for some time,[2] but its

Manuscript received April 18, 1967; revised November 7, 1967. This work was supported by the Defense Department of the Federal Republic of Germany, under Contract T-563-L-203.
The author is at 2 Lise Meitner Weg, 7501 Leopoldshafen, West Germany.

[1] Circuits of filters, modulators, Walsh function generators, etc., for telephony multiplex systems are shown in previous papers.[28] More general filters are discussed in a paper by Pichler.[29] Radio transmission by Walsh carriers is treated in "Über Funkverkehr mit Sequenzteilung an Stelle von Frequenz- oder Zeitteilung," Defense Department, Federal Republic of Germany Project Rept. T-675-L-203. Speech analysis was done by M. Tasto in a thesis, "Analyse von Zeitfunktionen durch Mäandertransformation und durch Fourier-Transformation," Technische Hochschule Darmstadt, Institut für allgemeine Nachrichtentechnik. The results of this thesis are presently used by M. Boesswetter of the same institute for development of a sequency channel vocoder. Experimental work on filters is further done by Prof. H. Lueg, Technische Hochschule Aachen. Filter and modulator circuits are also shown in Harmuth.[26] A thelphony multiplex system using Walsh functions as carriers has been developed by H. Lüke and R. Maile of AEG-Telefunken Research Institute. Schemes for applying the dyadic correlation function $\int F(t)G(t + T)\, dt$ instead of the usual one, for producing Walsh-shaped instead of sine-shaped electromagnetic waves in the visible-light region, for high-resolution and shape-recognizing Walsh-wave radar, etc., are in the theoretical stage.

[2] The Walsh–Fourier transform, in a different form from the one used here, is due to N. J. Fine. The term "Fine transform" is sometimes used.

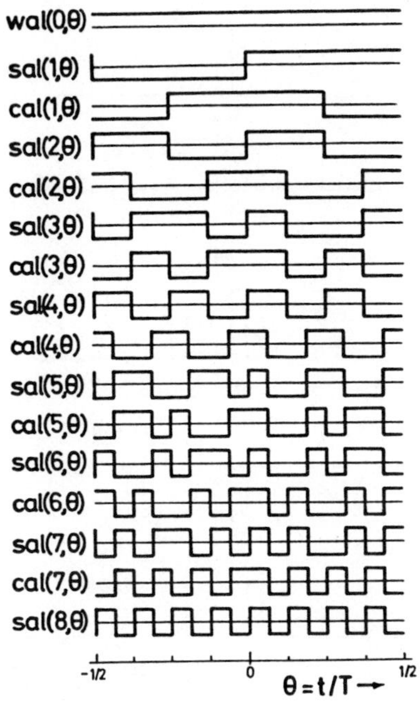

Fig. 1. Walsh functions wal $(0, \theta)$, cal (i, θ), and sal (i, θ).

form was not suitable for a generalization of the concept of frequency since the Walsh functions were ordered in a sequence that followed from their original definition by products of Rademacher functions.[20] The difference equation yields them in the sequence shown by Fig. 1. In this figure they are ordered according to the number of sign changes or zero crossings in the half-open interval $-\frac{1}{2} \leqq \theta < \frac{1}{2}$. The functions cal (i, θ) and sal (i, θ) have $2i$ zero crossings in this interval; $i = 1, 2, \cdots$. Furthermore, all functions cal (i, θ) equal $+1$ and all functions sal (i, θ) change from -1 to $+1$ for $\theta = 0$.[3]

Using the notation[4]

$$\text{wal}\,(2i, \theta) = \text{cal}\,(i, \theta),$$
$$\text{wal}\,(2i - 1, \theta) = \text{sal}\,(i, \theta), \quad -\tfrac{1}{2} \leqq \theta < \tfrac{1}{2}$$
$$\text{wal}\,(2i, \theta) = \text{wal}\,(2i - 1, \theta) \equiv 0, \quad \theta < -\tfrac{1}{2}, \quad \theta \geqq +\tfrac{1}{2}$$

one may write the difference equation of the Walsh functions in the following form:[5]

$$\text{wal}\,(2k + q, \theta) = \text{wal}\,(-1)^{[k/2]+q}[k, 2\theta + \tfrac{1}{2}) + (-1)^{k+q}\,\text{wal}\,(k, 2\theta - \tfrac{1}{2})]$$
$$\text{wal}\,(0, \theta) = 1, \quad -\tfrac{1}{2} \leq \theta < +\tfrac{1}{2}$$
$$= 0, \quad \theta < -\tfrac{1}{2}, \quad \theta \geq +\tfrac{1}{2}$$
$$q = 0 \text{ or } 1; \quad k = 0, 1, 2, \cdots.$$

[3] The functions $-\text{sal}\,(1, \theta)$ and $+\text{sal}\,(2^k, \theta)$, $k = 1, 2, \cdots$, are the Rademacher functions.

[4] The notations sal (i, θ) and cal (i, θ) have the advantage of showing the similarity to sine and cosine. The notation wal (k, θ) is frequently better suited for computation. For instance, the multiplication theorems (22) reduce to one theorem wal $(i \oplus k, \theta) =$ wal (i, θ) wal (k, θ). In analogy, the multiplication theorems (21) for sine and cosine reduce to one theorem if complex notation is used.

[5] $[k/2]$ means the largest integer smaller than or equal to $k/2$.

For explanation consider the Walsh function wal $(0, \theta)$ of Fig. 1. Shifting it by $\frac{1}{2}$ to the left into the interval $-1 \leqq \theta < 0$ yields wal $(0, \theta + \frac{1}{2})$ and compressing it by a factor 2 into the interval $-\frac{1}{2} \leqq \theta < 0$ yields wal $(0, 2\theta + \frac{1}{2})$. Similarly, wal $(0, 2\theta - \frac{1}{2})$ is obtained by shifting wal $(0, \theta)$ to the right and compressing it into the interval $0 \leqq \theta < \frac{1}{2}$. For $q = 0$ we obtain the sum $(-1)^{0+0}$[wal $(0, 2\theta + \frac{1}{2}) + (-1)^{0+0}$ wal $(0, 2\theta - \frac{1}{2})$] which evidently equals wal $(0, \theta)$. For $q = 1$ we obtain the sum $(-1)^{0+1}$ [wal $(0, >\theta + \frac{1}{2}) + (-1)^{0+1}$ wal $(0, 2\theta - \frac{1}{2})$] which equals wal $(1, \theta) =$ sal $(1, \theta)$. Similarly, one obtains from wal $(1, \theta)$ the shifted and compressed functions wal $(1, 2\theta + \frac{1}{2})$ and wal $(1, 2\theta - \frac{1}{2})$. The sum $(-1)^{0+0}$[wal $(1, 2\theta + \frac{1}{2}) + (-1)^{1+0}$ wal $(1, 2\theta - \frac{1}{2})$] yields wal $(2, \theta) =$ cal $(1, \theta)$, and the sum $(-1)^{0+1}$[wal $(1, 2\theta + \frac{1}{2}) + (-1)^{1+1}$ wal $(1, 2\theta - \frac{1}{2})$] yields wal $(3, \theta) =$ sal $(2, \theta)$.

A Walsh–Fourier series expansion of a function $F(\theta)$ defined in the interval $-\frac{1}{2} \leqq \theta < \frac{1}{2}$ has the following form:

$$F(\theta) = a(0)\,\text{wal}\,(0, \theta) + \sum_{i=1}^{\infty} [a_c(i)\,\text{cal}\,(i, \theta) + a_s(i)\,\text{sal}\,(i, \theta)], \tag{2}$$

where

$$a(0) = \int_{-1/2}^{1/2} F(\theta)\,\text{wal}\,(0, \theta)\,d\theta = \int_{-1/2}^{1/2} F(\theta)\,d\theta,$$
$$a_c(i) = \int_{-1/2}^{1/2} F(\theta)\,\text{cal}\,(i, \theta)\,d\theta,$$
$$a_s(i) = \int_{-1/2}^{1/2} F(\theta)\,\text{sal}\,(i, \theta)\,d\theta.$$

The set of functions required for the Walsh–Fourier transform may be derived by stretching the functions of Fig. 1 by a factor ξ and denoting the stretched functions cal (i, θ) and sal (i, θ) by cal $(i/\xi, \theta)$ and sal $(i/\xi, \theta)$. If ξ and i approach infinity in such a way that the limit

$$\lim_{\xi, i \to \infty} i/\xi = \mu \tag{3}$$

exists, one obtains the system of Walsh meander functions {cal (μ, θ), sal (μ, θ)} which are defined in the interval $-\infty < \theta < +\infty$ for all real non-negative values of μ. It is useful to extend the definition to negative values of μ by making cal (μ, θ) a symmetric and sal (μ, θ) a skew symmetric function of μ:

$$\text{cal}\,(-\mu, \theta) = \text{cal}\,(\mu, \theta), \text{sal}\,(-\mu, \theta) = -\text{sal}\,(\mu, \theta). \tag{4}$$

The derivation of the functions cal (μ, θ), sal (μ, θ) from the Walsh functions cal (i, θ), sal (i, θ) is discussed in a more mathematical fashion by Pichler.[15]

The Walsh–Fourier transform of a function $F(\theta)$ and its inverse have the following form:

$$a_c(\mu) = \int_{-\infty}^{\infty} F(\theta)\,\text{cal}\,(\mu, \theta)\,d\theta,$$
$$a_s(\mu) = \int_{-\infty}^{\infty} F(\theta)\,\text{sal}\,(\mu, \theta)\,d\theta, \tag{5}$$

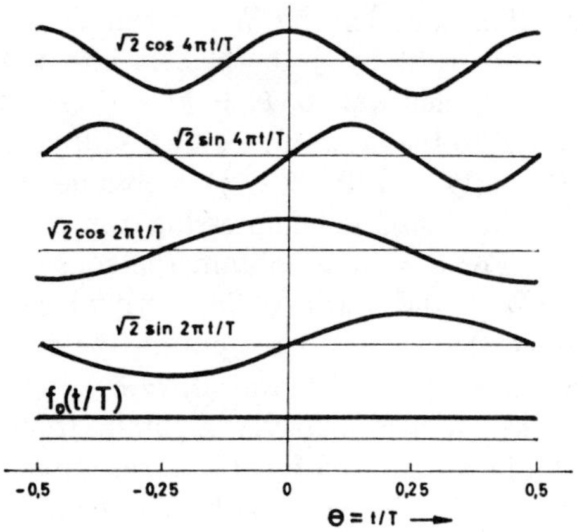

Fig. 2. Sine and cosine functions.

where

$$F(\theta) = \int_0^\infty [a_c(\mu) \text{ cal } (\mu, \theta) + a_s(\mu) \text{ sal } (\mu, \theta)] \, d\mu.$$

It has been shown[15],[19] that in the interval $-\frac{1}{2} \leqq \theta < \frac{1}{2}$ the following relations hold between the functions cal (μ, θ), sal (μ, θ) and cal (i, θ), sal (i, θ):

$$\left.\begin{aligned} \text{cal } (\mu, \theta) &= \text{wal } (0, \theta) = 1 \quad 0 \leqq \mu < 1 \\ \text{cal } (\mu, \theta) &= \text{cal } (i, \theta) \quad i \leqq \mu < i + 1 \\ \text{sal } (\mu, \theta) &= \text{sal } (i, \theta) \quad i - 1 < \mu \leqq i \end{aligned}\right\} \quad i = 1, 2, \cdots. \tag{6}$$

Hence, for a time limited function $F(\theta)$, that is identically zero outside the interval $-\frac{1}{2} \leqq \theta < \frac{1}{2}$, the Walsh–Fourier transform reduces to the simple form

$$a_c(\mu) = a(0) = \int_{-1/2}^{1/2} F(\theta) \, d\theta \quad 0 \leqq \mu < 1$$

$$a_c(\mu) = a_c(i) = \int_{-1/2}^{1/2} F(\theta) \text{ cal } (i, \theta) \, d\theta \quad i \leqq \mu < i + 1$$

$$a_s(\mu) = a_s(i) = \int_{-1/2}^{1/2} F(\theta) \text{ sal } (i, \theta) \, d\theta \quad i - 1 < \mu \leqq i \tag{7}$$

The inverse Walsh–Fourier transform is in this case given by

$$F(\theta) = a(0) \text{ wal } (0, \theta) + \sum_{i=1}^{\infty} [a_c(i) \text{ cal } (i, \theta) + a_s(i) \text{ sal } (i, \theta)] \quad -\tfrac{1}{2} \leqq \theta < \tfrac{1}{2} \tag{8}$$

Fig. 2 shows some sine and cosine pulses that vanish outside the interval $-\frac{1}{2} \leqq \theta < \frac{1}{2}$. Fig. 3 shows their Walsh–Fourier transforms $a_c(\mu)$ and $a_s(\mu)$. These transforms are shown for positive as well as negative values of μ. It follows from the definition (4) that $a_c(\mu)$ is a symmetric and $a_s(\mu)$ a skew symmetric function of μ.

Fig. 4 shows the coefficients $a(0)$, $a_c(i)$, and $a_s(i)$ of the Walsh–Fourier series of the same sine and cosine pulses. For engineering applications it is usual to interpret them as the coefficients of the series expansion of the periodic sine and cosine functions by periodic Walsh functions.

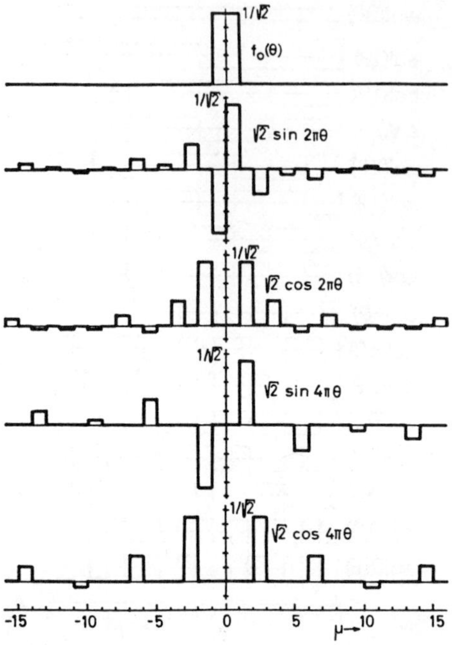

Fig. 3. Walsh–Fourier transforms $a_c(\mu)$ and $a_s(\mu)$ of sine and cosine pulses according to Fig. 2.

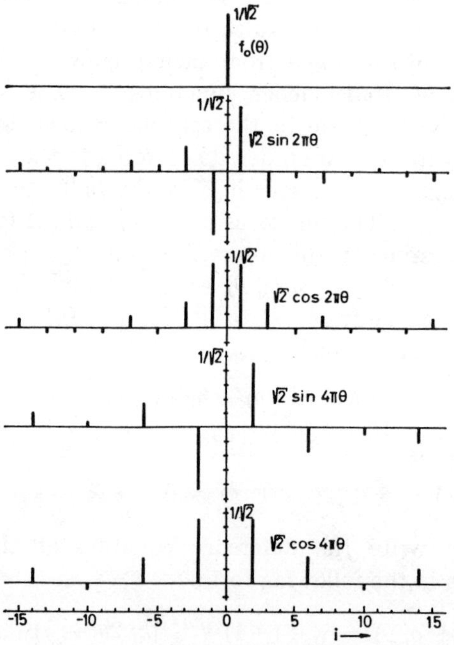

Fig. 4. Coefficients $a(0)$, $a_c(i)$, and $a_s(i)$ of the Walsh–Fourier series expansion of sine and cosine pulses according to Fig. 2.

The Generalized Frequency

Frequency is defined as the parameter f that distinguishes the individual functions of the system $\{\cos 2\pi ft, \sin 2\pi ft\}$. The generally used practical interpretation of f is the "number of cycles per unit of time." Another possible interpretation[21]–[24] is "one-half the average number of zero crossings per unit of time." For instance, a sine wave with 100 cycles per second (cps) has 200 zero crossings per second; one-half of this is 100 "per second" which is the same value and dimension as 100 cps. We call f the *sequency* rather than the frequency if measured by the number of zero crossings or sign changes and denote it by ϕ. As unit of sequency we use "zps" (one half the average number of zero crossings per second) in analogy to "cps". Sequency and frequency are identical for sine and cosine functions, but the concept of sequency can be applied to functions for which "cycle" has no obvious meaning and which do not even have to be periodic.[6]

We have seen that the Walsh functions cal (i, θ) and sal (i, θ) have $2i$ zero crossings in the interval $-\frac{1}{2} \leq \theta < \frac{1}{2}$. If the functions are periodically continued to infinity one may thus assign to them the normalized sequency i or the non-normalized sequency $\phi = i/T$; T is the time unit used to normalize the time variable t in cal (i, θ) and sal (i, θ)

$$\theta = t/T. \tag{9}$$

When the Walsh functions cal (i, θ) and sal (i, θ) are stretched by a factor ξ to obtain the Walsh meander functions cal (μ, θ) and sal (μ, θ), the time unit T is replaced by ξT and $2i$ becomes the number of zero crossings in the interval $-\xi T/2 \leq t < \xi T/2$. It follows from (3) that

$$\lim_{i,\xi \to \infty} i/T\xi = \mu/T = \phi. \tag{10}$$

If μ is not binary rational the Walsh meander functions cal (μ, θ) and sal (μ, θ) are not periodic. However, the limit $\mu/T = \phi$ exists and the definition of sequency is applicable.

The sequency of other functions may be defined in the same way. For instance, the Legendre polynomials $P_i(2\theta)$, $-1 \leq 2\theta < 1$, may be divided into the constant $P_0(\theta)$, the symmetric functions $P_{ci}(\theta)$, and the skew symmetric functions $P_{si}(\theta)$. Using a normalization that makes these functions orthonormal in the interval $-\frac{1}{2} \leq \theta < \frac{1}{2}$ and $P_{ci}(\theta) > 0$, $dP_{si}(\theta)/d\theta > 0$ for $\theta = 0$ one obtains

$$P_0(\theta) = 1, \qquad P_{ci}(\theta) = (-1)^i(4i+1)^{1/2}P_{2i}(2\theta),$$

$$P_{si}(\theta) = (-1)^{i-1}(4i-1)^{1/2}P_{2i-1}(2\theta); \qquad \theta = t/T.$$

$P_{ci}(\theta)$ and $P_{si}(\theta)$ have $2i$ zero crossings in the interval $-\frac{1}{2} \leq \theta < \frac{1}{2}$.

The functions $P_{c\mu}(\theta)$ and $P_{s\mu}(\theta)$ required for the Legendre-Fourier transform may again be obtained from $P_{ci}(\theta)$ and $P_{si}(\theta)$ by stretching them by a factor ξ, replacing i by i/ξ, and letting i and ξ increase towards infinity. It turns out[19] that the limits (3) and (10) exist so that $\mu/T = \phi$ is the sequency of $P_{c\mu}(\theta) = P_{c\mu}(t/T)$ and $P_{s\mu}(\theta) = P_{s\mu}(t/T)$.

If sine and cosine functions are written with normalized frequency $\nu = fT$ and normalized time $\theta = t/T$ and the non-normalized frequency and time are then substituted one obtains

$$\sin 2\pi\nu\theta = \sin 2\pi(fT)(t/T) = \sin 2\pi ft.$$

For these functions the time base T drops out. This is generally not so for other functions. Substitution of $\mu = \phi T$ and $\theta = t/T$ into the Walsh meander functions yields, e.g.,

$$\text{cal}(\mu, \theta) = \text{cal}(\phi T, t/T), \text{sal}(\mu, \theta) = \text{sal}(\phi T, t/T). \tag{11}$$

Hence, functions other then sine and cosine require the time base T in addition to the sequency ϕ for their definition. This additional parameter complicates the theory but it gives an additional degree of freedom that has resulted in some extremely simple filters based on Walsh meander functions rather than on sine and cosine functions.

The concepts of oscillation period $\tau = 1/f$ and wavelength $\lambda = v/f$ are replaced by the following more general concepts if frequency f is replaced by sequency ϕ:

average oscillation period $\tau = 1/\phi$—average time between two zero crossings multiplied by 2

average wavelength $\lambda = v/\phi$—average distance between two zero crossings multiplied by 2; v is the velocity of propagation of the zero crossings.

Power Spectra and Amplitude Spectra

If one squares the Walsh–Fourier series expansion (2) and integrates over the interval of orthogonality, $-\frac{1}{2} \leq \theta < \frac{1}{2}$, one obtains

$$\int_{-1/2}^{1/2} F^2(\theta)\, d\theta = \frac{1}{T}\int_{-1/2}^{1/2} F^2(t/T)\, dt$$
$$= a^2(0) + \sum_{i=1}^{\infty} [a_c^2(i) + a_s^2(i)]. \tag{12}$$

The left side of this equation may be interpreted as the average power of a signal $F(\theta)$. $a^2(0)$ and $a_c^2(i) + a_s^2(i)$ may then be interpreted as power spectrum of the discrete variable i. This interpretation will also hold if some other system of orthogonal functions is used instead of the Walsh functions in (2), provided such an expansion exists for $F(\theta)$ in the sense of a vanishing mean square deviation.

A power spectrum or, more precisely, a power density spectrum can also be defined for the Walsh–Fourier transform by squaring and integrating (5) over the interval of orthogonality $-\infty < \theta < +\infty$:

$$\int_{-\infty}^{\infty} F^2(\theta)\, d\theta = \frac{1}{T}\int_{-\infty}^{\infty} F^2(\theta)\, dt$$
$$= \int_0^{\infty} [a_c^2(\mu) + a_s^2(\mu)]\, d\mu; \qquad \theta = t/T. \tag{13}$$

$a_c^2(\mu) + a_s^2(\mu)$ may be interpreted as the power density spectrum of the continuous variable μ. Again, this definition will hold if other functions than the Walsh meander functions are used in (5). In particular one may substitute the complete orthonormal system $\{\sqrt{2} \cos 2\pi i\theta, \sqrt{2} \sin 2\pi i\theta\}$ for the system $\{\text{cal}(i, \theta), \text{sal}(i, \theta)\}$ to

[6] The need for using a new term "sequency" rather than a more general meaning of "frequency" is not evident here. However, Walsh waves in three-dimensional space may have a sequency as well as a (rotational) frequency and the distinction becomes necessary there.

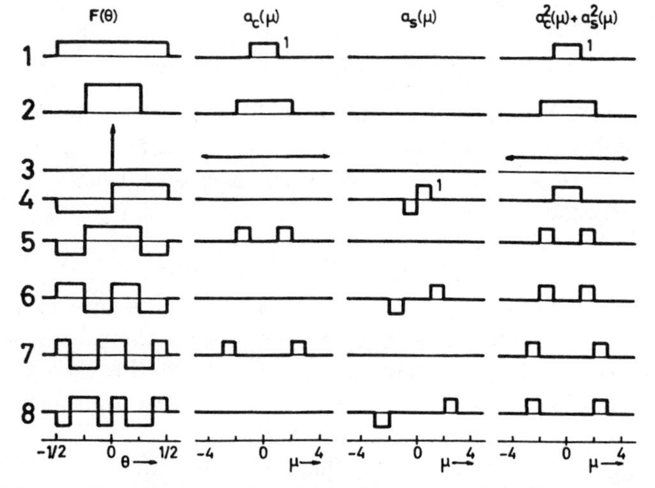

Fig. 5. Some time functions $F(\theta)$, their Walsh–Fourier transforms $a_c(\mu)$ and $a_s(\mu)$, and their power spectra $a_c^2(\mu) + a_s^2(\mu)$.

obtain the usual power spectrum from (12), or $\{\sqrt{2}\cos 2\pi\mu\theta,\ \sqrt{2}\sin 2\pi\mu\theta\}$ for $\{\mathrm{cal}\,(\mu, \theta),\ \mathrm{sal}\,(\mu, \theta)\}$ to obtain the usual power density spectrum from (13).

The amplitude spectrum of the ordinary Fourier transform is defined as the square root of the power spectrum. This definition is based on the relation

$$A \sin x + B \cos x = (A^2 + B^2)^{1/2} \cos (x - \arctan A/B). \tag{14}$$

Since such a relation does not generally hold for other functions there is no general analog to the amplitude spectrum. However, $a_c(\mu)$ and $a_s(\mu)$ may always be interpreted individually as amplitude spectra of the symmetric and the skew symmetric part of $F(\theta)$. The Wiener–Chintchin relation between the autocorrelation function of $F(\theta)$ and its power spectrum also requires the existence of (14). Hence, there is in general no direct analog to this relation.

Fig. 5 shows some time functions $F(\theta)$, their Walsh–Fourier transforms $a_c(\mu)$ and $a_s(\mu)$, and their power spectra $a_c^2(\mu) + a_s^2(\mu)$. It may be seen that the extension of the definition of μ to negative values is useful for $a_c(\mu)$ and $a_s(\mu)$; there is little advantage in plotting the power spectrum for negative values of μ since it is always symmetrical.

A striking feature of Fig. 5 is that all power spectra except the third one are sequency limited—that is, they vanish outside a finite sequency interval—although the corresponding time functions are time limited. This is in contrast to the ordinary Fourier transform where a time-limited function cannot have a frequency-limited power spectrum. A class of functions that are time and sequency limited follows from Fig. 5: the Walsh pulses wal $(0, \theta)$, cal (i, θ), and sal (i, θ) vanish outside the time interval $-\frac{1}{2} \leqq \theta < \frac{1}{2}$. Their Walsh–Fourier transforms vanish outside the sequency intervals $-1 \leqq \mu \leqq +1$, $-(i+1) \leqq \mu \leqq i+1$, and $-i \leqq \mu \leqq +i$. Hence, any time function $F(\theta)$, consisting of a superposition of Walsh pulses having finite values of i only, is time and sequency limited. Such functions occupy only a finite area of a "time–sequency domain" rather then the infinite area occupied by any function in the time–frequency domain.

Filters

The electric properties of linear time-invariant filters are usually stated by the frequency response of attenuation and phase shift. It is assumed that a voltage $V \cos 2\pi\mu\theta$ is applied to the input of such a filter and that amplitude and phase angle of the voltage $V_c(\mu) \cos [2\pi\mu\theta + \phi_c(\mu)]$ are measured at the output after the steady state has been reached. If a signal $F(\theta)$ at the input of the filter has the ordinary Fourier transforms $a_c(\mu)$ and $a_s(\mu)$ it will appear at the output as the signal $F_a(\theta)$:

$$F_a(\theta) = \sqrt{2} \int_0^\infty \{a_c(\mu)K_c(\mu) \cos [2\pi\mu\theta + \phi_c(\mu)] + a_s(\mu)K_c(\mu) \sin [2\pi\mu\theta + \phi_c(\mu)]\}\, d\mu; \tag{15}$$

where

$$K_c(\mu) = V_c(\mu)/V.$$

Linear[7] time-dependent filters have been devised and developed recently for an experimental sequency multiplex telephony system whose electric properties are best described by Walsh meander functions.[7] If a Walsh meander function V cal (μ, θ) or V sal (μ, θ) is applied to the input of such a filter, the steady-state output voltages $V_c(\mu)$ cal $[\mu, \theta - \theta_c(\mu)]$ and $V_s(\mu)$ sal $[\mu, \theta - \theta_s(\mu)]$ are obtained. A signal $F(\theta)$ with the Walsh–Fourier transforms $a_c(\mu)$ and $a_s(\mu)$ applied to the input of the filter will appear as the signal $F_a(\theta)$ at the output:

$$F_a(\theta) = \int_0^\infty \{a_c(\mu)K_c(\mu)\,\mathrm{cal}\,[\mu, \theta - \theta_c(\mu)] + a_s(\mu)K_s(\mu)\,\mathrm{sal}\,[\mu, \theta - \theta_s(\mu)]\}\, d\mu \tag{16}$$

where

$$K_c(\mu) = V_c(\mu)/V, \qquad K_s(\mu) = V_s(\mu)/V.$$

Since any signal may be considered to consist of a sequence of time limited signals, one may rewrite (16) into a simpler form that holds for a time interval $-\frac{1}{2} \leqq \theta < \frac{1}{2}$ rather than $-\infty < \theta < \infty$. In analogy to (6) one may choose

$$\begin{aligned} K_c(\mu) &= K(0) && 0 \leqq \mu < 1 \\ K_c(\mu) &= K_c(i) && i \leqq \mu < i+1 \\ K_s(\mu) &= K_s(i) && i-1 < \mu \leqq i. \end{aligned} \tag{17}$$

The simplest choice for $\theta_c(\mu)$ and $\theta_s(\mu)$ would be $\theta_c(\mu) = \theta_s(\mu) = 0$. However, this cannot be implemented practically. The simplest choice that can be implemented is

$$\theta_c(\mu) = \theta_s(\mu) = 1. \tag{18}$$

Using (6), (7), (8), (17), and (18), one may bring (16) into the following form:

[7] "Linear" means that the superposition and the proportionality law hold for the complete system of orthogonal functions used.

$$F_a(\theta) = a(0)K(0) \text{ wal } (0, \theta - 1) + \sum_{i=1}^{\infty} [a_c(i)K_c(i) \text{ cal } (i, \theta - 1) + a_s(i)K_s(i) \text{ sal } (i, \theta - 1)], \quad -\tfrac{1}{2} \leqq \theta < \tfrac{1}{2}. \tag{19}$$

Equations (15) and (16) hold generally if functions other then sine and cosine or Walsh meander functions are inserted. However, the condition that these functions will be attenuated and delayed only are hard to meet for other functions at the present time. These conditions can be met for sine and cosine due to the inherent linearity and time invariance of resistors, capacitors, and inductors. For Walsh meander functions they can be met by semiconductor circuits if the restrictions imposed by (17) and (18) are acceptable. Furthermore, synchronization has to be used to mark the beginning and end of the successive time intervals of duration 1 or T. While there is no practical way known to implement (16), the implementation of (19) is often trivially simple. It is worth noting that (18) means that the delay is constant for all values of the sequency μ; there are thus inherently no delay distortions in these "sequency filters."

For a practical example of a sequency filter based on (19), let us consider a "sequency low pass" having the following values of $K(0)$, $K_c(i)$ and $K_s(i)$:

$$K(0) = 1, \quad K_c(i) = K_s(i) = 0 \qquad i = 1, 2, \cdots .$$

Equation (19) is then reduced to

$$F_a(\theta) = F_a(t/T) = a(0) \text{ wal } (0, \theta - 1) = a(0) \text{ wal } (0, t/T - 1)$$
$$-\tfrac{1}{2} \leqq \theta = t/T < \tfrac{1}{2}, \quad -T/2 \leqq t < T/2. \tag{20}$$

$a(0)$ is obtained according to (7) by integrating the input signal $F(\theta)$ during an interval of duration T. The voltage obtained by this integration is then stored for the following time interval of duration T and this voltage is $F_a(\theta)$. Hence, the sequency low pass transforms $F(\theta)$ into a least-mean-square deviation step function, the duration of the steps being T. For $T = 125\ \mu s$ one has 8000 steps per second or 8000 independent amplitudes per second. This is the same information flow as that passing through a 4-kHz-wide "frequency low pass."

A number of "sequency bandpasses" can be implemented by sequency shifting the signal by means of amplitude modulation to the baseband, filtering it by a sequency low pass, and shifting it back to its original position by amplitude modulation. This principle is sometimes used for "frequency bandpasses" too. However, the amplitude modulation of sine and cosine functions produces two frequency (or sequency) sidebands and this causes complications. On the other hand, amplitude modulation of Walsh functions produces only one sequency sideband; hence, sequency shifting is easier to implement in this case. Beyond that, multiplication of a signal by Walsh functions means multiplication by +1 or −1 only. Hence, multiplication is done by switching.

The necessity of providing synchronization to mark beginning and end of successive time intervals of duration T is inconvenient in a number of applications. There are, however, applications such as telephony multiplexing where synchronization is used already for various methods of pulse modulation and time division. Synchronization of filters does not mean an additional burden in those cases.

Modulation

Sine and cosine carriers are obtained by periodic continuation of the sine and cosine functions of Fig. 2. Similarly, Walsh carriers are obtained by periodically continuing the Walsh functions of Fig. 1. It is evident that Walsh carriers and carriers constructed in this way from many other functions can be amplitude modulated like sine and cosine carriers. There are, however, some interesting differences. Amplitude modulation of sine and cosine carriers yields a double (frequency) sideband modulation due to the multiplication theorems of these functions:

$$\begin{aligned} 2 \cos i\theta \cos k\theta &= \cos (i - k)\theta + \cos (i + k)\theta \\ 2 \sin i\theta \cos k\theta &= \sin (i - k)\theta + \sin (i + k)\theta \\ 2 \cos i\theta \sin k\theta &= -\sin (i - k)\theta + \sin (i + k)\theta \\ 2 \sin i\theta \sin k\theta &= \cos (i - k)\theta - \cos (i + k)\theta. \end{aligned} \tag{21}$$

The functions $\cos k\theta$ and $\sin k\theta$ may be considered to represent carriers; $\cos i\theta$ and $\sin i\theta$ may represent components of a Fourier series or Fourier transform. The splitting of each component into two by multiplication with a carrier results in two sidebands if bands of components are considered.

Analogous multiplication theorems[15] exist for Walsh functions:

$$\begin{aligned} \text{cal } (i, \theta) \text{ cal } (k, \theta) &= \text{cal } (i \oplus k, \theta) \\ \text{sal } (i, \theta) \text{ cal } (k, \theta) &= \text{sal } \{[k \oplus (i - 1)] + 1, \theta\} \\ \text{cal } (i, \theta) \text{ sal } (k, \theta) &= \text{sal } \{[i \oplus (k - 1)] + 1, \theta\} \\ \text{sal } (i, \theta) \text{ sal } (k, \theta) &= \text{cal } [(i - 1) \oplus (k - 1), \theta] \\ &[\text{cal } (0, \theta) = \text{wal } (0, \theta)]. \end{aligned} \tag{22}$$

The sign $\oplus$ indicates an addition modulo 2. i and k are written as binary numbers and added as follows: $1 + 0 = 0 + 1 = 1, 0 + 0 = 1 + 1 = 0$ (no carry).

Let cal (k, θ) and sal (k, θ) represent Walsh carriers. cal (i, θ) and sal (i, θ) shall be components of a Walsh–Fourier series or a Walsh–Fourier transform. There is no splitting of components and as a consequence there is no double (sequency) sideband modulation for Walsh functions.

The output voltage of a sequency low pass according to (20) may be used to amplitude modulate Walsh carriers cal (k, θ) = cal $(k, t/T)$ and sal (k, θ) = sal $(k, t/T)$, $k = 1, 2, \cdots$ and several modulated carriers may be added without waste of sequency bandwidth. The need for single-sideband equipment is eliminated by the multiplication theorems (22). Other problems are of course created. Their discussion is left to more technical papers by other engineers.

Demodulation of amplitude modulated Walsh functions by rectification is possible. However, the superior method is synchronous demodulation. Let $F_a(\theta)$ be the output voltage of a sequency low pass. Suppressed carrier amplitude modulation of a Walsh carrier cal (k, θ) yields $F_a(\theta)$ cal (k, θ). Demodulation by a synchronous carrier cal (k, θ) yields

$$[F_a(\theta) \text{ cal } (k, \theta)] \text{ cal } (k, \theta) = F_a(\theta) \text{ wal } (0, \theta) = F_a(\theta), \quad (23)$$

due to the multiplication theorems (22). If a composite signal

$$F_a^{(1)}(\theta) \text{ cal } (1, \theta) + F_a^{(2)}(\theta) \text{ sal } (1, \theta) + F_a^{(3)}(\theta) \text{ cal } (2, \theta) + F_a^{(4)}(\theta) \text{ sal } (2, \theta) \cdots$$

is demodulated synchronously it can be shown that crosstalk from signals transmitted by the other carriers can be eliminated by feeding the demodulated signal through a sequency low-pass filter that equals the one used before modulation on the transmitter side. This is all in complete analogy to the procedure in frequency multiplex systems, except that only one sequency sideband is produced and synchronization of carriers and filters is required.

A check of multiplication theorems for other systems of functions found in mathematical handbooks shows that most of the better known ones have infinitely many terms rather than two terms or one term as on the right-hand side of (21) and (22). Hence, they would yield infinitely many sequency sidebands if used for amplitude modulation the way sine and cosine carriers are used.

Phase modulation of a sine carrier sin $2\pi\nu\theta$ by a signal $F(\theta)$ yields sin $2\pi\nu[\theta + (\Delta\nu/\nu)F(\theta)]$. The analog for a Walsh function cal (k, θ) is cal $[k, \theta + MF(\theta)]$, where M is a modulation index. This may be interpreted as a time position modulation of the carrier cal (k, θ). The extension of time position modulation to other carrier functions is evident. If the carrier function consists of periodic block pulses, one generally uses the term pulse position modulation rather than time position modulation.

Frequency modulation has two analogs for general carrier functions. It is known that frequency modulation may be interpreted as variation of the number of zero crossings of a sine wave per unit of time.[21],[23],[24] The number of zero crossings per unit of time of a Walsh function,

$$\text{cal } (i, \theta) = \text{cal } (\phi T, t/T); \qquad \theta = t/T, \quad i = \phi T, \qquad (24)$$

may be changed by varying the time base T and keeping $i = \phi T$ constant. This may be called time base modulation in the general case, while the term pulsewidth modulation is used if block pulses are modulated.

A second way is to modulate i or ϕ in (24) and keep the time base T constant. This may be called a code modulation for the following reason: in pulse code modulation, one uses a finite number $i = 1, 2, \cdots, n$ of functions $f_i(\theta)$ constructed from binary block pulses. Pulse code modulation may be considered to be a discontinuous change or modulation of the parameter i. The similarity to the modulation of the parameter i of the system of functions {cal (i, θ), sal (i, θ)} or other systems of functions is evident. Code modulation does not have to be discontinuous. For instance, the integer parameter i of the Walsh functions cal (i, θ) and sal (i, θ) is replaced by the real, continuous parameter μ in the case of Walsh meander functions cal (μ, θ) and sal (μ, θ). The index μ may be varied continuously in theory if not in practice.[8]

Acknowledgment

The author wishes to thank Dr. E. Schulze and Dr. M. Scholz of the Defense Department of the Federal Republic of Germany for many years of support. He is also grateful to the following for discussions, suggestions, and other forms of help: Prof. F. Fischer of Technische Hochschule Darmstadt; Prof. H. Friedburg, Prof. von Sanden, and F. Rühmann of Karlsruhe University; Prof. G. Lochs, Dr. P. Weiss, Dr. R. Liedl and Dr. F. Pichler of Innsbruck University; Prof. F. Lange of Rostock University; Prof. G. Ulrich of Technische Hochschule Ilmenau; Prof. H. Lueg of Technische Hoschschule Aachen; Dr. J. Piesch of the Austrian Post Office; Prof. D. Olson of St. Olaf College; Prof. E. Skudrzyk of Pennsylvania State University; Dr. H. F. Schlicke and D. P. Henkel of Allen-Bradley Co; Prof. Medina of Universidad Nacional de Mexico; Prof. Kato of Tokyo University; and L. Tirkschleit of Battelle Institute.

References

[1] A. H. Ballard, "Orthogonal multiplexing," *Aerospace Electronics, Advanced Communications*, vol. 3, pp. 51–60, November 1962.

[2] R. W. Sanders, "The DIGILOCK orthogonal modulation system," *Advanced Communication Systems*, vol. 1, pp. 57–75, 1965.

[3] R. M. Jaffe, "DIGILOCK telemetry system for the Air Force Special Weapons Center's Blue Scout, Jr.," *IRE Trans. Space Electronics and Telemetry*, vol. SET-8, pp. 44–50, March 1962.

[4] S. W. Golomb, L. D. Baumert, M. F. Easterling, J. J. Stiffler, and A. J. Viterbi, *Digital Communications*. Englewood Cliffs, N. J.: Prentice-Hall, 1964.

[5] P. Hsieh and M. Y. Hsiao, "Several classes of codes generated from orthogonal functions," *IEEE Trans. Information Theory*, vol. IT-10, pp. 88–91, January 1964.

[6] R. H. Aronstein, "Comparison of orthogonal and block codes," *Proc. IEE* (London), vol. 110, pp. 1965–1967, 1963.

[7] H. Harmuth, "Grundzüge einer Filtertheorie für die Mäanderfunktionen $A_n(\theta)$," *Arch. Elek. Übertragung*, vol. 18, pp. 544–554, 1964.

[8] J. L. Walsh, "A closed set of orthogonal functions," *Am. J. Math.*, vol. 55, pp. 5–24, 1923.

[9] N. J. Fine, "On the Walsh functions," *Trans. Am. Math. Soc.*, vol. 65, pp. 372–414, 1949.

[10] —, "The generalized Walsh functions," *Trans. Am. Math. Soc.*, vol. 69, pp. 66–77, 1950.

[11] G. W. Morgenthaler, "On Walsh–Fourier series," *Trans. Am. Math. Soc.*, vol. 84, pp. 472–507, 1957.

[12] R. Liedl, "Über eine spezielle Klasse von stark multiplikativ orthogonalen Funktionensystemen," *Monatsh. Math.*, vol. 68, pp. 130–137, 1964.

[13] P. Weiss, "Zusammenhang von Walsh–Fourier–Reihen mit Polynomen," *Monatsh. Math.*, vol. 71, pp. 165–179, 1967.

[14] K. W. Henderson, "Some notes on the Walsh functions," *IEEE Trans. Electronic Computers*, vol. EC-13, pp. 50–52, February 1964.

[8] One of the advantages of Walsh carriers is that a peak-power limited amplifier delivers 3 dB more average power when amplifying Walsh carriers than when amplifying sine carriers; another is the insensitivity to amplitude clipping.

[15] F. Pichler, "Das System der sal- und cal-Funktionen als Erweiterung des Systems der Walsh-Funktionen und die Theorie der sal- und cal-Fourier-Transformation," thesis, Mathematical Institute, Innsbruck University, Austria, 1967.

[16] N. Wiener, *Nonlinear Problems in Random Theory*. Cambridge, Mass.: M. I. T. Press, and New York: Wiley, 1958, p. 21.

[17] R. Bracewell, *The Fourier Transform and Its Applications*. New York: McGraw-Hill, 1965.

[18] P. M. Morse and H. Feshbach, *Methods of Theoretical Physics*, vol. 1. New York: McGraw-Hill, 1952, pp. 942–945.

[19] H. Harmuth, "Verallgemeinerung des Fourier-Integrals und des Begriffes Frequenz," *Arch. Elek. Uebertragung*, vol. 18, pp. 439–451, 1964.

[20] H. Rademacher, "Einige Sätze von allgemeinen Orthogonalfunktionen," *Math. Ann.*, vol. 87, pp. 122–138, 1922.

[21] P. L. Stumpers, "Theory of frequency-modulation noise," *Proc. IRE*, vol. 36, pp. 1081–1092, September 1948.

[22] P. A. Mann, "Der Zeitablauf von Rauschspannungen," *Elek. Nachrtech.*, vol. 20, pp. 183–189, 1943.

[23] P. F. Panter, *Modulation, Noise, and Spectral Analysis*. New York: McGraw-Hill, 1965.

[24] H. B. Voelcker, "Toward a unified theory of modulation," *Proc. IEEE*, vol. 54, pp. 340–353, March 1966; pp. 735–755, May 1966.

[25] F. F. Fowle, "The transposition of conductors," *AIEE Trans.*, vol. 23, pp. 659–687, 1905, particularly Fig. 11 and text on p. 675. This seems to be the oldest application of Walsh functions in communications. J. A. Barrett is credited for introducing them.

[26] H. Harmuth, "Sequency multiplex systems for telephony and data transmission," *Preconf. Digest, Internat'l Electronics Conf.* (Toronto, Canada, 1967), pp. 48–49. "Zeitbasis- und Zeitlagemodulation als Verallgemeinerung der Frequenz- und Phasenmodulation," *Proc. 10th Internat'l Wiss. Kolloq.* (Technischen Hochschule Ilmenau, East Germany,) 1965, pp. 1–5.

[27] R. F. Filipowsky, "Multiorthogonal data transmission systems," *Digest, IEEE Internat'l Conf. on Communications* (Minneapolis, Minn., 1967), p. 36.

[28] H. Harmuth, "Sequenz-Multiplex-Systeme für Telephonie- und Datenübertragung," *Arch. Elek. Übertragung*, vol. 22., pp. 27–35, January 1968; pp. 102–108, February 1968.

[29] F. Pichler, "Synthese linearer periodisch zeitvariabler Filter mit vorgeschriebenem Sequenzverhalten," *Arch. Elek. Übertragung*, vol. 22, pp. 150–161, March 1968.

Reprinted with permission from *IEEE Trans. Electromagn. Compat.* **EMC-15:**75–80 (1973)

On Notation and Definition of Terms Related to a Class of Complete Orthogonal Functions

NASIR AHMED, HEINZ H. SCHREIBER, AND PHILIP V. LOPRESTI

Abstract—This paper recommends a set of standards for notation and nomenclature that would assist prospective authors interested in submitting papers in the area of nonsinusoidal, complete, orthogonal sets of functions and their applications.

I. INTRODUCTION

The "Sequency Union" is a Specialist Working Group of the IEEE Electromagnetic Compatability Group. The objectives of this group are to "advance understanding in the uses of nonsinusoidal complete, orthonormal sets of functions; to diffuse the knowledge of these uses; and to assist the members in these activities." To this end, the main objective of this paper is to recommend a set of terms and related definitions that would assist authors interested in submitting papers in the area of nonsinusoidal, complete, orthogonal sets of functions.

II. DEFINITION OF SEQUENCY

Frequency is a term applied to the set of sinusoidal (periodic) functions whose zero-crossings are uniformly spaced over an interval. It is a parameter that distinguishes the individual functions of the sets $\{\cos 2\pi ft\}$ and $\{\sin 2\pi ft\}$, and it has the physical interpretation of the number of complete cycles (or one-half the number of zero-crossings) generated by the sinusoidal functions per unit time.

The generalization of frequency may be effected by defining generalized frequency as one-half the average number of zero-crossings per unit time. Harmuth [1] introduced the term sequency to describe generalized frequency, and applied it to distinguish functions whose zero-crossings are not uniformly spaced over an interval and which need not be periodic. This definition of sequency coincides with that of frequency when applied to sinusoidal functions. Applying this definition to periodic and aperiodic functions, one obtains:

1) the sequency of a periodic function equals one-half the number of sign changes per period;
2) the sequency of an aperiodic function equals one-half the number of sign changes per unit time, if this limit exists.

For the purposes of illustration, consider the continuous functions $f_1(t)$ and $f_2(t)$ in Fig. 1 which are defined in the half-open interval $[-0.5, 0.5)$. Each function has 4 zero-crossings in the interval; hence, the sequency of each of these functions equals 2. Analogous to frequency which is expressed in cycles per second or Hertz, sequency can be expressed in zero-crossings per second for which one may use the abbreviation "zps."

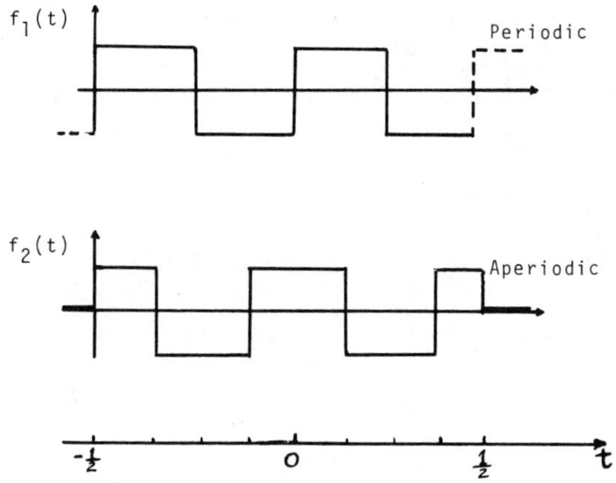

Fig. 1. Pertaining to definition of sequency.

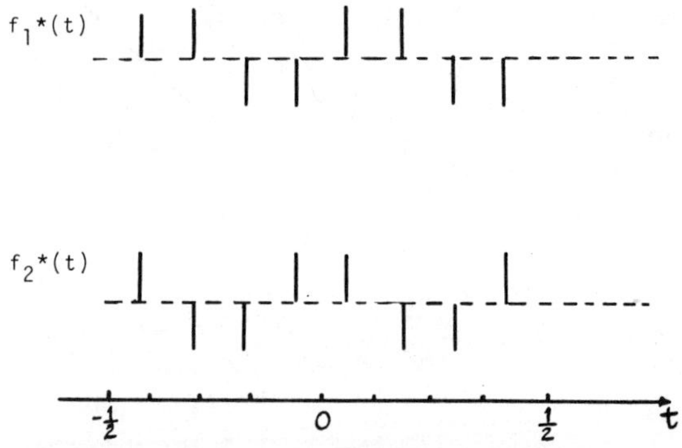

Fig. 2. Pertaining to definition of sequency.

The preceding definition of sequency can be applied with minor modification to the corresponding discrete function $f^*(t)$, which is obtained by sampling $f(t)$ at equidistant points. If the number of sign changes per unit time of $f^*(t)$ equals η, then the sequency of $f^*(t)$ is defined as $\eta/2$ or $(\eta + 1)/2$ when η is even or odd, respectively. For example, consider $f_1^*(t)$ and $f_2^*(t)$ shown in Fig. 2. These discrete functions are obtained by sampling $f_1(t)$ and $f_2(t)$ in Fig. 1, at a set of equidistant points. From Fig. 2 it follows that $\eta_1 = 3$ and $\eta_2 = 4$. Thus the sequency of each of the functions $f_1^*(t)$ and $f_2^*(t)$ equals 2, as before.

Manuscript received January 25, 1973.
N. Ahmed is with the Departments of Electrical Engineering and Computer Science, Kansas State University, Manhattan, Kan. 66502.
H. H. Schreiber is with the Grumman Aerospace Corporation, Bethpage, N.Y. 11714.
P. V. Lopresti is with the Engineering Research Center of Western Electric, Princeton, N.J. 08540.

TABLE I
NOTATION FOR CONTINUOUS AND DISCRETE FUNCTIONS

Name of Function	Notation: Continuous Functions	Notation: Discrete Functions
Rademacher	rad	Rad
Haar	har	Har
Walsh	wal	Wal
"cosine Walsh"	cal	Cal
"sine Walsh"	sal	Sal

III. SOME CONSIDERATIONS OF NOTATION

We observe that in the case of trignometric, exponential, and logarithmic functions, many of the notations accepted as "standard" consist of three letters. Examples of these include sin, cos, exp, erf, and log. Thus it is reasonable to introduce a similar scheme to denote nonsinusoidal, complete orthogonal functions. In addition, it is necessary to distinguish continuous functions from the corresponding discrete functions. A possible notation is proposed in Table I as suggested by Harmuth [1]. The set of Rademacher functions, although not a complete set, is also included in Table I because these functions are useful in the derivation of simple recurrence relations which enable one to generate Walsh functions [2]. The complete set of Walsh functions, defined on the unit interval, can be divided into two groups of even and odd functions about the point $t = 0.5$. These even and odd functions are analogous to the cosine and sine functions, respectively; hence they are denoted by sal (*s*ine W*al*sh) and cal (*c*osine W*al*sh), respectively. A rigorous mathematical development of the sal–cal system has been carried out by Pichler [3].

IV. RADEMACHER AND HAAR FUNCTIONS [4], [5]

Rademacher functions are an incomplete set of orthonormal functions which were developed in 1922. The Rademacher function of index m, denoted by $\text{rad}(m, t)$, is a train of rectangular pulses with 2^{m-1} cycles in the half-open interval $[0, 1)$, taking the values $+1$ or -1, with the exception of $\text{rad}(0, t)$ which is the unit pulse (see Fig. 3). If the Rademacher functions are periodically extended, then they satisfy the relation

$$\text{rad}\,(m, t + n2^{1-m}) = \text{rad}\,(m, t), \qquad m = 1, 2, 3, \cdots$$
$$n = \pm 1, \pm 2, \pm 3, \cdots.$$

Rademacher functions can be generated using the recurrence relation

$$\text{rad}\,(m, t) = \text{rad}\,(1, 2^{m-1} t)$$

with

$$\text{rad}\,(1, t) = \begin{cases} 1, & t \in [0, 1/2) \\ -1, & t \in [1/2, 1) \end{cases}. \tag{1}$$

The set of Haar functions $\{\text{har}\,(g, m, t)\}$ is periodic, orthonormal, and complete [5], and was proposed in 1910 by Haar. Fig. 4(a) shows the set of first 8 Haar functions. A recurrence relation that enables one to generate $\{\text{har}\,(g, m, t)\}$ is given by [5]

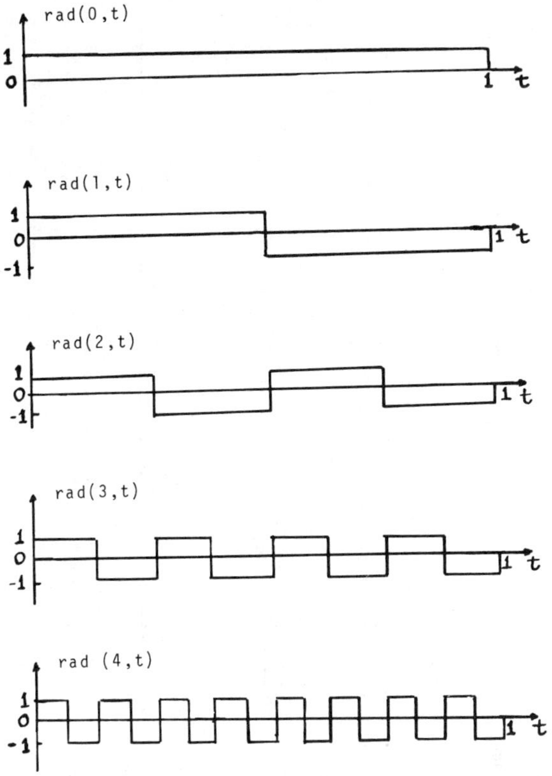

Fig. 3. Rademacher functions.

$$\text{har}\,(0, 0, t) = 1, \qquad t \in [0, 1)$$

$$\text{har}\,(g, m, t) = \begin{cases} 2^{g/2}, & \dfrac{m-1}{2^g} \leqslant t < \dfrac{m - 1/2}{2^g} \\ -2^{g/2}, & \dfrac{m - 1/2}{2^g} \leqslant t < \dfrac{m}{2^g} \\ 0, & \text{elsewhere for } t \in (0, 1) \end{cases}$$

where $0 \leqslant g \leqslant \log_2 N$ and $1 \leqslant m \leqslant 2^g$.

Sampling the set of Haar functions in Fig. 4(a) results in the array shown in Fig. 4(b), each row of which is a discrete Haar function $\text{Har}(g, m, t)$. Arrays so obtained are used in connection with the Haar transform [6] and may be denoted by $H_N^*, H^*(g)$ or $[H^*(g)]$ where $g = \log_2 N$.

V. WALSH FUNCTIONS

The incomplete set of Rademacher functions was completed by Walsh [7] in 1923, to form the complete orthnormal set of rectangular functions we now call the Walsh functions. The set of Walsh functions is generally classified into three groups. These groups differ from one another in that the *order* in which individual functions appear is different. The three types of orderings are: 1) sequency or Walsh ordering, 2) dyadic or Paley ordering, and, 3) natural or Hadamard ordering. In what follows, we discuss some aspects of each of these orderings.

Sequency or Walsh Ordering

This is the ordering which was originally employed by Walsh [7]. We denote the set of Walsh functions belonging to this

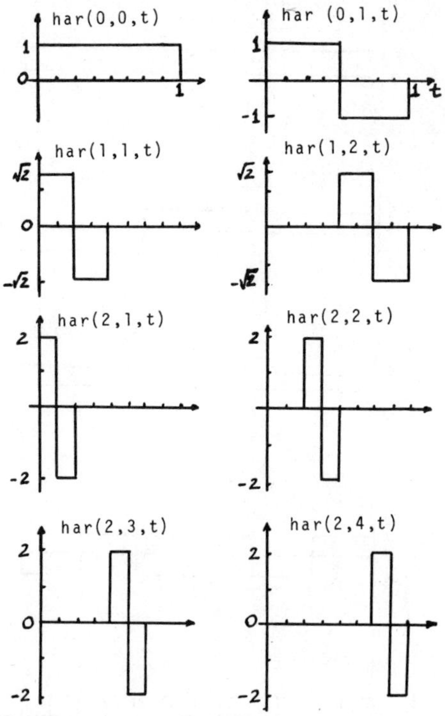

(a)

$$H^*(3)=\begin{bmatrix} 1 & 1 & 1 & 1 & 1 & 1 & 1 & 1 \\ 1 & 1 & 1 & 1 & -1 & -1 & -1 & -1 \\ \sqrt{2} & \sqrt{2} & -\sqrt{2} & -\sqrt{2} & 0 & 0 & 0 & 0 \\ 0 & 0 & 0 & 0 & \sqrt{2} & \sqrt{2} & -\sqrt{2} & -\sqrt{2} \\ 2 & -2 & 0 & 0 & 0 & 0 & 0 & 0 \\ 0 & 0 & 2 & -2 & 0 & 0 & 0 & 0 \\ 0 & 0 & 0 & 0 & 2 & -2 & 0 & 0 \\ 0 & 0 & 0 & 0 & 0 & 0 & 2 & -2 \end{bmatrix}$$

(b)

Fig. 4. (a) Continuous Haar functions, $N = 8$. (b) Discrete Haar functions, $N = 8$.

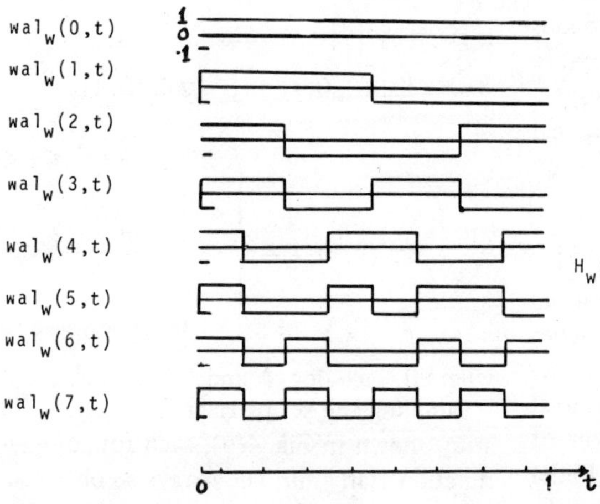

(a)

$$H_w(3)=\begin{bmatrix} 1 & 1 & 1 & 1 & 1 & 1 & 1 & 1 \\ 1 & 1 & 1 & 1 & -1 & -1 & -1 & -1 \\ 1 & 1 & -1 & -1 & -1 & -1 & 1 & 1 \\ 1 & 1 & -1 & -1 & 1 & 1 & -1 & -1 \\ 1 & -1 & -1 & 1 & 1 & -1 & -1 & 1 \\ 1 & -1 & -1 & 1 & -1 & 1 & 1 & -1 \\ 1 & -1 & 1 & -1 & -1 & 1 & -1 & 1 \\ 1 & -1 & 1 & -1 & 1 & -1 & 1 & -1 \end{bmatrix}$$

(b)

Fig. 5. (a) Walsh-ordered continuous Walsh functions, $N = 8$. (b) Walsh-ordered discrete Walsh functions, $N = 8$.

set by

$$S_w = \{\mathrm{wal}_w\,(i, t), \quad i = 0, 1, \cdots, (N-1)\} \tag{2}$$

where $N = 2^g$, $g = 1, 2, 3 \cdots$, the subscript w denotes Walsh ordering, and, i denotes the ith member of S_w.

If s_i represents the sequency of $\mathrm{wal}_w\,(i, t)$, then we have

$$s_i = \begin{cases} 0, & i = 0 \\ i/2, & i \text{ even} \\ (i+1)/2, & i \text{ odd}. \end{cases} \tag{3}$$

The cal and sal functions corresponding to $\mathrm{wal}_w(i, t)$ may be denoted as

$$\mathrm{cal}\,(s_i, t) = \mathrm{wal}_w(i, t), \quad i \text{ even}$$

and

$$\mathrm{sal}\,(s_i, t) = \mathrm{wal}_w(i, t), \quad i \text{ odd}. \tag{4}$$

Using the preceding notation, the first 8 of the Walsh functions $\mathrm{wal}_w(i, t)$ are shown in Fig. 5(a). From Fig. 5(a) it is apparent that the sequency of a Walsh function is greater than or equal to that of the preceding Walsh function and has precisely one more zero-crossing in $0 < t < 1$. Hence the alternate name, "sequency ordering." The set S_w can be generated using recurrence relations in [1], [2].

Discrete Case: By sampling the set of Walsh functions in Fig. 5(a), we obtain the (8×8) matrix shown in Fig. 5(b). In general one would obtain an $(N \times N)$ matrix. We shall show later while discussing the natural or Hadamard ordering, that such matrices can be obtained by reordering the rows of a class of Hadamard matrices. Thus a reasonable notation to denote the matrices which result from sampling the set of functions S_w would be $H_w(g)$, $[H_w(g)]$, or $H_{w,N}$ where, $g = \log_2 N$.

Let u_i and v_i be the ith bit in the binary representations of integers u and v, respectively. Thus, for example,

$$(u)_{\text{decimal}} = (u_{g-1} u_{g-2} \cdots, u_1 u_0)_{\text{binary}}.$$

Then the elements of $H_{w,N}$ can be generated using the relation [8]

$$h_{uv}^{(w)} = (-1)^{\sum_{i=0}^{g-1} r_i(u) v_i}, \quad u, v = 0, 1, \cdots, (N-1) \tag{5}$$

where

$$r_0(u) = u_{g-1}$$
$$r_1(u) = u_{g-1} + u_{g-2}$$
$$r_2(u) = u_{g-2} + u_{g-3}$$
$$\vdots$$
$$r_{g-1}(u) = u_1 + u_0.$$

Dyadic or Paley Ordering

The dyadic type of ordering was introduced by Paley [9]. Walsh functions are elements of the dyadic group and can be ordered using a Gray code [2], [10]. This set of N Walsh functions may be denoted as

$$S_p = \{\text{wal}_p(i, t), \quad i = 0, 1, \cdots, (N-1)\} \tag{6}$$

where p denotes Paley ordering and i denotes the ith member of S_p. This set of functions is related to the Walsh-ordered functions by the relation

$$\text{wal}_p(i, t) = \text{wal}_w(b(i), t) \tag{7}$$

where $b(i)$ is the *Gray code-to-binary conversion* of i. This conversion process is discussed in introductory computer texts such as [11]. An interesting algorithm for this transformation along with its implementation is discussed in [10]. For the purposes of illustration, we consider the case $N = 8$ and evaluate (7) as summarized in Table II. Using Table II and the functions $\text{wal}_w(i, t), i = 0, 1, \cdots, 7$ in Fig. 5(a), we obtain the first 8 Walsh functions $\text{wal}_p(i, t)$ shown in Fig. 6(a).

Discrete Case: Sampling the Walsh functions in Fig. 6(a) we obtain the (8×8) matrix shown in Fig. 6(b). As in the case of the Walsh ordering, this matrix can also be obtained by rearranging the rows of an (8×8) Hadamard matrix which will be discussed in the natural ordering case. Thus the class of Hadamard matrices associated with this type of ordering can be denoted by $H_p(g)$, $[H_p(g)]$, or $H_{p,N}$.

The elements of $H_p(g)$ can be generated using the relation [10]

$$h_{uv}^{(p)} = (-1)^{\sum_{i=0}^{g-1} u_{g-1} v_i}, \quad u, v = 0, 1, \cdots, (N-1). \tag{8}$$

Natural or Hadamard Ordering

This set of Walsh functions may be denoted by $\text{wal}_h(i, t)$ where h denotes Hadamard ordering and i denotes the ith member of the set S_h, such that

$$S_h = \{\text{wal}_h(i, t), \quad i = 0, 1, \cdots, (N-1)\}. \tag{9}$$

This set of functions are related to the Walsh-ordered functions by the relation

$$\text{wal}_h(i, t) = \text{wal}_w(b(\langle i \rangle), t) \tag{10}$$

where $\langle i \rangle$ is obtained by the *bit-reversal* of i and, $b(\langle i \rangle)$ is the *Gray code-to-binary conversion* of $\langle i \rangle$. For example, when $N = 8$, an evaluation of (10) yields Table III. Table III along with the functions $\text{wal}_w(i, t), i = 0, 1, \cdots, 7$ in Fig. 5(a), yield the first 8 Walsh functions $\text{wal}_h(i, t)$ shown in Fig. 7(a).

Discrete Case: Sampling of the Walsh functions in Fig. 7(a) results in the (8×8) Hadamard matrix shown in Fig. 7(b). In general one would obtain an $(N \times N)$ matrix, a reasonable notation for which is $H_h(g)$, $[H_h(g)]$, or $H_{h,N}$, where, $g = \log_2 N$. This class of Hadamard matrices can be partitioned in the form

$$H_h(g) = \begin{bmatrix} H_h(g-1) & H_h(g-1) \\ H_h(g-1) & -H_h(g-1) \end{bmatrix} \tag{11}$$

and are considered to be in "natural" form [8]. Hence the name, "natural ordering."

The elements of $H_h(g)$ can be generated using the relation [8], [12]

$$h_{uv}^{(h)} = (-1)^{\sum_{i=0}^{g-1} u_i v_i}, \quad u, v = 0, 1, \cdots, (N-1). \tag{12}$$

VI. DISCRETE WALSH TRANSFORMS

Using the three classes of Hadamard matrices discussed in Section V, one can define corresponding transformations as follows:

$$Y_N = [H_h(g)] X_N \tag{13}$$

$$Y_N = [H_w(g)] X_N \tag{14}$$

$$Y_N = [H_p(g)] X_N \tag{15}$$

where X_N and Y_N are the input and transform vectors, respectively. In the literature, (13) and (14) have been referred to as the Walsh-Hadamard [13], [14], Hadamard [8], Fourier-Hadamard [15], Walsh [16], or BIFORE transform [17], while (15) has been referred to as the dyadic-ordered Walsh transform. In all these cases, the subscripts h, w, or p were not used. Hence the need for using a variety of names to identify each transform with the particular Hadamard matrix. Clearly, if the subscript h, w, or p *is included* as part of the definition of the transform, the type of Hadamard matrix is automatically specified. Then, one can refer to the resulting transformation as: Hadamard-ordered Walsh-Hadamard transform $(\text{WHT})_h$; Walsh-ordered Walsh-Hadamard transform $(\text{WHT})_w$; and Paley-ordered Walsh-Hadamard transform $(\text{WHT})_p$.

TABLE II

$i_{decimal}$	i_{binary}	$b(i)_{binary}$	$b(i)_{decimal}$	Equation (7)
0	000	000	0	$\text{wal}_p(0,t) = \text{wal}_w(0,t)$
1	001	001	1	$\text{wal}_p(1,t) = \text{wal}_w(1,t)$
2	010	011	3	$\text{wal}_p(2,t) = \text{wal}_w(3,t)$
3	011	010	2	$\text{wal}_p(3,t) = \text{wal}_w(2,t)$
4	100	111	7	$\text{wal}_p(4,t) = \text{wal}_w(7,t)$
5	101	110	6	$\text{wal}_p(5,t) = \text{wal}_w(6,t)$
6	110	100	4	$\text{wal}_p(6,t) = \text{wal}_w(4,t)$
7	111	101	5	$\text{wal}_p(7,t) = \text{wal}_w(5,t)$

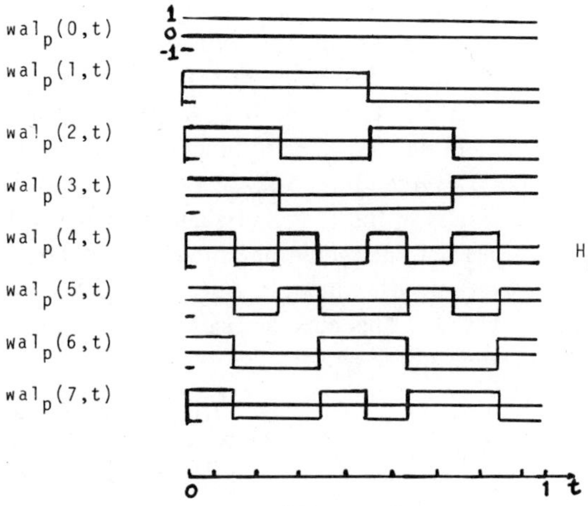

(a)

$$H_p(3) = \begin{bmatrix} 1 & 1 & 1 & 1 & 1 & 1 & 1 & 1 \\ 1 & 1 & 1 & 1 & -1 & -1 & -1 & -1 \\ 1 & 1 & -1 & -1 & 1 & 1 & -1 & -1 \\ 1 & 1 & -1 & -1 & -1 & -1 & 1 & 1 \\ 1 & -1 & 1 & -1 & 1 & -1 & 1 & -1 \\ 1 & -1 & 1 & -1 & -1 & 1 & -1 & 1 \\ 1 & -1 & -1 & 1 & 1 & -1 & -1 & 1 \\ 1 & -1 & -1 & 1 & -1 & 1 & 1 & -1 \end{bmatrix}$$

(b)

Fig. 6. (a) Paley-ordered continuous Walsh functions, $N = 8$. (b) Paley-ordered discrete Walsh functions, $N = 8$.

TABLE III

i	i_{binary}	$\langle i \rangle_{binary}$	$b(\langle i \rangle)_{binary}$	$b(\langle i \rangle)_{decimal}$	Equation (8)
0	000	000	000	0	$\text{wal}_h(0,t) = \text{wal}_w(0,t)$
1	001	100	111	7	$\text{wal}_h(1,t) = \text{wal}_w(7,t)$
2	010	010	011	3	$\text{wal}_h(2,t) = \text{wal}_w(3,t)$
3	011	110	100	4	$\text{wal}_h(3,t) = \text{wal}_w(4,t)$
4	100	001	001	1	$\text{wal}_h(4,t) = \text{wal}_w(1,t)$
5	101	101	110	6	$\text{wal}_h(5,t) = \text{wal}_w(6,t)$
6	110	011	010	2	$\text{wal}_h(6,t) = \text{wal}_w(2,t)$
7	111	111	101	5	$\text{wal}_h(7,t) = \text{wal}_w(5,t)$

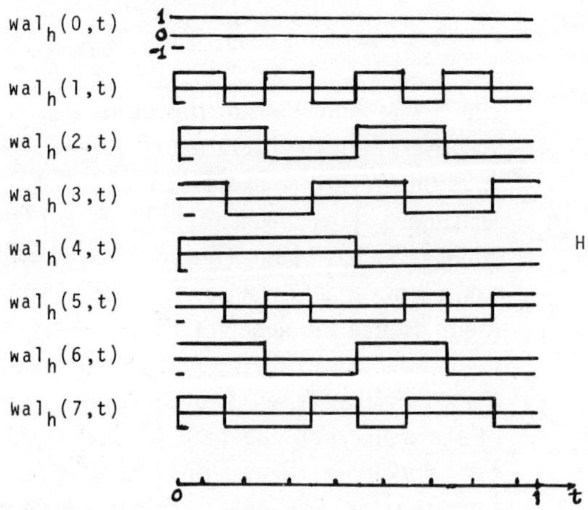

(a)

$$H_h(3) = \begin{bmatrix} 1 & 1 & 1 & 1 & 1 & 1 & 1 & 1 \\ 1 & -1 & 1 & -1 & 1 & -1 & 1 & -1 \\ 1 & 1 & -1 & -1 & 1 & 1 & -1 & -1 \\ 1 & -1 & -1 & 1 & 1 & -1 & -1 & 1 \\ 1 & 1 & 1 & 1 & -1 & -1 & -1 & -1 \\ 1 & -1 & 1 & -1 & -1 & 1 & -1 & 1 \\ 1 & 1 & -1 & -1 & 1 & 1 & -1 & -1 \\ 1 & -1 & -1 & 1 & -1 & 1 & 1 & -1 \end{bmatrix}$$

(b)

Fig. 7. (a) Hadamard-ordered continuous Walsh functions, $N = 8$. (b) Hadamard-ordered discrete Walsh functions, $N = 8$.

TABLE IV
NOTATION FOR A CLASS OF CONTINUOUS AND DISCRETE FUNCTIONS

Name of Function	Notation	
	Continuous Functions	Discrete Functions
A class of Haar functions	her	Her
A class of Haar functions	ter	Ter
Block pulses	blo	Blo
Paley functions	pal	Pal
A class of Hadamard functions	had	Had

VII. MISCELLANEOUS CONSIDERATIONS

In this paper we have considered the nonsinusoidal orthogonal functions which are referred to most often in the current literature. However, there are other sets of nonsinusoidal orthogonal functions in addition to modified versions of those considered in this paper. For our purposes, it is sufficient to list some of these as shown in Table IV. The interested reader may consult [1] for details pertaining to the same.

Only one-dimensional functions were considered throughout the paper. The extension to the multi-dimensional case is obvious. For example, $\{\mathrm{wal}_w(i, x), \mathrm{wal}_w(j, y), i, j = 0, 1, \cdots, N\}$ with $0 \leqslant x < 1$ and $0 \leqslant y < 1$ represents a set of N^2 two-dimensional Walsh-ordered Walsh functions.

VIII. CONCLUDING REMARKS

Since the interest in applying nonsinusoidal, complete, orthogonal sets of functions has grown rapidly in recent years, it is natural that a variety of notation, terminology, and related definitions have arisen. It is hoped that the material presented in this paper will assist authors in using a more compact and standard notation along with a set of terms whose definitions are not ambiguous as they have been in the past. Indeed, since this is the first attempt in this direction, there would obviously be room for additions, deletions, and improvements in the future. Forthcoming work should include topics pertaining to the definition of generalized Walsh functions and the structure of Walsh functions, related to their Rademacher product equivalence.

ACKNOWLEDGMENT

The authors are grateful to Dr. H. F. Harmuth of the Catholic University of America, Washington, D.C., for his valuable suggestions and comments in connection with preparing the paper.

REFERENCES

[1] H. F. Harmuth, *Transmission of Information by Orthogonal Functions*, Springer-Verlag, New York, Second Edition, pp. 10–89.
[2] R. B. Lackey and D. Meltzer, "A Simplified Definition of Walsh Functions," IEEE Trans. Computers, vol. C-20, pp. 211–213, February 1971.
[3] F. Pichler, "Das System der sal- und cal-Funktionen als Erweiterung des Systems der Walsh-Funktionen und die Theorie der sal- und cal-Fourier Transformation," thesis, Mathematical Institute, Innsbruck University, Austria, 1967.
[4] M. S. Corrington and R. N. Adams, "Advanced Analytical and Signal Processing Techniques," Rome Airforce Development Center, RADC-RDR-62-200, ASTIA Document N. AD-277942.
[5] B. S. Nagy, *Introduction to Real Functions and Orthogonal Expansions*, Oxford Univ. Press, New York (1965), pp. 338–340.
[6] C. H. Andrews and W. K. Pratt, "Digital Image Transform Processing," Proc. Symp. Walsh Function Applications, Washington, D.C., pp. 183–194, April 1970.
[7] J. L. Walsh, "A Closed Set of Normal Orthogonal Functions," Amer. Jour. Math., vol. 45, pp. 5–24, 1923.
[8] W. K. Pratt and H. C. Andrews, "Hadamard Transform Coding," Proc. IEEE, vol. 57, No. 1, pp. 58–67, January 1969.
[9] R. E. A. C. Paley, "A Remarkable Series of Orthogonal Functions," Proc. London Math. Soc., vol. 2, No. 24, pp. 241–279, 1952.
[10] C. Yuen, "Walsh Functions and Gray Code," Proc. Symp. Walsh Function Applications, Washington, D.C., pp. 68–73, April 1971.
[11] L. Nashelsky, *Digital Computer Theory*, John Wiley and Sons, Inc., 1966, pp. 60–74.
[12] J. E. Pearl, "Application of Walsh Transform to Statistical Analysis," IEEE Trans. Systems, Man, and Cybernetics, vol. SMC-1, No. 2, April 1971.
[13] F. R. Ohnsorg, "Spectral Modes of the Walsh-Hadamard Transform," Proc. Symp. Walsh Function Applications, Washington, D.C., pp. 55–59, April 1971.
[14] G. S. Robinson and R. Granger, "A Design Procedure for Non-Recursive Digital Filters Based on Walsh Functions," Proc. Symp. Walsh Function Applications, Washington, D.C., pp. 95-100, April 1971.
[15] J. E. Whelchel Jr. and G. F. Guinn, "The Fast Fourier-Hadamard Transform and Its Uses in Signal Representation and Classification," Aerospace Electronics Conference, EASCON Record, pp. 561–573, Sept. 9–11, 1968.
[16] J. E. Gibbs and F. R. Pichler, "Comments on Transformation of 'Fourier' Power Spectra Into 'Walsh' Power Spectra," Proc. Symp. Walsh Function Applications, Washington, D.C., pp. 51–54, April 1971.
[17] N. Ahmed, A. L. Abdussattar, and K. R. Rao, "BIFORE or Hadamard Transform," IEEE Trans. Audio and Electroacoustics, vol. AU-19 pp. 225–235, September 1971.
[18] C. K. Yuen, "Remarks on the Ordering of Walsh Functions," IEEE Trans. Computers, vol. C-21, pp. 1452–1453, Dec. 1972.

3

Reprinted with permission from *IEEE Trans. Electromagn. Compat.* **EMC-16:**134–136 (1974)

A Programmable Walsh Function Generator for Orthogonal Sequency Pairs

DONALD A. GAUBATZ MEMBER, IEEE, AND
REUVEN KITAI, SENIOR MEMBER, IEEE

Abstract—**Logic expressions are developed for a Walsh function generator for highly-orthogonal sal and cal sequency pairs, using clocked T-type flip-flop outputs. It is shown that the flip-flop toggle logic functions are generated fully by the rate multiplier type 7497, leading to economical implementation. Thus a generator for up to 63 sequency pairs consists of only seven integrated circuits.**

Walsh function generators for use in multiplexers, spectral analyzers and synthesizers often require close tolerance on orthogonality over wide sequency ranges. Programmability of the sequency order is also often required, as is adaptability to array generation. Peterson [1], Lebert [2], and Kitai and Siemens [3] reported designs of generators featuring a programmable Walsh output. A multiple output generator developed by Boesswetter [4] minimizes delay differences by using synchronous clocking so that orthogonality errors are due only to differences in flip-flop switching times. This principal has been extended by Besslich [5] who demonstrated the suitability of T-type flip-flops in Walsh array generators. The manipulation of toggle information leading to programmable and array generation has also been discussed by Yuen [6].

In sequential-type spectral analyzers the requirement is to produce orthogonal sal and cal functions of common sequency [7]. Sequency pairs are programmed by a binary coded input either directly, or by a decimally coded switch via a decimal to binary converter. Besslich's and Yuen's methods are well suited to this end. The Walsh pairs so generated are also hazard free. The logical design and implementation of such generators is considered here.

DERIVATION OF TOGGLE FUNCTIONS

To develop the sequency-pair sal and cal flip-flop toggle functions T_{sal} and T_{cal}, it suffices to consider sequency pairs up to $S = 7$. In Fig. 1(a), sal and cal values are given for the sixteen intervals r in one cycle, including Wal 0 and sal 8. A Walsh value $+1$ is denoted 1 and a value -1 is denoted 0. In the toggle table of Fig. 1(b) the values of r for which T is *HI* is denoted 1. For example, in generating sal 3, T_{sal} is *HI* immediately prior to clocking for intervals $r = 0, 2, 6, 8, 10$, and 14.

Inspection of the sal toggle table shows that, apart from the set of all integer sequencies (labelled D), there are only three sets of sequencies within the entire range of r. These are labelled A, B, C. For example, the sal toggle is *HI* for sequencies 3, 4, 7, 8 in intervals $r = 1, 5, 9, 13$. These sequencies comprise set B. Similarly, for the cal functions, the corresponding three sequency sets are labelled P, Q, R, the fourth set being the null set V. It is also seen that the cycling of toggle-function sets in the intervals r from zero to 7 is repeated in the intervals 8 to 15.

Manuscript received June 20, 1973.
The authors are with McMaster University, Hamilton, Ont., Canada.

TABLE 1

sal set	S	cal set	S
A	5,6,7,8	P	4,5,6,7
B	3,4,7,8	Q	2,3,6,7
C	2,4,6,8	R	1,3,5,7
D	$1 \rightarrow 8$	V	...

The toggle functions in the sets are listed in Table I. Comparing sets A, B, C with P, Q, R and noting their respective common intervals r, sequencies S in set P have corresponding sequencies S–1 in set A, and similarly for sets B, Q and C, R.

Let the programmed sequency S be binary abc as in Fig. 1, the least significant bit (LSB) being c. The expressions for the cal sets P, Q, R in terms of a, b, c are readily shown by standard reduction methods (and by inspection of Fig. 1(b) to be

$$P(a,b,c) = a \qquad Q(a,b,c) = b \qquad R(a,b,c) = c.$$

For the sal sets A, B, C the reduced expressions are not so simple; for example

$$A(a,b,c) = a(b + c).$$

In practical implementation it is more convenient to subtract 1 from $S = abc$ to form $S\text{–}1 = \alpha\beta\gamma$ binary, so that $A(\alpha,\beta,\gamma) = \alpha$, $B(\alpha,\beta,\gamma) = \beta$, $C(\alpha,\beta,\gamma) = \gamma$. Subtracting 1 from S is readily achieved by feeding the binary number S to a parallel full adder, with its other input word bits in the *HI* state. Ignoring the carry out, the output is S–1 binary.

Cycling through intervals r is conveniently implemented using a synchronous binary counter, with output bits x, y, z and complement bits $\bar{x}$, $\bar{y}$, $\bar{z}$, the LSB being z. Due to the repetitive cycling in the toggle functions, the binary counter required has a number of states equal to half the number of intervals r in the cycle (e.g., an 8-state or 3-bit counter for a 16-interval cycle). Referring to Fig. 1(b), the cal toggle set P occurs in even-numbered intervals r, including $r = 0$. If the xyz counter state is 000 during $r = 0$, then toggling for set P occurs in the reduced counter state $r(x,y,z) = \bar{z}$. Similarly set Q occurs in the counter state $r(x,y,z) = \bar{y}z$ and set R in $r(x,y,z) = \bar{x}yz$. The complete toggle function for the cal T-type flip-flop is therefore

$$T_{cal} = a\bar{z} + b\bar{y}z + c\bar{x}yz. \tag{1}$$

The sal toggle function includes set D so that

$$T_{sal} = \alpha\bar{z} + \beta\bar{y}z + \gamma\bar{x}yz + xyz. \tag{2}$$

The general toggle expressions are therefore

$$T_{cal} = a\bar{z} + b\bar{y}z + c\bar{x}yz + \cdots \tag{3}$$

$$T_{sal} = \alpha\bar{z} + \beta\bar{y}z + \gamma\bar{x}yz + \cdots + (\cdots xyz). \tag{4}$$

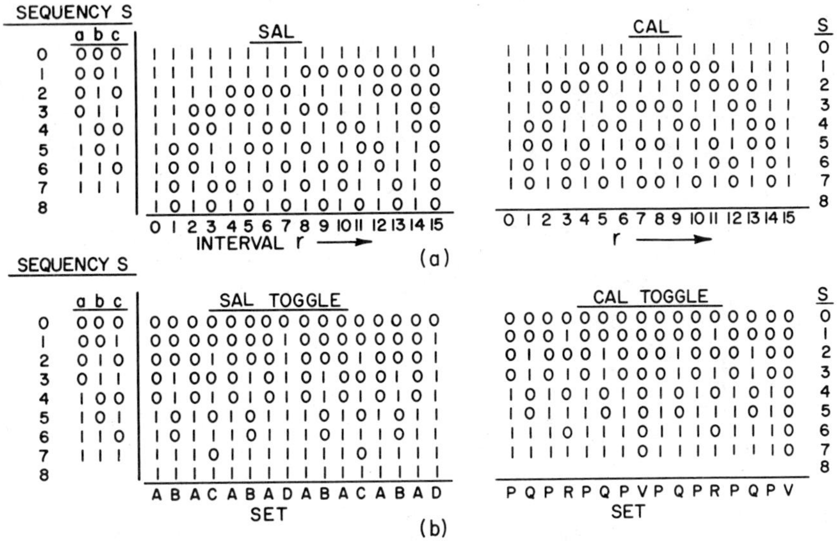

Fig. 1. (a) Discrete sal and cal values versus interval r. (b) Sal and cal toggle values for a T-type flip-flop.

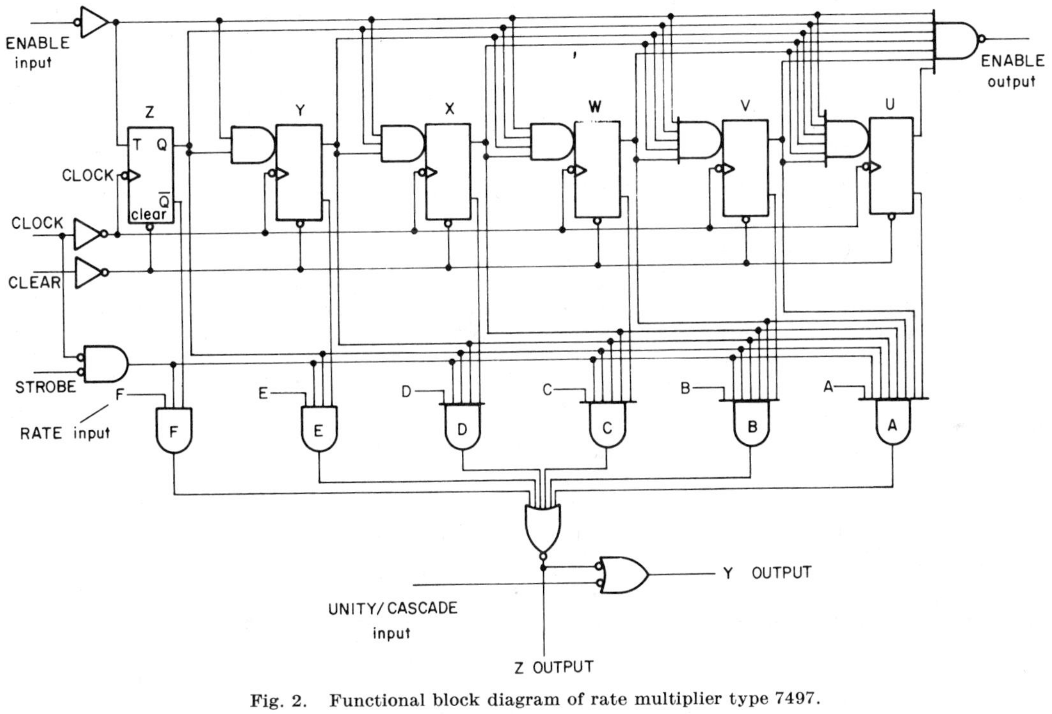

Fig. 2. Functional block diagram of rate multiplier type 7497.

IMPLEMENTATION

The sal and cal toggle logic functions for sequencies up to 63 are implemented fully by the Texas Instruments 6-bit binary rate multiplier type SN7497. A functional circuit diagram of this IC is shown in Fig. 2. It consists of a synchronous 6-bit binary counter and its associated decoding logic. The decoding circuitry can be strobed only when the positive-acting clock input has returned to the LO state, thus precluding erroneous rate multiplier outputs during counter-state changes.

The circuit diagram of a complete programmable Walsh function generator is illustrated in Fig. 3. The sequency address *abcdef* is applied to the RATE INPUT terminals of the 7497, the most significant bit (MSB) being applied to RATE INPUT F and the LSB f being applied to RATE INPUT A, for the cal toggle function. The UNITY/CASCADE terminal is wired *HI* and the STROBE INPUT LO. For the sal toggle (4) the address bits are $\alpha\beta\gamma\delta\epsilon\zeta$ and the additional counter state product term *uvwxyz*, i.e., the last term in (4) is provided by connecting the ENABLE OUTPUT to the UNITY/CASCADE INPUT. The subtraction of 1 from the sequency address bits to form S–1 = $\alpha\beta\gamma\delta\epsilon\zeta$ is performed by a 6-bit parallel full adder. The 6-adder inputs which are normally *HI* are also made accessible as the generator's RESET control so that when taken LO, the adder carry-out is LO, resetting the two rate multipliers and all the output flip-flops. All flip-flop outputs are *HI* under RESET, as is the sal 64 flip-flop output and the end-of-cycle flag which indicates the last interval in the cycle. The latter two outputs are useful for synchronising purposes.

No special "start" procedures are necessary; the generator is initialized completely by the RESET control. The negative-going clock pulse train has its first pulse occuring at the end of the first interval τ, as shown in Fig. 3. When the full 128 pulses of the cycle have passed, the generator does not have to be re-initiated or reset, but may be clocked continuously to provide periodic Walsh waves. A total of 7 integrated circuits is used.

The use of the 7497 binary rate multiplier is not restricted to Walsh function generators having S_{max} = 63. One needs only to recall, for example, that a cal 8 Walsh wave in a cycle of 128 inter-

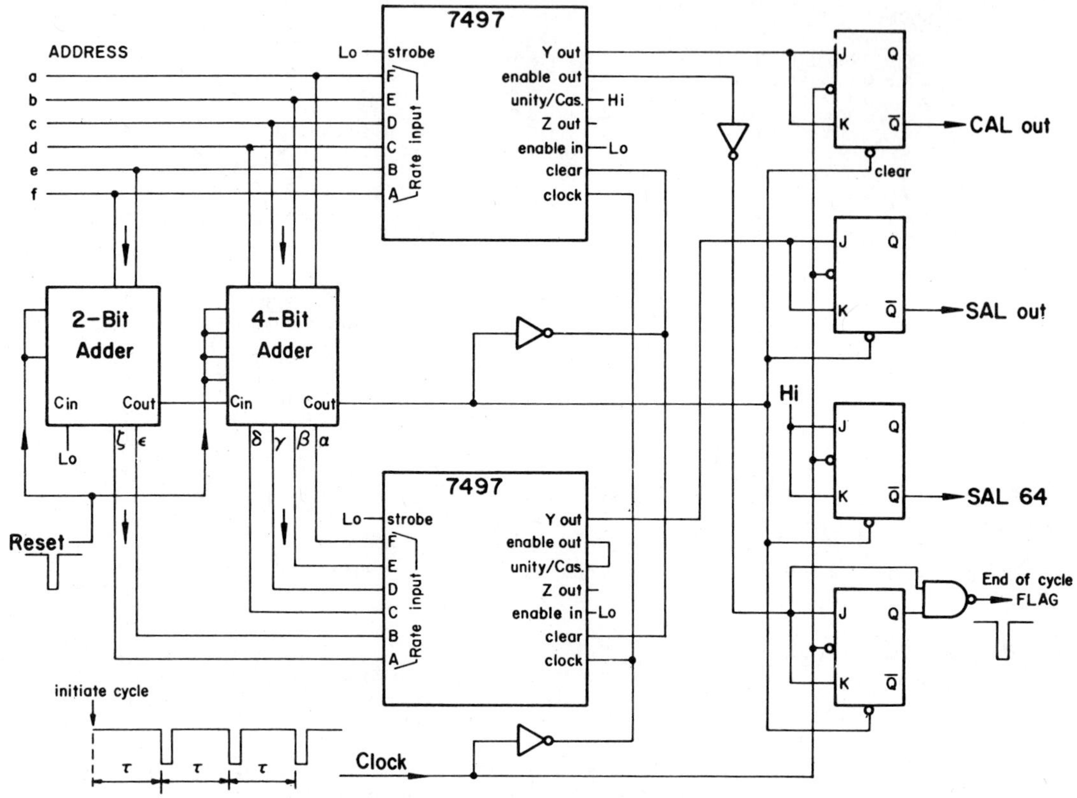

Fig. 3. Implementation of the generator for $S_{max} = 63$.

vals is identical to two consecutive cal 4 Walsh waves with cycles of 64 intervals. Fig. 1(a) may be inspected for similar relationships between Walsh waves having 16, 8, and 4 interval cycles. Thus a programmable Walsh function generator having a 6-bit input address capacity ($S_{max} = 63$) can be used to produce the set of Walsh wave pairs terminating in $S_{max} = 31$. This is accomplished by connecting the input address to the five most significant bit terminals a to e in Fig. 2(b) and tying the unused LSB f of the input address LO. In the case of a 4-bit address, for $S_{max} = 15$, the four most significant lines are used and the remaining two least significant input lines e, f are tied LO, etc. For a 5-bit input address the FLAG output pulse occurs at the end of every alternate cycle, for a 4-bit address at the end of every fourth cycle, etc.

If programmable Walsh function generators having S_{max} greater than 63 are desired, a number of binary rate multipliers may be cascaded synchronously via their ENABLE INPUT and ENABLE OUTPUT terminals. The addition of each 7497 increases the capacity of the sequency address input word by 6-bits. Thus $S_{max} = 2^{12} - 1 = 4095$ for two cascaded rate multipliers, etc. A programmable generator for $S_{max} = 4095$ consists of only ten integrated circuits, using three 4-bit parallel adders. Thus the rate multiplier provides the means of implementing programmable generators for any closed set of Walsh functions in an economical way.

Since the T-type flip-flops, which produce the Walsh function outputs, and the rate multipliers, which produce the toggle states, are clocked synchronously, input clock rates up to 15 MHz are typical. Any orthogonality error, due solely to differences in the switching times of the output flip-flops, can be reduced by the use of Schottky flip-flips to less than 5 ns, using present-day devices. Alternatively, the entire Walsh function generator circuit can be implemented with Schottky devices, offering this nanosecond orthogonality error at clock rates of the order of 50 MHz.

REFERENCES

[1] H. L. Peterson, "Generation of Walsh Functions," *Proc. Symp. Applications of Walsh Functions*, 1970, Washington, pp. 55–57.
[2] P. J. Lebert, "Walsh Function Generation For a Million Different Functions," *Proc. Symp. Applications of Walsh Functions*, 1970, Washington, pp. 52–55.
[3] R. Kitai and K. H. Siemens, "A Hazard-Free Walsh-Function Generator," *IEEE Trans. Instrum. Meas.*, vol. IM-21, Feb. 1972, pp. 80–83.
[4] C. Boesswetter, "Die Erzeugung Von Walsh Funktionen," *NTZ-Communications J.*, vol. 23, April 1970, pp. 201–207.
[5] P. W. Besslich, "Walsh Function Generator For Minimum Orthogonality Error," *IEEE Trans. Electromagn. Compat.*, vol. EMC-15, pp. 177–180, 1973.
[6] C. K. Yuen, "Walsh Function Generation Using Gray Code," *Proc. Symp. Applications of Walsh Functions*, 1973, Washington, pp. 284–289.
[7] K. H. Siemens and R. Kitai, "Digital Walsh-Fourier Analysis of Periodic Waveforms," *IEEE Trans. Instrum. Meas.*, vol. IM-18, 1969, pp. 316–321.

Part II

WALSH TRANSFORM COMPUTATION

Editor's Comments on Papers 4, 5, and 6

4 **CARL and SWARTWOOD**
A Hybrid Walsh Transform Computer

5 **FRANGAKIS and TZAFESTAS**
A Digital Walsh Function Analyser

6 **MUNIAPPAN and KITAI**
Microprocessor-Based Walsh-Fourier Spectral Analyzer

A great deal of research in the Walsh function field has been devoted to the study and development of easy and fast algorithms for computing Walsh transforms (see Selfridge, 1955; Crittenden, 1970). The sampling principle in the framework of Walsh signals and transforms was studied by Kak (1970) and Cheng and Johnson (1973). Their sequency-based sampling theorem states that: "A causal time function $x(t)$ sequency-band limited to $B = 2^k$ zeros per second, can be uniquely reconstructed from its samples at every $T = 1/B$ seconds for all positive time" (Cheng and Johnson, 1973, p. 675). The simplest version of Walsh transform, which provides the coefficients in natural (or Hadamard) ordering, was given by Whechel and Guinn (1968). They actually derived the so-called fast Hadamard transform (FHT) by using the recursive structure of the Hadamard matrix and applying a modified form of the Cooley-Tukey (1965) fast Fourier transform (FFT). In 1969 Shanks provided an efficient Walsh-transform computation algorithm analogous to the Cooley-Tukey FFT algorithm and based on the multiplicative, iterative equation generating the discrete Walsh functions. In 1970 Henderson showed how to circumvent the difficulty of obtaining the Walsh coefficients in bit-reversed order when the data are arranged in the natural forward order. Andrews and Caspari (1970) presented a generalized approach for fast-digital Walsh-Hadamard spectral analysis using the Kronecker-product matrix formulation. Another method for computing the Walsh-Hadamard transform and the R-transform was presented in 1970 by Ulman, where the transform coefficients were produced in increasing sequency order.

A real-time technique for computing the digital Walsh-Hadamard transform of two-dimensional pictures was developed by Alexandridis and Klinger (1972) by using the matrix formulation. A hardware implementation for real-time, parallel Walsh-Hadamard transformation was also provided. In a paper by Manz (1972) another sequency-ordered fast Walsh transform (FWT) computation algorithm was presented that was based on the Cooley-Tukey-type FHT and required an equal computational effort as FHT. The conversion from dyadic to sequency ordering was made by Yuen (1970) via a Gray code to binary decoder that slows down the FWT.

Usually sequency-ordered FWT involves three steps, namely bit reversing, transforming via the FHT, and extra Gray-code ordering by modifying the FHT structure. Fontaine (1970) provided a sequency-ordered transform which, instead of the bit reversing, involves an extra $N\log_2 N$ data transfer and thus does not actually have any computational advantage over Manz's algorithm. A procedure complementary to Manz's algorithm was given by Larsen (1976). Its basic difference is that it has a decimation-in-time structure and accepts data in normal order, returning the coefficients in bit-reversed sequency order. Then after bit reversal the Walsh coefficients are obtained in sequency order. The computational time of this algorithm in a *Zerox Sigma 6* computer was found to be the same as that of Manz's routine (1.16 sec), and much less than the computational time (2.16 sec) of Kunt's (1975) routine. An alternative algorithm was developed by Brown (1977) for computing sequency-ordered FWT using an additive recursion formula, which for an N-dimensional sampled-data set needs $N\log_2 N$ summations and can be applied to spectral analysis of nonstationary signals such as speech signals.

In Paper 4 a special-purpose hybrid-computing circuit for discrete Walsh transform computation is provided, and in Paper 5 a simple digital circuit is given for continuous Walsh function analysis. In Paper 6 a microprocessor-based Walsh-Fourier spectral analyzer is designed, based on an 8-bit microprocessor and 8-bit analog to digital (A/D) converter. Another Walsh spectral analyzer was developed in 1979 by Venkatesam and Prasada Rao. All the the Walsh spectral analyzers just discussed can be used in real-time applications.

REFERENCES

Alexandridis, N. A., and A. Klinger, 1972, Real-Time Walsh-Hadamard Transformation, *IEEE Trans. Comput.* **C-21:**288–292.

Andrews, H. C., and K. L. Caspari, 1970, A Generalized Technique for Spectral Analysis, *IEEE Trans. Comput.* **C-19:**16–25.

Brown, R. D., 1977, A Recursive Algorithm for Sequency-Ordered Fast Walsh Transforms, *IEEE Trans. Comput.* **C-26:**819–822.

Cheng, D. K., and D. L. Johnson, 1973, Walsh Transform of Sampled Time Functions and the Sampling Principle, *IEEE Proc.* **61:**674–675.

Cooley, J. W., and J. W. Tukey, 1965, An Algorithm for the Machine Computation of Complex Fourier Series, *Math. Comput.* **19:**297–302.

Crittenden, R. B., 1970, Walsh-Fourier Transforms, in *Proceedings of the Symposium on Walsh Functions Applications,* U.S. Naval Research Lab, AD707-431, U.S. Defense Documentation Center, Washington, D.C., pp. 170–174.

Fontaine, A. B., 1970, Single Dyadic and Sequency FORTRAN Mechanization of the Fast Walsh Transform, *Natl. Electron. Conf. Proc.* **28:**271–273.

Henderson, K. W., 1970, Comment on "Computation of the Fast Walsh-Fourier Transform," *IEEE Trans. Comput.* **C-19:**850.

Kak, S. C., 1970, Sampling Theorem in Walsh-Fourier Analysis, *Electron. Lett.* **6:**447–448.

Kunt, M., 1975, On Computation of the Hadamard Transform and the R-Transform in Ordered Form, *IEEE Trans. Comput.* **C-24:**1120–1121.

Larsen, H., 1976, An Algorithm to Compute the Sequency Ordered Walsh Transform, *IEEE Trans. Acoust. Speech, Signal Process.* **24:**335–336.

Manz, J. W., 1972, A Sequency-Ordered Fast Walsh Transform, *IEEE Trans. Audio Electroacoust.* **AU-20:**204–205.

Selfridge, R. G., 1955, Generalized Walsh Functions, *Pacific J. Math.* **5:**451–480.

Shanks, J. L., 1969, Computation of the Fast Walsh-Fourier Transform, *IEEE Trans. Comput.* **C-18:**457–459.

Ulman, L. J., 1970, Computation of the Hadamard Transform and the R-Transform in Ordered Form, *IEEE Trans. Comput.* **C-19:**359–360.

Venkatesam, S., and G. Prasada Rao, 1979, A Walsh Spectral Analyzer, *Int. J. Electron.* **46:**413–415.

Whechel, J. E., Jr., and D. F. Guinn, 1968, The Fast Fourier-Hadamard Transform and Its Use in Signal Representation and Classification, in *EASCON'68 Rec.*, pp. 561–573.

Yuen, C., 1970, Walsh Functions and Gray Code, in *Proceedings of the Symposium on Walsh Functions Applications,* U.S. Naval Research Lab, AD707-431, U.S. Defense Documentation Center, Washington, D.C., pp. 68–73.

4

A Hybrid Walsh Transform Computer

JOSEPH W. CARL AND RICHARD V. SWARTWOOD

Abstract—**A hybrid special-purpose computer based on a recursive algorithm for discrete Walsh transform computations is described. The device uses feedback in time to reduce the required number of summing junctions to N, the number of data points to be transformed. Computations are completed in 700 μs when $N = 256$.**

Index Terms—**Hybrid computer, Kronecker product, special-purpose computer, Walsh transform.**

Introduction

THE discrete Walsh transform (DWT), or Hadamard transform, has found application in many areas, including signal processing, pattern recognition, and communication theory (see [1], [2], and [6]). Much of the interest in this transform results from the computation advantages it offers over the more conventional Fourier transform, and from the simplicity of solid-state electronic hardware transform computers or hardware Walsh filters [7], [8]. The advantages of DWT hardware compared to discrete Fourier transform hardware have been pointed out quite effectively by Wishner [13].

The purposes of this paper are: 1) to present an often overlooked DWT algorithm that leads to an efficient hardware implementation; and 2) to describe a hybrid transform computer that has been constructed by the Air Force Institute of Technology and the Aerospace Medical Research Laboratory.

Manuscript received April 26, 1972; revised September 1, 1972. This work was supported in part by the Aerospace Medical Research Laboratory and in part by the Air Force Institute of Technology. The detailed circuit design was accomplished under Contract F33601-71-0765 by Systems Research Laboratories, Inc., Dayton, Ohio.

J. W. Carl is with the Mathematics and Analysis Branch, 6570 Aerospace Medical Research Laboratory, Wright-Patterson Air Force Base, Ohio 45433.

R. V. Swartwood is with the Support Branch, 6514 Test Squadron, Edwards Air Force Base, Calif. 93523.

Recursive Algorithm

The simplest method of defining the DWT, for our purposes, is in matrix notation. Let $N = 2^p$, where p is a positive integer. Then the Kronecker product of p matrices of the form

$$W_2 = \begin{bmatrix} 1 & 1 \\ 1 & -1 \end{bmatrix}$$

is the Walsh matrix of order N in "natural" order. For example,

$$W_8 = \begin{bmatrix} 1 & 1 \\ 1 & -1 \end{bmatrix} \times \begin{bmatrix} 1 & 1 \\ 1 & -1 \end{bmatrix} \times \begin{bmatrix} 1 & 1 \\ 1 & -1 \end{bmatrix}$$

$$= \begin{bmatrix} 1 & 1 & 1 & 1 \\ 1 & -1 & 1 & -1 \\ 1 & 1 & -1 & -1 \\ 1 & -1 & -1 & 1 \end{bmatrix} \times \begin{bmatrix} 1 & 1 \\ 1 & -1 \end{bmatrix}$$

$$= \begin{bmatrix} 1 & 1 & 1 & 1 & 1 & 1 & 1 & 1 \\ 1 & -1 & 1 & -1 & 1 & -1 & 1 & -1 \\ 1 & 1 & -1 & -1 & 1 & 1 & -1 & -1 \\ 1 & -1 & -1 & 1 & 1 & -1 & -1 & 1 \\ 1 & 1 & 1 & 1 & -1 & -1 & -1 & -1 \\ 1 & -1 & 1 & -1 & -1 & 1 & -1 & 1 \\ 1 & 1 & -1 & -1 & -1 & -1 & 1 & 1 \\ 1 & -1 & -1 & 1 & -1 & 1 & 1 & -1 \end{bmatrix}.$$

Now, let $[F]$ be a column vector of dimension N that represents sampled values of an input signal, and let the column vector of dimension N, $[A]$, represent the desired DWT of $[F]$. Then $[A]$ is defined by the matrix equation

$$[A] = W_N\,[F]. \tag{1}$$

The brute force computation of the Walsh coefficients according to (1) requires $N(N-1)$ additions or subtractions. However, Good [5] developed a matrix factorization technique that leads to a fast transform algorithm. Good's technique can be used to factor matrices derived from Kronecker products into p factors of order N, where N is the order of the original matrix, and p is the number of matrices in the Kronecker product. If the original matrix is the result of the Kronecker product of p identical matrices, as is the case for the Walsh matrices, then the p factors will also be identical.

The result of Good's factorization technique applied to Walsh matrices has the following form ($N = 8$).

$$\begin{bmatrix} 1 & 1 & 1 & 1 & 1 & 1 & 1 & 1 \\ 1 & -1 & 1 & -1 & 1 & -1 & 1 & -1 \\ 1 & 1 & -1 & -1 & 1 & 1 & -1 & -1 \\ 1 & -1 & -1 & 1 & 1 & -1 & -1 & 1 \\ 1 & 1 & 1 & 1 & -1 & -1 & -1 & -1 \\ 1 & -1 & 1 & -1 & -1 & 1 & -1 & 1 \\ 1 & 1 & -1 & -1 & -1 & -1 & 1 & 1 \\ 1 & -1 & -1 & 1 & -1 & 1 & 1 & -1 \end{bmatrix} = \begin{bmatrix} 1 & 1 & 0 & 0 & 0 & 0 & 0 & 0 \\ 0 & 0 & 1 & 1 & 0 & 0 & 0 & 0 \\ 0 & 0 & 0 & 0 & 1 & 1 & 0 & 0 \\ 0 & 0 & 0 & 0 & 0 & 0 & 1 & 1 \\ 1 & -1 & 0 & 0 & 0 & 0 & 0 & 0 \\ 0 & 0 & 1 & -1 & 0 & 0 & 0 & 0 \\ 0 & 0 & 0 & 0 & 1 & -1 & 0 & 0 \\ 0 & 0 & 0 & 0 & 0 & 0 & 1 & -1 \end{bmatrix}^3 \tag{2}$$

This result has been previously reported by Andrews and Caspari [4], and by Yuen [15], as well as Good. Since the factors of the Walsh matrices are rich in zeros, the number of required computations for the DWT is reduced to $N \log_2 N$. The preceding factorization results in a flow diagram for a computation algorithm that has the form of Fig. 1. Note the recursive structure of the algorithm: the interconnections of successive layers are identical. The recursive structure of this algorithm can be formalized in the following way.

Let $f(n)$ be a real function of the positive integers less than or equal to $N = 2^p$, where p is a positive integer. The DWT (in natural order) of $f(n)$ is obtained by

$$y_0(n) = f(n)$$

$$y_{i+1}(n) = \begin{cases} y_i(2n-1) + y_i(2n), & \text{for } 0 < n \leqslant N/2 \\ y_i(2n-N-1) - y_i(2n-N), & \text{for } N/2 < n \leqslant N \end{cases}$$

$$\text{DWT}(n) = y_p(n).$$

These results are easily extended to higher dimensions, but it is simpler to note that higher order transforms can be expressed as one-dimensional transforms with the inputs and outputs relabeled.

This recursive algorithm has been programmed on a Digital Equipment Corporation PDP-12 computer in Linc-mode assembly language using fixed-point arithmetic. The program requires a data storage area, a working area half the size of the data area, and 45 locations for the instructions. A 256-point DWT takes 71 ms to execute once the data is in core memory.

A prototype DWT special-purpose computer has been built to demonstrate the feasibility of speeding up DWT computations by a factor of more than 100. This prototype will now be described.

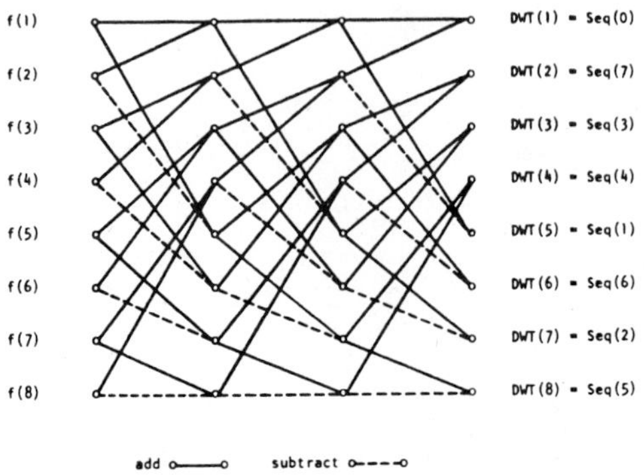

Fig. 1. Flow diagram for the recursive algorithm ($N = 8$), Seq (i) = "ordered" form.

Hybrid Walsh Transform Computer

A 256-point DWT was selected for hardware implementation because it has practical application, but is not too large for a prototype effort. The device was labeled to conform to a two-dimensional 16 X 16 configuration.

Three methods of implementation are available: analog, digital, and hybrid. An analog implementation was desired due to the speed, efficiency, and flexibility of analog circuits. (For example, in a bioengineering application it might be useful to investigate the effect of changing the gains of certain amplifiers, and thereby evaluate the discrepancies from an ideal that could be tolerated, say, in a model of biological function.) However, an entirely analog implementation of the recursive algorithm would require 2048 amplifiers if the interconnections were of the form shown in Fig. 2. This number could be reduced by increasing the number of inputs at each summing junction, but a fan-in fan-out problem quickly arises. Another way to reduce the number of summing junctions is to exploit the recursive nature of the algorithm. If each summing junction is used $\log_2 N$ times through a feedback loop, the number of amplifiers is reduced to 256 when each amplifier has two inputs. This feedback requires sample-and-hold circuitry at the input and output of each amplifier, and digital logic to control the sample-and-hold circuits. The realization is therefore a hybrid device. Note that an entirely digital device would also be an efficient realization.

The prototype DWT hardware was designed to be compatible with a flying-spot scanning system and with a PDP-12 general-purpose digital computer. Therefore, sample-and-hold circuits are already required at the input to each amplifier in the first level of computation to perform serial-to-parallel conversion of the video samples from the flying-spot scanner. Thus, the savings in the number of required amplifiers comes at a cost of sample-and-hold circuits at the output of each amplifier, some feedback clocking logic, and the time required for the feedback.

The block diagram of the hybrid realization is shown in Fig. 2. It implements the recursive algorithm described earlier in a slightly modified form. The modification consists of expressing the recursive factor matrix (for $N = 8$) as

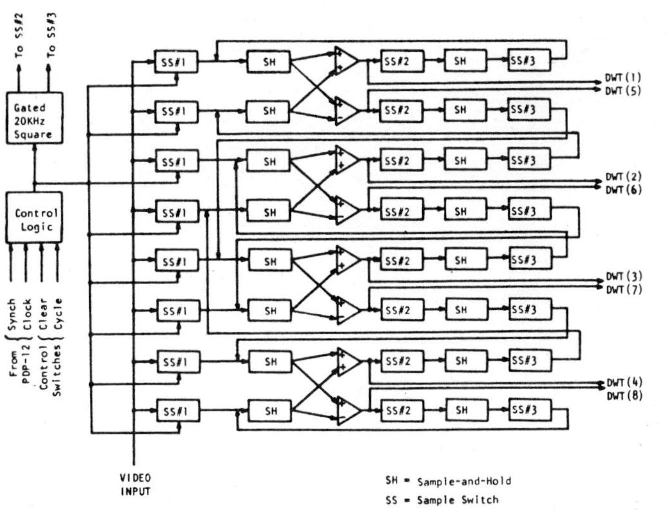

Fig. 2. Block diagram of the hybrid computer for $N = 8$.

$$\begin{bmatrix} 1 & 1 & 0 & 0 & 0 & 0 & 0 & 0 \\ 0 & 0 & 1 & 1 & 0 & 0 & 0 & 0 \\ 0 & 0 & 0 & 0 & 1 & 1 & 0 & 0 \\ 0 & 0 & 0 & 0 & 0 & 0 & 1 & 1 \\ 1 & -1 & 0 & 0 & 0 & 0 & 0 & 0 \\ 0 & 0 & 1 & -1 & 0 & 0 & 0 & 0 \\ 0 & 0 & 0 & 0 & 1 & -1 & 0 & 0 \\ 0 & 0 & 0 & 0 & 0 & 0 & 1 & -1 \end{bmatrix} = \begin{bmatrix} 1 & 0 & 0 & 0 & 0 & 0 & 0 & 0 \\ 0 & 0 & 1 & 0 & 0 & 0 & 0 & 0 \\ 0 & 0 & 0 & 0 & 1 & 0 & 0 & 0 \\ 0 & 0 & 0 & 0 & 0 & 0 & 1 & 0 \\ 0 & 1 & 0 & 0 & 0 & 0 & 0 & 0 \\ 0 & 0 & 0 & 1 & 0 & 0 & 0 & 0 \\ 0 & 0 & 0 & 0 & 0 & 1 & 0 & 0 \\ 0 & 0 & 0 & 0 & 0 & 0 & 0 & 1 \end{bmatrix} \begin{bmatrix} 1 & 1 & 0 & 0 & 0 & 0 & 0 & 0 \\ 1 & -1 & 0 & 0 & 0 & 0 & 0 & 0 \\ 0 & 0 & 1 & 1 & 0 & 0 & 0 & 0 \\ 0 & 0 & 1 & -1 & 0 & 0 & 0 & 0 \\ 0 & 0 & 0 & 0 & 1 & 1 & 0 & 0 \\ 0 & 0 & 0 & 0 & 1 & -1 & 0 & 0 \\ 0 & 0 & 0 & 0 & 0 & 0 & 1 & 1 \\ 0 & 0 & 0 & 0 & 0 & 0 & 1 & -1 \end{bmatrix}. \quad (3)$$

This slight modification allows the grouping of channels into pairs on the circuit boards, and results in reducing by half the number of signal interconnections between boards.

The prototype DWT computer was designed to accept a maximum data rate of 20 kHz, which is well above the upper limit of the available flying-spot scanner. The maximum data rate is determined by the sampling time of the input sample-and-hold circuits (50 μs in this device). The lowest anticipated data rate is 1 kHz, and this imposes a constraint on the maximum holding time of the input sample-and-hold circuits of 255 ms. These constraints were met by all the input sample-and-hold circuits so that the circuit boards would be interchangeable. Even with the additional constraint that the sample-and-hold circuits not droop more than 5 percent, the design criteria are not too strict, and relatively inexpensive circuitry can be used.

When the feedback operation proceeds at a rate of 20 kHz, the output sample-and-hold circuits need only hold for 50 μs. This short holding time allows considerable simplification in the output sample-and-hold circuits.

The only other major design consideration is the serial-to-parallel conversion of the video signal from the flying-spot scanner. This was accomplished by using a 256-bit shift register, with each bit used to gate an input sampling switch. When a "1" is entered into the register and shifted through it, the input switches are gated sequentially. A similar method can be used to convert the output back to a serial stream. This was not done in this prototype. Rather, the DWT coefficients are brought out in parallel, and fed to a 24-channel multiplexed analog-to-digital converter in the PDP-12. This arrangement is completely satisfactory for the application intended for the prototype.

Once these design considerations are complete, the functional operation of the device is easy to understand. In the block diagram of Fig. 2, the control logic and the bank of sample switches labeled SS #1 implement the serial-to-parallel conversion. The input bank of sample-and-hold circuits has a 50-μs sampling time, but a 255-ms holding time. The sums and differences of the sampled input appear at the output of the operational amplifiers as soon as the last video datum is sampled. The multiplication of the input vector by the right-most matrix appearing in (3) is now complete. This result is transferred to the output sample-and-hold circuits by gating the bank of sample switches labeled SS #2 on. (Note that this intermediate result also appears at the output of the device, but this is of no consequence.) The permutation required by the middle matrix of (3) is accomplished by routing the outputs of the output sample-and-hold circuits to the appropriate input sample-and-hold circuits. The permuted result is transferred back to the input sample-and-hold circuits by gating the SS #3's on. Since the sample-and-hold circuits all have a 50-μs sampling time, the switches SS #2 and SS #3 must be on for 50 μs at the proper time. This is accomplished by the 20-kHz square-wave generator that feeds SS #2 and SS #3. When SS #2 is on, SS #3 is off, and conversely, when SS #2 is off, SS #3 is on. One cycle of the square wave implements a multiplication by the factor matrix of (2) in the form shown in (3).

Note that the SS #1's do not enter the DWT computation; they merely provide the serial-to-parallel conversion. Also note that the desired output of the device appears after the seventh

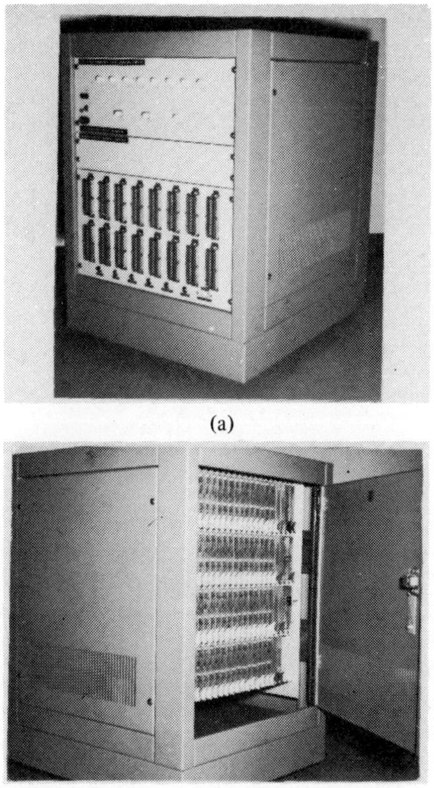

(a)

(b)

Fig. 3. The prototype transform computer. (a) Front. (b) Back.

iteration; this is 700 μs after the last datum is sampled. Since this output must be taken from the sample-and-hold circuits that have the long (255 ms) holding time, one permutation–the final one–is not done. The outputs are a permutation of the usual "natural" order of the DWT coefficients; this permutation is also indicated in Fig. 2 by the labels on the output leads.

The physical realization of the device is pictured in Fig. 3.

DISCUSSION

A hardware special-purpose DWT computer has been described that achieves a speedup factor of more than 100 compared to computations in the PDP-12. Even more important, the PDP-12 is freed to perform other functions during the transform computation. The device is much more efficient in terms of size, weight, and power requirements than a brute force device, or even a device with $N \log_2 N$ summing junctions, and it compares favorably in these respects to other proposed DWT hardware devices [3], [13].

ACKNOWLEDGMENT

The authors wish to thank Dr. T. G. Purnhagen, who pointed out the permutation that resulted in reduced interconnections between boards, and who provided many helpful discussions. They also wish to thank M. Kabrisky, C. F. Hall, R. Durham, and C. Skinn for their assistance.

REFERENCES

[1] *Proc. 1970 Symp. Appl. Walsh Functions*, Naval Res. Lab., Washington, D.C., ASTIA AD707,431.

[2] *Proc. 1971 Symp. Appl. Walsh Functions*, Naval Res. Lab., Washington, D.C., ASTIA AD727,000.

[3] N. A. Alexandridis and A. Klinger, "Real-time Walsh-Hadamard transformation," *IEEE Trans. Comput.* (Short Notes), vol. C-21, pp. 288-292, Mar. 1972.

[4] H. C. Andrews and K. L. Caspari, "A generalized technique for spectral analysis," *IEEE Trans. Comput.*, vol. C-19, pp. 16-25, Jan. 1970.

[5] I. J. Good, "The interaction algorithm and practical Fourier analysis," *J. Roy. Statist. Soc.* (London), vol. B20, 1958.

[6] H. F. Harmuth, *Transmission of Information by Orthogonal Functions*. New York: Springer, 1969.

[7] —, "Sequency filters based on Walsh functions for signals with two space variables," in *Proc. 4th Hawaii Int. Conf. Syst. Sci.*, Univ. Hawaii, Honolulu, 1971, pp. 414-416.

[8] —, "Survey of analog sequency filters based on Walsh functions," in *Proc. 1970 Symp. Appl. Walsh Functions*, Naval Res. Lab., Washington, D.C., ASTIA AD707,431, pp. 208-219.

[9] W. K. Pratt, J. K. Kane, and H. C. Andrews, "Hadamard transform image coding," *Proc. IEEE*, vol. 57, pp. 58-68, Jan. 1969.

[10] J. L. Shanks, "Computation of the fast Walsh-Fourier transform," *IEEE Trans. Comput.* (Short Notes), vol. C-18, pp. 457-459, May 1969.

[11] R. V. Swartwood, "A two-dimensional Walsh transform computer," M.S. thesis, Air Force Inst. Technol., Wright-Patterson AFB, Ohio, GE/EE/72-25, Mar. 1972.

[12] J. E. Whetchel and D. F. Guinn, "The fast Fourier-Hadamard transform and its use in signal representation and classification," in *EASCON 1968 Rec.*, pp. 561-573.

[13] H. D. Wishner, "Designing a special-purpose digital image processor," *Comput. Design*, vol. 11, pp. 71-76, Feb. 1972.

[14] F. Yates, *The Design and Analysis of Factorial Experiments*. Harpenden: Imperial Bureau of Soil Analysis, 1937.

[15] C. Yuen, "Some computational techniques in the applications of Walsh functions and their relation to Gray code ordering," in *Proc. 1971 Symp. Appl. Walsh Functions*, Naval Res. Lab., Washington, D.C., ASTIA AD727,000, pp. 68-73.

5

Reprinted from *Electron. Eng.*, July, pp. 89–91 (1979)

A DIGITAL WALSH FUNCTION ANALYSER

G. Frangakis and S. Tzafestas

Walsh functions have wonderful analogue-digital properties and are currently being used in a variety of engineering applications, which include two-dimensional image processing, sequence filtering, speech synthesis, simulation, and control. In this framework many attempts have been made to design special type computers for evaluating the Walsh transforms of signals and images. The present contribution provides a fully digital Walsh function analyzer, which accepts the signal in continuous-time form, and gives the Walsh coefficients in binary form.

The Walsh functions constitute a complete set of two-valued orthonormal functions $\Phi_k(t)$, $k=0,1,\ldots, n-1$, $n=2^m$ in the interval t (0,1), and they can be defined in several equivalent ways.[10] According to the Paley definition, they are formed by means of the Rademacher functions $R_0(t)=1$, $R_s(t)$, $s=1,2,\ldots,m$, as:

$$\Phi_0(t)=R_0(t), \Phi_k(t)=R_m{}^m(t)R_{m-1}^{m-1}(t) \quad R_1(t)$$

for $k=0,1,2,\ldots,n-1 (n=2^m)$, where $_{m\,m-1\,\cdots\,1}$ is the 8-4-2-1 binary expression of the decimal number k. The Rademacher functions are a set of square waves for t(0,1), of unit height and repetition rate (sequency) equal to 2^m, which can be generated by a BCD 8-4-2-1 counter. The Rademacher and Walsh-Paley functions for $n=8 (m=3)$ are shown in Fig. 1.

A continuous-time signal x(t), which is absolutely integrable in (0,1), has the Walsh series expansion

$$x(t) = {}^{n-1}_{k=0}\, a_k \Phi_k(t), a_k = {}^1{}_0 x(t)\Phi_k(t)dt,$$

which, in discrete time takes the form

$$x_i=\sum_{k=0}^{n-1} a_k\Phi_{ki},\quad a_k=\frac{1}{n}\sum_{i=0}^{n-1} x_i\Phi_{ki},$$

where $\overline{x}_i$ is the mean value of x(t) in the ith subinterval, and Φ_{ki} is the value of $\Phi_k(t)$ in the same subinterval. Some typical Walsh series expansions are the following:

The authors are at the University of Patras in Greece.

Ramp signal: $x(t)=t$

$$x(t)=t=\tfrac{1}{2}\Phi_0(t)-\tfrac{1}{4}\Phi_1(t)-\tfrac{1}{8}\Phi_2(t)-\tfrac{1}{16}\Phi_4(t)-\tfrac{1}{32}\Phi_8(t)-\tfrac{1}{64}\Phi_{16}(t)+\ldots \quad (3)$$

Triangular signal: $x(t)=t, t\epsilon[0,\tfrac{1}{2})$, $x(t)=1-t$, $t\epsilon[\tfrac{1}{2},1)$

$$x(t)=\tfrac{1}{4}\Phi_0(t)-\tfrac{1}{8}\Phi_3(t)+\ldots \quad (4)$$

Gate signal: $x(t)=0, t\epsilon[0,\tfrac{1}{4})$, $t\epsilon[\tfrac{1}{2},1)$, $x(t)=1$, $t\epsilon[\tfrac{1}{4},\tfrac{1}{2})$

$$x(t)=\tfrac{1}{4}\Phi_0(t)+\tfrac{1}{4}\Phi_1(t)-\tfrac{1}{4}\Phi_2(t)-\tfrac{1}{4}\Phi_3(t) \quad (5)$$

Sinusoidal signal: $x(t)=\sin(\pi t)$

Fig. 1: Rademacher and Walsh functions for m=3 and $n=2_m=8$

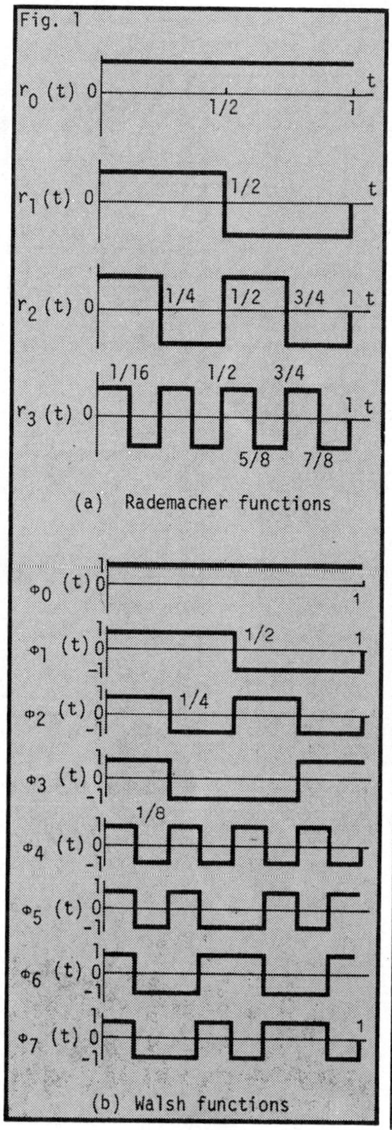

$$x(t)=0.637\Phi_0(t)-0.263\Phi_3(t)-0.126\Phi_5(t)-0.0573\Phi_6(t)+\ldots (6)$$

The structure of the present digital circuit is shown in Fig. 2. The signal x(t), to be analyzed is Walsh series, is fed to a sample-hold and linear voltage-to-frequency converter combination, the output of which is passed to the clock lines of the up-down counters. The outputs of the Walsh generator (i.e. the Walsh functions) are fed to the Up/Down control lines of the counters. This actually corresponds to the multiplication of x(t) and $\Phi_i(t)$, and time integration of the product, as required by eqn. 1. The Walsh coefficients are given by the contents of the counters at the end of the expansion period. A description of the functioning of each unit follows.

A parallel Walsh-Paley generator fo $n=4$ is shown in Fig. 3.

Sample and Hold

For the sample and hold operation the arrangement of Fig. 4 is used. The signal x(t) is sampled at the middle of each time subinterval (of duration 1/n) of the Walsh train, and its value is held up to the end of this subinternal, where it is reset to zero. At the middle of the next subinterval a new value of x(t) is sampled, and so on (see Fig. 6). In the present case the Motorola MC14016 Analogue switch and the MC1456 operational amplifier were used.

V-F Converter

This circuit converts a voltage V to a pulse train of frequency f. This conversion must be linear, with no bias. In our case the universal DATEL VFV converter was employed, which is available in two modes, namely VFV-10kHz for slow varying inputs, and VFV-100kHz for fast frequencies. The number N_t of output pulses, obtained up to time 5, is equal to

$N_t=KVt$

where V is the input voltage, and K is a constant, which takes the value K=1 for n=4, K=2 for n=8, etc. It is remarked that special care is required for adjusting the VF converter.

For example, we may adjust it

such that with $V=V_{max}$ (the maximum value of x(t)), the number of output pulses at the end of the expansion time t=T=1 is $N_T=64$ for K=1 (n=4), $N_T=128$ for K=2 (n=8), and so on.

Up-Down Counters

The length L of each counter depends on K and is determined by $s_{L-}{}^{1}=64K$, i.e. $L=7+\log_2 K$. Thus, for K=1 we need L=7 bits, the most significant bit (MSB) of which is used as sign bit, for K=2 we have L=8, etc. If MSB+0, the binary number is positive, and if MSB=1 then the binary number is negative (2's complement arithmetic). The output N_t of VF converter is driven to the clock channel of each counter and the Walsh function generator outputs $\Phi_0,\Phi_1,\Phi_2,\ldots$ are connected to the control channels of the corresponding up-down counters. When Φ_k+1 (logic 1) the counter counts up the N_t pulses, whereas when $\Phi_k=-1$ (logic 0) it counts down. At the end of the normalized time period (0,1), the content of counter C_k gives the Walsh coefficient a_k (k=0,1, ... n−1). In our case the MOTOROLA 14516 up-down counter was used.

Control Unit

This unit provides all required control signals. The START signal resets all counters as well as the Walsh Generator flip-flops. The END signal is the overflow of the counter of the Walsh generator. The sample and reset command signals are taken from the outputs of a flip-flop Q via monostables (Fig. 5). The output Q of this flip-flop also drives the Walsh generator.

Examples

Ramp Signal (Fig. 6): Consider first the Walsh expansion with four terms (n=4), in which case K=1 (eqn. 7) and L=7. The VF converter gives one pulse in the first subinterval (O,T/4) since $x(T/8)=V_{max}/8$ and $N_{T\,4}=(V_{max}/8)\ (T/4-T/8)=V_{max}T/64=1$. At the second subinterval (T/4,T/2) the VF converter gives $N_{T\,2}=(3V_{max}/8)\ (T/2-3T/8)=3V_{max}T/64=3$ pulses, at the third subinterval gives $N_{3T\,4}=(5V_{max\,8})\ (3T/4-5T/8)=5$ pulses, and at the final subinterval gives $N_T=(7V_{max\,8})\ (T-7T/8)=7$ pulses. Due to the form of $\Phi_0(t)$ (Fig. 1) the counter C_0 counts always up and so during (0,t) it counts 16 pulses and its binary content is 0010000. The counter C_1 counts up 4 pulses (from 0 to T/2) and down 12 pulses (from T/2 to T) and so its total result is −8 pulses, i.e. its content is 1111000. Similarly, the counter C_2 counts up 1+7=8 pulses and down 3+5=8 pulses, i.e. total result zero pulses or 0000000.

Now, due to the sample and hold scheme used, the VF converter emits pulses only at half of the time T, and the 16 pulses of C_0 correspond to 1/2 (16-1/2). Hence

$a_0=\frac{1}{2}$ (i.e. 16 pulses),

$a_1=\frac{1}{4}$ (i.e. −8 pulses),

$a_2=-\frac{1}{8}$ (i.e. −4 pulses),

$a_3=0$ (i.e. 0 pulses).

Fig. 2: Structure of digital Walsh analyser
Fig. 3: Walsh-Paley generator
Fig. 4: Sample and hold circuit

In Fig. 6 the ramp function, the pulses N_t, and all control signals are shown.

Now, support that n=8 in which case K=2, L=8, and the number of pulses, which the VF converter would give, with $x(t)=V_{max}=1$ for t (0,T=1), is 128. Hence:

$a_0=\frac{1}{2}$ (64 pulses,

$a_2=-\frac{1}{8}$ (−16 pulses),

$a_3=a_5=a_6=a_7=0$ (0 pulses)

$a_1=-\frac{1}{4}$(−32 pulses),

$a_4=-_{16}$

Clearly in this case a Walsh generator with three flip-flops must be used.

Triangular Signal: This case is depicted in Fig. 7 for n=4. The con-

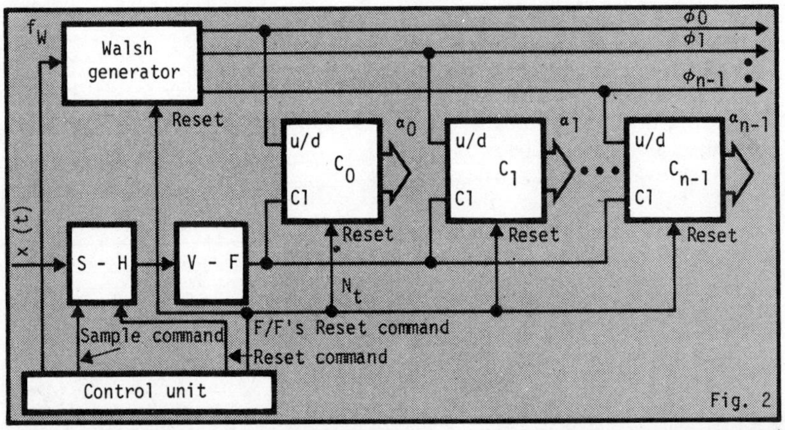

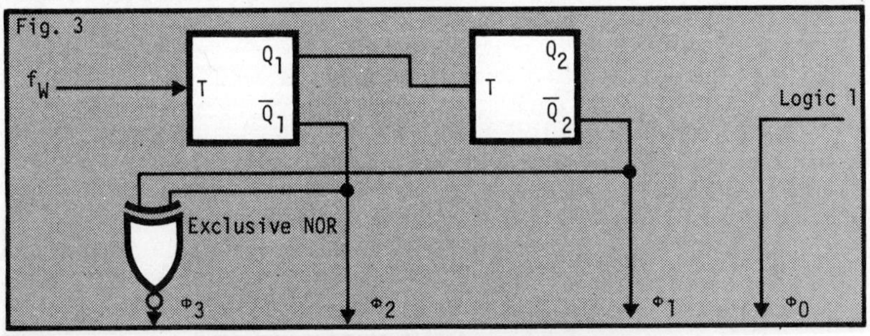

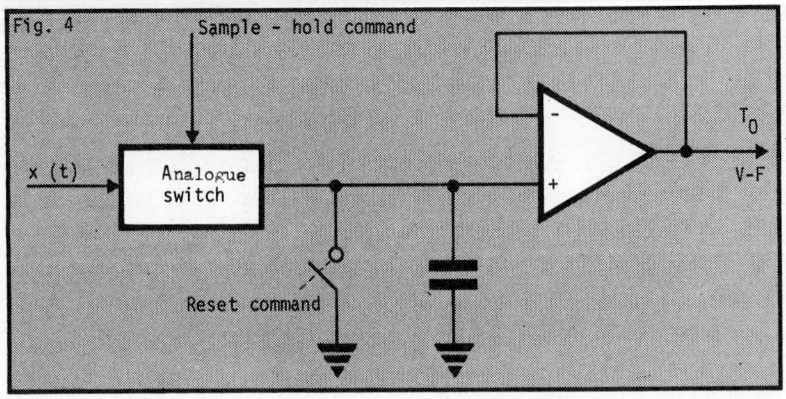

tents of the four up-down counters at the end of the period T are (in 2's complement form):

$C_0=0001000$,
$C_1=0000000$
$C_2=0000000$,
$C_3=1111100$

Hence, since in this case 16 pulses correspond to 1/2, we obtain $a_0=\frac{1}{4}$, $a_1=a_2=0$, $a_3=\frac{1}{8}$

Gate Signal: This case is illustrated in Fig. 8 for n=4. At the end of the period T, the counters contain: $C_0=0001000$, $C_1=0001000$, $C_2=1111000$, $C_3=1111000$. Hence $a_0=\frac{1}{4}$, $a_1=\frac{1}{4}$ $a_2=-\frac{1}{4}$ and $a_3=-\frac{1}{4}$

Sinusoidal Signal: This is a signal which is a nonlinear function of time. For n=8, (K=2, and L=8) we obtain the situation of Fig. 9., where the numbers 3,9,13, etc at the line of N_t represent the numbers of pulses emitted by the VF at the corresponding subintervals. The contents of the eight counters at the end of the expansion period are:

$C_0=3+9+13+15+15+13+9+3=80$
$C_1=3+9+13+15-(15+13+9+3)=0$
$C_2=(3+9)-(13+15)+(15+13)-(9+3)=0$
$C_3=(3+9)-(13+15+15+13)+(9+3)=-32$
$C_4=3-9+13-15+15-13+9-3=0$
$C_5=3-9+13-15-15+13-9+3=-16$
$C_6=3-(9+13)+(15+15)-(13+9)+3=-8$
$C_7=3=(9+13)+15-15+(13+9)+3=0$

Taking into account that for K=2, $N_T=128$ the above figures give $C_0=0.625$, $C_1=0$, $C_2=0$, $C_3=-0.25$, $C_4=0$, $C_5=-0.125$, $C_6=-0.0625$, $C_7=0$.

The circuit can be readily extended to compute the Walsh transform of vectorial signals $x(t)=x_1(t), x_2(t), \ldots, x_p(t)_T$.

The main restriction in using it, is due to the finite duration of the VF converter pulses, which may limit the subdivision of T into as many subintervals as desired. As we see in the examples, for signals which are piece-wise linear functions of t, the resulting coefficients coincide with those obtained analytically by using eqn.1 (or the approximate eqn.2). This is because the sampled value of x(t) at the middle of each subinterval coincides with the mean value x_i in this subinterval. In the general case, it yields coefficient values which are approximations to the exact ones. □

Fig. 5: Structure of control unit

Fig. 6: Analyser operation in ramp signal case

Fig. 7: Triangular signal case

Fig. 8: Gate signal case

Fig. 9: Sinusoidal signal case

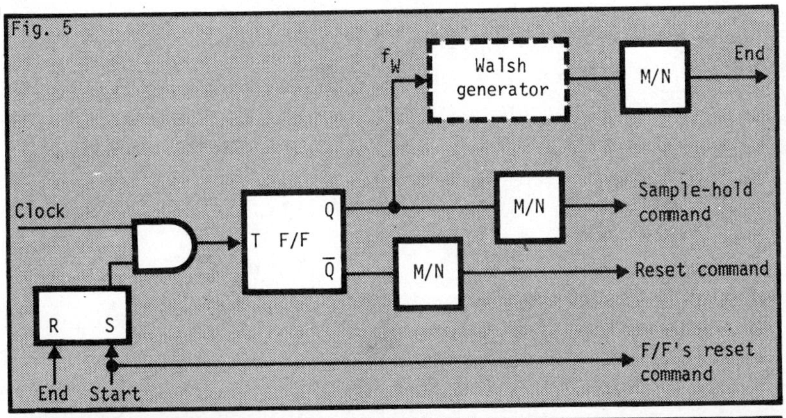

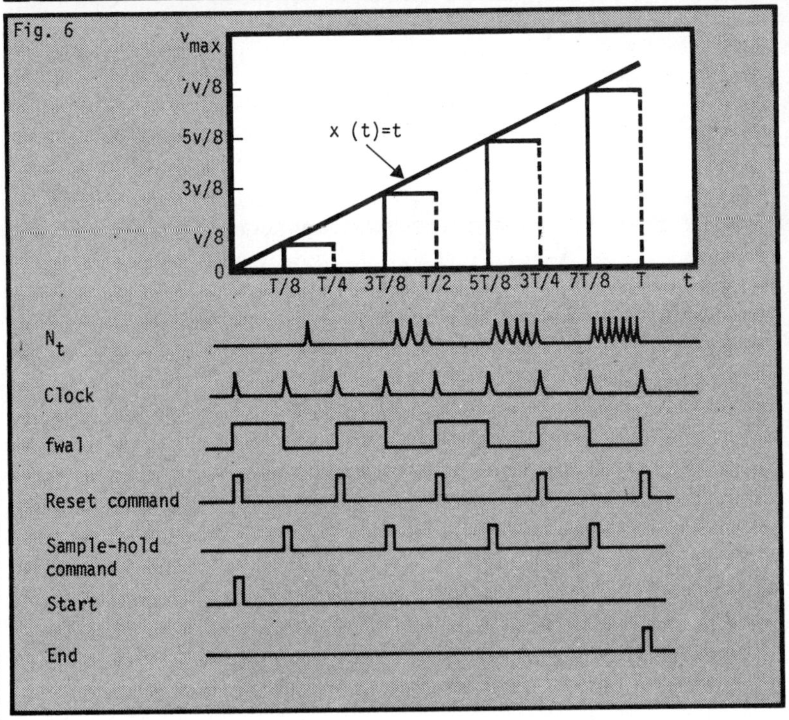

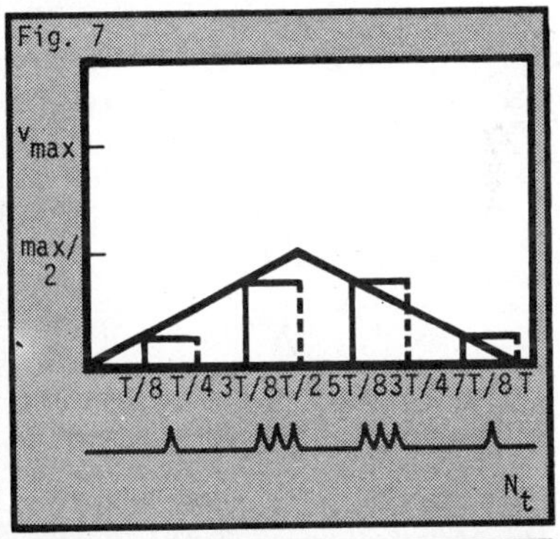

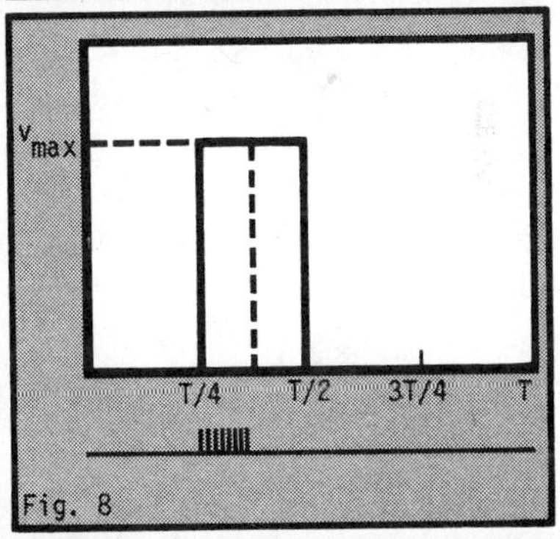

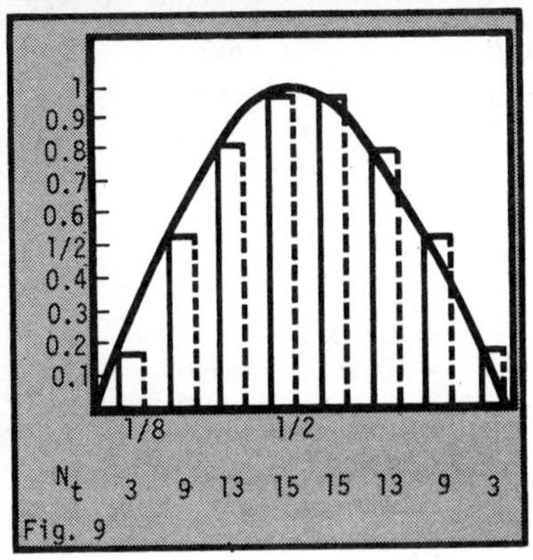

6

Reprinted with permission from *IEEE Trans. Instrum. Meas.* **IM-28**:295–299 (1979)

Microprocessor-Based Walsh–Fourier Spectral Analyzer

K. MUNIAPPAN AND REUVEN KITAI, SENIOR MEMBER, IEEE

Abstract—**A microprocessor-based Walsh–Fourier spectral analyzer is described. It includes the sampling of the incoming signal at 64 times the input signal frequency, using a special purpose frequency multiplier module (FMM), storing the digital data in a permuted sequence in the system memory under the control of a direct memory access (DMA) controller, and performing a fast Walsh–Hadamard transform (FWHT) of the permuted data sequence. The system uses a single board computer SBC80/10 in an Intel System 80/10 and a special purpose board which includes FMM, DMA, and A/D conversion circuits. Fourier coefficients are obtained via a Walsh to Fourier conversion algorithm; the total process is then faster than the Cooley–Tukey FFT algorithm for a data length of 64 or less.**

Introduction

THE LITERATURE abounds with algorithms and machine designs for the computation of the discrete Fourier transform [1]–[3], and digital spectral analyzers that are based on fast Fourier transform (FFT) algorithms are commercially available [4], [5]. While a Fourier analyzer permits the characterization of signals and systems in the frequency domain, an alternative representation in terms of bivalued orthogonal Walsh functions has led to the Walsh spectral analyzer (WSA) which finds application for the following reasons.

1) It can be efficiently implemented on either general-purpose or special-purpose computers due to the binary nature of Walsh functions.

2) In certain applications, e.g., image processing [6], fast and computationally efficient Walsh analysis may be preferred over Fourier analysis.

3) For band-limited signals, a measured Walsh spectrum may be transformed to yield the corresponding Fourier spectrum. The total process of Walsh spectrum measurement followed by conversion to Fourier spectrum is faster than the Cooley–Tukey FFT algorithm in situations where up to 64 data samples are used [17]. For data lengths greater than 64, this method is superior to the FFT method in applications where the number L of Fourier components is relatively small compared with N. The number of multiplications in that case is approximately NL/6. A 64-point data analysis suffices in many applications such as amplifier distortion measurement, biomedical signal analysis, power-system spectral measurements, and some areas of pattern recognition.

A set of N Walsh functions may be written as

$$\{\text{Wal}\,(k,t)\}_{k=0}^{N-1}$$

where $N = 2^m$, m is an integer, and k is the index [12]. Fig. 1 shows the set for $m = 3, 0 \leq t < 1$. Walsh functions permit the characterization of signals in the sequency domain. Sequency is defined as half the number of sign changes in the unit time interval, and is analogous to the frequency of sinusoidal functions. A series expansion of a signal in either domain converges to the signal in the least mean-square-error sense.

The purpose of the WSA is to compute the Walsh coefficients $\{B(k)\}_{k=0}^{N-1}$ of a signal $x(t)$ defined in a finite interval zero to T as the inner products of a set of orthogonal Walsh functions and the signal by

$$B(k) = \frac{1}{T}\int_0^T \text{Wal}\,(k,t)x(t)\,dt, \qquad k = 0, 1, 2, \cdots, N-1. \tag{1}$$

In the case of a sampled input signal, the integration in (1) is replaced by summation. Assuming N samples within a period T and taking the sampling interval as unity, (1) can be modified to

$$B(k) = \frac{1}{N}\sum_{i=0}^{N-1} \text{Wal}\,(k,i)X(i) \tag{2}$$

Manuscript received May 7, 1979; revised June 25, 1979. This work was supported by the National Research Council of Canada.

The authors are with the Department of Electrical Engineering, McMaster University, Hamilton, Ont. L8S 4L7, Canada.

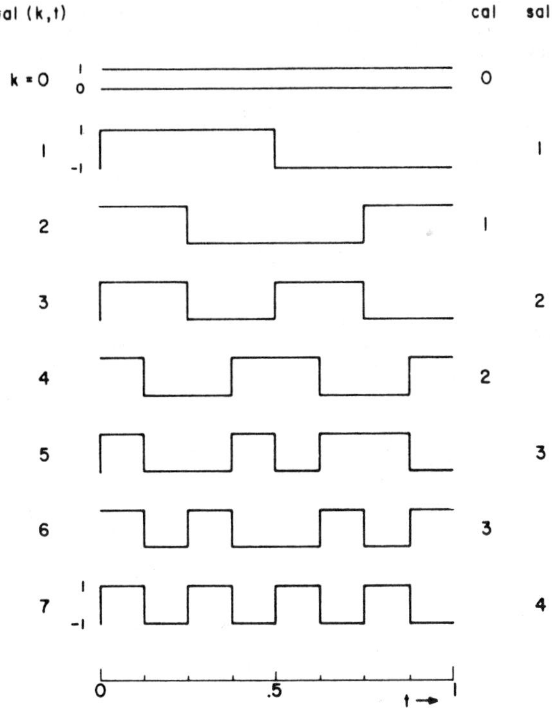

Fig. 1. The set of eight Walsh functions.

where $X(i)$ denotes the sampled value of the signal at the ith sampling instant. Wal (k, t) is called sal (z, t) or cal (z, t), by analogy with sine or cosine functions, depending on whether the index k is odd or even. The sequency is z, which is $(k + 1)/2$ if k is odd and $k/2$ if k is even. Correspondingly $B(k)$ is a sal or a cal coefficient, depending on the index k.

Varied approaches have been followed to obtain a finite number of Walsh coefficients, but all can be classified into the direct type or the fast Walsh transform (FWT) type. The direct type WSA [7], [8] is a straightforward implementation of (2) and uses N^2 arithmetic operations (additions or subtractions only), whereas the transform type WSA [9]–[11] uses $N \log_2 N$ arithmetic operations. The increased hardware in the direct type of WSA is sometimes not unreasonable, by virtue of the simplicity of (2), and is justified in applications where all Walsh coefficients are needed immediately after the Nth sample. In the present paper we are concerned with the design of a microprocessor-based WSA using a fast transform algorithm. The approach followed is different from earlier instrumentation for this purpose [9]–[11] in that the input sequence is permuted using minimal LSI hardware, and a fast Walsh–Hadamard transform (FWHT) is implemented.

Apart from output display, the instrument is contained entirely within an off-the-shelf Intel System 80/10 which consists of a cabinet containing an 8080-based single-board computer SBC 80/10, a card cage with ready-wired bus and power supply lines for up to three additional boards, and a power supply. The cabinet dimensions are 43.2 cm × 8.9-cm height, 50.8-cm depth (17 × 3.5 × 20 in). Only one additional specially designed board is used: It contains circuits for trigger-level adjustment, A/D conversion, 1K of READ–WRITE memory, a frequency multiplier module (FMM) and a direct memory access (DMA) controller.

The FMM circuit has been described elsewhere [15]; it generates 64 equally spaced pulses within one period of the input signal, so that the data window spans one cycle. The leakage error in the Fourier spectrum measurement is therefore negligibly small. Another feature of the circuit is that it does not require a high-speed clock for its proper functioning. The signal frequency may lie anywhere within the range 0.2 Hz to 10 kHz without range-change switching. A measurement uses two cycles of input signal, the first cycle being used for frequency determination and the second cycle for data acquisition. This is followed by processing, of about 24-ms duration, to yield the Walsh spectrum. The corresponding Fourier spectrum is available about 1 s later, most of this time being devoted to software multiplication (using Booth's algorithm). Less than 2K words of program memory (ROM) and 1K words of RAM are used. The System 80/10 monitor is not made use of so that space is left for an additional 2K words of ROM. Also two additional boards may be plugged into the cardcage. The system therefore lends itself to extension to include other measurements at low cost. The use of a System 80/10 package and LSI controller type 8257 for DMA were found to reduce development costs and time in a significant way.

Design Choices

An earlier approach was to first compute the FWHT using N sampled values; this is simple in software and fast by virtue of the recursive nature of the Hadamard matrix. The Hadamard-ordered coefficients $B_h(k)$ were then reordered [10] in increasing sequency-ordered coefficients $B_w(k)$ according to

$$B_h(k) = B_w(g\langle k\rangle) \tag{3}$$

where $\langle k\rangle$ is obtained by bit reversal of k, and $g\langle k\rangle$ is the Gray-to-binary code conversion of $\langle k\rangle$ [12]. Gray-to-binary code conversion and bit reversal is time-consuming using software. Alternately, this procedure calls for more hardware and additional processing time than are required in the method we have adopted.

A second possibility is to use the FWT algorithm due to Manz [13] to obtain $B_w(k)$. This algorithm requires a bit reversal operation on input data and employs two different kinds of butterfly operations. However, bit reversal can be achieved easily through hardware during data acquisition.

Thirdly, Berauer has described a fast-in-place algorithm to perform the sequency order Walsh transform [14]. His approach was to permute the incoming data according to (4) and (5) below, and perform the FWHT of the permuted data sequence. We have replaced the time-consuming software permutation of input data by specially designed hardware using a microprocessor DMA pheripheral chip 8257 which results in a substantial saving over other hardware schemes. Only three chips are required; the other two comprise two-input EXCLUSIVE-OR gates for $m = \log_2 N$ least significant memory address bits during data acquisition. The advantage of this method over Manz's is that the two

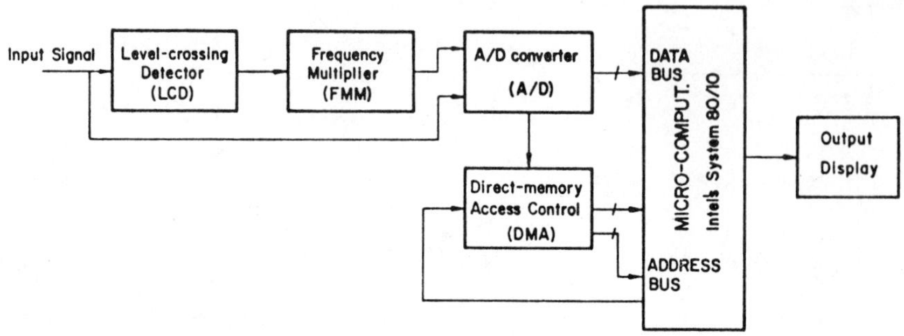

Fig. 2. Organization of a microcomputer based Walsh–Fourier spectral analyzer.

kinds of butterfly operations are reduced to one. This reduces program length, and also improves on the processing time by reducing the bookkeeping operations. The permutation operation for a data length N is a binary to Gray-code conversion followed by bit reversal

$$X(i) \rightarrow \bar{X}(\hat{i}) \tag{4}$$

where $X(i) = X(0), X(1), \cdots, X(N-1)$ is the input data sequence, $X(0)$ first, and $\bar{X}(\hat{i})$ is the permuted data in the computer memory. If the binary representation of i and $\hat{i}$ are

$$\begin{aligned} i &= i_{m-1} i_{m-2} \cdots i_r \cdots i_0 \\ \hat{i} &= \hat{i}_{m-1} \hat{i}_{m-2} \cdots \hat{i}_r \cdots \hat{i}_0 \end{aligned} \tag{5}$$

we have

$$\begin{aligned} \hat{i}_0 &= i_{m-1} \\ \hat{i}_r &= i_{m-r} \oplus i_{m-r-1}, \qquad r \in \{1, 2, \cdots, m-1\}. \end{aligned}$$

System Organization

The organization of the WSA is shown in Fig. 2. The level-crossing detect circuit and FMM determine the period T_S (digitally) during the first signal cycle after starting the measurement. Using this, 64 uniformly spaced pulses are generated in the next cycle. In the case of nonperiodic signals, external marker pulses which define the measuring interval are supplied to FMM.

The measurement is started by the RESET switch of the system 80/10 which begins program execution from memory address 0000 (hexadecimal). The address register and the terminal count register of the DMA controller are programmed with the starting address of the data and the number of samples to be acquired. Then a "DMA REQUEST" pulse, generated by software, is applied via an I/O port of system 80/10. DMA takes control of the system bus, and provides 16-bit address data in increasing order, starting from a base address which is programmable. To obtain the permuted address of (4) and (5), the six lsbs of the DMA output are modified as shown in Fig. 3. Note that the base address, i.e., the address for the first data $X(0)$, is chosen to have its six lsbs as zeros so that no separate adder is required to obtain the actual address information for data storage. (In a practical system the memory space for data storage may not begin from zero.) DMA releases the system bus after 64 samples have been obtained and the software

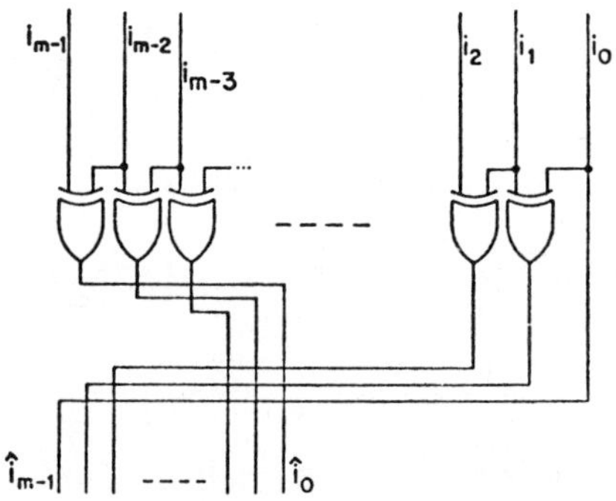

Fig. 3. Circuit for modifying the DMA address output for permuted data storage.

computation of Walsh coefficients begins. This software is based on the signal-flow graph for the FWHT [12] which is similar to the Cooley–Tukey signal-flow graph for FFT, except that multipliers associated with FFT are absent. The Walsh coefficients are computed using 16-bit arithmetic. The results are converted into signed BCD numbers and transmitted to a Hewlett-Packard graphics terminal type 2647A with processing capability for displaying numerical and graphic data.

Given the Walsh coefficients of a band-limited signal, the corresponding Fourier coefficients can be obtained by way of a matrix multiplication so that both spectra are available. The Walsh-to-Fourier conversion process uses the matrix multiplication [16],

$$[a] = [K][F_a][C] \tag{6}$$

$$[b] = [K][F_b][S] \tag{7}$$

where

$[a] = (N/2 \times 1)$ cosine coefficient matrix
$[b] = (N/2 \times 1)$ sine coefficient matrix
$[S] = (N/2 \times 1)$ sal coefficient matrix
$[C] = (N/2 \times 1)$ cal coefficient matrix
$[K] =$ diagonal $(N/2 \times N/2)$ compensation matrix
$[F_a], [F_b] = (N/2 \times N/2)$ conversion matrices for cosine and sine components, respectively.

The kth row of F_a or F_b corresponds to the first $N/2$ cal or sal coefficients of sinusoids cos $2\pi kt$ and sin $2\pi kt$, respectively. The corresponding elements in F_a and F_b may differ only in signs but have the same magnitudes. Only odd-numbered rows in F_a and F_b are independent; any other row of

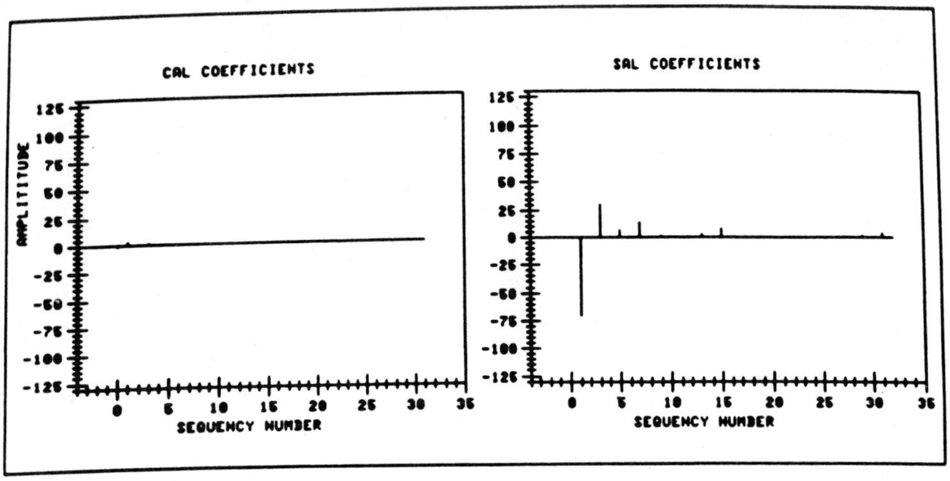

(a)

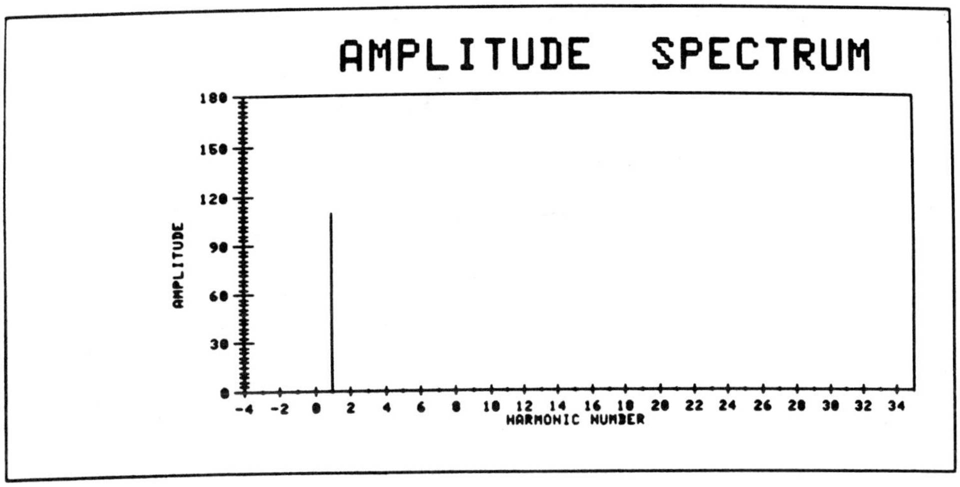

(b)

Fig. 4. (a) Display of Walsh spectrum of 1-kHz sinusoid. (b) Fourier amplitude spectrum of the same signal after Walsh-to-Fourier conversion.

elements is a subset of some odd numbered row. As a result, in software implementation, only $(N/4)$ odd-numbered rows need to be stored. The signs of conversion matrix elements of F_a in a row either differ or do not differ from the signs of corresponding F_b matrix elements alternately. As a result, only F_b elements in two's complement form are stored, while during the conversion process for cosine elements, successive results of multiplication corresponding to a particular cosine component are either added or subtracted during the accumulation process alternately. The compensation matrix elements are given by

$$\begin{aligned} k(f, f) &= \text{sinc}^{-2}\,(f/N), \qquad f = 1, 2, \cdots, (N/2) - 1 \\ &\tfrac{1}{2}\,\text{sinc}^{-2}\,\tfrac{1}{2}, \qquad f = N/2. \end{aligned} \tag{8}$$

The Fourier coefficients are transmitted to the graphics terminal for the display of the amplitude spectrum both in numeric and graphic form. Also the phase information of Fourier components is made available, using software in the terminal. The instrument therefore also serves as a transfer-function analyzer. Circuit and software details of the instrument are too lengthy to be included here; they are available on request.

Discussion

The present design is based on an 8-bit microprocessor and uses an 8 bit A/D converter. This limits the dynamic range of the signal as well as the output coefficients to 42 dB (range is defined as the ratio of maximum allowable input signal to the resolution of the A/D converter). For better dynamic range, an A/D converter of 12 bits may be used and the use of a 16-bit microprocessor is desirable to keep the software simple and to reduce the computation time. The maximum input signal frequency, in the system built, is limited to 4.5 kHz by the 2.8-μs conversion time of the A/D converter. The Walsh/Fourier Spectrum of a sinusoidal signal as displayed on the graphics terminal is shown in Fig. 4.

References

[1] L. R. Rabiner and B. Gold, *Theory and Applications of Digital Signal Processing*. Englewood Cliffs, NJ: Prentice-Hall, 1975.
[2] H. L. Gorginsky and G. A. Works, "A pipeline fast Fourier trans-

form," *IEEE Trans. Comput.*, vol. C-19, pp. 1015–1019, Nov. 1970.

[3] M. J. Corinthios, "The design of a class of fast Fourier transform computers," *IEEE Trans. Comput.*, vol. C-20, pp. 617–623, June 1971.

[4] "Spectrum analysis-theory, implementations and applications," Rockland System Corporation, 1977.

[5] N. A. Pendergrass and J. S. Farnhoch, "A high-resolution, low frequency spectrum analyzer," *Hewlett-Packard J.*, pp. 2–13, Sept. 1978.

[6] H. C. Andrews, *Computer Techniques in Image Processing*. New York: Academic Press, 1970.

[7] K. H. Siemens and R. Kitai, "Digital Walsh–Fourier analysis of periodic waveforms," *IEEE Trans. Instrum. Meas.*, vol. IM-18, pp. 316–321, Dec. 1969.

[8] Y. Tanada and H. Sano, "A hybrid Walsh transform analyzer," *Syst., Comput. and Contr.*, vol. 7, no. 1, pp. 80–89, 1976.

[9] M. R. Ashouri and A. G. Constantinides, "A pipeline fast Walsh Fourier transform," in *Proc. IEEE Int. Conf. ASSP* Hartford, CT, May 9–11), pp. 515–518, 1977.

[10] A. R. Elliott and Y. Y. Shum, "A parallel array handware implementation, of the fast Hadamard and Walsh transforms," in *Proc. Symp. Walsh Functions*, pp. 181–183, Mar. 1972.

[11] Y. A. Geadah and M. J. G. Corinthios, "Natural, dyadic, and sequency order algorithms and processors for Walsh Hadamard transform," *IEEE Trans. Comput.*, vol. C-26, pp. 435–442, May 1977.

[12] N. Ahmed and K. R. Rao, *Orthogonal Transforms for Digital Signal Processing*. New York: Springer-Verlag, 1975.

[13] J. W. Manz, "A sequency-ordered fast Walsh transform," *IEEE Trans. Audio Electroacoust.*, vol. AU-20, pp. 204–205, Aug. 1972.

[14] G. Berauer, "Fast 'in-place' computation of the discrete Walsh transform in sequency order," in *Proc. Symp. Walsh Functions*, pp. 272–279, 1972.

[15] K. Muniappan and R. Kitai, "Improved frequency multiplier," in *Dig. Conf. Precision Electromagnetic Meas.* (Ottawa, Ont., Canada), pp. 81–83, 1978.

[16] R. Kitai, "Walsh-to-Fourier spectral conversion for periodic waves," *IEEE Trans. Electromag. Compat.*, vol. EMC-17, pp. 266–269, Nov. 1975.

[17] Y. Tadokoro and T. Higuchi, "Discrete Fourier transform computation via Walsh transform," *IEEE Trans. Acoust., Speech, Signal Processing*, vol. ASSP-26, pp. 236–240, June 1978.

Part III

DYNAMIC SYSTEMS ANALYSIS

Editor's Comments on Papers 7 Through 10

The initiation of the analysis of dynamic systems in the time domain via Walsh functions was made by Corrington (1973) in his paper on the solution of differential and integral equations. The key idea was the observation that successive integrals of Walsh functions are expressed as Walsh series with well-defined, tabulated coefficients. The differential equation is solved for the highest derivative, and the result is then integrated as many times as required to give the solution. Two years later C. F. Chen and Hsiao (1975) presented the solution of dynamic systems in state space formulation by a more systematic use of the Walsh function integration property expressed by an operational equation $\int \phi dt = P\phi$, where P is a well-defined operational matrix. Using this operational equation, the state-space differential system is converted to a linear algebraic system, which has to be solved for a set of unknown Walsh series coefficients.

The application of Walsh function analysis to distributed parameter systems is made in Paper 7 using a generalized set of orthogonal functions (block pulse functions) along with the associated, generalized operational matrices, and in Paper 8 using double Walsh series expansions. Delay dynamic systems were studied by W. L. Chen and Shih (1978) using the shift Walsh matrix obtained from the m-dimensional Walsh matrix by shifting the columns of the Walsh matrix to the right, dropping the last $k \leqq m$ columns and selecting all elements of the first k columns of the new matrix as zero.

In Paper 9, W. L. Chen provides an analysis of multi delay systems, which is based on the concept of delay operational matrix. This analysis is actually an extension of the single delay analysis also by W. L. Chen. Paper 10 discusses the analysis of nonlinear systems via Walsh series by using a characteristic function technique for a Walsh-transformable nonlinearity or by a matrix method for the case of inputs represented by a truncated Walsh series expansion.

A useful method for extending the computation (i.e., the solution) beyond the limit of the initial normal interval of the Walsh series analysis was provided by Prasada Rao, Palanisamy, and Srinivasan (1980). Using this method the solution can be extended over contiguous, normalized intervals of time to any length.

Regarding the analysis of stochastic dynamic systems, the work of Frick (1979) is worth mentioning. He constructed a noise process with piece-wise constant sample functions in the form of a finite random Walsh series, studied its properties, and explored its applicability to the analysis of stochastic dynamic systems. Prasada Rao and Palanisamy (1982) presented and discussed the properties of an operational Walsh matrix for "Stretch," which is useful in solving certain functional differential equations.

REFERENCES

Chen, C. F., and C. H. Hsiao, 1975, A State Space Approach to Walsh Series Solution of Linear Systems, *Int. J. Syst. Sci.* **6:**833–858.

Chen, W. L., and Y. P. Shih, 1978, Shift Walsh Matrix and Delay-Differential Equations, *IEEE Trans. Autom. Control* **AC-23:**1023–1028.

Corrington, M. S., 1973, Solution of Differential and Integral Equations with Walsh Functions, *IEEE Trans. Circuit Theory* **CT-20:**470–476.

Frick, P. A., On Walsh Noise: Its Properties and Use in Dynamic Stochastic Systems, *IEEE Trans. Syst. Man, Cybern.* **SMC-9:**411–419.

Prasada Rao, G., and K. R. Palanisamy, 1982, Walsh Stretch Matrices and Functional Differential Equations, *IEEE Trans. Autom. Control* **AC-27:**272–276.

Prasada Rao, G., K. R. Palanisamy, and T. Srinivasan, 1980, Extension of Computation Beyond the Limit of Initial Normal Interval in Walsh Series Analysis of Dynamical Systems, *IEEE Trans. Autom. Control,* **AC-25:**317–319.

7

Reprinted from *J. Franklin Inst.* **303**:267–284 (1977)

Walsh Operational Matrices for Fractional Calculus and Their Application to Distributed Systems

by C. F. CHEN *and* Y. T. TSAY

Department of Electrical Engineering, University of Houston, Houston, Texas

and T. T. WU

Engineering and Applied Physics Department, Harvard University, Cambridge, Massachusetts

ABSTRACT: *The Walsh operational matrix for performing integration and solving state equations is generalized to fractional calculus for investigating distributed systems. A new set of orthogonal functions is derived from Walsh functions. By using the new functions, the generalized Walsh operational matrices corresponding to $\sqrt{s}$, $\sqrt{(s^2+1)}$, e^{-s} and $e^{-\sqrt{s}}$ etc. are established. Several distributed parameter problems are solved by the new approach.*

I. Introduction

Since Corrington (**1**) constructed Walsh tables for solving high-order differential equations and Chen and Hsiao (**2**) developed the Walsh operational matrix for solving state equations, the Walsh operational method has been successfully applied to various problems such as time domain synthesis (**3**) in system theory; piecewise gain determination in optimal control (**4**) and the direct method of solution of variational problems (**5**).

The unique feature of the operational matrix of Chen and Hsiao is that it performs like an integrator in the time domain or $1/s$ in the Laplace domain. We naturally ask: can this operational matrix be generalized to fractional calculus? In other words, are there Walsh operational matrices corresponding to irrational functions such as $\sqrt{s}$, $\sqrt{(s^2+1)}$ or transcendental functions such as e^{-s}, $e^{-\sqrt{s}}$ etc.?

This paper will investigate this problem and develop several Walsh matrices for fractional calculus, then apply these matrices for solving distributed parameter systems.

II. Review of Operational Matrix for Integer Integration

It is known that Walsh functions $\phi_i(t)$ are a set of square waves which are orthonormal. Figure 1 shows the functions from ϕ_0 to ϕ_{15} in the dyadic order.

For an arbitrary function $f(t)$, we can expand it into Walsh series, if the

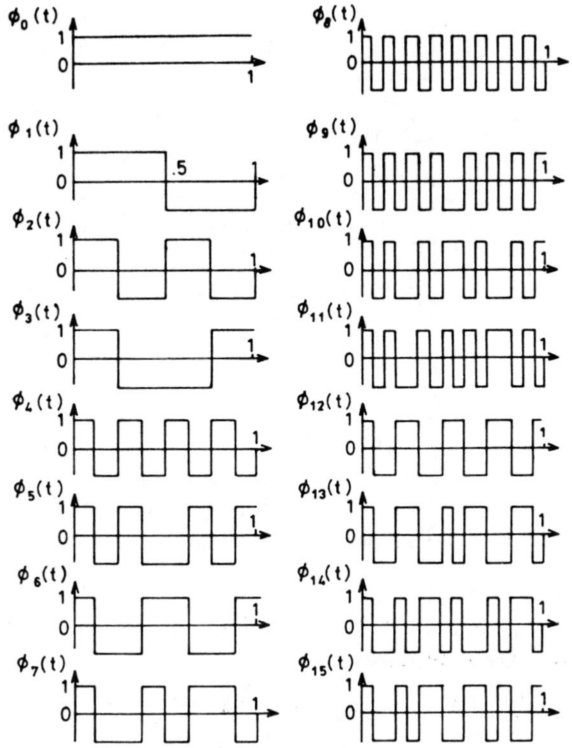

FIG. 1. Walsh functions.

function is absolutely integrable in [0, 1). We write:

$$f(t)=\sum_{n=0}^{N} C_n\phi_n(t) \tag{1}$$

where

$$C_n=\int_0^1 \phi_n(t)f(t)\,\mathrm{d}t \tag{2}$$

are determined such that the following integral square error is minimized, i.e.

$$\varepsilon=\int_0^1 [f(t)-\sum_{n=0}^{N} C_n\phi_n(t)]^2\,\mathrm{d}t. \tag{3}$$

For example, if we expand $f(t)=t$, the result is

$$f(t)=\tfrac{1}{2}\phi_0(t)-\tfrac{1}{4}\phi_1(t)-\tfrac{1}{8}\phi_2(t)-\tfrac{1}{16}\phi_4(t)+\dots \tag{4}$$

Partial sums of (4) form stairwise waves that approximate the original function t, as shown in Fig. 2.

If we perform integration on the Walsh functions, various triangular waves are obtained. Evaluating the Walsh coefficients for these triangular waves,

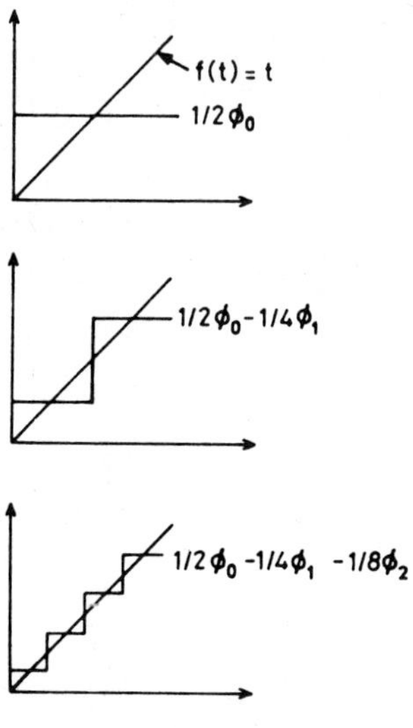

FIG. 2. Decomposing a ramp into Walsh series.

gives the following formula: (Typically, with no loss of generality, only the first eight Walsh waves are considered.)

$$\begin{bmatrix} \int\phi_0\,dt \\ \int\phi_1\,dt \\ \int\phi_2\,dt \\ \int\phi_3\,dt \\ \int\phi_4\,dt \\ \int\phi_5\,dt \\ \int\phi_6\,dt \\ \int\phi_7\,dt \end{bmatrix} = \left[\begin{array}{cccc|cccc} \frac{1}{2} & -\frac{1}{4} & -\frac{1}{8} & 0 & -\frac{1}{16} & 0 & 0 & 0 \\ \frac{1}{4} & 0 & 0 & -\frac{1}{8} & 0 & -\frac{1}{16} & 0 & 0 \\ \frac{1}{8} & 0 & 0 & 0 & 0 & 0 & -\frac{1}{16} & 0 \\ 0 & \frac{1}{8} & 0 & 0 & 0 & 0 & 0 & -\frac{1}{16} \\ \hline \frac{1}{16} & 0 & 0 & 0 & 0 & 0 & 0 & 0 \\ 0 & \frac{1}{16} & 0 & 0 & 0 & 0 & 0 & 0 \\ 0 & 0 & \frac{1}{16} & 0 & 0 & 0 & 0 & 0 \\ 0 & 0 & 0 & \frac{1}{16} & 0 & 0 & 0 & 0 \end{array}\right] \begin{bmatrix} \phi_0 \\ \phi_1 \\ \phi_2 \\ \phi_3 \\ \phi_4 \\ \phi_5 \\ \phi_6 \\ \phi_7 \end{bmatrix} \tag{5}$$

or

$$\int \boldsymbol{\phi}_{(8)}\,dt \cong \mathbf{G}_{(8\times8)}\boldsymbol{\phi}_{(8)}. \tag{5a}$$

The subscript means the dimension taken. It is preferable to take 2^{Ω}, where Ω is an integer, as a dimension number. Note that the ith row of the square matrix shows the Walsh series coefficients of the function $\int\phi_{i-1}\,dt$, $i=1, 2, \ldots$. For example, the first row of (5) gives the Walsh coefficients of $\int\phi_0\,dt$. The eight coefficients may be found in the first eight terms of (4), of course.

The square matrix **G** performs as an integrator and is called the Walsh operational matrix for integer integration. It is interesting to note that if **G** is partitioned into four equal parts as shown; the upper left corner of $G_{(8\times8)}$ is identical to $\mathbf{G}_{(4\times4)}$; and the upper left part of $\mathbf{G}_{(4\times4)}$ is $\mathbf{G}_{(2\times2)}$. Therefore, the regularity of the structure of the **G** matrix enables us to enlarge the **G** matrix.

We write the general operational matrix for integration as follows:

$$\mathbf{G}_{(m\times m)}=\left[\begin{array}{c:c:c:c}
\frac{1}{2}\;\ddots & -\frac{2}{m}\mathbf{I}_{(\frac{m}{8})} & -\frac{1}{m}\mathbf{I}_{(\frac{m}{4})} & -\frac{1}{2m}\mathbf{I}_{(\frac{m}{2})} \\ \hdashline
\frac{2}{m}\mathbf{I}_{(\frac{m}{8})} & \mathbf{O}_{(\frac{m}{8})} & & \\ \hdashline
\frac{1}{m}\mathbf{I}_{(\frac{m}{4})} & & \mathbf{O}_{(\frac{m}{4})} & \\ \hdashline
\frac{1}{2m}\mathbf{I}_{(\frac{m}{2})} & & & \mathbf{O}_{(\frac{m}{2})}
\end{array}\right] \tag{6}$$

The operational matrix (6) perform integration for integer calculus nicely. Therefore, it can be used to solve linear differential equations effectively. Chen and Hsiao demonstrate its power in solving state equations, in synthesizing transfer functions in the time domain, and in finding time-varying gains for optimal control.

We will generalize this matrix for fractional integration or non-integer calculus.

III. A Set of New Functions

Since $\mathbf{G}_{(m\times m)}$ corresponds to $1/s$, if we want to study $\sqrt{s}$ for example, what we have to effect is an extension of the operational matrix idea: (1) to invert the operational matrix in order to find a matrix corresponding to the differentiator; (2) to take the square root of the differentiator matrix thus obtained for finding a new matrix equivalent to $\sqrt{s}$. However, difficulties arise: The inversion of a large matrix is not a trivial matter if m is large! Secondly, for taking the square root of a matrix, we naturally think of diagonalization of the matrix in question. It is easy to find that all the eigenvalues of the operational matrix are clustered together. There are m repeated roots and the value is $1/2m$. Therefore, diagonalization of this problem is not a routine matter.

For overcoming these difficulties, we develop a new set of orthogonal functions. There is no loss of generality if we use $\mathbf{G}_{(4\times 4)}$ for deriving the new set. Consider:

$$\begin{bmatrix}\int\phi_0\,dt\\ \int\phi_1\,dt\\ \int\phi_2\,dt\\ \int\phi_3\,dt\end{bmatrix}\cong\begin{bmatrix}\frac{1}{2} & -\frac{1}{4} & -\frac{1}{8} & 0\\ \frac{1}{4} & 0 & 0 & -\frac{1}{8}\\ \frac{1}{8} & 0 & 0 & 0\\ 0 & \frac{1}{8} & 0 & 0\end{bmatrix}\begin{bmatrix}\phi_0\\ \phi_1\\ \phi_2\\ \phi_3\end{bmatrix} \tag{7}$$

Or, in general,

$$\left[\int \mathbf{\Phi}_{(m)}\,dt\right]=\mathbf{G}_{(m\times m)}\psi_{(m)}. \tag{8}$$

We define a set of new functions which relates the set of Walsh functions by the following formula:

$$\begin{bmatrix} \phi_0 \\ \phi_1 \\ \phi_2 \\ \phi_3 \end{bmatrix} = \begin{bmatrix} 1 & 1 & 1 & 1 \\ 1 & 1 & -1 & -1 \\ 1 & -1 & 1 & -1 \\ 1 & -1 & -1 & 1 \end{bmatrix} \begin{bmatrix} \psi_0 \\ \psi_1 \\ \psi_2 \\ \psi_3 \end{bmatrix} \tag{9}$$

or, in general,

$$\boldsymbol{\phi}_{(m)} = \mathbf{W}_{(m \times m)} \boldsymbol{\psi}_{(m)} \tag{9a}$$

where the ψ_i are called the block pulse functions with unity height and $1/m$ width. They are shown in Fig. 3 for the first four functions. The square matrix in (9), denoted by $W_{(4\times 4)}$, is called the Walsh matrix. The relation between the matrix and Walsh wave forms is shown in Fig. 4. From the figure we see that it is very easy to construct the Walsh matrix because the elements of the matrix correspond directly to the magnitude of the wave forms.

One of the properties of the W matrix is that

$$\mathbf{W}^2_{(m \times m)} = m\mathbf{I}_m \tag{10}$$

or

$$\mathbf{W}^{-1}_{(m \times m)} = \frac{1}{m} \mathbf{W}_{(m \times m)}. \tag{11}$$

Substituting (9a) into (8), yields

$$\int_0^t \mathbf{W}_{(m \times m)} \boldsymbol{\psi}_{(m)} \, dt = \mathbf{G}_{(m \times m)} \mathbf{W}_{(m \times m)} \boldsymbol{\psi}_{(m)}. \tag{12}$$

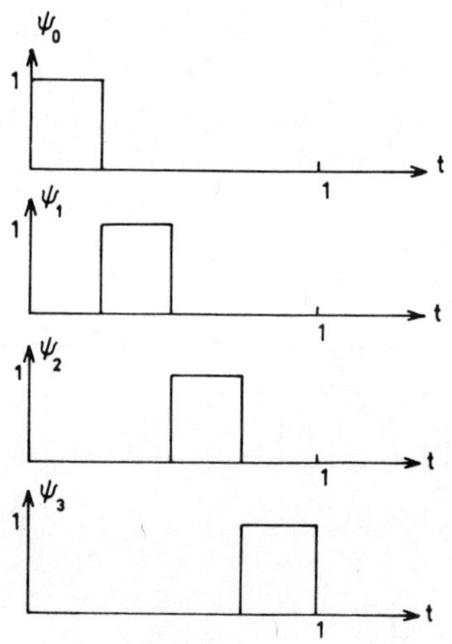

FIG. 3. Block pulse functions.

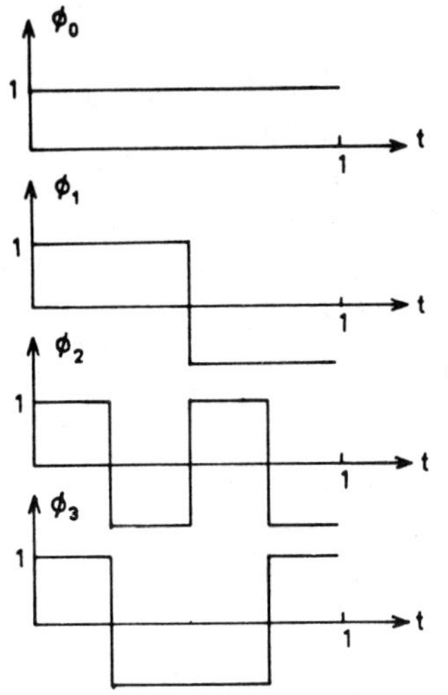

$$\Rightarrow \begin{bmatrix} 1 & 1 & 1 & 1 \\ 1 & 1 & -1 & -1 \\ 1 & -1 & 1 & -1 \\ 1 & -1 & -1 & 1 \end{bmatrix}$$

FIG. 4. Walsh functions and Walsh matrix.

Therefore,

$$\int_0^t \boldsymbol{\psi}_{(m)}\,\mathrm{d}t = \mathbf{W}^{-1}_{(m\times m)}\mathbf{G}_{(m\times m)}\mathbf{W}_{(m\times m)}\boldsymbol{\psi}_{(m)}. \tag{13}$$

We define the following:

$$\mathbf{W}^{-1}_{(m\times m)}\mathbf{G}_{(m\times m)}\mathbf{W}_{(m\times m)} \triangleq \mathbf{H}_{(m\times m)}. \tag{14}$$

Combining it with (11) gives

$$\mathbf{H}_{(m\times m)} = \frac{1}{m}\mathbf{WGW}. \tag{15}$$

Evaluation of the similarity transformation yields

$$\mathbf{H}_{(m\times m)} = \frac{1}{m}\begin{bmatrix} \frac{1}{2} & 1 & 1 & 1 & \dots & 1 \\ 0 & \frac{1}{2} & 1 & 1 & \dots & 1 \\ 0 & 0 & \frac{1}{2} & 1 & \dots & 1 \\ \cdot & \cdot & \cdot & \cdot & \cdots & \cdot \\ \cdot & \cdot & \cdot & \cdot & \cdots & \cdot \\ \cdot & \cdot & \cdot & \cdot & \cdots & \cdot \\ 0 & 0 & 0 & 0 & \dots & \frac{1}{2} \end{bmatrix} \tag{16}$$

Matrix H is the operational matrix as expressed in the block pulse function domain. Of course, this matrix can be directly obtained from integration of Fig. 3, by reading the magnitudes as shown in Fig. 5.

It is seen that the set of the block pulse functions is more fundamental than the Walsh functions: The operational matrix for integration shown in (16) is in

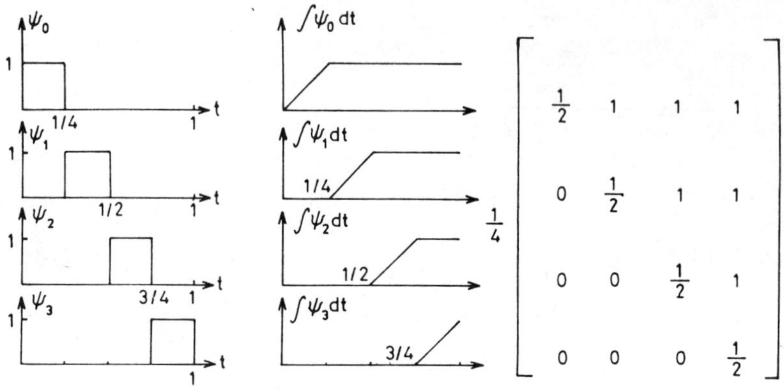

FIG. 5. Block pulse functions and their integrations.

terms of block pulse functions and is simpler than the operational matrix derived from Walsh.

Inspecting the H matrix, we make the following decomposition:

$$\mathbf{H}_{(m\times m)}=\frac{1}{m}\left(\tfrac{1}{2}\mathbf{I}_m+\mathbf{Q}_{(m\times m)}+\mathbf{Q}^2_{(m\times m)}+\cdots+\mathbf{Q}^{m-1}_{(m\times m)}\right) \tag{17}$$

where

$$\mathbf{Q}_{(m\times m)}=\begin{bmatrix}0&1&0&0&\dots&0\\0&0&1&0&\dots&0\\0&0&0&1&\dots&0\\ \cdot&\cdot&\cdot&\cdot&\dots&\cdot\\ \cdot&\cdot&\cdot&\cdot&\dots&\cdot\\ \cdot&\cdot&\cdot&\cdot&\dots&\cdot\\0&0&0&0&\dots&0\end{bmatrix}\triangleq\left[\begin{array}{c|ccc}0&&&\\ \vdots&&\mathbf{I}_{(m-1)}&\\0&&&\\ \hline 0&0&\dots&0\end{array}\right] \tag{18}$$

and it is easy to see the following important property:

$$\mathbf{Q}^i_{(m\times m)}=\left[\begin{array}{c|c}\mathbf{O}&\mathbf{I}_{m-i}\\ \hline \mathbf{O}_i&\mathbf{O}\end{array}\right]\quad \text{for}\quad i<m \tag{19}$$

and

$$\mathbf{Q}^i_{(m\times m)}=\mathbf{O}_{(m)}\qquad \text{for}\quad i\geq m. \tag{20}$$

Using (19) and (20), we write (17) as

$$\begin{aligned}\mathbf{H}_{(m\times m)}&=\frac{1}{m}\left(\tfrac{1}{2}\mathbf{I}_m+\sum_{i=1}^{\infty}\mathbf{Q}^i_{(m\times m)}\right)\\&=\frac{1}{m}\left(-\tfrac{1}{2}\mathbf{I}_m+(\mathbf{I}_m-\mathbf{Q}_{(m\times m)})^{-1}\right)\\&=\frac{1}{2m}(\mathbf{I}_m+\mathbf{Q}_{(m\times m)})(\mathbf{I}_m-\mathbf{Q}_{(m\times m)})^{-1}.\end{aligned} \tag{21}$$

We rewrite (17) by expressing $\mathbf{H}_{(m\times m)}$ as a polynomial of $\mathbf{Q}_{(m\times m)}$

$$\mathbf{H}_{(m\times m)} = h_m(\mathbf{Q}_{m\times m}) \tag{22}$$

where

$$h_m(x) = \frac{1}{m}\left(\tfrac{1}{2} + x + x^2 + \ldots x^{m-1}\right). \tag{23}$$

If q is an eigenvalue of $\mathbf{Q}_{(m\times m)}$, it is known (6) that the corresponding eigenvalue for $\mathbf{H}_{(m\times m)}$ is

$$h = h_m(q) = \frac{1}{2m}\frac{1+q}{1-q}. \tag{24}$$

However, the eigenvalues of $\mathbf{Q}_{(m\times m)}$ in (18) are all zeros. From (24), we obtain

$$h = h_m(0) = \frac{1}{2m}$$

Therefore, we state that the eigenvalues of $\mathbf{H}_{(m\times m)}$ is $1/2m$ with multiplicity m.

Equations (21) through (24) will be used repeatedly to derive various generalized operational matrices.

IV. Operational Matrix for Differentiation

Theoretically, it should not be very difficult to find the operational matrix for differentiation, when that for integration is known. However, it is much simpler to find the inversion in the block pulse domain.

Let us denote the operational matrix for differentiation as $\mathbf{B}_{(m\times m)}$. We have

$$\begin{aligned}\mathbf{B}_{(m\times m)} &= \mathbf{H}^{-1}_{(m\times m)} = 2m(\mathbf{I}_m + \mathbf{Q}_{(m\times m)})(\mathbf{I}_m - \mathbf{Q}_{(m\times m)})\\ &= 2m(\mathbf{I}_m - 2\mathbf{Q}_{(m\times m)} + 2\mathbf{Q}^2_{(m\times m)} + \ldots (-1)^{m-1}\mathbf{Q}^{m-1}_{(m\times m)}\\ &= 4m\left(\tfrac{1}{2}\mathbf{I}_m + \sum_{i=1}^{m-1}(-1)^i\mathbf{Q}^i_{(m\times m)}\right).\end{aligned} \tag{26}$$

For $m = 4$, we have

$$\mathbf{B}_{(4\times 4)} = 16\begin{bmatrix} \frac{1}{2} & -1 & 1 & -1\\ 0 & \frac{1}{2} & -1 & 1\\ 0 & 0 & \frac{1}{2} & -1\\ 0 & 0 & 0 & \frac{1}{2}\end{bmatrix} \tag{27}$$

Similarly transformating back to the Walsh domain yields the operational matrix for differentiation, denoted by $\mathbf{D}_{(m\times m)}$:

$$\mathbf{D}_{(m\times m)} \triangleq \mathbf{G}^{-1}_{(m\times m)} = \frac{1}{m}\mathbf{W}_{(m\times m)}\mathbf{H}^{-1}_{(m\times m)}\mathbf{W}_{(m\times m)} \tag{28}$$

For $m=4$, we have

$$\mathbf{D}_{(4\times 4)}=8\begin{bmatrix} 0 & 0 & 1 & 0\\ 0 & 0 & 0 & 1\\ -1 & 0 & 4 & -2\\ 0 & -1 & 2 & 0 \end{bmatrix}. \tag{29}$$

In general, the formula is

$$\mathbf{D}_{(m\times m)}=2m\left[\begin{array}{c:cc:c} \mathbf{O}_{(\frac{m}{2})} & \multicolumn{3}{c}{\mathbf{I}_{(\frac{m}{2})}} \\ \hdashline & \begin{array}{c:c} m & \\ \hdashline & \ddots \end{array} & -4\mathbf{I}_{(\frac{m}{8})} & \\ \cdashline{2-3} -\mathbf{I}_{(\frac{m}{2})} & 4\mathbf{I}_{(\frac{m}{8})} & \mathbf{O}_{(\frac{m}{8})} & -2\mathbf{I}_{(\frac{m}{4})} \\ \cdashline{2-4} & \multicolumn{2}{c:}{2\mathbf{I}_{(\frac{m}{4})}} & \mathbf{O}_{(\frac{m}{4})} \end{array}\right] \tag{30}$$

From (26) the eigenvalue, h^{-1}, of matrix $\mathbf{H}^{-1}_{(m\times m)}$ can be expressed as the eigenvalue, q, of $\mathbf{Q}_{(m\times m)}$

$$b=4m\left(\tfrac{1}{2}+\sum_{i=1}^{m-1}(-1)^i q^i\right) \tag{31}$$

or

$$b=2m\frac{1-q}{1+q}. \tag{32}$$

We could use (30) directly to evaluate the operator of differentiation or use (26) and (28) indirectly to deal with it. When the dimension number m is higher, the former is not as simple as the latter.

V. Operational Matrices for Fractional Calculus

Now we try to find the operational matrix for fractional differentiation. The general form of (32) could be written as follows:

$$b^{1/l}=\left(2m\frac{1-q}{1+q}\right)^{1/l} \tag{33}$$

where l is an integer. Equation (33) can be expanded into polynomial of q and terminated at q^{m-1}. As the result, Eq. (33) becomes

$$b^{1/l}=(2m)^{1/l}\rho_{l,m}(q) \tag{34}$$

where $\rho_{l,m}$ is the polynomial of order $m-1$ for $1/l$ differentiation. Thus the operational matrix for $1/l$ differentiation from (26) is given by

$$\mathbf{B}^{1/l}_{(m\times m)} = (2m)^{1/l}\rho_{l,m}(\mathbf{Q}_{(m\times m)}). \tag{35}$$

In the Walsh domain, the corresponding $1/l$ differentiation operational matrix is

$$\mathbf{D}^{1/l}_{(m\times m)} = (2m)^{1/l}\,\mathbf{W}^{-1}_{(m\times m)}\rho_{l,m}(\mathbf{Q}_{(m\times m)})\mathbf{W}_{(m\times m)}. \tag{36}$$

To illustrate the procedure, we try to find the operational matrix of $\sqrt{s}$. By choosing $m=4$, we have

$$\rho_{l,m}(q) = \rho_{2,4}(q) = \left(\frac{1-q}{1+q}\right)^{\frac{1}{2}} = 1 - q + \tfrac{1}{2}q^2 - \tfrac{1}{2}q^3 + \ldots \tag{37}$$

Then (35) becomes

$$\mathbf{B}^{\frac{1}{2}}_{(4\times 4)} = \sqrt{(2\times 4)}(\mathbf{I}_{(4\times 4)} - \mathbf{Q}_{(4\times 4)} + \tfrac{1}{2}\mathbf{Q}^2_{(4\times 4)} - \tfrac{1}{2}\mathbf{Q}^3_{(4\times 4)} + \ldots). \tag{38}$$

However,

$$\mathbf{Q}^i = \mathbf{O} \quad \text{for} \quad i \geq 4.$$

Therefore, we have

$$\mathbf{B}^{\frac{1}{2}}_{(4\times 4)} = \sqrt{(8)}[\mathbf{I}_{(4\times 4)} - \mathbf{Q}_{(4\times 4)} + \tfrac{1}{2}\mathbf{Q}^2_{(4\times 4)} - \tfrac{1}{2}\mathbf{Q}^3_{(4\times 4)}]$$

$$= \sqrt{8}\begin{bmatrix} 1 & -1 & \frac{1}{2} & -\frac{1}{2} \\ 0 & 1 & -1 & \frac{1}{2} \\ 0 & 0 & 1 & -1 \\ 0 & 0 & 0 & 1 \end{bmatrix} \tag{39}$$

In the Walsh domain, using (36) gives

$$\mathbf{D}^{\frac{1}{2}}_{(4\times 4)} = \frac{\sqrt{2}}{4}\begin{bmatrix} 3 & 1 & 3 & 1 \\ -1 & 5 & -1 & 5 \\ -3 & -1 & 17 & -5 \\ 1 & -5 & 5 & 7 \end{bmatrix} \tag{40}$$

For finding the operational matrix of fractional integration, we can use the reasoning similar to the fractional differentiation case.

Generalizing (24), yields

$$h^{1/l} = \left[\frac{1-q}{2m(1+q)}\right]^{1/l} \tag{41}$$

$$= \sqrt[l]{\left(\frac{1}{2m}\right)}\gamma_{l,m}(q) \tag{42}$$

where $\gamma_{l,m}$ is the polynomial of order $m-1$ for $1/l$ integration.

The operational matrix for $1/l$-integration in terms of the block pulse function is given by

$$\mathbf{H}^{1/l}_{(m\times m)} = \frac{1}{(2m)^{1/l}}\gamma_{l,m}(\mathbf{Q}_{(m\times m)}) \tag{43}$$

and the corresponding $1/l$-integration operational matrix in the Walsh domain is easily found as

$$\mathbf{G}^{1/l}_{(m\times m)} = \frac{1}{(2m)^{1/l}}\mathbf{W}^{-1}_{(m\times m)}\gamma_{l,m}(\mathbf{Q}_{(m\times m)})\mathbf{W}_{(m\times m)} \tag{44}$$

$$= \frac{1}{m(2m)^{1/l}}\mathbf{W}_{(m\times m)}\gamma_{l,m}(\mathbf{Q}_{(m\times m)})\mathbf{W}_{(m\times m)} \tag{45}$$

Illustratively, we try to find the operational matrix for $1/\sqrt{s}$, assuming $m=4$, i.e.

$$\gamma_{l,m}(g) = \gamma_{2,4}(q) = 1 + q + \tfrac{1}{2}q^2 + \tfrac{1}{2}q^3. \tag{46}$$

Then

$$\mathbf{H}^{\frac{1}{2}}_{(4\times 4)} = \frac{1}{\sqrt{8}}(\mathbf{I}_{(4\times 4)} + \mathbf{Q}_{(4\times 4)} + \tfrac{1}{2}\mathbf{Q}^2_{(4\times 4)} + \tfrac{1}{2}\mathbf{Q}^3_{(4\times 4)})$$

$$= \frac{1}{\sqrt{8}}\begin{bmatrix} 1 & 1 & \frac{1}{2} & \frac{1}{2} \\ 0 & 1 & 1 & \frac{1}{2} \\ 0 & 0 & 1 & 1 \\ 0 & 0 & 0 & 1 \end{bmatrix}$$

From (45), we have

$$\mathbf{G}^{\frac{1}{2}}_{(4\times 4)} = \frac{1}{4\cdot\sqrt[2]{8}}\mathbf{W}_{(4\times 4)}\mathbf{H}^{\frac{1}{2}}_{(4\times 4)}\mathbf{W}_{(4\times 4)}$$

$$= \frac{1}{16/2}\begin{bmatrix} 17 & -5 & -3 & -1 \\ 5 & 7 & 1 & -5 \\ 3 & 1 & 3 & 1 \\ -1 & 5 & -1 & 5 \end{bmatrix} \tag{47}$$

Of course, this result can be obtained directly from inverting the matrix of the first example.

VI. Application to Distributed Systems

For many engineering systems described by partial differential equations or time-varying differential equations, when evaluated in the Laplace domain, fractional functions or transcendental functions of s result. For example, problems in thermal processes, in hole diffusion of transistors, in electromagnetic devices, in transmission lines and in percolation processes often have mathematical models involving $\sqrt{(s)}$, $\sqrt{(s^2+1)}$, $e^{-\sqrt{s}}$ etc. To find their inverse

Laplace transform is not a trivial matter. There is a long history in attacking these types of problems. To mention a few: Pade's approximation (**6**); Carlson and Halijak's approach (**7**) using regular Newton's process; Lerner's work on partial analogue approximation (**8**); Kiloneitreva and Netushil's technique (**9**) developing special functions; Chen and Chiu's use the Fast Fourier transform (10).

The operational matrix method for fractional calculus developed in this paper is a general and powerful approach to the distributed problem. We now apply the method to various distributed systems.

Bessel functions

One of the well known functions

$$F(s)=\frac{1}{\sqrt{(s^2+1)}} \tag{48}$$

can be constructed as the Laplace transform of the solution of a time-varying system. We would like to find its inverse Laplace transform as the first application. Rewrite

$$F(s)=\frac{1}{\sqrt{(s^2+1)}}=\frac{s}{\sqrt{(s^2+1)}}\cdot\frac{1}{s}\triangleq F_1(s)\frac{1}{s}. \tag{49}$$

Suppose we are interested in the solution for

$$0\leq t<8.$$

Time scaling must be used.

Note that the general rule of scaling is based on the following

$$F(s)\to\frac{1}{\alpha}F\left(\frac{s}{\alpha}\right)\quad\text{for}\quad t\to\alpha t \tag{50}$$

where α is the scaling factor.

For this case, $\alpha=8$

$$F(s)\to\tfrac{1}{8}F\left(\frac{s}{8}\right)=\tfrac{1}{8}F_1\left(\frac{s}{8}\right)\frac{1}{(s/8)} \tag{51}$$

where

$$F_1\left(\frac{s}{8}\right)=\frac{s/8}{\sqrt{[(s/8)^2+1]}}. \tag{52}$$

For $m=8$, substituting (32) into (51), yields

$$\begin{aligned}F_1(q)&=\frac{2\left(\frac{1-q}{1+q}\right)}{\sqrt{\left[2\left(\frac{1-q}{1+q}\right)^2+1\right]}}\\&=\frac{2(1-q)}{\sqrt{(5-6q+5q^2)}}.\end{aligned} \tag{53}$$

Taylor series expansion at $q=0$ gives

$$F_1(q)=0.8944-0.3578q-0.5009q^2-0.3578q^3 -0.0429q^4+0.2284q^5+0.2905q^6+0.1346q^7+\dots \quad (54)$$

Replacing q by Q gives the operational matrix of this problem in the ψ domain. Note that $1/s$ in (51) should be expressed by the ψ functions, i.e.

$$\frac{1}{s}\to[1\ 1\ 1\ 1\ 1\ 1\ 1\ 1]\psi_{(8)}. \quad (55)$$

Consequently, (48), becomes

$$\mathscr{L}^{-1}\frac{1}{\sqrt{(s^2+1)}}=\mathscr{L}^{-1}\frac{1}{s}F_1\left(\frac{s}{8}\right)$$

$$=[1\ 1\ 1\ 1\ 1\ 1\ 1\ 1]\begin{bmatrix} 0.8944 & -0.3578 & \dots & 0.1346 \\ & \ddots & \ddots & \vdots \\ 0 & & \ddots & -0.3578 \\ & & & 0.8944 \end{bmatrix}\psi_{(8)}$$

$$=0.8944\psi_0+0.5367\psi_1+0.03584\psi_2-0.3220\psi_3 -0.3650\psi_4-0.1365\psi_5+0.1721\psi_6+0.3067\psi_7. \quad (56)$$

The exact solution and (56) are compared in Fig. 6.

If we use the W matrix (9a), we can transform (56) in terms of the Walsh functions. Of course, the two approaches yield the same answer.

Generalized exponential function

Consider the following function

$$F(s)=\frac{s^{-\frac{1}{2}}}{\sqrt{(s)}+1} \quad (57)$$

We rewrite it as

$$F(s)=\frac{1}{1+\dfrac{1}{\sqrt{s}}}\cdot\frac{1}{s}\triangleq F_1(s)\cdot\frac{1}{s}.$$

Assume $m=4$, we are interested in the interval $0\le t<2.4$. The scaling factor is $s\to s/2.4$. Then we have

$$F_1(s)=\frac{1}{1+\dfrac{\sqrt{2.4}}{\sqrt{s}}}.$$

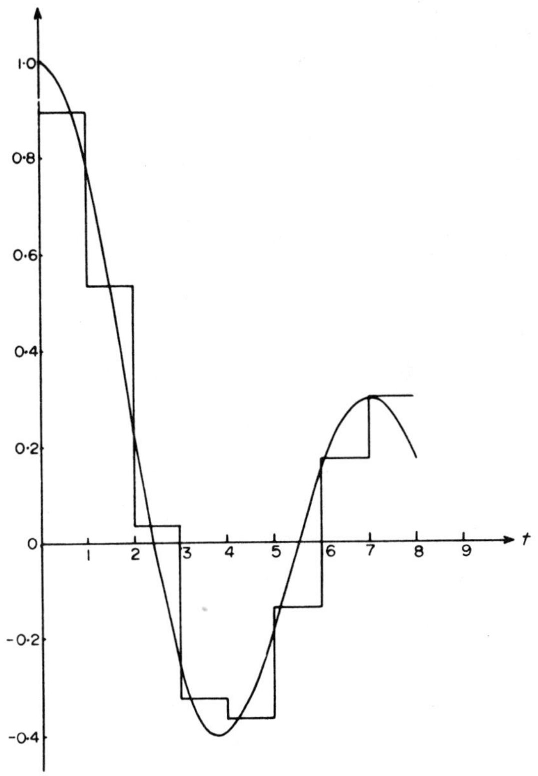

FIG. 6. Solution of Bessel function.

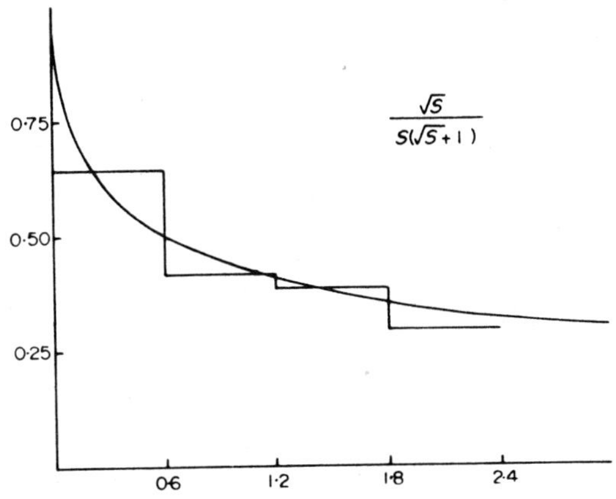

FIG. 7. Solution of generalized exponential function.

Using (32), gives

$$F_1(q) = \frac{1}{1+0.5477(1+q+\frac{1}{2}q^2+\frac{1}{2}q^3+\ldots)}$$
$$= 0.6461 - 0.2287q - 0.03341q^2 - 0.09602q^3 \qquad (58)$$

Then replacing q by $\mathbf{Q}$, gives

$$\mathbf{F}_{1(4\times4)} = 0.6461\mathbf{I}_{(4\times4)} - 0.2287\mathbf{Q}_{(4\times4)} - 0.02341\mathbf{Q}^2_{(4\times4)} - 0.09602\mathbf{Q}^3_{(4\times4)}.$$

Finally, we obtain

$$\mathscr{L}^{-1}F(s) = [1\ \ 1\ \ 1\ \ 1]\begin{bmatrix} 0.6461 & -0.2287 & -0.03341 & -0.09602 \\ 0 & 0.6461 & -0.2287 & -0.03341 \\ 0 & 0 & 0.6461 & -0.2283 \\ 0 & 0 & 0 & 0.6461 \end{bmatrix}(4) \qquad (59)$$

$$= [0.6461,\ 0.41074,\ 0.3840,\ 0.2880]\psi_{(4)}.$$

Figure 7 shows the exact solution and the approximation calculated by the new method.

Heat conduction

For the heat conduction problem, we easily arrive at the following function:

$$F(s) = \frac{1}{s}\,e^{-\sqrt{s}}.$$

Find the inverse Laplace transform for $0 \le t < 8$.

By using the scale $s \rightarrow s/8$ and letting $m = 4$, we have

$$F(s) = \frac{1}{s/8}\cdot\frac{1}{8}\exp\left(-\sqrt{\frac{s}{8}}\right) = \frac{1}{s}\exp\left(-\sqrt{\frac{1-q}{1+q}}\right) \triangleq \frac{1}{s}F_1(q). \qquad (60)$$

Taylor expanding and evaluating the coefficients, give

$$\frac{\partial F_1(q)}{\partial q} = \exp\left(-\sqrt{\frac{1-q}{1+q}}\right)(1+q)^{-3/2}(1-q)^{-1/2}$$

$$\frac{\partial^2 F_1(q)}{\partial q^2} = \exp\left(-\sqrt{\frac{1-q}{1+q}}\right)[(1+q)^{-3}(1-q)^{-1} + (-1+2q)(1+q)^{-5/2}(1-q)^{-3/2}]$$

$$\frac{\partial^3 F_1(q)}{\partial q^3} = \exp\left(-\sqrt{\frac{1-q}{1+q}}\right)[(1+q)^{-q/2}(1-q)^{-3/2} + (2q-1)(1+q)^{-4}(1-q)^{-2}$$
$$-3(1+q)^{-4}(1-q)^{-1} + (1-q)^{-2}(1+q)^{-3} - 2(1+q)^{-5/2}(1-q)^{-3/2}$$
$$+(2q-1)(-5/2)(1+q)^{-7/2}(1-q)^{-3/2} + (2q-1)(1+q)^{-5/2}(3/2)(1-q)^{-5/2}].$$

Then Taylor series

$$F_1(q)=F_1(0)+\frac{\partial}{\partial q}F_1(0)\cdot q+\frac{1}{2!}\frac{\partial^2}{\partial^2 q}F_1(0)\cdot q^2+\frac{1}{3!}\frac{\partial^3}{\partial q^3}F_1(0)\cdot q^3$$

becomes

$$F_1(q)=\left(1+q+\frac{1}{3!}q^3\right)e^{-1}.$$

Replacing q by Q, we finally have

$$\mathscr{L}^{-1}\left(\frac{1}{s}\cdot e^{-\sqrt{s}}\right)=[1\quad 1\quad 1\quad 1]\begin{bmatrix} e^{-1} & e^{-1} & 0 & \frac{e^{-1}}{6} \\ 0 & e^{-1} & e^{-1} & 0 \\ 0 & 0 & e^{-1} & e^{-1} \\ 0 & 0 & 0 & e^{-1} \end{bmatrix}\psi_{(4)} \tag{61}$$

$$=0.3679\psi_0+0.7358\psi_1+0.73584\psi_2+0.79714\psi_3.$$

Figure 8 shows the exact solution and the approximation of the heat conduction problem.

Percolation problem

It is well known that the percolation problem is difficult to solve. We use the same procedure as above.

Consider the function

$$F(s)=\frac{1}{s}\exp\left(-\frac{1}{s+1}\right)=\frac{1}{s}F_1(s).$$

If $m=4$ and $0\leq t<8$, we substitute s by $s/8$ for scaling. Then we obtain

$$F_1(q)=\exp\left[-\left(\frac{1-q}{1+q}\right)\Big/\left[\frac{1-q}{1+q}+1\right]\right]$$

$$=e^{-\frac{1}{2}}\left[1+\frac{q}{2}+\frac{(q/2)^2}{2!}+\frac{(q/2)^3}{3!}+\ldots\right]$$

$$=e^{-\frac{1}{2}}[1+0.5q+0.125q^2+0.0185q^3+\ldots].$$

Consequently, we have

$$\mathscr{L}^{-1}\left[\frac{1}{s}\exp\left(-s/(s+1)\right)\right]=[1\quad 1\quad 1\quad 1]\begin{bmatrix} 0.6065 & 0.3033 & 0.07582 & 0.01123 \\ 0 & 0.6065 & 0.3033 & 0.07582 \\ 0 & 0 & 0.6065 & 0.3033 \\ 0 & 0 & 0 & 0.6065 \end{bmatrix}\psi_{(4)}$$

$$=0.6065\psi_0+0.90984\psi_1+0.9856\psi_2+0.9969\psi_3. \tag{63}$$

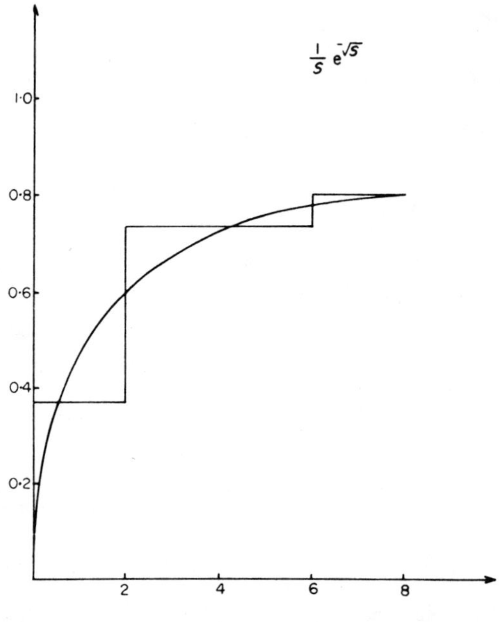

FIG. 8. Solution of heat conduction problem.

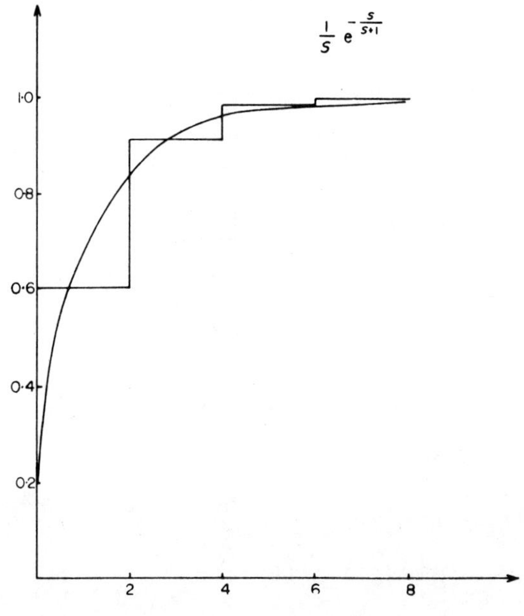

FIG. 9. Solution of percolation problem.

The comparison between the exact solution and the approximation is shown in Fig. 9.

VII. Conclusion

Fractional functions in general, and the functions $\sqrt{s}$, $\sqrt{(s^2+1)}$, e^{-s} and $e^{-\sqrt{s}}$ in particular, are inverted by the Walsh operational matrix approach. A new set of orthogonal functions $\psi_i(t)$, which is closely related to the Walsh set $\phi_i(t)$, is derived. Various operational matrices for the typical functions are established, both in the ψ domain and the ϕ domain. The application of these operational matrices for the solution of distributed-parameter systems is demonstrated.

Acknowledgements

The first author wishes to acknowledge the support of the National Science Foundation grant HES75-20759. To one of the reviewers the authors wish to express their gratitude for valuable suggestions.

References

(1) M. S. Corrington, "Solution of differential and integral equations with Walsh functions", *IEEE Trans. on Circuit Theory*, Vol. CT-20, No. 5, pp. 470–476, Sept. 1973.

(2) C. F. Chen and C. H. Hsiao, "A state space approach to Walsh series solution of linear systems", *Int. J. Systems Sci.* Vol. 6, No. 9, pp. 833–858, 1975.

(3) C. F. Chen and C. H. Hsiao, "Time-domain synthesis via Walsh functions", *IEE Proc.* Vol. 122, pp. 565–570, May 1975.

(4) C. F. Chen and C. H. Hsiao, "Design of piecewise constant gains for optimal control via Walsh functions", *IEEE Trans. on Automatic control*, Vol. AC-20, No. 5, pp. 596–603, Oct. 1975.

(5) C. F. Chen and C. H. Hsiao, "Walsh series direct method for solving variational problems" *J. Franklin Inst.*, Vol. 300, No. 4, pp. 265–280, Oct. 1975.

(6) J. L. Stewart, "Generalized Pade approximation", *IRE Proc.* pp. 2003–2008, 1960.

(7) G. E. Carlson and C. A. Halijak, "Approximation of fractional Capacitors $(1/S)^{1/n}$ by the regular Newton Process", *IEEE Trans. on Circuit Theory*, Vol. CT-11, pp. 210–213, 1964.

(8) R. M. Lerner, "The design of a constant angle or power-law magnitude impedance", *IEEE Trans. on Circuit Theory*, Vol. CT-10, pp. 98–107, 1963.

(9) M. B. Kilomeitreva and A. V. Netushil, "Transients in an automatic control systems with irrational transfer functions", *Automation and Remote Control*, Vol. 203, pp. 359–364, 1965.

(10) C. F. Chen and R. F. Chiu, "Evaluation of irrational and transcendental transfer functions via the fast Fourier transform", *Int. J. Electron.* Vol. 35, No. 2, pp. 267–278, 1973.

8

Reprinted from *Int. J. Syst. Sci.* **9**:569–578 (1978)

Double Walsh series solution of first-order partial differential equations

YEN-PING SHIH† and JOUNG-YI HAN†

A double Walsh series is introduced to represent approximately functions of two independent variables, and is then applied to analyse single as well as simultaneous first-order partial differential equations. The solutions for the coefficient matrices can be obtained directly from Kronecker product formulae, which are suitable for computer computation. An example for a single first-order partial differential equation is solved by a double Walsh series approximation with satisfactory results.

1. Introduction

The applications of Walsh functions have been given much attention (Proceedings 1970, 1971, 1972, 1973, 1974). Most recently, Walsh functions have been applied in the analysis, optimization, and time-domain synthesis of linear systems (Chen and Hsiao 1975 a, b, c, d, Corrington 1973, Burkhardt 1975). Chen and Shih (1977) first introduced a Walsh product matrix to extend the Walsh series approximation to time-varying linear systems.

In this paper, a double Walsh series approximation of functions of two independent variables is first introduced. Therefore, a Walsh function can be applied to analyse single and simultaneous first-order partial differential equations.

2. Double Walsh series

First, consider the double Walsh series approximation.

A function $y(x, t)$ which is absolutely integrable in $x \in [0, 1)$ and $t \in [0, 1)$ can be approximately represented by a Walsh series with respect to t as follows :

$$y(x, t) = \sum_{i=0}^{\infty} y_i(x)\phi_i(t) \tag{1}$$

where $\phi_i(t)$ is the ith-order Walsh function. The elementary properties of Walsh functions can be found elsewhere (Chen and Hsiao 1975 a, b, c, d). A Walsh series approximation is obtained by the truncation of eqn. (1) to the first n terms :

$$y(x, t) = \sum_{i=0}^{n-1} y_i(x)\phi_i(t) \tag{2}$$

Using the orthonormal property of Walsh functions,

$$\int_0^1 \phi_i(t)\phi_j(t)\,dt = \begin{cases} 0, & i \neq j \\ 1, & i = j \end{cases} \tag{3}$$

Received 13 July 1977.

† Department of Chemical Engineering, National Cheng Kung University, Tainan, Taiwan, China.

The coefficient functions $y_i(x)$ of eqn. (2) become

$$y_i(x) = \int_0^1 y(x, t)\phi_i(t)\,dt \tag{4}$$

Likewise, a Walsh series approximation of $y_i(x)$ gives

$$y_i(x) = \sum_{j=0}^{m-1} \phi_j(x) y_{ji} \tag{5}$$

where $\phi_j(x)$ are Walsh functions with respect to x, and y_{ji} are coefficients obtained by

$$y_{ji} = \int_0^1 \phi_j(x) y_i(x)\,dx = \int_0^1 \int_0^1 \phi_j(x) y(x, t)\phi_i(t)\,dx\,dt \tag{6}$$

Combination of eqns. (2) and (5) yields

$$y(x, t) = \sum_{i=0}^{n-1} \sum_{j=0}^{m-1} \phi_j(x) y_{ji} \phi_i(t) \tag{7}$$

Letting the Walsh vectors,

$$\boldsymbol{\phi}_{(n)}(t) = [\phi_0(t), \phi_1(t), \ldots, \phi_{n-1}(t)]^{\mathrm{T}} \tag{8}$$

$$\boldsymbol{\phi}_{(m)}(x) = [\phi_0(x), \phi_1(x), \ldots, \phi_{m-1}(x)]^{\mathrm{T}} \tag{9}$$

and coefficient matrix $\mathbf{Y}$ of $m \times n$ dimension

$$\mathbf{Y} = \begin{bmatrix} y_{00} & y_{01} & \cdots & y_{0,n-1} \\ y_{10} & y_{11} & \cdots & y_{1,n-1} \\ \cdots & \cdots & \cdots & \cdots \\ y_{m-1,0} & y_{m-1,1} & \cdots & y_{m-1,n-1} \end{bmatrix} \tag{10}$$

eqn. (7) is written in matrix form as

$$y(x, t) = \boldsymbol{\phi}_{(m)}^{\mathrm{T}}(x) \mathbf{Y} \boldsymbol{\phi}_{(n)}(t) \tag{11}$$

This is the double Walsh series approximation of $y(x, t)$.

For completeness in the series expansion (Corrington 1973) choose $m = 2^d$, $n = 2^c$, c, d = positive integers. Integration of Walsh vectors of eqns. (8) and (9) gives, respectively,

$$\int_0^t \boldsymbol{\phi}_{(n)}(t')\,dt' = \mathbf{P}_{(n \times m)} \boldsymbol{\phi}_{(n)}(t) \tag{12}$$

$$\int_0^x \boldsymbol{\phi}_{(m)}(x')\,dx' = \mathbf{P}_{(m \times m)} \boldsymbol{\phi}_{(m)}(x) \tag{13}$$

where $\mathbf{P}_{(n\times n)}$ and $\mathbf{P}_{(m\times m)}$ are operational matrices of integration (Chen and Hsiao 1975 a). The general form of $\mathbf{P}_{(n\times n)}$ is

$$\mathbf{P}_{(n\times n)}=\left[\begin{array}{c:c:c:c} \begin{array}{c:c} \frac{1}{2} & \\ \hdashline & \ddots \end{array} & -\frac{1}{n/2}\mathbf{I}_{(n/8)} & -\frac{1}{n}\mathbf{I}_{(n/4)} & -\frac{1}{2n}\mathbf{I}_{(n/2)} \\ \hdashline \frac{1}{n/2}\mathbf{I}_{(n/8)} & \mathbf{0}_{(n/8)} & & \\ \hdashline \multicolumn{2}{c:}{\frac{1}{n}\mathbf{I}_{(n/4)}} & \mathbf{0}_{(n/4)} & \\ \hdashline \multicolumn{3}{c:}{\frac{1}{2n}\mathbf{I}_{(n/2)}} & \mathbf{0}_{(n/2)} \end{array}\right] \tag{14}$$

where $\mathbf{I}$ and $\mathbf{0}$ are identity and null matrices, respectively. For example :

$$\mathbf{P}_{(4\times 4)}=\begin{bmatrix} \frac{1}{2} & -\frac{1}{4} & -\frac{1}{8} & 0 \\ \frac{1}{4} & 0 & 0 & -\frac{1}{8} \\ \frac{1}{8} & 0 & 0 & 0 \\ 0 & \frac{1}{8} & 0 & 0 \end{bmatrix} \tag{15}$$

Applications of eqns. (10), (12) and (13) to the solution of a single first-order partial differential equation (PDE) is given first. Then, the method is extended to the solution of simultaneous first-order PDE.

3. Single first-order PDE

Consider a simple single first-order PDE with constant coefficients

$$\frac{\partial y}{\partial t}+\frac{\partial y}{\partial x}=ay+bu \tag{16}$$

$$t=0, \quad y(x, 0)=g(x) \tag{17}$$

$$x=0, \quad y(0, t)=f(t) \tag{18}$$

where a and b are constants, $u=u(x, t)$ is the forcing function. We try to obtain an approximate solution by Walsh functions for $t\in[0, 1)$ and $x\in[0, 1)$. Integration of eqn. (16) with respect to t and x gives

$$\underset{(1)}{\int_0^t\int_0^x \frac{\partial y}{\partial t'}\,dx'\,dt'}+\underset{(2)}{\int_0^t\int_0^x \frac{\partial y}{\partial x'}\,dx'\,dt'}=\underset{(3)}{\int_0^t\int_0^x ay\,dx'\,dt'}+\underset{(4)}{\int_0^t\int_0^x bu\,dx'\,dt'} \tag{19}$$

The double Walsh series approximation of $y(x, t)$ is given by eqn. (11). Likewise,

$$u(x, t)=\boldsymbol{\phi}_{(m)}{}^{\mathrm{T}}(x)\,\mathbf{U}\boldsymbol{\phi}_{(n)}(t) \tag{20}$$

where $\mathbf{U}$ is an $m \times n$ matrix, with components u_{ji}:

$$u_{ji} = \int_0^1 \int_0^1 u(x, t)\phi_j(x)\phi_i(t)\, dx\, dt \tag{21}$$

A Walsh series approximation of $g(x)$ of eqn. (17) is

$$g(x) = \boldsymbol{\phi}_{(m)}{}^{\mathrm{T}}(x)\mathbf{g} \tag{22}$$

where $\mathbf{g} = [g_0, g_1, \ldots, g_{m-1}]^{\mathrm{T}}$ and

$$g_j = \int_0^1 g(x)\phi_j(x)\, dx \tag{23}$$

Since $\phi_0(t) = 1$ for $t \in [0, 1)$, eqn. (22) can be written as

$$g(x) = \boldsymbol{\phi}_{(m)}{}^{\mathrm{T}}(x)[\mathbf{g}, \underbrace{\mathbf{0}, \ldots, \mathbf{0}}_{(n-1)\text{ columns}}]\boldsymbol{\phi}_{(n)}(t) = \boldsymbol{\phi}_{(m)}{}^{\mathrm{T}}(x)\mathbf{G}\boldsymbol{\phi}_{(m)}(t) \tag{24}$$

Similarly a Walsh series approximation of $f(t)$ of eqn. (18) can be obtained:

$$f(t) = \mathbf{f}^{\mathrm{T}}\boldsymbol{\phi}_{(n)}(t) = \boldsymbol{\phi}_{(m)}{}^{\mathrm{T}}(x)\begin{bmatrix} \mathbf{f}^{\mathrm{T}} \\ \mathbf{0}^{\mathrm{T}} \\ \vdots \\ \mathbf{0}^{\mathrm{T}} \end{bmatrix} \left.\begin{matrix} \\ \\ \\ \end{matrix}\right\} (m-1)\text{ rows} \;\boldsymbol{\phi}_{(n)}(t) = \boldsymbol{\phi}_{(m)}{}^{\mathrm{T}}(x)\mathbf{F}\boldsymbol{\phi}_{(n)}(t) \tag{25}$$

where $\mathbf{f}^{\mathrm{T}} = [f_0, f_1, \ldots, f_{n-1}]$ and

$$f_i = \int_0^1 f(t)\phi_i(t)\, dt \tag{26}$$

Using the Walsh series approximations of eqns. (11), (20), (24) and (25), and eqns. (12) and (13), the four terms of eqn. (19) can be evaluated, respectively, as follows:

$$\begin{aligned} \text{term (1)} &= \int_0^x [y(x,'t) - y(x', 0)]\, dx' \\ &= \int_0^x [\boldsymbol{\phi}_{(m)}{}^{\mathrm{T}}(x')\mathbf{Y}\boldsymbol{\phi}_{(n)}(t) - \boldsymbol{\phi}_{(m)}{}^{\mathrm{T}}(x')\mathbf{G}\boldsymbol{\phi}_{(n)}(t)]\, dx' \\ &= \boldsymbol{\phi}_{(m)}{}^{\mathrm{T}}(x)\mathbf{P}_{(m\times m)}{}^{\mathrm{T}}\mathbf{Y}\boldsymbol{\phi}_{(n)}(t) - \boldsymbol{\phi}_{(m)}{}^{\mathrm{T}}(x)\mathbf{P}_{(m\times m)}{}^{\mathrm{T}}\mathbf{G}\boldsymbol{\phi}_{(n)}(t) \end{aligned} \tag{27}$$

where eqn. (17) has been used:

$$\begin{aligned} \text{term (2)} &= \int_0^t [y(x, t') - y(0, t')]\, dt' \\ &= \int_0^t [\boldsymbol{\phi}_{(m)}{}^{\mathrm{T}}(x)\mathbf{Y}\boldsymbol{\phi}_{(n)}(t') - \boldsymbol{\phi}_{(m)}{}^{\mathrm{T}}(x)\mathbf{F}\boldsymbol{\phi}_{(n)}(t')]\, dt' \\ &= \boldsymbol{\phi}_{(m)}{}^{\mathrm{T}}(x)\mathbf{Y}\mathbf{P}_{(n\times n)}\boldsymbol{\phi}_{(n)}(t) - \boldsymbol{\phi}_{(m)}{}^{\mathrm{T}}(x)\mathbf{F}\mathbf{P}_{(n\times n)}\boldsymbol{\phi}_{(n)}(t) \end{aligned} \tag{28}$$

where eqn. (18) has been used:

$$\text{term (3)} = a\boldsymbol{\phi}_{(m)}{}^{\mathrm{T}}(x)\mathbf{P}_{(m\times m)}{}^{\mathrm{T}}\mathbf{Y}\mathbf{P}_{(n\times n)}\boldsymbol{\phi}_{(n)}(t) \tag{29}$$

$$\text{term (4)} = b\boldsymbol{\phi}_{(m)}{}^{\mathrm{T}}(x)\mathbf{P}_{(m\times m)}{}^{\mathrm{T}}\mathbf{U}\mathbf{P}_{(n\times n)}\boldsymbol{\phi}_{(n)}(t) \tag{30}$$

Substitution of eqns. (27)–(30) into eqn. (19) gives an equation of the following form :

$$\mathbf{\Phi}_{(m)}^{\mathrm{T}}(x)[\ \dots\ ]\mathbf{\Phi}_{(n)}(t)=0 \tag{31}$$

Since eqn. (31) is valid for any x and t in the domain of consideration, the quantity in the brackets should be equal to zero. That is

$$\begin{aligned}\mathbf{P}_{(m\times m)}^{\mathrm{T}}\,\mathbf{Y}+\mathbf{YP}_{(n\times n)}-a\mathbf{P}_{(m\times m)}^{\mathrm{T}}\,\mathbf{YP}_{(n\times n)}&=\mathbf{P}_{(m\times m)}^{\mathrm{T}}\,\mathbf{G}\\&\quad+\mathbf{FP}_{(n\times n)}+b\mathbf{P}_{(m\times m)}^{\mathrm{T}}\,\mathbf{UP}_{(n\times n)}\end{aligned} \tag{32}$$

The components of $\mathbf{Y}$ are y_{ji}, $j=0, 1, \dots, m-1$, $i=0, 1, \dots, n-1$, which can be solved from the set of $m\times n$ linear algebraic equations of eqn. (32) in matrix form.

The explicit form of the solution of $\mathbf{Y}$ is of practical importance for computer computation. Derivation of the explicit form of the solution of Y is as follows. First, multiply eqn. (32) by $\mathbf{P}_{(n\times n)}^{-1}$ from the right :

$$\begin{aligned}\mathbf{P}_{(m\times m)}^{\mathrm{T}}\,\mathbf{YP}_{(n\times n)}^{-1}+\mathbf{Y}-a\mathbf{P}_{(m\times m)}^{\mathrm{T}}\,\mathbf{Y}&=\mathbf{P}_{(m\times m)}^{\mathrm{T}}\,\mathbf{GP}_{(n\times n)}^{-1}\\&\quad+\mathbf{F}+b\mathbf{P}_{(m\times m)}^{\mathrm{T}}\,\mathbf{U}\triangleq\mathbf{Q}\end{aligned} \tag{33}$$

where $\mathbf{Q}$ is an $m\times n$ matrix. The elements of $\mathbf{Q}$ are known. The first column of $\mathbf{Y}$ is defined as $\mathbf{y}_0$, the second column, $\mathbf{y}_1$, etc. Similarly, the first column of $\mathbf{Q}$ is defined as $\mathbf{q}_0$, the second column, $\mathbf{q}_1$, etc. Hence, eqn. (33) can be rewritten as

$$\begin{aligned}[\mathbf{y}_0, \mathbf{y}_1, \dots, \mathbf{y}_{n-1}]+\mathbf{P}_{(m\times m)}^{\mathrm{T}}[\mathbf{y}_0, \mathbf{y}_1, \dots, \mathbf{y}_{n-1}]\mathbf{P}_{(n\times n)}^{-1}\\-a\mathbf{P}_{(m\times m)}^{\mathrm{T}}[\mathbf{y}_0, \mathbf{y}_1, \dots, \mathbf{y}_{n-1}]=[\mathbf{q}_0, \mathbf{q}_1, \dots, \mathbf{q}_{n-1}]\end{aligned} \tag{34}$$

The matrix $\mathbf{Y}$ is now rearranged into a column vector of mn elements by changing its first column $\mathbf{y}_0$ into the first m components of the vector, and then the second column $\mathbf{y}_1$ into the second m components, etc. The $m\times n$ matrix $\mathbf{Q}$ is rearranged by the same manner. Finally, we have

$$\begin{aligned}\begin{bmatrix}\mathbf{y}_0\\ \mathbf{y}_1\\ \vdots\\ \mathbf{y}_{n-1}\end{bmatrix}+\mathbf{P}_{(m\times m)}^{\mathrm{T}}\otimes(\mathbf{P}_{(n\times n)}^{-1})^{\mathrm{T}}\begin{bmatrix}\mathbf{y}_0\\ \mathbf{y}_1\\ \vdots\\ \mathbf{y}_{n-1}\end{bmatrix}\\-a\mathbf{P}_{(m\times m)}^{\mathrm{T}}\otimes\mathbf{I}_{(n\times n)}\begin{bmatrix}\mathbf{y}_0\\ \mathbf{y}_1\\ \vdots\\ \mathbf{y}_{n-1}\end{bmatrix}=\begin{bmatrix}\mathbf{q}_0\\ \mathbf{q}_1\\ \vdots\\ \mathbf{q}_{n-1}\end{bmatrix}\end{aligned} \tag{35}$$

Here $\mathbf{A}\otimes\mathbf{B}^{\mathrm{T}}$ is the Kronecker product defined as

$$\mathbf{A}\otimes\mathbf{B}^{\mathrm{T}}=\begin{bmatrix}b_{11}\mathbf{A} & b_{21}\mathbf{A} & \dots & b_{n1}\mathbf{A}\\ b_{12}\mathbf{A} & b_{22}\mathbf{A} & \dots & b_{n2}\mathbf{A}\\ \dots & \dots & \dots & \dots\\ b_{1n}\mathbf{A} & b_{2n}\mathbf{A} & \dots & b_{nn}\mathbf{A}\end{bmatrix}$$

where b_{ji} are the components of $n \times n$ matrix **B**. Therefore, the solution for **Y** comes directly from eqn. (35) :

$$\begin{bmatrix} \mathbf{y}_0 \\ \mathbf{y}_1 \\ \vdots \\ \mathbf{y}_{n-1} \end{bmatrix} = [\mathbf{I} + \mathbf{P}_{(m\times m)}{}^{\mathrm{T}} \otimes (\mathbf{P}_{(n\times n)}{}^{-1})^{\mathrm{T}} - a\mathbf{P}_{(m\times m)}{}^{\mathrm{T}} \otimes \mathbf{I}_{(n\times n)}]^{-1} \begin{bmatrix} \mathbf{q}_0 \\ \mathbf{q}_1 \\ \vdots \\ \mathbf{q}_{n-1} \end{bmatrix} \tag{36}$$

Equation (36) is the solution of **Y**. It is convenient to be adopted to computer computation.

4. Example

Consider a simple example

$$\frac{\partial y}{\partial t} + \frac{\partial y}{\partial x} = -y + 1 \tag{37}$$

$$x = 0, \quad y(0, t) = 0$$

$$t = 0, \quad y(x, 0) = 1 - \exp(-x)$$

The exact solution is

$$y(x, t) = 1 - \exp(-x) \tag{38}$$

Notice that $y(x, t)$ is independent of t.

Choosing $m = n = 4$, the computation result from eqn. (36) is

$$\mathbf{Y} = \begin{bmatrix} 0{\cdot}36691 & 0{\cdot}00049 & 0{\cdot}00025 & -0{\cdot}00002 \\ -0{\cdot}15627 & 0{\cdot}00030 & 0{\cdot}00013 & 0{\cdot}00046 \\ -0{\cdot}07939 & 0{\cdot}00008 & 0{\cdot}00005 & -0{\cdot}00002 \\ -0{\cdot}01967 & -0{\cdot}00043 & 0{\cdot}00005 & 0{\cdot}00008 \end{bmatrix} \tag{39}$$

The values of $y(x, t)$ obtained from eqn. (39) is

(40)

	$y(x, t)$			
	$0 \leqslant t < \frac{1}{4}$	$\frac{1}{4} \leqslant t < \frac{1}{2}$	$\frac{1}{2} \leqslant t < \frac{3}{4}$	$\frac{3}{4} \leqslant t < 1$
$0 \leqslant x < \frac{1}{4}$	0·11303	0·11099	0·11111	0·11111
$\frac{1}{4} \leqslant x < \frac{1}{2}$	0·31152	0·31017	0·30845	0·30865
$\frac{1}{2} \leqslant x < \frac{3}{4}$	0·46368	0·46464	0·46350	0·46203
$\frac{3}{4} \leqslant x < 1$	0·58232	0·58283	0·58372	0·58275

The double Walsh series approximation of the exact solution of eqn. (38) is

$$\mathbf{Y}_{\text{exact}}=\begin{bmatrix} 0{\cdot}36787 & 0 & 0 & 0 \\ -0{\cdot}15481 & 0 & 0 & 0 \\ -0{\cdot}07860 & 0 & 0 & 0 \\ -0{\cdot}01925 & 0 & 0 & 0 \end{bmatrix} \tag{41}$$

The values of $y(x, t)$ obtained from eqn. (41) are

	$y(x, t), \quad 0 \leqslant t < 1$
$0 \leqslant x < \frac{1}{4}$	0·11520
$\frac{1}{4} \leqslant x < \frac{1}{2}$	0·31092
$\frac{1}{2} \leqslant x < \frac{3}{4}$	0·46267
$\frac{3}{4} \leqslant x < 1$	0·58338

(42)

The agreement between eqns. (40) and (42) is excellent.

5. Simultaneous first-order PDE

Consider simultaneous first-order partial differential equations of k equations as follows :

$$\frac{\partial y_i}{\partial t}+\sum_{j=1}^{k} c_{ij}\frac{\partial y_j}{\partial x}=\sum_{j=1}^{k} a_{ij}y_j+\sum_{j=1}^{p} b_{ij}u_j(x, t), \quad i=1, 2, \ldots, k \tag{43}$$

The boundary conditions are

$$t=0, \quad y_i(x, 0)=g_i(x), \quad i=1, 2, \ldots, k \tag{44}$$

$$x=0, \quad y_i(0, t)=f_i(t), \quad i=1, 2, \ldots, k \tag{45}$$

Letting the double Walsh series approximations of $y_i(x, t)$, $u_i(x, t)$, $g_i(x)$, $f_i(t)$ be

$$\left.\begin{aligned} y_i(x, t) &= \boldsymbol{\Phi}_{(m)}{}^{\mathrm{T}}(x)\mathbf{Y}_i\boldsymbol{\Phi}_{(n)}(t) \\ u_i(x, t) &= \boldsymbol{\Phi}_{(m)}{}^{\mathrm{T}}(x)\mathbf{U}_i\boldsymbol{\Phi}_{(n)}(t) \\ g_i(x) &= \boldsymbol{\Phi}_{(m)}{}^{\mathrm{T}}(x)\mathbf{G}_i\boldsymbol{\Phi}_{(n)}(t) \\ f_i(t) &= \boldsymbol{\Phi}_{(m)}{}^{\mathrm{T}}(x)\mathbf{F}_i\boldsymbol{\Phi}_{(n)}(t) \end{aligned}\right\} \tag{46}$$

$\mathbf{Y}_i$, $\mathbf{U}_i$, $\mathbf{G}_i$ and $\mathbf{F}_i$ are $m \times n$ matrices defined similar to eqns. (20) and (21). By the same procedure for the derivation of eqn. (33), we obtain

$$\mathbf{P}_{(m\times m)}{}^{\mathrm{T}}\,\mathbf{Y}_i\mathbf{P}_{(n\times n)}{}^{-1} + \sum_{j=1}^{k} c_{ij}\mathbf{Y}_j - \sum_{j=1}^{k} a_{ij}\mathbf{P}_{(m\times m)}{}^{\mathrm{T}}\,\mathbf{Y}_j$$

$$= \mathbf{P}_{(m\times m)}{}^{\mathrm{T}}\,\mathbf{G}_i\mathbf{P}_{(n\times n)}{}^{-1} + \sum_{j=1}^{k} c_{ij}\mathbf{F}_j + \sum_{j=1}^{p} b_{ij}\mathbf{P}_{(m\times m)}{}^{\mathrm{T}}\,\mathbf{U}_j$$

$$\triangleq \mathbf{Q}_i, \quad i=1, 2, \ldots, k \tag{47}$$

where the $\mathbf{Q}$s are $m \times n$ matrices. Equation (47) is a k set of algebraic equations for the unknown matrices $\mathbf{Y}_i$. The solution of $\mathbf{Y}_i$, $i=1, 2, \ldots, k$ is derived as follows.

Let the components of $\mathbf{Y}_i$ be rearranged into an mn vector $\mathbf{z}_i$:

$$\mathbf{z}_i = \begin{bmatrix} \text{first column of } \mathbf{Y}_i \\ \text{second column of } \mathbf{Y}_i \\ \cdots\cdots \\ n\text{th column of } \mathbf{Y}_i \end{bmatrix} \tag{48}$$

Similarly, the mn vector $\mathbf{w}_i$ is

$$\mathbf{w}_i = \begin{bmatrix} \text{first column of } \mathbf{Q}_i \\ \text{second column of } \mathbf{Q}_i \\ \cdots\cdots \\ n\text{th column of } \mathbf{Q}_i \end{bmatrix} \tag{49}$$

Notice that the columns of $\mathbf{Y}_i$ and $\mathbf{Q}_i$ are m-dimensional. Following the procedure of eqns. (33)–(36), eqn. (47) becomes

$$\mathbf{P}_{(m\times m)}{}^{\mathrm{T}} \otimes (\mathbf{P}_{(n\times n)}{}^{-1})^{\mathrm{T}}\,\mathbf{z}_i + \sum_{j=1}^{k} c_{ij}\mathbf{z}_j$$

$$- \sum_{j=1}^{k} a_{ij}(\mathbf{P}_{(m\times m)}{}^{\mathrm{T}} \otimes \mathbf{I}_{(n\times n)})\mathbf{z}_j = \mathbf{w}_i, \quad i=1, 2, \ldots, k \tag{50}$$

where $\mathbf{I}_{(n\times n)}$ is an $n \times n$ identity matrix. Taking the transpose of eqn. (50), we have

$$\mathbf{z}_i{}^{\mathrm{T}}[\mathbf{P}_{(m\times m)}{}^{\mathrm{T}} \otimes (\mathbf{P}_{(n\times n)})^{\mathrm{T}}]^{\mathrm{T}} + \sum_{j=1}^{k} c_{ij}\mathbf{z}_j{}^{\mathrm{T}}$$

$$- \sum_{j=1}^{k} a_{ij}\mathbf{z}_j{}^{\mathrm{T}}[\mathbf{P}_{(m\times m)}{}^{\mathrm{T}} \otimes \mathbf{I}_{(n\times n)}]^{\mathrm{T}} = \mathbf{w}_i{}^{\mathrm{T}}, \quad i=1, 2, \ldots, k \tag{51}$$

Notice that $\mathbf{z}_i{}^{\mathrm{T}}$ and $\mathbf{w}_i{}^{\mathrm{T}}$ are $1 \times mn$ vectors (row vectors). Letting

$$\mathbf{Z} = \begin{bmatrix} \mathbf{z}_1{}^{\mathrm{T}} \\ \mathbf{z}_2{}^{\mathrm{T}} \\ \vdots \\ \mathbf{z}_k{}^{\mathrm{T}} \end{bmatrix}, \quad \mathbf{W} = \begin{bmatrix} \mathbf{w}_1{}^{\mathrm{T}} \\ \mathbf{w}_2{}^{\mathrm{T}} \\ \vdots \\ \mathbf{w}_k{}^{\mathrm{T}} \end{bmatrix} \tag{52}$$

$\mathbf{Z}$ and $\mathbf{W}$ are $k\times mn$ matrices. The k equations of (51) are now combined together :

$$\mathbf{I}_{(k\times k)}\mathbf{Z}[\mathbf{P}_{(m\times m)}{}^{\mathrm{T}}\otimes(\mathbf{P}_{(n\times n)}{}^{-1})^{\mathrm{T}}]^{\mathrm{T}}+\mathbf{C}_{(k\times k)}\mathbf{Z}$$
$$-\mathbf{A}_{(k\times k)}\mathbf{Z}\ [\mathbf{P}_{(m\times m)}{}^{\mathrm{T}}\otimes\mathbf{I}_{(n\times n)}]^{\mathrm{T}}=\mathbf{I}_{(k\times k)}\mathbf{W} \tag{53}$$

where

$$\mathbf{C}_{(k\times k)}=[c_{ij}]$$
$$\mathbf{A}_{(k\times k)}=[a_{ij}]$$

are the coefficient matrices of the partial differential equations of eqn. (43).

Letting the ith column of $\mathbf{Z}$ be $(\mathbf{Z})_i$, and the ith column of $\mathbf{W}$ be $(\mathbf{W})_i$, eqn. (53) can be rewritten as

$$\mathbf{I}_{(k\times k)}[(\mathbf{Z})_1,(\mathbf{Z})_2,\ldots,(\mathbf{Z})_{mn}][\mathbf{P}_{(m\times m)}{}^{\mathrm{T}}\otimes(\mathbf{P}_{(n\times n(}{}^{-1})^{\mathrm{T}}]^{\mathrm{T}}$$
$$+\mathbf{C}_{(k\times k)}[(\mathbf{Z})_1,\ldots,(\mathbf{Z})_{mn}]\mathbf{I}_{(mn\times mn)}-\mathbf{A}_{(k\times k)}[(\mathbf{Z})_1,\ldots,(\mathbf{Z})_{mn}]$$
$$[\mathbf{P}_{(m\times m)}{}^{\mathrm{T}}\otimes\mathbf{I}_{(n\times n)}]^{\mathrm{T}}=\mathbf{I}_{(k\times k)}[(\mathbf{W})_1,(\mathbf{W})_2,\ldots,(\mathbf{W})_{mn}]\mathbf{I}_{(mn\times mn)} \tag{54}$$

Notice that $(\mathbf{Z})_i$ and $(\mathbf{W})_i$ are $k\times 1$ vectors. Rearrangement of $\mathbf{Z}$ and $\mathbf{W}$ into kmn vectors gives

$$\mathbf{I}_{(k\times k)}\otimes[\mathbf{P}_{(m\times m)}{}^{\mathrm{T}}\otimes(\mathbf{P}_{(n\times n)}{}^{-1})^{\mathrm{T}}]\begin{bmatrix}(\mathbf{Z})_1\\(\mathbf{Z})_2\\\vdots\\(\mathbf{Z})_{mn}\end{bmatrix}+[\mathbf{C}_{(k\times k)}\otimes\mathbf{I}_{(mn\times mn)}]\begin{bmatrix}(\mathbf{Z})_1\\(\mathbf{Z})_2\\\vdots\\(\mathbf{Z})_{mn}\end{bmatrix}$$
$$-\mathbf{A}_{(k\times k)}\otimes[\mathbf{P}_{(m\times m)}{}^{\mathrm{T}}\otimes\mathbf{I}_{(n\times n)}]\begin{bmatrix}(\mathbf{Z})_1\\(\mathbf{Z})_2\\\vdots\\(\mathbf{Z})_{mn}\end{bmatrix}=[\mathbf{I}_{(k\times k)}\otimes\mathbf{I}_{(mn\times mn)}]\begin{bmatrix}(\mathbf{W})_1\\(\mathbf{W})_2\\\vdots\\(\mathbf{W})_{mn}\end{bmatrix} \tag{55}$$

Therefore,

$$\begin{bmatrix}(\mathbf{Z})_1\\(\mathbf{Z})_2\\\vdots\\(\mathbf{Z})_{mn}\end{bmatrix}=\{\mathbf{I}_{(k\times k)}\otimes[\mathbf{P}_{(m\times m)}{}^{\mathrm{T}}\otimes(\mathbf{P}_{(n\times n)}{}^{-1})^{\mathrm{T}}]+\mathbf{C}_{(k\times k)}\otimes\mathbf{I}_{(mn\times mn)}$$
$$-\mathbf{A}_{(k\times k)}\otimes[\mathbf{P}_{(m\times m)}{}^{\mathrm{T}}\otimes\mathbf{I}_{(n\times n)}]\}^{-1}[\mathbf{I}_{(k\times k)}\otimes\mathbf{I}_{(mn\times mn)}]\begin{bmatrix}(\mathbf{W})_1\\(\mathbf{W})_2\\\vdots\\(\mathbf{W})_{mn}\end{bmatrix} \tag{56}$$

Equation (56) is the solution of $\mathbf{Y}_i$, $i=1,2,\ldots,k$. Once $\mathbf{Y}_i$ are determined, the solutions $y_i(x,t)$ are obtained.

6. Conclusion

A double Walsh series approximation is introduced. Application of a double Walsh series approximation, the solution of simultaneous first-order differential equations is reduced to the solution of linear simultaneous algebraic equations. Using the Kronecker product formula, the key step is the inversion of a $kmn \times kmn$ matrix.

The application of a double Walsh series should not be limited to the example of this paper. In fact the wide applications of a Walsh series, in the research field for multiple Walsh series should be of interest.

Acknowledgments

The financial support of the National Science Foundation is appreciated.

References

Chen, C. F., and Hsiao, C. H., 1975 a, *Int. J. Systems Sci.*, **6,** 833 ; 1975 b, *Int. J. Control*, **21,** 881 ; 1975 c, *Proc. Instn elect. Engrs*, **122,** 565 ; 1975 d, *I.E.E.E. Trans. autom. Control*, **20,** 596.

Chen, W. L., and Shih, Y. P., 1978, *Int. J. Control* (in the press).

Corrington, M. S., 1973, *I.E.E.E. Trans. Circuit Theory*, **20,** 470.

Proceedings of Walsh Function Symposium, 1970, 1971, 1972, 1973, 1974, Naval Research Laboratories, Washington, D.C., U.S.A.

9

Reprinted from *J. Franklin Inst.* **313**:207–217 (1982)

Walsh Series Analysis of Multi-delay Systems

by WEN-LIANG CHEN

Department of Power Mechanical Engineering, National Tsing Hua University, Hsinchu, Taiwan, Republic of China

ABSTRACT: *A matrix, called the "delay operational matrix", is constructed from the Walsh matrix. This matrix, together with some matrices obtained from the delay operational matrix after performing right-shift operations, is used to solve multi-delay linear dynamic systems. A simple example is given to compare the actual solution and the solution obtained by the techniques of this paper.*

I. Introduction

During the past few years, many authors have given much attention to the study of single delay dynamic systems (**1–5**). Among the methods used were direct integration, Walsh series expansion, and block pulse series expansion, respectively. In this paper, the Walsh series expansion method is chosen as the tool to solve multi-delay dynamic systems. Although the authors (**3**) have presented the same method to solve single delay systems, they will, however, have difficulty in choosing the reasonable number m, the number of terms of Walsh series, in multi-delay systems. For example, if a system contains two delays, 0.25 and 0.3, it cannot find a finite number m such that the technique (**3**) can be applied. Therefore, some modifications are given in this paper. A matrix called the *delay operational matrix* is introduced first from the well-known Walsh matrix. This matrix, together with some modified matrices formed from the delay operational matrix, is used to analyze multi-delay systems.

II. Delay Operational Matrix

Consider a set of Walsh functions $\phi_0(t), \phi_1(t), \ldots, \phi_{m-1}(t)$, where $m = 2^d$ for some positive integer d. It is well known that the Walsh matrix corresponding to these Walsh functions is

$$W_{m\times m} = \begin{bmatrix} \phi_0(0) & \phi_0(\frac{1}{m}) & \cdots & \phi_0\left(\frac{m-1}{m}\right) \\ \phi_1(0) & \phi_1\left(\frac{1}{m}\right) & \cdots & \phi_1\left(\frac{m-1}{m}\right) \\ \vdots & \vdots & & \vdots \\ \phi_{m-1}(0) & \phi_{m-1}\left(\frac{1}{m}\right) & \cdots & \phi_{m-1}\left(\frac{m-1}{m}\right) \end{bmatrix} \tag{1}$$

and the shift Walsh matrices $W(m; -k)$ for $k = 1, 2, \ldots, m-1$ given in Ref. (3) are

$$W(m; -k) = \begin{bmatrix} 0\cdots 0 & \phi_0(0) & \phi_0\left(\frac{1}{m}\right) & \cdots \phi_0\left(\frac{m-1-k}{m}\right) \\ 0\cdots 0 & \phi_1(0) & \phi_1\left(\frac{1}{m}\right) & \cdots \phi_1\left(\frac{m-1-k}{m}\right) \\ \vdots \quad \vdots & \vdots & \vdots & \\ 0\cdots 0 & \phi_{m-1}(0) & \phi_{m-1}\left(\frac{1}{m}\right) & \cdots \phi_{m-1}\left(\frac{m-1-k}{m}\right) \end{bmatrix}. \tag{2}$$

For example, let $m = 4$ and $k = 1$

$$W_{(4\times4)} = \begin{bmatrix} 1 & 1 & 1 & 1 \\ 1 & 1 & -1 & -1 \\ 1 & -1 & 1 & -1 \\ 1 & -1 & -1 & 1 \end{bmatrix} \quad \text{and} \quad W(4; -1) = \begin{bmatrix} 0 & 1 & 1 & 1 \\ 0 & 1 & 1 & -1 \\ 0 & 1 & -1 & 1 \\ 0 & 1 & -1 & -1 \end{bmatrix}.$$

A matrix $Q_{(m\times m)}$, which is called a *delay operational matrix*, is formed from the Walsh matrix $W_{(m\times m)}$ according to the following rule: Let q_{ij} denote the element of $Q_{(m\times m)}$ in the ith row and j^{th} column and w_{ij} denote the element of $W_{(m\times m)}$ in the ith row and j^{th} column. Define q_{ij}, $i, j = 1, 2, \ldots, m$, which is related to the w_{ij} as

$$q_{ij} = \begin{cases} w_{ij} & \text{for } i = 1, 2, \ldots, m;\ j = 1 \\ -2 & \text{for } j \neq i,\ w_{i,j-1} = 1 \text{ and } w_{ij} = -1 \\ 2 & \text{for } j \neq 1,\ w_{i,j-1} = -1 \text{ and } w_{ij} = 1 \\ 0 & \text{otherwise.} \end{cases} \tag{3}$$

For example, let $m = 4$. Then the delay operational matrix $Q_{(4\times4)}$ is

$$Q_{(4\times4)} = \begin{bmatrix} 1 & 0 & 0 & 0 \\ 1 & 0 & -2 & 0 \\ 1 & -2 & 2 & -2 \\ 1 & -2 & 0 & 2 \end{bmatrix}. \tag{4}$$

This example shows that $Q_{(m\times m)}$ can be obtained from $W_{(m\times m)}$ directly.

If a shift-right operation is performed on $Q_{(m\times m)}$ which is the same operation as $W(m; -k)$ to $W_{(m\times m)}$, then a set of matrices, denoted by $Q(m; -k)$, $k = 1, 2, \ldots, m-1$, is obtained. For example, if $m = 4$,

$$Q_{(4\times4)} = \begin{bmatrix} 1 & 0 & 0 & 0 \\ 1 & 0 & -2 & 0 \\ 1 & -2 & 2 & -2 \\ 1 & -2 & 0 & 2 \end{bmatrix} \quad Q(4; -1) = \begin{bmatrix} 0 & 1 & 0 & 0 \\ 0 & 1 & 0 & -2 \\ 0 & 1 & -2 & 2 \\ 0 & 1 & -2 & 0 \end{bmatrix}$$

$$Q(4;-2)=\begin{bmatrix}0&0&1&0\\0&0&1&0\\0&0&1&-2\\0&0&1&-2\end{bmatrix}\qquad Q(4;-3)=\begin{bmatrix}0&0&0&1\\0&0&0&1\\0&0&0&1\\0&0&0&1\end{bmatrix}.$$

The matrices $Q(m;-k)$, $k=1,2,\ldots,m-1$, together with $Q_{(m\times m)}$, are used to introduce delay into the Walsh functions which is not a multiple of $(1/m)$.

Let $\phi_i[t-(k/m+\tau)]\,u[t-(k/m+\tau)]$ be the $(i+1)^{\text{th}}$ Walsh function with time delay $(k/m)+\tau$, where $u(t)$ is unit step function, k is a positive integer, $0\leq k<m$, and $0\leq\tau<(1/m)$. If this function is examined carefully, it will be found that $\phi_i[t-(k/m+\tau)]\,u[t-(k/m+\tau)]$ can be decomposed into the summation of Walsh function $\phi_i(t-k/m)\,u(t-k/m)$ and some pulse functions. Let $p_\tau(t)$ be defined as a unit pulse function with pulse duration τ. Then from the piecewise-constant property of Walsh functions

$$\phi_i\left[t-\left(\frac{k}{m}+\tau\right)\right]u\left[t-\left(\frac{k}{m}+\tau\right)\right]=\phi_i\left(t-\frac{k}{m}\right)u\left(t-\frac{k}{m}\right)-q_{i+1}(m;\,-k)P_\tau(t) \tag{5}$$

where $q_{i+1}(m;\,-k)$ is the $(i+1)^{\text{th}}$ row of $Q(m;\,-k)$ and

$$P_\tau(t)\triangleq\left[p_\tau(t)\quad p_\tau\left(t-\frac{1}{m}\right)\ldots p_\tau\left(t-\frac{m-1}{m}\right)\right]^T. \tag{6}$$

For example,

$$\begin{aligned}\phi_2\left[t-\left(\frac{1}{4}+\tau\right)\right]u\left[t-\left(\frac{1}{4}+\tau\right)\right]&=\phi_2\left(t-\frac{1}{4}\right)u\left(t-\frac{1}{4}\right)-q_3(4;\,-1)P_\tau(t)\\&=\phi_2\left(t-\frac{1}{4}\right)u\left(t-\frac{1}{4}\right)-[0\quad 1\quad -2\quad 2]\begin{bmatrix}p_\tau(t)\\p_\tau\left(t-\frac{1}{4}\right)\\p_\tau\left(t-\frac{2}{4}\right)\\p_\tau\left(t-\frac{3}{4}\right)\end{bmatrix}\\&=\phi_2\left(t-\frac{1}{4}\right)u\left(t-\frac{1}{4}\right)-p_\tau\left(t-\frac{1}{4}\right)+2p_\tau\left(t-\frac{2}{4}\right)-2p_\tau\left(t-\frac{3}{4}\right).\end{aligned} \tag{7}$$

Equation (7) is shown in Fig. 1.

If $\phi_i[t-(k/m+\tau)]\,u[t-(k/m+\tau)]$ is expanded into m-term Walsh series, i.e.

$$\phi_i\left[t-\left(\frac{k}{m}+\tau\right)\right]u\left[t-\left(\frac{k}{m}+\tau\right)\right]=C_{i0}\phi_0(t)+C_{i1}\phi_1(t)+\cdots+C_{i,\,m-1}\phi_{m-1}(t) \tag{8}$$

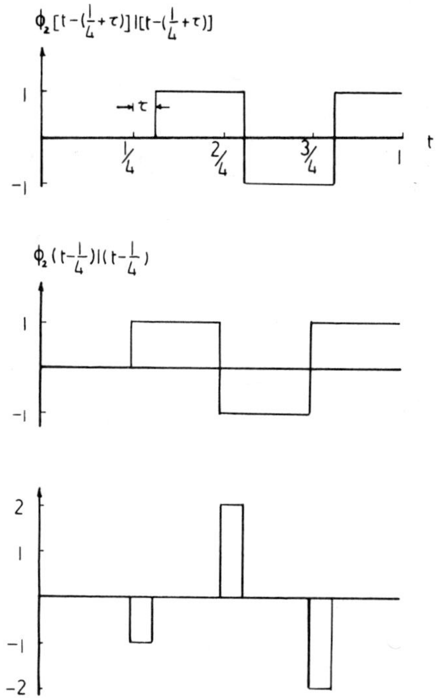

FIG. 1. Decomposition of delay Walsh functions.

the coefficients $C_{i0}, C_{i1}, \ldots, C_{i,m-1}$ can be evaluated as follows:

$$\begin{aligned}
C_{ij} &= \int_0^1 \phi_i\left[t-\left(\frac{k}{m}+\tau\right)\right] u\left[t-\left(\frac{k}{m}+\tau\right)\right]\phi_j(t)\,\mathrm{d}t \\
&= \int_0^1 \left[\phi_i\left(t-\frac{k}{m}\right)u\left(t-\frac{k}{m}\right)-q_{i+1}(m;\,-k)P_\tau(t)\right]\phi_j(t)\,\mathrm{d}t \\
&= \int_0^1 \phi_i\left(t-\frac{k}{m}\right)u\left(t-\frac{k}{m}\right)\phi_j(t)\,\mathrm{d}t - \int_0^1 q_{i+1}(m;\,-k)P_\tau(t)\phi_j(t)\,\mathrm{d}t \\
&= \left\{\frac{1}{m}\times[(i+1)^{\text{th}}\text{ row of } W(m;\,-k)]\times[(j+1)^{\text{th}}\text{ column of } W_{(m\times m)}]\right\} \\
&\quad - \tau q_{i+1}(m;\,-k)\times[(j+1)^{\text{th}}\text{ column of } W_{(m\times m)}] \\
&= \left\{\frac{1}{m}\times[(i+1)^{\text{th}}\text{ row of } W(m;\,-k)] - \tau q_{i+1}(m;\,-k)]\right\} \\
&\quad \times\{(j+1)^{\text{th}}\text{ column of } W_{(m\times m)}\}
\end{aligned} \tag{9}$$

for $j=0,1,\ldots,m-1$.

Substituting (9) into (8), yields

$$\phi_i\left[t-\left(\frac{k}{m}+\tau\right)\right]u\left[t-\left(\frac{k}{m}+\tau\right)\right]$$
$$=\left\{\frac{1}{m}[(i+1)^{\text{th}}\text{ row of }W(m;\,-k)]-\tau q_{i+1}(m;\,-k)\right\}W_{(m\times m)}\phi_{(m)}(t) \quad (10)$$

where

$$\phi_{(m)}(t)\triangleq[\phi_0(t)\quad \phi_1(t)\ldots\phi_{m-1}(t)]^T.$$

Because (10) is valid for $i=0,1,\ldots,m-1$, therefore

$$\phi_{(m)}\left[t-\left(\frac{k}{m}+\tau\right)\right]u\left[t-\left(\frac{k}{m}+\tau\right)\right]$$
$$=\left\{\frac{1}{m}W(m;\,-k)-\tau Q(m;\,-k)\right\}W_{(m\times m)}\phi_{(m)}(t). \quad (11)$$

Equation (11) is the general form of the Walsh series expansion of delay Walsh functions and will be used to solve the multi-delay systems in the next section.

Example 1

For $m=4$ and $k=1$, the shift Walsh matrix $W(4;\,-1)$ and $Q(4;\,-1)$ are, respectively,

$$W(4;\,-1)=\begin{bmatrix}0&1&1&1\\0&1&1&-1\\0&1&-1&1\\0&1&-1&-1\end{bmatrix}\quad\text{and}\quad Q(4;\,-1)=\begin{bmatrix}0&1&0&0\\0&1&0&-2\\0&1&-2&2\\0&1&-2&0\end{bmatrix}.$$

Therefore

$$\phi_{(4)}\left[t-\left(\frac{1}{4}+\tau\right)\right]u\left[t-\left(\frac{1}{4}+\tau\right)\right]=\left\{\frac{1}{4}W(4;\,-1)-\tau Q(4;\,-1)\right\}W_{(4\times4)}\phi_{(4)}(t)$$
$$=\begin{bmatrix}\frac{3}{4}-\tau & -\frac{1}{4}-\tau & -\frac{1}{4}+\tau & -\frac{1}{4}\tau\\ \frac{1}{4}+\tau & \frac{1}{4}-3\tau & \frac{1}{4}-\tau & -\frac{3}{4}+3\tau\\ \frac{1}{4}-\tau & \frac{1}{4}-\tau & -\frac{3}{4}+5\tau & \frac{1}{4}-3\tau\\ -\frac{1}{4}\tau & \frac{3}{4}-3\tau & -\frac{1}{4}+3\tau & -\frac{1}{4}-\tau\end{bmatrix}\phi_{(4)}(t). \quad (12)$$

Although the result in (12) can also be obtained through direct expansion, the method used in this paper is a very convenient matrix manipulation for computer computation.

III. Multi-delay Systems

Consider the multi-delay linear dynamic system

$$\begin{aligned} \dot{X}(t) &= AX(t) + \sum_{i=1}^{r} B_i X\left[t - \left(\frac{k_i}{m} + \tau_i\right)\right] + DU(t) \\ X(t) &= F(t) \text{ for } t < 0 \\ X(0) &= X_0 \end{aligned} \tag{13}$$

where $k_1, k_2, \ldots, k_r$ are positive integers with $k_1 \leq k_2 \leq \ldots \leq k_r < m$, $0 \leq \tau_i < (1/m)$ for $i = 1, 2, \ldots, r$, and $m = 2^d$ for positive integer d. The dimensions of (13) are $X(t) \in R^n$ and $U(t) \in R^p$.

Because the function $F(t)$ is given, the following data can be calculated beforehand, i.e.

$$F\left[t - \left(\frac{k_i}{m} + \tau_i\right)\right] = V_i \phi_{(m)}(t) \tag{14}$$

$$\int_0^{k_i/m+\tau_i} F\left[t - \left(\frac{k_i}{m} + \tau_i\right)\right] \mathrm{d}t = Z_i \tag{15}$$

for $i = 1, 2, \ldots, r$, where V_i is an $n \times m$ matrix and this matrix is the coefficients of Walsh series of $F[t - (k_i/m + \tau_i)]$. Without loss of generality, we assume

$$\frac{k_1}{m} + \tau_1 < \frac{k_2}{m} + \tau_2 < \ldots < \frac{k_r}{m} + \tau_r.$$

Then the solution of (13) for $0 \leq t < 1$ can be derived as follows:

Integrating, (13) becomes

$$\begin{aligned} X(t) = &\int_0^t AX(t')\,\mathrm{d}t' + \int_0^t \sum_{i=1}^{r} B_i X\left[t' - \left(\frac{k_i}{m} + \tau_i\right)\right] \mathrm{d}t' \\ &+ \int_0^t DU(t')\,\mathrm{d}t' + X_0. \end{aligned} \tag{16}$$

Assume that the Walsh series expansions of $X(t)$ and $U(t)$ are, respectively,

$$X(t) \triangleq H\phi_{(m)}(t)$$

and

$$U(t) \triangleq G\phi_{(m)}(t) \tag{17}$$

where H and G are $n \times m$ and $p \times m$ matrices, respectively. Then (16) can be solved by the Walsh series method of this paper. The solution is partitioned into several sections, each calculated separately as follows:

For $0 \le t < (k_1/m) + \tau_1$,

$$X(t) = \int_0^t AX(t')\,dt' + \int_0^t \sum_{i=1}^{r} B_i F\left[t' - \left(\frac{k_i}{m} + \tau_i\right)\right] dt' + \int_0^t DU(t')\,dt' + X_0. \tag{18}$$

Substituting (14) and (17) into (18), one has

$$H\phi_{(m)}(t) = AHP\phi_{(m)}(t) + \sum_{i=1}^{r} B_i V_i P\phi_{(m)}(t) + DGP\phi_{(m)}(t) + E\phi_{(m)}(t) \tag{19}$$

where P is called operational matrix in Ref. (**6**) and

$$E \triangleq [X_0 \quad \underbrace{0 \quad 0 \ldots 0}_{m-1 \text{ columns}}].$$

Thus

$$H = AHP + \sum_{i=1}^{r} B_i V_i P + DGP + E. \tag{20}$$

This is a set of $n \times m$ linear algebraic equations which can be solved through algebraic computation.

For $\quad \dfrac{k_1}{m} + \tau_1 \le t < \dfrac{k_2}{m} + \tau_2,$

$$\begin{aligned} X(t) = {} & \int_0^t AX(t')\,dt' + \int_0^{k_1/m+\tau_1} B_1 F\left[t' - \left(\frac{k_1}{m} + \tau_1\right)\right] dt' \\ & + \int_0^t B_1 X\left[t' - \left(\frac{k_1}{m} + \tau_1\right)\right] u\left[t' - \left(\frac{k_1}{m} + \tau_1\right)\right] dt' \\ & + \int_0^t \sum_{i=2}^{r} B_i F\left[t' - \left(\frac{k_i}{m} + \tau_i\right)\right] dt' \\ & + \int_0^t DU(t')\,dt' + X_0. \end{aligned} \tag{21}$$

Substituting (14), (15) and (17) into (21), one obtains

$$
\begin{aligned}
H\phi_{(m)}(t) &= AHP\phi_{(m)}(t) + B_1\bar{Z}_1\phi_{(m)}(t) \\
&\quad + B_1H\int_0^t \phi_{(m)}\left[t' - \left(\frac{k_1}{m} + \tau_1\right)\right] u\left[t' - \left(\frac{k_1}{m} + \tau_1\right)\right] \mathrm{d}t' \\
&\quad + \sum_{i=2}^{r} B_iV_iP\phi_{(m)}(t) + DGP\phi_{(m)}(t) + E\phi_{(m)}(t) \\
&= AHP\phi_{(m)}(t) + B_1\bar{Z}_1\phi_{(m)}(t) \\
&\quad + B_1H\left\{\frac{1}{m}W(m; -k_1) - \tau_1Q(m; -k_1)\right\} W_{(m\times m)}P\phi_{(m)}(t) \\
&\quad + \sum_{i=2}^{r} B_iV_iP\phi_{(m)}(t) + DGP\phi_{(m)}(t) + E\phi_{(m)}(t)
\end{aligned}
$$

where

$$
\bar{Z}_1 \triangleq [Z_1 \quad \underbrace{0 \quad 0 \ldots 0}_{m-1 \text{ columns}}].
$$

Thus

$$
\begin{aligned}
H = AHP + B_1\bar{Z}_1 + B_1H\left\{\frac{1}{m}W(m; -k_1) - \tau_1Q(m; -k_1)\right\} W_{(m\times m)}P \\
+ \sum_{i=2}^{r} B_iV_iP + DGP + E.
\end{aligned} \tag{22}
$$

Equation (22) is also a set of $n \times m$ algebraic equations and can be solved through algebraic computation.

In general, the Walsh series coefficients of the solution $X(t)$ can be calculated from the following $n \times m$ algebraic equations:

$$
\begin{aligned}
H = AHP + \sum_{j=1}^{i} B_j\bar{Z}_j + \sum_{j=1}^{i} B_jH\left\{\frac{1}{m}W(m; -k_j) - \tau_jQ(m; -k_j)\right\} W_{(m\times m)}P \\
+ \sum_{j=i+1}^{r} B_jV_jP + DGP + E
\end{aligned} \tag{23}
$$

for

$$
\frac{k_i}{m} + \tau_i \leqslant t < \frac{k_{i+1}}{m} + \tau_{i+1}, \quad i = 1, 2, \ldots, r-1,
$$

and

$$
\begin{aligned}
H = AHP + \sum_{j=1}^{r} B_j\bar{Z}_j + \sum_{j=1}^{r} B_jH\left\{\frac{1}{m}W(m; -k_j) - \tau_jQ(m; -k_j)\right\} \\
W_{(m\times m)}P + DGP + E
\end{aligned} \tag{24}
$$

for $(k_r/m)+\tau_r \leq t < 1$. Therefore, the solutions of (23) for $i = 1, 2, \ldots, r-1$ together with the solutions of (20) and (24) form the piecewise constant solution of the multi-delay system.

Example 2
Let the delay dynamic system be

$$\dot{X}(t) = X(t-0.35) + X(t-0.7) + 1$$
$$X(t) = 0 \text{ for } t \leq 0. \tag{25}$$

Equation (25) is rewritten to the form given in (13) and if $m = 4$ is chosen,

$$\dot{X}(t) = X\left[t - \left(\frac{1}{4} + 0.1\right)\right] + X\left[t - \left(\frac{2}{4} + 0.2\right)\right] + 1. \tag{26}$$

For $0 \leq t < 0.35$, (20) becomes

$$H = GP \tag{27}$$

where

$$H \triangleq [h_0 \quad h_1 \quad h_2 \quad h_3],$$
$$G = [1 \quad 0 \quad 0 \quad 0]$$

and

$$P = \begin{bmatrix} 0.5 & -0.25 & -0.125 & 0 \\ 0.25 & 0 & 0 & -0.125 \\ 0.125 & 0 & 0 & 0 \\ 0 & 0.125 & 0 & 0 \end{bmatrix}.$$

The solution of (27) is

$$H = [0.5 \quad -0.25 \quad -0.125 \quad 0]$$

i.e. $X(t) = 0.5\phi_0(t) - 0.25\phi_1(t) - 0.125\phi_2(t)$.
For $0.35 \leq t < 0.7$, (23) becomes

$$H = B_1 H\left\{\frac{1}{4}\, W(4; -1) - 0.1Q(4; -1)\right\} W_{(4\times4)}P + DGP \tag{28}$$

where

$$W_{(4\times4)} = \begin{bmatrix} 1 & 1 & 1 & 1 \\ 1 & 1 & -1 & -1 \\ 1 & -1 & 1 & -1 \\ 1 & -1 & -1 & 1 \end{bmatrix}, \qquad W(4; -1) = \begin{bmatrix} 0 & 1 & 1 & 1 \\ 0 & 1 & 1 & -1 \\ 0 & 1 & -1 & 1 \\ 0 & 1 & -1 & -1 \end{bmatrix}$$

$$Q(4;\,-1)=\begin{bmatrix}0&1&0&0\\0&1&0&-2\\0&1&-2&2\\0&1&-2&0\end{bmatrix},\qquad D=B_1=1.$$

Solving (28), one obtains

$$H=[0.555579\quad -0.300081\quad -0.153612\quad 0.023953]$$

i.e. $X(t)=0.555579\phi_0(t)-0.300081\phi_1(t)-0.153612\phi_2(t)+0.023953\phi_3(t)$.
For $0.7 \leqslant t < 1$, (24) becomes

$$H=B_1H\left\{\frac{1}{4}W(4;\,-1)-0.1Q(4;\,-1)\right\}W_{(4\times 4)}P + B_2H\left\{\frac{1}{4}W(4;\,-2)-0.2Q(4;\,-2)\right\}W_{(4\times 4)}P+DGP \tag{29}$$

where $B_1=B_2=1$,

$$W(4;\,-2)=\begin{bmatrix}0&0&1&1\\0&0&1&1\\0&0&1&-1\\0&0&1&-1\end{bmatrix}\quad\text{and}\quad Q(4;\,-2)=\begin{bmatrix}0&0&1&0\\0&0&1&0\\0&0&1&-2\\0&0&1&-2\end{bmatrix}.$$

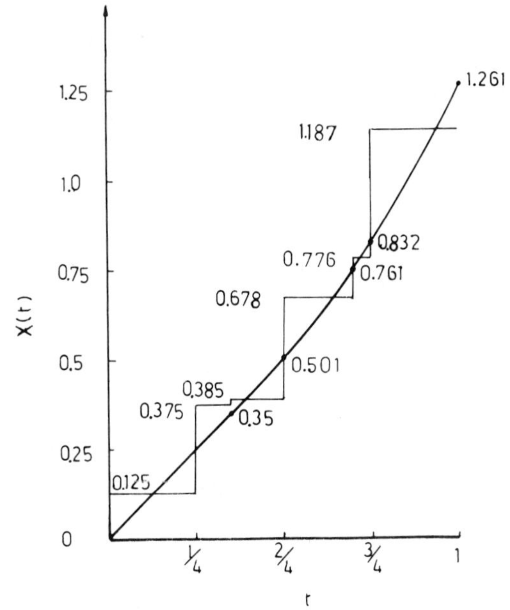

FIG. 2. Comparison of solutions.

Solving (29), one has

$$H = [0.655315 \quad -0.326205 \quad -0.169165 \quad 0.036774]$$

i.e. $X(t) = 0.655315\phi_0(t) - 0.326205\phi_1(t) - 0.169165\phi_2(t) + 0.036774\phi_3(t)$.
Figure 2 compares the approximate solution obtained above and the exact solution

$$X(t) = \begin{cases} t & \text{for } 0 \leq t < 0.35 \\ \frac{1}{2}(t-0.35)^2 + t & \text{for } 0.35 \leq t < 0.7 \\ \frac{1}{6}(t-0.7)^3 + \frac{1}{2}(t-0.7)^2 + \frac{1}{2}(t-0.35) + t & \text{for } 0.7 \leq t < 1. \end{cases}$$

From Fig. 2, the result we have obtained is quite satisfactory.

IV. Conclusions

A general formula was derived to analyze multi-delay dynamic systems. Although the solution obtained by this method is not exact, it has several advantages:

(1) It is eminently suitable for computer computation.

(2) The number of time intervals to be partitioned is equal to the number of delays plus one.

(3) The number of terms in Walsh series expansion is independent of the values of delays.

In this paper, only time-invariant systems were considered. However, the method presented here can also be applied to time-varying systems.

References

(1) A. C. Tsoi, "Explicit solution of a class of delay-differential equations", *Int. J. Control,* Vol. 21, pp. 39–48, 1975.

(2) A. C. Tsoi, "Explicit solution to a class of neutral delay differential equations", *Int. J. Control,* Vol. 22, pp. 787–799, 1975.

(3) W. L. Chen and Y. P. Shih, "Shift Walsh matrix and delay-differential equations", *IEEE Trans. Automat. Control,* Vol. AC-23, pp. 1023–1028, 1978.

(4) Y. P. Shih, "Time-partition solution for delay-differential equations", *Trans. Aeronaut. Astronaut. Soc.* Vol. 10, pp. 80–94, 1977.

(5) G. P. Rao and T. Srinivasan, "Analysis and synthesis of dynamic systems containing time delays via block pulse functions", *Proc. IEE,* Vol. 125, pp. 1064–1068, 1978.

(6) C. F. Chen and C. H. Hsiao, "Walsh series analysis in optimal control", *Int. J. Control,* Vol. 21, pp. 881–897, 1975.

10

On the Walsh Analysis of Nonlinear Systems

MOHAMMAD MAQUSI, MEMBER, IEEE

***Abstract*–This paper considers some aspects of the analysis of nonlinear systems via Walsh functions. Two basic approaches are investigated. These are 1) a characteristic-function method which assumes a Walsh-transformable nonlinearity and 2) a matrix method that is suitable when the input can be represented by a truncated Walsh series expansion. Illustrative examples are included.**

***Key Words:* Nonlinear systems, Walsh analysis, statistical dyadic, matrix.**

I. INTRODUCTION

In recently published work related to Walsh functions, most attention has been devoted to the analysis of linear problems. The nonlinear case has received little attention. Thus in this paper we devote our attention to the nonlinear case.

Corrington [1] was the first to outline a graphical procedure for specifying the output of a nonlinear system when excited by deterministic Walsh-type signals. Weiser [2] then studied

Manuscript received April 15, 1977, revised March 15, 1978.

The author is with the Electrical Engineering Department, University of Jordan, Amman, Jordan. 65111, ext. 1678.

the problem of analyzing nonlinear systems when subjected to stochastic inputs. However, neither of these efforts present results that are convenient for performing computations.

This paper presents two different approaches for Walsh analysis of nonlinear systems. These are 1) a characteristic-function method for deriving an expression for the output stayadic (statistical dyadic) correlation function and 2) a matrix method that is suitable when the input can be represented by a truncated Walsh series expansion.

II. WALSH CHARACTERISTIC-FUNCTION METHOD

In general, for a random variable x with a probability density function $f(x)$, a Walsh characteristic function [3] is defined as the expectation of a Walsh function (kernel) $\psi(\sigma, x)$. The quantity σ is referred to as a sequency parameter, and $\psi(\sigma, x)$ denotes a Walsh-Fine-type function [4], [5]. Both σ and x are taken as nonnegative.

Hence

$$S_x(\sigma) = E[\psi(\sigma, x)] = \int_0^\infty f(x)\psi(\sigma, x)dx \tag{1}$$

which implies that $S_x(\sigma)$ is simply the Walsh transform of the density function $f(x)$. This characteristic function is such that

$$S_x(0) = 1$$

and

$$|S_x(\sigma)| \leqslant 1.$$

These properties can be established easily by noting that $\psi(0, x) = 1$ and $|\psi(\sigma, x)| \leqslant 1$ for all x. It is possible to derive moments of the random variable x from $S_x(\sigma)$ via the dyadic derivative operator [6].

Our main objective here is to derive the output stayadic correlation function for a Walsh-transformable nonlinearity. We consider a zero-memory nonlinearity described by

$$y = g(x), \qquad x \geqslant 0. \tag{2}$$

The nonlinearity is assumed to be such that it can be specified by the integral

$$g(x) = \int_0^\infty G(\sigma)\psi(\sigma, x)d\sigma \tag{3}$$

where $G(\sigma)$ is the Walsh transform of $g(x)$; i.e.,

$$G(\sigma) = \int_0^\infty g(x)\psi(\sigma, x)dx. \tag{4}$$

We also assume that the inputs to the nonlinear system are sample functions of a weakly dyadic stationary (WDS) process. A process $z(t)$ is WDS, if

$$E[z^2(t)] < \infty$$

and

$$E[z(t_1)z(t_2)] = C_z(\tau), \qquad t_2 = t_1 \oplus \tau$$

where the operator $\oplus$ denotes dyadic summation (mod-2 addition), and $C_z(\tau)$ is called a stayadic correlation function. Two important properties of $C_z(\tau)$ are:

$$C_z(0) = E[z^2(t)]$$

and

$$|C_z(\tau)| \leqslant C_z(0).$$

Thus the output stayadic correlation function is given by

$$C_y(\tau; t_1) = E[y(t_1)y(t_1 \oplus \tau)]. \tag{5}$$

If $f(x_1, x_2; \tau)$ denotes the joint probability density function for the random variables $x_1 = x(t_1)$ and $x_2 = x(t_2) = x(t_1 \oplus \tau)$, then (5) can be rewritten as

$$C_y(\tau) = \int_0^\infty\int_0^\infty g(x_1)g(x_2)f(x_1, x_2; \tau)dx_1dx_2. \tag{6}$$

Again, the Walsh characteristic function for the random variables x_1 and x_2 is given by

$$S_x(\sigma, z; \tau) = E[\psi(\sigma, x_1)\psi(z, x_2)]$$

$$= \int_0^\infty\int_0^\infty f(x_1, x_2; \tau)\psi(\sigma, x_1)\psi(z, x_2)dx_1dx_2. \tag{7}$$

Multiplying both sides of (7) by $\psi(\sigma, x_1')\ \psi(z, x_2')$ followed by integrating with respect to σ and z, and using the notation of a delta function as

$$\delta(x) = \int_0^\infty \psi(\sigma, x)d\sigma \tag{8}$$

we obtain

$$\int_0^\infty\int_0^\infty S_x(\sigma, z; \tau)\psi(\sigma, x_1')\psi(z, x_2')d\sigma dz = f(x_1', x_2'; \tau). \tag{9}$$

In (9) we multiply both sides by $g(x_1')\ g(x_2')$, integrate with respect to x_1' and x_2', and use (6) to obtain the end result

$$C_y(\tau) = \int_0^\infty\int_0^\infty S_x(\sigma, z; \tau)G(\sigma)G(z)d\sigma dz. \tag{10}$$

Although we have written $C_y(\tau)$ as a function of τ only, in some cases it may be a function of t_1 also.

This method is now illustrated by two examples.

A. *Example 1: Half-Wave Rectifier/Detector*

Let

$$g(x) = \begin{cases} 1, & x \geqslant 0 \\ 0, & x < 0. \end{cases}$$

Then

$$G(\sigma) = \int_0^\infty \psi(\sigma, x)dx = \delta(\sigma).$$

This device represents a rectifier hard-limiter system which is used in communication problems. For instance, this device may be followed by a bandpass filter to form a nonlinear amplifier (also called a bandpass half-wave rectifier) which may be employed in FM receivers. On the other hand, if it is followed by a low-pass filter, the combination yields a detector [7].

Now,

$$C_y(\tau) = \int_0^\infty \int_0^\infty S_x(\sigma, z; \tau)\delta(\sigma)\delta(z)d\sigma dz.$$

Assuming that $x_1 = x(t_1)$, and $x_2 = x(t_1 \oplus \tau)$ are independent and uniform on $[0, a]$ for all τ, we obtain

$$S_x(\sigma, z; \tau) = \int_0^a \int_0^a \frac{1}{a^2}\psi(\sigma, x_1)\psi(z, x_2)dx_1dx_2$$

$$= \frac{1}{a^2}J(\sigma, a)J(z, a)$$

where $J(\sigma, a)$ is an integral originally defined by Fine [4], [8] as

$$J(\sigma, a) = \int_0^a \psi(\sigma, x)dx. \tag{11}$$

Evaluation of $C_y(\tau)$ yields

$$C_y(\tau) = \frac{1}{a^2}J(0, a)J(0, a) = 1, \qquad \tau \geqslant 0$$

since $J(0, a) = a$. For this simple example, the above result may be verified by direct computation, giving

$$C_y(\tau) = \int_0^\infty \int_0^\infty g(x_1)g(x_2)f(x_1, x_2; \tau)dx_1dx_2$$

$$= \int_0^\infty \int_0^\infty f(x_1)f(x_2)dx_1dx_2 = 1, \qquad \tau \geqslant 0.$$

B. Example 2: Spatial Filter

Let

$$g(x) = \begin{cases} 1, & 0 \leqslant x < 2^m \text{ where } m \text{ is a positive} \\ & \text{a positive integer} \\ 0, & \text{otherwise.} \end{cases}$$

Then

$$G(\sigma) = \int_0^{2^m} \psi(\sigma, x)dx = J(\sigma, 2^m)$$

$$= 2^m J(2^m\sigma, 1) = \begin{cases} 2^m, & 0 \leqslant 2^m\sigma < 1 \\ 0, & 1 \leqslant 2^m\sigma. \end{cases}$$

This nonlinearity can be viewed as an amplitude sensitive filter which yields an output if the input amplitude is in the range $(0, 2^m)$. Such a filter may be employed to detect the presence or absence of a signal with amplitude variations in the range $(0, 2^m)$. We assume that $S_x(\sigma, z; \tau)$ is as given in Example 1. With $a = 2^m$, we get

$$C_y(\tau) = (2^m)^2 \int_0^{1/2m} \int_0^{1/2m} \frac{1}{(2^m)^2} J(\sigma, 2^m) \cdot J(z, 2^m)d\sigma dz.$$

But

$$J(\sigma, 2^m) = \sum_{k=0}^{2^m-1} \psi(k, \sigma) + J([\sigma], 0)\psi(2^m, \sigma)$$

due to the following property of $J(\sigma, w)$:

$$J(\sigma, w) = C_1(\sigma) \sum_{k=0}^{[w]-1} \psi(k, \sigma) + \psi([w], \sigma)J([\sigma], w - [w])$$

where $[\sigma]$ is the greatest integer in σ; $C_1(\sigma) = 1$, $0 \leqslant \sigma < 1$, and is zero otherwise.

Hence

$$J(\sigma, 2^m) = \sum_{k=0}^{2^m-1} \psi(k, \sigma) = D_{2^m}(\sigma) = 2^m$$

where $D_{2^m}(\sigma)$ is a Walsh-Dirichlet kernel, which implies that

$$C_y(\tau) = 1, \qquad \tau \geqslant 0.$$

III. MATRIX METHOD

Here we develop a useful output expression for the output of a nonlinear system in cases where the input can be approximated by a truncated Walsh series, i.e.,

$$x(t) \doteq x_n(t) = \sum_{n=0}^{N-1} A_n \text{ wal}(n, t) \tag{12}$$

where wal (n, t) is a sequency-ordered Walsh function [5], [9], and A_n is a Walsh coefficient given by

$$A_n = \int_0^1 x(t) \text{ wal}(n, t)\, dt. \tag{13}$$

It is convenient to represent Walsh functions by corresponding block orthogonal functions defined by

$$b(i, t) = \begin{cases} 1, & \frac{i-1}{N} \leqslant t < \frac{i}{N} \\ 0, & \text{otherwise,} \end{cases} \quad i = 1, 2, \cdots, N. \tag{14}$$

The first N of these functions are used to represent the first N Walsh functions. For convenience, N is considered to be of the form 2^m.

Now

$$\text{wal}(n, t) = \sum_{i=1}^{N} a(i, n)b(i, t) \tag{15}$$

where

$$a(i, n) = N\int_0^1 \text{wal}(n, t)b(i, t)\, dt$$
$$= \pm 1, \quad \begin{array}{l} i = 1, 2, \cdots, N \\ n = 0, 1, \cdots, N-1. \end{array} \tag{16}$$

In matrix notation, (15) is written as

$$[w(t)] = [a][b(t)] \quad w(t) = \text{wal}(\cdot, t). \tag{17}$$

The matrix $[a]$ is an orthogonal Walsh matrix W [9] of order N. Thus

$$[a] = W; \ [a]^{-1} = W^{-1} = \frac{1}{N}[a]. \tag{18}$$

The input in matrix form is

$$x(t) \doteq [A]^T[w(t)] \tag{19}$$

where $[A]^T$ denotes the transpose of $[A]$. We obtain the output as

$$\begin{aligned} y(t) &\doteq g([A]^T[a][b(t)]) \\ &= [g([A]^T[a])][b(t)] \\ &= [B]^T[b(t)] \end{aligned} \tag{20}$$

where $[B]^T$ is a row matrix whose components are a function of $g(\cdot)$. Thus

$$\begin{aligned} y(t) &\doteq [B]^T[a]^{-1}[a][b(t)] \\ &= [B]^T[a]^{-1}[w(t)] \\ &= [C]^T[w(t)]. \end{aligned} \tag{21}$$

The representation in (21) may be quite useful in certain types of computations, e.g., dyadic correlation [3], [9]. Moreover, such an expression can be computed efficiently via the fast Walsh transform (FWT).

A. Example 3

To illustrate the previous matrix method, we present a rather simple example. Assume that the nonlinearity is specified by

$$g(x) = \exp(-x)$$

and let

$$x(t) = t, \qquad 0 \leqslant t < 1.$$

Thus

$$y(t) = g[x(t)] = e^{-t}, \qquad 0 \leqslant t < 1.$$

Let the input be approximated as

$$x(t) \doteq \sum_{n=0}^{7} A_n \text{wal}(n, t).$$

The Walsh coefficients $\{A_n\}$ are evaluated to obtain

$$[A]^T = [0.5, -0.25, 0, -0.125, 0, 0, 0, -0.0625].$$

The 8 × 8 Walsh matrix of interest is given by [5], [9] as

$$W = [a] = \begin{bmatrix} + & + & + & + & + & + & + & + \\ + & + & + & + & - & - & - & - \\ + & + & - & - & - & - & + & + \\ + & + & - & - & + & + & - & - \\ + & - & - & + & - & + & + & - \\ + & - & + & - & - & + & - & + \\ + & - & + & - & + & - & + & - \end{bmatrix}$$

where + and − denote +1 and −1, respectively, and

$$[a]^{-1} = \frac{1}{8}W.$$

The related computations yield:

$$[A]^T[a] = [0.0625, 0.1875, 0.3125, 0.4375, 0.5625, 0.6875, 0.8125, 0.9375].$$

$[B]^T$ is obtained by operating (component wise) on the elements of $[A]^T[a]$. Thus we get

$$[B]^T = [0.9394, 0.8290, 0.7316, 0.6457, 0.5698, 0.5028, 0.4438, 0.3916]$$

where, for instance, the second entry is obtained as $\exp(-0.1875) = 0.8290$. Hence

$$[C]^T = [B]^T[a]^{-1} = [0.6317, 0.1547, 0.0192, 0.0786, 0.0049, 0.0012, 0.0097, 0.0394].$$

Thus the output is given by (21) to be

$$\begin{aligned} y(t) \doteq\ & 0.6317\,\text{wal}(0, t) + 0.1547\,\text{wal}(1, t) + 0.0192\,\text{wal}(2, t) \\ & + 0.0786\,\text{wal}(3, t) + 0.0049\,\text{wal}(4, t) + 0.0012 \\ & \text{wal}(5, t) + 0.0097\,\text{wal}(6, t) + 0.0394\,\text{wal}(7, t). \end{aligned}$$

Table I lists the values of $y(t)$ so obtained. For the purpose of comparison, the exact output $y(t) = e^{-t}$ values (by direct evaluation) are also included. Of course, if the truncated input Walsh series contains further terms, we expect the error indicated in Table I to decrease further.

B. Error Considerations

We now estimate the truncation error associated with Walsh expansions. Assume that the input is a WDS process with a stayadic correlation function $C_x(\tau)$. Then the resulting Walsh coefficients $\{A_n\}$can be shown to be orthogonal, i.e.,

$$\langle A_n A_m \rangle = \Gamma_x(n)\delta_{nm} \tag{22}$$

TABLE I
COMPARISON OF OUTPUT VALUES

t	x(t)=t	ŷ(t)=exp(-t)	y(t)
0.0	0.0	1.000	0.939
0.1	0.1	0.905	0.939
0.2	0.2	0.819	0.829
0.3	0.3	0.741	0.732
0.4	0.4	0.670	0.646
0.55	0.55	0.577	0.570
0.6	0.6	0.549	0.503
0.7	0.7	0.497	0.444
0.9	0.9	0.407	0.392

where $\langle \cdot \rangle$ denotes ensemble averaging, and δ_{nm} denotes a Kronecker delta. $\Gamma_x(n)$ may be referred to as a finite Walsh (sequency) power spectral density. It is defined accordingly as

$$\Gamma_x(n) = \int_0^1 Cx(t) \text{ wal } (n, t) dt. \tag{23}$$

The corresponding truncation mean-squared error is given by

$$e_{ms}(t) = \langle [x(t) - x_n(t)]^2 \rangle$$
$$= C_x(0) + \sum_{n=0}^{n-1} \Gamma_x(n)[1 - 2 \text{ wal } (n, t)]. \tag{24}$$

IV. CONCLUDING REMARKS

We have presented two methods that enable a Walsh analysis of nonlinear systems. These methods are general enough to be applicable to an appreciably wide class of problems. Furthermore, one of the methods is amenable to machine computation in contrast to other currently available techniques.

REFERENCES

[1] M. S. Corrington, "Advanced analytical and signal processing techniques," ASTIA Document AD-277942, Apr. 1962.

[2] F. E. Weiser, "Walsh function analysis of instantaneous nonlinear stochastic problems," Ph.D. Dissertation, Polytechnic Institute of New York, New York, NY, 1964.

[3] M. Maqusi, "Walsh analysis of linear dyadic-invariant and instantaneous nonlinear systems," Sc.D. Dissertation, New Mexico State University, Las Cruces, NM, 1973.

[4] N. J. Fine, "The generalized Walsh functions," *Trans. Amer. Math. Soc.*, vol. 69, pp. 66-77, 1950.

[5] N. Ahmed, H. Schreiber, and P. Lopresti, "On notation and definition of terms related to a class of complete orthogonal functions," *IEEE Trans. Electromag. Compat.*, vol. EMC-15, pp. 75-80, May 1973.

[6] M. Maqusi, "On moments and Walsh characteristics functions," *IEEE Trans. Commun. Technol.*, vol. COM-21, pp. 768-770, June 1973.

[7] J. B. Thomas, *An Introduction to Statistical Communication Theory.* New York: Wiley, 1969.

[8] N. J. Fine, "On the Walsh functions," *Trans. Amer. Math. Soc.*, vol. 65, pp. 373-414, 1949.

[9] H. F. Harmuth, *Transmission of Information by Orthogonal Functions,* 2nd ed. New York: Springer, 1972.

Part IV

SYSTEM IDENTIFICATION

Editor's Comments on Papers 11 Through 15

Owing to their salient properties, Walsh functions proved to be very powerful in solving the identification (or synthesis) problem of dynamic systems from given input-output records. Part IV contains the basic papers treating the identification problem of a variety of dynamic systems such as linear, bilinear, nonlinear, and distributed parameter.

Paper 11 appears to be the first paper in which the problem of identifying dynamic systems is solved with the aid of Walsh functions. The key idea is the application of repeated integration together with the Walsh operational matrix employed for determining the system response. In Paper 12 bilinear system identification is considered and solved by using the same tools as covered in Paper 11, namely the Walsh operational matrix and the group properties of Walsh functions. The same type of systems were also researched by W. L. Chen and Shih (1978). Lumped parameter and distributed parameter systems in differential, integral, and state equation form are discussed in Paper 13. Non-constant parameters of polynomial type are also considered.

Nonlinear systems are considered by Maqusi in Paper 14. The nonlinear system under investigation was assumed to consist of a linear-dyadic invariant system followed by a single valued nonlinearity and a second linear-dyadic invariant system. Two methods were developed, one based on direct output measurements with Walsh function input and one assuming ergodic processes for the input. Nonlinear-distributed parameter systems were considered in a paper by M. S. P. Sinha, Rajamani, and A. K. Sinha (1980) and treated by extending the linear-distributed parameter identification results of Paraskevopoulos and Bounas (1978).

Paper 15 provides a methodology for improving the identification accuracy of continuous systems through the so-called one-shot operational matrices for repeated integration via Walsh functions. The paper also gives a multistep parameter estimation algorithm for systems with large, unknown time delays. Some additional works in the area of systems identification via Walsh functions are described in Prasada Rao and Sivakumar (1975), Gopalsami and Deekshatulu (1976), Tzafestas and Chrysochoides (1977), and Tzafestas, Papastergiou, and Anoussis (1984). In Prasada Rao and Sivakumar (1975) the initial states of the system are included in the identification problem, and in Gopalsami and Deekshatulu (1976) it is shown that the method of Chen and Hsiao (Paper 11) coincides with the repeated trapezoidal rule of numerical integration.

Tzafestas and Chrysochoides (1977) treat the problem of reconstructing (identifying) the time-varying reactivity input of a nuclear reactor point- and space-dependent integral equation model via Walsh functions. The solution provides the formulas for computing the coefficients of the Walsh series representation of the reactivity. The same problem is also considered in Tzafestas, Papastergiou, and Anoussis (1984) through block pulse functions.

Cohn-Sfetcu and Nichols (1975) reveal the analogies between linear time-invariant systems and linear-dyadic invariant systems, and, using spectral analysis (Walsh-Fourier transform), solve the identification problem of multi-input and multi-output dyadic systems. Prasada Rao and Sivakumar (1981) deal with the transfer-function identification problem of multi-input and multi-output systems, and present an iterative Walsh-based algorithm for parameter identification in piece-wise linear systems (1982).

REFERENCES

Chen, W. L., and Y. P. Shih, 1978, Parameter Estimation of Bilinear Systems Via Walsh Functions, *J. Franklin Inst.* **305:**249–257.

Cohn-Sfetcu, S., and S. T. Nichols, 1975, On the Identification of Linear Dyadic Invariant Systems, *IEEE Trans. Electromagn. Compat.* **EMC-17:**111–117.

Gopalsami, N., and B. L. Deekshatulu, 1976, Time-Domain Synthesis via Walsh Functions, *IEE Proc.* **123:**461–462.

Paraskevopoulos, P. N., and A. C. Bounas, 1978, Distributed Parameter System Identification Via Walsh Functions, *Int. J. Syst. Sci.,* **9:**75–83.

Prasada Rao, G., and L. Sivakumar, 1975, System Identification via Walsh Functions, *IEE Proc.* **122:**1160–1161.

Prasada Rao, G., and L. Sivakumar, 1981, Transfer Function Matrix Identification in MIMO systems via Walsh Functions, *IEEE Proc.* **69:**465–466.

Prasada Rao, G., and L. Sivakumar, 1982, Piecewise Linear System Identification via Walsh Functions, *Int. J. Syst. Sci.* **13:**525–530.

Sinha, M. S. P., V. S. Rajamani, and A. K. Sinha, 1980, Identification of Non-linear Distributed System Using Walsh Functions, *Int. J. Control* **32:**669–676.

Tzafestas, S. G., and N. Chrysochoides, 1977, Time-Varying Reactivity Reconstruction via Walsh Functions, *IEEE Trans. Autom. Control* **AC-22:**886–888.

Tzafestas, S. G., C. N. Papastergiou, and J. N. Anoussis, 1984, Dynamic Reactivity Computation in Nuclear Reactors Using Block-Pulse Function Expansion, *Int. J. Modelling Simulation* **4:**73–76.

Reprinted from *IEE Proc.* **122:**565–570 (1975)

TIME-DOMAIN SYNTHESIS VIA WALSH FUNCTIONS

C. F. Chen and C. H. Hsiao

List of symbols

$\phi_n(t)$ = Walsh function
$r_q(t)$ = Rademacher function
$b_q b_{q-1}$ = binary expression of n
$f(t)$ = an arbitrary function
c_n = Walsh-series coefficient
f_k = discrete value of $f(t)$ at k
W = Walsh matrix
$P(\sigma \times \sigma)$ = operational matrix with $\sigma \times \sigma$ dimensions
$\boldsymbol{\phi}_\sigma$ = truncated Walsh vector with σ elements
m = order of operational matrix
$y(t)$ = output variable
$u(t)$ = input variable
h_i = input variable Walsh-series coefficients
c = output-variable Walsh-coefficient vector
h = input-variable Walsh-coefficient vector
$\boldsymbol{\Phi}'$ = rectangular matrix $2n \times m$
a = coefficient vector to be determined
b = coefficient vector to be determined
$Y(s)$ = Laplace transform of $y(t)$
$U(s)$ = Laplace transform of $u(t)$
$\underline{C}$ = $[c_1 c_2, \ldots, c_n]'$
$\bar{x}_1$ = average value of x_1
$\bar{x}_2$ = average value of x_2

1 Introduction

Finding the simplest system that will realise a prescribed input/output behaviour has been a fundamental problem in systems theory since Guillemin's time.[1] Only a few techniques are generally known, and they, including the one developed by Guillemin himself, are based on repeated differentiation and contain all the inherent disadvantages involved therein.[2, 3]

In searching for a new direction, the question naturally arises: why not use a principle based on integration instead? Walsh-function theory,[4] when used in the solving of differential equations, is such a principle. Corrington[5] used this approach on the solution problem. This paper proceeds in the inverse direction, attacking the time-domain-synthesis problem via Walsh functions.

Let us first briefly review Walsh functions.

2 Rademacher and Walsh functions

In recent years, Walsh-function theory has been innovated and applied to various fields in engineering and science.[6, 7] The original papers were published in 1922 and 1923 by Rademacher[8] and Walsh,[9] respectively.

Rademacher's function is a set of square waves of unit height with periods equal to $1, \frac{1}{2}, \frac{1}{4}, \frac{1}{8}, \ldots, 2^{(1-k)}$, respectively. The first four square waves are shown in Fig. 1. It is noted that the set involves only odd functions, and therefore it is not complete. In 1923, Walsh independently developed a complete set known as Walsh functions. The set of Walsh functions and the set of Rademacher functions have the following relations:

Paper 7389 C, first received 27th September and in revised form 17th December 1974

Prof. Chen is, and Mr. Hsiao was formerly, with the Department of Electrical Engineering, University of Houston, Houston, Texas 77004, USA. Mr. Hsiao is now with Cheng Kung University and Chung Shan Institute, Taiwan

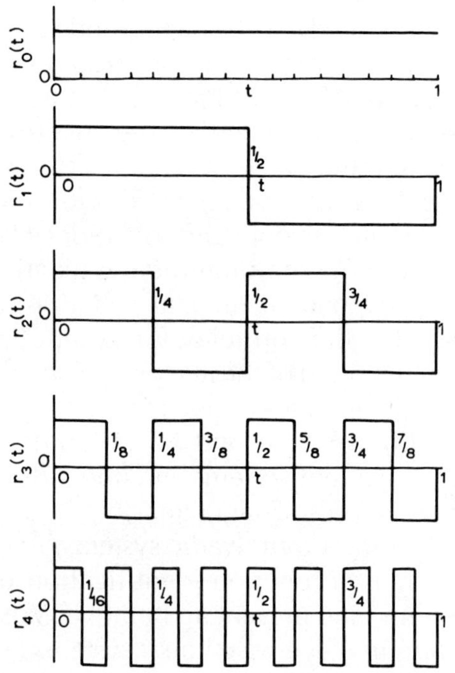

Fig. 1
Rademacher functions

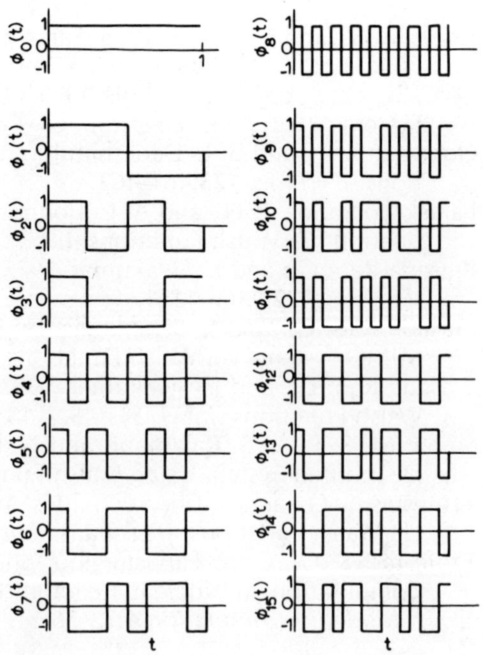

Fig. 2
Walsh functions

$$\left.\begin{aligned}
\phi_0(t) &= r_0(t)\\
\phi_1(t) &= r_1(t)\\
\phi_2(t) &= \{r_2(t)\}^1\,\{r_1(t)\}^0\\
\phi_3(t) &= \{r_2(t)\}^1\,\{r_1(t)\}^1\\
\phi_4(t) &= \{r_3(t)\}^1\,\{r_2(t)\}^0\,\{r_1(t)\}^0\\
\phi_5(t) &= \{r_3(t)\}^1\,\{r_2(t)\}^0\,\{r_1(t)\}^1\\
\phi_6(t) &= \{r_3(t)\}^1\,\{r_2(t)\}^1\,\{r_1(t)\}^0\\
\phi_7(t) &= \{r_3(t)\}^1\,\{r_2(t)\}^1\,\{r_1(t)\}^1
\end{aligned}\right\} \quad (1)$$

$$\vdots$$

$$\phi_n(t) = \{r_q(t)\}^{b_q}\,\{r_{q-1}(t)\}^{b_{q-1}}\,\{r_{q-2}(t)\}^{b_{q-2}}\ldots \quad (2)$$

where

$$q = [\log_2 n] + 1 \quad (3)$$

in which $[\cdot]$ means 'taking the greatest integer of'. Therefore,

$$n = b_q 2^{q-1} + b_{q-1} 2^{q-2} + \ldots b_1 2^0 \quad (4)$$

where $b_q b_{q-1} \ldots b_1$ is the binary expression of n.

Therefore, if a particular Walsh function $\phi_n(t)$ is given and its Rademacher function components are required, we simply change n into binary form and then substitute in eqn. 2. For example, to find the Rademacher-function components of Walsh function $\phi_{10}(t)$,

$$q = [\log_2 10] + 1$$
$$= 4$$

$$\begin{array}{ccccccccc}
10 = & 1 \times 2^3 & + & 0 \times 2^2 & + & 1 \times 2^1 & + & 0 \times 2^0\\
& \uparrow & & \uparrow & & \uparrow & & \uparrow\\
& b_4 & & b_3 & & b_2 & & b_1
\end{array}$$

and therefore

$$\phi_{10}(t) = \{r_4(t)\}^1\,\{r_3(t)\}^0\,\{r_2(t)\}^1\,\{r_1(t)\}^0$$

The Walsh functions, like the Rademacher functions, are easy to draw.

3 Walsh-coefficient determination

A function $f(t)$ that is absolutely integrable in $[0, 1)$ may be expanded into a Walsh series:

$$f(t) = c_0\phi_0(t) + c_1\phi_1(t) + \ldots + c_n\phi_n(t) + \ldots \quad (5)$$

where c_n are coefficients of the Walsh series of $f(t)$.

It is desirable to determine the coefficients c_n such that the integral-square error satisfies the following relation:

$$\lim_{N\to\infty} \int_0^1 |f(t) - \sum_{n=0}^{N} c_n\phi_n(t)|^2 dt = 0 \quad (6)$$

Multiplying by $\phi_n(t)$ on both sides of eqn. 5 and then integrating each term from 0 to 1, we obtain

$$c_n = \int_0^1 \phi_n(t) f(t)\, dt \quad (7)$$

This is because of the orthonormal property of Walsh functions.

Let us illustrate the Walsh-series expansion by the following simple ramp-function example:

$$f(t) = t$$

Substituting $f(t)$ into eqn. 7 and taking only four terms, we obtain

$$c_0 = \int_0^1 \phi_0(t)\, t\, dt = \tfrac{1}{2}$$

$$c_1 = \int_0^1 \phi_1(t)\, t\, dt = -\tfrac{1}{4}$$

$$c_2 = \int_0^1 \phi_2(t)\, t\, dt = -\tfrac{1}{8}$$

$$c_3 = \int_0^1 \phi_3(t)\, t\, dt = 0$$

After substituting these coefficients into eqn. 5, we have

$$t = \tfrac{1}{2}\phi_0(t) - \tfrac{1}{4}\phi_1(t) - \tfrac{1}{8}\phi_2(t) + 0\,\phi_3(t)$$

Which is the four-term Walsh-series expansion of the ramp function.

3.1 Discrete formula

If the given function is not in its analytic form but in tabulated-data or graphical form, and if its Walsh-series expansion is desired, we would modify eqns. 5 and 7 into discrete forms:

$$f_k = \sum_{n=0}^{m-1} \phi_{kn} c_n \qquad k = 0, 1, 2, \ldots, m-1 \quad (5a)$$

$$c_n = \sum_{k=0}^{m-1} \phi_{nk} f_k \frac{1}{m} \qquad n = 0, 1, 2, \ldots, m-1 \quad (7a)$$

where f_k is the average value of the function in question in the kth subinterval, $\phi_{n,k}$ is the value of the nth Walsh function in the kth subinterval, and m is the total number of subintervals.

To illustrate the use of a discrete formula, let us evaluate the Walsh series again for the ramp function in its tabulated form. Given

k	0	1	2	3
f_k	$\frac{1}{8}$	$\frac{3}{8}$	$\frac{5}{8}$	$\frac{7}{8}$

The corresponding graphical form is the ramp function.

Eqn. 7a in its expansion form for $m = 4$ is as follows:

$$\begin{bmatrix} c_0\\ c_1\\ c_2\\ c_3 \end{bmatrix} = \begin{bmatrix} \phi_{00} & \phi_{01} & \phi_{02} & \phi_{03}\\ \phi_{10} & \phi_{11} & \phi_{12} & \phi_{13}\\ \phi_{20} & \phi_{21} & \phi_{22} & \phi_{23}\\ \phi_{30} & \phi_{31} & \phi_{32} & \phi_{33} \end{bmatrix} \begin{bmatrix} f_0\\ f_1\\ f_2\\ f_3 \end{bmatrix} \times \tfrac{1}{4} \quad (8)$$

Substituting the tabulated data of the ramp function into eqn. 8, we have

$$\begin{bmatrix} c_0\\ c_1\\ c_2\\ c_3 \end{bmatrix} = \begin{bmatrix} 1 & 1 & 1 & 1\\ 1 & 1 & -1 & -1\\ 1 & -1 & 1 & -1\\ 1 & -1 & -1 & 1 \end{bmatrix} \begin{bmatrix} \frac{1}{8}\\ \frac{3}{8}\\ \frac{5}{8}\\ \frac{7}{8} \end{bmatrix} \times \tfrac{1}{4} \quad (9a)$$

$$= \begin{bmatrix} \frac{1}{2}\\ -\frac{1}{4}\\ -\frac{1}{8}\\ 0 \end{bmatrix} \quad (9b)$$

The square matrix defined in eqn. 8 and the numerical values shown in eqn. 9a are easily recognised from the definition of Walsh functions.

It is seen that the Walsh series of a unit ramp function obtained from a discrete formula, or from eqns. 9, and that obtained from the analytic formula are, of course, the same.

Eqn. 8 can be written into a general compact form:

$$c = Wf\frac{1}{m} \quad (10)$$

where W is called the Walsh matrix.

3.2 Operational matrix

In the preceding Section, we showed that the ramp function can be expressed by a Walsh series, or

$$t \cong \tfrac{1}{2}\phi_0(t) - \tfrac{1}{4}\phi_1(t) - \tfrac{1}{8}\phi_2(t) \quad (11)$$

However, a ramp function can be considered as the first integral of a unit step function, or $\phi_0(t)$. Therefore, we write the following

$$\int_0^t \phi_0(x)\,dx \cong [\tfrac{1}{2}, -\tfrac{1}{4}, -\tfrac{1}{8}, 0] \begin{bmatrix} \phi_0\\ \phi_1\\ \phi_2\\ \phi_3 \end{bmatrix} \quad (12)$$

The first integral of $\phi_1(t)$ is a triangular function, and if we expand the triangular function into a Walsh series by using discrete formula with $m = 4$, we have

$$\int_0^t \phi_1(t)\,dx \cong [\tfrac{1}{4}, 0, 0, -\tfrac{1}{8}] \begin{bmatrix} \phi_0\\ \phi_1\\ \phi_2\\ \phi_3 \end{bmatrix} \quad (13)$$

Similarly, we can evaluate the Walsh-series coefficients of the first integration of $\phi_2(t)$ and $\phi_3(t)$ and easily obtain

$$\int_0^t \phi_2(x)\,dx \cong [\tfrac{1}{8},\ 0,\ 0,\ 0]\begin{bmatrix}\phi_0\\ \phi_1\\ \phi_2\\ \phi_3\end{bmatrix} \tag{14}$$

and

$$\int_0^t \phi_3(x)\,dx \cong [0,\ \tfrac{1}{8},\ 0,\ 0]\begin{bmatrix}\phi_0\\ \phi_1\\ \phi_2\\ \phi_3\end{bmatrix} \tag{15}$$

Combining eqns. 12–15, we have

$$\begin{bmatrix}\int_0^t \phi_0(x)\,dx\\ \int_0^t \phi_1(x)\,dx\\ \int_0^t \phi_2(x)\,dx\\ \int_0^t \phi_3(x)\,dx\end{bmatrix} \cong \begin{bmatrix}\frac{1}{2} & -\frac{1}{4} & -\frac{1}{8} & 0\\ \frac{1}{4} & 0 & 0 & -\frac{1}{8}\\ \frac{1}{8} & 0 & 0 & 0\\ 0 & \frac{1}{8} & 0 & 0\end{bmatrix}\begin{bmatrix}\phi_0(t)\\ \phi_1(t)\\ \phi_2(t)\\ \phi_3(t)\end{bmatrix} \tag{16}$$

or in compact form

$$\int_0^t \boldsymbol{\phi}_{(4)}(x)\,dx \cong \boldsymbol{P}_{(4\times 4)}\boldsymbol{\phi}_{(4)}(t) \tag{17}$$

$\boldsymbol{P}_{(4\times 4)}$ is called the operational matrix of dimension 4 which relates Walsh functions and their integrals. It is chosen as a square matrix for the reason of convenient calculation.

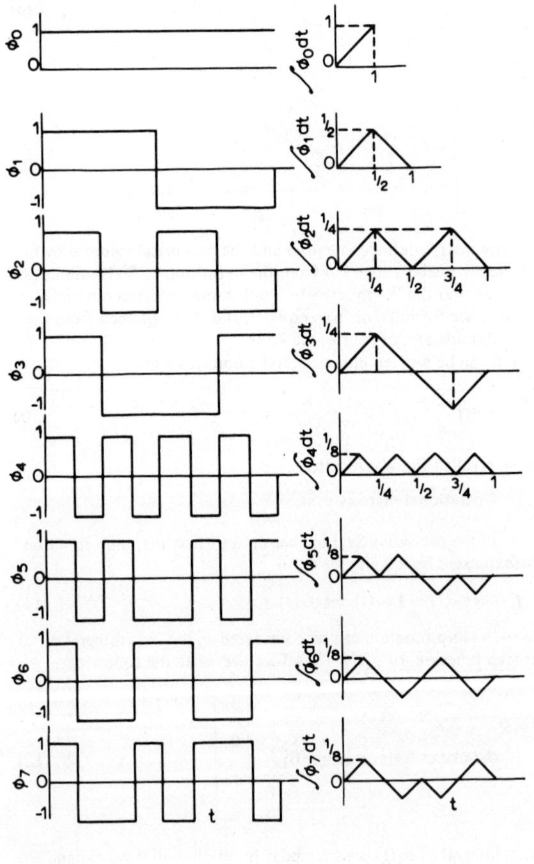

Fig. 3
Walsh functions and their first integrals

By the use of eqn. 17, integration becomes multiplication, therefore, we consider $\boldsymbol{P}$ as an operational matrix.

If we divided the unit [0, 1) into eight subintervals instead of four, and evaluated $\int \phi_0\,dt, \int \phi_1\,dt, \ldots, \int \phi_7\,dt$ by either an analytic method or a discrete formula, we would obtain a group of triangular waves as shown in Fig. 3. Then we could expand the triangular waves into Walsh functions, arriving at the following formula:

$$\begin{bmatrix}\int\phi_0\,dt\\ \int\phi_1\,dt\\ \int\phi_2\,dt\\ \int\phi_3\,dt\\ \hline \int\phi_4\,dt\\ \int\phi_5\,dt\\ \int\phi_6\,dt\\ \int\phi_7\,dt\end{bmatrix} \cong \left[\begin{array}{cccc|cccc}\frac{1}{2} & -\frac{1}{4} & -\frac{1}{8} & 0 & -\frac{1}{16} & 0 & 0 & 0\\ \frac{1}{4} & 0 & 0 & -\frac{1}{8} & 0 & -\frac{1}{16} & 0 & 0\\ \frac{1}{8} & 0 & 0 & 0 & 0 & 0 & -\frac{1}{16} & 0\\ 0 & \frac{1}{8} & 0 & 0 & 0 & 0 & 0 & -\frac{1}{16}\\ \hline \frac{1}{16} & 0 & 0 & 0 & 0 & 0 & 0 & 0\\ 0 & \frac{1}{16} & 0 & 0 & 0 & 0 & 0 & 0\\ 0 & 0 & \frac{1}{16} & 0 & 0 & 0 & 0 & 0\\ 0 & 0 & 0 & \frac{1}{16} & 0 & 0 & 0 & 0\end{array}\right]\begin{bmatrix}\phi_0\\ \phi_1\\ \phi_2\\ \phi_3\\ \hline \phi_4\\ \phi_5\\ \phi_6\\ \phi_7\end{bmatrix} \tag{18}$$

which is

$$\int_0^t \boldsymbol{\phi}_{(8)}(x)\,dx \cong \boldsymbol{P}_{(8\times 8)}\boldsymbol{\phi}_{(8)}(t) \tag{18a}$$

It is interesting to note that the upper left corner of $\boldsymbol{P}_{(8\times 8)}$ is exactly $\boldsymbol{P}_{(4\times 4)}$ in eqn. 17, the upper right corner and the lower corner are unit matrices multiplied by $-\frac{1}{16}$ and $\frac{1}{16}$, respectively, and the lower right corner is simply a null matrix.

Following a similar reasoning line, we can write a general expression for the operational matrix $\boldsymbol{P}$ of order m (which is a positive integer power of 2) as follows:

$$\begin{bmatrix} \begin{bmatrix} \begin{bmatrix} \frac{1}{2}\ddots & -\frac{2}{m}I_{\left(\frac{m}{8}\right)} \\ \frac{2}{m}I_{\left(\frac{m}{8}\right)} & O_{\left(\frac{m}{8}\right)} \end{bmatrix} & -\frac{1}{m}I_{\left(\frac{m}{4}\right)} \\ \frac{1}{m}I_{\left(\frac{m}{4}\right)} & O_{\left(\frac{m}{4}\right)} \end{bmatrix} & -\frac{1}{2m}I_{\left(\frac{m}{2}\right)} \\ \frac{1}{2m}I_{\left(\frac{m}{2}\right)} & O_{\left(\frac{m}{2}\right)} \end{bmatrix} \tag{19}$$

This operational matrix will play an important role in the time-domain-synthesis problems.

4 Principle of transfer-function synthesis

Consider the following differential equation:

$$y^{(n)} + a_1 y^{(n-1)} + \ldots a_{n-1}y^1 + a_n y = b_1 u^{(n-1)} + \ldots + b_{n-1}u + b_n u \tag{20}$$

where $a_1, a_2, \ldots, a_n, b_1, b_2, \ldots, b_n$ are unknowns while input u and output y are given analytically or numerically. Also, assume all initial conditions are equal to zero.

Integrating both sides of eqn. 20 n times, we have

$$y(t) + a_1 \int_0^t y(t)\,dt + \ldots a_n \underbrace{\int \ldots \int y(t)\,dt}_{n} = b_1 \int_0^t u(t)\,dt + \ldots + b_n \underbrace{\int \ldots \int u(t)\,dt}_{n} \tag{21}$$

Both given $y(t)$ and $u(t)$ may be expanded into Walsh series. We write

$$y(t) = \boldsymbol{c}'\boldsymbol{\phi}(t) = c_0\phi_0 + c_1\phi_1 + c_2\phi_2 + \ldots \tag{22}$$

and

$$u(t) = \boldsymbol{h}'\boldsymbol{\phi}(t) = h_0\phi_0 + h_1\phi_1 + h_2\phi_2 + \ldots \tag{23}$$

As explained in Section 3, the first integration of a Walsh function may be expressed approximately by

$$\int_0^t \phi(x)\,dx \cong \boldsymbol{P}\boldsymbol{\phi}(t) \tag{24}$$

Substituting eqns. 22, 23 and 24 into eqn. 21, we have

$$\boldsymbol{c}'[\boldsymbol{I} + a_1\boldsymbol{P} + a_2\boldsymbol{P}^2 + \ldots + a_n\boldsymbol{P}^n]\,\boldsymbol{\phi}(t) = \boldsymbol{h}'[b_1\boldsymbol{P} + b_2\boldsymbol{P}^2 + \ldots + b_n\boldsymbol{P}^n]\,\boldsymbol{\phi}(t) \tag{25}$$

Eqn. 25 must be satisfied for any value of t. Let us take, say, $2n$ samples, at $t_1, t_2, \ldots, t_{2n}$. We have

$$c'[I+a_1P+a_2P^2+\dots a_nP^n]\boldsymbol{\phi}(t_1)=h'[b_1P+b_2P^2+\dots+b_nP^n]\boldsymbol{\phi}(t_1)$$
$$c'[I+a_1P+a_2P^2+\dots a_nP^n]\boldsymbol{\phi}(t_2)=h'[b_1P+b_2P^2+\dots+b_nP^n]\boldsymbol{\phi}(t_2)$$
$$\vdots$$
$$c'[I+a_1P+a_2P^2+\dots a_nP^n]\boldsymbol{\phi}(t_{2n})=h'[b_1P+b_2P^2+\dots+b_nP^n]\boldsymbol{\phi}(t_{2n}) \qquad (26)$$

We can solve the $2n$ equations for the $2n$ unknowns $\boldsymbol{a}$ and $\boldsymbol{b}$. Then let us define a matrix $\boldsymbol{\Phi}$ of $2n \times m$ as

$$\boldsymbol{\Phi}' \triangleq \begin{bmatrix} \boldsymbol{\phi}'(t_1) \\ \boldsymbol{\phi}'(t_2) \\ \vdots \\ \boldsymbol{\phi}'(t_{2n}) \end{bmatrix} \qquad (27)$$

Eqn. 26 is then simplified to

$$\{\boldsymbol{\Phi}'[-P'c, -P'^2c, \dots, -P'^nc, P'h \dots P'^nh]\}\begin{bmatrix} a \\ b \end{bmatrix} = \boldsymbol{\Phi}'c \qquad (26a)$$

Performing inversions on the $2n \times 2n$ matrix in the $\{\cdot\}$, we obtain $\boldsymbol{a}$ and $\boldsymbol{b}$:

$$\begin{bmatrix} a \\ b \end{bmatrix} = \{\boldsymbol{\Phi}'[-P'c \dots -P'^nc, P'h \dots P'^nh]\}^{-1}\boldsymbol{\Phi}c \qquad (28)$$

Eqn. 28 is the basic formula to fit eqn. 20, if input/output behaviour is given.

4.1 Illustrative example

Suppose the input to a system is a unit step and the output data are as follows:

t	0	$\frac{1}{4}$	$\frac{1}{2}$	$\frac{3}{4}$	1
$y(t)$	0	0·632	0·865	0·95	0·982

It is required to find the differential equation or the transfer function of the system.

Let us average $y(t)$ of each subinterval first.

k	0	1	2	3
y_k	0·316	0·7485	0·9075	0·966

Applying the discrete formula shown in eqns 9 with $n = 4$, we obtain

$$c = W\bar{y}\frac{1}{m}$$
$$= \begin{bmatrix} 1 & 1 & 1 & 1 \\ 1 & 1 & -1 & -1 \\ 1 & -1 & 1 & -1 \\ 1 & -1 & -1 & 1 \end{bmatrix}\begin{bmatrix} 0·316 \\ 0·7485 \\ 0·9075 \\ 0·966 \end{bmatrix} \times \tfrac{1}{4} = \begin{bmatrix} 0·7345 \\ -0·20225 \\ -0·12275 \\ -0·0935 \end{bmatrix}$$

Suppose the linear system is defined as

$$y + ay = bu$$
$$u = [1, 0, 0, 0]\,\boldsymbol{\phi}(t)$$

and choose $t_1 = \frac{1}{8}$ and $t_2 = \frac{3}{8}$. Then

$$\boldsymbol{\Phi}' = \begin{bmatrix} \boldsymbol{\phi}'(\frac{1}{8}) \\ \boldsymbol{\phi}'(\frac{3}{8}) \end{bmatrix} = \begin{bmatrix} 1 & 1 & 1 & 1 \\ 1 & 1 & -1 & -1 \end{bmatrix}$$

Applying eqn. 28 with $\boldsymbol{P}$ defined in eqn. 16 yields

$$\begin{bmatrix} a \\ b \end{bmatrix} = \begin{bmatrix} 3·48 \\ 3·63 \end{bmatrix}$$

The required differential equation is therefore

$$\dot{y} + 3·48y = 3·63u \qquad (29)$$

or the required transfer function is

$$\frac{Y(s)}{U(s)} = \frac{3·63}{s + 3·48} \qquad (29a)$$

For the same problem, if we took more samples we would obtain a better result. To demonstrate this statement, let us take nine samples

t	0	$\frac{1}{8}$	$\frac{2}{8}$	$\frac{3}{8}$	$\frac{4}{8}$	$\frac{5}{8}$	$\frac{6}{8}$	$\frac{7}{8}$	1
$y(t)$	0	0·394	0·632	0·789	0·865	0·918	0·950	0·970	0·982
$\bar{y}_k$	0·197	0·513	0·710	0·826	0·889	0·934	0·960	0·976	

Using eqn. 9 again with $m = 8$, we have

$$c = Wy\tfrac{1}{8} = \tfrac{1}{8}[6·063, -1·511, -0·941, -0·711, -0·495, -0·369, -0·231, -0·169]$$

choosing $t_1 = 13/16$ and $t_2 = 15/16$, we find

$$\boldsymbol{\Phi} = \begin{bmatrix} \boldsymbol{\phi}'(\frac{13}{16}) \\ \boldsymbol{\phi}'(\frac{15}{16}) \end{bmatrix} = \begin{bmatrix} 1 & -1 & -1 & 1 & 1 & -1 & -1 & 1 \\ 1 & -1 & -1 & 1 & -1 & 1 & 1 & -1 \end{bmatrix}$$

Applying eqn 28 with $P = P_{(8\times 8)}$, we finally obtain

$$\begin{bmatrix} a \\ b \end{bmatrix} = \begin{bmatrix} 3·9287759 \\ 3·9243553 \end{bmatrix}$$

Therefore the differential equation of the system is

$$\dot{y} + 3·9287759y = 3·9243553u \qquad (30)$$

When we took five samples, the differential equation fitted was eqn. 29; for nine samples, we obtain eqn. 30. The unit step response is actually taken from the following system:

$$\dot{y} + 4y = 4u$$

It is interesting to note that the result obtained by the five-point approach is acceptable, and that by the nine-point approach is quite satisfactory.

5 Principle of state-equation synthesis

Suppose we are interested in realising the following system·

$$\dot{x} = Ax \qquad (31)$$

from a set of zero-input response data $x(t_i), i = 0, 1, \dots, m$.

First of all, we expand $x(t)$ into a Walsh series with m terms

$$x(t) = \begin{bmatrix} c_1 \\ c_2 \\ \vdots \\ c_n' \end{bmatrix}\begin{bmatrix} \phi_0 \\ \phi_1 \\ \vdots \\ \phi_{m-1} \end{bmatrix} = C\boldsymbol{\phi}(t) \qquad (32)$$

where C is an $n \times m$ matrix; $\boldsymbol{\phi}(t)$, an $m \times 1$ vector.

Integrating both sides of eqn. 31 and using the expression of eqn. 32 and the operational matrix $\boldsymbol{P}$, we have

$$\int_0^t \dot{x}(\lambda)d\lambda = A\int_0^t x(\lambda)d\lambda = AC\int_0^t \boldsymbol{\phi}(\lambda)d\lambda \qquad (33)$$
$$x(t) - x(0) = ACP\boldsymbol{\phi}(t)$$
$$C\boldsymbol{\phi}(t) - [x(0), 0, \dots, 0]\,\boldsymbol{\phi}(t) = ACP\boldsymbol{\phi}(t) \qquad (33a)$$

Eqn. 33a must hold for any value of t. Let us take n samples.

Define

$$\boldsymbol{\Phi} \triangleq [\boldsymbol{\phi}(t_1), \boldsymbol{\phi}(t_2) \dots \boldsymbol{\phi}(t_n)] \qquad (34)$$
$$\{C - [x(0), 0, 0, \dots, 0]\}\boldsymbol{\Phi} = ACP\boldsymbol{\Phi} \qquad (35)$$

Then the system matrix A can be evaluated by

$$A = \{C - [x(0), 0, 0, \dots, 0]\}\boldsymbol{\Phi}\{CP\boldsymbol{\Phi}\}^{-1} \qquad (36)$$

5.1 Illustrative example

A zero-input response of a system is tabulated as follows:

t	$x_1(t)$	$x_2(t)$	$\bar{x}_1$	$\bar{x}_2$
0	1	−1		
			0·8772	−0·9737
$\frac{1}{4}$	0·7545	−0·9494		
			0·6433	−0·8852
$\frac{1}{2}$	0·5321	−0·8231		
			0·4388	−0·7453
$\frac{3}{4}$	0·3454	−0·6676		
			0·2720	−0·5879
1	0·1985	−0·0508		

Using the discrete formula, we find

$$c_1 \cong W\bar{x}_1\frac{1}{m} = \begin{bmatrix} 1 & 1 & 1 & 1 \\ 1 & 1 & -1 & -1 \\ 1 & -1 & 1 & -1 \\ 1 & -1 & -1 & 1 \end{bmatrix} \begin{bmatrix} 0{\cdot}8772 \\ 0{\cdot}6433 \\ 0{\cdot}4388 \\ 0{\cdot}2720 \end{bmatrix} \times \tfrac{1}{4} = \begin{bmatrix} 2{\cdot}23153 \\ 0{\cdot}809797 \\ 0{\cdot}4007092 \\ 0{\cdot}0671214 \end{bmatrix} \times \tfrac{1}{4}$$

$$c_2 \cong W\bar{x}_2\frac{1}{m} = \begin{bmatrix} 1 & 1 & 1 & 1 \\ 1 & 1 & -1 & -1 \\ 1 & -1 & 1 & -1 \\ 1 & -1 & -1 & 1 \end{bmatrix} \begin{bmatrix} -0{\cdot}8737 \\ -0{\cdot}8852 \\ -0{\cdot}7453 \\ -0{\cdot}5879 \end{bmatrix} \times \tfrac{1}{4} = \begin{bmatrix} -3{\cdot}192311 \\ -0{\cdot}525631 \\ -0{\cdot}2458685 \\ 0{\cdot}0689903 \end{bmatrix} \times \tfrac{1}{4}$$

$$\text{Let } \boldsymbol{\Phi} = [\boldsymbol{\psi}(\tfrac{1}{8}), \boldsymbol{\psi}(\tfrac{3}{8})] = \begin{bmatrix} 1 & 1 \\ 1 & -1 \\ 1 & -1 \\ 1 & 1 \end{bmatrix}$$

The initial conditions are given as

$$\boldsymbol{x}(0) = \begin{bmatrix} 1 \\ -1 \end{bmatrix}$$

We use $\boldsymbol{P}_{(4\times 4)}$ formula, or

$$\boldsymbol{P}_{(4\times 4)} = \left[\begin{array}{c:c|cc} \frac{1}{2} & -\frac{1}{4} & -\frac{1}{8} & 0 \\ \hdashline \frac{1}{4} & 0 & 0 & -\frac{1}{8} \\ \hline \frac{1}{8} & 0 & 0 & 0 \\ 0 & \frac{1}{8} & 0 & 0 \end{array}\right]$$

Substituting $\boldsymbol{C}$, $x(0)$, $\boldsymbol{P}$ and $\boldsymbol{\Phi}$ into eqn. 36, we have

$$\{\boldsymbol{C} - [x(0), 0, 0, 0]\}\,\boldsymbol{\Phi} = \tfrac{1}{4}\begin{bmatrix} -0{\cdot}4908416 & -2{\cdot}911856 \\ 0{\cdot}1051796 & 1{\cdot}6481788 \end{bmatrix}$$

$$\boldsymbol{CP\Phi} = \begin{bmatrix} 3{\cdot}5091584 & 16{\cdot}764101 \\ -3{\cdot}8948204 & -23{\cdot}186666 \end{bmatrix} \times \tfrac{1}{32}$$

Therefore

$$\boldsymbol{A} = \tfrac{32}{4}\begin{bmatrix} -0{\cdot}4908416 & -2{\cdot}911854 \\ 0{\cdot}1051796 & 1{\cdot}6481788 \end{bmatrix} \begin{bmatrix} 3{\cdot}5091584 & 16{\cdot}764101 \\ -3{\cdot}8948204 & -23{\cdot}186666 \end{bmatrix}^{-1}$$

$$= \begin{bmatrix} 0{\cdot}0198261 & 0{\cdot}9903525 \\ -1{\cdot}9962949 & -2{\cdot}0061511 \end{bmatrix} \tag{37}$$

The samples actually are taken from the zero-input response of a system

$$\dot{\boldsymbol{x}} = \boldsymbol{Ax}$$

where

$$\boldsymbol{A} = \begin{bmatrix} 0 & 1 \\ -2 & -2 \end{bmatrix} \tag{38}$$

Comparing eqn. 37 with eqn. 38, we see the power of the method.

6 Laboratory test

The common excitations usually used in a laboratory are neither unit-step-input nor initial conditions, but rather gate functions. When the input is a given gate function and the output is measured, it is desired to find the transfer function of the system. The classical approaches for the domain synthesis are not easily applied. The new Walsh-function method, however, can fit the transfer function as usual. The following example will demonstrate the procedure and the accurate results.

Suppose a gate function is applied to a linear system, and the output is recorded as follows:

t	0		$\frac{1}{4}$		$\frac{1}{2}$		$\frac{3}{4}$		1
$y(t)$	0		0·245		0·222		0·1867		0·1469
$\bar{y}_k$		0·1225		0·2335		0·20435		0·1668	

The gate input is defined as

$$u(t) = \begin{cases} 1 & \text{or } 0 \leqslant t < \frac{1}{4} \\ 0 & \text{elsewhere} \end{cases}$$

Using a discrete formula, we can find the Walsh series for the output.

$$y(t) \cong \boldsymbol{c}'\boldsymbol{\psi} = [c_0, c_1, c_2, c_3]\boldsymbol{\psi}$$

$$\boldsymbol{c}' = W\bar{y}\frac{1}{m} = \tfrac{1}{4}[0{\cdot}72715, -0{\cdot}01515, -0{\cdot}07345, -0{\cdot}14855]$$

The Walsh-series expansion for the gate input can be easily shown as

$$u(t) = [\tfrac{1}{4}, \tfrac{1}{4}, \tfrac{1}{4}, \tfrac{1}{4}]\,\boldsymbol{\psi} = \boldsymbol{h}'\boldsymbol{\psi}$$

$$\boldsymbol{h}' = \tfrac{1}{4}[1, 1, 1, 1]$$

Using operational matrix $\boldsymbol{P}$, we can find the first and second integrations of $y(t)$ and $u(t)$. For a second-order system, there are four unknowns, two as and two bs, and $m = 4$ for five discrete data. Therefore

$$\boldsymbol{\Phi} = \boldsymbol{W} = \begin{bmatrix} 1 & 1 & 1 & 1 \\ 1 & 1 & -1 & -1 \\ 1 & -1 & 1 & -1 \\ 1 & -1 & -1 & 1 \end{bmatrix}$$

Substituting $\boldsymbol{c}$, $\boldsymbol{h}$, $\boldsymbol{P}$ and $\boldsymbol{\Phi}$ we obtain

$$\begin{bmatrix} a \\ b \end{bmatrix} = \begin{bmatrix} 1{\cdot}8173251 \\ 1{\cdot}8531946 \\ 1{\cdot}0107953 \\ 1{\cdot}7734002 \end{bmatrix}$$

That means the system may be approximately described as

$$\ddot{y} + 1{\cdot}8173251\,\dot{y} + 1{\cdot}8531946\,y = 1{\cdot}0107953\,\dot{u} + 1{\cdot}7734002\,u$$

To obtain better results, we may take more data, say 17 points as follows:

t	0	$\frac{1}{16}$	$\frac{1}{8}$	$\frac{3}{16}$	$\frac{1}{4}$	$\frac{5}{16}$	$\frac{3}{8}$	$\frac{7}{16}$	$\frac{1}{2}$
$y(t)$	0	0·0624	0·1244	0·1855	0·245	0·241	0·236	0·230	0·222

t	$\frac{9}{16}$	$\frac{5}{8}$	$\frac{11}{16}$	$\frac{3}{4}$	$\frac{13}{16}$	$\frac{7}{8}$	$\frac{15}{16}$	1
$y(t)$	0·214	0·205	0·1962	0·1866	0·1768	0·1669	0·1569	0·1469

And choose $t_1 = 25/32$, $t_2 = 27/32$, $t_3 = 29/32$, $t_4 = 31/32$, namely

$$\boldsymbol{\Phi}' = \begin{bmatrix} \boldsymbol{\psi}'(25/32) \\ \boldsymbol{\psi}'(27/32) \\ \boldsymbol{\psi}'(29/32) \\ \boldsymbol{\psi}'(31/32) \end{bmatrix} =$$

$$= \begin{bmatrix} 1 & -1 & -1 & 1 & 1 & -1 & -1 & 1 & 1 & -1 & -1 & 1 & 1 & -1 & -1 & 1 \\ 1 & -1 & -1 & 1 & 1 & -1 & -1 & 1 & -1 & 1 & 1 & -1 & -1 & 1 & 1 & -1 \\ 1 & -1 & -1 & 1 & -1 & 1 & 1 & -1 & 1 & -1 & -1 & 1 & -1 & 1 & 1 & -1 \\ 1 & -1 & -1 & 1 & -1 & 1 & 1 & -1 & -1 & 1 & 1 & -1 & 1 & -1 & -1 & 1 \end{bmatrix}$$

Then we obtain

$$\begin{bmatrix} a \\ b \end{bmatrix} = \begin{bmatrix} 2{\cdot}050781 \\ 2{\cdot}039063 \\ 0{\cdot}9980469 \\ 2{\cdot}042969 \end{bmatrix}$$

$$\ddot{y} + 2{\cdot}050781\,\dot{y} + 2{\cdot}039063\,y = 0{\cdot}9980469\,\dot{u} + 2{\cdot}041969\,u$$

The actual system is

$$\ddot{y} + 2\dot{y} + 2y = \dot{u} + 2u.$$

7 Conclusions

A new time-domain synthesis method based on the Walsh functions has been established. With zero-state response given and the transfer function of the system desired, by using the Walsh-function principle, we derived eqn. 28 to fit the transfer function. On the other hand, with zero-input response given and the state equation desired, we derived eqn. 36 to fit. Three examples are included for illustration.

If the order of the transfer function assumed is too high, the inverse of eqn. 28 does not exist. Then we can decrease the order and solve the problem again. Similarly, if the dimension of the state equation is assumed too high initially, the inversion of the matrix in eqn. 36 does not exist, and we should lower the dimension.

If what we fitted is a high-order transfer function and what we want is a lower order one, then we face a new problem, namely, the model reduction problem. The first author[10] has suggested continued-fraction methods to solve this problem effectively.

It is seen that the Walsh-function approach belongs to the school of least-squares identification. In general, however, least-squares identification is to use orthogonal polynomials or

$$C_i = \frac{\sum_{t=0}^{n} f(t)\, \psi_{ni}(t)}{\sum_{t=0}^{n} \psi_{ni}^2(t)} \qquad i = 1, ..., m$$

where $\psi(t)$ is an orthogonal polynomial, and C_i and $f(t)$ are coefficients and the given data, respectively, as defined before.

The advantages of the Walsh-function approach are apparent: (*a*) it is still least-squares identification, (*b*) the denominator becomes 1, because Walsh functions are not only orthogonal, but also orthonormal, (*c*) the fast Walsh algorithm[11, 12] is available.

8 References

1 GUILLEMIN, E.A.: 'Synthesis of passive networks' (Wiley, 1957)
2 KALMAN, R.E., and DECLARIS, N.: 'Aspects of network and system theory' (Holt, Rinehart & Winston, 1971), pp. 311–325
3 SHIEH, L.S., CHEN, C.F., and HUANG, C.J.: 'On identifying transfer functions and state equations for linear systems', *IEEE Trans.*, 1972, **AES-8**, pp. 811–820
4 PICHER, F.: 'Walsh-function and optimal linear systems'. Proceedings of the Walsh-function symposium, Washington DC, 1970, pp. 17–22
5 CORRINGTON, M.S.: 'Solution of differential and integral equations with Walsh functions', *IEEE Trans.*, 1973, **CT-20**, pp. 470–475
6 LEE, J.D.: 'Review of recent work on applications of Walsh functions in communications', Proceedings of the Walsh-function symposium, Washington, DC, 1970, pp. 26–35
7 GIBBS, J.E., and GEBBIE, H.A.: 'Application of Walsh function to transform spectroscopy', *Nature*, 1969, **224**, pp. 1021–1013
8 RADEMACHER, H.: 'Einige Salze uber Reiben von allgemeinen Orthogonal functionen', *Math. Ann.*, 1922, **87**, pp. 712–718
9 WALSH, J.L.: 'A closed set of orthogonal functions', *Am. J. Math*, 1923, **45**, pp. 5–24
10 CHEN, C.F.: 'Model reduction of multivariable systems via matrix continued fractions. Presented at the 5th International Federation of Automatic Control, Paris, France, 1972 and published in *Int. J. of Contr.*, 1974, **20**, pp. 225–238
11 MANZ, J.H.: 'A sequency-ordered fast Walsh transform,' *IEEE Trans.*, 1972, **AU-20**, pp. 204–205
12 SHANKS, J.L.: 'Computation of the fast Walsh-Fourier transform,' *ibid.*, 1969, **C-18**, pp. 457–459

12

BILINEAR SYSTEM IDENTIFICATION BY WALSH FUNCTIONS

V. R. Karanam, P. A. Frick, and R. R. Mohler

***Abstract*—A new technique, which uses a finite orthonormal expansion to approximate input and output functions, is proposed for the identification of a class of nonlinear systems. Walsh functions form the basis of the expansions as a consequence of their useful properties here. That is, the parameter estimation problem is reduced to algebraic form by Fine's relationship for the integral of Walsh functions [1], by the operational matrix concept of Chen and Hsiao [2], and by the very desirable group properties of Walsh functions [3]. Computational examples are presented to illustrate the utility of this method. The availability of Walsh-function programs makes the method particularly amenable to practice.**

I. Introduction

In recent years Walsh functions have been used in the identification and optimal control of linear systems [2], [4], [5]. Chen and Hsiao [2] deal with the problem of determining a suitable internal structure for a linear time-invariant system from its input–output data. Such a structure is obtained using the method of repeated integration and the operational matrix concept relating Walsh functions to their integrals. A similar identification problem is considered in [4] with unknown initial states. The design of piecewise constant gains for optimal control of linear systems is examined in [5].

The results of Section III make it possible to extend the use of Walsh functions to identification problems for a class of nonlinear systems with perfect information, and specifically to variable-structure systems and systems with polynomial-type nonlinearities by exploiting the nice group properties of Walsh functions. For purposes of discussion and to illustrate the developed method, bilinear system [6], [7] is studied as an example. Specifically, consider the system

$$\dot{x}(t)=A(t)x(t)+\sum_{k=1}^{q} N_k(t)x(t)u_k(t)+B(t)u(t) \tag{1.1}$$

$$y(t)=d(t)x(t) \tag{1.2}$$

where $x,\ y\in R^n$, $u\in R^q$; $A(t)$, $N_k(t)$, $B(t)$ are Lebesque-measurable matrix-valued functions of appropriate dimension, and $d(t)$ is a known nonzero scalar-valued function. The bilinear system of (1.1), (1.2) can therefore be represented by

$$\begin{aligned} y(t)=y(0)&+\int_0^t\sum_{i=1}^{n} a_i(\tau)y_i(\tau)\,d\tau \\ &+\int_0^t\sum_{k=1}^{q}\sum_{i=1}^{n} n_{k_i}(\tau)y_i(\tau)u_k(\tau)\,d\tau \\ &+\int_0^t\sum_{j=1}^{q} b_j(\tau)d(\tau)u_j(\tau)\,d\tau \end{aligned} \tag{1.3}$$

where a_i, n_{k_i}, and b_j are the column vectors of matrices A, N_k, and B, respectively, and y_i is the ith component of the vector y.

In section III a method of identifying the a_i, n_{k_i}, and b_j of (1.3) is developed employing Walsh functions. Some computational results are presented in section IV. The next section contains a brief review of the Walsh functions and their relevant properties.

II. Walsh Functions

Following Fine [1] it is convenient to define Walsh functions in terms of Rademacher functions. The Rademacher functions $\{\phi_n\}$ form an incomplete set of orthogonal functions on $[0,1)$ and are defined by

$$\begin{aligned} &\phi_0(t)=\begin{cases} 1 & 0\leqslant t<1/2 \\ -1 & 1/2\leqslant t<1 \end{cases} \\ &\phi_0(t+1)=\phi_0(t) \\ &\phi_n(t)=\phi_0(2^n t) \qquad (n=1,2,\cdots) \end{aligned} \tag{2.1}$$

and Walsh functions $\{\psi_n\}$ are then obtained by defining

$$\begin{aligned} &\psi_0(t)=1 \\ &\psi_n(t)=\phi_{n_1}(t)\phi_{n_2}(t)\cdots\phi_{n_r}(t) \end{aligned} \tag{2.2}$$

for $n=2^{n_1}+2^{n_2}+\cdots+2^{n_r}$, where the nonnegative integers n_i are uniquely determined by $n_{i+1}<n_i$.

The Walsh functions, being a complete orthornormal set, allows a representation of every Lebesgue-measurable function $f(t)$ on $[0,1)$ in the form

$$f(t)=a_0\psi_0(t)+a_1\psi_1(t)+a_2\psi_2(t)+\cdots \tag{2.3}$$

where the coefficients a_n are given by

$$a_n=\int_0^1\psi_n(t)f(t)\,dt \qquad (n=0,1,2,\cdots). \tag{2.4}$$

Two properties of Walsh functions particularly attractive for our purposes are the following:

1) The Walsh functions form an Abelian group [3]. Specifically, they are closed under multiplication, with the multiplication for Walsh functions expressed as

$$\psi_m(t)\psi_n(t)=\psi_{m\oplus n}(t). \tag{2.5}$$

The sign $\oplus$ represents no-carry modulo-2 addition [3].

2) Fine [1] also showed that the integral of a Walsh function can be represented in terms of Walsh functions. For an integral of the form

$$J_k(t)=\int_0^t\psi_k(x)\,dx \qquad (k=0,1,2,\cdots). \tag{2.6}$$

Fine showed that, with $k=2^n+k'$, $0\leqslant k'\leqslant 2^n$,

$$J_0(t)=1/2\psi_0(t)-1/4\sum_{p=0}^{\infty}2^{-p}\psi_{2^p}(t) \tag{2.7}$$

and

$$J_k(t)=2^{-(n+2)}\left\{\psi_{k'}(t)-\sum_{p=1}^{\infty}2^{-p}\psi_{2^{n+p}+k}(t)\right\},\qquad \text{for } k\geqslant 1.$$

(2.7) may be manipulated to a more compact form (see Appendix A) as

$$\int_0^t\Psi(x)\,dx=P\Psi(t). \tag{2.8}$$

Here, $\Psi(x)$ is a column vector consisting of Walsh functions and the matrix P is termed "operational matrix" in [2]. Ψ and P in (2.8) are of infinite dimension, but finite dimensional approximations are obtained by restricting Ψ to contain only 2^n Walsh functions for some integer $n>1$. The resulting square operational matrix P is therefore of order $m=2^n$ and

$$\int_0^t\Psi_{(m)}(x)\,dx\cong P_{(m)}\Psi_{(m)}(t). \tag{2.9}$$

For the sake of completeness, the derivation and structure of $P_{(m)}$ is included in Appendix A. Although Corrington [8] is generally credited with the development of the integral representation of Walsh functions [4], [5], it is apparently due to Fine [1] in view of (2.7).

III. The Identification Problem

In this section a method is developed for identifying the parameter vector-valued functions $a_i(t)$, $n_{k_i}(t)$ for $i=1,2,\cdots,n$; and $b_j(t)$ for $j=1,2,\cdots,q$ of (1.3).

Manuscript received September 21, 1976; revised August 29, 1977 and April 10, 1978. Paper recommended by R. K. Mehra, Past Chairman of the Identification Committee. This work was supported in part by the National Science Foundation under Grant ENG 77-07027.

The authors are with the Department of Electrical and Computer Engineering, Oregon State University, Corvallis, OR 97331.

Employing $m=2^n$ (for some integer n) Walsh functions in Ψ, $a_i(t)$, $b_j(t)$, $n_{k_i}(t)$, and $d(t)$ are approximated by their Walsh function expansions on $[0,1)$ by

$$\begin{aligned} a_i(t) &\simeq F_i\Psi_{(m)}(t) \qquad b_j(t)\simeq H_j\Psi_{(m)}(t)\\ n_{k_i}(t) &\simeq G_{k_i}\Psi_{(m)}(t) \qquad d(t)\simeq \nu^T\Psi_{(m)}(t) \end{aligned} \tag{3.1}$$

thus reducing the identification problem to identifying F_i, G_{k_i}, and H_j.

The known input u and the output y functions similarly are approximated by their Walsh function expansions

$$\begin{aligned} y_i(t) &\simeq e_i^T\Psi_{(m)}(t)=\Psi_{(m)}^T(t)e_i\\ u_k(t) &\simeq z_k^T\Psi_{(m)}(t)=\Psi_{(m)}^T(t)z_k. \end{aligned} \tag{3.2}$$

Substitution of (3.1) and (3.2) into (1.3) leads to

$$\begin{aligned} E\Psi_{(m)}(t) \simeq y(0) &+ \sum_{i=1}^{n} F_i\int_0^t \Psi_{(m)}(\tau)\Psi_{(m)}^T(\tau)\,d\tau e_i\\ &+\sum_{k=1}^{q}\sum_{i=1}^{n} G_{k_i}\int_0^t \Psi_{(m)}(\tau)\Psi_{(m)}^T(\tau)e_i z_k^T\Psi_{(m)}(\tau)\,d\tau\\ &+\sum_{k=1}^{q} H_k\int_0^t \Psi_{(m)}(\tau)\Psi_{(m)}^T(\tau)\nu z_k^T\Psi_{(m)}(\tau)\,d\tau \end{aligned} \tag{3.3}$$

where E represents the matrix formed by e_i^T as row vectors.

In order to render (3.3) useful for computations, it is now shown that the matrix

$$M_{(m)}(t)=\Psi_{(m)}(t)\Psi_{(m)}^T(t) \qquad t\in[0,1) \tag{3.4}$$

is reducible to a simple algebraic form. $M_{(m)}$ is a symmetric matrix with $\psi_i(t)\psi_j(t)$ as a typical element $(i,j=0,1,\cdots,m-1)$, and employing the multiplication rule for Walsh functions it can be represented in the form

$$M_{(m)}(t)=\begin{bmatrix} \psi_0(t) & \psi_1(t) & \cdots & \cdots & \psi_{m-1}(t)\\ \psi_1(t) & \psi_0(t) & \cdots & \cdots & \psi_{m-2}(t)\\ \cdots & & \cdots & \cdots & \cdots\\ \cdots & & \psi_{i\oplus j}(t) & \cdots & \cdots\\ \cdots & & \cdots & \cdots & \cdots\\ \psi_{m-1}(t) & \psi_{m-2}(t) & \cdots & \cdots & \psi_0(t) \end{bmatrix}. \tag{3.5}$$

It is seen that the first column of the matrix $M_{(m)}(t)$ is $\Psi_{(m)}(t)$ and the remaining columns are permutations of the first column elements. $M_{(m)}(t)$ can be represented in the composite matrix notation

$$M_{(m)}(t)=[\Psi_{(m)}(t),\Lambda_1^{(m)}\Psi_{(m)}(t),\Lambda_2^{(m)}\Psi_{(m)}(t),\cdots,\Lambda_{m-1}^{(m)}\Psi_{(m)}(t)] \tag{3.6}$$

where, for example, when $m=4$,

$$\Lambda_1(4)=\begin{bmatrix}0&1&0&0\\1&0&0&0\\0&0&0&1\\0&0&1&0\end{bmatrix};\quad \Lambda_2(4)=\begin{bmatrix}0&0&1&0\\0&0&0&1\\1&0&0&0\\0&1&0&0\end{bmatrix};\quad \text{etc.}$$

In general, the $\Lambda_i^{(m)}$ matrices can be obtained from

$$\Lambda_i^{(m)}=\left[\begin{array}{c|c}\Lambda_i\left(\frac{m}{2}\right) & 0\left(\frac{m}{2}\right)\\ \hline 0\left(\frac{m}{2}\right) & \Lambda_i\left(\frac{m}{2}\right)\end{array}\right] \quad\text{and}\quad \Lambda_{i+(m/2)}^{(m)}=\left[\begin{array}{c|c}0\left(\frac{m}{2}\right) & \Lambda_i\left(\frac{m}{2}\right)\\ \hline \Lambda_i\left(\frac{m}{2}\right) & 0\left(\frac{m}{2}\right)\end{array}\right], \qquad m>1 \tag{3.7}$$

for $i=0,1,2,\cdots,(m/2-1)$ and $\Lambda_0^{(m/2)}=I_{(m/2)}$. Some additional interesting properties of Λ_i matrices are discussed in Appendix B.

To fully exploit the simple algebraic structure outlined above in the system representation of (3.3), the following two lemmas are established:

Lemma 3.1: $\Psi_{(m)}\Psi_{(m)}^T\alpha=\Lambda_\alpha\Psi_{(m)}$ where $\Lambda_\alpha=[\alpha,\Lambda_1^{(m)}\alpha,\cdots,\Lambda_{m-1}^{(m)}\alpha]$.

Proof: From (3.6)

$$\Psi_{(m)}\Psi_{(m)}^T=[\Psi_{(m)},\Lambda_1^{(m)}\Psi_{(m)},\cdots,\Lambda_{m-1}^{(m)}\Psi_{(m)}]$$

and since the matrices $\Psi_{(m)}\Psi_{(m)}^T$ and $\Lambda_i^{(m)}$'s are symmetric,

$$\Psi_{(m)}\Psi_{(m)}^T\alpha=\begin{bmatrix}\Psi_{(m)}^T\alpha\\ \Psi_{(m)}^T\Lambda_1^{(m)}\alpha\\ \vdots\\ \Psi_{(m)}^T\Lambda_{m-1}^{(m)}\alpha\end{bmatrix}.$$

Noting that $\Psi_{(m)}^T\Lambda_i^{(m)}\alpha$ is a scalar,

$$\Psi_{(m)}^T\Lambda_i^{(m)}\alpha=\left(\Psi_{(m)}^T\Lambda_i^{(m)}\alpha\right)^T=\alpha^T\Lambda_i^{(m)}\Psi_{(m)}, \qquad \text{for } i=1,2,\cdots,(m-1).$$

Hence,

$$\Psi_{(m)}\Psi_{(m)}^T\alpha=\begin{bmatrix}\alpha^T\Psi_{(m)}\\ \alpha^T\Lambda_1^{(m)}\Psi_{(m)}\\ \vdots\\ \alpha^T\Lambda_{m-1}^{(m)}\Psi_{(m)}\end{bmatrix}=[\alpha,\Lambda_1^{(m)}\alpha,\cdots,\Lambda_{m-1}^{(m)}\alpha]\Psi_{(m)}=\Lambda_\alpha\Psi_{(m)}$$

since Λ_α is symmetric.

Lemma 3.2: $\Psi_{(m)}\Psi_{(m)}^T\alpha\beta^T\Psi_{(m)}=\Lambda_\alpha\Lambda_\beta\Psi_{(m)}$.

Proof: Using Lemma 3.1, $\Psi_{(m)}\Psi_{(m)}^T\alpha\beta^T\Psi_{(m)}=\Lambda_\alpha\Psi_{(m)}\beta^T\Psi_{(m)}$ and since $\beta^T\Psi_{(m)}$ is a scalar, and again using Lemma 3.1, $\Psi_{(m)}\beta^T\Psi_{(m)}=\Psi_{(m)}(\Psi_{(m)}^T\beta)=\Psi_{(m)}\Psi_{(m)}^T\beta=\Lambda_\beta\Psi_{(m)}$, thus proving the lemma.

From (2.9) and Lemmas 3.1 and 3.2, (3.3) can be written as

$$\begin{aligned} E\Psi_{(m)}(t)\simeq y(0) &+ \sum_{i=1}^{n} F_i\Lambda_{e_i}P_{(m)}\Psi_{(m)}(t)\\ &+\sum_{k=1}^{q}\sum_{i=1}^{n} G_{k_i}\Lambda_{e_i}\Lambda_{z_k}P_{(m)}\Psi_{(m)}(t)\\ &+\sum_{k=1}^{q} H_k\Lambda_\nu\Lambda_{z_k}P_{(m)}\Psi_{(m)}(t). \end{aligned} \tag{3.8}$$

It is to be noted that the above derivation is performed for arbitrary $t\in[0,1)$ and, therefore, holds for all t in this interval. It is now a relatively simple computational task to identify F_i, G_{k_i}, and H_k for known E, z_k and ν from input–output data. m linearly independent equations that contain all the parameters can be generated from (3.8) by sampling the Walsh functions at $t=(2i-1)/2m$ for $i=1,2,\cdots,m$. These equations can be solved for the unknowns F_i, G_{k_i}, and H_k. Situations may arise where the number of unknowns is greater than m. For example, in the identification, using eight Walsh functions of a second-order time-invariant bilinear system with a two-input vector, the unknown components of matrices A, N_k, and B total 16. Such situations may be handled by using additional input–output data with a different initial condition or with a different input vector. Another possible solution to such a problem is to increase the number of Walsh functions being used for the identification.

IV. Computational Results

Four computational examples are presented to illustrate the method proposed in Section III.

Example 4.1: Consider the following variable-structure system

$$\dot y(t)=ay(t)+by^2(t)u(t)+cy(t)u^2(t)+du(t) \tag{4.1}$$

with $a=-1$, $b=0.5$, $c=0.3$, $d=1.5$, $u(t)=e^{-0.5t}$ and zero initial condition. For simulation purposes the output $y(t)$ is generated using a fourth-order Runge–Kutta method with $t=j\Delta$ where $j=0,1,2,\cdots,8$ and $\Delta=0.125$. Using the method developed in Section III, (4.1) can be written as

$$e^T\Psi(t)=ae^TP\Psi(t)+be^T\Lambda_e\Lambda_zP\Psi(t)+cz^T\Lambda_e\Lambda_zP\Psi(t)+dz^TP\Psi(t) \tag{4.2}$$

where e and z are the Walsh coefficient vectors of $y(t)$ and $u(t)$,

TABLE I

Parameter	Total number of Walsh functions used		True Value
	m = 4	m = 8	
a	-0.9983	-0.9996	-1.0
b	0.4963	0.4991	0.5
c	0.2987	0.2997	0.3
d	1.4951	1.4988	1.5

respectively. To solve for the four parameters a, b, c, d, four equations are generated from (4.2) using four arbitrarily chosen but distinct time points. For the results tabulated in Table I, time points $t=3/16$, $7/16$, $11/17$, and $15/16$ were used.

Example 4.2: Consider the second-order bilinear system described by

$$\dot{y}(t)=Ay(t)+\sum_{k=1}^{2} N_k y(t)u_k(t)+Bu(t) \tag{4.3}$$

where $u(t)$ and $y(t)$ are input and output vectors, respectively, and A, N_1, N_2, and B are constant 2×2 matrices. Again, using a fourth-order Runge–Kutta method for generating the output data with

$$A=\begin{bmatrix}0 & 1\\ -0.5 & -1\end{bmatrix}\ N_1=\begin{bmatrix}0 & 1\\ 0.5 & 0.2\end{bmatrix}\ N_2=\begin{bmatrix}1 & 0\\ 0.2 & 0.3\end{bmatrix}\ B=\begin{bmatrix}2 & 1\\ 1 & 1\end{bmatrix}$$

$u_1(t)=e^{-0.5t}$, $u_2(t)=e^{-t}$, and $t=j\Delta$, where $j=0,1,2,\cdots,8$ and $\Delta=0.125$. Two initial conditions $y^T(0)=(0,0)$ and $y^T(0)=(0.2,0.3)$ are used.

Using the technique of Section III, (4.3) may be manipulated to yield two equations:

$$C\alpha_1 \cong \gamma_1$$

and

$$C\alpha_2 \cong \gamma_2 \tag{4.4}$$

where

$$\alpha_i^T=\left(a_{i1}a_{i2}n_{i1}^{(1)}n_{i2}^{(1)}n_{i1}^{(2)}n_{i2}^{(2)}b_{i1}b_{i2}\right)$$

$$\gamma_i^T=\left(e_i^T\Psi_{(m)}(t_1),e_i^T\Psi_{(m)}(t_2),\cdots,e_i^T\Psi_{(m)}(t_m)\right)$$

for $i=1,2$, and C is a $m\times 8$ matrix with an ith row as

$$\big[e_1^TP\Psi_{(m)}(t_i),e_2^TP\Psi_{(m)}(t_i),e_1^T\Lambda_{z_1}P\Psi_{(m)}(t_i),e_2^T\Lambda_{z_1}P\Psi_{(m)}(t_i),$$

$$e_1^T\Lambda_{z_2}P\Psi_{(m)}(t_i),e_2^T\Lambda_{z_2}P\Psi_{(m)}(t_i),z_1^TP\Psi_{(m)}(t_i),z_2^TP\Psi_{(m)}(t_i)\big].$$

With $m=8$ Walsh functions and since (4.4) is an approximate equation, to smooth out any errors, 12 equations were chosen arbitrarily from the 16 generated with two different initial conditions to solve for the eight unknowns in each equation of (4.4) to yield the following values:

$$A=\begin{bmatrix}-.005 & .979\\ -.500 & -1.004\end{bmatrix}\ N_1=\begin{bmatrix}.016 & 1.033\\ .500 & .206\end{bmatrix}$$

$$N_2=\begin{bmatrix}.991 & -.019\\ .200 & .296\end{bmatrix}\ B=\begin{bmatrix}2.022 & .969\\ 1.001 & .992\end{bmatrix}.$$

Example 4.3: Consider the bilinear system with time-varying parameters:

$$\dot{y}(t)=a(t)y(t)+n(t)y(t)u(t)+b(t)u(t). \tag{4.5}$$

Using the technique of Section III, (3.8) corresponding to (4.5), with

$$a(t)=f^T\Psi(t)$$
$$n(t)=g^T\Psi(t)$$
$$b(t)=h^T\Psi(t)$$

can be written as

$$e^T\Psi(t)=y(0)+f^T\Lambda_eP\Psi(t)+g^T\Lambda_e\Lambda_zP\Psi(t)+h^T\Lambda_zP\Psi(t) \tag{4.6}$$

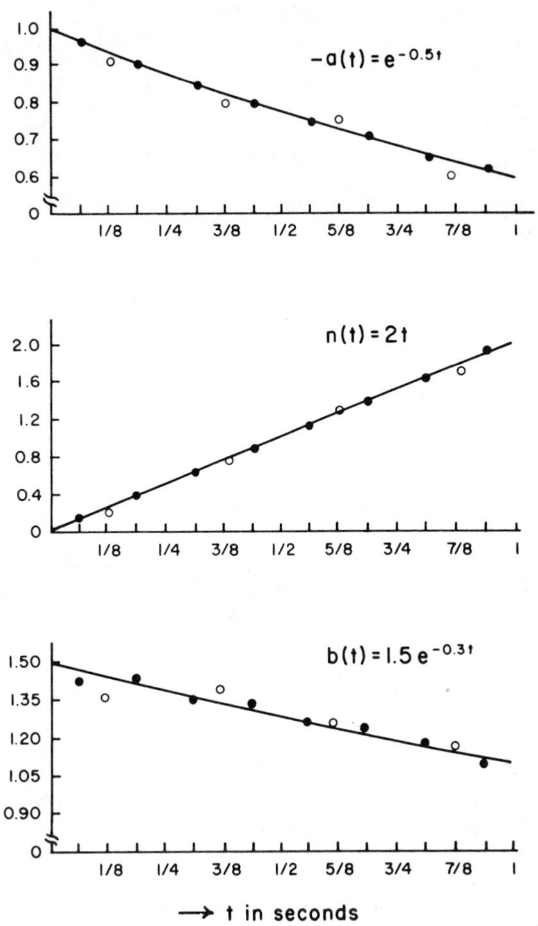

Fig. 1. Computed results for example 4.3. The solid circles denote the values obtained using 8 Walsh functions and the open circles using 4 Walsh functions.

where e and z are the Walsh coefficient vectors of $y(t)$ and $u(t)$, respectively. Using m Walsh functions, it is required to find $3m$ components of f, g, and h. Four different inputs e^{-2t} (0.2), $2t$ (0.1), te^{-t} (4.1), and $1/(t+0.5)$ (2.6), (the numbers in parentheses being the initial conditions) are used in (4.6) to generate $4m$ equations to solve for the $3m$ unknowns. The plots of $a(t)$, $n(t)$, and $b(t)$ for $m=4$ and $m=8$ Walsh functions are illustrated in Fig. 1.

Example 4.4: This example illustrates the effect of modeling and measurement noise. Consider the scalar bilinear system

$$\dot{x}(t)=ax(t)+nx(t)u(t)+bu(t)+\xi(t) \tag{4.7}$$

$$y(t)=x(t)+\eta(t) \tag{4.8}$$

where $\xi(t)$ and $\eta(t)$ are independent Gaussian white noises with zero-mean and variances given by

$$E\left[\eta(t)\eta(\tau)\right]=r\delta(t-\tau)$$
$$E\left[\xi(t)\xi(\tau)\right]=q\delta(t-\tau)$$

with the parameter values $a=-2$, $n=1$, $b=3$, $u(t)=2\exp(2(t-0.5))$, $x(0)=5.0$, and $\Delta=1/64$. Output data is generated using Euler's method for several values of r and q.

In order to apply the technique of Section III, $x(t)$ is to be estimated. With $x(t)=\Psi_{(m)}^T(t)e$, (4.8) becomes

$$y(t)=\Psi_{(m)}^Te+\eta(t). \tag{4.9}$$

The least-squares estimate of e[9] for m observations in (4.9) is given by

$$e=\left(V_m^Tr^{-1}V_m\right)^{-1}V_m^Tr^{-1}y_m \tag{4.10}$$

TABLE II

Noise Covariance	Parameter Estimates		
	a	n	b
q = 0.01, r = 0.09	-2.047	0.912	3.509
q = 0.01, r = 0.16	-2.057	0.897	3.634
q = 0.01, r = 0.25	-2.066	0.882	3.756
q = 0.04, r = 0.09	-2.040	0.913	3.479
True value	-2.00	1.00	3.000

where

$$V_m = (\Psi_{(m)}(t_1), \cdots, \Psi_{(m)}(t_m))^T$$

and

$$y_m = (y(t_1), \cdots, y(t_m))^T.$$

With $t_i = (2i-1)/2m$ for $i = 1, 2, \cdots, m$, it is easy to see that $V_m = W_m$, the Walsh matrix of order m. Noting that $W_m W_m^{-1} = mI_m$ where I_m is the identity matrix of order m, and $W_m = W_m^{-1}$, (4.10) becomes

$$e = \left(\frac{1}{m}\right) W_m y_m = f$$

where f is the Walsh coefficient vector of $y(t)$ [3]. This implies that, since the Walsh coefficient vector for a function is unique, the least-squares estimate of $x(t)$ of (4.8) is $y(t)$. Thus, using the output generated for $x(t)$ and neglecting $\xi(t)$ in the parameter estimation scheme, the technique of Section III, with $m = 16$, is applied to (4.7). In order to average out the effect of noise, all 16 equations generated are used to solve for the three unknown parameters. The results are tabulated in Table II.

In Example 4.1, the parameter estimation error with a 4-Walsh function approximation is less than 0.8 percent while for eight Walsh functions the error was less than 0.2 percent. Employing the norm $\|A\| = \sup_{\|x\|=1} \|Ax\|$, the maximum estimation error (that of the matrix N_1) is 0.034 in Example 4.2. By employing the error definition

$$\frac{1}{m} \sum_{i=1}^{m} [p(t_i) - \hat{p}(t_i)]^2$$

to the results of Example 4.3, the error is less than 0.3 percent for $m = 4$ and less than 0.1 percent for the 8-Walsh function approximation.

V. Concluding Remarks

While the derivation of the proposed method in Section III is carried through for bilinear systems, the method can be applied to a very large class of nonlinear systems. For example, the polynomial type nonlinear term x^n, (x, a scalar) can be handled using Lemmas 3.1 and 3.2 repeatedly. Let $x(t) = e^T \Psi(t)$. Then $x^n(t) = e^T \Lambda_e^{n-1} \Psi(t)$. Fractional powers in nonlinearities, however, cannot be handled in this framework because Lemmas 3.1 and 3.2 do not apply and it is doubtful if analogous results could be obtained to handle such cases.

Problems with unknown initial conditions pose no additional difficulties and can be handled in an analogous way. The proposed method, however, cannot be used in cases where $\dim(y) \neq \dim(x)$; a shortcoming of a large class of parameter estimation methods. A very recent result by Friedland [10], however, indicates that this difficulty can be avoided in cases where

$$y(t) = H(t)x(t)$$

and H is a known time-varying matrix-valued function.

The remarkable accuracy of the proposed Walsh-function method in the computational examples of Section IV surprised even the authors. Moreover, since Walsh functions are relatively easy to handle in computations, the proposed technique is quite simple to implement and has, therefore, obvious advantages in practical situations. The only computational disadvantage is that 2^n (n, an integer) Walsh functions have to be employed, and as a consequence the computer storage requirements for higher dimensional problems might become excessive if more than 8–16 functions are used. Example 4.4 indicates that acceptable parameter estimates can be obtained even in cases where the bilinear system is subject to small modeling and measurement errors.

Appendix A

In this Appendix it is shown how (2.7) may be written in compact form as (2.9). Taking the terms containing the first four Walsh functions (that is, ψ_0, ψ_1, ψ_2, and ψ_3) $J_0(t)$ is approximated by

$$J_0(t) \cong 1/2\psi_0(t) - 1/4\psi_1(t) - 1/8\psi_2(t). \tag{A1}$$

Let $\Psi_{(4)}(t) = [\psi_0(t)\psi_1(t)\psi_2(t)\psi_3(t)]$. Then, (A1) may be written as

$$J_0(t) \cong [1/2 - 1/4 - 1/8 \quad 0]\Psi_{(4)}(t). \tag{A2}$$

Similarly, from (2.7) it can be seen that

$$\begin{aligned} J_1(t) &\cong [1/4 \quad 0 \quad 0 \quad -1/8]\Psi_{(4)}(t) \\ J_2(t) &\cong [1/8 \quad 0 \quad 0 \quad 0]\psi_{(4)}(t) \end{aligned} \tag{A3}$$

and

$$J_3(t) \cong [0 \quad 1/8 \quad 0 \quad 0]\psi_{(4)}(t).$$

Combining (A2) and (A3), and writing J_i's explicitly,

$$\begin{bmatrix} \int_0^t \psi_0(x)\,dx \\ \int_0^t \psi_1(x)\,dx \\ \int_0^t \psi_2(x)\,dx \\ \int_0^t \psi_3(x)\,dx \end{bmatrix} \cong \begin{bmatrix} 1/2 & -1/4 & -1/8 & 0 \\ 1/4 & 0 & 0 & -1/8 \\ 1/8 & 0 & 0 & 0 \\ 0 & 1/8 & 0 & 0 \end{bmatrix} \begin{bmatrix} \psi_0(t) \\ \psi_1(t) \\ \psi_2(t) \\ \psi_3(t) \end{bmatrix}$$

or in matrix form

$$\int_0^t \Psi_{(4)}(x)\,dx \cong P_{(4)}\Psi_{(4)}(t). \tag{A4}$$

A general expression for the operational matrix P of an arbitrary order m can be obtained and is given by

$$P_{(1)} = \frac{1}{2}; \quad P_{(m)} = \left[\begin{array}{c|c} P_{(m/2)} & -\frac{1}{2m} I_{(m/2)} \\ \hline \frac{1}{2m} I_{(m/2)} & 0_{(m/2)} \end{array}\right] \qquad \begin{array}{l} m = 2^n \\ n = 1, 2, \cdots. \end{array} \tag{A5}$$

Here $I_{(m/2)}$ and $0_{(m/2)}$ denote, respectively, the identity and null matrices of order $m/2$.

Appendix B

Three crucial properties of Λ_i matrices are established in the following three lemmas.

Lemma B.1: $\Lambda_i^{(m)} \Lambda_j^{(m)} = \Lambda_{i \oplus j}^{(m)}$.

Proof by induction: The lemma is trivially true for $m = 1$. For $m = 2$ four cases arise depending on whether $i = 0$ or 1, and $j = 0$ or 1. Consider, for example, the case $i = j = 0$ giving

$$\Lambda_i^{(2)} \Lambda_j^{(2)} = \begin{bmatrix} \Lambda_i^{(1)} & 0_{(1)} \\ 0_{(1)} & \Lambda_i^{(1)} \end{bmatrix} \begin{bmatrix} \Lambda_j^{(1)} & 0_{(1)} \\ 0_{(1)} & \Lambda_j^{(1)} \end{bmatrix} = \begin{bmatrix} \Lambda_{i \oplus j}^{(1)} & 0_{(1)} \\ 0_{(1)} & \Lambda_{i \oplus j}^{(1)} \end{bmatrix} = \Lambda_{i \oplus j}^{(2)} \quad \text{by (3.7).}$$

A similar simple calculation can be performed in the other three cases. Assume the lemma holds for $m/2$, $m=2^n$ for some $n>1$ (having already established this for $m=2$) then, for $i<m/2$ and $j>m/2$, say $j=l+m/2$ where $l \geqslant 0$,

$$\Lambda_i^{(m)}\Lambda_j^{(m)} = \begin{bmatrix} \Lambda_i^{(m/2)} & 0_{(m/2)} \\ 0_{(m/2)} & \Lambda_i^{(m/2)} \end{bmatrix} \begin{bmatrix} 0_{(m/2)} & \Lambda_l^{(m/2)} \\ \Lambda_l^{(m/2)} & 0_{(m/2)} \end{bmatrix}$$

$$= \begin{bmatrix} 0_{(m/2)} & \Lambda_{i\oplus l}^{(m/2)} \\ \Lambda_{i\oplus l}^{(m/2)} & 0_{(m/2)} \end{bmatrix} = \Lambda_{i\oplus j}^{(m)} \quad \text{by (3.7).}$$

Thus, the lemma is true for all integer powers of 2.

Lemma B.2: $\Lambda_i^{(m)}$ $(i=0,1,2,\cdots,(m-1))$ form an Abelian group under the multiplication operation.

Proof: Associativity holds for matrix multiplication. The identity element $\Lambda_0^{(m)}$ exists such that $\Lambda_i^{(m)}\Lambda_0^{(m)}=\Lambda_0^{(m)}\Lambda_i^{(m)}=\Lambda_0^{(m)}$ for all i. The inverse of each $\Lambda_i^{(m)}$ is itself, since

$$\Lambda_i^{(m)}\Lambda_i^{(m)}=\Lambda_{i\oplus i}^{(m)}=\Lambda_0^{(m)} \quad \text{using Lemma B.1.}$$

This proves Lemma B.2.

Lemma B.3:

$$\prod_{i-1}^{2^k-1} \Lambda_i^{(m)}=\Lambda_0^{(m)}, \qquad \text{for all } k>1.$$

The proof follows trivially from Lemma B.1 and the fact that $1\oplus 2\oplus 3\cdots\oplus(2^k-1)=0$ for all $k>1$.

References

[1] N. J. Fine, "On the Walsh functions," *Trans. Amer. Math. Soc.*, vol. 65, pp. 372–414, May 1949.

[2] C. F. Chen and C. H. Hsiao, "Time domain synthesis via Walsh functions," *Proc. IEE*, vol. 122, pp. 565–570, May 1975.

[3] H. F. Harmuth, *Transmission of Information by Orthogonal Functions*. New York: Springer-Verlag, 1972.

[4] G. Prasada Rao and L. Sivakumar, "System identification via Walsh functions," *Proc. IEE*, vol. 122, pp. 1160–1161, Oct. 1975.

[5] C. F. Chen and C. H. Hsiao, "Design of piecewise constant gains for optimal control via Walsh functions," *IEEE Trans. Automat. Contr.*, vol. AC-20, pp. 596–603, Oct. 1975.

[6] R. R. Mohler, *Bilinear Control Processes*. New York: Academic, 1973.

[7] A. Ruberti and R. R. Mohler, Eds., *Variable Structure Systems with Applications to Economic and Biology*. New York: Springer-Verlag, 1975.

[8] M. S. Corrington, "Solution of differential and integral equations with Walsh functions," *IEEE Trans. Circuit Theory*, vol. CT-20, pp. 470–475, Sept. 1973.

[9] A. P. Sage and J. L. Melsa, *Estimation Theory with Applications to Communications and Control*. New York: McGraw-Hill, 1971.

[10] B. Friedland, "On the calibration problem," *IEEE Trans. Automat. Contr.*, vol. AC-22, pp. 899–905, Dec. 1977.

13

Reprinted from *J. Franklin Inst.* **305:**199–220 (1978)

Walsh Series Approach to Lumped and Distributed System Identification

by S. G. TZAFESTAS

Control Systems Laboratory, University of Patras, Patras, Greece

ABSTRACT: *This paper considers the problem of identifying the parameters of dynamic systems from input–output records. Both lumped-parameter and distributed-parameter systems, deterministic and stochastic, are studied. The approach adopted is that of expanding the system variables in Walsh series. The key point is an operational matrix P which relates the coefficient matrix Γ of the Walsh series of a given function with the coefficient matrix of its first derivative. Using this operational matrix P one overcomes the necessity to use differentiated data, a fact that usually is avoided either by integration of the data or by using discrete-time models. Actually, the original differential input–output model is converted to a linear algebraic (or regression) model convenient for a direct (or a least squares) solution. A feature of the method is that it permits the identification of unknown initial conditions simultaneously with the parameter identification. The results are first derived for single-input single-output systems and then are extended to multi-input multi-output systems. The case of non-constant parameters is treated by assuming polynomial forms. Some results are also included concerning the identification of state-space and integral equation models. The theory is supported by two examples, which give an idea of how effective the method is expected to be in the real practice.*

I. Introduction

Walsh functions constitute a complete family of binary functions with interesting properties, and have recently found application in a variety of fields such as two-dimensional image processing, sequency filtering, spectroscopy, multiplexing, speech synthesis and analysis, statistical analysis, system simulation, control theory etc. (**1**)–(**9**).

This paper presents solutions to some identification problems by using the approach of expanding the system variables in Walsh series. The paper was motivated by the work of Chen and Hsiao concerning the Walsh series approach to state space simulation, time domain synthesis, and optimal control (**6**)–(**8**).

The paper is organized as follows. Section II treats the identification of single-input single-output systems, starting with the basic identification formula which is based on the results of (**8**), and covers both the zero-initial condition and the non-zero unknown initial condition cases (**16**). Section III extends the results of Section II to multi-input multi-output systems, and Section IV treats stochastic systems via the least-squares method. In Section V we consider time varying systems in which the parameters are polynomials of the time variable. In Section VI the state space model identification is considered for the cases of full-state and partial-state measurements (time-invariant and time-varying cases). Section VII treats the identification of integral equation models, and Section VIII is devoted to the case of distributed-parameter systems treating the problem through the usual way of transforming the system into appropriate lumped-parameter forms. Finally Section IX provides two illustrative computational examples, and the two appendices give a brief review of the Walsh function expansion and the proof of a recursive formula utilized in the time-varying case.

II. Single-Input Single-Output Systems

1. *Zero-initial conditions: The basic identification formula*

Consider a deterministic time-invariant single-input single-output (SISO) system with zero initial conditions of the type

$$A_n(D)y(t) = B_n(D)u(t), \quad t \geq 0 \tag{1}$$

where

$$\begin{aligned} A_n(D) &= D^n + a_{n-1}D^{n-1} + \ldots + a_1 D + a_0 \\ B_n(D) &= b_n D^n + b_{n-1}D^{n-1} + \ldots + b_1 D + b_0. \end{aligned} \tag{2}$$

The problem is to determine (compute or estimate) the $2n+1$ parameters $a_0, a_1, \ldots, a_{n-1}, b_0, b_1, \ldots, b_n$ by using a record of input-output measurements $\{u(t), y(t), t\varepsilon T\}$ over a certain time interval T. Without loss of generality we can assume that $T = \{t: 0 \leq t \leq 1\}$. If $T = \{t: 0 \leq t \leq t_f\}$ we normalize the time as $\sigma = t/t_f$ to obtain the normalized interval $T = \{\sigma: 0 \leq \sigma \leq 1\}$.

To solve this problem, fundamental in our approach, the differential model (1), which involves the derivatives of the available input–output data, must first be converted to an algebraic model not involving differentiated data. To this end, the nth derivatives of $y(t)$ and $u(t)$ are expanded in Walsh series and the Walsh operational formula (A3) of the Appendix A is successively used, together with the fact that the initial conditions are all zero, to yield

$$D^n y(t) = \Phi^T(t)c, \quad D^{n-1}y(t) = \Phi^T(t)P^T c, \ldots, y(t) = \Phi^T(t)(P^n)^T c \tag{3a}$$

$$D^n u(t) = \Phi^T(t)\Theta, \quad D^{n-1}u(t) = \Phi^T(t)P^T\Theta, \ldots, u(t) = \Phi^T(t)(P^n)^T\Theta \tag{3b}$$

where c and Θ are the Walsh coefficient vectors of $D^n y(t)$ and $D^n u(t)$ respectively.

Then denoting by γ and ζ the Walsh coefficient vectors of $y(t)$ and $u(t)$ which are at least $(2n+1)$ dimensional, we obtain the relationships

$$c^T P^n = \gamma^T, \quad c^T P^{n-1} = \gamma^T P^{-1}, \ldots, c^T = \gamma^T P^{-n} \tag{4a}$$

$$\Theta^T P^n = \zeta^T, \quad \Theta^T P^{n-1} = \zeta^T P^{-1}, \ldots, \Theta^T = \zeta^T P^{-n}. \tag{4b}$$

Using Eqs. (3a–b) and (4a–b) the model (1) becomes

$$\gamma^T A_n(P^{-1})\Phi(t) = \zeta^T B_n(P^{-1})\Phi(t). \tag{5}$$

Equation (5) is the algebraic equivalent of model (1) sought. Since the data coefficient vectors γ and ζ can be computed from

$$\gamma=\int_0^1 y(t)\Phi(t)\,dt,\quad \zeta=\int_0^1 u(t)\Phi(t)\,dt \tag{6}$$

and the Walsh function vector $\Phi(t)$ is available for all t in the interval $0\leq t\leq 1$, to compute the parameters a_i and b_i we define the vectors

$$\Theta=\begin{bmatrix}\Theta_a\\ \cdots\\ \Theta_b\end{bmatrix},\quad \Theta_a=\begin{bmatrix}a_{n-1}\\ a_{n-2}\\ \cdot\\ \cdot\\ \cdot\\ a_0\end{bmatrix},\quad \Theta_b=\begin{bmatrix}b_n\\ b_{n-1}\\ \cdot\\ \cdot\\ \cdot\\ b_0\end{bmatrix} \tag{7}$$

and write (5) in the regression form

$$\Phi^T(t)Z\Theta=\Phi^T(t)(P^T)^{-n}\gamma,\quad 0\leq t\leq 1 \tag{8a}$$

where

$$Z=[-(P^T)^{-n+1}\gamma,\ldots,-\gamma;\ (P^T)^{-n}\zeta,\ (P^T)^{-n+1}\zeta,\ldots,\zeta]. \tag{8b}$$

Now, applying (8a) for $t=t_1, t_2, t_3,\ldots, t_{2n+1}$, and defining the $(2n+1)\times 2n+1)$ square matrix $\tilde{\Phi}$ as

$$\tilde{\Phi}^T=\begin{bmatrix}\Phi^T(t_1)\\ \Phi^T(t_2)\\ \cdot\\ \cdot\\ \cdot\\ \Phi^T(t_{2n+1})\end{bmatrix} \tag{9}$$

we obtain

$$(\tilde{\Phi}^T Z)\Theta=\tilde{\Phi}^T(P^T)^{-n}\gamma$$

which, upon the assumption that $\tilde{\Phi}^T Z$ is nonsingular, gives

$$\Theta=(\tilde{\Phi}^T Z)^{-1}\tilde{\Phi}^T(P^T)^{-n}\gamma. \tag{10}$$

Equation (10) provides the values of the unknown parameters, and holds when the measured data of $y(t)$ and $u(t)$ do not contain any errors or noise. The case where there exists stochastic errors will be treated in Section IV.

The formula (10) involves P^{-n}, $P^{-n+1},\ldots,P^{-1}$ which require the time consuming inversion of $P, P^2,\ldots,P^n$. However, an equivalent formula containing $P, P^2,\ldots,P^n$ and not $P^{-1}, P^{-2},\ldots,P^{-n}$ can be derived as follows. Apply (5) for $t=t_1, t_2,\ldots,t_{2n+1}$ and obtain

$$\gamma^T A_n(P^{-1})\tilde{\Phi}=\zeta^T B_n(P^{-1})\tilde{\Phi} \tag{11}$$

where care is taken for the matrix

$$\tilde{\Phi} = [\Phi(t_1), \quad \Phi(t_2), \ldots, \Phi(t_{2n+1})] \tag{12}$$

to be non-singular. Then (11) reduces to

$$\gamma^T A_n(P^{-1}) = \zeta^T B_n(P^{-1})$$

which if post multiplied by $P^n \tilde{\Phi}$ yields

$$\gamma^T(I + a_{n-1}P + \ldots + a_0 P^n)\tilde{\Phi} = \zeta^T(b_n I + b_{n-1}P + \ldots + b_0 P^n)\tilde{\Phi}. \tag{13}$$

Equation (13) is now written in the regression form

$$(\tilde{\Phi}^T H)\Theta = \tilde{\Phi}\gamma$$

from which we obtain the basic identification formula (**10**)

$$\Theta = (\tilde{\Phi}^T H)^{-1} \tilde{\Phi}^T \gamma \tag{14a}$$

where

$$H = [-P^T\gamma, \ldots, -(P^T)^n\gamma; \zeta, \quad (P^T)\zeta, \ldots, (P^T)^n\zeta] = (P^T)^n Z. \tag{14b}$$

It is noted that actually the model (13) could be obtained directly by successively integrating n times from $t=0$ up to $t=t$ the model (1) and applying the Walsh function operational formula (A3) of the Appendix A (**8**).

2. *Non-zero initial conditions*

Consider the SISO system (1) and assume that the initial conditions

$$[y]_0 = Y_0, \quad [Dy]_0 = Y_1, \ldots, [D^{n-1}y]_0 = Y_{n-1} \tag{15a}$$

of the response function are unknown and are to be determined simultaneously with the system parameters a_i and b_i. The initial conditions

$$[u]_0 = U_0, \quad [Du]_0 = U_1, \ldots, [D^{n-1}u]_0 = U_{n-1} \tag{15b}$$

of the input function are assumed to be known.

Successively integrating the model (1) n times from $t=0$ up to $t=t$ and assuming that the initial conditions of the integrals $\int u\,\mathrm{d}t$, $\int y\,\mathrm{d}t$, $\iint u\,\mathrm{d}t\,\mathrm{d}t$, $\iint y\,\mathrm{d}t\,\mathrm{d}t$ etc. are zero, one obtains

$$\begin{aligned} &y(t) + a_{n-1}D^{-1}y(t) + \ldots + a_1 D^{-(n-1)}y(t) + a_0 D^{-n}y(t) \\ &+ Y_0 + Y_1 t + \frac{1}{2!}Y_1 t^2 + \ldots + \frac{1}{(n-1)!}Y_{n-1}t^{n-1} \\ &+ a_{n-1}\left[Y_0 t + \frac{1}{2!}Y_1 t^2 + \ldots + \frac{1}{(n-1)!}Y_{n-2}t^{n-1}\right] \\ &+ \ldots + a_2\left[\frac{1}{(n-2)!}Y_0 t^{n-2} + \frac{1}{(n-1)!}Y_1 t^{n-1}\right] + a_1 \frac{1}{(n-1)!}Y_0 t^{n-1} \end{aligned}$$

$$
\begin{aligned}
&= b_n u(t) + b_{n-1} D^{-1} u(t) + \ldots + b_1 D^{-(n-1)} u(t) + b_0 D^{-n} u(t) \\
&\quad + b_n\left[U_0 + U_1 t + \frac{1}{2!} U_2 t^2 + \ldots + \frac{1}{(n-1)!} U_{n-1} t^{n-1}\right] \\
&\quad + b_{n-1}\left[U_0 t + \frac{1}{2!} U_1 t^2 + \ldots + \frac{1}{(n-1)!} U_{n-2} t^{n-1}\right] \\
&\quad + \ldots + b_2\left[\frac{1}{(n-2)!} U_0 t^{n-2} + \frac{1}{(n-1)!} U_1 t^{n-1}\right] + b_1 \frac{1}{(n-1)!} U_0 t^{n-1}
\end{aligned} \tag{16}
$$

where $D^{-1}y = \int_0^t y\, dt$, $D^{-2}y = \int_0^t \int_0^t y\, dt\, dt$, etc.

Clearly, in the present case we have $(3n+1)$ unknown parameters and so the Walsh expansion coefficient vectors γ and ζ of $y(t)$ and $u(t)$ must be at least $(3n+1)$ dimensional. Hence using the expansions

$$
\begin{aligned}
y(t) &= \gamma^T \Phi(t), \quad D^{-1} y(t) = \gamma^T P \Phi(t), \ldots, D^{-n} y(t) = \gamma^T P^n \Phi(t) \\
u(t) &= \zeta^T \Phi(t), \quad D^{-1} u(t) = \zeta^T P \Phi(t), \ldots, D^{-n} u(t) = \zeta^T P^n \Phi(t) \\
t^0 &= 1 = \varepsilon_0^T \Phi(t), \quad t = \varepsilon_1^T \Phi(t), \ldots, t^{n-1} = \varepsilon_{n-1}^T \Phi(t),
\end{aligned}
$$

Eq. (16) becomes

$$
\begin{aligned}
\Phi^T(t)\gamma &= \Phi^T(t)[-P^T\gamma, \ldots, -(P^T)^n\gamma;\ \zeta, P^T\zeta, \ldots, (P^T)^n\zeta]\Theta \\
&\quad + \Phi^T(t)[\varepsilon_0, \varepsilon_1, \ldots, \varepsilon_{n-1}]\Theta_c
\end{aligned} \tag{17}
$$

where $\Theta_c = \Theta_u - \Theta_y$,

$$
\Theta_y = \begin{bmatrix} Y_0 \\ Y_1 + a_{n-1} Y_0 \\ \cdot \\ \cdot \\ \cdot \\ \frac{1}{(n-1)!}(Y_{n-1} + a_{n-1} Y_{n-2} + \ldots + a_1 Y_0) \end{bmatrix}
\tag{17a}
$$

$$
\Theta_u = \begin{bmatrix} b_n U_0 \\ b_n U_1 + b_{n-1} U_0 \\ \cdot \\ \cdot \\ \cdot \\ \frac{1}{(n-1)!}(b_n U_{n-1} + b_{n-1} U_{n-2} + \ldots + b_1 U_0) \end{bmatrix}.
$$

Defining the matrices

$$
f = \begin{bmatrix} \Theta \\ \Theta_c \end{bmatrix}, \quad E = [\varepsilon_0, \varepsilon_1, \ldots, \varepsilon_{n-1}] \tag{18}
$$

and applying (17) at $t=t_1, t_2, \ldots, t_{3n+1}, \ldots, t_m$ yields

$$\tilde{\Phi}^T\gamma = \tilde{\Phi}Qf, \quad Q=[H \,\vdots\, E] \tag{18a}$$

from which it follows that

$$f=(\tilde{\Phi}Q)^{-1}\tilde{\Phi}^T\gamma \tag{19}$$

where the matrix H is defined as in (14b).

Equation (19) determines f, i.e. Θ and Θ_c. To determine the initial condition vector

$$Y=[Y_0, Y_1, \ldots, Y_{n-1}]^T$$

we write Θ_c in (17a) as

$$\Theta_c = L_bU - L_aY \tag{20}$$

where

$$U=[U_0, U_1, \ldots, U_{n-1}]^T$$

$$L_a = \begin{bmatrix} 1 & & & \\ a_{n-1} & 1 & 0 & \\ \cdots & \cdots & \cdots & \cdots \\ \dfrac{a_1}{(n-1)!} & \dfrac{a_2}{(n-1)!} & \cdots & \dfrac{1}{(n-1)!} \end{bmatrix}, \quad L_b = \begin{bmatrix} b_n & & & \\ b_{n-1} & b_n & 0 & \\ \cdots & \cdots & \cdots & \cdots \\ \dfrac{b_1}{(n-1)!} & \dfrac{b_2}{(n-1)!} & \cdots & \dfrac{b_n}{(n-1)!} \end{bmatrix} \tag{21}$$

and solve for Y, i.e.

$$Y=L_a^{-1}(L_bU-\Theta_c). \tag{22}$$

III. Multi-Input Multi-Output Systems

In this section the identification technique of the previous section will be extended to the canonical multi-input multi-output (MIMO) system:

$$(D^n + A_{n-1}D^{n-1} + \ldots + A_1D + A_0)\mathbf{y}(t) = (B_nD^n + B_{n-1}D^{n-1} + \ldots + B_1D + B_0)\mathbf{u}(t) \tag{23}$$

where

$$A_k = [a_{ij}^k] = \begin{bmatrix} \mathbf{a}_1^{k^T} \\ \mathbf{a}_2^{k^T} \\ \cdot \\ \cdot \\ \cdot \\ \mathbf{a}_p^{k^T} \end{bmatrix}, \quad B_k = [b_{ij}^k] = \begin{bmatrix} \mathbf{b}_1^{k^T} \\ \mathbf{b}_2^{k^T} \\ \cdot \\ \cdot \\ \cdot \\ \mathbf{b}_p^{k^T} \end{bmatrix}, \quad \mathbf{y} = \begin{bmatrix} y_1 \\ y_2 \\ \cdot \\ \cdot \\ \cdot \\ y_p \end{bmatrix}, \quad \mathbf{u} = \begin{bmatrix} u_1 \\ u_2 \\ \cdot \\ \cdot \\ \cdot \\ u_p \end{bmatrix}$$

$$\mathbf{a}_i^{k^T} = [a_{i1}^k, a_{i2}^k, \ldots, a_{ip}^k], \quad \mathbf{b}_i^{k^T} = [b_{i1}^k, b_{i2}^k, \ldots, b_{ip}^k].$$

The problem is to determine the parameters a_{ij}^k, b_{ij}^k and the initial conditions of the vector function $\mathbf{y}(t)$.

Expanding $\mathbf{y}(t)$ and $\mathbf{u}(t)$ in Walsh series, we obtain

$$\begin{aligned} \mathbf{y}(t) &= \Gamma^T\Phi(t), \quad D^{-1}\mathbf{y}(t) = \Gamma^T P\Phi(t), \ldots, D^{-n}\mathbf{y}(t) = \Gamma^T P^n\Phi(t) \\ \mathbf{u}(t) &= \Delta^T\Phi(t), \quad D^{-1}\mathbf{u}(t) = \Delta^T P\Phi(t), \ldots, D^{-n}\mathbf{u}(t) = \Delta^T P^n\Phi(t) \end{aligned} \tag{24}$$

where Γ^T and Δ^T are the matrix Walsh coefficients.

Consider the ith row of (23):

$$(D^n + \mathbf{a}_i^{n-1^T}D^{n-1} + \ldots + \mathbf{a}_i^{0^T})\mathbf{y}(t) = (\mathbf{b}_i^{n^T}D^n + \mathbf{b}_i^{n-1^T}D^{n-1} + \ldots + \mathbf{b}_i^{0^T})\mathbf{u}(t)$$

which, when integrated, from 0 up to t, n times yields

$$\Phi^T(t)\Gamma = \Phi^T(t)Qf^i, \quad Q = [H \,\vdots\, E], \quad f^i = \begin{bmatrix} \Theta_a^i \\ \Theta_b^i \\ \Theta_u^i - \Theta_y^i \end{bmatrix}$$

whence as usual

$$f^i = (\tilde{\Phi}^T Q)^{-1}\tilde{\Phi}^T\Gamma, \quad i = 1, 2, \ldots, p. \tag{25}$$

As in the SISO case, we further have

$$\Theta_y^i = L_a^i Y, \quad \Theta_u^i = L_b^i U, \quad i = 1, 2, \ldots, p \tag{26}$$

where L_a^i and L_b^i have the structure shown in (21). From (25) we obtain f^i, i.e. Θ_a^i, Θ_b^i and the difference $\Theta_u^i - \Theta_y^i = \Theta_c^i$. To determine Y, we write

$$\boldsymbol{\Theta}_c = \begin{bmatrix} \Theta_c^1 \\ \Theta_c^2 \\ \cdot \\ \cdot \\ \cdot \\ \Theta_c^p \end{bmatrix} = \begin{bmatrix} \Theta_u^1 \\ \Theta_u^2 \\ \cdot \\ \cdot \\ \cdot \\ \Theta_u^p \end{bmatrix} - \begin{bmatrix} \Theta_y^1 \\ \Theta_y^2 \\ \cdot \\ \cdot \\ \cdot \\ \Theta_y^p \end{bmatrix} = L_b U - L_a Y$$

whence

$$Y = L_a^{-1}(L_b U - \boldsymbol{\Theta}_c), \quad L_a = \begin{bmatrix} L_a^1 \\ L_a^2 \\ \cdot \\ \cdot \\ \cdot \\ L_a^p \end{bmatrix}, \quad L_b = \begin{bmatrix} L_b^1 \\ L_b^2 \\ \cdot \\ \cdot \\ \cdot \\ L_b^p \end{bmatrix}.$$

IV. Stochastic Systems

We shall study only SISO stochastic systems. Multivariable systems can be treated analogously. Consider a stochastic model of the type

$$A_n(D)y(t) = B_n(D)u(t) + e(t) \tag{28}$$

where $e(t)$ represents the errors in the input and output measurements, or, equivalently, a stochastic noise or disturbance affecting the system. All other quantities are as in the model (1).

The problem is to determine a best estimate of the parameters a_i, b_i and the initial conditions Y_i of the output variables.

Here the least squares approach will be employed which does not require to know the stochastic nature of $e(t)$.

Following the procedure of Section II.2, we arrive at the model

$$\Phi^T(t)\gamma = \Phi^T(t)\{H\Theta + E\Theta_c\} + \varepsilon(t) \tag{29}$$

which is the stochastic analogue of (17). Here $\varepsilon(t)$ is the n-fold time integral of $e(t)$.

Now, we apply (29) for $t = t_{1,k}, t_{2,k}, \ldots, t_{3n+1,k}, \ldots, t_{m,k}$, $k = 1, 2, \ldots, N$ and obtain*

$$r_k = R_k f + \varepsilon_k \tag{30}$$

where

$$r_k = \tilde{\Phi}_k^T \gamma, \quad R_k = \tilde{\Phi}_k^T Q, \quad Q = [H \,\vdots\, E]$$

$$\tilde{\Phi}^T = \begin{bmatrix} \Phi^T(t_{1,k}) \\ \Phi^T(t_{2,k}) \\ \cdot \\ \cdot \\ \cdot \\ \Phi^T(t_{m,k}) \end{bmatrix}, \quad f = \begin{bmatrix} \Theta \\ \\ \Theta_c \end{bmatrix} \tag{30a}$$

Then, assuming that R_k and ε_k are such that

$$\lim_{N\to\infty} \sum_{k=1}^{N} R_k^T \varepsilon_k = 0 \tag{31}$$

the conventional least squares estimate of f is given by†

$$\hat{f} = \left(\sum_{k=1}^{N} R_k^T R_k\right)^{-1} \left(\sum_{k=1}^{N} R_k^T r_k\right). \tag{32}$$

*We can choose the $t_{i,k}$'s such that $t_{1,k} < t_{2,k} < \ldots < t_{m,k} < t_{1,k+1} < t_{2,k+1} < \ldots < t_{m,k+1} < t_{1,k+2} < \ldots$.

†This is found by premultiplying both sides of (30) by R_k^T, summing from $k = 1$ up to $k = N$, and taking into account (31), (**10**). Equivalently the same result is obtained by minimizing $\frac{1}{2}\sum_{k=1}^{N} \varepsilon_k^2$ with respect to f.

The condition (31) means that R_k and ε_k are uncorrelated. When ε_k is correlated with R_k one can use either the generalized least-squares method, if the statistical properties of ε_k are known, or the modified least squares, if the properties of ε_k are not known (**10**).

V. *Time-Varying Systems*

Consider a SISO system of the form (1) with time-varying parameters of the polynomial type (**11**):

$$\begin{aligned} a_i(t) &= \alpha_i^0 + \alpha_i^1 t + \ldots + \alpha_i^{\sigma_i} t^{\sigma_i} \\ b_i(t) &= \beta_i^0 + \beta_i^1 t + \ldots + \beta_i^{s_i} t^{s_i}. \end{aligned} \tag{33}$$

For simplicity we assume zero initial conditions. The extension to unknown nonzero initial conditions is straight forward.

Integrating n times from $t=0$ up to $t=t$, yields

$$\begin{aligned} y(t) &+ \alpha_{n-1}^0 I_n(D^{n-1}y) + \ldots + \alpha_{n-1}^{\sigma_{n-1}} I_n(t^{\sigma_{n-1}} D^{n-1} y) \\ &+ \ldots + \alpha_1^0 I_n(Dy) + \ldots + \alpha_1^{\sigma_1} I_n(t^{\sigma_1} Dy) \\ &+ \alpha_0^0 I_n(y) + \ldots + \alpha_0^{\sigma_0} I_n(t^{\sigma_0} y) \\ &= \beta_n^0 u + \beta_n^1 J_n(tD^n u) + \ldots + \beta_n^{s_n} J_n(t^{s_n} D^n u) \\ &+ \ldots + \beta_1^0 J_n(Du) + \ldots + \beta_1^{s_1} J_n(t^{s_1} Du) \\ &+ \beta_0^0 J_n(u) + \ldots + \beta_0^{s_0} J_n(t^{s_0} u) \end{aligned} \tag{34}$$

where

$$I_n(t^k D^\sigma y) = \int_0^t \int_0^t \ldots \int_0^t t^k D^\sigma y \, dt^n \triangleq I_n(k, \sigma) \tag{35a}$$

$$J_n(t^k D^\sigma u) = \int_0^t \int_0^t \ldots \int_0^t t^k D^\sigma u \, dt^n \triangleq J_n(k,\sigma). \tag{35b}$$

The integrals $I_n(k, \sigma)$ and $J_n(k, \sigma)$ can be determined recursively by using the following relationships: (for the proof see Appendix B)

$$I_n(k, \sigma) = \gamma^T S_n(k, \sigma)\Phi(t) \tag{36a}$$

$$J_n(k, \sigma) = \zeta^T S_n(k, \sigma)\Phi(t) \tag{36b}$$

where

$$S_n(k, \sigma+1) = S_{n-1}(k, \sigma) - kS_n(k-1, \sigma) \tag{37a}$$

with boundary conditions

$$S_n(k, 0) = R_k^n(t), \quad S_n(0, \sigma) = P^{n-\sigma} \tag{37b}$$

$$R_k^n(t) = P[R_k^{n-1}(t) - kR_{k-1}^n(t)] \tag{37c}$$

$$R_k^0(t) = t^k I, \quad R_0^n(t) = P^n \tag{37d}$$

for $k = 1, 2, 3, \ldots$ and $\sigma = 0, 1, 2, 3, \ldots$.

Using (36a–b) and (37a–d), (34) gives

$$\Phi^T(t)\gamma = \Phi^T(t)\tilde{H}\Theta \tag{38}$$

where

$$\Theta^T = [\alpha_{n-1}^0, \ldots, \alpha_{n-1}^{\sigma_{n-1}}, \ldots, \alpha_0^0, \ldots, \alpha_0^{\sigma_0}, \quad \beta_n^0, \ldots, \beta_n^{s_n}, \ldots, \beta_0^0, \ldots, \beta_0^{s_0}] \tag{39a}$$

$$\begin{aligned}\tilde{H} = [&-S_n^T(0, n-1)\gamma, \ldots, -S_n^T(\sigma_{n-1}, n-1)\gamma, \ldots, -S_n^T(0, 0)\gamma, -S_n^T(\sigma_0, 0)\gamma, \\ &\zeta, S_n^T(1, n)\zeta, \ldots, S_n^T(s_n, n)\zeta, \ldots, S_n^T(0, 0)\zeta, \ldots, S_n^T(s_0, 0)\zeta].\end{aligned} \tag{39b}$$

Hence applying (38) at $t = t_1, t_2, \ldots, t_m$, as usual we obtain

$$\Theta = (\tilde{\Phi}^T\tilde{H})^{-1}\tilde{\Phi}^T\gamma. \tag{40}$$

VI. State-Space Model Identification

1. *Full state measurements*

Consider first the case where the entire state vector is perfectly measured:

$$\dot{x} = Ax + Bu, \quad x(0) = x_0 \tag{41}$$

The problem is to determine the matrices A and B and the initial condition x_0. The methodology is the same as that described for MIMO systems (sec. III). We take the ith line of (41), namely

$$\dot{x}_i = a_i^T x + b_i^T u \tag{42}$$

and integrate from $t = 0$ up to $t = t$, using the Walsh expansions

$$x(t) = \Gamma^T\Phi(t) \quad \text{and} \quad u(t) = \Delta^T\Phi(t). \tag{43}$$

The result is

$$\begin{aligned}\Phi^T(t)(\gamma_i - X_{0,i}) &= \Phi^T(t)P^T\Gamma a_i + \Phi^T(t)P^T\Delta b_i \\ &= \Phi^T(t)H\Theta_i\end{aligned} \tag{44}$$

where

$$H = P^T[\Gamma \,\vdots\, \Delta], \quad \Theta_i = \begin{bmatrix} a_i \\ b_i \end{bmatrix}, \quad X_0 = [x_0, 0, 0, \ldots, 0]. \tag{45}$$

Hence as usual

$$\Theta_i = (\tilde{\Phi}^T H)^{-1}\tilde{\Phi}(\gamma_i - X_{0,i}), \quad i = 1, 2, \ldots, n \tag{46}$$

where γ_i is the ith column of Γ.

2. *Partial state measurements*

Now consider the case where one has available observations of the p-dimensional output vector

$$y(t) = Cx(t) + \text{errors} \tag{47}$$

where the matrix C is known but not square since $p < n$.

In this case, under the assumption that the system (41) and (47) is observable, one has to minimize a least squares performance functional of the type

$$J(\Theta) = \tfrac{1}{2}\int_0^1 \tilde{y}^T(t)Q\tilde{y}(t)\,dt$$

$$\tilde{y}(t) = y(t) - Cx(t)$$

with respect to the parameter vector $\Theta^T = [a_1^T, a_2^T, \ldots, a_n^T, b_1^T, b_2^T, \ldots, b_n^T]$.

This minimization can be performed in two ways: (i) by gradient methods; and (ii) by non-linear filtering. In either case one needs an initial guess of the system parameters which is improved by the algorithm. The solution of (41), or of the resulting filter equations, can be performed with the aid of the Walsh functions as described by Chen and Hsiao (**6**).

3. *Non-constant matrices*

Consider the case where the matrices A and B in (41) are non-constant, and can be expressed in the polynomial form

$$\begin{aligned} A &= A_0 + A_1 t + A_2 t^2 + \ldots + A_\sigma t^\sigma \\ B &= B_0 + B_1 t + B_2 t^2 + \ldots + B_s t^s. \end{aligned} \tag{48}$$

Suppose that the state $x(t)$ is perfectly measured. The problem is to determine all matrices A_i, B_i.

By taking the ith row of (41), and integrating it with respect to t from $t = 0$ up to $t = t$, yields

$$\Phi^T(t)(\gamma_i - X_{0,i}) = \Phi^T(t)H\Theta_i \tag{49}$$

where

$$H_i = [R_0^{1^T}(t)\Gamma, R_1^{1^T}(t)\Gamma, \ldots, R_\sigma^{1^T}(t)\Gamma, R_0^{1^T}(t)\Delta, R_1^{1^T}(t)\Delta, \ldots, R_s^{1^T}(t)\Delta]$$

$$\Theta_i^T = [a_i^{0^T}, a_i^{1^T}, \ldots, a_i^{\sigma^T}, b_i^{0^T}, b_i^{1^T}, \ldots, b_i^{s^T}]$$

$a_i^{k^T}$, $b_i^{k^T}$ are the ith rows of A_k and B_k respectively, and the $R_i^1(t)$ are given by (37c–d). From (49) we obtain Θ_i as usual, i.e.

$$\Theta_i = (\tilde{\Phi}^T H)^{-1}\tilde{\Phi}(\gamma_i - X_{0,i}). \tag{50}$$

If the state vector is measured through the output relation $y(t) = C(t) \times (t) + \text{error}$, then one has to apply the techniques mentioned in Section VI. 2.

VII. Integral Equation Model Identification

Consider first the SISO model

$$y(t)=y^0(t)+\int_0^t u(t-s)g(s)\,ds. \tag{51}$$

The problem is to identify the weighting function $g(s)$ on the basis of an input–output record $\{u(t), y(t), 0\leq t\leq 1\}$.

To solve the problem we expand $g(s)$ in Walsh series

$$g(s)=\sum_{j=0}^{m-1}\Theta_j\Phi_j(s)=\Phi^T(s)\Theta,\quad \Theta=\begin{bmatrix}\Theta_0\\ \Theta_1\\ \cdot\\ \cdot\\ \cdot\\ \Theta_{m-1}\end{bmatrix} \tag{52}$$

and introduce it in (51). i.e.

$$y(t)=y^0(t)+\sum_{j=0}^{m-1}\Theta_j\int_0^t u(t-s)\Phi_j(s)\,ds. \tag{53}$$

Now define $\tilde{y}(t)=y(t)-y^0(t)$ and introduce the approximation (which is good for large m)

$$\int_0^t u(t-s)\Phi_j(s)\,ds=\frac{1}{m}\sum_{k=0}^{p}\bar{\mathbf{u}}_{p-k}\Phi_{jk}$$

where the indexes p and k represent discrete t and s, and $\bar{\mathbf{u}}_{p-k}$ is the mean value of $u(t-s)$ in the kth subinterval of s. Then, Eq. (53) takes the form

$$\begin{aligned}\tilde{y}_p&=\frac{1}{m}\sum_{k=0}^{p}\bar{u}_{p-k}\left(\sum_{j=0}^{m-1}\Theta_j\Phi_{jk}\right)\\ &=\left(\frac{1}{m}\sum_{k=0}^{p}\bar{u}_{p-k}\Phi_k\right)\Theta=\boldsymbol{v}_p^T\Theta\end{aligned} \tag{54}$$

where

$$\boldsymbol{v}_p^T=\frac{1}{m}\sum_{k=0}^{p}\bar{u}_{p-k}\Phi_k. \tag{54a}$$

Applying (54) for $p=0, 1, 2, \ldots, m-1$ yields

$$\tilde{Y}=U\Theta \tag{55}$$

where

$$\tilde{Y}=\begin{bmatrix}\tilde{y}_0\\ \tilde{y}_1\\ \cdot\\ \cdot\\ \cdot\\ \tilde{y}_{m-1}\end{bmatrix},\quad U=\begin{bmatrix}\boldsymbol{v}_0^T\\ \boldsymbol{v}_1^T\\ \cdot\\ \cdot\\ \cdot\\ \boldsymbol{v}_{m-1}^T\end{bmatrix}. \tag{55a}$$

Hence

$$\Theta = U^{-1}\tilde{Y}. \tag{56}$$

The value of $g(s)$ in the kth subinterval of s is given by

$$g_k = \Phi_k^T\Theta = \Phi_k^T U^{-1}\tilde{Y}. \tag{57}$$

Now consider the multi-input and multi-output system

$$y(t) = y^0(t) + \int_0^t G(s)u(t-s)\,ds \tag{58}$$

where $y(t)$ is a p-dimensional, $u(t)$ a q-dimensional vector, and $G(s)$ is a $p \times q$ dimensional matrix with rows

$$g_i^T(s) = [g_{i1}(s), \ldots, g_{iq}(s)].$$

Taking the ith row of (58) and working as above, we obtain

$$\tilde{y}_{i,p} = \tilde{\boldsymbol{v}}_p\boldsymbol{\Theta}_i, \quad \tilde{y}_{i,p} = y_{i,p} - y_{i,p}^0 \tag{59}$$

where

$$\tilde{\boldsymbol{v}}_p = [\boldsymbol{v}_{1,p}^T, \boldsymbol{v}_{2,p}^T, \ldots, \boldsymbol{v}_{q,p}^T], \quad \boldsymbol{\Theta}_i = \begin{bmatrix} \boldsymbol{\Theta}_{i1} \\ \cdot \\ \cdot \\ \boldsymbol{\Theta}_{iq} \end{bmatrix}, \quad \boldsymbol{\Theta}_{ir} = \begin{bmatrix} \Theta_{0,ir} \\ \cdot \\ \cdot \\ \Theta_{m-1,ir} \end{bmatrix}$$

$$\boldsymbol{v}_{r,p}^T = \frac{1}{m}\sum_{l=1}^{p} \bar{u}_{r,p-l}\Phi_l^T, \quad g_{ir}(s) = \sum_{k=0}^{m-1} \Theta_{k,ir}\Phi_k(s)$$

and $\bar{u}_{r,p-l}$ is the mean value of $u_r(\sigma - s)$ in the lth time subinterval of s.

Applying (59) for $p = 0, 1, \ldots$ we obtain $\tilde{Y}_i = \tilde{U}\boldsymbol{\Theta}_i$ as in (55), whence

$$\boldsymbol{\Theta}_i = \tilde{U}^{-1}\tilde{Y}_i, \quad i = 1, 2, \ldots, p. \tag{60}$$

VIII. Distibuted-Parameter Systems

The most promising approach to the identification of constant parameters in distributed-parameter systems (DPS) is by transforming the system into lumped form and then using existing lumped-parameter identification methods.

The principal methods of carrying out this conversion are the following:

(i) Modulating function method (**12**);
(ii) Method of lines (**12**);
(iii) Method of characteristics (**12**);
(iv) Galerkin expansion method (**13**);
(v) Space–time integration method (**14**), (**15**).

Here we shall apply the Galerkin expansion method combined with the present Walsh expansion approach.

Consider a SISO DPS of the type

$$A_n(D)y(x, t) = c\mathcal{L}_x y(x, t) + B_n(D)u(x, t), \quad x \in D \tag{61}$$

defined on a spatial domain D, where $A_n(D)$ and $B_n(D)$ are as in (2), $\mathscr{L}_x$ is a linear spatial operator over D, $y(x, t)$ is the output function, and $u(x, t)$ is the input function over D. The initial conditions are

$$\begin{aligned}&[y(x,t)]_0 = Y_0(x), \quad [Dy(x,t)]_0 = Y_1(x), \ldots, [D^{n-1}y(x,t)]_0 = Y_{n-1}(x)\\ &[u(x,t)]_0 = U_0(x), \quad [Du(x,t)]_0 = U_1(x), \ldots, [D^{n-1}u(x,t)]_0 = U_{n-1}(x).\end{aligned} \tag{62}$$

The boundary condition has the form

$$\beta_x y(x,t) = u_b(x,t), \quad x \in \vartheta D \tag{63}$$

where ϑD denotes the boundary surface of D, β_x is a linear boundary spatial operator such that the pair $(\mathscr{L}_x, \beta_x)$ is well posed in the sense of Hadamard, and $u_b(x, t)$ is the boundary input function.

The problem is to determine the parameters a_i, b_i and c as well as the unknown initial conditions of the output function on the basis of an appropriate set of input–output measurements.

Concerning the measurements, in practice it is desirable to use point (or small region) measurements at a limited number of points. The choice of the number of point measurements and their location is closely related to the observability property as has been defined by Goodson and Klein (G–K observability) (**17**). According to the G–K definition, a *DPS* is observable if the variable $y(x, t)$ can be uniquely reconstructed over D by using the available measurements. A system is called N-mode observable if the uniqueness of the first N modes (eigenfunctions) is guaranteed by the measurement data.

The measurements are assumed of the form $z(x^i, t) = \mu(x^i, t)y(x^i, t)$ where x^i, $i = 1, 2, \ldots, s$ are the measurement points. Thus we can write

$$\mathbf{z}_s(t) = \mathbf{M}_s(t)\mathbf{y}_s(t) \tag{64}$$

where

$$\mathbf{z}_s(t) = \begin{bmatrix} z(x^1,t) \\ z(x^2,t) \\ \vdots \\ z(x^s,t) \end{bmatrix}, \quad \mathbf{y}_s(t) = \begin{bmatrix} y(x^1,t) \\ y(x^2,t) \\ \vdots \\ y(x^s,t) \end{bmatrix},$$

$$\mathbf{M}_s(t) = \begin{bmatrix} \mu(x^1,t) & & & \\ & \mu(x^2,t) & & 0 \\ & & \ddots & \\ 0 & & & \mu(x^s,t) \end{bmatrix}.$$

To convert the system (61) into lumped-parameter form we assume that $y(x, t)$ is approximated by

$$\hat{y}(x,t) = n_0(x,t) + \sum_{j=1}^{N} \psi_j(x)\xi_j(t) \tag{65}$$

where $n_0(x, t)$ is a function chosen so as to satisfy the non-homogeneous boundary and/or initial conditions.

The Galerkin approximating functions $\psi_j(x)$ are free to be selected for better convergence rate, and must satisfy the homogeneous boundary conditions. Usually, the $\psi_j(x)$'s are polynomials, trigonometric functions, or eigenfunctions. Although they are not required to be orthogonal, choosing the $\psi_j(x)$'s orthonormal over D, i.e. such that

$$\int_D \psi_i(x)\psi_j(x)\,dx = \delta_{ij} \quad (\delta_{ij} = \text{Kronecker's delta})$$

yields a decoupled set of lumped-parameter equations. Introducting (65) into (61), multiplying by $\psi_i(x)$, and integrating over D yields

$$A_n(D)\boldsymbol{\xi}(t) = cG\boldsymbol{\xi}(t) + B_n(D)\mathbf{v}(t) + c\mathbf{d}(t) - A_n(D)\boldsymbol{\zeta}(t) \tag{66}$$

where

$$\boldsymbol{\xi}(t) = \begin{bmatrix} \xi_1(t) \\ \xi_2(t) \\ \vdots \\ \xi_N(t) \end{bmatrix}, \quad \mathbf{v}(t) = \begin{bmatrix} v_1(t) \\ v_2(t) \\ \vdots \\ v_n(t) \end{bmatrix}, \quad \mathbf{d} = \begin{bmatrix} d_1(t) \\ d_2(t) \\ \vdots \\ d_N(t) \end{bmatrix}, \quad \boldsymbol{\zeta}(t) = \begin{bmatrix} \zeta_1(t) \\ \zeta_2(t) \\ \vdots \\ \zeta_N(t) \end{bmatrix}$$

$$v_i(t) = \int_D u(x, t)\psi_i(x)\,dx, \quad \zeta_i(t) = \int_D n_0(x, t)\psi_i(x)\,dx \tag{67}$$

$$\xi_i(t) = \int_D \{y(x, t) - n_0(x, t)\}\psi_i(x)\,dx, \quad d_i(t) = \int_D \{\mathscr{L}_x n_0(x, t)\}\psi_i(x)\,dx$$

$$G_{ij} = \int_D \{\mathscr{L}_x \psi_j(x)\}\psi_i(x)\,dx, \quad G = [G_{ij}].$$

Similarly, the boundary condition (63) yields

$$\beta_x n_0(x, t) = u_b(x, t), \quad x \in \vartheta D \tag{68}$$

since all the $\psi_j(x)$'s satisfy the homogeneous boundary condition $\beta_x\psi_j(x) = 0$. Actually, solving (68) provides the function $n_0(x, t)$ required in (65).

If the functions $\psi_i(x)$, $i = 1, 2, \ldots$ are not orthonormal, i.e.

$$\int_D \psi_i(x)\psi_j(x)\,dx = F_{ji}, \quad i, j = 1, 2, \ldots, N \tag{69}$$

then in place of (66), we obtain

$$A_n(D)F(\boldsymbol{\xi} + \boldsymbol{\zeta}) = cG(\boldsymbol{\xi} + \boldsymbol{\zeta}) + B_n(D)F\mathbf{v} \tag{70}$$

where $F = [F_{ij}]$. Note that if $\{\psi_i(x)\}$ is an orthonormal family then F is the unit matrix.

Clearly (66) or (70) is in the form of the MIMO system (28) and so the results of section IV apply directly. To determine the initial conditions $Y_i(x)$ we use the expansions

$$Y_k(x) = \sum_{j=1}^{N} \Xi_j^k \psi_j, \quad U_k(x) = \sum_{j=1}^{N} U_j^k \psi_j(x) \tag{71}$$

ınd work as in Section II.2.

The only problem remaining is to reconstruct $\boldsymbol{\xi}(t)$ and $\mathbf{v}(t)$ by using the available measurements of $y(x, t)$ and $u(x, t)$, $x \in D$.

Firstly, suppose that it is possible to obtain measurements at a sufficient number of spatial points $x^1, x^2, \ldots, x^N$. Then we can write the measurement vector $\mathbf{z}_N(t)$ as

$$\mathbf{z}_N(t) = M_N(t)\mathbf{y}_N(t) \tag{72}$$

where

$$\mathbf{y}_N(t) = \begin{bmatrix} y(x^1, t) \\ y(x^2, t) \\ \vdots \\ y(x^N, t) \end{bmatrix}, \quad M_N(t) = \begin{bmatrix} \mu(x^1, t) & & & \\ & \mu(x^2, t) & & 0 \\ & & \ddots & \\ 0 & & & \mu(x^N, t) \end{bmatrix}.$$

Now

$$\mathbf{y}_N(t) = \begin{bmatrix} \sum_{j=1}^{N} \psi_j(x^1)\{\xi_j(t) + \gamma_j(t)\} \\ \vdots \\ \sum_{j=1}^{N} \psi_j(x^N)\{\xi_j(t) + \gamma_j(t)\} \end{bmatrix} = \boldsymbol{\psi}_N(\boldsymbol{\xi} + \boldsymbol{\xi})$$

where

$$\boldsymbol{\psi}_N = \begin{bmatrix} \psi_1(x^1) & \psi_2(x^1) & \ldots & \psi_N(x^1) \\ \psi_1(x^2) & \psi_2(x^2) & \ldots & \psi_N(x^2) \\ \cdots & \cdots & \cdots & \cdots \\ \psi_1(x^N) & \psi_2(x^N) & \ldots & \psi_N(x^N) \end{bmatrix}, \tag{73}$$

Thus we have $\mathbf{z}_N(t) = M_N(t)\boldsymbol{\psi}_N(\boldsymbol{\xi} + \boldsymbol{\zeta})$, whence

$$\boldsymbol{\xi} = \{M_N(t)\boldsymbol{\Psi}_N\}^{-1}\mathbf{z}_N(t) - \boldsymbol{\zeta}. \tag{74}$$

In the same way we have $\mathbf{u}_N(t) = \boldsymbol{\psi}_N\mathbf{v}(t)$, whence

$$\mathbf{v}(t) = \boldsymbol{\Psi}_N^{-1}\mathbf{u}_N(t) \quad \text{where} \quad \mathbf{u}_N(t) = \begin{bmatrix} u(x^1, t) \\ \vdots \\ u(x^N, t) \end{bmatrix}. \tag{75}$$

Next consider the case where the system is observed at only a few points $x^1, x^2, \ldots, x^s$, $s < N$ which implies that $\boldsymbol{\psi}_s$ is nonsquare. This case can be treated as illustrated in Section VI.2. for the lumped parameter situation where one has partial state measurements, i.e. by minimizing the least squares performance functional

$$J(\Theta)=\tfrac{1}{2}\int_0^1 \tilde{\mathbf{z}}_s^T(t)Q\tilde{\mathbf{z}}_s^T(t)\,dt$$

$$\tilde{\mathbf{z}}(t)=\mathbf{z}_s(t)-M_s(t)\mathbf{y}_s(t)$$

with respect to the unknown parameter vector Θ.

Again one can either use a gradient technique or a *DP* non-linear filter as e.g. that described in (**13**), (**18**) and (**19**).

IX. Examples

To illustrate the theoretical results we shall give two numerical examples.

Example 1.

The system under consideration is a point nuclear reactor in the integral form

$$y(t)=\int_0^t h(t-s)y(s)\delta k(s)\,ds+y^0(t)$$

where the output $y(t)=n(t)$ is the neutron level, $\delta k(t)=\rho(t)$ the reactivity input, and, for one group of delayed neutrons, $g(t)$ and $y^0(t)$ are given by

$$g(t)=\frac{1}{l^*}\left\{\lambda+\frac{\beta}{l^*}e^{-\gamma t}\right\},\quad y^0(t)=\frac{1}{\gamma}\{r_1+r_2e^{-\gamma t}\}$$

with $\gamma=\lambda+\beta/l^*$, $r_1=(n_0+C_0)\lambda$ and $r_2=\beta n_0/l^*-\lambda C_0$.

Here, λ is the decay constant, l^* is the prompt neutrón generation time, β is the delayed neutron fraction, and n_0, C_0 are the initial values of $n(t)$ and $C(t)$ respectively. The parameter values used are $\lambda=0.1\ \text{sec}^{-1}$, $l^*=10^{-3}$ sec, $\delta k=+0.003$ and $\beta=0.0075$, and the system was assumed initially in a steady-state condition ($dn_0/dt=dC_0/dt=0$).

The data were generated by the analytic expression of the response. Applying the formulas (56) and (57) of Section VII with the understanding that here $u(t-s)$ of the model (51) is equal to $u(t-s)=h(t-s)y(s)$ we have found (for $m=64$) the following representative values of δk_s

s	$\delta k_s\times 10^2$
8	0.2893972592
16	0.2917671721
24	0.2933708125
32	0.2946927741
40	0.2957822694
48	0.2966802534
56	0.2974203677

Example 2

The system under consideration is selected to be the one dimensional diffusion equation studied in (**13**), i.e.

$$\frac{\vartheta y}{\vartheta t}-c\frac{\vartheta^2 y}{\vartheta x^2}=0,\quad x\in D=(0,1)$$

$$y(0,t)=0,\quad \vartheta y(1,t)/\vartheta x=0,\quad y(x,0)=1.$$

Since the boundary conditions are homogeneous the term $n_0(x,t)$ in Eq. (65) must be zero.

The Galerkin approximating functions are chosen to be the orthogonalized polynomials

$$\psi_1(x)=12x^2-24x,\quad \psi_2(x)=8x^3-\frac{61}{4}x^2+\frac{13}{2}x$$

$$\psi_3(x)=\frac{17290}{2717}x^4-\frac{39368}{2717}x^3+\frac{1431}{143}x^2-2x$$

$$\psi_4(x)=\frac{576}{11}x^5-\frac{17190}{121}x^4+\frac{14952}{121}x^3-\frac{6051}{121}x^2+6x,\ \text{etc.}$$

Then equation (66) gives

$$\dot{\boldsymbol{\xi}}(t)=cG\boldsymbol{\xi}(t),\quad \boldsymbol{\xi}(0)=\boldsymbol{\xi}_0$$

where

$$G_{ij}=\int_0^1\{\vartheta^2\psi_j(x)/\vartheta x^2\}\psi_i(x)\,\mathrm{d}x,\quad \xi_{0,i}=\int_0^1 y(x,0)\psi_i(x)\,\mathrm{d}x.$$

The analytical expression of the response is

$$y(x,t)=\frac{4}{\pi}\sum_{i=1}^{\infty}\frac{1}{2i-1}\exp\left[-c(2i-1)^2(\pi/2)^2t\right]\sin\left(\frac{2i-1}{2}\pi x\right). \tag{76}$$

Application of the method of Section VIII with data generated by using (76), with $c=0.9$ and 20 terms, gave the following results:

m | N	1	2	3
4	0.918766	0.868534	0.838910
8	0.914813	0.881835	0.872427
16	0.908300	0.912955	0.893216
32	0.903856	0.902445	0.898339

These results indicate that actually an approximation with $N=2$ and $m=32$ is sufficient.

X. Conclusions

This paper has presented a complete study of the utilization of Walsh function expansions for the identification of both lumped- and distributed-parameter systems. The basic advantage of using Walsh functions is the easy implementation of the data due to the integration property of Walsh functions and their orthonormality. The paper has covered most practical problems namely, the unknown initial conditions case, the MIMO case, the stochastic case, and the time varying case.

It is hoped that the results derived provide a useful tool for treating any linear identification problem. Work is in progress for treating nonlinear situations. Concerning the solution of state observer and estimator designs one may refer to (**9**).

References

(**1**) H. Harmuth, "Transmission of Information by Orthogonal Functions", Berlin, Springer Verlag, 1969.

(**2**) R. Lackey, "The wonderful World of Walsh Functions", *Proc.* 1972 *Symp. on Applic. of Walsh Functions, The Catholic Univ. of America,* Washington D.C. NTIS, pp. 2–7, 1972.

(**3**) J. Pearl, "Application of Walsh Transform to Statistical Analysis", *IEEE Trans.,* Vol. SMC-1, No. 2, pp. 111–119, 1971.

(**4**) M. Corrington, "Solution of Differential and Integral Equations with Walsh Functions", *IEEE Trans.,* Vol. CT-20, No. 5, pp. 470–476, 1973.

(**5**) W. Pratt *et al.*: "Hadamard Transform Image Coding", *Proc. IEEE,* Vol. 57, No. 1, pp. 58–68, 1969.

(**6**) C. Chen and C. H. Hsiao, "State-Space Approach to Walsh Series Solution of Linear Systems", *Int. J. Systems Sci.,* Vol. 6., pp. 833–858, 1975.

(**7**) C. Chen and C. H. Hsiao, "Design of Piece wise Constant Gains for Optimal Control Via Walsh Functions", *IEEE Trans.,* Vol. AC-20, pp. 596–602, 1975.

(**8**) C. Chen and C. H. Hsiao, "Time-Domain Synthesis via Walsh Functions", *Proc. IEE,* Vol. 122, pp. 565–570, 1975.

(**9**) P. Stavroulakis and S. Tzafestas, "Walsh Series Approach to Observer and Filter Design in Optimal Control Systems", *Int. J. Control,* Vol. 26, pp. 721–736, 1977.

(**10**) S. Tzafestas, "Some Computer-Aided Estimators in Stochastic Control Systems Identification", *Int. J. Control,* Vol. 12, pp. 385–399, 1970.

(**11**) M. Budin, "Parameter Estimation Using Least Squares Polynomial Smoothing", *IEEE Trans.,* Vol. SMC-3, pp. 371–381, 1973.

(**12**) S. Tzafestas, "Parameter Estimation in Distributed-Parameter Dynamic Models", Proc. European Federation of Chem. Engrg: *Decision, Design and the Computer,* London, Sept. 1972 (Also *Trans. I. Chem. E.,* No. 35. pp. 5: 43–5: 51, 1973).

(**13**) M. Polis, E. Goodson and M. Wozny "On Parameter Identification for Distributed Systems Using Galerkin Criterion", *Automatica,* Vol. 9, pp. 53–64, 1973.

(**14**) S. Diamessis, "The *n*th-Order Distributed-Parameter System: Parameter Identification and Boundary Condition Determination", *Proc. NATO Advanced Institute on Network and Signal Theory,* pp. 275–279, Bournemouth, Sept. 1972.

(**15**) J. Diamessis, "A New Method for Determining the Parameters of Physical Systems", *Proc. IEEE,* pp. 205–206, Feb. 1965.

(**16**) A. Mathew and F. Fairman, "Identification in the Presence of Initial Conditions", *IEEE Trans.*, Vol. AC-17, pp. 394–396, 1972.

(**17**) R. Goodson and R. Klein, "A Definition and Some Results for Distributed System Observability", *IEEE Trans.*, Vol. AC-15, pp. 165–174, 1970.

(**18**) D. Detchmendy and R. Sridhar, "Sequential Estimation of States and Parameters in Noisy Nonlinear Systems", *Trans. ASME: J. Basic Engng*, Vol. 99 (Series D), pp. 362–368, 1966.

(**19**) M. Hwang, J. Seinfeld and G. Gavalas, "Optimal Least-Squares Filtering and Interpolation in Distributed-Parameter Systems", *J. Appl. Analysis and Applications*, Aug. 1972.

Appendix A: Review of Walsh Function Expansion

The Walsh functions $\Phi_k(t)$, $k = 0, 1, \ldots, m$, $m = 2^n$ are two-valued functions which are orthonormal in the interval $0 \leq t < 1$. Denoting the value $+1$ by "+" and the value -1 by "−" the functions $\Phi_k(t)$, $k = 0, 1, 2, \ldots, 15$ have the following form

$\Phi_0(t)$	+	+	+	+	+	+	+	+	+	+	+	+	+	+	+	+
$\Phi_1(t)$	+	+	+	+	+	+	+	+	−	−	−	−	−	−	−	−
$\Phi_2(t)$	+	+	+	+	−	−	−	−	+	+	+	+	−	−	−	−
$\Phi_3(t)$	+	+	+	+	−	−	−	−	−	−	−	−	+	+	+	+
$\Phi_4(t)$	+	+	−	−	+	+	−	−	+	+	−	−	+	+	−	−
$\Phi_5(t)$	+	+	−	−	+	+	−	−	−	−	+	+	−	−	+	+
$\Phi_6(t)$	+	+	−	−	−	−	+	+	+	+	−	−	−	−	+	+
$\Phi_7(t)$	+	+	−	−	−	−	+	+	−	−	+	+	+	+	−	−
$\Phi_8(t)$	+	−	+	−	+	−	+	−	+	−	+	−	+	−	+	−
$\Phi_9(t)$	+	−	+	−	+	−	+	−	−	+	−	+	−	+	−	+
$\Phi_{10}(t)$	+	−	+	−	−	+	−	+	+	−	+	−	−	+	−	+
$\Phi_{11}(t)$	+	−	+	−	−	+	−	+	−	+	−	+	+	−	+	−
$\Phi_{12}(t)$	+	−	−	+	+	−	−	+	+	−	−	+	+	−	−	+
$\Phi_{13}(t)$	+	−	−	+	+	−	−	+	−	+	+	−	−	+	+	−
$\Phi_{14}(t)$	+	−	−	+	+	−	−	−	+	+	+	−	−	+	+	−
$\Phi_{15}(t)$	+	−	−	+	−	+	+	−	−	+	+	−	+	−	−	+

We see that the regularity of the $\Phi_k(t)$'s is not clear. However, one way of constructing the above Walsh matrix of any order is by decomposing each Walsh function $\Phi_k(t)$ into Rademacher functions $R_s(t)$, $s = 1, 2, \ldots, n$ which are a set of orthonormal square waves for $0 \leq t < 1$ with unit height and repetition rate (sequency) equal to 2^{s-1}.

The decomposition formula is

$$\Phi_k(t) = R_s^{d_s}(t) R_{s-1}^{d_{s-1}}(t) \ldots R_1^{d_1}(t), \quad k = 0, 1, \ldots, 2^s - 1$$

with initial condition $\Phi_0(t) = R_0(t) = 1$, where "$d_s d_{s-1} \ldots d_1$" is the binary expression of the decimal number k.

The Walsh functions constitute a complete set of functions, and so one can expand any given function $f(t)$, which is absolutely integrable in the interval $[0, 1]$ in a Walsh–Fourier series as

$$f(t) = \sum_{k=0}^{\infty} \xi_k \Phi_k(t), \quad \xi_k = \int_0^1 \Phi_k(t) f(t) \, dt. \tag{A1}$$

Consider the first four Walsh functions $\Phi_k(t)$, $k=0,1,2,3$, and integrate them from 0 up to t. The result is

$$\int_0^t \begin{bmatrix}\Phi_0(\tau)\\ \Phi_1(\tau)\\ \Phi_2(\tau)\\ \Phi_3(\tau)\end{bmatrix} d\tau = \left[\begin{array}{cc:cc} \frac{1}{2} & -\frac{1}{4} & -\frac{1}{8} & 0\\ \frac{1}{4} & 0 & 0 & -\frac{1}{8}\\ \hdashline \frac{1}{8} & 0 & 0 & 0\\ 0 & \frac{1}{8} & 0 & 0 \end{array}\right] \begin{bmatrix}\Phi_0(t)\\ \Phi_1(t)\\ \Phi_2(t)\\ \Phi_3(t)\end{bmatrix}. \tag{A2}$$

Equation (A2) can be written in the general form

$$\int_0^t \mathbf{\Phi}(\tau)\, d\tau = P\mathbf{\Phi}(t) \tag{A3}$$

where

$$\mathbf{\Phi}(t) = \begin{bmatrix}\Phi_0(t)\\ \Phi_1(t)\\ \cdot\\ \cdot\\ \cdot\\ \Phi_{m-1}(t)\end{bmatrix}, \quad P = P_m = \left[\begin{array}{c:c} P_{m/2} & -\dfrac{1}{2m}I_{m/2}\\ \hdashline \dfrac{1}{2m}I_{m/2} & 0_{m/2} \end{array}\right], \quad P_1 = \tfrac{1}{2} \tag{A4}$$

for $m=2^q$, $q=1,2,\ldots$. The operational formula (A3) shows that the integration of $\mathbf{\Phi}(\tau)$ over the interval $[0,t]$ is performed by simply multiplying $\mathbf{\Phi}(t)$ by the matrix P_m. This is the key of the Walsh series approach to system identification.

Appendix B: Proof of the Formulas (36a-b)

Define the operator $R_k^\sigma(t)$ by

$$R_k^\sigma(t)\mathbf{\Phi}(t) = \underbrace{\int_0^t\int_0^t \cdots \int_0^t}_{\sigma} t^k \mathbf{\Phi}(t)\, dt \ldots dt. \tag{A5}$$

Then, the following recursive formula holds

$$R_{k+1}^{\sigma+1}(t) = P\{R_{k+1}^{\sigma}(t) - (k+1)R_k^{\sigma+1}(t)\} \tag{A6}$$

for $k=0,1,2,\ldots$, $\sigma=0,1,2,\ldots$, with initial conditions

$$R_{k+1}^{0}(t) = t^{k+1}I, \quad R_0^{\sigma+1} = P^{\sigma+1} \tag{A7}$$

where $R_0^1(t) = P$ and $R_1^0(t) = tI$.

This recursive formula can be established by induction. In fact, one can easily see that

$$\int_0^t t^k\Phi(t)\,dt = R_k^1(t)\Phi(t) \quad \text{where} \quad R_{k+1}^1(t) = P\{I\cdot t^{k+1} - (k+1)R_k^1(t)\}$$

$$\int_0^t\int_0^t t^k\Phi(t)\,dt\,dt = R_k^2(t)\Phi(t) \quad \text{where} \quad R_{k+1}^2(t) = P\{R_{k+1}^1(t) - (k+1)R_k^2(t)\}$$

$$\vdots \quad \vdots \quad \vdots$$

$$\underbrace{\int_0^t\int_0^t\cdots\int_0^t}_{\sigma} t^k\Phi(t)\,dt\ldots dt = R_k^\sigma(t)\Phi(t) \quad \text{where} \quad R_k^\sigma(t) \quad \text{satisfies (A6).}$$

Introduce the definition of $I_n(k,\sigma)$ given by Eq. (35a). Then, we have

$$I_n(t^k y) = I_n(k,0) = \gamma^T \underbrace{\int_0^t\cdots\int_0^t}_{n} t^k\Phi(t)\,dt\ldots dt = \gamma^T R_k^n(t)\Phi(t) \tag{A8}$$

where $R_k^n(t)$ satisfies Eq. (A6).

Now

$$I_n(D^k y) = I_{n-k}(y)$$

and so one can verify that

$$I_n(t^k D^{\sigma+1} y) = I_{n-1}(t^k D^\sigma y) - kI_n(t^{k-1}D^\sigma y) \tag{A9}$$

or

$$I_n(k,\sigma+1) = I_{n-1}(k,\sigma) - kI_n(k-1,\sigma) \tag{A10}$$

with boundary conditions

$$I_n(t^k D^0 y) = I_n(k,0) = \gamma^T R_k^n(t)\Phi(t) \tag{A10a}$$

$$I_n(0,\sigma) = \gamma^T P^{n-\sigma}\Phi(t). \tag{A10b}$$

Thus defining $S_n(k,\sigma)$ by

$$I_n(k,\sigma) = \gamma^T S_n(k,\sigma)\Phi(t) \tag{A11}$$

Eqs. (A10) to (A10b) imply that

$$S_n(k,\sigma+1) = S_{n-1}(k,\sigma) - kS_n(k-1,\sigma) \tag{A12}$$

with boundary conditions

$$S_n(k,0) = R_k^n(t) \quad \text{and} \quad S_n(0,\sigma) = P^{n-\sigma} \tag{A13}$$

where $R_k^n(t)$ satisfies Eqs. (A6) and (A7). Q.E.D.

14

Reprinted from *J. Franklin Inst.* **310**:65–75 (1980)

Identification of Voltera Kernels of a Class of Nonlinear Systems by Walsh Function Techniques

by MOHAMMAD MAQUSI*

Electrical Engineering Department, University of Texas at El Paso, El Paso, TX 79968, U.S.A.

ABSTRACT: *This paper discusses the identification problem for a class of nonlinear systems. A member of this class may be represented by a single-valued power-law type nonlinearity preceded and succeeded by linear dyadic invariant systems. Such an arrangement allows for a Voltera functional series representation. The identification problem is then concerned with the specification of the associated Voltera kernels.*

Two approaches are presented for dealing with this problem. Both approaches are, however, based on Walsh function techniques. The first approach relies on direct output measurements when the input is a Walsh function. This approach is suitable for a deterministic case. The second approach assumes ergodic processes for the input. Based on measurements drawn from an input–output dyadic correlation function, determination of the Voltera kernels is made.

I. Introduction

Several authors (**1–5**), have discussed the problem of the identification of certain classes of nonlinear systems. A number of these works consider the identification problem in the case where the nonlinear system is representable by a Voltera functional series expansion. Indeed, such a representation provides an adequate description for a large class of nonlinear systems. However, the computations associated with identification schemes based on this representation are often burdensome.

In this paper we study the identification of a class of nonlinear systems by Walsh function techniques. As shown in Fig. 1, the nonlinear system under study consists of a linear dyadic invariant (LDI) system followed by a single-valued nonlinearity and another LDI system. In general, analysis of nonlinear systems with Walsh-type signals yields useful and computationally convenient expressions (**6, 7**).

II. General Analysis

We consider the space of real, positive-coordinate signals. Thus the input $x(t)$ is defined for $t \geq 0$. In general, an input–output relation for a casual LDI

* At present on leave from the Electrical Engineering Department, University of Jordan, Amman, Jordan.

FIG. 1. A class of nonlinear systems.

system (**8, 9**) is given by

$$q(t) = \int_0^\infty h(\tau)x(t \oplus \tau)\, d\tau = \int_0^\infty h(t \oplus \tau)\, x(\tau)\, d\tau, \tag{1}$$

where $x(t)$, $q(t)$ and $h(t)$ denote the system input, output, and impulse response, respectively. The operator $\oplus$ denotes modulo-two addition (without-carry).

The two forms indicated in (1) are made possible by the use of a dyadic translation invariant integral property (**10, 11**), which states that for a Lebesgue-integrable function $f(t)$ on $[0, \infty)$, we have

$$\int_0^\infty f(t \oplus \tau)\, dt = \int_0^\infty f(t)\, dt, \quad \tau \in [0, \infty). \tag{2}$$

Furthermore, relations (1) represent dyadic convolution integrals for the functions $x(t)$ and $h(t)$.

Identification of linear dyadic invariant systems has been studied elsewhere, and follows after the case of linear time invariant systems to a large extent (**12, 13**).

Referring to Fig. 1, we have

$$q(t) = \int_0^\infty x(\tau)h(t \oplus \tau)\, d\tau = \int_0^\infty x(t \oplus \tau)h(\tau)\, d\tau. \tag{3}$$

We consider the no-memory nonlinear element to be represented by a transfer characteristic as

$$y(t) = \gamma_1 q(t) + \gamma_2 q^2(t) + \ldots + \gamma_n q^n(t). \tag{4}$$

Employing (3) in (4) gives

$$\begin{aligned} y(t) = {} & \gamma_1 \int_0^\infty h_1(\tau_1)x(t \oplus \tau_1)\, d\tau_1 + \gamma_2 \int_0^\infty \int_0^\infty h_1(\tau_1)h_1(\tau_2) \\ & \times x(t \oplus \tau_1)x(t \oplus \tau_2)\, d\tau_1 d\tau_1 + \ldots \\ & + \gamma_n \int_0^\infty \ldots \int_0^\infty h_1(\tau_1) \ldots h_1(\tau_n)x(t \oplus \tau_1) \ldots x(t \oplus \tau_n)\, d\tau_1 \ldots d\tau_n. \end{aligned} \tag{5}$$

The final system output is given by

$$z(t) = \int_0^\infty h_2(\alpha)y(t \oplus \alpha)\, d\alpha. \tag{6}$$

Substitution of (5) in (6), and utilization of the dyadic translation invariant integral property (2) give the measured system output as

$$z(t) = \gamma_1 \int_0^{\infty}\int_0^{\infty} h_1(\tau_1 \oplus \alpha) h_z(\alpha) x(t \oplus \tau_1)\, d\tau_1\, d\alpha$$
$$+ \gamma_2 \int_0^{\infty}\int_0^{\infty}\int_0^{\infty} h_1(\tau_1 \oplus \alpha) h_1(\tau_2 \oplus \alpha) h_2(\alpha)$$
$$\times x(t \oplus \tau_1) x(t \oplus \tau_2)\, d\tau_1 d\tau_2\, d\alpha + \dots$$
$$+ \gamma_n \int_0^{\infty} \dots \int_0^{\infty} h_1(\tau_1 \oplus \alpha) \dots h_1(\tau_n \oplus \alpha) h_2(\alpha)$$
$$\times x(t \oplus \tau_1) \dots x(t \oplus \tau_n)\, d\tau_1 \dots d\tau_n\, d\alpha. \tag{7}$$

In a more compact form, we may rewrite (7) as

$$z(t) = \sum_{i=1}^{n} \int_i g_i(\tau_1, \dots, \tau_i) \prod_{p=1}^{i} x(t \oplus \tau_p)\, d\tau_p, \tag{8}$$

where

$$g_i(\tau_1, \dots, \tau_i) = \gamma_i \int_0^{\infty} h_2(\alpha) \left[\prod_{p=1}^{i} h_1(\tau_p \oplus \alpha) \right] d\alpha. \tag{9}$$

The function $g_i(\tau_1, \dots, \tau_i)$ is referred to as the ith-order Voltera kernel or the ith-order impulse response of the nonlinear system, and the integral indicated in (8) denotes a multiple (i times) integral ranging from 0 to ∞. The series expansion in (8) is called a Voltera series expansion, and according to the more dictated form in (7) it affords an output expression which is similar in form to that used in the case of linear dyadic invariant systems.

Furthermore, the Voltera kernels may be specified in the sequency domain by the use of Walsh transform. Indeed, applying a multidimensional Walsh transform to (9), gives

$$G_i(\sigma_1, \dots, \sigma_i) = \int_0^{\infty} \dots \int_0^{\infty} g_i(\tau_1, \dots, \tau_i) \psi(\sigma_i, \tau_i) \dots \psi(\sigma_i, \tau_i)\, d\tau_1 \dots d\tau_i, \tag{10}$$

where the transform kernel $\psi(\sigma, \tau)$ denotes a generalized Walsh–Fine type function (**11**). The nonnegative real numbers σ and τ denote sequency and spatial (e.g., time) (**14**) variables, respectively. In general, the Walsh transform of a suitable (e.g., square-integrable) function $f(t)$ is given by

$$F(\sigma) = \int_0^{\infty} f(t) \psi(\sigma, t)\, dt = \mathcal{W}[f(t)], \tag{11}$$

and the inverse Walsh transform is given by

$$f(t) = \int_0^{\infty} F(\sigma) \psi(\sigma, t)\, d\sigma = \mathcal{W}^{-1}[F(\sigma)]. \tag{12}$$

For illustration, we compute $G_1(\sigma)$ using (9) and (10). Hence,

$$G_1(\sigma)=\gamma_1\int_0^\infty\left[\int_0^\infty h_2(\alpha)h_1(\tau_1\oplus\alpha)\,d\alpha\right]\psi(\sigma,\tau_1)\,d\tau_1. \tag{13}$$

Interchanging orders of integration and using the (dyadic time-domain) shift property of Walsh transform; i.e. for a positive real number b,

$$\mathscr{W}[f(t\oplus b)]=F(\sigma)\psi(\sigma,b), \tag{14}$$

we obtain

$$\begin{aligned}G_1(\sigma)&=\gamma_1\int_0^\infty h_2(\alpha)H_1(\sigma)\psi(\sigma,\alpha)\,d\alpha\\&=\gamma_1H_1(\sigma)H_2(\sigma).\end{aligned} \tag{15}$$

In a similar manner, we obtain

$$G_2(\sigma_1,\sigma_2)=\gamma_2H_1(\sigma_1)H_1(\sigma_2)H_2(\sigma_1\oplus\sigma_2), \tag{16}$$

and in general we get

$$G_i(\sigma_1,\ldots,\sigma_i)=\gamma_iH_1(\sigma_1)\ldots H_1(\sigma_i)H_2(\sigma_1\oplus\ldots\oplus\sigma_i). \tag{17}$$

By a multidimensional inverse Walsh transform we have

$$\begin{aligned}g_i(\tau_1,\ldots,\tau_i)=\gamma_i\int_0^\infty\ldots\int_0^\infty &H_1(\sigma_1)\ldots H_1(\sigma_i)H_2(\sigma_1\oplus\ldots\oplus\sigma_i)\\&\times\psi(\sigma_1,\tau_1)\ldots\psi(\sigma_1,\tau_i)\,d\sigma_1\ldots d\sigma_i.\end{aligned} \tag{18}$$

III. Determination of the Voltera Kernels

In this section we discuss two methods for determining the Voltera kernels, defined earlier in Section II. Two approaches are used. One approach is based on direct output measurements, while the other approach depends on computations of an input–output dyadic crosscorrelation function. However, both methods utilize Walsh signals for the input.

A. *Output measurements approach*

In this case we assume an input $x(t)$ given by

$$x(t)=A\psi(r,t),\quad r\text{ is real}, \tag{19}$$

where A and r represent the amplitude and sequency values of the input, respectively.

Under assumption (19), we obtain

$$q(t)=\int_0^\infty Ah_1(\tau)\psi(r,t\oplus\tau)\,d\tau=AH_1(r)\psi(r,\tau). \tag{20}$$

In (20) we have used a multiplicative identity for Walsh functions: $\psi(r,t\oplus\tau)=\psi(r,t)\psi(r,\tau)$, and the definition of the Walsh transform of $h_1(t)$. We note from

(20) that $q(t)$ is in essence a Walsh function, save for a scaling factor, with the same sequency as that of the input $x(t)$.

Now

$$[\psi(r,t)]^i = \begin{cases} \psi(r,t), & i = \text{odd integer} \\ \psi(0,t), & i = \text{even integer} \end{cases}$$

$$= \psi\left(\sum_{j=1}^{i} \oplus\, r, t\right), \tag{21}$$

where

$$\sum_{j=1}^{i} \oplus\, r = r \oplus \ldots \oplus r, \; i \text{ times}.$$

Thus, the output of the nonlinear element is given by

$$y(t) = \sum_{i=1}^{n} \gamma_i q^i(t) = \sum_{i=1}^{n} \gamma_i A^i H_1^i(r)[\psi(r,t)]^i$$

$$= \sum_{i=1}^{n} \gamma_i A^i H_1^i(r) \psi\left(\sum_{j=1}^{i} \oplus\, r, t\right). \tag{22}$$

The quantity $H_1(r)$ represents the steady state gain of the linear dyadic invariant system at the sequency r. The measured output of the system is now specified by

$$z(t) = \int_0^\infty h_2(\alpha) y(t \oplus \alpha) d\alpha = \sum_{i=1}^{n} \gamma_i A^i H_1^i(r)$$

$$\times \int_0^\infty h_2(\alpha) \psi\left(\sum_{j=1}^{i} \oplus r, t \oplus \alpha\right) d\alpha. \tag{23}$$

Applying the same procedure to the integral indicated in (23) as was done in (20), we then obtain

$$z(t) = \sum_{i=1}^{n} \gamma_i A^i H_1^i(r) H_2\left(\sum_{j=1}^{i} \oplus\, r\right) \psi\left(\sum_{j=1}^{i} \oplus\, r, t\right). \tag{24}$$

Since $\sum_{j=1}^{i} \oplus\, r$ gives either r or 0, for all $i = 1, 2, \ldots, n$, we may group terms of like sequency and rewrite (24) as

$$z(t) = \sum_{i=1}^{[(n+1)/2]} \{\gamma_{2i} A^{2i} H_2(0) H_1^{2i}(r) + \gamma_{2i-1} A^{2i-1} H_2(r) H_1^{2i-1}(r)\, \psi(r,t)\}. \tag{25}$$

According to (24), or equivalently (25), and (8) we observe that if the linear dyadic subsystems $h_1(t)$ and $h_2(t)$ are isolated and identified, then identification of the Voltera kernels is made complete by determining the γ-set, $\{\gamma_1, \gamma_2, \ldots, \gamma_n\}$. If we now take measurements of the output at times t_m,

$m = 1, 2, \ldots, n$, we obtain, from (25), equations for determining the γ-unknowns. Hence

$$z(t_m) = [A^2H_2(0)H_1^2(r)\gamma_2 + A^4H_2(0)H_1^4(r)\gamma_4 + \ldots] + [AH_2(r)H_1(r)\gamma_1 + A^3H_2(r)H_1^3(r)\gamma_3 + \ldots]\psi(r, t_m),$$
$$m = 1, 2, \ldots, n, \tag{26}$$

where $\psi(r, t_m) = +1$ or -1, for all $m = 1, 2, \ldots, n$.

Without loss of generality, we assume that n is even. By (26), the terms of interest in determining the Voltera kernels are defined by

$$\begin{aligned}
\beta_1 &= \gamma_1 H_1(r)H_2(r) = G_1(r)\\
\beta_2 &= \gamma_2 H_1^2(r)H_2(0) = G_2(r, r)\\
&\vdots\\
\beta_{n-1} &= \gamma_{n-1}H_1^{n-1}(r)H_2(r) = G_{n-1}(r, \ldots, r)\\
\beta_n &= \gamma_n H_1^n(r)H_2(0) = G_n(r, \ldots, r).
\end{aligned} \tag{27}$$

In terms of (27), (26) becomes

$$z(t_m) = [\beta_2 + \beta_4 + \ldots + \beta_n] + [\beta_1 + \beta_3 + \ldots + \beta_{n-1}]\psi(r, t_m),$$
$$m = 1, 2, \ldots, n. \tag{28}$$

As mentioned previously, if the two linear dyadic subsystems are identified, then (28) specifies the β-set which in turn specifies the Voltera kernels, and this completes the identification problems. A method for decoupling the identification of the linear subsystems from the identification of the nonlinear element is presented in (**1**) for the case of linear time invariant systems. This method may be applied to Fig. 1 with minor modifications to suit the dyadic invariant systems. Furthermore, we point out that in general we may evaluate the terms $\beta_1, \ldots, \beta_n$ for each test sequency (i.e., vary r) which renders a set of magnitude response characteristics from which the sequency transfer function expressions for the two linear dyadic invariant systems may be derived by a least-squares fitting technique, for instance.

B. *Dyadic Correlation Function Approach*

Now let

$$z(t) = \sum_{m=1}^{n} w_m(t), \tag{29}$$

where

$$w_m(t) = \gamma_m \int_{m+1} h_1(\tau_1)\ldots h_1(\tau_m)h_2(\alpha) \times x(t \oplus \tau_1 \oplus \alpha)\ldots x(t \oplus \tau_m \oplus \alpha)\, d\tau_1 \ldots d\tau_m\, d\alpha. \tag{30}$$

Before proceeding with this approach we digress to state a few pertinent concepts. For a process $x(t)$, we define a finite dyadic correlation function (**8, 15**) by

$$D_x(\tau; T) = \frac{1}{T}\int_0^T x(t)x(t \oplus \tau)\,dt, \tag{31}$$

and a dyadic correlation function by

$$D_x(\tau) = \lim_{T\to\infty} \frac{1}{T}\int_0^T x(t)x(t \oplus \tau)\,dt. \tag{32}$$

A dyadic crosscorrelation function for two processes $x(t)$ and $y(t)$ is defined by

$$D_{xy}(\tau) = \lim_{T\to\infty} \frac{1}{T}\int_0^T x(t)y(t \oplus \tau)\,dt. \tag{33}$$

Similarly, we may define a finite dyadic crosscorrelation function without the limit operation. For random processes, these functions are random functions of the parameter τ. The description of random processes may be made more appropriately by defining a stayadic correlation function (**16**), for such a process $g(t)$, by

$$C_g(\tau) = E[g(t)g(t \oplus \tau)]. \tag{34}$$

A corresponding sequency power spectral density is given by

$$\Gamma_g(\sigma) = \mathcal{W}[C_g(\tau)] = \int_0^\infty C_g(\tau)\psi(\sigma, \tau)\,d\tau. \tag{35}$$

For ergodic processes, dyadic and stayadic correlation functions become interchangeable in the limit as the time interval under consideration becomes infinite. This result is modeled after Birkoff's theorem for the interchange of time and ensemble averages for such processes, and is discussed in Section IV. We thus assume that the random processes we consider are of the ergodic type.

According to (29), an input-output finite dyadic crosscorrelation function is given by

$$\begin{aligned} D_{xz}(\tau; T) &= \frac{1}{T}\int_0^T x(t)z(t \oplus \tau)\,dt \\ &= \sum_{m=1}^{n} \frac{1}{T}\int_0^T x(t)w_m(t \oplus \tau)\,dt \\ &= \sum_{m=1}^{n} D_{xw_m}(\tau; T). \end{aligned} \tag{36}$$

Employing (30), we obtain

$$\begin{aligned} D_{xw_m}(\tau; T) = \gamma_m \int_{m+1} & h_1(\tau_1)\ldots h_1(\tau_m)h_2(\alpha) \\ & \times\left[\frac{1}{T}\int_0^T x(t)x(t \oplus \tau \oplus \tau_1 \oplus \alpha)\ldots x(t \oplus \tau \oplus \tau_m \oplus \alpha)\,dt\right] \\ & \times d\tau_1 \ldots d\tau_m\,d\alpha. \end{aligned} \tag{37}$$

Denoting the bracketed integral in (37) by $D_x^m(\tau; T)$, we obtain for $x(t)=A\psi(r, t)$,

$$D_x^m(\tau; T)=A^{m+1}I\psi\left(\sum_{i=1}^{m}\oplus r, \alpha\right)\psi\left(\sum_{i=1}^{m}\oplus r, \tau\right) \\ \times\psi(r, \tau_1)\ldots\psi(r, \tau_m), \tag{38}$$

where

$$I=\frac{1}{T}\int_0^T\psi\left(\sum_{i=1}^{m+1}\oplus r, t\right)dt=\begin{cases}\frac{1}{T}\int_0^T\psi(r, t)\,dt, & \text{even } m\\ 1, & \text{odd } m\end{cases}. \tag{39}$$

For integral r and T, we apply orthogonality of Walsh functions to reduce (39) to

$$I=\begin{cases}1, & \text{odd } m\\ 0, & \text{even } m\end{cases}. \tag{40}$$

However, for real r and T, we have (39) which gives an upper bound on I, as

$$|I|\leqslant\frac{1}{T}\int_0^T\left|\psi\left(\sum_{i=1}^{m+1}\oplus r, t\right)\right|dt=1. \tag{41}$$

Evaluating $D_{xw_m}(\tau; T)$ according to (37) and (38), we have

$$D_{xw_m}(\tau; T)=\gamma_m A^{m+1}IH_1^m(r)H_2\left(\sum_{i=1}^{m}\oplus r\right)\psi\left(\sum_{i=1}^{m}\oplus r, \tau\right). \tag{42}$$

In view of (42), the input–output dyadic crosscorrelation, as given by (36), now reduces to

$$D_{xz}(\tau; T)=D_{xz}(\tau) \\ =\sum_{m=1}^{n}\gamma_m A^{m+1}IH_1^m(r)H_2\left(\sum_{i=1}^{m}\oplus r\right)\psi\left(\sum_{i=1}^{m}\oplus r, \tau\right). \tag{43}$$

The application of this result is similar to the result obtained earlier for identification by direct output measurements. If the two linear dyadic subsystems are identified, then measurements of $D_{xz}(\tau)$ at times τ_k, $k=1, 2, \ldots, n$ determine the γ-set by application of (43). On the other hand, by applying different values for the input sequency r, and drawing measurements from $D_{xz}(\tau)$, we get different values for the terms of interest: $\gamma_1 A^2IH_1(r)H_2(r)\psi(r, \tau)$, $\gamma_2 A^3IH_1^2(r)H_2(0), \ldots,$. From these latter values, the Voltera kernels may be determined as indicated in part (A) of this section.

IV. Discussion of Results

In the previous sections we considered a certain class of nonlinear systems composed essentially of nonlinear elements and LDI systems. Section II discusses a Voltera series expansion representation for such systems. The representation hinges primarily on the specification of associated Voltera kernels. Consequently, Section III is concerned with the identification of these kernels by the use of Walsh type signals for the excitations. A method of identification based on system output measurements is viewed appropriate for a deterministic case. For random cases, if we assume ergodic signals at the input, then identification of the kernels may still be made using Walsh techniques. But computations are now based on input–output dyadic crosscorrelation functions. However, either approach will ultimately benefit from the various computational advantages rendered to Walsh functions.

The assumption of ergodicity in the random case facilitates the interchange of statistical and dyadic correlation functions. In an earlier work, and after Lee (**17**), Weiser (**18**) produced such a result by heuristic arguments. Due to the rather difficult accessibility of Weiser's work, this result is reproduced in this paper.

Consider an ensemble of sample functions of a random process, $x(t)$, placed parallel in time as shown in Fig. 2. Stayadic correlation for such a process is then defined by

$$C_x(\tau) = E[x(t)x(t \oplus \tau)], \tag{44}$$

and dyadic correlation is defined by

$$D_x(\tau) = \lim_{T\to\infty} \frac{1}{T}\int_0^T x(t)x(t \oplus \tau)\,dt. \tag{45}$$

Now suppose that each single function is divided into increments of infinitesimal duration Δt as shown in Fig. 2(b). Since the ensemble members (sample functions) are generated by sources of indentical nature (i.e. ergodic), we make the assumption that all members have the same member aggregate of amplitude values, although the order of these values may be different. Next, we

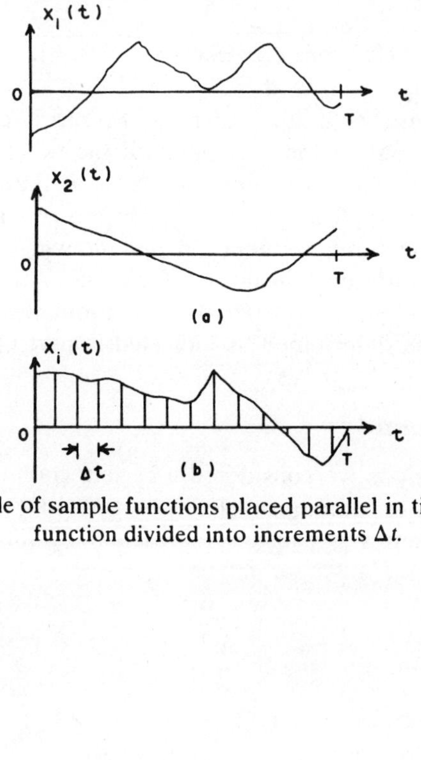

FIG. 2. (a) An ensemble of sample functions placed parallel in time. (b) A single sample function divided into increments Δt.

examine the nature of the dyadic time shift. Since the dyadic transformation is measure-preserving (**10**), then the aggregate of amplitude values consists of the same members as the aggregate of amplitudes formed by a single sample function. By the ergodic hypothesis, this aggregate is the same as the ensemble aggregate for a fixed time, the elements merely being taken in a different order. Note also that the ordering of the elements in the $x(t \oplus \tau)$ aggregate will differ from the ordering of the elements in the $x(t)$ aggregate. Finally, consider the aggregate of pairs formed by first selecting the amplitude value $x(t)$ and then the amplitude value $x(t \oplus \tau)$ over the single ensemble. Since it is assumed that aggregates contain the same elements whether formed from an ensemble or from a single sample function over all time, the aggregate of pairs should also contain the same members whether formulated by scanning the ensemble of sample functions or a single sample function. By the same previous reasoning, it is then possible to say that the mean of the product of the two elements, in the limit as T goes to infinity, is the same no matter how it is expressed whether as a mean with respect to the member aggregate or as a mean with respect to the ensemble aggregate.
Thus,

$$\begin{aligned} C_x(\tau) &= E[x(t)x(t \oplus \tau)] \\ &= \lim_{T\to\infty} E[x_T(t)x_T(t \oplus \tau)] \\ &= \lim_{T\to\infty} \frac{1}{T}\int_0^T x(t)x(t \oplus \tau)\,\mathrm{d}t \\ &= D_x(\tau). \end{aligned} \tag{46}$$

This latter result indicates the possible interchange of stayadic and dyadic correlation functions associated with an ergodic process.

V. Conclusions

This paper presents two methods for the identification of Voltera kernels for a certain class of nonlinear systems. Both methods utilize Walsh signals as inputs. According to the first method, identification is accomplished through output measurements at different input sequencies. The second method relies on measurements taken for input–output dyadic crosscorrelation functions. A primary advantage of employing Walsh signals is in general attributed to their computational advantages.

References

(**1**) S. A. Billings and S. Y. Fakhouri, "Identification of a class of nonlinear systems using correlation analysis", *Proc. IEEE*, Vol. 125, pp. 691–697, 1978.

(**2**) S. A. Billings and S. Y. Fakhouri, "Identification of nonlinear systems using the Wiener model", *Electron. Letters*, Vol. 13, pp. 502–504, 1977.

(3) B. Cooper and A. H. Falkner, "Identification of simple nonlinear systems", *Proc. IEEE*, Vol. 122, pp. 753–755, 1975.

(4) R. V. Webb, "Identification of the Voltera kernels of a process containing single-valued nonlinearities", *Electron. Letters*, Vol. 10, pp. 344–346, 1974.

(5) E. Economakos, "Identification of a group of internal signals of zero-memory nonlinear systems", *Electron. Letters*, Vol. 7, pp. 99–100, 1971.

(6) M. Maqusi, "Walsh analysis of power-law systems", *IEEE Trans. Information Theory*, Vol. IT–23, pp. 144–146, 1977.

(7) M. Maqusi, "On the Walsh analysis of nonlinear systems", *IEEE Trans. electromag. Compat.*, Vol. EMC–20, pp. 519–523, 1978.

(8) F. Pichler, "Walsh functions and optimal linear systems", *Proc. Symp. Walsh Function Applications*, pp. 17–22, Washington, D.C., April 1970.

(9) M. Maqusi, "Walsh analysis of linear dyadic invariant and instantaneous nonlinear systems", Sc.D. Dis., New Mexico State Univ., 1973.

(10) N. J. Fine, "On the Walsh functions", *Trans. Am. math. Soc.*, Vol. 65, pp. 373–414, 1949.

(11) N. J. Fine, "On generalized Walsh functions", *Trans. Am. math. Soc.*, Vol. 69, pp. 66–77, 1950.

(12) S. Cohn-Sfetcu and S. T. Nichols, "On the identification of dyadic invariant systems", *IEEE Trans. electramag. Compat.* Vol. EMC–17, pp. 111–117, 1975.

(13) G. P. Rao and L. Sivakumar, "Identification of time-lag systems via Walsh functions", *IEEE Trans. autom. Control*, Vol. AC–24, pp. 806–808, 1979.

(14) N. Ahmed, H. H. Schreider and P. V. Lopresti, "On notation and definition of terms related to a class of complete orthogonal functions", *IEEE Trans. electromag. Compat.*, Vol. EMC–15, pp. 76–80, 1973.

(15) F. Pichler, "Walsh functions and linear system theory", *Proc. Symp. Walsh Function Applications*, pp. 175–182, Washington, D.C., April 1970.

(16) M. Maqusi, "A sampling theorem for dyadic stationary processes", *IEEE Trans. Acoust., Speech, Signal Processing*, Vol. ASSP–26, pp. 265–267, 1978.

(17) Y. W. Lee, "Statistical theory of communication", John Wiley, New York, 1960.

(18) F. W. Weiser, "Walsh function analysis of instantaneous nonlinear stochastic problems", Ph.D. Dis., Polytechnic Institute of New York, 1964.

Reprinted from *IEE Proc.* **130:**9–16 (1983)

Improved algorithms for parameter identification in continuous systems via Walsh functions

Prof. G. Prasada Rao, B.E., M.Tech., Ph.D., M.I.S.T.E., and K.R. Palanisamy, B.Sc., B.E., M.Sc. (Eng.), Ph.D., M.I.S.T.E., Mem. I.E.E.E.

Indexing terms: *Control theory, Mathematical techniques*

Abstract: The paper first reviews the status of the Walsh-functions (WF) approach to the problem of parameter identification in continuous dynamical systems. The WF method of analysis and parameter estimation relies on the use of the so-called operational matrix for integration. For higher-order systems, the existing operational matrix has so far been applied after being raised to the required power. This procedure results in an accumulation of error at each stage of integration. To improve this situation, the paper introduces one-shot operational matrices for repeated integration (OSOMRI) via Walsh functions. The superiority of OSOMRI is demonstrated in problems of analysis and identification of higher-order systems. The paper then presents a direct noniterative method of parameter estimation for lumped linear systems in the presence of small unknown delays employing Walsh delay matrices developed recently by the authors. When the delay terms are not small, the results of this noniterative method will be considerably in error. However, these results can be taken as good initial values for the inevitable refining iterative shift algorithm of Prasada Rao and Sivakumar. The paper also presents an efficient multistep algorithm for parameter estimation in systems with large unknown time delays.

1 Introduction

This paper is concerned with parameter estimation in continuous dynamical systems from continuous input-output data. This problem has received considerable attention in recent years [1–20]. There have been mainly two different approaches to this problem. The first one is a direct procedure in which continuous process data are fitted into the chosen continuous model by carefully handling the derivative terms. The other [9–12], in which process data are initially sampled, consists of two major steps. In the first step, a discrete model is estimated from the sampled data. The discrete model is then converted into continuous form in the second stage. The conversion of discrete models into continuous (Laplace transform) form is not without difficulties. The difficulties encountered in the sampled-data approach are mainly due to the loss of intersample information, and attempts to reduce these rely on *a priori* knowledge of the signal nature between sampling instants. At the beginning of the effort, sampling has to be arbitrarily fast if one does not know initially the approximate range of values of the various modes of the system under indentification. Some guidelines may be drawn from the theory of digital filters to aid the procedure to convert the estimated discrete model into a continuous one. Sinha [12] suggests bilinear transformation as a method suitable for a large number of practical situations. Discussions on the two-stage approach with sampled data [9–12] as an alternative to the direct approach with continuous data do not establish superiority of one approach over the other. The choice of an approach is usually based on suitability to the actual situation.

In the meantime, efforts to avoid noise-accentuating derivative operations directly on process data in estimating continuous models, making the direct approach more attractive, have been successful and on the increase. These efforts include the following:

(*a*) Integration of process signals either directly [13] or by first modulating them with a set of 'method functions' [14] before fitting into the model differential equation.

(*b*) Use of Poisson moment functionals (PMF) of process data [15–20]. The PMFs of process signals may be physically measured on-line as the response of a Poisson filter chain to the respective signals. Off-line computation of integrals as required in the method function technique is avoided.

(*c*) Walsh spectral characterisation of process data [1–8]. The Walsh spectral characterisation of signals reduces the calculus of continuous systems to an elegant optimally (in the sense of least squares) approximate algebra through the so-called operational matrices. Models which are linear in their parameters give rise to linear algebraic equations which can be solved for the parameters and unknown initial conditions. Walsh spectral analysers [21] may be connected physically to the system under investigation for a practically attractive implementation of the WF technique. Apart from avoiding the direct derivative operations on process data, the Walsh-function (WF) method is computationally attractive and inherently resistive to zero-mean additive noise [7]. Hardware for Walsh function processing is simple and rugged as it involves standard switching elements; because of these, the WF technique has received much attention in recent years.

The WF technique is based on the operational matrices suggested by Prasada Rao and Sivakumar [1] and Chen and Hsiao [2] in identifying lumped linear systems. Tzafestas [3], Karanam *et al.* [4] and Chen and Shih [5] gave algorithms based on the operational matrices for other kinds of lumped models. Tzafestas [3] and Paraskevopoulos and Bounas [6] apply these ideas in estimating models described by partial differential equations. Prasada Rao and Sivakumar extend the original algorithm to handle models with unknown time delays [7] and multiple inputs and multiple outputs [8].

While the existing methods are sufficiently accurate for models of order not greater than one, the use of higher powers of the operational matrix to approximate repeated integration in higher-order models results in an accumulation of error at every stage of the integration. Consequently the resulting algorithms become increasingly inaccurate as the system order increases. This paper first attempts to correct this state of affairs by limiting the error of approximation to one stage only by suggesting one-shot operational matrices for repeated integration (OSOMRI). Sections 2 and 3 present the development of the OSOMRI and clearly demonstrate their efficiency in problems of analysis of continuous dynamical systems. Section 4 presents improved identification algorithms by OSOMRI for lag-free systems and demonstrates their superiority over the existing ones. Section 5 describes a

Paper 2284D, received 7th December 1981

Prof. Prasada Rao is presently on sabbatical leave at the Lehrstuhl für Elektrische Steuerung und Regelung, Ruhr Universität Bochum, 4630 Bochum 1, W. Germany from the Department of Electrical Engineering, Indian Institute of Technology, Kharagpur, 721302, India, and Dr. Palanisamy is with the Department of Electronics & Communication Engineering, Government College of Technology, Coimbatore 641013, India direct noniterative algorithm for parameter identification in systems containing small unknown time lags. The usefulness of the approximate results of the direct algorithm as good initial values for the inevitable iterative shift algorithm of Prasada Rao and Sivakumar [7], for the case when delay is not small, is illustrated in Section 6. Section 7 proposes an algorithm for the case of systems containing several unknown time delays.

2 Walsh-functions approach

Consider an m-set of Walsh functions $\{\psi_i(t),\ i=0,1,\ldots,(m-1)\}$ defined in the interval (0, 1) such that $m=2^{k_w}$, k_w an integer. A process signal $f(t)$ on (0, 1) may be approximated in terms of the Walsh functions as

$$\left.\begin{aligned} f(t) &\simeq F^T\Psi(t) \\ \text{where}& \\ F &= [f_0, f_1, \ldots, f_{m-1}]^T \\ \Psi(t) &= [\psi_0(t), \psi_1(t), \ldots, \psi_{m-1}(t)]^T \\ f_i &= \int_0^1 f(t)\,\psi_i(t)\,dt \end{aligned}\right\} \quad (1)$$

minimising the integral of the squared error over the interval.

Here we consider the normal interval without loss of generality. The treatment can be generalised for any interval or arbitrary length.

The integral

$$\underbrace{\int_0^t\int_0^t\cdots\int_0^t}_{j\text{ times}} f(t)\,dt^j$$

may be approximated as

$$\underbrace{\int_0^t\int_0^t\cdots\int_0^t}_{j\text{ times}} F^T\Psi(t)\,dt^j$$

The existing practice is to write

$$\left.\begin{aligned} &\underbrace{\int_0^t\int_0^t\cdots\int_0^t}_{j\text{ times}} \Psi(t)\,dt^j = E^j\Psi(t) \\ &E_{(m\times m)} = \left[\begin{array}{c|c} E_{(m/2\times m/2)} & -\dfrac{1}{2m} I_{(m/2\times m/2)} \\ \hline \dfrac{1}{2m} I_{(m/2\times m/2)} & 0_{(m/2\times m/2)} \end{array}\right] \end{aligned}\right\} \quad (2)$$

with $E_{(1\times 1)} = 1/2$.

Although approximate realisation of

$$f_{(j)}(t) \triangleq \underbrace{\int_0^t\int_0^t\cdots\int_0^t}_{j\text{ times}} f(t)\,dt^j$$

as $F^TE^j\Psi(t)$ is algebraically a simple approach, it leads to an accumulation of error at each of the j stages of integration. Therefore we consider alternatively

$$\Psi_{(j)}(t) \triangleq \underbrace{\int_0^t\int_0^t\cdots\int_0^t}_{j\text{ times}} \Psi(t)\,dt^j$$

in its exact form and represent it in Walsh series at the final stage only. The result is in the form

$$\Psi_{(j)}(t) \simeq E_j\Psi(t) \quad (3)$$

where E_j is the intended OSOMRI. The elements of E_j naturally minimise the integral of the squared error between $\Psi_{(j)}(t)$ and $E_j\Psi(t)$. It can be shown that

$$E_j = \frac{1}{m} WP_jW \quad (4a)$$

where W is the orthogonal Walsh matrix (with + 1s and − 1s as its elements). The corresponding multistage operator E^j may be written as

$$E^j = \frac{1}{m} WP^jW \quad (4b)$$

The upper triangular matrix P_j is given by

$$P_j = \frac{1}{m^j}\left[\frac{1}{(j+1)!}I + \sum_{r=1}^{m-1}\left\{\sum_{q=0}^{j-1}\frac{r^{q+1}-(r-1)^{q+1}}{(q+1)!(j-q)!}\right\}\nabla^r\right] \quad (5)$$

in which, for instance,

$$\nabla^1_{(4\times 4)} = \begin{bmatrix} 0 & 1 & 0 & 0 \\ 0 & 0 & 1 & 0 \\ 0 & 0 & 0 & 1 \\ 0 & 0 & 0 & 0 \end{bmatrix}$$

Table 1 lists the matrices P^j (power matrix) and P_j (the OSOMRI) for certain values of j.

In the case when $m=4$, we get the following, using eqns. 4 and eqns. 6a–h from Table 1:

$$E^1 = E_1 = \begin{bmatrix} \frac{1}{2} & -\frac{1}{4} & -\frac{1}{8} & 0 \\ \frac{1}{4} & 0 & 0 & -\frac{1}{8} \\ \frac{1}{8} & 0 & 0 & 0 \\ 0 & \frac{1}{8} & 0 & 0 \end{bmatrix} \quad (6i)$$

$$E^2 = \frac{1}{64}\begin{bmatrix} 11 & -8 & -4 & 2 \\ 8 & -5 & -2 & 0 \\ 4 & -2 & -1 & 0 \\ 2 & 0 & 0 & -1 \end{bmatrix} \quad (6j)$$

$$E_2 = \frac{1}{64}\begin{bmatrix} \frac{32}{3} & -8 & -4 & 2 \\ 8 & -\frac{16}{3} & -2 & 0 \\ 4 & -2 & -\frac{4}{3} & 0 \\ 2 & 0 & 0 & -\frac{4}{3} \end{bmatrix} \quad (6k)$$

Table 1: Matrices P^j and P_j

j	P^j		P_j	
1	$\frac{1}{m}\left[\frac{I}{2}+\sum_{r=1}^{m-1}\nabla^r\right]$	(6*a*)	$\frac{1}{m}\left[\frac{I}{2}+\sum_{r=1}^{m-1}\nabla^r\right]$	(6*b*)
2	$\frac{1}{m^2}\left[\frac{I}{4}+\sum_{r=1}^{m-1}r\nabla^r\right]$	(6*c*)	$\frac{1}{m^2}\left[\frac{I}{3!}+\sum_{r=1}^{m-1}r\nabla^r\right]$	(6*d*)
3	(for $m = 4$, say) $\frac{1}{64}\left[\frac{I}{8}+\frac{3}{4}\nabla+\frac{9}{4}\nabla^2+\frac{19}{4}\nabla^3\right]$	(6*e*)	$\frac{1}{m^3}\left[\frac{I}{4!}+\sum_{r=1}^{m-1}\frac{1+6r^2}{12}\nabla^r\right]$	(6*f*)
4	(for $m = 4$, say) $\frac{1}{256}\left[\frac{I}{16}+\frac{1}{2}\nabla+2\nabla^2+\frac{11}{2}\nabla^3\right]$	(6*g*)	$\frac{1}{m^4}\left[\frac{I}{5!}+\sum_{r=1}^{m-1}\frac{2r^3+r}{12}\nabla^r\right]$	(6*h*)

$$E^3 = \frac{1}{256}\begin{bmatrix} 12 & -10 & -\frac{11}{2} & 4 \\ 10 & -8 & -4 & \frac{5}{2} \\ \frac{11}{2} & -4 & -2 & 1 \\ 4 & -\frac{5}{2} & -1 & 0 \end{bmatrix} \tag{6l}$$

$$E_3 = \frac{1}{256}\begin{bmatrix} \frac{32}{3} & -\frac{28}{3} & -\frac{31}{6} & 4 \\ \frac{28}{3} & -8 & -4 & \frac{17}{6} \\ \frac{31}{6} & -4 & -2 & 1 \\ 4 & -\frac{17}{6} & -1 & 0 \end{bmatrix} \tag{6m}$$

and so on. From these, we note that

$$E^2 - E_2 = \frac{1}{m^2}\frac{I}{12} \tag{6n}$$

$$E^3 - E_3 = \frac{1}{m^3}\frac{2E}{3} \tag{6o}$$

$$E^4 - E_4 = \frac{1}{m^4}\left(4E^2 - \frac{I}{60}\right) = \frac{1}{m^4}\left(4E_2 + \frac{I}{80}\right) \tag{6p}$$

3 Superiority of E_j over E^j

Consider the Walsh-series representation of $f(t)$ given by eqn. 1:

$$f(t) = F^T\Psi(t) + e(t)$$

in which the L_2-norm of the 'representation error' [22] is given by

$$\|e\| \leq \frac{1}{2^{(K_w+1)}\sqrt{3}}\left|\sup_{(0,1)}\frac{df}{dt}\right| \tag{6q}$$

Prasada Rao and Srinivasan [22] have discussed other errors. Now let us consider the use of E^j and E_j in integrating $f(t)$ over the interval (0, 1):

$$f_{(j)}(t) = F^T E^j \Psi(t) + \Gamma_r(t) + e_{(j)}(t) \tag{6r}$$

and alternatively

$$f_{(j)}(t) = F^T E_j \Psi(t) + \Gamma_0(t) + e_{(j)}(t) \tag{6s}$$

where $\Gamma_r(t)$ and $\Gamma_0(t)$ are errors resulting from the use of E^j and E_j, respectively. To compare $\|\Gamma_r\|$ with $\|\Gamma_0\|$, without loss of generality, we invoke the single-term Walsh-series (STWS) approach [23] to integration. The STWS method [23] asserts that it would suffice if we consider a single (the first) Walsh function $\psi_0(t)$, $t \in (0, 1/m)$, i.e. on the first subinterval, and extend the computation over the successive subintervals $((i-1)/m,\ i/m)$, $i = 2, 3, \ldots, m$. The result is the same as obtained by considering m Walsh functions in (0, 1) and handling them together. Eqns. 6*r* and *s* become scalar. Thus

$$f(t) \simeq F_0\psi_0(t) \qquad t \in (0, 1/m)$$

Consequently,

$$E^j = \frac{1}{m^j 2^j}$$

and

$$E_j = \frac{1}{m^j(j+1)!}$$

Also, the exact jth integral of the single Walsh term

$$F_0\psi_{0(j)}(t) = F_0\frac{t^j}{j!}$$

The errors resulting from the application of E^j and E_j are, respectively,

$$\gamma_r(t) = F_0\left(\frac{t^j}{j!} - \frac{1}{2^j m^j}\right)$$

and

$$\gamma_0(t) = F_0 \left(\frac{t^j}{j!} - \frac{1}{(j+1)!m^j} \right)$$

It is possible to show that

$$\epsilon_j \triangleq \frac{\|\gamma_r\|}{\|\gamma_0\|} = \frac{[\int_0^{1/m} \gamma_r^2(t)\,dt]^{1/2}}{[\int_0^{1/m} \gamma_0^2(t)\,dt]^{1/2}}$$

$$= \frac{(j+1)^{1/2}}{j2^j} \times$$

$$[(j+1)2^{2j} + j!(2j+1)\{(j+1)! - 2^{j+1}\}]^{1/2}$$

$$\geqslant 1 \qquad \text{for all } j$$

Table 2 shows ϵ_j for some values of j.

Table 2: Norm ratio ϵ_j

j	ϵ_j
1	1.0000000
2	1.1456439
3	2.0275875
4	4.9765073
5	14.2881060

Example 1
Consider the solution of a second-order initial-value problem over (0, 1):

$$\ddot{x}(t) = -\pi^2 x(t) \qquad x(0) = 0, \quad \dot{x}(0) = -\pi \tag{7}$$

The exact solution is

$$x(t) = -\sin \pi t$$

Consider an m-size Walsh series in (0, 1):

$$x(t) = X^T \Psi(t) \qquad X^T = [X_0, X_1, \ldots, X_{m-1}]$$

$$X_i = \int_0^1 x(t)\psi_i(t)\,dt$$

The solution may be computed:
(i) using E^2, from

$$X^T\Psi(t) = -\pi^2 X^T E^2 \Psi(t) + \underbrace{[-\pi, 0, 0, \ldots, 0]}_{m-1 \text{ columns}} E\Psi(t) \tag{8a}$$

(ii) using E_2, from

$$X^{*T}\Psi(t) = -\pi^2 X^{*T} E_2 \Psi(t) + \underbrace{[-\pi, 0, 0, \ldots, 0]}_{m-1 \text{ columns}} E\Psi(t) \tag{8b}$$

We distinguish the result of eqn. 8*b* with an asterisk. X and X^* may be computed using the extension method with STWS [23]. Choosing $m = 16$, the STWS solutions in the first subinterval $(0, 1/m)$, for instance, are

$$X_1 = \frac{-2m\pi}{(4m^2 + \pi^2)} = -0.0972375$$

$$X_1^* = \frac{-3m\pi}{(6m^2 + \pi^2)} = -0.097549$$

These results, after rounding up to four decimal places, are shown in Table 3 as Y_k and Y_k^*, the piecewise-constant values in the subintervals. Note that Y_k^* is closer to the exact solution. The solution is shown up to $t = 1/2$, the point of symmetry in (0, 1). There is an interesting feature of the solution of eqn. 7. Eqns. 8*a* and *b* imply that

Table 3: Comparison of solutions of eqn. 7 using E^2 and E_2 with its exact solution

k	Piecewise-constant subinterval values		
	Y_k using E^2	Y_k^* using E_2	exact subinterval averages
1	−0.0972	−0.0975	−0.0979
2	−0.2880	−0.2889	−0.2898
3	−0.4676	−0.4691	−0.4706
4	−0.6293	−0.6313	−0.6334
5	−0.7668	−0.7693	−0.7718
6	−0.8749	−0.8777	−0.8805
7	−0.9493	−0.9523	−0.9554
8	−0.9872	−0.9904	−0.9936

$$X^T(I + \pi^2 E^2) = X^{*T}(I + \pi^2 E_2)$$

In view of eqn. 6*n*,

$$X^T \left[I + \frac{\pi^2}{12m^2}(I + \pi^2 E_2)^{-1} \right] = X^{*T}$$

This relationship, being free from initial-condition terms, applies to piecewise-constant subinterval values also; i.e.

$$\frac{Y_k^*}{Y_k} = 1 + Z$$

where

$$Z = \frac{\pi^2}{12m^2}\left(1 + \frac{\pi^2}{6m^2}\right)$$

In the present case, with $m = 16$, $Z = 3.2334036 \times 10^{-3}$. The results in Table 3 confirm this. Having thus established the superiority of E_j over E^j in the solution problem, we turn to the problem of parameter identification and discuss the use and performance of OSOMRI therein.

4 Improved algorithm for identification of lag-free systems

Consider a lumped lag-free system modelled by

$$y(t) + \sum_{i=1}^{n} a_i \frac{d^i y(t)}{dt^i} = b_0 x(t) + \sum_{j=1}^{n-1} b_j \frac{d^j x(t)}{dt^j} \tag{9}$$

with $x(t)$ and $y(t)$ available in an arbitrary but active (and normalised) period of time

$$\{a_i, i = 1, 2, 3, \ldots, n\}$$

$$\{b_j, j = 0, 1, 2, \ldots, (n-1)\}$$

are the essential parameters to be determined. Simultaneously, the presence of other unknowns

$$\left\{ \alpha_{dj} \triangleq \left. \frac{d^{j-1}y}{dt^{j-1}} \right|_{t=0} \qquad j = 1, 2, 3, \ldots, n-1 \right\}$$

$$\left\{ \beta_{dj} \triangleq \left. \frac{d^{j-1}x}{dt^{j-1}} \right|_{t=0} \qquad j = 1, 2, 3, \ldots \right\}$$

should also be considered. Although $y(0)$ and $x(0)$ are known, we will deliberately ignore the information and treat them also as unknowns for reasons to be given later.

Integrating both sides of eqn. 9 n times in terms of Walsh spectra, equating like sequency components on either side, and rearranging, we get [2, 3]

$$\mathscr{L}\mathscr{M}=\mathscr{N} \tag{10}$$

where

$$\mathscr{L} = [\mathscr{P} \vdots \mathscr{I}^T] \qquad \mathscr{M} = \begin{bmatrix} \mathscr{R} \\ \mathscr{H} \end{bmatrix}$$

$$\mathscr{R}_{(2n \times 1)} = [a_1, a_2, \ldots, a_n, b_0, b_1, \ldots, b_{n-1}]^T \triangleq [A^T \vdots B^T]$$

$$\mathscr{H}_{(n \times 1)} = [f_1, f_2, \ldots, f_n]^T$$

$$f_i = \alpha_d^T \nabla^{n-i} A + \beta_d^T \nabla_B^{n-i-1}$$

$$\alpha_d = -[\alpha_{d1}, \alpha_{d2}, \ldots, \alpha_{dn}]^T$$

$$\beta_d = [\beta_{d1}, \beta_{d2}, \ldots, \beta_{dn}]^T$$

$$\mathscr{P} = [E_{n-1}^T Y_d \vdots E_{n-2}^T Y_d \vdots \ldots \vdots Y_d \vdots -E_n^T X_d \vdots -E^T X_d]$$

$$\mathscr{N} = -E_n^T Y_d$$

$$y(t) = Y_d^T \Psi(t)$$

$$x(t) = X_d^T \Psi(t)$$

$$Y_d = [y_{d0}, y_{d1}, \ldots, y_{dm-1}]^T$$

$$X_d = [x_{d0}, x_{d1}, \ldots, x_{dm-1}]^T$$

$$x_{dj} = \int_0^1 x(t)\, \psi_j(t)\, dt$$

$$y_{dj} = \int_0^1 y(t) \psi_j(t)\, dt$$

$\mathscr{I}$ has elements which are known to be Walsh spectra of terms of the form $t^j/j!$. In fact,

$$\mathscr{I}_{(n \times m)} = \begin{bmatrix} \text{1st row of } E_0 \\ \text{1st row of } E_1 \\ \cdot \\ \cdot \\ \cdot \\ \text{1st row of } E_{n-1} \end{bmatrix}$$

with $E_0 = E^0 = I$, the identity matrix.

Example 2

Consider the problem of identification in a second-order lag-free model given by

$$\ddot{y}(t) + a_1 \dot{y}(t) + a_2 y(t) = x(t)$$

with input-output data in (0, 1) simulated in two cases with unit step input and zero initial conditions: case 1: $a_1 = 0.3$, $a_2 = 0.02$ and case 2: $a_1 = 0.12$, $a_2 = 0.002$. The identification algorithm described above is applied by letting $n = 2$ with a set of eight Walsh functions ($m = 8$). Results using E_j and E^j are shown together against the actual results in Table 4.

The mean square error (MSE) is $\sum_i (p_i - \bar{P}_i)^2$, where p_i and P_i denote, respectively, the true and the calculated values of system parameters. This is a reasonable basis for comparison of the results from E_j and E^j with the actual values. In the present case, E_2 clearly gives better results than E^2.

5 Identification in the presence of small unknown time delays

When the model happens to contain time-delay terms, as far as the authors are aware, the iterative shift algorithm (ISA) [7] seems to be the only way in which the problem can be handled via Walsh functions. The Walsh delay matrices (WDMs) [24] render the method of identification direct for small delays. The ISA which is inevitable for large delays, may be improved in general if WDMs are employed. A brief account of WDMs is given here to aid understanding of the proposed schemes.

Let a function $f(t)$ and its Walsh spectral representation F be available in (0, 1). Then $f(t-\tau)$, $0 \leqslant \tau \leqslant 1$, may be written as

$$f(t-\tau) \simeq F^T \Psi(t-\tau) \qquad t \in (0, 1)$$

Let us further assume that the initial function is zero for t $(-\tau, 0)$. In particular, if $\tau = \alpha/m$, $0 \leqslant \alpha \leqslant 1$, we can write

$$\Psi(t-\tau) = D_{(m)}^{\alpha} \Psi(t) \tag{11}$$

where $D_{(m)}$ is called the Walsh delay matrix.

For a given α we have

$$D_{(m=2,\alpha)} \triangleq D_{(2)}^{\alpha} = \frac{1}{2}\begin{bmatrix} 2-\alpha & -\alpha \\ \alpha & 2-3\alpha \end{bmatrix}$$

$$= I_{(2)} + \frac{\alpha}{2}\begin{bmatrix} -1 & -1 \\ 1 & -3 \end{bmatrix}$$

$$\triangleq I_{(2)} + \alpha\, \Delta_{(2)} \qquad \text{(say)}$$

Table 4: Identification of parameters in example 2

Parameters	a_1	a_2	$y(0)$	$\dot{y}(0)$	Mean square error (MSE)
Case 1					
True value	0.3	0.02	0.0	0.0	–
Result using E^2	0.30025	0.019442	0.000774	−0.001302	2.69266×10^{-6}
Result using E_2	0.300405	0.019227	0.000380	0.000001	9.05955×10^{-7}
Case 2					
True value	0.12	0.002	0.0	0.0	–
Result using E^2	0.120061	0.001881	0.000311	−0.001302	1.8098×10^{-6}
Result using E_2	0.120072	0.001860	0.000154	0.000000	4.85×10^{-8}

$$D_{(m=4,\alpha)} \triangleq D^{\alpha}_{(4)} = \frac{1}{4}\begin{bmatrix} 4-\alpha & -\alpha & -\alpha & -\alpha \\ \alpha & 4-3\alpha & \alpha & -3\alpha \\ \alpha & \alpha & 4-7\alpha & \alpha \\ -\alpha & 3\alpha & -\alpha & 4-5\alpha \end{bmatrix}$$

$$= I_{(4)} + \alpha\,\Delta_{(4)}$$

and so on.

If $\tau > 1/m$, i.e. if $\tau = (N+\alpha)/m$, $0 \leqslant N \leqslant m$, $0 \leqslant \alpha \leqslant 1$, the corresponding WDMs can be generated. With $N = 0$ and $\alpha = 1$, we have

$$D^{1}_{(2)} = \frac{1}{2}\begin{bmatrix} 1 & -1 \\ 1 & -1 \end{bmatrix}$$

$$D^{1}_{(4)} = \frac{1}{4}\begin{bmatrix} 3 & -1 & -1 & -1 \\ 1 & 1 & 1 & -3 \\ 1 & 1 & -3 & 1 \\ -1 & 3 & -1 & -1 \end{bmatrix}$$

and so on.

For the various properties and methods of generation the reader may refer to Prasada Rao and Palanisamy [24, 25]. In order to understand the proposed noniterative technique of identification, consider a model of the form

$$y(t) + \sum_{i=1}^{n} a_i \frac{d^i y(t)}{dt^i} = b_0 x(t-\tau_x) \tag{12}$$

with an unknown small delay τ_x in the input process. We now integrate eqn. 12 n times and insert the operational matrices E_j. The delay term

$$x(t-\tau_x) = X_d^T D^{\epsilon_x}_{(m)} \Psi(t) = [X_d^T + \epsilon_x X_d^T \Delta_{(m)}]\,\Psi(t) \tag{13}$$

where $\tau_x = \epsilon_x/m$, leading to

$$[E_{n-1}^T Y_d \,\vdots\, E_{n-2}^T Y_d \,\vdots \ldots \vdots\, Y_d \,\vdots\, -E_n^T X_d \,\vdots\, -(\Delta_{(m)}E_n)^T X_d \,\vdots\, \mathscr{J}^T] \begin{bmatrix} \mathscr{R} \\ \mathscr{H} \end{bmatrix} = -E_n^T Y_d \tag{14}$$

where

$$\mathscr{R} = [a_1, a_2, \ldots, a_n, b_0, \epsilon_x b_0]$$

and

$$\mathscr{H} = [f_1, f_2, \ldots, f_n]^T$$

Example 3

Consider a process described by

$$\dot{y}(t) + a_1 y(t) = x(t-\tau_x) \tag{15}$$

with a small unknown delay τ_x in the input process. Data generated with zero initial conditions and unit step input are employed in the algorithm.

Table 5 summarises the results obtained in different numerical experiments. Note how the unknown initial condition also is obtained along with the other parameters.

Table 5: Identification of parameters in a first-order model (example 3)

		Parameters			
		a	$y(0)$	delay	ϵ_x
1	True value	−0.5	0.0	0.0292968	0.9375
	Result	−0.510261	−0.000388	0.0298809	0.956189
2	True value	−1.0	0.0	0.0292968	0.9375
	Result	−1.006355	−0.001077	0.0287801	0.920966
3	True value	−2.0	1.0	0.0273437	0.875
	Result	−1.99977	0.99735	0.0248665	0.795731
4	True value	−2.0	0.0	0.0292968	0.9375
	Result	−2.000475	−0.001772	0.027408	0.877059

Example 4

Consider a second-order model of the form

$$\ddot{y}(t) + a_1 \dot{y}(t) + a_2 y(t) = x(t-\tau_x) \tag{16}$$

with $\tau_x = \epsilon_x/m$, $0 \leqslant \epsilon_x \leqslant 1$. Data generated with zero initial conditions and unit step input are employed in the algorithm, and the results are summarised in Table 6. Note the automatically evaluated initial conditions and the superiority of E_2.

6 Identification of systems with larger time delays

When τ_x is not small, expressing it as fraction of $1/m$ requires a small m. This will influence the numerical accuracy of the representation and lead to large errors in the results. We should therefore let m remain as large as required and correspondingly modify our approach. In what follows, we present such an algorithm. The delayed term is written as

$$x(t-\tau_x) = X_d^T D^{N+\epsilon_x}_{(m)} \Psi(t)$$

$$= [X_d^T D^N_{(m)} + \epsilon_x X_d^T \Delta_{(m)} D^N_{(m)}]\,\Psi(t)$$

where $\tau_x = (N+\epsilon_x)/m$, $0 \leqslant \epsilon_x \leqslant 1$.

Table 6: Identification of parameters in a second-order model (example 4)

	Parameters	a_1	a_2	ϵ_x	$\dot{x}(0)$	$x(0)$	MSE $\sum_i (p_i - \hat{p}_i)^2$
1	True value	0.3	0.02	0.875	0.0	0.0	–
	Result using E_2	0.299940	0.02139	0.675562	−0.006217	0.000045	0.0398159
	Result using E^2	0.299924	0.021395	0.450359	−0.013258	0.000073	0.1804977
2	True value	0.25	0.01	0.875	0.0	0.0	–
	Result using E_2	0.249953	0.01131	0.675716	−0.006214	0.000045	0.0397544
	Result using E^2	0.249944	0.011346	0.450469	−0.013257	0.000073	0.1804041

We will first try to determine N and then apply the direct algorithm to find ϵ_x. The form of the identification equation for eqn. 12 is the same as eqn. 14 with one difference: the term $\vdots -(E_n D_{(m)}^N) X_d \vdots$ will appear in place of $\vdots -(\Delta_{(m)} E_n)^T X_d \vdots$. Given the input-output record over (t_0, t_f), the algorithm may be applied in the following steps:

(*a*) Choose a WF defined on the base (t_1, t_{d1}), $t_1, t_{d1} \in (t_0, t_f)$

(*b*) Form eqn. 14 with $D_{(m)}^N$ in place of $D_{(m)}^{\epsilon_x}$.

(*c*) If $m > 2n + 1$, the number of parameters, apply the principle of least squares to normalise eqn. 14.

(*d*) Calculate the parameter vector

$$\mathscr{P}_1 \triangleq \begin{bmatrix} \mathscr{R}_1 \\ \hline \mathscr{H}_1 \end{bmatrix}$$

(*e*) Repeat steps (*b*)–(*d*) with $N = 0, 2, \ldots, m-3$.

(*f*) Repeat steps (*a*)–(*e*) choosing a base (t_1, t_{d2}) for the WF, where $t_1, t_{d2} \in (t_0, t_f)$, and obtain the corresponding parameter vector $\mathscr{P}_2$.

(*g*) Compute the disparity between $\mathscr{P}_1$ and $\mathscr{P}_2$ as

$$\mathscr{W}(N) = \| \mathscr{P}_1 - \mathscr{P}_2 \|$$

For a value of N corresponding to the neighbourhood of τ_x, $\mathscr{W}(N)$ is a minimum.

(*h*) Obtain N by direct search.

(*i*) Employing the derived value of N, compute ϵ_x following the direct algorithm on the lines of Section 4.

Example 5

Consider a delay process

$$\dot{y}(t) = ay(t) + x(t - \tau_x) \tag{17}$$

With a unit step input, the process is simulated with $a = -0.5$, $y(0) = 1$, and $\tau_x = 0.0527343$. For the chosen value of $1/m$, i.e. $1/16$, $\epsilon_x = 0.84375$. The function $\mathscr{W}(N)$ is shown in Fig. 1. It may be seen that $N = 0$ has the least $\mathscr{W}(N)$. This means that the actual value of lag lies between $N = 0$ and 1. Now we apply the direct algorithm and get the parameters including ϵ_x. The final results are shown in Table 7. The second set of results in Table 7 belongs to another situation in which the actual delay is fairly large such that, for $N = 4$, $\mathscr{W}(N)$ is minimum. The function $\mathscr{W}(N)$ is shown in Fig. 2.

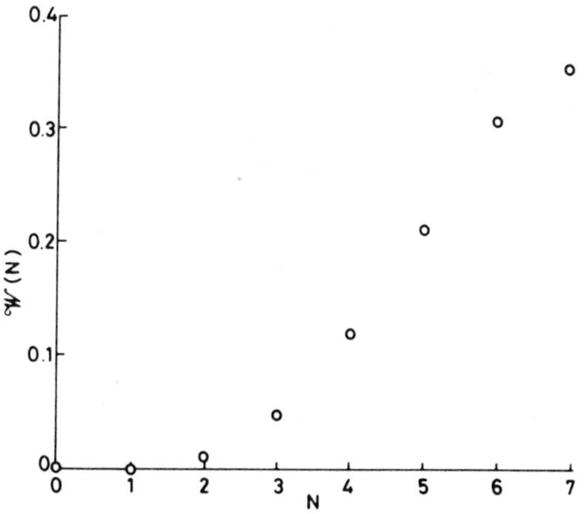

Fig. 1 *Loss function of example 5 (m = 16)*

7 Identification in the presence of several small delays

Consider a system with two delays:

$$y(t - \tau_y) + \sum_{i=1}^{n} a_i \frac{d^i y(t)}{dt^i} b_0 x(t - \tau_x) \tag{18}$$

where τ_x and τ_y are small. We will write

$$y(t - \tau_y) = Y_d^T D_{(m)}^{\epsilon_y} \Psi(t) = [Y_d^T + \epsilon_y Y_d^T \Delta_{(m)}] \Psi(t)$$

and

$$x(t - \tau_x) = X_d^T D_{(m)}^{\epsilon_x} \Psi(t) = [X_d^T + \epsilon_x X_d^T \Delta_{(m)}] \Psi(t)$$

The identification equation then becomes

$$[E_{n-1}^T Y_d \vdots E_{n-2}^T Y_d \vdots \cdots \vdots Y_d \vdots (\Delta_{(m)} E_n)^T Y_d \vdots -E_n^T X_d \vdots -(\Delta_{(m)} E_n z)^T X_d \vdots \mathscr{F}^T] \begin{bmatrix} \mathscr{R} \\ \hline \mathscr{H} \end{bmatrix} = -(E_n^T Y_d) \tag{19}$$

where

$$\mathscr{R} = [a_1, a_2, \ldots, a_n, \epsilon_y, b_0, \epsilon_x b_0]^T$$

Example 6

Consider a process with two delay terms:

$$\dot{y}(t) = ay(t - \tau_y) + x(t - \tau_x) \tag{20}$$

With $a = -0.5$, $\tau_x = \tau_y = 0.046875$, i.e. $\epsilon_x = \epsilon_y = 0.75$, the input-output data are generated with $x(t)$ as a unit step. With $m = 16$, the results of computation are presented in Table 8. The data are used from the record beginning arbitrarily such that $y(0) = 1.0$.

Example 7

Consider a process with two delayed inputs each with a small delay:

$$\dot{y}(t) = ay(t) + x_1(t - \tau_{x1}) + x_2(t - \tau_{x2}) \tag{21}$$

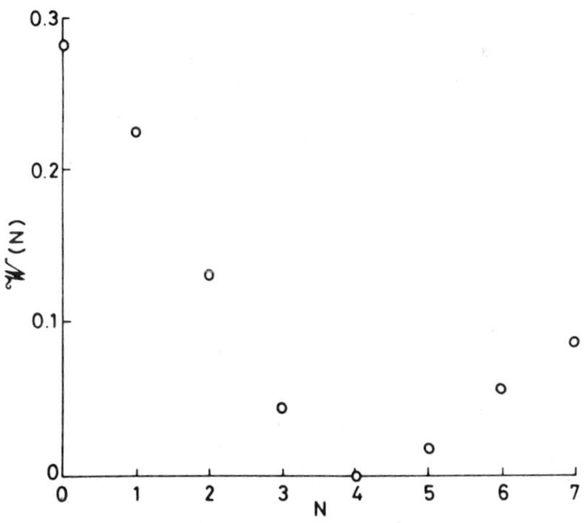

Fig. 2 *Loss function of example 5 (m = 16), larger delay*

Table 7: Results of the multistep algorithm for systems suspected to have large delays

	Parameters	a	$y(0)$	ϵ_x	N	τ_x
1	True value	−0.5	1.0	0.84375	0	0.0527343
	Result	−0.500082	0.991784	0.710999	0	0.044434
2	True value	−0.5	1.0	0.84375	4	0.5527343
	Result	−0.500264	0.999750	0.839935	4	0.5524959

Table 8: Results of parameter identification in eqn. 20

Parameters	a	$y(0)$	ϵ_x	ϵ_y	τ_x	τ_y
True value	−0.5	1.0	0.75	0.75	0.046875	0.046875
Result	−0.500283	0.994009	0.663679	0.773617	0.0414799	0.048351

Table 9: Results of parameter identification in eqn. 21

Parameters	a	$y(0)$	ϵ_{x1}	ϵ_{x2}	τ_{x1}	τ_{x2}
True value	−0.5	0.0	0.75	0.9375	0.0234375	0.0292968
Result	−0.501620	−0.001635	0.884831	0.948916	0.0276509	0.0296536

Input-output data have been simulated with $a = 0.5$, $\tau_{x1} = 0.0234375$, $\tau_{x2} = 0.0292968$ and $y(0) = 0.0$. One of the inputs is a delayed unit ramp and the other is a delayed unit step.

8 Discussion and conclusion

The error in the Walsh-series approximation of repeated integrals is limited to only one stage as a result of the newly proposed one-shot operational matrices for repeated integration. The algebraically simple idea of using higher powers of E results in an accumulation of errors at each stage of integration. Moreover, generation of E^j from the basic E-matrix by successive matrix multiplications is computationally less economical than the generation of E_j through eqn. 5, in which ∇^r, the power of the shift matrix, can be affected by simply shifting the diagonal of unity elements appropriately. The superiority of the operational matrix E_j is clearly demonstrated in algorithms of analysis and identification. A direct noniterative algorithm is suggested for systems containing small time delays. A multistep algorithm is proposed for problems involving large delays. The algorithms can handle data on an arbitrary but active interval of time. The unknown initial conditions are calculated simultaneously with the system parameters. The initial conditions $y(0)$ and $x(0)$ are actually known but this fact is deliberately ignored in the treatment. Acceptance and direct use of this information, which may be noisy in actual practice, will affect the elements of the information matrix $\mathscr{M}$, leading to erroneous results. On the other hand, if we treat $x(0)$ and $y(0)$ as unknowns, $\mathscr{M}$ will contain only Walsh spectral information on input-output data, which is known to be considerably immune to zero-mean additive noise [7].

9 References

1 PRASADA RAO, G., and SIVAKUMAR, L.: 'System identification via Walsh functions', *Proc. IEE*, 1975, **122**, (10), pp. 1160–1161

2 CHEN, C.F., and HSIAO, C.H.: 'Time domain synthesis via Walsh functions', *ibid.*, 1975, **122**, (5), pp. 565–570

3 TZAFESTAS, S.: 'Walsh series approach to lumped and distributed system identification', *J. Franklin Inst.*, 1978, **305**, p. 199

4 KARANAM, V.R., FRICK, P.A., and MOHLER, R.R.: 'Bilinear system identification by Walsh functions', *IEEE Trans.*, 1978, **AC-23**, pp. 709–713

5 CHEN, W.L., and SHIH, Y.P.: 'Parameter estimation of Bilinear systems via Walsh functions', *J. Franklin Inst.*, 1978, **305**, pp. 249–259

6 PARASKEVOPOULOS, P.N., and BOUNAS, A.C.: 'Distributed parameter system identification via Walsh functions', *Int. J. Syst. Sci.*, 1978, **9**, pp. 75–83

7 PRASADA RAO, G., and SIVAKUMAR, L.: 'Identification of time-lag systems via Walsh functions', *IEEE Trans.*, 1979, **24**, pp. 806–808

8 PRASADA RAO, G., and SIVAKUMAR, L.: 'Transfer function matrix identification in MIMO systems via Walsh functions', *Proc. IEEE*, 1981, **69**, pp. 465–466

9 SMITH, F.W.: 'System Laplace transform estimation from sampled data', *IEEE Trans.*, 1968, **AC-13**, pp. 37–44

10 HSIA, T.C.: 'A discrete method for parameter identification in linear systems with transport lags', *ibid.*, 1969, **AES-5**, pp. 236–239

11 SINHA, N.K.: 'Estimation of transfer function of continuous system from sampled data', *Proc. IEE*, 1972, **119**, (5), pp. 612–614

12 HSIA, T.C.: 'On sampled data approach to parameter identification of continuous linear systems', *IEEE Trans.*, 1972, **AC-17**, pp. 247–249

13 DIAMESSIS, J.E.: 'A new method of determining the parameters of a physical system', *Proc. IEEE*, 1965, **53**, pp. 205–206

14 SHINBROT, M.: 'On the analysis of linear and nonlinear systems', *Trans. ASME*, 1957, **79**, pp. 547–552

15 FAIRMAN, F.W., and SHEN, D.W.C.: 'Parameter identification for linear time varying dynamic processes', *Proc. IEE*, 1970, **117**, ( ), pp. 2025–2029

16 SAHA, D.C., and PRASADA RAO, G.: 'Time domain synthesis via Poisson moment functionals', *Int. J. Control*, 1979, **30**, pp. 417–426

17 SAHA, D.C., and PRASADA RAO, G.: 'Identification of lumped linear systems in the presence of unknown initial conditions via Poisson moment functionals', *ibid.*, 1980, **31**, pp. 637–644

18 SAHA, D.C., and PRASADA RAO, G.: 'Identification of lumped linear time-varying parameter systems via Poisson moment functionals', *ibid.*, 1980, **32**, pp. 709–721

19 SAHA, D.C., and PRASADA RAO, G.: 'Identification of lumped linear systems in the presence of small unknown time delays via Poisson moment functionals', *ibid.*, 1981, **33**, p. 945

20 SAHA, D.C., and PRASADA RAO, G.: 'A general algorithm for parameter identification in lumped continuous systems. – The Poisson moment functional approach', *IEEE Trans.*, 1982, **AC-27**, pp. 223–225

21 VENKATESAM, S., and PRASADA RAO, G.: 'A Walsh spectral analyser', *Int. J. Electron.*, 1979, **46**, pp. 413–415

22 PRASADA RAO, G., and SRINIVASAN, T.: 'Analysis and synthesis of dynamic systems containing time delays via block-pulse functions', *Proc. IEE*, 1978, **125**, (9), pp. 1064–1068

23 PRASADA RAO, G., PALANISAMY, K.R., and SRINIVASAN, T.: 'Extension of computation beyond the initial normal interval in Walsh series analysis dynamical systems', *IEEE Trans.*, 1980, **AC-25**, pp. 317–319

24 PRASADA RAO, G., and PALANISAMY, K.R.: 'A new operational matrix for delay via Walsh functions and some aspects of its algebra and applications'. Proceedings of national systems conference NSC 78, PAU Ludbiana, Sept. 1978, pp. 1/60–61

25 PRASADA RAO, G., and PALANISAMY, K.R.: 'Walsh stretch matrices and functional differential equations', *IEEE Trans.*, 1982, **AC-27**, pp. 272–276

Part V

OPTIMAL OBSERVER, FILTER, AND CONTROLLER DESIGN

Editor's Comments on Papers 16 Through 22

In addition to being applied to system analysis and identification problems, Walsh function expansions have also been applied with success to the design and implementation of optimal filters and controllers, naturally providing piece-wise constant approximations of the optimal feedback gains. Previously such approximations were determined by prespecifying the structural form of the time varying gains (Kleinman, Fortmann, and Athans, 1968).

The idea of using Walsh function series in optimal control problems was first employed by C. F. Chen and C. H. Hsiao (Paper 16). Essentially the method belongs to the direct variational approach and is very powerful and easily implementable. Some comments on Chen and Hsiao's paper are provided in the note of Gopalsami and Deekshatulu (1976) where it is shown that the same solution can be easily obtained by working directly with piece-wise constant approximations of the rate variables. In his reply Chen indicates that the

Σ matrix, directly derived by Gopalsami and Deekshatulu, is actually the integration operational matrix of the block pulse functions employed in the work of Chen, Tsay, and Wu (1977) (see Paper 7).

In the three papers by Tzafestas and Stavroulakis (Papers 17, 18, and 19) the finite Walsh series expansion method is used for designing approximate (suboptimal) observers and filters incorporated in a closed-loop optimal controller. Both continuous and discrete time systems, time-invariant and time-varying with or without delays are considered. The solution provides a computational algorithm that gives the Walsh expansion coefficients of the state and observer output. Paper 17 actually extends the results of Paper 19 to time delay systems, thus developing a computational algorithm for the determination of the coefficients in the Walsh expansion. It is shown that the given initial conditions of the state and the observer can be used iteratively for computing the state and the observer at any subsequent time. An attempt to extend these results for implementing observer-based controllers of distributed parameter systems using finite spatial measurements is made in Paper 20. The results are derived by using double Walsh series expansions for approximately reconstructing the system state that depends on space and time.

Paper 21 uses the Walsh function variational-synthesis method (see Chen and Hsiao, 1975) for solving the optimal control problem of point reactors as well as space-dependent nuclear reactors. The nonlinearity due to the product of the reactivity input and the power level output is bypassed by employing "power $\times$ reactivity" penalty terms rather than pure reactivity terms in the cost functional.

Mahapatra (1980) considers the case of a distributed-parameter control problem and solves the resulting partial differential Riccati equation using Walsh function expansions. A more complete attempt in this direction was made by Tzafestas (1978) where a combination of Galerkin (in space) and Walsh (in time) expansions was used to implement the solution of the optimal filtering and control problems for a general class of distributed parameter systems. Also the control-like least squares technique of Orner, Salamon, and Yu (1975) was applied to obtain improved stability and accuracy specifications, and the direct Walsh variational technique of Chen and Hsiao (1975) was employed for treating distributed parameter systems described by their integral equation model.

In Paper 22 an embedding approach is followed for systems with delays in state and control, and a computational algorithm via Walsh functions is developed. This algorithm utilizes the concept of Walsh operational matrices for delay and advance. See Palanisamy and Prasada Rao (1983) for a paper in which the minimum energy control of linear systems with time delay satisfying some terminal constraints is treated by means of Walsh functions.

REFERENCES

Chen, C. F., and C. H. Hsiao, 1975, A Walsh Series Direct Method for Solving Variational Problems, *J. Franklin Inst.* **300:**265–280.

Chen, C. F., Y. T. Tsay, and T. T. Wu, 1977, Walsh Operational Matrices for Fractional Calculus and Their Application to Distributed Systems, *J. Franklin Inst.* **303:**267–284.

Gopalsami, N., and B. L. Deekshatulu, 1976, Comments on "Design of Piecewise Constant Gains for Optimal Control via Walsh Functions," *IEEE Trans. Autom. Control* **AC-21:**634–636.

Kleinman, D. L., T. Fortmann, and M. Athans, 1968, On the Design of Linear Systems with Piece-Wise Constant Feedback Gains, *IEEE Trans. Autom. Control* **AC-13:** 354–361.

Mahapatra, G. B., 1980, Solution of Optimal Control Problem of Linear Diffusion Equations via Walsh Functions, *IEEE Trans. Autom. Control* **AC-25:**319–321.

Orner, P., P. Salamon, Jr., and W. Yu, 1975, Least Squares Simulation of Distributed Systems, *IEEE Trans. Autom. Control* **AC-20:**75–83.

Palanisamy, K. R., and G. Prasada Rao, 1983, Minimum Energy Control of Time-Delay Systems via Walsh Functions, *Optim. Control Appl. Methods* **4:**213–226.

Tzafestas, S. G., 1978, Design of Distributed-Parameter Optimal Controllers and Filters via Walsh Galerkin Expansions, in *Control of Distributed Parameter Systems,* (S. Banks and A. Pritchard, eds.,) Pergamon Press, Oxford, England, pp. 201–217.

Reprinted with permission from *IEEE Trans. Autom. Control* **AC-20:**596–602 (1975)

DESIGN OF PIECEWISE CONSTANT GAINS FOR OPTIMAL CONTROL VIA WALSH FUNCTIONS

C. F. Chen and C. H. Hsiao

***Abstract*—This paper presents a technique for determinating time-varying feedback gains of linear systems with quadratic performance criteria. The gains are approximated by the piecewise constants which are naturally determined by Walsh functions. After introducing Walsh functions in the beginning we develop an operational matrix for solving state equations. Then using the operational matrix we solve the piecewise constant gains problem.**

I. Introduction

IT IS known that the control of a linear system with respect to quadratic performance criteria often involves time varying gain design. Kleinman, Fortmann, and Athans proposed a very elegant approach [1]–[3] to the problem by taking practical engineering constraints into consideration: it was done by prespecifying the structural form of time varying feedback gains, while leaving various free parameters to be chosen in an optimal fashion. Regarding the theoretical investigation, their thinking showed certain similarities between their approach and the direct method of Ritz in calculus of variation.

This paper presents a new approach to the optimal problem by using the Walsh functions [4]–[6]. The approach should be classified as a direct method, but is more powerful than the direct methods of Ritz [7], Euler [7], or Gelerkin [8] on the one hand and is simpler than the procedure proposed by Kleinman, Fortmann, and Athans on the other. The piecewise constant gains obtained by the approach are naturally formed, equally distributed and therefore can be easily implemented.

As a start, we will briefly introduce Walsh functions.

II. Fourier Series and Walsh Series

In direct methods of calculus of variation, we often use Fourier series, power series, etc., with coefficients to be determined as the initial step. Our new approach starts with Walsh series.

It is well known that a function which is periodic may be expanded into Fourier series. Analogously speaking, a function, $f(t)$, which is absolutely integrable in $(0,1]$ may be expanded into Walsh series.

Manuscript received August 30, 1974; revised January 29, 1975. Paper recommended by D. L. Kleinman, Chairman of the IEEE S-CS Optimal Systems Committee.
C.-F. Chen is with the Department of Electrical Engineering, University of Houston, Houston, Tex. 77004.
C.-H. Hsiao was with the Department of Electrical Engineering, University of Houston, Houston, Tex. 77004. He is now with Cheng Kung University, Taiwan, China.

$$f(t)=c_0\phi_0(t)+c_1\phi_1(t)+c_2\phi_2(t)+\cdots+c_n\phi_n(t)+\cdots \tag{1}$$

where

$$c_n=\int_0^1 \phi_n(t)f(t)\,dt \tag{2}$$

are determined such that the following integral square error is minimized:

$$\epsilon=\int_0^1\left[f(t)-\sum_{n=0}^{N}c_n\phi_n(t)\right]^2 dt. \tag{3}$$

The Walsh functions $\phi_0(t)$, $\phi_1(t),\cdots,\phi_n(t)$ are a set of square waves which are orthonormal. Fig. 1. shows the Walsh functions from ϕ_0 to ϕ_{11} in the dyadic order. However, the regularity of the set of Walsh functions in the dyadic order is not very easily seen. Each Walsh function can be decomposed into more elementary square waves, or Rademacher functions [10].

Rademacher functions, $r_k(t)$, are a set of square waves of unit height with periods equal to 1, $\frac{1}{2}$, $\frac{1}{4}$, $\frac{1}{8},\cdots,2^{(1-k)}$. Alternatively, we can state that the number of cycles of the square waves of $r_k(t)$ is 2^{k-1}. The first six waves of Rademacher are shown in Fig. 2. It is seen that

$$\begin{aligned}\phi_0(t)&=r_0(t)\\ \phi_1(t)&=r_1(t)\\ &\vdots\\ \phi_i(t)&=(r_k(t))^{b_k}(r_{k-1}(t))^{b_{k-1}}(r_{k-2}(t))^{b_{k-2}}\cdots\end{aligned} \tag{4}$$

where

$$k=[\log_2 i]+1 \tag{5}$$

where [•] means taking the integer part of "•" and $b_k,b_{k-1},\cdots,b_1$ is the binary number expression of i.

To draw the wave form of any Walsh function becomes a trivial matter if we use the above mentioned decomposition technique.

Let us then return to the Walsh coefficient evaluation for a function. Consider a given function $f(t)=t$. It is desired to expand it into Walsh series. Substituting $f(t)=t$

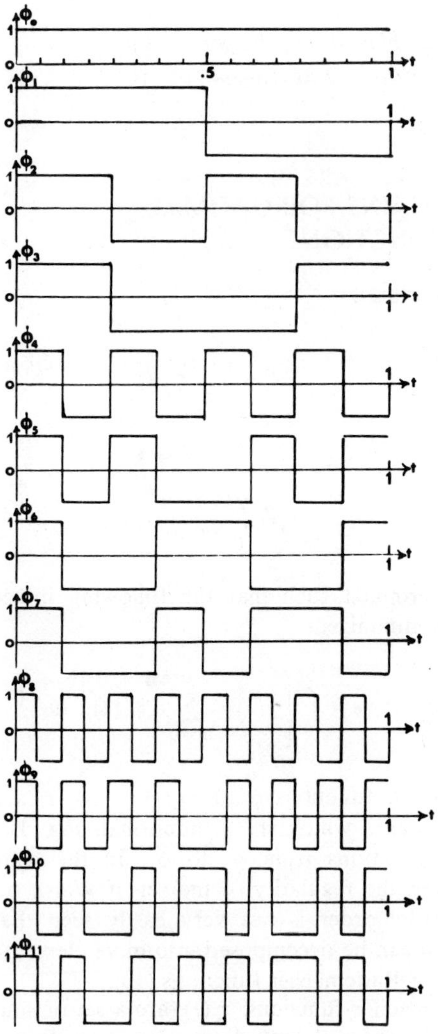

Fig. 1. Dyadic ordered Walsh functions.

into (2), we have

$$f(t)=t=\tfrac{1}{2}\phi_0(t)-\tfrac{1}{4}\phi_1(t)-\tfrac{1}{8}\phi_2(t)$$
$$-\tfrac{1}{16}\phi_4(t)-\tfrac{1}{32}\phi_8(t)-\tfrac{1}{64}\phi_{16}(t)\cdots. \quad (6)$$

The original curve $f(t)=t$ and its Walsh series approximations are shown in Fig. 3. They are stairwise waves. The crudest representation is obtained by taking one term of the Walsh series, or $\frac{1}{2}\phi_0$; the second one consists of two terms $\frac{1}{2}\phi_0-\frac{1}{4}\phi_1$. The figure shows up to a four term approximation which is

$$f(t)=t\cong\tfrac{1}{2}\phi_0-\tfrac{1}{4}\phi_1-\tfrac{1}{8}\phi_2-\tfrac{1}{16}\phi_4. \quad (6a)$$

From the coefficient evaluation process, we easily see that the similarities between Fourier series and Walsh series are obvious.

III. Integration and Operational Matrix

In this section we will derive a method by which we can perform any integration by multiplying a constant matrix.

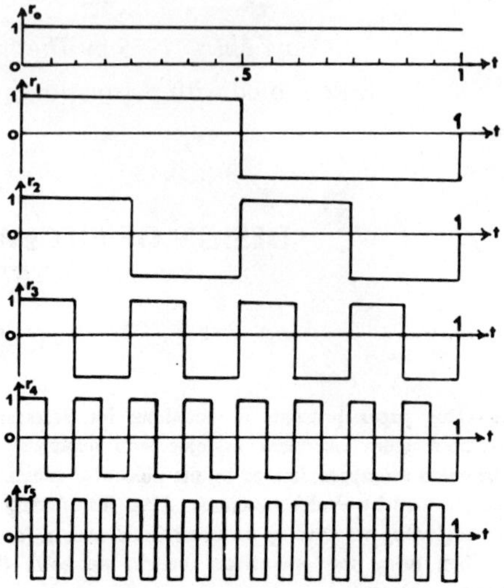

Fig. 2. Rademacher functions.

Let us take $\phi_0,\phi_1,\cdots,\phi_7$ and integrate them; we will have various triangular waves [6]. If we evaluate the Walsh coefficients for these triangular waves, we will easily arrive at the following formula for approximation:

$$\begin{bmatrix}\int\phi_0 dt\\ \int\phi_1 dt\\ \int\phi_2 dt\\ \int\phi_3 dt\\ \int\phi_4 dt\\ \int\phi_5 dt\\ \int\phi_6 dt\\ \int\phi_7 dt\end{bmatrix} = \begin{bmatrix}\frac{1}{2} & \frac{-1}{4} & \frac{-1}{8} & 0 & \frac{-1}{16} & 0 & 0 & 0\\ \frac{1}{4} & 0 & 0 & \frac{-1}{8} & 0 & \frac{-1}{16} & 0 & 0\\ \frac{1}{8} & 0 & 0 & 0 & 0 & 0 & \frac{-1}{16} & 0\\ 0 & \frac{1}{8} & 0 & 0 & 0 & 0 & 0 & \frac{-1}{16}\\ \frac{1}{16} & 0 & 0 & 0 & 0 & 0 & 0 & 0\\ 0 & \frac{1}{16} & 0 & 0 & 0 & 0 & 0 & 0\\ 0 & 0 & \frac{1}{16} & 0 & 0 & 0 & 0 & 0\\ 0 & 0 & 0 & \frac{1}{16} & 0 & 0 & 0 & 0\end{bmatrix}\begin{bmatrix}\phi_0\\ \phi_1\\ \phi_2\\ \phi_3\\ \phi_4\\ \phi_5\\ \phi_6\\ \phi_7\end{bmatrix} \quad (7)$$

or

$$\int \boldsymbol{\phi}_{(8)}dt=\boldsymbol{P}_{(8\times 8)}\boldsymbol{\phi}_{(8)}. \quad (7a)$$

The subscript means the dimension taken. It is preferable to take 2^{Ω}, where Ω is an integer, as a dimension number. Making this choice will enable us to obtain simpler results and to have a much easier calculation.

It is noted that

$$\int\phi_0 dt=t;$$

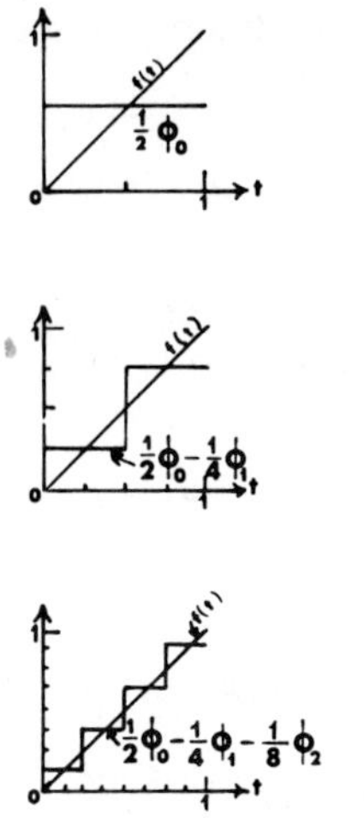

Fig. 3. Expanding a ramp function into Walsh functions.

therefore the Walsh coefficients of $\int \phi_0 dt$ is found in (6a) which is simply the first row of (7) times the $\phi(t)$ vector.

Equation (7) is for $m=8$, a general formula $P_{(m\times m)}$ can be written as follows:

$$P_{(m\times m)} = \left[\begin{array}{cc|c|c} \begin{array}{cc} \frac{1}{2} & \\ & \ddots \end{array} & \frac{-2}{m}I_{(m/8)} & \frac{-1}{m}I_{(m/4)} & \\ \frac{2}{m}I_{(m/8)} & 0_{(m/8)} & & \frac{-1}{2m}I_{(m/2)} \\ \hline \multicolumn{2}{c|}{\frac{1}{m}I_{(m/4)}} & 0_{(m/4)} & \\ \hline \multicolumn{3}{c|}{\frac{1}{2m}I_{(m/2)}} & 0_{(m/2)} \end{array}\right]. \quad (8)$$

It is interesting to note that if (8) is partitioned into four parts as shown, the upper left part of $P_{(m\times m)}$ is identical to $P_{(\overline{m/2}\times\overline{m/2})}$; and the upper left corner of $P_{(\overline{m/2}\times\overline{m/2})}$ is $P_{(\overline{m/4}\times\overline{m/4})}$. Therefore, this regularity of the structure of the P matrix enables us to write the mth enlarged matrix to any dimension, if the dimension number is restricted to 2^{Ω} where Ω is an integer number.

IV. State Equation Solution by Kronecker Product Formula

Consider the following state equation:

$$\dot{x} = Ax + Bu, \qquad x(0) = x_0 \qquad (9)$$

where x is a state vector of n components and u is an input vector of l components. A and B are $n\times n$ and $n\times l$ matrices, respectively. We would like to establish a procedure for solving the state equations via Walsh series.

First of all, we assume the rate variable vector $\dot{x}$ to be a set of m-term Walsh series whose $n\times m$ coefficients are to be determined. Let

$$\begin{bmatrix} \dot{x}_1 \\ \dot{x}_2 \\ \vdots \\ \dot{x}_n \end{bmatrix} = \begin{bmatrix} c_{10} & c_{11} & c_{12} & \cdots & c_{1(m-1)} \\ c_{20} & c_{21} & c_{22} & \cdots & c_{2(m-1)} \\ \vdots & \vdots & \vdots & \cdots & \vdots \\ c_{n0} & c_{n1} & c_{n2} & \cdots & c_{n(m-1)} \end{bmatrix} \begin{bmatrix} \phi_0 \\ \phi_1 \\ \vdots \\ \phi_{(m-1)} \end{bmatrix}. \qquad (10)$$

We write each column as a vector and define the rectangular matrix as C. Then (10) becomes

$$\dot{x} = [c_0, c_1, \cdots, c_{m-1}]\phi \triangleq C\phi. \qquad (10a)$$

It should be noted that this initial step is quite different than that of the regular procedure of the series solution of a differential equation. We assume the rate variable $\dot{x}$ as an undetermined vector series, instead of assuming the state variable itself.

The state variable x may be obtained by integration:

$$x(t) = C\int_0^t \phi(s)ds + x_0. \qquad (11)$$

The integration can be performed approximately by using the P matrix.

$$\int_0^t \phi(s)ds = P\phi(t). \qquad (7a)$$

Also the input vector can also be expressed by Walsh series:

$$u = H\phi \qquad (12)$$

where H is a $l\times m$ matrix; the determination of which can be achieved by the techniques illustrated in Section II.

Substituting (10a), (11), (7a), and (12) into (9) yields

$$C\phi = A(CP\phi + x_0) + BH\phi. \qquad (13)$$

Ax_0 can be written as the product of a matrix G and the vector ϕ.

$$Ax_0 = Ax_0\phi_0 = [Ax_0, \underbrace{0, 0, \cdots, 0}_{(m-1)\text{columns}}]\phi \triangleq G\phi. \qquad (14)$$

Finally we have

$$C = ACP + G + BH. \qquad (15)$$

The last two terms of (15) can be combined by letting

$$G + BH \triangleq K. \qquad (16)$$

Then

$$C = ACP + K. \qquad (15a)$$

If we rearrange the $n \times m$ matrix C as an nm-vector c by changing its first column into the first n components of the vector; the second column, the second n components of the vector, etc.; and rearrange K in the same manner; we finally obtain an even simpler form in terms of a Kronecker product for (15a);

$$c = [A \otimes P']c + k \tag{17}$$

where $A \otimes P'$ is a Kronecker product defined as

$$A \otimes P' \triangleq \begin{bmatrix} p_{11}A & p_{21}A & \cdots & p_{m1}A \\ p_{12}A & p_{22}A & \cdots & p_{m2}A \\ \cdots & \cdots & \cdots & \cdots \\ p_{1m}A & p_{2m}A & \cdots & p_{mm}A \end{bmatrix} \tag{18}$$

and P' is the transposition of P.

The solution of c comes from (17) directly,

$$c = [I - A \otimes P']^{-1}k. \tag{19}$$

Once c has been decided, the Walsh series representation for the rate variable is determined. The state variable vector is then found by substitution.

$$x = CP\phi + x_0. \tag{20}$$

When m is large, the computation of (19) becomes a difficult problem, we take advantage of the special pattern of matrix P and develop an algorithm for calculation. The details are shown in the Appendix.

A high-order differential equation with constant coefficients can always be written as a state equation. Therefore (19) and (20) enable us to solve linear time invariant systems elegantly and completely. Of course, the answer is in terms of Walsh functions.

A subroutine WALDE has been written in Fortran language for solving an nth-order differential equation.

V. Optimal Problem

The optimal control of a linear time-invariant system

$$\dot{x} = Ax + Bu \tag{21}$$

with quadratic performance index

$$\delta = \tfrac{1}{2}\int_0^{t_f}(x'Qx + u'Ru)\,dt \tag{22}$$

is well known to be

$$u^* = R^{-1}B'p(t) \tag{23}$$

where $p(t)$ satisfies the following canonical equation:

$$\begin{bmatrix} \dot{x} \\ \dot{p} \end{bmatrix} = \begin{bmatrix} A & BR^{-1}B' \\ Q & -A' \end{bmatrix}\begin{bmatrix} x \\ p \end{bmatrix} \tag{24}$$

and the boundary conditions are specified as

$$x(0) = x_0 \tag{25}$$

$$p(t_f) = 0. \tag{26}$$

Equation (26) is the transversality condition.

It is more convenient to change the independent variable by defining

$$\tau = t_f - t; \tag{27}$$

then (24) becomes

$$\begin{bmatrix} \dot{x}(\tau) \\ \dot{p}(\tau) \end{bmatrix} = \begin{bmatrix} -A & -BR^{-1}B' \\ -Q & A' \end{bmatrix}\begin{bmatrix} x(\tau) \\ p(\tau) \end{bmatrix} \triangleq -M\begin{bmatrix} x(\tau) \\ p(\tau) \end{bmatrix}. \tag{28}$$

The transition matrix of (28) is

$$e^{-Mt} = \begin{bmatrix} \eta_{11}(\tau) & \eta_{12}(\tau) \\ \eta_{21}(\tau) & \eta_{22}(\tau) \end{bmatrix}. \tag{29}$$

Since $p(\tau = 0) = 0$, the solution of (28) can be written as

$$x(\tau) = \eta_{11}(\tau)x(\tau = 0) \tag{30}$$

$$p(\tau) = \eta_{21}(\tau)x(\tau = 0). \tag{31}$$

From (30), we have

$$x(\tau = 0) = \eta_{11}^{-1}(\tau)x(\tau). \tag{30a}$$

Substituting (30a) into (31) yields

$$p(\tau) = \eta_{21}(\tau)\eta_{11}^{-1}(\tau)x(\tau). \tag{31a}$$

Then optimal control is reduced to

$$\begin{aligned} u^*(t) &= R^{-1}B'\eta_{21}(t_f - t)\eta_{11}^{-1}(t_f - t)x(t_f - t) \\ &\triangleq -L(t_f - t)x(t_f - t) \end{aligned} \tag{32}$$

where $L(t_f - t)$ is the optimal feedback gain matrix.

VI. Walsh Series Solution to the Problem

Because Walsh series is defined in the 0 to 1 interval we normalize the problem first by using

$$\lambda = \tau / t_f; \tag{33}$$

then (28) becomes

$$\begin{bmatrix} \dot{x}(\lambda) \\ \dot{p}(\lambda) \end{bmatrix} = -t_f M\begin{bmatrix} x(\lambda) \\ p(\lambda) \end{bmatrix}, \quad 0 \leqslant \lambda < 1. \tag{34}$$

Next, assume $\dot{x}(\lambda)$ and $\dot{p}(\lambda)$ to be expanded into Walsh series whose coefficients are to be determined.

$$\begin{bmatrix} \dot{x}(\lambda) \\ \dot{p}(\lambda) \end{bmatrix} = C\phi(\lambda) \tag{35}$$

where C is an $2n \times m$ matrix, and $\phi(\lambda)$, an m-vector.

Then (7a) is applied to perform integration on (35):

$$\begin{bmatrix} x(\lambda) \\ p(\lambda) \end{bmatrix} = CP\phi(\lambda) + \begin{bmatrix} x(\lambda=0) \\ 0_n \end{bmatrix}. \quad (36)$$

Substituting (36) and (35) into (34) gives

$$C\phi(\lambda) = -t_f M \left\{ CP + \left[\begin{matrix} x(\lambda=0) \\ 0_n \end{matrix}, \; 0_{2n}, \cdots, 0_{2n} \right] \right\} \phi(\lambda). \quad (37)$$

Defining k as

$$k = \begin{bmatrix} -t_f A x(\lambda=0) \\ -t_f Q x(\lambda=0) \\ 0_{2n} \\ \cdot \\ \cdot \\ \cdot \\ 0_{2n} \end{bmatrix} \quad (38)$$

then (37) is simplified into

$$c = \left[I + t_f M \otimes P' \right]^{-1} k. \quad (37a)$$

Solving (37a) for c, we obtain the Walsh coefficients of the rate variable $\dot{x}(\lambda)$ and the rate co-state variable $\dot{p}(\lambda)$. Then substitute them into (36). The answer of $x(\lambda)$ and $p(\lambda)$ in terms of Walsh function are finally found.

VII. Example of Feedback Gain Determination

Let us consider Kleinman's example [1], [2],

$$\dot{x}(t) = Ax(t) + Bu(t)$$

$$= \begin{bmatrix} -1 & 0 & 0 \\ 0 & 0 & -2 \\ 0 & 2 & 0 \end{bmatrix} x(t) + \begin{bmatrix} 2 \\ 2 \\ 1 \end{bmatrix} u(t), \qquad x(0) = x_0. \quad (39)$$

The performance index is specified as

$$\mathcal{S} = \frac{1}{2} \int_0^2 (x'Qx + u'Ru)\,dt$$

$$= \frac{1}{2} \int_0^2 \left\{ x' \begin{bmatrix} 2 & -2 & 0 \\ -2 & 2 & 0 \\ 0 & 0 & 0 \end{bmatrix} x + 2u^2 \right\} dt. \quad (40)$$

Since $t_f = 2 \neq 1$, we need to normalize the time scale.

$$\lambda = (t_f - t)/t_f = 1 - 0.5t. \quad (41)$$

The canonical equation then becomes

$$\begin{bmatrix} \dot{x}(\lambda) \\ \dot{p}(\lambda) \end{bmatrix} = \begin{bmatrix} 2 & 0 & 0 & -4 & -4 & 2 \\ 0 & 0 & 4 & -4 & -4 & 2 \\ 0 & -4 & 0 & 2 & 2 & -1 \\ -4 & 4 & 0 & -2 & 0 & 0 \\ 4 & -4 & 0 & 0 & 0 & -4 \\ 0 & 0 & 0 & 0 & 4 & 0 \end{bmatrix} \begin{bmatrix} x(\lambda) \\ p(\lambda) \end{bmatrix}. \quad (42)$$

Assuming that the Walsh expansion of $\dot{x}(\lambda)$ and $\dot{p}(\lambda)$ may be approximately expressed with 16 terms, we have

$$\begin{bmatrix} \dot{x}(\lambda) \\ \dot{p}(\lambda) \end{bmatrix} = C\phi(\lambda) \quad (43)$$

where C is a 6×16 matrix to be determined and $\phi(\lambda)$ a Walsh vector of 16 components. Letting $x_1(\lambda=0)=1$ and all other initial conditions $(\lambda=0)$ equal to zero, and applying (37a), we obtain the first column of $\eta_{11}(\lambda)$ and $\eta_{21}(\lambda)$. Then, letting $x_2(\lambda=0)=1$, we obtain the second column of $\eta_{11}(\lambda)$ and $\eta_{21}(\lambda)$; and letting $x_3(\lambda=0)=1$, we obtain the third column. The optimal gain is evaluated with (32), which is a multiplication of matrices. The waveform is shown at the lower right hand corner of Fig. 4.

Taking the average for each pair of consecutive values of the above obtained optimal gains with 16 subintervals, we obtain a new set of piecewise feedback gain for 8 subintervals. Continuing this procedure, we obtain constant feedback gain for 4, 2, and 1 subintervals. All of them are shown in Fig. 4.

The cost matrix associated with the gain matrix $L(t)$ is a unique solution of the following:

$$\dot{V}(t) = -V(t)\left[A - BL(t)\right] - \left[A - BL(t)\right]'V(t) - Q - L'(t)RL(t) \quad (44)$$

$$V(t_f) = 0 \quad (44a)$$

and μ is defined as the trace of $V(0)$, namely

$$\mu = \mathrm{tr}\left[V(0)\right]. \quad (45)$$

For a detailed explanation of $V(t)$ and μ, the reader is referred to [1].

We evaluate $V(0)$ and μ for each control law we have previously derived. These μ's are tabulated below.

Number of Subintervals	16	8	4	2	1
μ	1.7125	1.7115	1.7114	1.7471	1.9063

These values are slightly larger than those of Kleinman, but our method is much simpler.

VIII. Conclusion

A Walsh function method for determining the optimal piecewise constant gains for a linear system is established. The basic formula is (37a). Compared with Kleinman, Fortman and Athans method, the proposed approach is much simpler in analysis and easier in implementation.

Appendix

Algorithm for Solving C Via the Kronecker Product Formula

In this Appendix we derive a recursive algorithm to solve C from (19) instead of inversing $[I - A \otimes P']$ directly. Let us illustrate the procedures for $m = 2^3 = 8$.

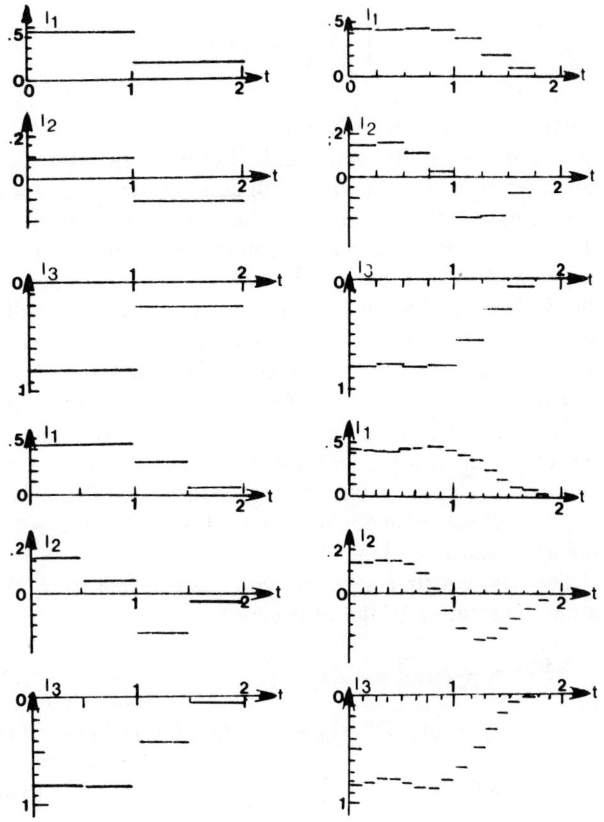

Fig. 4. Solving Kleinman's problem via the Walsh function method.

Step 1: Equation (15a) may be rewritten explicitly for $m=8$.

$$\begin{bmatrix} \mathbf{c}_0 \\ \mathbf{c}_1 \\ \mathbf{c}_2 \\ \mathbf{c}_3 \\ \mathbf{c}_4 \\ \mathbf{c}_5 \\ \mathbf{c}_6 \\ \mathbf{c}_7 \end{bmatrix} = \begin{bmatrix} (\frac{1}{2})A & (\frac{1}{4})A & (\frac{1}{8})A & 0 & (\frac{1}{16})A & 0 & 0 & 0 \\ (\frac{-1}{4})A & 0 & 0 & (\frac{1}{8})A & 0 & (\frac{1}{16})A & 0 & 0 \\ (\frac{-1}{8})A & 0 & 0 & 0 & 0 & 0 & (\frac{1}{16})A & 0 \\ 0 & (\frac{-1}{8})A & 0 & 0 & 0 & 0 & 0 & (\frac{1}{16})A \\ (\frac{-1}{16})A & 0 & 0 & 0 & 0 & 0 & 0 & 0 \\ 0 & (\frac{-1}{16})A & 0 & 0 & 0 & 0 & 0 & 0 \\ 0 & 0 & (\frac{-1}{16})A & 0 & 0 & 0 & 0 & 0 \\ 0 & 0 & 0 & (\frac{-1}{16})A & 0 & 0 & 0 & 0 \end{bmatrix} \cdot \begin{bmatrix} \mathbf{c}_0 \\ \mathbf{c}_1 \\ \mathbf{c}_2 \\ \mathbf{c}_3 \\ \mathbf{c}_4 \\ \mathbf{c}_5 \\ \mathbf{c}_6 \\ \mathbf{c}_7 \end{bmatrix} + \begin{bmatrix} \mathbf{k}_0 \\ \mathbf{k}_1 \\ \mathbf{k}_2 \\ \mathbf{k}_3 \\ \mathbf{k}_4 \\ \mathbf{k}_5 \\ \mathbf{k}_6 \\ \mathbf{k}_7 \end{bmatrix}. \tag{A1}$$

It is obvious that $\mathbf{c}_4$, $\mathbf{c}_5$, $\mathbf{c}_6$, and $\mathbf{c}_7$ may be expressed in terms of $\mathbf{c}_0$, $\mathbf{c}_1$, $\mathbf{c}_2$, and $\mathbf{c}_3$.

$$\begin{bmatrix} \mathbf{c}_4 \\ \mathbf{c}_5 \\ \mathbf{c}_6 \\ \mathbf{c}_7 \end{bmatrix} = \begin{bmatrix} (\frac{-1}{16})A & 0 & 0 & 0 \\ 0 & (\frac{-1}{16})A & 0 & 0 \\ 0 & 0 & (\frac{-1}{16})A & 0 \\ 0 & 0 & 0 & (\frac{-1}{16})A \end{bmatrix} \cdot \begin{bmatrix} \mathbf{c}_0 \\ \mathbf{c}_1 \\ \mathbf{c}_2 \\ \mathbf{c}_3 \end{bmatrix} + \begin{bmatrix} \mathbf{k}_4 \\ \mathbf{k}_5 \\ \mathbf{k}_6 \\ \mathbf{k}_7 \end{bmatrix}. \tag{A2}$$

In order to keep a consistent notation, we may define

$$G_3 \triangleq I \tag{A3}$$

$$R_3 \triangleq -(1/2^{3+1})G_3^{-1}A; \tag{A4}$$

then

$$\mathbf{c}_{i+4} = R_3 \mathbf{c}_i + \mathbf{k}_{i+4}, \qquad i=0,1,2,3. \tag{A5}$$

Substituting (A2) into (A1), we have

$$\begin{bmatrix} \mathbf{c}_0 \\ \mathbf{c}_1 \\ \mathbf{c}_2 \\ \mathbf{c}_3 \end{bmatrix} = \begin{bmatrix} (\frac{1}{2})A + (\frac{-1}{256})A^2 & (+\frac{1}{4})A & (\frac{1}{8})A & 0 \\ (\frac{-1}{4})A & (\frac{-1}{256})A^2 & 0 & (\frac{1}{8})A \\ (\frac{-1}{8})A & 0 & (\frac{-1}{256})A^2 & 0 \\ 0 & (\frac{-1}{8})A & 0 & (\frac{-1}{256})A^2 \end{bmatrix} \cdot \begin{bmatrix} \mathbf{c}_0 \\ \mathbf{c}_1 \\ \mathbf{c}_2 \\ \mathbf{c}_3 \end{bmatrix} + \begin{bmatrix} \mathbf{k}_0 + (\frac{1}{16})A\mathbf{k}_4 \\ \mathbf{k}_1 + (\frac{1}{16})A\mathbf{k}_5 \\ \mathbf{k}_2 + (\frac{1}{16})A\mathbf{k}_6 \\ \mathbf{k}_3 + (\frac{1}{16})A\mathbf{k}_7 \end{bmatrix}. \tag{A6}$$

Now the diagonal elements are no longer null matrices. We may define the new diagonal matrix as F_3, and the new inputs as $\mathbf{k}_{i,\text{III}}$.

$$F_3 \triangleq \tfrac{-1}{256}A^2 = \frac{1}{2^{3+1}}AR_3 \tag{A7}$$

$$\mathbf{k}_{i,\text{III}} \triangleq \mathbf{k}_i + \frac{1}{2^{3+1}}AG_3^{-1}\mathbf{k}_{i+4}, \qquad i=0,1,2,3. \tag{A8}$$

Step 2: Eliminating $\mathbf{c}_2$, $\mathbf{c}_3$ from the lower half of (A6), we have

$$\begin{bmatrix} \mathbf{c}_2 \\ \mathbf{c}_3 \end{bmatrix} = \begin{bmatrix} (\frac{-1}{8})(I + \frac{1}{256}A^2)^{-1}A & 0 \\ 0 & (-\frac{1}{8})(I + \frac{1}{256}A^2)^{-1}A \end{bmatrix} \cdot \begin{bmatrix} \mathbf{c}_0 \\ \mathbf{c}_1 \end{bmatrix} + (I + \tfrac{1}{256}A^2)^{-1} \begin{bmatrix} \mathbf{k}_{2,\text{III}} \\ \mathbf{k}_{3,\text{III}} \end{bmatrix}. \tag{A9}$$

This may be put into recursive form by defining

$$G_2 \triangleq I - F_3 = I + \tfrac{1}{256}A^2 \tag{A10}$$

$$R_2 \triangleq \frac{1}{2^{2+1}}G_2^{-1}A = (-\tfrac{1}{8})(I + \tfrac{1}{256}A^2)^{-1}A; \tag{A11}$$

then

$$c_{i+2} = R_2 c_i + G_2^{-1}\mathbf{k}_{(i+2),\text{III}}, \qquad i=0,1. \tag{A12}$$

Substituting (A9) and (A12) into (A6), we have

$$\begin{bmatrix}\mathbf{c}_0\\ \mathbf{c}_1\end{bmatrix} = \begin{bmatrix} (\frac{1}{2})\boldsymbol{A} + \boldsymbol{F}_3\frac{1}{64}\boldsymbol{A}\left(\boldsymbol{I}+\frac{1}{256}\boldsymbol{A}^2\right)^{-1}\boldsymbol{A} & (\frac{1}{4})\boldsymbol{A} \\ (\frac{-1}{4})\boldsymbol{A} & \boldsymbol{F}_3\frac{-1}{64}\boldsymbol{A}\left(\boldsymbol{I}+\frac{1}{256}\boldsymbol{A}^2\right)^{-1}\boldsymbol{A}\end{bmatrix} \cdot \begin{bmatrix}\mathbf{c}_0\\ \mathbf{c}_1\end{bmatrix} + \begin{bmatrix}\mathbf{k}_{0,\mathrm{III}} + \frac{1}{8}\boldsymbol{A}\boldsymbol{G}_2^{-1}\boldsymbol{k}_{2,\mathrm{III}} \\ \mathbf{k}_{1,\mathrm{III}} + \frac{1}{8}\boldsymbol{A}\boldsymbol{G}_2^{-1}\boldsymbol{k}_{3,\mathrm{III}}\end{bmatrix}. \tag{A13}$$

For recursive operation, we should define again

$$\boldsymbol{F}_2 \triangleq \boldsymbol{F}_3 + \frac{1}{2^{2+1}}\boldsymbol{A}\boldsymbol{R}_2 = -\frac{1}{256}\boldsymbol{A}^2 - \frac{1}{64}\boldsymbol{A}\left(\boldsymbol{I}+\frac{1}{256}\boldsymbol{A}^2\right)^{-1}\boldsymbol{A} \tag{A14}$$

$$\mathbf{k}_{i,\mathrm{II}} \triangleq \mathbf{k}_{i,\mathrm{III}} + \frac{1}{2^{2+1}}\boldsymbol{A}\boldsymbol{G}_2^{-1}\mathbf{k}_{i+2,\mathrm{III}}, \qquad i=0,1. \tag{A15}$$

Step 3: Eliminating $\mathbf{c}_1$ from (A13) and (A14), we have

$$\mathbf{c}_1 = \left(\frac{-1}{4}\right)(\boldsymbol{I}-\boldsymbol{F}_2)^{-1}\boldsymbol{A}\mathbf{c}_0 + (\boldsymbol{I}-\boldsymbol{F}_2)^{-1}\mathbf{k}_{i,\mathrm{II}} \triangleq \boldsymbol{R}_1\mathbf{c}_0 + \boldsymbol{G}_1^{-1}\mathbf{k}_{i,\mathrm{II}} \tag{A16}$$

where $\boldsymbol{G}_1$, $\boldsymbol{R}_1$ are defined in the same form as $\boldsymbol{G}_2$, $\boldsymbol{R}_2$, $\boldsymbol{G}_3$, $\boldsymbol{R}_3$.

$$\boldsymbol{G}_1 \triangleq \boldsymbol{I} - \boldsymbol{F}_2 \tag{A17}$$

$$\boldsymbol{R}_1 \triangleq -\frac{1}{2^{i+1}}\boldsymbol{G}_1^{-1}\boldsymbol{A}. \tag{A18}$$

Substituting (A16) and (A14) into the upper half of (A13), we have

$$\mathbf{c}_0 = \left[(\tfrac{1}{2})\boldsymbol{A} + \boldsymbol{F}_2 + (\tfrac{1}{4})\boldsymbol{A}\boldsymbol{R}_1\right]\mathbf{c}_0 + \mathbf{k}_{0,\mathrm{II}} + (\tfrac{1}{4})\boldsymbol{A}\boldsymbol{G}_1^{-1}\mathbf{k}_{i,\mathrm{II}}.$$

Therefore

$$\mathbf{c}_0 = \boldsymbol{G}_0^{-1}\mathbf{k}_{0,\mathrm{I}} \tag{A19}$$

where

$$\boldsymbol{G}_0 \triangleq \boldsymbol{I} - (\tfrac{1}{2})\boldsymbol{A} - \boldsymbol{F}_1 \tag{A20}$$

$$\boldsymbol{F}_1 \triangleq \boldsymbol{F}_2 + \frac{1}{2^{1+1}}\boldsymbol{A}\boldsymbol{R}_1 \tag{A21}$$

$$\mathbf{k}_{0,\mathrm{I}} \triangleq \mathbf{k}_{0,\mathrm{II}} + \frac{1}{2^{1+1}}\boldsymbol{A}\boldsymbol{G}_1^{-1}\mathbf{k}_{1,\mathrm{II}}. \tag{A22}$$

Step 4: $\mathbf{c}_1$ is obtained by substituting $\mathbf{c}_0$ into (A16). Similarly, $\mathbf{c}_2$, $\mathbf{c}_3$, and $\mathbf{c}_4$, $\mathbf{c}_5$, $\mathbf{c}_6$, $\mathbf{c}_7$ are obtained by applying (A12) and (A5), respectively. This completes the derivation of the algorithm for solving $\boldsymbol{c}$ from (28), with $m=8$.

In general, for $m=2^\alpha$, α is any positive integer, we start with

$$\boldsymbol{G}_\alpha = \boldsymbol{I}, \quad \boldsymbol{R}_a = -\frac{1}{2^{\alpha+1}}\boldsymbol{A}, \quad \boldsymbol{F}_\alpha = \frac{1}{2^{\alpha+1}}\boldsymbol{A}\boldsymbol{R}_\alpha$$

$$\boldsymbol{F}_{\alpha+1} = \mathbf{0}, \quad k_{i,\alpha+1} = k_i. \tag{A23}$$

Then, we can calculate $\boldsymbol{R}_\beta$ and $\mathbf{k}_{i,\beta}$, $\beta=\alpha,\alpha-1,\cdots,1$, $i=0,1,\cdots,2^{\beta-1}$, from the following recursive formulas:

$$\boldsymbol{G}_\beta = \boldsymbol{I} - \boldsymbol{F}_{\beta+1}$$

$$\boldsymbol{R}_\beta = -2^{-\beta-1}\boldsymbol{G}_\beta^{-1}\boldsymbol{A}$$

$$\boldsymbol{F}_\beta = \boldsymbol{F}_{\beta+1} + 2^{-\beta-1}\boldsymbol{A}\boldsymbol{R}_\beta$$

$$\mathbf{k}_{i,\beta} = \mathbf{k}_{i,\beta+1} + 2^{-\beta-1}\boldsymbol{A}\boldsymbol{G}_\beta^{-1}\mathbf{k}_{(i+2^{\beta-1}),\beta+1}. \tag{A24}$$

Then we obtain $\mathbf{c}_0$:

$$\mathbf{c}_0 = \boldsymbol{G}_0^{-1}\mathbf{k}_{0,\mathrm{I}}, \boldsymbol{G}_0 = \mathbf{1} - \tfrac{1}{2}\boldsymbol{A} - \boldsymbol{F}_1. \tag{A25}$$

All the other vectors $\mathbf{c}_j, j=0,1,\cdots,(m-1)$ are found by substituting them with

$$\mathbf{c}_{i+2}\beta - 1 = \boldsymbol{R}_\beta\mathbf{c}_i + \boldsymbol{G}_\beta^{-1}\mathbf{k}_{(i+2^{\beta-1}),\beta+1},$$

$$\beta = 1,2,\cdots,\alpha \quad i=0,1,\cdots,(2^{\beta-1}-1). \tag{A26}$$

In the above algorithm, we always work with matrices $(\boldsymbol{G},\boldsymbol{F},\boldsymbol{R})$ of $n\times n$. Therefore there is no need to operate with larger matrices. For instance, if we wish to apply 128 Walsh functions to solve a 6th-order differential equation, we only work with matrices 6×6. The calculation of matrix inverse of $[\boldsymbol{I}-\boldsymbol{A}\otimes\boldsymbol{P}']$, which is of 768×768, is avoided. Therefore, we have saved computing time and storage. In addition, we have reduced round-off errors significantly.

References

[1] D. L. Kleinman, T. Fortmann, and M. Athans, "On the design of linear systems with piecewise-constant feedback gains," *IEEE Trans. Automat. Contr.*, vol. AC-13, pp. 354–361, Aug. 1968.

[2] D. L. Kleinman, and M. Athans, "The design of suboptimal linear time-varying systems," *IEEE Trans. Automat. Contr.*, vol. AC-13, pp. 150–159, Apr. 1968.

[3] T. Fortmann, "Optimal piece constant solutions of the linear regulator problems," MIT Electron. Syst. Lab. Rep. ESL-326, Oct. 1967.

[4] *Proc. Walsh Functions Symp.*, Nav. Res. Lab., Washington, D. C., 1970, 1971, 1972.

[5] N. F. Harmuth, "Application of Walsh functions in communication," *IEEE Spectrum*, pp. 82–91, Nov. 1969.

[6] M. S. Corrington, "Solution of differential and integral equations with Walsh functions," *IEEE Trans. Circuit Theory*, vol. CT-20, pp. 470–476., Sept. 1973.

[7] L. E. Elsgoe, *Calculus of Variations*. New York: Pergamon, 1961.

[8] R. S. Schechter, *The Variation Method in Engineering*. New York: McGraw-Hill, 1967, pp. 23–26.

[9] J. L. Walsh, "A closed set of orthogonal functions," *Amer. J. Math.*, vol. 45, pp. 5–24, 1923.

[10] H. Rademacher, "Einge Satze Uber Reihen Von Allgemen Orthogonal Function," *Ann. Math.*, vol. 87, pp. 712–738, 1922.

[11] M. Athans, "The matrix minimum principle," *Inform. Contr.*, Nov. 1967.

17

Reprinted from *Int. J. Syst. Sci.* **9**:287–299 (1978)

A Walsh series approach to time-delay control system observer design

P. STAVROULAKIS† and S. TZAFESTAS‡

The optimal control problem of time-delay systems with partial state measurements is studied through the application of a Walsh function expansion. The main purpose of the paper is to provide a computational technique for determining the expansion coefficient matrices for both the system state and the observer output. It is shown that the known initial conditions of the state and the observer can be used iteratively to compute the state and the observer at any subsequent time and thus implement the optimal controller. Also, by subdividing the time interval of interest, the procedure is simplified such that it does not involve any matrix inversion.

1. Introduction

Time-delay differential equations have been used over the years in modelling various economic, biological, traffic-flow, chemical, and man–machine systems.

The optimal control of this type of system (a system in which the time derivative of the state depends on present as well as past values of the process) by a quadratic cost functional has been extensively studied by many authors in the vast literature of the subject. The most significant results used in this paper are those of Krasovskii (1962), Ross and Flügge-Lotz (1969), Koivo (1972) and Schmitendorf (1973). Ross and Flügge-Lotz show that there exists a direct parallel between the results for ordinary differential equations and those for time-delay systems. That is, the optimal control for time-delay systems with a quadratic cost functional is linear with respect to the state of the system, and includes compensation for the hereditary effects of the systems (linear functional). This ' state feedback ' feature of the optimal control has created the impetus for developing an observer theory for time-delay control systems (Bhat and Koivo 1976, Hewer and Nazaroff 1973) similar to that developed by Luenberger (1964) for finite dimensional linear systems. As shown in by Bhat and Koivo, the special decomposition of the infinitesimal generator of the semi-group corresponding to the delay differential equations of the delay system can be used to develop an asymptotically stable observer. It is also shown that the spectral separation property of the observer is also valid for delay systems. The spectral separation property states that the eigenvalues of the closed-loop system implemented via the observer is the union of the eigenvalues of the system with state feedback and the eigenvalues of the observer.

In a previous paper (Stavroulakis and Tzafestas 1977) the authors presented a method for developing computational algorithms using Walsh functions to study observer and filter designs in optimal control problems. The objective of this paper is to extend these results to time-delay systems, developing computational techniques for studying the optimal control of time, delay systems with partial state observations.

Received 22 February 1977
† Bell Laboratories, Holmdel, New Jersey 07733, U.S.A.
‡ Control Systems Laboratory, University of Patras, Patras, Greece, and NRC ' Demokritos ' Aghia Paraskevi, Athens, Greece.

2. Preliminaries

Consider the time-invariant time-delay system given by

$$\dot{\mathbf{x}}(t)=A_0\mathbf{x}(t)+A_1\mathbf{x}(t-h)+B\mathbf{u}(t), \quad t_0\leqslant t\leqslant T \tag{1 a}$$

$$\mathbf{y}(t)=C\mathbf{x}(t) \tag{1 b}$$

with

$$\mathbf{x}(\xi)=\mathbf{x}_0(\xi), \quad \xi\in[t_0-h,t_0] \tag{1 c}$$

where $\mathbf{x}(t)$ is an n-vector A_0, A_1 are $n\times n$ constant matrices, $\mathbf{y}$ is an m-vector which symbolizes the output of the system, C is an $m\times n$ constant matrix, and $h>0$ is the constant delay. The function $\mathbf{x}_0(\xi)$ is a known continuous function in the interval $[-h,0]$.

In the interval $[t_0,T]$ we obtain $\mathbf{x}(t)$ from

$$\mathbf{x}(t)=\int_{t_0}^{t}\dot{\mathbf{x}}(\tau')\,d\tau'+\mathbf{x}(t_0), \quad t\in[t_0,T] \tag{2}$$

and

$$\lim_{t\to t_0+h}\mathbf{x}(t-h)=\mathbf{x}_0(t_0) \tag{3}$$

If we define

$$\dot{\mathbf{x}}(t)=\begin{cases}0, & t\in[t_0-h,t_0)\\ \dot{\mathbf{x}}(t), & t\in[t_0,T]\end{cases} \tag{4}$$

then letting

$$\dot{\mathbf{x}}(t)=X\mathbf{w}\left(\frac{t-t_0+h}{T-t_0+h}\right), \quad t\in[t_0-h,T] \tag{5}$$

where

$$X=\int_{t_0-h}^{T}\dot{\mathbf{x}}(\tau)\mathbf{w}'\left(\frac{\tau-t_0+h}{T-t_0+h}\right)d\tau \tag{6}$$

we observe that, as t varies in the interval $[t_0-h,T]$, $\mathbf{w}\left(\frac{t-t_0+h}{T-t_0+h}\right)$ varies in the interval $[0,1]$, where $\mathbf{w}(\cdot)$ are the first k Walsh functions and X is an $n\times k$ matrix. If we make a change of variable

$$\sigma=\frac{\tau-t_0+h}{T-t_0+h}, \quad \tau\in[t_0-h,T]$$

eqn. (6) becomes

$$X=\frac{1}{T-t_0+h}\int_0^1\dot{\mathbf{x}}((T-t_0+h)\sigma+t_0-h)\mathbf{w}'(\sigma)\,d\sigma \tag{7}$$

Using eqn. (5), we obtain an approximation for $\mathbf{x}(t)$:

$$\mathbf{x}(t) = \int_{t_0}^{t} \dot{\mathbf{x}}(\tau)\, d\tau + \mathbf{x}(t_0) = \int_{t_0}^{t} X\mathbf{w}\left(\frac{\tau - t_0 + h}{T - t_0 + h}\right) d\tau + \mathbf{x}(t_0) \tag{8}$$

It is known that in general (Chen and Hsiao 1975)

$$\int_0^t \mathbf{w}(\tau)\, d\tau = P\mathbf{w}(t) \tag{9}$$

where P is a constant square matrix. Hence eqn. (8) takes the form

$$\mathbf{x}(t) = XP\mathbf{w}\left(\frac{t - t_0 + h}{T - t_0 + h}\right) + \mathbf{x}(t_0), \quad t \in [t_0 - h, T] \tag{10}$$

Similarly, making the following definitions

$$\begin{aligned} \hat{\mathbf{x}}(t) &\equiv 0, & t &< t_0 \\ \hat{\mathbf{x}}(t) &\equiv \mathbf{x}_0(t - h), & t &\in [t_0, t_0 + h] \\ \hat{\mathbf{x}}(t) &\equiv \mathbf{x}(t - h), & t &\in [t_0 + h, T] \end{aligned} \tag{11}$$

we have

$$\hat{\mathbf{x}}(t) = \hat{X}\mathbf{w}\left(\frac{t - t_0 + h}{T - t_0 + h}\right)$$

where

$$\begin{aligned} \hat{X} &\equiv \int_{t_0}^{T} \hat{\mathbf{x}}(t)\mathbf{w}'\left(\frac{t - t_0 + h}{T - t_0 + h}\right) dt \\ &= \int_{t_0}^{t_0 + h} \mathbf{x}_0(t - h)\mathbf{w}'\left(\frac{t - t_0 + h}{T - t_0 + h}\right) dt + \int_{t_0 + h}^{T} \mathbf{x}(t - h)\mathbf{w}'\left(\frac{t - t_0 + h}{T - t_0 + h}\right) dt \end{aligned} \tag{12 a}$$

When h is a variable such that $\xi \in [0, h]$, eqn. (12 a) becomes

$$\hat{X}(\xi) = \int_{t}^{t_0 + \xi} \mathbf{x}_0(t - \xi)\mathbf{w}'\left(\frac{t - t_0 + h}{T - t_0 + h}\right) dt + \int_{t_0 + \xi}^{T} \mathbf{x}(t - \xi)\mathbf{w}'\left(\frac{t - t_0 + h}{T - t_0 + h}\right) dt \tag{12 b}$$

where $\mathbf{x}(t - \xi)$ can now be taken from eqn. (10). Hence, in general, we can write

$$\mathbf{x}(t - \xi) = \hat{X}(\xi)\mathbf{w}\left(\frac{t - t_0 + h}{T - t_0 + h}\right), \quad t \in [t_0, T] \tag{12 c}$$

Defining the variable $s \equiv \dfrac{t-t_0+h}{T-t_0+h}$, eqns. (5), (10) and (12 *c*) take the form

$$\dot{\mathbf{x}}(t) = X\mathbf{w}(s), \quad t \in [t_0, T] \tag{12d}$$

$$\mathbf{x}(t) = \tilde{X}\mathbf{w}(s), \quad t \in [t_0, T] \tag{12e}$$

where

$$\tilde{X} \equiv [XP + X^0], \quad X^0 \equiv [\mathbf{x}(t_0) \vdots \mathbf{0} \vdots \mathbf{0} \vdots \dots \vdots \mathbf{0}]$$

since

$$x(t_0) = x(t_0)w_0\left(\frac{t-t_0+h}{T-t_0+h}\right) + \mathbf{0}w_1\left(\frac{t-t_0+h}{T-t_0+h}\right) + \dots + \mathbf{0}w_k\left(\frac{t-t_0+h}{T-t_0+h}\right) = X^0\mathbf{w}(s)$$

Then we can write

$$\mathbf{x}(t-\xi) = \hat{X}(\xi)\mathbf{w}(s), \quad t \in [t_0, T] \tag{12f}$$

Introducing eqn. (12 *e*) into eqn. (12 *b*), we obtain

$$\hat{X}(\xi) = \int_{t_0}^{t_0+\xi} \mathbf{x}_0(t-\xi)\mathbf{w}'\left(\frac{t-t_0+h}{T-t_0+h}\right) dt + \int_{t_0+\xi}^{T} \tilde{X}\mathbf{w}\left(\frac{t-\xi-t_0+h}{T-t_0+h}\right)\mathbf{w}'\left(\frac{t-t_0+h}{T-t_0+h}\right) dt$$

or

$$\hat{X}(\xi) = \hat{X}_1(\xi) + \tilde{X}\hat{X}_2(\xi)$$

where

$$\hat{X}_1(\xi) \equiv \int_{t_0}^{t_0+\xi} \mathbf{x}_0(t-\xi)\mathbf{w}'\left(\frac{t-t_0+h}{T-t_0+h}\right) dt$$

and

$$\hat{X}_2(\xi) \equiv \int_{t_0+\xi}^{T} \mathbf{w}\left(\frac{t-\xi-t_0+h}{T-t_0+h}\right)\mathbf{w}'\left(\frac{t-t_0+h}{T-t_0+h}\right) dt$$

The definitions we have made so far will prove to be very helpful in the subsequent problem formulation and the development of the computational algorithm.

3. Problem formulation

Let the time-delay system be described by the eqns. (1 *a*)–(1 *c*), which are repeated here for convenience:

$$\dot{\mathbf{x}}(t) = A_0\mathbf{x}(t) + A_1\mathbf{x}(t-h) + B\mathbf{u}(t) \tag{13a}$$

$$\mathbf{y}(t) = C\mathbf{x}(t) \tag{13b}$$

$$\mathbf{x}(\xi) = \mathbf{x}_0(\xi), \quad \xi \in [t_0 - h, t_0] \tag{13c}$$

where $\mathbf{x}_0(\xi)$ is a known vector function.

Given the quadratic cost functional

$$J = \tfrac{1}{2} \int_{t_0}^{t_f} (\mathbf{x}'Q\mathbf{x} + \mathbf{u}'R\mathbf{u})\, dt \tag{14}$$

we seek to find an optimal control law $\mathbf{u}^*(t)$ based on the measurement vector $\mathbf{y}(t)$, such that eqns. (13) are satisfied and the cost functional eqn. (14) is minimized. The solution of this problem, when the entire vector $\mathbf{x}(t)$ is available for measurement, is well established and it is given either in the form of a set of canonical equations or as a set of Riccati-like equations. In the first case, we have (Schmitendorf 1973)

$$\mathbf{u}^*(t) = -R^{-1}B'\mathbf{p}(t) \tag{15}$$

where

$$\left.\begin{array}{ll} \dot{\mathbf{x}}(t) = A_0\mathbf{x}(t) - BR^{-1}B\mathbf{p}(t) + A_1\mathbf{x}(t-h), & t_0 \leqslant t \leqslant T \\ \dot{\mathbf{p}}(t) = Q\mathbf{x}(t) - A_0'\mathbf{p}(t) - A_1'\mathbf{p}(t+h), & t_0 \leqslant t \leqslant T-h \\ \dot{\mathbf{p}}(t) = -Q\mathbf{x}(t) - A_0'\mathbf{p}(t), & T-h \leqslant t \leqslant T \\ \mathbf{p}(T) = 0 \text{ and } \mathbf{p}(T-h^+) = \mathbf{p}(T-h^-) \to & \text{continuity condition} \end{array}\right\} \tag{16}$$

The direct coupling of the above equations indicates that the optimal control law $\mathbf{u}^*(t)$ given by eqn. (15) depends on the entire state $(\mathbf{x}t)$. To prove this, we adopt the same procedure used in optimal control problems without delays, decoupling eqns. (16) by setting

$$\mathbf{p}(t) = K_1\mathbf{x}(t) + \int_{-h}^{0} K_2(\xi)\mathbf{x}(t+\xi)\, d\xi \tag{17}$$

It is known (Ross and Flügge-Lotz 1969) that the matrices K_1 and $K_2(\xi)$ satisfy the set of equations

$$A_0'K_1 + K_1A_0 - K_1BR^{-1}B'K_1 + K_2'(0) + K_2(0) + Q = 0$$

$$\frac{dK_2}{d\xi} = (A_0' - K_1BR^{-1}B')K_2(\xi) + K_3(0,\xi), \quad -h \leqslant \xi \leqslant 0$$

$$\frac{\partial K_3}{\partial \theta}(\theta,\xi) + \frac{\partial K_3}{\partial \xi}(\theta,\xi) = -K_2'(\theta)BR^{-1}B'K_2(\xi) \quad \left\{\begin{array}{l} -h \leqslant \xi \leqslant 0 \\ -h \leqslant \theta \leqslant 0 \end{array}\right. \tag{18}$$

with the boundary conditions

$$\left.\begin{array}{l} K_2(-1) = K_1A_1 \\ K_3(-1,\xi) = A_1K_2(\xi) \end{array}\right\} \tag{19}$$

Substituting eqn. (17) into eqn. (15), we obtain

$$\mathbf{u}^*(t) = -R^{-1}B'\left(K_1\mathbf{x}(t) + \int_{-h}^{0} K_2(\xi)\mathbf{x}(t+\xi)\, d\xi\right) \tag{20}$$

We observe that the optimal control law $\mathbf{u}^*(t)$ can be expressed in the 'feedback' form via the linear functional of the state $\mathbf{x}(t)$. A natural question to ask at this point is whether it is possible to develop an observer which, using the measurement $\mathbf{y}(t)$, asymptotically approximates the optimal control $\mathbf{u}^*(t)$ given by eqn. (20). Bhat and Koivo (1976) showed that, by using spectral decomposition of the infinitesimal generator of the semi-group corresponding to the delay differential equations of the system, an asymptotically stable observer can be developed for which the error decay can be made as fast as desired. The form of this observer is

$$\dot{\mathbf{z}}(t) = A_0\mathbf{z}(t) + A_1\mathbf{z}(t-h) + \Phi(0)K[C\mathbf{z}(t) - \mathbf{y}(t)] + A_1 \int_0^h \Phi(\xi - h)K[C\mathbf{z}(t-\xi) - \mathbf{y}(t-\xi)]\,d\xi + B\mathbf{u}(t) \quad (21)$$

where $\Phi(\xi)$ is the matrix of all eigenvectors which form a basis for the subspace of all unstable modes of the system (1 a) (Bhat and Koivo 1976, Hale 1971), and K is chosen to stabilize the unstable modes of the system. When $\mathbf{u}^*(t)$ is implemented through the use of the observer (state estimator), then

$$\mathbf{u}^*(t) = -R^{-1}B'\left(K_1\mathbf{z}(t) + \int_{-h}^{0} K_2(\xi)\mathbf{z}(t+\xi)d\xi\right) \quad (22)$$

Combining eqns. (22) and (13 a) and introducing the definition

$$-R^{-1}B'K_1 = K_1^*, \quad -R^{-1}B'K_2(\xi) = K_2^*(\xi)$$

we obtain

$$\dot{\mathbf{x}}(t) = (A_0 + BK_1^*)\mathbf{x}(t) + A_1\mathbf{x}(t-h) + \int_{-h}^{0} BK_2^*(\xi)\mathbf{x}(t+\xi)\,d\xi + BK_1^*\mathbf{e}(t) + B\int_{-h}^{0} K_2^*(\xi)\mathbf{e}(t+\xi)\,d\xi \quad (23)$$

$$\dot{\mathbf{e}}(t) = (A_0 + \Phi(0)KC)\mathbf{e}(t) + A_1\mathbf{e}(t-h) + A_1 \int_0^h \Phi(\xi - a)K\mathbf{e}(t-\xi)\,d\xi \quad (24\,a)$$

where

$$\mathbf{e}(t) \equiv \mathbf{z}(t) - \mathbf{x}(t) \quad (24\,b)$$

and

$$\lim_{t\to\infty} \mathbf{e}(t) = 0$$

Introducing eqns. (12 *d*), (12 *e*) and (12 *f*), and the approximate Walsh expressions for $\mathbf{e}(t)$†, into eqns. (23) and (24 *a*) yields

$$X\mathbf{w}(s) = (A_0 + BK_1^*)(XP + X^0)\mathbf{w}(s) + A_1(\hat{X}_1(h) + \tilde{X}\hat{X}_2(h))\mathbf{w}(s) + \int_h^0 BK_2^*(-\xi)(\hat{X}_1(-\xi) + \tilde{X}\hat{X}_2(-\xi))\,d\xi\mathbf{w}(s) + BK_1^*E\mathbf{w}(s) + B\int_h^0 K_2^*(-\xi)(\hat{E}_1(-\xi) + \tilde{E}\hat{E}_2(-\xi))\,d\xi\mathbf{w}(s) \quad (25)$$

$$E\mathbf{w}(s) = (A_0 + \Phi(0)KC)(EP + E^0)\mathbf{w}(s) + A_1(\hat{E}_1(h) + \tilde{E}\hat{E}_2(h))\mathbf{w}(s) + A_1\int_0^h \Phi(\xi - h)K(\hat{E}_1(\xi) + \tilde{E}\hat{E}_2(\xi))\,d\xi\mathbf{w}(s) \quad (26)$$

The above equation can be put into a more convenient form:

$$E\mathbf{w}(s) = (A_0 + \Phi(0)KC)EP\mathbf{w}(s) + A_1EP\hat{E}_2(h)\mathbf{w}(s) + A_1\left(\int_0^h \Phi(\xi - h)KEP\hat{E}_2(\xi)\,d\xi\right)\mathbf{w}(s) + (A_0 + \Phi(0)KC)E^0\mathbf{w}(s) + A_1(\hat{E}_1(h) + E^0\hat{E}_2(h))\mathbf{w}(s) + A_1\left(\int_0^h \Phi(\xi - h)K(\hat{E}_1(\xi) + E^0\hat{E}_2(\xi))\,d\xi\right)\mathbf{w}(s) \quad (27)$$

Equation (27) must hold for any value of $s \in \left[\dfrac{h}{T - t_0 + h}, 1\right]$, and thus we obtain

$$E = (A_0 + \Phi(0)KC)EP + A_1EP\hat{E}_2(h) + A_1\int_0^h \Phi(\xi - h)KEP\hat{E}_2(\xi)\,d\xi + E_c \quad (28)$$

where

$$E_c \equiv (A_0 + \Phi(0)KC)E^0 + A_1(\hat{E}_1(h) + E^0\hat{E}_2(h)) + A_1\int_0^h \Phi(\xi - h)K(\hat{E}_1(\xi) + E^0\hat{E}_2(\xi))\,d\xi \quad (29)$$

Since all the matrices in the defining expression for E_c are known, E_c is a known constant matrix. By the same reasoning, eqn. (25) can be expressed as

$$X = (A_0 + BK_1^*)XP + A_1XP\hat{X}_2(h) + \int_{-h}^0 BK_2^*(-\xi)XP\hat{X}_2(-\xi)\,d\xi + BK_1^*EP + B\int_{-h}^0 K_2^*(-\xi)EP\hat{E}_2(-\xi)\,d\xi + X_c \quad (30)$$

where again

$$X_c \equiv (A_0 + BK_1^*)X^0 + A_1(\hat{X}_2(h) + X^0\hat{X}_2(h)) + BK_1^*E^0 + \int_{-h}^0 BK_2^*(-\xi)(\hat{X}_1(-\xi) + X^0\hat{X}_2(-\xi))\,d\xi + \int_{-h}^0 BK_2^*(-\xi)(\hat{E}_1(-\xi) + E^0\hat{E}_2(-\xi))\,d\xi$$

† The approximate expression for $\mathbf{e}(t)$, equivalent to $\mathbf{x}(t)$ in eqn. (10), is

$$\mathbf{e}(t) = EP\mathbf{w}\left(\frac{t - t_0 + h}{T - t_0 + h}\right) + \mathbf{e}(t_0). \quad (24\,c)$$

Using the procedure developed by Stavroulakis and Tzafestas (1977), we obtain the following equivalent system of equations for the solution of eqns. (28) and (30):

$$\left.\begin{aligned}\hat{\mathbf{x}} &= \hat{A}_1\hat{\mathbf{x}} + \hat{A}_2\hat{\mathbf{x}} + \hat{B}\hat{\mathbf{e}} + \hat{\mathbf{x}}_0 \\ \hat{\mathbf{e}} &= \hat{\hat{A}}_1\hat{\mathbf{e}} + \hat{\hat{A}}_2\hat{\mathbf{e}} + \hat{\mathbf{e}}_0\end{aligned}\right\} \tag{31}$$

where

$$\hat{\mathbf{x}} \equiv \begin{bmatrix}\mathbf{x}_1\\ \mathbf{x}_2\\ \vdots \\ \mathbf{x}_n\end{bmatrix}, \quad \hat{\mathbf{e}} \equiv \begin{bmatrix}\mathbf{e}_1\\ \mathbf{e}_2\\ \vdots \\ \mathbf{e}_n\end{bmatrix} \tag{32}$$

with $\mathbf{x}_1 \ldots \mathbf{x}_n$ and $\mathbf{e}_1 \ldots \mathbf{e}_n$ being the columns of X and E respectively. Also

$$\hat{A}_1 \equiv (A_0 + BK_1{}^*) \circledast P'$$

$$\hat{A}_2 \equiv A_1 \circledast (\hat{X}_2{}'(h)P') + \int_{-h}^{0} [BK_2{}^*(-\xi) \circledast (X_2{}'(-\xi)P')]\, d\xi$$

$$\hat{B} \equiv BK_1{}^* \circledast P'$$

$$\hat{\hat{A}}_1 \equiv (A_0 + \Phi(0)KC) \circledast P'$$

$$\hat{\hat{A}}_2 \equiv A_1 \circledast (E_2{}'(h)P') + \int_{-h}^{0} (\Phi(\xi - h)K \circledast (E_2{}'(\xi)P'))\, d\xi$$

and $\hat{\mathbf{x}}_0$ and $\hat{\mathbf{e}}_0$ are the vectors formed by the columns of X_c and E_c expressed in the vector form of eqn. (32).

The notation $A \circledast B$ for any two matrices A and B denotes the Kronecker product of these matrices:

$$A \circledast B = \begin{bmatrix} b_{11}A & b_{12}A & \ldots & b_{1n}A \\ b_{21}A & b_{22}A & \ldots & b_{2n}A \\ \vdots & \vdots & & \vdots \\ b_{n1}A & b_{n2}A & \ldots & b_{nn}A \end{bmatrix}$$

Having stated the observer design problem in the format of eqn. (31), and using a finite Walsh function expansion, we can now proceed to develop a computational algorithm similar to that developed by Stavroulakis and Tzafestas (1977) for the case of non-delay systems. The approximations used avoid the necessity of finding the inverse of matrices at the expense of an iterative algorithm when a large study period is involved.

4. Computational algorithm

It is shown in Stavroulakis and Tzafestas (1977) that a system similar to that of eqn. (31) can be solved iteratively due to the special structure of the matrix P. When the first eight Walsh functions are used for the approximation of the state vector $\mathbf{x}(t)$, the matrix P takes the form (Chen and Hsaio 1975)

$$P' = \left[\begin{array}{cccc|cccc} & (\alpha) & & & & (\beta) & & \\ \frac{1}{2} & \frac{1}{4} & \frac{1}{8} & 0 & \frac{1}{16} & 0 & 0 & 0 \\ -\frac{1}{4} & 0 & 0 & \frac{1}{8} & 0 & \frac{1}{16} & 0 & 0 \\ -\frac{1}{8} & 0 & 0 & 0 & 0 & 0 & \frac{1}{16} & 0 \\ 0 & -\frac{1}{8} & 0 & 0 & 0 & 0 & 0 & \frac{1}{16} \\ & (\gamma) & & & & (\delta) & & \\ \hline -\frac{1}{16} & 0 & 0 & 0 & 0 & 0 & 0 & 0 \\ 0 & -\frac{1}{16} & 0 & 0 & 0 & 0 & 0 & 0 \\ 0 & 0 & -\frac{1}{16} & 0 & 0 & 0 & 0 & 0 \\ 0 & 0 & 0 & -\frac{1}{16} & 0 & 0 & 0 & 0 \end{array}\right] \tag{33}$$

The matrix P can be expressed in a more compact form as

$$P = \left[\begin{array}{c|c} P^{\alpha} & P^{\beta} \\ \hline P^{\gamma} & P^{\delta} \end{array}\right]^{\dagger}$$

Introducing the definitions:

$$\tilde{A}_1 = (A_0 + BK_1^*)$$
$$\tilde{\tilde{A}}_1 = (A_0 + \Phi(0)KC)$$

enables eqn. (31) to be expressed as

$$\left.\begin{aligned} \hat{\mathbf{x}} &= (\tilde{A}_1 \circledast P')\hat{\mathbf{x}} + \hat{A}_2\hat{\mathbf{x}} + \hat{B}\hat{\mathbf{e}} + \hat{\mathbf{x}}_0 \\ \hat{\mathbf{e}} &= (\tilde{\tilde{A}}_1 \circledast P')\hat{\mathbf{e}} + \hat{\hat{A}}_2\hat{\mathbf{e}} + \hat{\mathbf{e}}_0 \end{aligned}\right\} \tag{34}$$

It is shown in the Appendix that eqns. (34) can be solved simultaneously for $\hat{\mathbf{e}}$ and $\hat{\mathbf{x}}$ as functions of $\hat{\mathbf{e}}_0$ and of $\hat{\mathbf{e}}_0$ and $\hat{\mathbf{x}}_0$ respectively. The solution, which avoids the inversion of matrices by making an approximation via the subdivision of the time interval of the problem study period, leads to a convenient form for $\hat{\mathbf{x}}$ and $\hat{\mathbf{e}}$. This procedure leads iteratively, as shown in the Appendix, to the following set of equations for $\hat{\mathbf{e}}$ and $\hat{\mathbf{x}}$:

$$\left.\begin{aligned} \hat{\mathbf{e}} &= \left[\begin{array}{c} E^u \\ \hline E^l \end{array}\right]\hat{\mathbf{e}}_0 \\ \hat{\mathbf{x}} &= \left[\begin{array}{c} X^w \\ \hline X^l \end{array}\right]\hat{\mathbf{x}}_0 + \left[\begin{array}{c} X_E{}^u \\ \hline X_E{}^l \end{array}\right]\hat{\mathbf{e}}_0 \end{aligned}\right\} \tag{35}$$

† The general form of P is also available for any order of the matrix.

where the matrices which multiply the vectors $\hat{\mathbf{x}}_0$ and $\hat{\mathbf{e}}_0$ can be determined iteratively using the known matrices $\hat{A}_1$, $\hat{A}_2$, $\hat{\hat{A}}_1$, $\hat{\hat{A}}_2$, $\hat{B}$ defined previously.

Having determined $\hat{\mathbf{e}}$ and $\hat{\mathbf{x}}$ we can now return to form the matrices X and E which can be used to determine $\mathbf{x}(t)$ and $\mathbf{e}(t)$ via eqns. (10) and (24*c*), which can in turn be used to determine $\mathbf{z}(t)$ via eqn. (24*b*). Finally, we can substitute $\mathbf{z}(t)$ into eqn. (22) to determine $\mathbf{u}^*(t)$. A block diagram for this algorithm is shown in the figure.

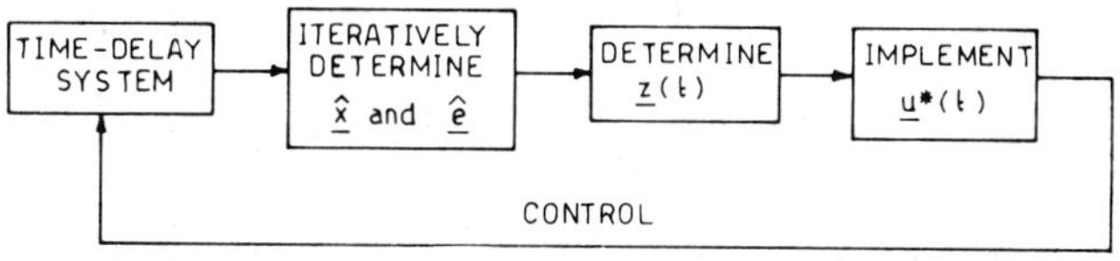

5. Conclusion

It is seen that Walsh functions can be used as an approximating set of functions to find a finite series expansion of both the state and the observer of a time-delay control system with partial state measurements. By subdividing the problem study period, it is shown that the known initial conditions of the state and the observer can be used iteratively to determine the optimal control $\mathbf{u}^*(t)$.

Appendix

Substituting eqn. (32) into eqn. (31) yields

$$\hat{\mathbf{x}} = \left[\begin{array}{cc|cc|cccc} \frac{1}{2}\tilde{A}_1 & \frac{1}{4}\tilde{A}_1 & \frac{1}{8}\tilde{A}_1 & 0 & \frac{1}{16}\tilde{A}_1 & 0 & 0 & 0 \\ -\frac{1}{4}\tilde{A}_1 & 0 & 0 & \frac{1}{8}\tilde{A}_1 & 0 & \frac{1}{16}\tilde{A}_1 & 0 & 0 \\ \hline -\frac{1}{8}\tilde{A}_1 & 0 & 0 & 0 & 0 & 0 & \frac{1}{16}\tilde{A}_1 & 0 \\ 0 & -\frac{1}{8}\tilde{A}_1 & 0 & 0 & 0 & 0 & 0 & \frac{1}{16}\tilde{A}_1 \\ \hline -\frac{1}{16}\tilde{A}_1 & 0 & 0 & 0 & 0 & 0 & 0 & 0 \\ 0 & -\frac{1}{16}\tilde{A}_1 & 0 & 0 & 0 & 0 & 0 & 0 \\ 0 & 0 & -\frac{1}{16}\tilde{A}_1 & 0 & 0 & 0 & 0 & 0 \\ 0 & 0 & 0 & -\frac{1}{16}\tilde{A}_1 & 0 & 0 & 0 & 0 \end{array}\right] \hat{\mathbf{x}} + \hat{A}_2\hat{\mathbf{x}} + \hat{B}\hat{\mathbf{e}} + \hat{\mathbf{x}}_0 \quad (A\,1)$$

$$
\hat{\mathbf{e}} = \left[\begin{array}{cc|cc|cccc}
\frac{1}{2}\tilde{\tilde{A}}_1 & \frac{1}{4}\tilde{\tilde{A}}_1 & \frac{1}{8}\tilde{\tilde{A}}_1 & 0 & \frac{1}{16}\tilde{\tilde{A}}_1 & 0 & 0 & 0 \\
-\frac{1}{4}\tilde{\tilde{A}}_1 & 0 & 0 & \frac{1}{8}\tilde{\tilde{A}}_1 & 0 & \frac{1}{16}\tilde{\tilde{A}}_1 & 0 & 0 \\
\hline
-\frac{1}{8}\tilde{\tilde{A}}_1 & 0 & 0 & 0 & 0 & 0 & \frac{1}{16}\tilde{\tilde{A}}_1 & 0 \\
0 & -\frac{1}{8}\tilde{\tilde{A}}_1 & 0 & 0 & 0 & 0 & 0 & \frac{1}{16}\tilde{\tilde{A}}_1 \\
\hline
-\frac{1}{16}\tilde{\tilde{A}}_1 & 0 & 0 & 0 & 0 & 0 & 0 & 0 \\
0 & -\frac{1}{16}\tilde{\tilde{A}}_1 & 0 & 0 & 0 & 0 & 0 & 0 \\
0 & 0 & -\frac{1}{16}\tilde{\tilde{A}}_1 & 0 & 0 & 0 & 0 & 0 \\
0 & 0 & 0 & -\tilde{\tilde{A}}_{\frac{1}{16}1} & 0 & 0 & 0 & 0
\end{array}\right] \hat{\mathbf{e}} + \hat{A}_2\hat{\mathbf{e}} + \hat{\mathbf{e}}_0 \qquad \text{(A 2)}
$$

As a further approximation, we make the following change of variable:

$$\lambda = at, \quad a \text{ is a positive constant}$$

and obtain

$$\dot{\mathbf{x}}(t) = \dot{\mathbf{x}}(\lambda)a \quad \text{and} \quad \dot{\mathbf{e}}(t) = \dot{\mathbf{e}}(\lambda)a$$

If we break the vector $\hat{\mathbf{e}}$ into two parts,

$$\hat{\mathbf{e}} = \begin{bmatrix} \mathbf{e}^u \\ \mathbf{e}^l \end{bmatrix}$$

and solve eqn. (A 2) for the last four vectors, we obtain

$$\mathbf{e}^l = \frac{1}{a}[(\tilde{\tilde{A}}_\mathrm{p})^\gamma \mathbf{e}^u + (\hat{\hat{A}}_2)^\gamma \mathbf{e}^u + (\hat{\hat{A}}_2)^\delta \mathbf{e}^l + \mathbf{e}_0{}^l] \qquad \text{(A 3)}$$

where $(\tilde{\tilde{A}}_\mathrm{p})^\gamma$, $(\hat{\hat{A}}_2)^\gamma$ and $(\hat{\hat{A}}_2)^\delta$ correspond to the equivalent partitions of the matrix P. Similarly, eqn. (A 1) can be solved for the last four vectors of the augmented vector $\hat{\mathbf{x}}$ to give

$$\mathbf{x}^l = \frac{1}{a}[(\tilde{A}_\mathrm{p})^\gamma \mathbf{x}^u + (\hat{A}_2)^\gamma \mathbf{x}^u + (\hat{A}_2)^\delta \mathbf{x}^l + (\hat{B})^\gamma \mathbf{e}^u + (\hat{B})^\delta \mathbf{e}^l + \mathbf{x}_0{}^l] \qquad \text{(A 4)}$$

where again $(\tilde{A}_\mathrm{p})^\gamma$, $(\hat{A}_2)^\gamma$, $(\hat{A}_2)^\delta$, $(\hat{B})^\gamma$ and $(\hat{B})^\delta$ are partitions of the matrices $\hat{A}_2$ and $\hat{B}$ corresponding to similar partitions of the matrix P. If we solve eqn. (A 3) for $\mathbf{e}^l$, we obtain

$$\mathbf{e}^l = \left(-I\frac{1}{a}(\hat{\hat{A}}_2)^\delta\right)^{-1}\left[\frac{1}{a}((\tilde{\tilde{A}}_\mathrm{p})^\gamma \mathbf{e}^u + (\hat{\hat{A}}_2)^\gamma \mathbf{e}^u + \mathbf{e}_0{}^l)\right] \qquad \text{(A 5)}$$

If a is chosen sufficiently large, and the approximation

$$\left(I-\frac{1}{a}A\right)^{-1}\approx I+\frac{1}{a}A \tag{A 6}$$

is made, eqn. (A 5) takes the form

$$\mathbf{e}^l=\left(I+\frac{1}{a}(\hat{A}_2)^\delta\right)\left[\frac{1}{a}((\tilde{A}_p)^\gamma\mathbf{e}^u+(\hat{A}_2)^\gamma\mathbf{e}^u+\mathbf{e}_0^l)\right] \tag{A 7}$$

In a more compact form, eqn. (A 7) can be written as

$$\mathbf{e}^l=R_1^{\delta\gamma}\mathbf{e}^u+R_2^\delta\mathbf{e}_0^l \tag{A 8}$$

where

$$R_1^{\delta\gamma}\equiv\frac{1}{a}\left(I+\frac{1}{a}(\hat{A}_2)^\delta\right)((\tilde{A}_p)^\gamma+(\hat{A}_2)^\gamma)$$

and

$$R_2^\delta\equiv\frac{1}{a}\left(I+\frac{1}{a}(\hat{A}_2)^\delta\right)$$

Solving eqn. (A 2) for the first four vectors of the augmented vector $\hat{\mathbf{e}}$, we obtain

$$\mathbf{e}^u=\frac{1}{a}[(\tilde{A}_p)^\alpha\mathbf{e}^u+(\tilde{A}_p)^\beta\mathbf{e}^l+(\hat{A}_2)^\alpha\mathbf{e}^u+(\hat{A}_2)^\beta\mathbf{e}^l+\mathbf{e}_0^u] \tag{A 9}$$

Substituting eqn. (A 8) into eqn. (A 9) we obtain

$$\begin{aligned}\mathbf{e}^u=\frac{1}{a}[(\tilde{A}\ )^\alpha\mathbf{e}^u&+(\tilde{A}_p)^\beta(R_1^{\delta\gamma})\mathbf{e}^u+(\hat{A}_2)^\alpha\mathbf{e}^u\\&+(\hat{A}_2)^\beta R_1^{\delta\gamma}\mathbf{e}^u+(\tilde{A}_p)^\beta R_2^\delta\mathbf{e}_0^l+(\hat{A}_2)^\beta R_2^\delta\mathbf{e}_0^l+\mathbf{e}_0^u]\end{aligned} \tag{A 10}$$

Equation (A 10) can be written as

$$\mathbf{e}^u=\frac{1}{a}[R_1^{\alpha\beta}\mathbf{e}^u+R_2^{\beta\delta}\mathbf{e}_0^l+\mathbf{e}_0^u] \tag{A 11}$$

where

$$R_1^{\alpha\beta}\equiv[(\tilde{A}_p)^\alpha+(\tilde{A}_p)^\beta R_1^{\delta\gamma}+(\hat{A}_2)^\alpha+(\hat{A}_2)^\beta R_1^{\delta\gamma}]$$

and

$$R_2^{\beta\delta}\equiv(\tilde{A}_p)^\beta R_2^\delta+(\hat{A}_2)^\beta R_2^\delta$$

Solving eqn. (A 11) for $\mathbf{e}^u$, we obtain

$$\mathbf{e}^u=\left(I-\frac{1}{a}R_1^{\alpha\beta}\right)^{-1}\frac{1}{a}(R_2^{\beta\delta}\mathbf{e}_0^l+\mathbf{e}_0^u) \tag{A 12}$$

and making use of the approximation (A 6), eqn. (A 12) becomes

$$\mathbf{e}^u = \frac{1}{a}\left(I + \frac{1}{a}R_1^{\alpha\beta}\right)(R_2^{\beta\delta}\mathbf{e}_0^l + \mathbf{e}_0^u) \tag{A 13}$$

Equation (A 13) is sufficient to determine $\mathbf{e}^u$, which subsequently can be used in eqn. (A 8) for the determination of $\mathbf{e}^l$ which is the other half of the vector $\hat{\mathbf{e}}$. This procedure yields

$$\mathbf{e}^l = R_1^{\delta\gamma}\left[\frac{1}{a}\left(I + \frac{1}{a}R_1^{\alpha\beta}\right)(R_2^{\beta\delta}\mathbf{e}_0^l + \mathbf{e}_0^u) + R_2^{\delta}\mathbf{e}_0^l\right] \tag{A 14}$$

Hence we observe that $\hat{\mathbf{e}}$ can be expressed as

$$\hat{\mathbf{e}} \equiv \begin{bmatrix} \mathbf{e}^u \\ --- \\ \mathbf{e}^l \end{bmatrix} = \begin{bmatrix} E^u \\ --- \\ E^l \end{bmatrix} \hat{\mathbf{e}}_0 \tag{A 15}$$

where

$$E^u \equiv \left[\frac{1}{a}\left(I + \frac{1}{a}R_1^{\alpha\beta}\right) \;\middle|\; \frac{1}{a}\left(I + \frac{1}{a}R_1^{\alpha\beta}\right)R_2^{\beta\delta}\right]$$

$$E^l \equiv \left[R_1^{\delta\gamma}\left(\frac{1}{a}\left(I + \frac{1}{a}R_1^{\alpha\beta}\right)\right) \;\middle|\; R_1^{\delta\gamma}\left(\frac{1}{a}\left(I + \frac{1}{a}R_1^{\alpha\beta}\right)R_2^{\beta\delta} + R_2^{\delta}\right)\right]$$

Having determined $\hat{\mathbf{e}}$ from eqn. (A 15), we can now introduce it into eqn. (A 4), and, following *exactly* the same procedure for the determination of $\hat{\mathbf{x}}$ as we did for $\mathbf{e}$, we can express $\hat{\mathbf{x}}$ as

$$\hat{\mathbf{x}} \equiv \begin{bmatrix} \mathbf{x}^u \\ \\ \mathbf{x}^l \end{bmatrix} = \begin{bmatrix} X^u \\ --- \\ X^l \end{bmatrix} \hat{\mathbf{x}}_0 + \begin{bmatrix} X_E^u \\ ---- \\ X_E^l \end{bmatrix} \hat{\mathbf{e}}_0 \tag{A 16}$$

where X^u, X^l, X_E^u and X_E^l are the matrices which will result from this procedure. These tedious steps are omitted here.

References

Bhat, K. P. M., and Koivo, H. N., 1976, Department of Electrical Engineering, University of Toronto, Technical Report No. 7502.

Chen, C. F., and Hsiao, C. H., 1975, *Int. J. Systems Sci.*, **6,** 833.

Hale, T. K., 1971, *Functional Differential Equations* (New York: Springer-Verlag).

Hewer, G. A., and Nazaroff, G. J., 1973, *Int. J. Control*, **18,** 1.

Koivo, H. N., 1972, *Appl. Mech. Rev.*, **25,** 239.

Krasovskii, N. N., 1962, *Appl. Math. Mech.*, **26,** 50.

Luenberger, D. G., 1964, *I.E.E.E. Trans. milit. Electron.*, **8,** 74.

Ross, D. W., and Flügge-Lotz, I., 1969, *J. SIAM Control*, **7,** 609.

Schmitendorf, W. E., 1973, *Automatica*, **9,** 633.

Stavroulakis, P., and Tzafestas, S., 1977, *Int. J. Control*, **26,** 5.

Reprinted from *J. Franklin Inst.* **305**:293–305 (1978)

A Method for Suboptimal Control of Time-varying Systems

by S. TZAFESTAS *and* P. STAVROULAKIS†
Control Systems Laboratory, University of Patras, Patras, Greece

ABSTRACT: *A finite set expansion method is presented for the suboptimal control of linear time-varying systems with both complete and incomplete state availability. By projecting the dynamical model of the system into a subspace of Walsh signals, we are able to implement the control by employing a computational method which involves constant coefficients. The method is illustrated by an example.*

I. Introduction

When modern control techniques are used in the design of control systems for dynamic processes, a feedback law often results which requires availability of the whole state of the plant.

For time varying or time invariant systems whose regulation time is finite, the optimal control is generated by a feedback law of the form $u^*(t) = -K(t)\mathbf{x}(t)$. The elements of the matrix $K(t)$ are time-varying feedback gains which are computed from the solution of a nonlinear matrix Riccati differential equation. The design approach to be taken in this paper is to replace the optimal gains $K(t)$ with their approximate values which are generaled using a finite set of Walsh functions. The problem of designing feedback gains for the case where the state is not completely known, is also studied using a similar procedure. The results reported here rest heavily on those previously reported by the authors in (**1, 2**) for the case of time invariant systems. For the sake of completeness we shall briefly present the basic results of optimal control of time varying systems with and without complete state availability.

II. Preliminary Considerations

Consider the linear time-varying system described by

$$\dot{x} = A(t)\mathbf{x}(t) + b(t)\mathbf{u}(t) \tag{1}$$

with initial condition

$$x(0) = \mathbf{x}_0. \tag{2}$$

The optimal linear regulator problem is to determine the control $u^*(t)$ in the time interval $[t_0, T]$ which minimizes the quadratic cost functional

$$J = \int_0^T \{\mathbf{x}'(t)Q(t)\mathbf{x}(t) + \mathbf{u}'(t)R(t)\mathbf{u}(t)\}\, dt \times \mathbf{x}'(T)Q_0\mathbf{x}(T) \tag{3}$$

† On leave of absence from Bell Labs., U.S.A.

where

$$Q_0 \geq 0 \quad (\text{constant } n\times n \text{ matrix})$$
$$Q(t) \geq 0, \quad t\varepsilon[0, T] \quad (n\times n \text{ matrix})$$
$$R(t) > 0, \quad t\varepsilon[0, T] \quad (n\times n \text{ matrix})$$

and $A(t)_{n\times n}$, $B(t)_{n\times p}$ are time-varying matrices of proper dimensions dependent on the dimensions of the vectors $\mathbf{x}(t)$ and $\mathbf{u}(t)$.

The above problem has been studied over the years (**3, 4, 9**) using modern control techniques such as the maximum principle, dynamic programming, leading to the Hamilton–Jacobi–Bellman partial functional differential equation, and general results of the projection theorem. It has been shown that the optimal control $\mathbf{u}^*(t)$ is of the form

$$\mathbf{u}^*(t) = K(t)\mathbf{x}(t) \tag{4}$$

where

$$K(t) = -R^{-1}(t)B'(t)\tilde{K}(t),$$

and

$$\frac{d\tilde{K}}{dt}(t) = \tilde{K}(t)A(t) + A'(t)\tilde{K}(t) - \tilde{K}(t)B(t)R^{-1}(t)B'(t)\tilde{K}(t) + Q(t) \tag{5}$$

with final condition $\tilde{K}(T) = Q_0$.

A different formulation of the problem leads to the set of canonical equations

$$\left.\begin{aligned} &(a) \quad \dot{\mathbf{x}}(t) = A(t)\mathbf{x}(t) + B(t)R^{-1}(t)B'(t)\lambda(t) \\ &(b) \quad \dot{\lambda}(t) = -Q(t)\mathbf{x}(t) - A'(t)\lambda(t) \\ &(c) \quad \mathbf{u}^*(t) = -R^{-1}(t)B'(t)\lambda(t) \\ &(d) \quad \lambda(T) = Q_0\mathbf{x}(T) \end{aligned}\right\}. \tag{6}$$

When only the output of the system, i.e. $\mathbf{y}(t) = C(t)\mathbf{x}(t)$, is available for measurement and the dimension of the output is $r < n$, then an *observer* is used for the asymptotic reconstruction of the state $\mathbf{x}(t)$. The basic concept of observer theory is to reconstruct the state vector $\mathbf{x}(t)$ such that

$$\lim_{t\to\infty}\left\{\mathbf{x}(t) - S(t)\begin{bmatrix}\mathbf{y}(t)\\ \mathbf{z}(t)\end{bmatrix}\right\} = 0 \tag{7}$$

where $\mathbf{z}(t)$ is the state of the observer and satisfies the following differential equation:

$$\dot{\mathbf{z}}(t) = F(t)\mathbf{z}(t) + G(t)\mathbf{y}(t) + H(t)\mathbf{u}(t). \tag{8}$$

It is shown (**1**) that, if we set

$$\mathbf{z}(t) = T(t)\mathbf{x}(t) - \mathbf{e}(t), \tag{9}$$

then

$$\left.\begin{aligned}
&\text{(a)}\quad I-S(t)\begin{bmatrix} C(t) \\ \hdashline T(t) \end{bmatrix}=0\\
&\text{(b)}\quad \dot{\mathbf{e}}(t)=F(t)\mathbf{e}(t)\\
&\text{(c)}\quad F(t)=T(t)A(t)N(t)+\dot{T}(t)N(t)\\
&\text{(d)}\quad G(t)=T(t)A(t)M(t)+\dot{T}(t)N(t)\\
&\text{(e)}\quad \hat{\hat{\mathbf{x}}}(t)=S(t)\begin{bmatrix} \mathbf{y}(t) \\ \hdashline \mathbf{z}(t) \end{bmatrix}\ \text{(asymptotic estimate of } \hat{x}(t))
\end{aligned}\right\}\tag{10}$$

where $M(t)$ and $N(t)$ are the matrices obtained by partitioning $S(t)$ as follows

$$S(t)=[\underbrace{M(t)}_{m}\ \vdots\ \underbrace{N(t)}_{r}]\tag{11}$$

and the control can be implemented via the expression

$$\mathbf{u}(t)=K(t)\hat{\hat{\mathbf{x}}}(t).\tag{12}$$

It can also be shown that the augmented vector $\begin{bmatrix} \mathbf{x}(t) \\ \hdashline \mathbf{e}(t) \end{bmatrix}$ satisfies the following equation

$$\begin{bmatrix} \dot{\mathbf{x}}(t) \\ \hdashline \dot{\mathbf{e}}(t) \end{bmatrix}=\begin{bmatrix} A(t)+B(t)K(t) & \vdots & -B(t)K(t)N(t) \\ \hdashline 0 & \vdots & F(t) \end{bmatrix}\begin{bmatrix} \mathbf{x}(t) \\ \hdashline e(t) \end{bmatrix}.\tag{13}$$

III. Solution of Problem: Walsh Series Approach

1. *Complete State Availability*

We observe that the set of Eqs. (6) takes the following form:

$$\dot{\hat{\mathbf{x}}}(t)=\begin{bmatrix} A(t) & \vdots & -B(t)R^{-1}(t)B'(t) \\ \hdashline -Q(t) & \vdots & -A'(t) \end{bmatrix}\hat{\mathbf{x}}(t)\tag{14}$$

$$\hat{\mathbf{x}}(0)=\begin{bmatrix} \mathbf{x}(0) \\ \hdashline \lambda(0) \end{bmatrix}\quad\text{and}\quad \hat{\mathbf{x}}(T)=\begin{bmatrix} I \\ \hdashline Q_0 \end{bmatrix}\mathbf{x}(T),\tag{14a}$$

where

$$\hat{\mathbf{x}}(t)=\begin{bmatrix} \mathbf{x}(t) \\ \hdashline \lambda(t) \end{bmatrix}.$$

Introducing normalized time $\tau=t/T$, then Eq. (14) can be put into a more compact form and the period of regulation of the system under consideration is normalized to [0, 1]. This normalization is necessary to make the interval [0, 1] coincide with the finite interval of completeness of the orthonormal Walsh functions we are considering. Hence Eq. (14) takes the following form

$$\dot{\hat{\mathbf{x}}}(\tau)=T\hat{A}(\tau)\hat{\mathbf{x}}(\tau),\quad \tau\varepsilon[0,1]\tag{15}$$

and $\hat{A}(\tau)$ is the composite matrix in Eq. (14). Using the procedure described in

(**1**) we assume that the derivative of the state vector $\hat{\mathbf{x}}(t)$ can be approximated by a finite set of Walsh functions, namely

$$\dot{\hat{\mathbf{x}}}(\tau) = X\mathbf{w}(\tau) \tag{16}$$

where X is a constant matrix with elements X_{ij} given by

$$X_{ij} = \int_0^1 \dot{\hat{x}}_i(\tau)\mathbf{w}_j(\tau)\,\mathrm{d}\tau \tag{17}$$

where $\dot{\hat{x}}_i(\tau)$ is the ith component of the vector $\dot{\hat{\mathbf{x}}}(\tau)$, $w_j(\tau)$ is the Walsh function of order j, and

$$\mathbf{w}(\tau) = \begin{bmatrix} w_0(\tau) \\ w_1(\tau) \\ \cdot \\ \cdot \\ \cdot \\ w_{m-1}(\tau) \end{bmatrix}. \tag{18}$$

For convergence considerations we assume $m = 2^q$, $q = 1, 2, \ldots,$ since, when the Walsh series that will be generated from (18) is truncated at the end of a group which includes all functions with a given minimum interval between transitions, then the partial sum of the first 2^q terms of the series will be equal to the average value of the function in each subinterval of length 2^{-q}(**10, 11**).

Equation (16) can be integrated to give $\hat{\mathbf{x}}(\tau)$

$$\hat{\mathbf{x}}(\tau) = \int_0^\tau \dot{\hat{\mathbf{x}}}(\Theta)\,\mathrm{d}\Theta + \hat{\mathbf{x}}(0). \tag{19}$$

But, it is known (**1, 6**) that

$$\int_0^\tau \mathbf{w}(\Theta)\,\mathrm{d}\Theta = P\mathbf{w}(\tau) \tag{20}$$

where P is a constant matrix of a special structure as shown in (**6**). In view of (20), eqs. (15) and (19) give

$$\hat{\mathbf{x}}(\tau) = XP\mathbf{w}(\tau) + \hat{\mathbf{x}}(0) \tag{21}$$

and

$$X\mathbf{w}(\tau) = T\hat{A}(\tau)[XP\mathbf{w}(\tau) + \hat{\mathbf{x}}(0)]. \tag{22}$$

Postmultiplying Eq. (22) by $\mathbf{w}'(\tau)$ and integrating over the interval [0, 1] we obtain

$$\int_0^1 X\mathbf{w}(\tau)\mathbf{w}'(\tau)\,\mathrm{d}\tau = T\int_0^1 \hat{A}(\tau)XP\mathbf{w}(\tau)w'(\tau)\,\mathrm{d}\tau + T\int_0^1 \hat{A}(\tau)\hat{\mathbf{x}}(0)w'(\tau)\,\mathrm{d}\tau. \tag{23}$$

But

$$\int_0^1 \mathbf{w}(\tau)\mathbf{w}'(\tau)\,\mathrm{d}\tau = I \tag{24}$$

due to the fact that the Walsh functions involved in the vector $\mathbf{w}(\tau)$ are orthonormal. Hence, Eq. (23) can be written as

$$X = T\int_0^1 \hat{A}(\tau)XP\mathbf{w}(\tau)w'(\tau)\,d\tau + T\int_0^1 \hat{A}(\tau)\hat{\mathbf{x}}(0)\mathbf{w}'(\tau)d\tau. \tag{25}$$

From Eq. (21), we get

$$\hat{\mathbf{x}}(1) = \begin{bmatrix}\mathbf{x}(1)\\ \lambda(1)\end{bmatrix} = XP\mathbf{w}(1) + \begin{bmatrix}\mathbf{x}(0)\\ \lambda(0)\end{bmatrix}. \tag{26}\dagger$$

But Eq. (14a) gives

$$\lambda(1) = Q_o\mathbf{x}(1). \tag{27}$$

Solving Eq. (26) with respect to $\mathbf{x}(1)$ and $\lambda(1)$, we get

$$\mathbf{x}(1) = X^aP\mathbf{w}(1) + \mathbf{x}(0) \tag{28}$$

$$\lambda(1) = X^bP\mathbf{w}(1) + \lambda(0) \tag{29}$$

where X^a and X^b are the upper and lower halves of the matrix X partitioned as follows:

$$X = \begin{bmatrix}X^a\\ X^b\end{bmatrix}. \tag{30}$$

Introducing Eq. (27) into (29) yields:

$$Q_0\mathbf{x}(1) = X^bP\mathbf{w}(1) + \lambda(0). \tag{31}$$

Similarly, (28) can be introduced into (31) to give

$$Q_0[X^aP\mathbf{w}(1) + \mathbf{x}(0)] = X^bP\mathbf{w}(1) + \lambda(0). \tag{32}$$

Hence

$$(Q_0X^aP - X^bP)\mathbf{w}(1) + Q_0\mathbf{x}(0) = \lambda(0). \tag{33}$$

Finally, introducing (33) into (25) we obtain

$$X = T\int_0^1 \hat{A}(\tau)XP\mathbf{w}(\tau)\mathbf{w}'(\tau)\,d\tau + T\int_0^1 \hat{A}(\tau)\begin{bmatrix}\mathbf{x}(0)\\ Q_0\mathbf{x}(0)\end{bmatrix}\mathbf{w}'(\tau)\,d\tau$$
$$+ T\int_0^1 \hat{A}(\tau)\begin{bmatrix}\mathbf{0}\\ Q_0(X^aP - X^bP)\mathbf{w}(1)\end{bmatrix}\mathbf{w}'(\tau)\,d\tau. \tag{34}$$

Now, if we partition the matrix equation (34) as shown in (30), and if we redefine $\hat{A}(\tau)$ as

$$\hat{A}(\tau) = \begin{bmatrix}A(\tau) & \tilde{B}(\tau)\\ -Q(\tau) & -A'(\tau)\end{bmatrix} \tag{35}$$

† To ensure convergence at $t = T$ we consider the expansion interval equal to $T+\epsilon$, where ε is a small positive constant. In this case the results will be valid over the interval $[0, T]$ which is considered to be our regulation interval. Clearly $t = T$ now corresponds to $\tau = T/(T+\varepsilon) < 1$.

where $\tilde{B}(\tau) = -B(\tau)R^{-1}(\tau)B'(\tau)$, we get

$$X^a = T\int_0^1 A(\tau)X^a P\mathbf{w}(\tau)w'(\tau)\,d\tau + T\int_0^1 \tilde{B}(\tau)X^b P\mathbf{w}(\tau)\mathbf{w}'(\tau)\,d\tau + T\int_0^1 A(\tau)\mathbf{x}(0)\mathbf{w}'(\tau)\,d\tau + T\int_0^1 \tilde{B}(\tau)Q_0\mathbf{x}(0)\mathbf{w}'(\tau)\,d\tau + T\int_0^1 \tilde{B}(\tau)Q_0(X^aP - X^bP)\mathbf{w}(1)\mathbf{w}'(\tau)\,d\tau \quad (36)$$

and

$$X^b = -T\int_0^1 Q(\tau)X^a P\mathbf{w}(\tau)\mathbf{w}'(\tau)\,d\tau - T\int_0^1 A'(\tau)X^b \mathbf{P}\mathbf{w}(\tau)\mathbf{w}'(\tau)\,d\tau - T\int_0^1 A'(\tau)Q_0(X^aP - X^bP)w(1)\mathbf{w}'(\tau)\,d\tau. \quad (37)$$

Then, writing X^a and X^b into vector forms and following exactly the procedure presented in (**1**, **2**), Eqs. (36) and (37) reduce to the following form:

$$\hat{\mathbf{x}}_a = A_1\hat{\mathbf{x}}_a + A_2\hat{\mathbf{x}}_b + \hat{\mathbf{c}}_a \quad (36a)$$

$$\hat{\mathbf{x}}_b = A_3\hat{\mathbf{x}}_a + A_4\hat{\mathbf{x}}_b + \hat{\mathbf{c}}_b \quad (37a)$$

where

$$\hat{\mathbf{x}}_a = \begin{bmatrix} \mathbf{x}_1^a \\ \mathbf{x}_2^a \\ \cdot \\ \cdot \\ \cdot \\ \mathbf{x}_m^a \end{bmatrix} \quad \text{and} \quad \hat{\mathbf{x}}_b = \begin{bmatrix} \mathbf{x}_1^b \\ \mathbf{x}_2^b \\ \cdot \\ \cdot \\ \cdot \\ \mathbf{x}_m^b \end{bmatrix}$$

with $\mathbf{x}_i^a$ and $\mathbf{x}_i^b$ being the column vectors of X^a and X^b, and

$$A_1 = T\int_0^1 [A(\tau)\times\{\mathbf{w}(\tau)\mathbf{w}'(\tau)P'\}]\,d\tau - T\int_0^1 [\tilde{B}(\tau)Q_0\times\{\mathbf{w}(\tau)\mathbf{w}'(1)P'\}]\,d\tau$$

$$A_2 = T\int_0^1 [\tilde{B}(\tau)\times\{\mathbf{w}(\tau)\mathbf{w}'(\tau)P'\}]\,d\tau - T\int_0^1 [\tilde{B}(\tau)Q_0\times\{\mathbf{w}(\tau)\mathbf{w}'(1)P'\}]\,d\tau\dagger$$

$\hat{\mathbf{c}}_a$ = column vector form of the matrix

$$T\int_0^1 A(\tau)\mathbf{x}(0)\mathbf{w}'(\tau)\,d\tau + T\int_0^1 \tilde{B}(\tau)Q_o\mathbf{x}(0)\mathbf{w}'(\tau)\,d\tau$$

$$A_3 = -T\int_0^1 [Q(\tau)\times\{\mathbf{w}(\tau)\mathbf{w}'(\tau)P'\}]\,d\tau - T\int_0^1 A'(\tau)Q_0\times\{\mathbf{w}(\tau)\mathbf{w}'(1)P'\}\,d\tau$$

$$A_4 = -T\int_0^1 [A'(\tau)\times\{\mathbf{w}(\tau)\mathbf{w}'(\tau)P'\}]\,d\tau + T\int_0^1 [A'(\tau)Q_0\times\{\mathbf{w}(\tau)\mathbf{w}'(1)P'\}]\,d\tau$$

† For the definition of the matrix Kronecker Product × see (**1**).

$\hat{\mathbf{c}}_b$ = column vector form of the matrix

$$-T\int_0^1 Q(\tau)\mathbf{x}(0)\mathbf{w}'(\tau)\,\mathrm{d}\tau - T\int_0^1 A'(\tau)Q_0\mathbf{x}(0)\mathbf{w}'(\tau)\,\mathrm{d}\tau.$$

Having calculated the matrices A_1, A_2, A_3, A_4 and the vectors $\hat{\mathbf{c}}_a$ and $\hat{\mathbf{c}}_b$, we then define

$$\hat{\mathbf{x}}_c = \begin{bmatrix}\hat{\mathbf{x}}_a\\ \hat{\mathbf{x}}_b\end{bmatrix},\quad \hat{\mathbf{c}} = \begin{bmatrix}\hat{\mathbf{c}}_a\\ \hat{\mathbf{c}}_b\end{bmatrix},\quad A_c = \left[\begin{array}{c|c}A_1 & A_2\\ \hline A_3 & A_4\end{array}\right]$$

and reduce Eqs. (36a) and (37a) into the compact form

$$\hat{\mathbf{x}}_c = A_c\hat{\mathbf{x}}_c + \hat{\mathbf{c}}. \tag{38}$$

For a further step in our approximation procedure we make the following change of the normalized time variable τ:

$$\lambda = \alpha\tau \quad (\alpha > 0 \text{ is a constant})$$

which implies that Eq. (38) takes the form

$$\hat{\mathbf{x}}_c = \frac{1}{\alpha}[A_c\hat{\mathbf{x}}_c + \hat{\mathbf{c}}]. \tag{39}$$

Solving Eq. (39), under the assumption that "α" is chosen sufficiently large so that the approximation

$$\left\{I - \frac{1}{\alpha}A_c\right\}^{-1} \approx I + \frac{1}{\alpha}A_c \tag{40}$$

is valid, yields

$$\hat{\mathbf{x}}_c = \left(I + \frac{1}{\alpha}A_c\right)\hat{\mathbf{c}}. \tag{41}$$

Knowledge of the vector $\hat{\mathbf{x}}_c$ is sufficient to determine the matrix X, and hence to determine the state vector $\mathbf{x}(\tau)$ and the adjoint vector $\lambda(\tau)$ (via Eq. (21)) which are required to implement the control $u^*(\tau)$ using Eq. (6c).

The implementation of the suboptimal controller by the computational procedure developed here is shown in a block diagram form in Fig. 1.

We observe that by expanding the rate of change of the state of a linear time-varying control system in a finite series of Walsh functions (which contains as many terms as it is necessary for the convergence to the average value of the

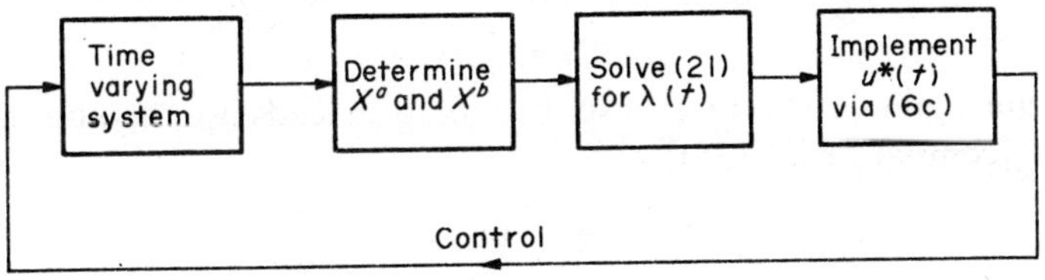

FIG. 1. Implementation of the suboptimal controller.

function in each subinterval of length 2^{-q}) and using the procedure of this paper we can implement the suboptimal control through the computation of only constant coefficients. The same procedure can be used for the case where the system state vector is partially available through the measurement of the output vector of smaller dimension. This is made in the next section.

2. *Partial State Measurement (Observer Case)*

We have seen that when the state is available only via the output $\mathbf{y}(t) = C(t)\mathbf{x}(t)$, the estimate $\hat{x}(t)$ of $\hat{\mathbf{x}}(t)$ can be used to implement the control via the observer $\mathbf{z}(t)$.

Setting

$$\begin{bmatrix} \dot{\mathbf{x}}(t) \\ \hdashline \dot{e}(t) \end{bmatrix} = E\mathbf{w}(t) \tag{42}$$

we then obtain [see Eqs. (16) and (21)]

$$\begin{bmatrix} \mathbf{x}(t) \\ \hdashline \mathbf{e}(t) \end{bmatrix} = EP\mathbf{w}(t) + \begin{bmatrix} \mathbf{x}(0) \\ \hdashline \mathbf{e}(0) \end{bmatrix}. \tag{43}$$

Introducing Eqs. (42) and (43) into (13) yields

$$E\mathbf{w}(t) = A_0(t)\left\{EP\mathbf{w}(t) + \begin{bmatrix} \mathbf{x}(0) \\ \mathbf{e}(0) \end{bmatrix}\right\} \tag{44}$$

where

$$A_0(t) = \left[\begin{array}{c:c} A(t)+B(t)K(t) & -B(t)K(t)N(t) \\ \hdashline 0 & F(t) \end{array}\right].$$

Postmultiplying (44) by $\mathbf{w}'(t)$ and integrating over the interval [0, 1] gives†

$$\int_0^1 E\mathbf{w}(t)\mathbf{w}'(t)\,\mathrm{d}t = \int_0^1 A_0(t)\left\{EP\mathbf{w}(t) + \begin{bmatrix} \mathbf{x}(0) \\ \hdashline \mathbf{e}(0) \end{bmatrix}\right\}\mathbf{w}'(t)\,\mathrm{d}t$$

which, due to the fact that the Walsh functions of the vector $\mathbf{w}(t)$ are orthonormal, reduces to

$$E = \int_0^1 A_0(t)EP\mathbf{w}(t)\mathbf{w}'(t)\,\mathrm{d}t + \int_0^1 A_0(t)\begin{bmatrix} \mathbf{x}(0) \\ \mathbf{e}(0) \end{bmatrix}\mathbf{w}'(t)\,\mathrm{d}t. \tag{45}$$

A procedure similar to that described before leads to the following vector equation [compare with (38)]

$$\hat{\mathbf{q}} = \tilde{\tilde{A}}\hat{q} + \hat{\mathbf{q}}_0 \tag{46}$$

† It is assumed that the time interval of control is normalized to [0, 1].

where

$$\hat{\mathbf{q}} = \begin{bmatrix} \mathbf{e}_1 \\ \mathbf{e}_2 \\ \vdots \\ \mathbf{e}_n \end{bmatrix}, \quad \hat{\mathbf{q}}_0 = \begin{bmatrix} \mathbf{e}_0^1 \\ \mathbf{e}_0^2 \\ \vdots \\ \mathbf{e}_0^n \end{bmatrix}$$

with $\mathbf{e}_1, \mathbf{e}_2, \ldots, \mathbf{e}_n$ and $\mathbf{e}_0^1, \mathbf{e}_0^2, \ldots, \mathbf{e}_0^n$ being the column vectors of the matrices E and

$$\int_0^1 A_0(t)\begin{bmatrix} \mathbf{x}(0) \\ \hdashline \mathbf{e}(0) \end{bmatrix}\mathbf{w}'(t)\,\mathrm{d}t$$

respectively, and

$$\tilde{\tilde{A}} = \int_0^1 [A_0(t) \times \{\mathbf{w}(t)\mathbf{w}'(t)P'\}]\,\mathrm{d}t.$$

We observe that by determining the vector $\hat{\mathbf{q}}$ from (46), which is a *linear algebraic equation with constant coefficients*, we can determine E. The vectors $\mathbf{x}(t)$ and $\mathbf{e}(t)$ can subsequently be determined from (43). Then, Eqs. (9), (10e), and ((12) can be used to implement the control $\mathbf{u}^*(t)$. Figure 2 shows the implementation in a block diagram form.

Clearly, the implementation of the control $\mathbf{u}^*(t)$ makes use of both the output $\mathbf{y}(t)$ and the observer $\mathbf{z}(t)$. The solution of Eq. (46), can be found by the procedure used for the solution of Eq. (38). Without repeating all the details, Eq. (46) yields

$$\hat{\mathbf{q}} = \left(I + \frac{1}{\alpha}\tilde{\tilde{A}}\right)\hat{\mathbf{q}}_0. \tag{47}$$

IV. Example

Let a system described by the following state space equation

$$\dot{\mathbf{x}} = A(t)\mathbf{x}(t) + B(t)u(t) \tag{48}$$

where

$$x = \begin{bmatrix} x_1 \\ x_2 \end{bmatrix}, \quad a = \begin{bmatrix} 0 & t \\ 0 & 0 \end{bmatrix} \quad \text{and} \quad B = \begin{bmatrix} 0 \\ \sqrt{t} \end{bmatrix}$$

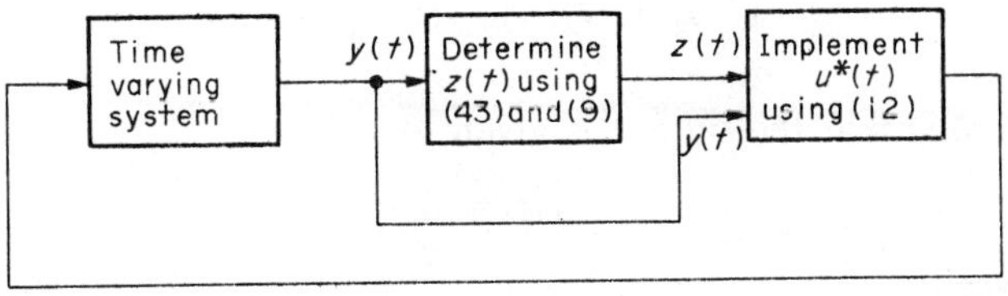

FIG. 2. Implementation of $u^*(t)$ via an observer.

with the cost functional

$$J=\mathbf{x}'(\tau)Q_0\mathbf{x}(\tau)+\int_0^T \mathbf{x}'(t)Q(t)\mathbf{x}(t)\,\mathrm{d}t+\int_0^T u^2(t)\,\mathrm{d}t.$$

where

$$Q_0=\begin{bmatrix}1 & 0\\ 0 & 1\end{bmatrix}\quad Q(t)=\begin{bmatrix}t & 0\\ 0 & t\end{bmatrix}.$$

The optimal control of the above system using standard techniques (4) is found by solving the following set of canonical equations

$$\begin{bmatrix}\dot{\mathbf{x}}\\ \hline \dot{\lambda}\end{bmatrix}=\left[\begin{array}{cc|cc}0 & t & 0 & 0\\ 0 & 0 & 0 & -t\\ \hline -t & 0 & 0 & 0\\ 0 & -t & -t & 0\end{array}\right]\begin{bmatrix}\mathbf{x}(t)\\ \hline \lambda(t)\end{bmatrix}\tag{49}$$

and

$$u^*(t)=[0,\sqrt{t}]\lambda(t).$$

Solution of the system of Eq. (49) gives, after tedious but standard manipulations

$$\begin{aligned}u^*(t)={}&-M\sqrt{t}[(\alpha(t)\gamma(t)+\beta(t)\,\delta(t)+2\beta(t)\gamma(t))(\alpha(t)\gamma(t)+\beta(t)\,\delta(t)\\&+\alpha(t)\,\delta(t)+\beta(t)\gamma(t))+(\alpha(t)\,\delta(t)+\beta(t)\gamma(t)+2\beta(t)\,\delta(t))\\&\times(-2\beta(t)\gamma(t)-\alpha(t)\gamma(t)+\beta(t)\,\delta(t))]x_1(t)\\&-M\sqrt{t}[(\alpha(t)\gamma(t)+\beta(t)\,\delta(t)+2\beta(t)\gamma(t))(2\beta(t)\gamma(t)\\&+\alpha(t)\gamma(t)+\beta(t)\,\delta(t))+(\alpha(t)\,\delta(t)+\beta(t)\gamma(t)+2\beta(t)\,\delta(t))\\&\times(-\alpha(t)\beta(t)+\beta(t)\,\delta(t)+\alpha(t)\,\delta(t)-\beta(t)\gamma(t))]x_2(t)\end{aligned}\tag{50}$$

where

$$\alpha(t)=\cosh\frac{\sqrt{3}}{4}(T-t)^2,\quad \beta(t)=\frac{1}{\sqrt{3}}\sinh\frac{\sqrt{3}}{4}(T-t)^2,\quad \gamma(t)=\sin\frac{(T-t)^2}{4},$$

$$\delta(t)=\cos\frac{(T-t)^2}{4},\quad M=\frac{1}{3\beta^2(0)\gamma^2(0)+2\alpha(0)\beta(0)+2\alpha^2(0)\,\delta^2(0)}.$$

Denoting by $K_1(t)$ the coefficient of $x_1(t)$ and $K_2(t)$ the coefficient of $x_2(t)$ in Eq. (50), we can write

$$u^*(t)=-K_1(t)x_1(t)-K_2(t)x_2(t).\tag{51}$$

Then, assuming that the output is given by

$$y(t)=x_1(t)$$

Eq. (8) gives

$$\dot{z}(t)=F(t)z(t)+G(t)y(t)+H(t)u(t).\tag{52}$$

Solving the observer equations we obtain:

$$F(t) = -t^2$$
$$G(t) = -1 - t^3$$
$$H(t) = \sqrt{t}.$$

Finally Eqs. (51) and (52) give

$$\begin{aligned} \dot{x}_1 &= t x_2 \\ \dot{x}_2 &= -K_1(t)x_1 - K_2(t)x_2 \\ \dot{z} &= -t^2 z - ((1+t^3) - \sqrt{t}\,K_1(t))x_1 - \sqrt{t}\,K_2(t)x_2. \end{aligned} \tag{53}$$

Hence

$$\dot{\hat{\mathbf{x}}}(t) = \hat{A}(t)\hat{\mathbf{x}}(t) \tag{54}$$

where

$$\hat{x}(t) = \begin{bmatrix} x_1 \\ x_2 \\ z \end{bmatrix} \quad \text{and} \quad \hat{A}(t) = \left[\begin{array}{c:c:c} 0 & t & 0 \\ -K_1(t) & K_2(t) & 0 \\ -(1+t^3)+\sqrt{t}K_1(t) & -\sqrt{t}K_2(t) & -t^2 \end{array}\right].$$

Now, setting

$$\dot{\hat{\mathbf{x}}}(t) = X\mathbf{w}(t) \tag{55}$$

where

$$\mathbf{w}(t) = \begin{bmatrix} w_1(t) \\ w_2(t) \\ \cdot \\ \cdot \\ \cdot \\ w_8(t) \end{bmatrix} \quad \text{(the first eight Walsh functions)}$$

and subsequently introducing (55) into (54), we obtain

$$X\mathbf{w}(t) = \hat{A}(t)XP\mathbf{w}(t) + \hat{\mathbf{x}}(0), \quad \hat{\mathbf{x}}(0) = \begin{bmatrix} x_1(0) \\ x_2(0) \\ z(0) \end{bmatrix}. \tag{56}$$

Postmultiplying (56) by $\mathbf{w}'(t)$ and integrating over the proper interval, as explained at the footnote of Eq. (26), we obtain

$$\int_0^T X\mathbf{w}(t)\mathbf{w}'(t)\,\mathrm{d}t = \frac{1}{\alpha}\int_0^T \hat{A}(t)XP\mathbf{w}(t)\mathbf{w}'(t)\,\mathrm{d}t + \int_0^T \hat{\mathbf{x}}(0)\mathbf{w}'(t)\,\mathrm{d}t \tag{57}$$

or

$$X = \frac{1}{\alpha}\int_0^T \hat{A}(t)XP\mathbf{w}(t)\mathbf{w}'(t)\,\mathrm{d}t + \int_0^T \hat{\mathbf{x}}(0)\mathbf{w}'(t)\,\mathrm{d}t$$

or

$$X = \frac{1}{\alpha}\int_0^T \hat{A}(t)XP\mathbf{w}(t)\mathbf{w}'(t)\,\mathrm{d}t + [\hat{\mathbf{x}}(0), \mathbf{0}, \ldots, \mathbf{0}]. \tag{58}$$

Clearly, the above equation can be put into the form [see Eq. (39)]

$$\hat{\mathbf{x}} = \frac{1}{\alpha}\left(\int_0^T \hat{A}(t)\times(\mathbf{w}(t)\mathbf{w}'(t)P')\,\mathrm{d}t\right)\hat{\mathbf{x}} + \begin{bmatrix}\hat{x}(0)\\ \mathbf{0}\\ \mathbf{0}\\ \mathbf{0}\end{bmatrix}$$

or

$$\hat{\mathbf{x}} = \frac{1}{\alpha}A^0\hat{\mathbf{x}} + \mathbf{c} \tag{59}$$

where

$$A_0 = \int_0^T \hat{A}(t)\times(\mathbf{w}(t)\mathbf{w}'(t)P')\,\mathrm{d}t$$

$$\mathbf{c} = \begin{bmatrix}\hat{\mathbf{x}}(0)\\ \mathbf{0}\\ \cdot\\ \cdot\\ \cdot\\ \mathbf{0}\end{bmatrix}.$$

Finally, Eq. (59) yields the approximate solution [see Eq. (40)]

$$\hat{\mathbf{x}} = \left(I + \frac{1}{\alpha}A^0\right)\mathbf{c}.$$

V. Conclusion

We have seen that time-varying control systems can be studied using a procedure originally proposed for time-invariant systems. The main idea is to expand the time varying components of the system into a finite set of Walsh functions grouped in powers of 2 for establishing convergence in the mean of the function at the subintervals 2^{-q} of the normalized time interval [0, 1]. The mathematical model of the time-varying system is projected to a finite Walsh function subspace in which we can use the methodology developed in (**1**, **2**). The reverse procedure enables us to implement the suboptimal control for the case with complete state feedback control and output feedback control via an observer.

References

(**1**) P. Stavroulakis and S. Tzafestas, "Walsh series approach to observer and filter design in optimal control systems", *Int. J. Control*, Vol. 26, pp. 721–736, 1977.

(2) P. Stavroulakis and S. Tzafestas, "Walsh series approach to time-delay control system observer design", *Int. J. System Sci.* In press.

(3) M. Athans and P. L. Falb, "optimal Control", McGraw-Hill, New York, 1966.

(4) A. E. Bryson Jr. and Y. E. Ho., "Applied Optimal Control", Blaisdell, Waltham, Mass., 1969.

(5) D. G. Luenbenger, "Observing the state of a linear system", *IEEE Trans. Mil. Electronics,* Vol. MIL-8, pp. 74–80, 1964.

(6) C. F. Chen and C. H. Hsiao, "Design of piecewise constant gains for optimal control via Walsh functions", *IEEE Trans. on Automatic Control,* Vol. AC-20, pp. 596–602, 1975.

(7) Y. O. Yuksel and J. J. Bongiorno, "Observers for linear multivariable systems with applications", *IEEE Trans. Automatic Control,* Vol. AC-16, No. 6, 1971.

(8) J. L. Walsh, "A closed set of orthogonal functions", *Am. J. Math.,* Vol. 45, pp. 5–24, 1923.

(9) D. G. Luenberger, "A new derivation of the quadratic loss control equation", *IEEE Trans. Automatic Control,* Vol. AC-10, pp. 202–203, 1965.

(10) R. E. A. C. Paley, "A remarkable series of orthogonal functions", *Proc. Lond. math. Soc.,* Vol. 34, pp. 241–279, 1932.

(11) N. J. Fine, "On the Walsh functions", *Trans. Am. math. Soc.,* Vol. 65, pp. 372–414, 1949.

(12) D. L. Kleinman, T. Fortmann and M. Athans, "On the design of linear systems with piecewise constant feedback gains", *IEEE Trans. Automatic Control,* Vol. AC-13, No. 4, 1968.

Reprinted from *Int. J. Control* **26**:721–736 (1977)

Walsh series approach to observer and filter design in optimal control systems

P. STAVROULAKIS† and S. TZAFESTAS‡

In this paper the observer design problem in control systems is studied using the Walsh functions as the approximating set of functions to find a finite series expansion of both the state of system and the observer. The basic objective is to develop a computational algorithm for the determination of the coefficients in the expansion. The special properties of the Walsh functions are exploited properly for both continuous and discrete-time systems and it is shown that the initial conditions on the state and the observer can be used iteratively to determine the column vectors of the matrix needed in the expansion of the state and observer output. The same procedure is applied to stochastic discrete-time systems.

1. Introduction

Recent work in using approximation procedures for studying optimal control problems (Chen and Hsiao 1975) has shown that the expansion of the system state and the control input into a finite series of Walsh functions leads to interesting results.

Recently, Walsh's series has been used widely in almost all areas of engineering science, such as communication theory, electromagnetic theory, bioscience, and the solution of differential equations. The advent of computers has made the method of approximating certain physical quantities by a finite Walsh series approximation a useful and well managed technique. The objective of this paper is to use Walsh functions to study the problem of designing observers in optimal control systems. The technique to be developed here is similar to that given in Chen and Hsiao (1975) for the case of optimal control systems.

2. Problem statement

Consider the following linear system in state space

$$\dot{\mathbf{x}} = A\mathbf{x} + B\mathbf{u} \tag{1}$$

$$\mathbf{y}(t) = C\mathbf{x}(t) \tag{2}$$

where $\mathbf{x}(t)$ is the state vector of dimension n, A, B, C are constant matrices, $\mathbf{u}(t)$ is the control vector of dimension $p < n$, and $\mathbf{y}(t)$ is the output vector of dimension $m < n$.

We seek to find the optimal control law $\mathbf{u}^*(t)$ which enters in (1) and minimizes the quadratic cost functional

$$J = \int_0^\infty [\mathbf{x}'(t)Q\mathbf{x}(t) + \mathbf{u}'(t)R\mathbf{u}(t)]\, dt \tag{3}$$

Received 22 June 1976.

† Visiting Assistant Professor, University of Athens, Athens, Greece. On leave of absence from Bell Laboratories.

‡ Professor of Control Systems, University of Patras, Patras, Greece, and NRC ‘ Demokritos ’, Aghia Paraskevi, Attiki, Greece.

under the constraints

(i) the output $\mathbf{y}(t)$ is only available for control,
(ii) a digital computational technique is to be used for the determination of the control $\mathbf{u}^*(t)$,

where Q is a symmetric positive semidefinite matrix and R is a symmetric positive definite matrix.

The preceding problem without the constraint (ii) has been extensively studied, e.g. by Luenberger (1964), Miller (1973) and Yüskel and Bongiorno (1971). It has been shown that it is possible to use the output $\mathbf{y}(t)$ as the input to another linear dynamic system called ' observer ' of the form

$$\dot{\mathbf{z}} = D\mathbf{z} + G\mathbf{y} + H\mathbf{u} \tag{4}$$

such that

$$\mathbf{z}(t) = T\mathbf{x}(t) + \mathbf{e}(t) \tag{5}$$

and

$$\hat{\mathbf{x}}(t) = \begin{bmatrix} C \\ \hline T \end{bmatrix}^{-1} \begin{bmatrix} \mathbf{y}(t) \\ \hline \mathbf{z}(t) \end{bmatrix} \tag{6}$$

where the vector $\hat{\mathbf{x}}(t)$ is an estimate of the state vector $\mathbf{x}(t)$ which is used with the same feedback control matrix K as when all the state $\mathbf{x}(t)$ is available for measurement, i.e.

$$\mathbf{u}^*(t) = K\hat{\mathbf{x}}(t) \tag{7}$$

where (B' denotes the transpose of B) :

$$K = -R^{-1}B'P \tag{8}$$

and P is the solution of the steady-state Riccati equation. Also, choosing $\mathbf{z}(0) = T\mathbf{x}(0)$ it follows that

$$\lim_{t\to\infty} (\hat{\mathbf{x}}(t) - \mathbf{x}(t)) = 0 \tag{9}$$

3. Solution of problem : Walsh series approach

We observe that if the equation $\mathbf{u}^* = K\mathbf{x}$ is introduced into (1), we obtain the following closed-loop system :

$$\dot{\mathbf{x}} = (A + BK)\mathbf{x} \tag{10}$$

Similarly from eqn. (6) we obtain

$$\hat{\mathbf{x}}(t) = M\mathbf{y}(t) + N\mathbf{z}(t) \tag{11}$$

where

$$[M \vdots N] = \begin{bmatrix} C \\ \hline T \end{bmatrix}^{-1}$$

Introducing the equation $\mathbf{u}^* = K\mathbf{x}$ into (4) gives

$$\dot{\mathbf{z}} = D\mathbf{z} + (GC + HK)\mathbf{x} \tag{12}$$

The solution of eqns. (10) and (12) is necessary for the determination of the optimal control, since from (11) and (7) we obtain

$$\mathbf{u}^*(t) = KM\mathbf{y}(t) + KN\mathbf{z}(t) \tag{13}$$

The solution of this system of equations will be carried out here using Walsh functions as the approximating set of functions for the state and the observer. Such an approximation will lead to an iterative technique amenable to digital computation.

To start with, we assume that the derivative of the state $\mathbf{x}(t)$ can be approximated by a finite summation of Walsh functions as

$$\dot{\mathbf{x}}(t) = X\mathbf{w}(t) \tag{14}$$

where X is a constant matrix with element X_{ij} given by

$$X_{ij} = \int_0^1 \dot{x}_i(t) w_j(t)\, dt \tag{15}$$

where $\dot{x}_i(t)$ is the ith component of the vector $\dot{\mathbf{x}}(t)$, $w_j(t)$ is the Walsh function of order j, and

$$\mathbf{w}(t) = \begin{bmatrix} w_1(t) \\ w_2(t) \\ \vdots \\ w_n(t) \end{bmatrix} \tag{16}$$

Equation (14) can be integrated to give $\mathbf{x}(t)$

$$\mathbf{x}(t) = \int_0^t X\mathbf{w}(\lambda)\, d\lambda + \mathbf{x}(0) \tag{17}$$

But it can be shown that (Chen and Hsiao 1975)

$$\int_0^t \mathbf{w}(\lambda)\, d\lambda = P\mathbf{w}(t) \tag{18}$$

where P is a matrix of constants which has the following particular form :

$$P_{N\times N} = \begin{bmatrix} P_{N/2\times N/2} & -\dfrac{1}{2N} I_{N/2\times N/2} \\ \dfrac{1}{2N} I_{N/2\times N/2} & 0_{N/2\times N/2} \end{bmatrix} \tag{19}$$

where N is preferably some power of 2, and

$$P_{2\times 2} = \begin{bmatrix} \frac{1}{2} & -\frac{1}{4} \\ \frac{1}{4} & 0 \end{bmatrix} \tag{20}$$

Using the definitions (19) and (20) one can determine the matrix P of any order and in turn $\mathbf{x}(t)$. Thus

$$\mathbf{x}(t) = XP\mathbf{w}(t) + \mathbf{x}(0) \tag{21}$$

Similarly, the state of the observer can be expanded in a finite Walsh series to give

$$\dot{\mathbf{z}}(t) = Z\mathbf{w}(t) \tag{22}$$

and

$$\mathbf{z}(t) = ZP\mathbf{w}(t) + \mathbf{z}(0) \tag{23}$$

Introducing (14) and (21)–(23) into (10) and (12) one obtains

$$X\mathbf{w}(t) = (A + BK)[XP\mathbf{w}(t) + \mathbf{x}(0)] \tag{24}$$

$$Z\mathbf{w}(t) = D(ZP\mathbf{w}(t) + \mathbf{z}(0)) + (GC + HK)(XP\mathbf{w}(t) + \mathbf{x}(0)) \tag{25}$$

The expression $XP\mathbf{w}(t) + \mathbf{x}(0) = \mathbf{x}(t)$ can be written as

$$[XP + (\mathbf{x}(0) \vdots \mathbf{0} \vdots \mathbf{0} \vdots \ldots \vdots \mathbf{0})]\mathbf{w}(t) = \mathbf{x}(t) \tag{25 a}$$

since $w_1(t) = 1$. Now, if we define X_0 as $X_0 = [\mathbf{x}(0) \vdots \mathbf{0} \vdots \mathbf{0} \vdots \ldots \vdots \mathbf{0}]$ the expression (25 a) can be written as

$$\mathbf{x}(t) = (XP + X_0)\mathbf{w}(t) \tag{26}$$

Similarly

$$\mathbf{z}(t) = (ZP + Z_0)\mathbf{w}(t) \tag{27}$$

Introduction of (26) and (27) into (24) and (25) gives

$$X = \hat{A}(XP + X_0) \tag{28}$$

$$Z = D(ZP + Z_0) + \hat{G}(XP + X_0) \tag{29}$$

where

$$\hat{A} = A + BK, \quad \hat{G} = GC + HK \tag{29 a}$$

Rearranging the columns $\mathbf{x}_1, \mathbf{x}_2, \ldots, \mathbf{x}_n$ and $\mathbf{z}_1, \mathbf{z}_2, \ldots, \mathbf{z}_n$ of the matrices X and Z into vector form, i.e.

$$\hat{\mathbf{x}} = \begin{bmatrix} \mathbf{x}_1 \\ \mathbf{x}_2 \\ \vdots \\ \mathbf{x}_n \end{bmatrix}, \quad \hat{\mathbf{z}} = \begin{bmatrix} \mathbf{z}_1 \\ \mathbf{z}_2 \\ \vdots \\ \mathbf{z}_n \end{bmatrix}$$

eqns. (28) and (29) take the form

$$\hat{\mathbf{x}} = (\hat{A} \circledast P')\hat{\mathbf{x}} + (\hat{A} \circledast I)\hat{\mathbf{x}}_0 \tag{30}$$

$$\hat{\mathbf{z}} = (D \circledast P')\hat{\mathbf{z}} + (\hat{G} \circledast P')\hat{\mathbf{x}} + (D \circledast I)\hat{\mathbf{z}}_0 + (\hat{G} \circledast I)\hat{\mathbf{x}}_0 \tag{31}$$

where for any two matrices A, B the operation $\circledast$ in $A \circledast B$, called the Kronecker product, is defined as

$$A \circledast B' = \begin{bmatrix} Ab_{11} & Ab_{21} & \cdots & Ab_{n1} \\ Ab_{12} & Ab_{22} & \cdots & Ab_{n2} \\ \vdots & \vdots & & \vdots \\ Ab_{1n} & Ab_{2n} & \cdots & Ab_{nn} \end{bmatrix}$$

and

$$\hat{\mathbf{x}}_0 = \begin{bmatrix} \mathbf{x}(0) \\ \mathbf{0} \\ \vdots \\ \mathbf{0} \end{bmatrix}, \quad \hat{\mathbf{x}}_0 = \begin{bmatrix} \mathbf{z}(0) \\ \mathbf{0} \\ \mathbf{0} \\ \vdots \\ \mathbf{0} \end{bmatrix}$$

Assuming that in the expansions (26) and (27), eight terms constitute a sufficient approximation of the state vectors $\mathbf{x}(t)$ and $\mathbf{z}(t)$, then it can be shown (Chen and Hsiao 1975) that a simple recursive form for the evaluation of $\mathbf{x}_i$, $i > 0$ with respect to $\mathbf{x}_1$ can be developed. For a further step in our approximation procedure we make the following change of the variable t:

$$\lambda = at, \quad a \text{ is a positive constant}$$

and obtain

$$\dot{\mathbf{x}}(t) = \dot{\mathbf{x}}(\lambda)\,.\,a, \quad \dot{z}(t) = \dot{z}(\lambda)\,.\,a \tag{32}$$

Therefore, upon introducing (32) in (14) and (22), eqns. (30) and (31) take the form

$$\hat{\mathbf{x}}(\lambda) = \frac{1}{a}\,[(\hat{A} \circledast P')\hat{\mathbf{x}}(\lambda) + (\hat{A} \circledast I)\hat{\mathbf{x}}_0] \tag{33}$$

$$\hat{\mathbf{z}}(\lambda) = \frac{1}{a}\,[(D \circledast P')\hat{\mathbf{z}}(\lambda) + (\hat{G} \circledast \hat{P}')\hat{\mathbf{x}}(\lambda) + (D \circledast I)\hat{\mathbf{z}}_0 + (\hat{G} \circledast I)\hat{\mathbf{x}}_0] \tag{34}$$

In order to obtain some indication about the structure of the Kronecker product, we give an example concerning the case $N = 2^3$. In this case

$$P'_{8\times 8} = \left[\begin{array}{cc|cc|cccc} \frac{1}{2} & \frac{1}{4} & \frac{1}{8} & 0 & \frac{1}{16} & 0 & 0 & 0 \\ -\frac{1}{4} & 0 & 0 & \frac{1}{8} & 0 & \frac{1}{16} & 0 & 0 \\ \hline -\frac{1}{8} & 0 & 0 & 0 & 0 & 0 & \frac{1}{16} & 0 \\ 0 & -\frac{1}{8} & 0 & 0 & 0 & 0 & 0 & \frac{1}{16} \\ \hline -\frac{1}{16} & 0 & 0 & 0 & 0 & 0 & 0 & 0 \\ 0 & -\frac{1}{16} & 0 & 0 & 0 & 0 & 0 & 0 \\ 0 & 0 & -\frac{1}{16} & 0 & 0 & 0 & 0 & 0 \\ 0 & 0 & 0 & -\frac{1}{16} & 0 & 0 & 0 & 0 \end{array}\right]_{8\times 8} \tag{35}$$

$$A \circledast P_{8\times 8}' = \left[\begin{array}{cc:cc:cccc}
\frac{1}{2}\hat{A} & \frac{1}{4}\hat{A} & \frac{1}{8}\hat{A} & 0 & \frac{1}{16}\hat{A} & 0 & 0 & 0 \\
-\frac{1}{4}\hat{A} & 0 & 0 & \frac{1}{8}\hat{A} & 0 & \frac{1}{16}\hat{A} & 0 & 0 \\
\hdashline
-\frac{1}{8}\hat{A} & 0 & 0 & 0 & 0 & 0 & \frac{1}{16}\hat{A} & 0 \\
0 & -\frac{1}{8}\hat{A} & 0 & 0 & 0 & 0 & 0 & \frac{1}{16}\hat{A} \\
\hdashline
-\frac{1}{16}\hat{A} & 0 & 0 & 0 & 0 & 0 & 0 & 0 \\
 & -\frac{1}{16}\hat{A} & 0 & 0 & 0 & 0 & 0 & 0 \\
0 & 0 & -\frac{1}{16}\hat{A} & 0 & 0 & 0 & 0 & 0 \\
0 & 0 & 0 & -\frac{1}{16}\hat{A} & 0 & 0 & 0 & 0
\end{array}\right]_{(N8)\times(N8)} \tag{36}$$

and

$$\hat{A} \circledast I = \begin{bmatrix}
\hat{A} & 0 & 0 & 0 & 0 & 0 & 0 & 0 \\
0 & \hat{A} & 0 & 0 & 0 & 0 & 0 & 0 \\
 & \vdots & & \vdots & & & \vdots & \\
0 & 0 & 0 & 0 & 0 & 0 & 0 & \hat{A}
\end{bmatrix} \tag{37}$$

The first iteration of our procedure gives

$$\mathbf{x}_{i+4} = -\frac{1}{a}\left(\frac{1}{2^4}\hat{A}\mathbf{x}_i + \hat{A}\mathbf{x}_{0(i+4)}\right), \quad i = 1, 2, 3, 4 \tag{38}$$

and

$$\mathbf{z}_{i+4} = -\frac{1}{a}\left(\frac{1}{2^4}(D\mathbf{z}_i + \hat{G}\mathbf{x}_i) + D\mathbf{z}_{0(i+4)}\right), \quad i = 1, 2, 3, 4 \tag{39}$$

where $\mathbf{x}_{0(i+4)}$ and $\mathbf{z}_{0(i+4)}$ are the $(i+4)$th elements of the augmented vectors

$$\hat{\mathbf{x}}_0 = \begin{bmatrix} \mathbf{x}_0 \\ \mathbf{0} \\ \vdots \\ \mathbf{0} \\ \vdots \\ \mathbf{0} \end{bmatrix} \begin{array}{l} \leftarrow \text{1st} \\ \leftarrow \text{2nd} \\ \vdots \\ \leftarrow (i+4)\text{th}, \\ \\ \leftarrow \text{8th} \end{array} \qquad \hat{\mathbf{z}}_0 = \begin{bmatrix} \mathbf{z}_0 \\ \mathbf{0} \\ \vdots \\ \mathbf{0} \\ \vdots \\ \mathbf{0} \end{bmatrix} \begin{array}{l} \leftarrow \text{1st} \\ \leftarrow \text{2nd} \\ \\ \leftarrow (i+4)\text{th} \\ \\ \leftarrow \text{8th} \end{array}$$

and in our case are zero for both $\hat{\mathbf{x}}_0$ and $\hat{\mathbf{z}}_0$.

If in the original observer problem we assume $\mathbf{z}(0) = T\mathbf{x}(0)$, then it is easy to show (Luenberger 1964) that

$$\mathbf{z}(t) = T\mathbf{x}(t) \tag{40}$$

for all t, and eqns. (26), (27) and (40) give

$$\mathbf{z}_{i+4} = T\mathbf{x}_{i+4}, \quad i = 1, 2, 3, 4 \tag{41}$$

Now, if we multiply eqn. (38) by T we obtain

$$T\mathbf{x}_{i+4} = -\frac{1}{a} T\left(\frac{1}{2^4}\hat{A}\mathbf{x}_i + \hat{A}\mathbf{x}_{0(i+4)}\right), \quad i=1, 2, 3, 4 \tag{42}$$

Similarly from (39) and (41) we obtain

$$T\mathbf{x}_{i+4} = -\frac{1}{a}\left(\frac{1}{2^4} DT\mathbf{x}_i + \hat{G}\mathbf{x}_i\right) + D\mathbf{z}_{0(i+4)} \tag{43}$$

Hence, since $\mathbf{x}_{i(0+4)} = 0$ and $\mathbf{z}_{0(i+4)} = 0$, the expressions (42) and (43) give

$$-\frac{1}{a} T\left(\frac{1}{2^4}\hat{A}\mathbf{x}_i\right) = -\frac{1}{a}\left(\frac{1}{2^4} DT\mathbf{x}_i + \hat{G}\mathbf{x}_i\right) \tag{44}$$

Thus (see (29 a)) :

$$T(A+BK) = DT + (GC+HK) \tag{45}$$

which gives for all i

$$\left.\begin{aligned} TA - DT &= GC \\ TB &= H \end{aligned}\right\} \tag{46}$$

The set of algebraic eqns. (46) are the well-known conditions for the regular observer/control problem. Thus the basic solution of the observer is also valid for the approximation used to produce the computational technique via the Walsh function expansion method.

Having determined $\mathbf{x}_{i+4}$ and $\mathbf{z}_{i+4}$ for $i=1, 2, 3, 4$ with respect to $\mathbf{x}_i$ and $\mathbf{z}_i$ from (38) and (39), and T from (46) and (11), we can, in turn, continue the iteration to calculate all other vector elements $\mathbf{x}_i$ and $\mathbf{z}_i$ with respect to $\mathbf{x}_1$ and $\mathbf{z}_1$. In fact, introducing eqns. (35)–(37) into (33) and (34) we can solve for the vectors $\mathbf{x}_{i+2}$ and $\mathbf{z}_{i+2}$, $i=1, 2$ with respect to $\mathbf{x}_i$ and $\mathbf{z}_i$, and obtain

$$\left.\begin{aligned} \mathbf{x}_{i+2} &= R_3{}^{\hat{A}}\mathbf{x}_i \\ \mathbf{z}_{i+2} &= R_3{}^{D}\mathbf{z}_i + R_3{}^{*}\mathbf{x}_i \end{aligned}\right\} i=1, 2 \tag{47}$$

$$\left.\begin{aligned} \mathbf{x}_{i+1} &= R_2{}^{\hat{A}}\mathbf{x}_i \\ \mathbf{z}_{i+1} &= R_2{}^{D}z_i + R_2{}^{*}\mathbf{x}_i \end{aligned}\right\} i=1, 2 \tag{48}$$

$$\left.\begin{aligned} \mathbf{x}_1 &= \left(I + \frac{\hat{A}}{2a}\right)\hat{A}\mathbf{x}(0) \\ \mathbf{z}_1 &= \left(I + \frac{D}{2a}\right) D\mathbf{z}(0) + \hat{G}\mathbf{x}(0) + R_1{}^{*}\mathbf{x}_1 \end{aligned}\right\} \tag{49}$$

More details about the derivation of (47)–(49) are given in the Appendix.

Example

Consider the optimal control problem, where

$$\begin{bmatrix} \dot{x}_1 \\ \dot{x}_2 \end{bmatrix} = \begin{bmatrix} 0 & 1 \\ 0 & -2 \end{bmatrix}\begin{bmatrix} x_1 \\ x_2 \end{bmatrix} + \begin{bmatrix} 0 \\ 1 \end{bmatrix} u \tag{50}$$

$$y=[1 \quad 0]\begin{bmatrix} x_1 \\ x_2 \end{bmatrix}=x_1 \tag{51}$$

$$J=\int_0^\infty \left\{[x_1 \quad x_2]\begin{bmatrix} 4 & 0 \\ 0 & 1 \end{bmatrix}\begin{bmatrix} x_1 \\ x_2 \end{bmatrix}+u^2\right\} dt \tag{52}$$

$$x_1(0)=x_{01}=1 \quad \text{and} \quad x_2(0)=0 \tag{53}$$

For this problem we can easily show that

$$u^*=R^{-1}B'K\mathbf{x}=[-2 \quad -1]\begin{bmatrix} x_1 \\ x_2 \end{bmatrix}=-2x_1-x_2$$

We observe that the optimal control u^* is a function of both $x_1(t)$ and $x_2(t)$, and that the output $y(t)=x_1(t)$ gives information about x_1 only. Therefore, the reconstruction of the whole state

$$\begin{bmatrix} x_1 \\ x_2 \end{bmatrix}$$

will require the utilization of an observer of dimension 1. Let the observer be given by

$$\dot{z}=dz+gy+hu \tag{54}$$

where, if we take

$$gC=[1, 0] \quad \text{and} \quad d=-3$$

we obtain from (46)

$$[\tau_1 \quad \tau_2]\begin{bmatrix} 0 & 1 \\ 0 & -2 \end{bmatrix}+3[\tau_1 \quad \tau_2]=[1 \quad 0] \tag{55}$$

where $T=[\tau_1, \tau_2]$. Solving eqn. (55) for τ_1 and τ_2 we obtain

$$\tau_1=\tfrac{1}{3}, \quad \tau_2=-\tfrac{1}{3}$$

Hence

$$T=[\tfrac{1}{3}, -\tfrac{1}{3}] \quad \text{and} \quad H=TB=-\tfrac{1}{3}$$

Equation (54) now becomes

$$\dot{z}=-3z+x_1-\tfrac{1}{3}u \tag{56}$$

Combining eqns. (50) and (56) we obtain

$$\begin{bmatrix} \dot{x}_1 \\ \dot{x}_2 \\ \dot{z} \end{bmatrix}=\begin{bmatrix} 0 & 1 & 0 \\ -2 & -3 & 0 \\ \frac{5}{3} & \frac{1}{3} & -3 \end{bmatrix}\begin{bmatrix} x_1 \\ x_2 \\ z \end{bmatrix} \tag{57}$$

Let

$$\mathbf{q} = \begin{bmatrix} x_1 \\ x_2 \\ z \end{bmatrix} \tag{58}$$

then

$$\dot{\mathbf{q}} = A\mathbf{q} \tag{59}$$

and

$$\mathbf{q}(0) = \begin{bmatrix} x_1(0) \\ x_2(0) \\ z(0) \end{bmatrix} = \begin{bmatrix} 1 \\ 0 \\ \frac{1}{3} \end{bmatrix}$$

If we choose $a = 1$, then eqns. (A 1), (A 12), (A 13), (A 17), and (A 20) of the Appendix give

$$\left.\begin{aligned}
\mathbf{q}_{i+4} &= -\begin{bmatrix} 0 & 0{\cdot}625 & 0 \\ -0{\cdot}187 & -0{\cdot}125 & 0{\cdot}187 \\ 0{\cdot}125 & 0 & 0{\cdot}25 \end{bmatrix}\mathbf{q}_i, \quad i = 1, 2, 3, 4 \\
\mathbf{q}_{i+2} &= -\begin{bmatrix} -0{\cdot}006 & 0{\cdot}124 & 0{\cdot}009 \\ -0{\cdot}356 & -0{\cdot}255 & 0{\cdot}338 \\ 0{\cdot}237 & 0{\cdot}005 & -0{\cdot}471 \end{bmatrix}\mathbf{q}_i, \quad i = 1, 2 \\
\mathbf{q}_{i+1} &= -\begin{bmatrix} -0{\cdot}046 & 0{\cdot}246 & 0{\cdot}070 \\ -0{\cdot}6 & -0{\cdot}54 & 0{\cdot}452 \\ 0{\cdot}35 & 0{\cdot}0156 & 0{\cdot}724 \end{bmatrix}\mathbf{q}_i, \quad i = 1 \\
\mathbf{q}_1 &= \begin{bmatrix} 0{\cdot}8 & 0{\cdot}2 & 0{\cdot}1 \\ -0{\cdot}4 & 0{\cdot}4 & 0{\cdot}2 \\ -0{\cdot}366 & 0{\cdot}066 & 0{\cdot}366 \end{bmatrix}\begin{bmatrix} 1 \\ 0 \\ \frac{1}{3} \end{bmatrix}
\end{aligned}\right\} \tag{60}$$

From eqns. (58) and (60) we obtain

$$z(t) = -0{\cdot}244w_1(t) - 0{\cdot}113w_2(t) - 0{\cdot}311w_3(t) - 0{\cdot}069w_4(t) - 0{\cdot}042w_5(t) + 0{\cdot}028w_6(t) + 0{\cdot}077w_7(t) + 0{\cdot}023w_8(t)$$

4. Discrete-time systems

The procedure described in § 3 can be used to study discrete-time systems by using the discrete-time Walsh functions (Pearl 1971). For the sake of completeness we start with a presentation of the basic properties of the discrete-time Walsh functions. The discrete-time Walsh function of order k at time n, where k and n are integers, is defined as

$$w_k(n) = (-1)^{\sum_{i=0}^{N-1} k_i n_i} \tag{61}$$

where

$$k = \sum_{i=0}^{N-1} k_i 2^i, \quad n = \sum_{i=0}^{N-1} n_i 2^i$$

i.e. k_i and n_i are the digits in the binary representation of k and n. It can be shown (Tse and Athans 1970) that

$$\left.\begin{aligned} \sum_n w_k(n) w_{k'}(n) &= 2^N \delta_{k,k'} \\ \sum_n w_k(n) w_k(n') &= 2^N \delta_{n,n'} \end{aligned}\right\} \tag{62}$$

These properties allow the expansion of any discrete-time function to be presented in the form

$$f(n) = \frac{1}{2^N} \sum_k a_k w_k(n) \tag{63}$$

where

$$a_k = \sum_n w_k(n) f(n) \tag{64}$$

On the basis of the above discussion the equivalent discrete-time observer problem takes the following form :

Given

$$\left.\begin{aligned} \mathbf{x}(n+1) &= A\mathbf{x}(n) + B\mathbf{u}(n) \\ \mathbf{y}(n) &= C\mathbf{x}(n) \end{aligned}\right\} \tag{65}$$

Find an observer of the type

$$\mathbf{z}(n+1) = F\mathbf{z}(n) + G\mathbf{y}(n) + H\mathbf{u}(n) \tag{66}$$

such that

$$\mathbf{z}(n) = T\mathbf{x}(n) \tag{67}$$

and

$$\mathbf{x}(n) = P\mathbf{z}(n) + v\mathbf{y}(n) \tag{68}$$

where

$$\begin{bmatrix} T \\ \hdashline C \end{bmatrix}^{-1} = [P \,\vdots\, V] \tag{69}$$

It can be shown (Tse and Athans 1970) that if we choose

$$\left.\begin{aligned} F&=TAP \\ G&=TAV \\ H&=TB \\ \mathbf{z}(0)&=T\mathbf{x}(0) \end{aligned}\right\} \tag{70}$$

then the observer solves the problem in the way it is stated here.

In order to be able to use the discrete Walsh functions for determining the control $\mathbf{u}(n)$ we work as follows.

Let

$$\mathbf{x}(n)=X\mathbf{w}(n) \tag{71}$$

where

$$\mathbf{w}(n)=\begin{bmatrix} w_1(n) \\ \vdots \\ w_j(n) \\ \vdots \\ w_N(n) \end{bmatrix} \tag{72}$$

and the elements X_{ij} of the matrix X are given by

$$X_{ij}=\sum_{n=0}^{N-1} x_i(n)w_j(n) \tag{73}$$

where $x_i(n)$ is the ith component of the state vector $\mathbf{x}(n)$. Due to the property (62), (73) can be written

$$X=\tfrac{1}{2}N\sum_{n=0}^{N-1} \mathbf{x}(n)\mathbf{z}(n) \tag{74}$$

Similarly we can write

$$\left.\begin{aligned} \mathbf{z}(n)&=Z\mathbf{w}(n) \\ \mathbf{x}(n+1)&=\hat{X}\mathbf{w}(n) \\ \mathbf{z}(n+1)&=\hat{Z}\mathbf{w}(n) \end{aligned}\right\} \tag{75}$$

Let the optimal control be given in the feedback form

$$\mathbf{u}^*(n)=K\mathbf{x}(n) \tag{76}$$

Then eqns. (65) and (66) can be written

$$\mathbf{x}(n+1)=(A+BK)\mathbf{x}(n) \tag{77}$$

$$\mathbf{z}(n+1)=F\mathbf{z}(n)+(GC+HK)\mathbf{x}(n) \tag{78}$$

Setting again

$$\hat{A}=A+BK, \quad \hat{G}=GC+HK$$

we can write (77) and (78) in the form

$$\mathbf{x}(n+1)=\hat{A}\mathbf{x}(n) \tag{79}$$

$$\mathbf{z}(n+1)=F\mathbf{z}(n)+\hat{G}\mathbf{x}(n) \tag{80}$$

Multiplying eqns. (79) and (80) by $w'(n)$ and summing over the time n $(0\leqslant n<2^N)$ we obtain

$$\sum_n \mathbf{x}(n+1)\mathbf{w}'(n)=\sum_n \hat{A}\mathbf{x}(n)\mathbf{w}'(n) \tag{81}$$

$$\sum_n \mathbf{z}(n+1)\mathbf{w}'(n)=\sum_n (F\mathbf{z}(n)\mathbf{w}'(n)+\hat{G}\mathbf{x}(n)\mathbf{w}'(n)) \tag{82}$$

But from (74) and (75) we observe that (81) and (82) give

$$\hat{X}=\hat{A}X \tag{83}$$

$$\hat{Z}=FZ+\hat{G}X \tag{84}$$

Successive application of (79) and (80) for $n=0, 1, \ldots$ by using the fact that $\mathbf{z}(0)=T\mathbf{x}(0)$, where $\mathbf{x}(0)$ is the known initial condition of the state vector, yields

$$\mathbf{x}(n)=\hat{A}^n\mathbf{x}(0) \tag{85}$$

$$\mathbf{z}(n)=\left[F^{n-1}(FT+\hat{G})+\sum_{j=0}^{n-2} F^j\hat{G}A^{n-j-1}\right]\mathbf{x}(0), \quad n\geqslant 1 \tag{86}$$

where

$$\sum_{j=0}^{-1} F^j\hat{G}A^{n-j-1}\equiv 0$$

Introduction of (83) and (84) into (74) and (75) gives

$$X=\frac{1}{2^N}\sum_{n=0}^{N-1}(\hat{A})^n\mathbf{x}(0)\mathbf{w}'(n) \tag{87}$$

$$Z=\frac{1}{2^N}\sum_{n=1}^{N-1}\left[F^{n-1}(FT+\hat{G})+\sum_{j=0}^{n-2} F^j\hat{G}A^{n-j-1}\right]\mathbf{x}(0)\mathbf{w}'(n)+T\mathbf{x}(0)\mathbf{w}'(0) \tag{88}$$

5. Stochastic control systems (discrete time)

Similarly in the case of stochastic discrete-time control systems which are described by the difference equations

$$\mathbf{x}(n+1)=A\mathbf{x}(n)+B\mathbf{u}(n)+\boldsymbol{\xi}(n) \tag{89}$$

$$\mathbf{y}(n)-C\mathbf{x}(n)+\boldsymbol{\eta}(n) \tag{90}$$

we can find an observer of the form

$$\mathbf{z}(n+1)=F\mathbf{z}(n)+G\mathbf{y}(n)+H\mathbf{u}(n) \tag{91}$$

such that the observer output $\mathbf{z}(n)$ together with the system output $\mathbf{y}(n)$ give a minimum variance estimate $\hat{\mathbf{x}}(n)$ of the state $\mathbf{x}(n)$, of the form

$$\hat{\mathbf{x}}(n)=P\mathbf{z}(n)+V\mathbf{y}(n) \tag{92}$$

For optimal control systems it is known that this minimum variance estimate can be used to construct the optimal control

$$\mathbf{u}^*(n) = K\hat{\mathbf{x}}(n) = KP\mathbf{z}(n) + KV\mathbf{y}(n) \tag{93}$$

Here, without loss of generality, we shall constrain ourselves to the discussion of state estimation.

It has been shown (Tse and Athans 1970) that the observer estimator of the system (89) is given by

$$\mathbf{z}(n+1) = F\mathbf{z}(n) + G\mathbf{y}(n) \tag{94}$$

where

$$\left.\begin{aligned} \mathbf{u}(n) &= 0 \\ F &= TAP \\ G &= TAV \\ V &= \Delta C'[C\Delta C' + Q]^{-1} \\ \Delta &= A\Sigma A' + R \\ \Sigma &= \Delta - \Delta C'[C\Delta C' + Q]^{-1}C\Delta \end{aligned}\right\} \tag{95}$$

and

$$E\{\boldsymbol{\xi}(n)\boldsymbol{\xi}'(k)\} = R\delta_{nk} \tag{96}$$

$$E\{\boldsymbol{\eta}(n)\boldsymbol{\eta}'(k)\} = Q\delta_{nk} \tag{97}$$

Introducing eqns. (90) into (94) we obtain

$$\mathbf{z}(n+1) = F\mathbf{z}(n) + G(C\mathbf{x}(n) + \boldsymbol{\eta}(n)) \tag{98}$$

Then for one realization of the stochastic processes $\boldsymbol{\xi}(n)$ and $\boldsymbol{\eta}(n)$, eqns. (89) and (98) give

$$\mathbf{x}(n) = A^n\mathbf{x}(0) + \sum_{i=1}^{n} A^{n-i}\boldsymbol{\xi}(i-1) \tag{99}$$

$$\begin{aligned} \mathbf{z}(n) = (TAP)^n\mathbf{z}(0) &+ \sum_{i=1}^{n} (TAP)^{n-i}(TAVC)A^{i-1}\mathbf{x}(0) \\ &+ \sum_{i-1}^{n} (TAP)^{i-1}(TAVC)\left[\sum_{j=1}^{i=1} A^{i-1-j}\boldsymbol{\xi}(j-1)\right] \\ &+ \sum_{i=1}^{n} (TAP)^{n-i}(TAV)\boldsymbol{\eta}(i-1) \end{aligned} \tag{100}$$

Then eqns. (74) and (75) give

$$X = \frac{1}{2^N}\left[\sum_{n=1}^{N-1} A^n\mathbf{x}(0)\mathbf{w}'(n) + \sum_{n=1}^{N-1}\left(\sum_{i=1}^{n} A^{n-1}\boldsymbol{\xi}(i-1)\mathbf{w}'(n)\right)\right] \tag{101}$$

$$Z=\frac{1}{2^N}\left[\sum_{n=1}^{N-1}(TAP)^n\mathbf{z}(0)\mathbf{w}'(n)+\sum_{n=1}^{N-1}\sum_{i=1}^{n}(TAP)^{n-i}(TAVC)A^{i-1}\mathbf{x}(0)\mathbf{w}'(n)\right.$$
$$+\sum_{n=1}^{N-1}\sum_{i=1}^{n}\sum_{j=1}^{i-1}(TAP)^{n-i}(TAVC)A^{i-1-j}\boldsymbol{\xi}(j-1)\mathbf{w}'(n)$$
$$\left.+\sum_{n=1}^{N-1}\sum_{i=1}^{n}(TAP)^{n-i}(TAV)\boldsymbol{\eta}(i-1)\mathbf{w}'(n)\right] \tag{102}$$

For the determination of the estimate $\hat{\mathbf{x}}(n)$ based on the measurement $\mathbf{y}(n)$, we use eqn. (94) and obtain

$$\mathbf{z}(n)=(TAP)^n\mathbf{z}(0)+\sum_{i=1}^{n}(TAP)^{n-i}TAV\mathbf{y}(i-1) \tag{103}$$

$$Z=\sum_{n=1}^{N-1}\left[(TAP)^n\mathbf{z}(0)+\sum_{i=1}^{n}(TAP)^{n-i}G\mathbf{y}(i-1)\right]\mathbf{w}'(n) \tag{104}$$

where for all summations $\sum_{j=k}^{N}\{\,.\,\}$, it is assumed that $N \geqslant k$, otherwise the sum is zero.

The knowledge of the matrix Z can be used to determine the vector $\mathbf{z}(n)$ using eqn. (75), which in turn can be used in (92) for the determination of the minimum variance estimate $\hat{\mathbf{x}}(n)$ of $\mathbf{x}(n)$.

6. Conclusion

We have seen that the Walsh functions can be used as an approximating set of functions to find a finite series expansion of both the state and the observer. The standard results of both discrete and continuous time systems are embedded in the procedure of the respective cases. It is shown that the initial conditions on the state and the observer can be used iteratively to determine the state and the observer at every case. By subdividing the time interval of interest, we show in the Appendix that the procedure is still simplified and it involves only multiplication of $(n \times n)$ matrices. The procedure is extended to the case of stochastic discrete-time systems for determining the minimum variance estimate $\hat{\mathbf{x}}(n)$ of the stochastic state $\mathbf{x}(n)$ via an observer which is fed by the output of the stochastic system.

Appendix

Solving eqns. (30) and (31) for the last four vectors $\mathbf{x}_{i+4}$, $i=1, 2, 3, 4$ and $\mathbf{z}_{i+4}$, $i=1, 2, 3, 4$ it is easy to show (due to the special structure of the matrices (35), (36) and (37)) that

$$\mathbf{x}_{i+4}=-\frac{1}{2^4a}\hat{A}\mathbf{x}_i, \qquad i=1, 2, 3, 4 \tag{A 1}$$

$$\mathbf{z}_{i+4}=-\frac{1}{2^4a}(D\mathbf{z}_i+\hat{G}x_i), \quad i=1, 2, 3, 4 \tag{A 2}$$

In (A 1) and (A 2) use was made of the fact that $Z_{0(i+4)}=0$ and $X_{0(i+4)}=0$, $i=1, 2, 3, 4$.

Introducing eqns. (A 1) and (A 2) into (30) and (31), and solving for the vectors $\mathbf{x}_{i+2}$, $i=1, 2$ and $\mathbf{z}_{i+2}$, $i=1, 2$ we obtain

$$\mathbf{x}_{i+2} = -\frac{1}{a}\left[\tfrac{1}{8}\left(I+\frac{1}{256a^2}\hat{A}^2\right)^{-1}\hat{A}\right]\mathbf{x}_i, \quad i=1, 2 \tag{A 3}$$

$$\mathbf{z}_{i+2} = -\frac{1}{a}\left[\left(I+\frac{1}{256a^2}D^2\right)^{-1}\left(\tfrac{1}{8}D\mathbf{z}_i+\tfrac{1}{8}\hat{G}\mathbf{x}_i+\frac{1}{256a}(D\hat{G}+\hat{G}\hat{A})\mathbf{x}_{i+2}\right)\right] \tag{A 4}$$

The value of a is chosen such that the following approximations can be made :

$$\left(I+\frac{1}{a}\hat{A}\right)^{-1} \simeq I-\frac{1}{a}\hat{A} \tag{A 5†}$$

$$\left(I+\frac{1}{a}D\right)^{-1} \simeq I-\frac{1}{a}D \tag{A 6}$$

Then eqns. (A 3) and (A 4) can be written as

$$\mathbf{x}_{i+2} = -\frac{1}{8a}\left(I-\frac{1}{256a^2}\hat{A}^2\right)\hat{A}\mathbf{x}_i, \quad i=1, 2 \tag{A 7}$$

$$\mathbf{z}_{i+2} = -\left(I-\frac{1}{256a^2}D^2\right)\left(\frac{1}{8a}D\mathbf{z}_i+\frac{1}{8a}\hat{G}\mathbf{x}_i+\frac{1}{256a^2}(D\hat{G}+\hat{G}\hat{A})\mathbf{x}_{i+2}\right) \tag{A 8}$$

Introduce the following definitions

$$R_4{}^{\hat{A}} = -\frac{1}{2^4a}\Theta_4{}^{\hat{A}}\hat{A}, \quad \Theta_4{}^{\hat{A}} \equiv I, \qquad \Lambda_4{}^{\hat{A}} = -\left(\frac{1}{2^4a}\hat{A}\right)^2 \tag{A 9}$$

$$R_3{}^{\hat{A}} = -\frac{1}{2^3a}\Theta_3{}^{\hat{A}}\hat{A}, \quad \Theta_3{}^{\hat{A}} = I-\Lambda_4{}^{\hat{A}}, \quad \Lambda_3{}^{\hat{A}} = -\left(\frac{1}{2^3a}\hat{A}\right)^2 \tag{A 10}$$

$$R_2{}^{\hat{A}} = -\frac{1}{2^2a}\Theta_2{}^{\hat{A}}\hat{A}, \quad \Theta_2{}^{\hat{A}} = I-\Lambda_3{}^{\hat{A}} \tag{A 11}$$

For any other matrix, say K, we write $R_i{}^K$, $\Theta_i{}^K$, $\Lambda_i{}^K$ to denote the equivalent expressions obtained if in (A 9), (A 10), and (A 11) the matrix $\hat{A}$ is replaced by K.

Hence, equations (A 1), (A 2), (A 7) and (A 8) can be written as

$$\left.\begin{aligned}\mathbf{x}_{i+4} &= R_4{}^{\hat{A}}\mathbf{x}_i \\ \mathbf{z}_{i+4} &= R_4{}^{D}z_i + R_4{}^{\hat{A}}\mathbf{x}_i\end{aligned}\quad i=1, 2, 3, 4\right\} \tag{A 12}$$

Similarly eqns. (A 7) and (A 8) can be written as

$$\mathbf{x}_{i+2} = R_3{}^{\hat{A}}\mathbf{x}_i, \quad i=1, 2 \tag{A 13}$$

$$\mathbf{z}_{i+2} = R_3{}^{D}\mathbf{z}_i + 2\Theta_3{}^{D}R_4{}^{\hat{G}}\mathbf{x}_i + \Theta_3{}^{D}(R_4{}^{D}R_4{}^{\hat{G}} + R_4{}^{\hat{G}}R_4{}^{\hat{A}})\mathbf{x}_{i+2} \tag{A 14}$$

† Of course, choosing a large-time normalization factor a may restrict the actual computation interval. This does not make any restriction of generality since one may use successive intervals to complete any desired computation time.

Combining (A 13) and (A 14) gives

$$\mathbf{z}_{i+2} = R_3{}^D\mathbf{z}_i + \Theta_3{}^D[2R_4{}^{\hat{G}} - (R_4{}^D R_4{}^{\hat{G}} + R_4{}^{\hat{G}} R_4{}^{\hat{A}})R_3{}^{\hat{A}}]\mathbf{x}_i \qquad \text{(A 15)}$$

Then defining

$$R_3{}^* = \Theta_3{}^D[2R_4{}^{\hat{G}} - (R_4{}^D R_4{}^{\hat{G}} + R_4{}^{\hat{G}} R_4{}^{\hat{A}})R_3{}^{\hat{A}}]$$

eqn. (A 15) becomes

$$\mathbf{z}_{i+2} = R_3{}^D\mathbf{z}_i + R_3{}^*\mathbf{x}_i, \quad i = 1, 2 \qquad \text{(A 16)}$$

Now, introducing eqns. (A 13) and (A 7) into (30) and solving for $\mathbf{x}_{i+1}$, $i=1$ we obtain

$$\mathbf{x}_{i+1} = -\frac{1}{2^2 a}\left(I - \frac{\hat{A}^2}{64a^2}\right)\hat{A}\mathbf{x}_i, \quad i = 1 \qquad \text{(A 17)}$$

which can be written as

$$\mathbf{x}_{i+1} = R_2{}^{\hat{A}}\mathbf{x}_i, \quad i = 1 \qquad \text{(A 18)}$$

A similar process gives

$$\mathbf{z}_{i+1} = R_2{}^D\mathbf{z}_i + R_2{}^*\mathbf{x}_i, \quad i = 1 \qquad \text{(A 19)}$$

where

$$R_2{}^* = \Theta_2{}^D[4R_4{}^{\hat{G}} - 2R_4{}^D R_3{}^* R_2{}^{\hat{A}} - (R_4{}^D R_4{}^{\hat{G}} + R_4{}^{\hat{G}} R_4{}^{\hat{A}})R_2{}^{\hat{A}} - 2R_4{}^{\hat{G}} R_3{}^{\hat{A}} R_2{}^{\hat{A}}]$$

Finally, solving eqns. (30) and (31) for the first vectors $\mathbf{x}_1$ and $\mathbf{z}_1$ and making use of the expressions derived so far, we obtain

$$\mathbf{x}_1 = \left(I + \frac{\hat{A}}{2a}\right)\hat{A}\mathbf{x}(0) \qquad \text{(A 20)}$$

and

$$\mathbf{z}_1 = \left(I + \frac{D}{2a}\right)[(-4R_4{}^D R_2{}^* - 2R_4{}^D R_3{}^* - (R_4{}^D R_4{}^{\hat{G}} + R_4{}^{\hat{G}} R_4{}^{\hat{A}}) - 8R_4{}^{\hat{G}} + 4R_4{}^{\hat{G}} R_2{}^{\hat{A}} - 2R_4{}^{\hat{G}} R_3{}^{\hat{A}})\mathbf{x}_1 + D\mathbf{z}(0) + \hat{G}\mathbf{x}(0)] \qquad \text{(A 21)}$$

Defining

$$R_1{}^* \equiv -4R_4{}^D R_2{}^* - 2R_4{}^D R_3{}^* - (R_4{}^D R_4{}^{\hat{G}} + R_4{}^{\hat{G}} R_4{}^{\hat{A}}) - 8R_4{}^{\hat{G}} + 4R_4{}^{\hat{G}} R_2{}^{\hat{A}} - 2R_4{}^{\hat{G}} R_3{}^{\hat{A}}$$

eqn. (A 21) is written as

$$\mathbf{z}_1 = \left(I + \frac{D}{2a}\right)[D\mathbf{z}(0) + \hat{G}\mathbf{x}(0) + R_1{}^*\mathbf{x}_1)] \qquad \text{(A 22)}$$

References

Blochman, N. M., 1974, *Proc. I.E.E.E.*, **62.**

Chen, C. F., and Hsiao, C. H., 1975, *Int. J. Systems Sci.*, **6,** 833.

Harmuth, N. F., 1969, *I.E.E.E. Spectrum*, November, p. 82.

Luenberger, D. G., 1964, *I.E.E.E. Trans. mil. Electron.*, **8,** 74.

Miller, R. A., 1973, *Int. J. Control.*, **18,** 139.

Pearl, J., 1971, *I.E.E.E. Trans. syst. Man. Cybernet.*, **1,** 111.

Tse, E., and Athans, M., 1970, *I.E.E.E. Trans. autom. Control*, **15,** 416.

Walsh, J. L., 1923, *Am. J. math.*, **45,** 5.

Yüskel, Y. O., and Bongiorno, Jr., J. J., 1971, *I.E.E.E. Trans. autom. Control*, **16,** 603.

20

Reprinted from *Math. Computers Simulation* **22:**373–379 (1980)

DISTRIBUTED–PARAMETER OBSERVER–BASED CONTROL IMPLEMENTATION USING FINITE SPATIAL MEASUREMENTS

P. STAVROULAKIS
School of Engineering, Oakland University, Rochester, MI 48063, U.S.A.

S. TZAFESTAS
Control Systems Laboratory, University of Patras, Patras, Greece

Distributed-parameter systems have infinite dimensional states and hence, in practical observer and control design applications, the problem of approximate finite-dimensional state reconstruction arises. In this paper a useful approach for implementing such an approximation is presented, which makes use of double Walsh series expansions and is based on finited output measurements at a specified set of spatial points.

1. INTRODUCTION

This paper examines the problem of the approximate reconstruction of the unknown state variables for a class of distributed parameter systems via an observer. Previously the subject of observers for distributed-parameter systems was studied in [1]- [4] from the point of view of approximately reconstructing the state (Galerkin method, eigenfunction expansion) or by examining conditions under which the identity and reduced order observers exist for systems represented by strongly continuous groups. In [1] the Lyapunov stability approach for observer design has been generalized to include important classes of linear and non-linear operator, partial differential and partial-integral equations in describing the distributed parameter systems. In the present paper, a double Walsh series is introduced to represent approximately functions of two independent variables and is then applied to approximately reconstruct the state of a distributed parameter system through the use of an observer. This procedure leads to a structure similar to the one used in [10] and [11] by which a computational algorithm can be developed for the determination of the coefficients in the expansion. The special properties of the Walsh functions [9-12] are exploited so that the initial conditions in the state and the observer can be used iteratively to determine the column vectors of the matrix required in the expansion of the state and observer output.

2. PROBLEM FORMULATION

Consider the important class of distributed parameter systems (DPS) described by the following mathematical model [6,7]:

$$\frac{\partial \underline{y}(x,t)}{\partial t} = A(x)\frac{\partial \underline{y}}{\partial x}(x,t)+B(x)\underline{y}(x,t)+C(x)\underline{u}(t) \qquad (1)$$

where A and B are nxn symmetric matrices, C is an nxm matrix $(m \leq n)$, $\underline{y}(x,t)$ is the state vector of dimensions nx1, and the spatial variable x is a scalar in the interval $0 \leq x \leq 1$. We shall assume that in general A,B,C are functions of the spatial variable x. For this system, we shall pose the following initial-boundary conditions.

Initial condition: $\underline{y}(x,0)=\underline{g}(x)$ (2)

Boundary condition: $\underline{y}(0,t)=\underline{y}(1,t)=0$ (3)

This class of systems is characterized by a finite dimensional control and the control criterion is the minimization of an energy-type cost functional [6]. For the case where the cost functional represents the total energy of the system at a given time after the exercise of control is begun, it is shown that the optimal control exists and satisfies the maximum principle.

The uniqueness of the optimal control function is obtained through the uniqueness of the terminal state of the system and the satisfaction of a certain maximum principle by the control function. In situations where the maximum principle does not tell as much as in problems decribed by ordinary differential equations, a special normality condition is developed which determines when the optimal control can fail to be a uniquely determined piece-wise function.

For the problem defined here the output of this system is defined by a set of measurements at a limited number of points. The choice of the number of point measurements and their locations is closely related to the observability property as has been defined by Goodson and Klein [8].

According to this definition, a DPS is observable if the variable $\underline{y}(x,t)$ can be uniquely reconstructed over the interval $0\leq x\leq 1$ by using the available measurements. A system is called N-mode observable if the uniqueness of the first N modes (eigenfunctions) is guaranteed by the measurement data.

The measurements are assumed of the form

$$\hat{\underline{y}}_i(t)=\mu(x^i,t)\underline{y}(x^i,t)$$

where $x^i, i=1,2,\ldots,s$ are the measurement points. Thus we can write:

$$\hat{\underline{y}}(t)=M_s(t)\underline{y}_s(t) \tag{4}$$

where

$$\hat{\underline{y}}(t)=\begin{bmatrix}\hat{\underline{y}}_1(t)\\ \hat{\underline{y}}_2(t)\\ \vdots\\ \hat{\underline{y}}_s(t)\end{bmatrix}_{ns\times 1},\quad \underline{y}_s(t)=\begin{bmatrix}\underline{y}(x^1,t)\\ \underline{y}(x^2,t)\\ \vdots\\ \underline{y}(x^s,t)\end{bmatrix}_{ns\times 1}$$

$$M_s(t)=\begin{bmatrix}\mu(x^1,t) & & & & & & & & 0\\ & \ddots & & & & & & & \\ & & \mu(x^1,t) & & & & & & \\ & & & \mu(x^2,t) & & & & & \\ & & & & \ddots & & & & \\ & & & & & \mu(x^2,t) & & & \\ & & & & & & \mu(x^s,t) & & \\ & & & & & & & \ddots & \\ 0 & & & & & & & & \mu(x^s,t)\end{bmatrix}$$

We seek to find the optimal control law $\underline{u}(t)$ which satisfies (1) and minimizes the quadratic cost functional

$$J=\int_0^T(\underline{y}'(x,t)Q\underline{y}(x,t)+\underline{u}'(t)R(x,t)\underline{u}(t)\,dxdt \tag{5}$$

where $Q(x,t)$ is a symmetric positive semidefinite matrix and $R(x,t)$ is a symmetric positive definite matrix, under the constraints:

i) the output vector $\hat{\underline{y}}(t)$ is the only available measurement for control.

ii) a digital computational technique is to be used for the determination of the control $\underline{u}(t)$.

The preceding problem without the constraint (ii) has been studied in [1] and [3]. It has been shown that it is possible to use the output $\hat{\underline{y}}(t)$ as the input to a lumped or distributed parameter observer in order to reconstruct the state $\underline{y}(x,t)$ required for the implementation of the control. Namely, for the distributed-parameter case we can define the observer system as follows:

$$\frac{\partial \underline{Z}(x,t)}{\partial t}= A(x)\frac{\partial \underline{Z}(x,t)}{\partial x}+B(x)\underline{Z}(x,t)$$
$$+D\{\hat{\underline{y}}(t)-M_s(t)\underline{Z}_s(t)\}+C(x)\underline{u}(t) \tag{6}$$

where $\underline{Z}(x,t)$ is the full-order observer state vector and $\hat{\underline{Z}}(t)$ is defined in a way similar to the one used to define $\hat{\underline{y}}(t)$, i.e. $\hat{\underline{Z}}(t)=M_s(t)\underline{Z}_s(t)$. Furthermore, let the reconstruction error state $\underline{e}(x,t)$ be defined by

$$\underline{e}(x,t)=\underline{y}(x,t)-\underline{Z}(x,t) \tag{7}$$

such that $\underline{e}(x,t)$ approaches zero [1] in a much shorter time than the control time period $[0,1]$.

Equation (6) can now be written as:

$$\frac{\partial \underline{Z}(x,t)}{\partial t}=A(x)\frac{\partial \underline{Z}(x,t)}{\partial x}-DM_s(t)\underline{Z}_s(t)$$
$$+B(x)\underline{Z}(x,t)+D\hat{\underline{y}}(t)+C(x)\underline{u}(t) \tag{8}$$

Using (7) in (6) and (1) one obtains:

$$\frac{\partial \underline{e}(x,t)}{\partial t}=A(x)\frac{\partial \underline{e}(x,t)}{\partial x}+B(x)\underline{e}(x,t)-DM_s(t)\underline{e}_s(t) \tag{9}$$

We observe that equations (6), (7) and (9) can be used to determine the state vector $\underline{y}(x,t)$ by using the measurements $\hat{\underline{y}}(t)$, i.e.

$$\underline{y}(x,t)=\underline{Z}(x,t)+\underline{e}(x,t)$$

The important question of observability with reference to the observer structure just presented has been examined in [1]. It is well known in system theory that the intrisic property of the original system as being observable is the necessary and sufficient condition for the existence of a physically realizable observer so that the system states can be determined from the observed output data through the use of the auxiliary dynamic observer. The observer gain operator D must not only stabilize the state reconstruction error dynamics, but it must also be compatible with the observability requirements. An important class of observer gain operators D are the constant linear operators. These classes of systems were studied in [1]. Actually a very important requirement is that the original system operator possesses a semigroup property and be completely observable. Using the results reported in [5] we can define the optimal control $\underline{u}^*(t)$ of the system described by (1) and (5) by

$$\underline{u}^*(t)=-K^{-1}(t)\int_0^1 C(x)\underline{p}(x,t)dx \tag{10}$$

where

$$K(t)=\int_0^1 R(x,t)dx \tag{11}$$

and $\underline{p}(x,t)$ is the equivalent costate variable vector.

Relating the costate vector $\underline{p}(x,t)$ with the state vector $\underline{y}(x,t)$ via the distributed parameter Riccati matrix [5] we can write equation (10) in the form

$$\underline{u}^*(t)=-K^{-1}(t)\int_0^1 C(x)\left(\int_0^1 M(x,\xi,t)\underline{y}(\xi,t)d\xi\right)dx \tag{12}$$

where

$$\underline{p}(x,t)=\int_0^1 M(x,\xi,t)\underline{y}(\xi,t)d\xi \tag{13}$$

We observe that for the determination of the control $\underline{u}^*(t)$, it is required to have the entire vector $\underline{y}(x,t)$ for all points in the region $0\leq x\leq 1$ and $0\leq t\leq 1$. This is achieved via the observer $\underline{Z}(x,t)$ which in the limit approximates the state vector $\underline{y}(x,t)$ as $\underline{e}(x,t)$ approaches zero.

3. SOLUTION OF PROBLEM: WALSH SERIES APPROACH

Suppose that the vector function $\underline{y}(x,t)$ is expandable in a double Walsh series, i.e.

$$\underline{y}(x,t)=\sum_{j=0}^{\infty}\sum_{i=0}^{\infty}\underline{y}_{ij}\Phi_i(t)\Psi_j(x) \tag{14}$$

where $\Psi_o(x)$, $\Psi_1(x),\ldots,$ are the Walsh functions with respect to the variable x, $\Phi_i(t)$ are the Walsh functions with respect to the variable t, and $\underline{y}_{ij}$ are the coefficient vectors of the double Walsh expansion given by:

$$\underline{y}_{ij}=\int_0^1\int_0^1 \underline{y}(x,t)\Phi_i(t)\Psi_j(x)dxdt \tag{15}$$

An approximation of $\underline{y}(x,t)$ will then be

$$\underline{y}(x,t)=\sum_{j=0}^{M-1}\sum_{i=0}^{N-1}\underline{y}_{ij}\Phi_i(t)\Psi_j(x) \tag{16}$$

Similarly we assume approximate expansions of $\underline{Z}(x,t)$ and $\underline{e}(x,t)$ of the form

$$\underline{Z}(x,t)=\sum_{j=0}^{M-1}\sum_{i=0}^{N-1}\underline{Z}_{ij}\Phi_i(t)\Psi_j(x) \tag{17}$$

and

$$\underline{e}(x,t)=\sum_{j=0}^{M-1}\sum_{i=0}^{N-1}\underline{E}_{ij}\Phi_i(t)\Psi_j(x) \tag{18}$$

Equations (16), (17) and (18) can be written as:

$$\left.\begin{aligned}\underline{y}(x,t)&=Y(t)\underline{\Psi}(x)\\ Z(x,t)&=Z(t)\underline{\Psi}(x)\\ \underline{e}(x,t)&=E(t)\underline{\Psi}(x)\end{aligned}\right\} \tag{19}$$

where

$$Y(t)=\left[Y_o\underline{\Phi}(t)\,\vdots\,Y_1\underline{\Phi}(t)\,\vdots\ldots\vdots\,Y_{M-1}\underline{\Phi}(t)\right] \tag{20}$$

$$Y_\ell=\left[\underline{y}_{o\ell}\,\vdots\,\underline{y}_{1\ell}\,\vdots\,\underline{y}_{2\ell}\,\vdots\cdots\vdots\,\underline{y}_{N-1\ell}\right] \tag{21}$$

$$\underline{\Phi}(t)=\begin{bmatrix}\Phi_o(t)\\ \Phi_1(t)\\ \vdots\\ \Phi_{N-1}(t)\end{bmatrix},\quad \underline{\Psi}(x)=\begin{bmatrix}\Psi_o(x)\\ \Psi_1(x)\\ \vdots\\ \Psi_{M-1}(x)\end{bmatrix} \tag{22}$$

Similar expressions hold for the definition of the matrices Z(t) and E(t).

In the following development, we will assume that matrices A and B are not functions of the spatial variable x.

Introducing expression (19) into (12) we can show after some manipulations that

$$\underline{u}^*(t)=-K^{-1}(t)\int_0^1\sum_{i=0}^{M-1}C_i\Psi_i(x)\left(\int_0^1\sum_{k=0}^{M-1}\sum_{\ell=0}^{M-1}M_{k\ell}(t)\cdot\Psi_k(x)\Psi_\ell(\xi)Y(t)\underline{\Psi}(\xi)d\xi\right)dx$$

or

$$\underline{u}^*(t)=-K^{-1}(t)\sum_{\ell=0}^{M-1}\sum_{i=0}^{M-1}C_iM_{i\ell}(t)Y(t)\underline{\epsilon}^\ell$$

$$=-\sum_{\ell=0}^{M-1}\sum_{i=0}^{M-1}K^{-1}(t)C_iM_{i\ell}(t)Y(t)\underline{\epsilon}^\ell \tag{23}$$

where

$$C_i=\int_0^1 C(x)\Psi_i(x)dx$$

$$M_{i\ell}(t)=\int_0^1\int_0^1 M(x,\xi,t)\Psi_i(x)\Psi_\ell(\xi)dxd\xi$$

$$\underline{\epsilon}^\ell=\begin{bmatrix}0\\0\\0\\ \vdots\\1\\0\\0\\ \vdots\end{bmatrix}\leftarrow \ell\text{th position}$$

Hence

$$\underline{u}^*(t)=-\left(\sum_{\ell=0}^{M-1}\sum_{i=0}^{M-1}K^{-1}(t)C_iM_{i\ell}(t)Y_\ell\right)\underline{\Phi}(t)=$$

$$=-\left(\sum_{\ell=0}^{M-1}\hat{M}_\ell(t)Y_\ell\right)\underline{\Phi}(t) \tag{24}$$

where

$$\hat{M}_\ell(t)=\sum_{i=0}^{M-1}K^{-1}(t)C_iM_{i\ell}(t) \tag{25}$$

Assuming that

$$\hat{M}_\ell(t)=\sum_{p=0}^{N-1}\hat{M}_{\ell p}\Phi_p(t)$$

where

$$\hat{M}_{\ell p} = \int_0^1 \hat{M}_\ell(t)\Phi_p(t)dt,$$

then equation (24) becomes

$$\underline{u}^*(t) = -\sum_{\ell=0}^{M-1}\left(\sum_{p=0}^{N-1}\hat{M}_{\ell p}\Phi_p(t)Y_\ell\right)\underline{\Phi}(t) \tag{26}$$

Using the multiplication properties of the Walsh functions [9], i.e.

$$\Phi_i(t)\Phi_j(t) = \Phi_{i\oplus j}(t)$$

where $i\oplus j$ is the modulo-2 addition, we can write

$$\Phi_p(t)\underline{\Phi}(t) = \Phi_p(t)\begin{bmatrix}\Phi_o(t)\\ \Phi_1(t)\\ \Phi_2(t)\\ \vdots\\ \Phi_{N-1}(t)\end{bmatrix} = \begin{bmatrix}\Phi_o(t)\Phi_p(t)\\ \Phi_1(t)\Phi_p(t)\\ \vdots\\ \vdots\\ \Phi_{N-1}(t)\Phi_p(t)\end{bmatrix} = \Phi^p\underline{\Phi}(t) \tag{27}$$

For example, in the situation where n=8 and p=5 the expression (27) becomes

$$\begin{bmatrix}\Phi_o(t)\Phi_5(t)\\ \Phi_1(t)\Phi_5(t)\\ \cdot\\ \cdot\\ \cdot\\ \cdot\\ \cdot\\ \Phi_7(t)\Phi_5(t)\end{bmatrix} = \begin{bmatrix}\Phi_5(t)\\ \Phi_4(t)\\ \Phi_7(t)\\ \Phi_6(t)\\ \Phi_1(t)\\ \Phi_o(t)\\ \Phi_3(t)\\ \Phi_2(t)\end{bmatrix} = \begin{bmatrix}0&0&0&0&0&1&0&0\\ 0&0&0&0&1&0&0&0\\ 0&0&0&0&0&0&0&1\\ 0&0&0&0&0&0&1&0\\ 0&1&0&0&0&0&0&0\\ 1&0&0&0&0&0&0&0\\ 0&0&0&1&0&0&0&0\\ 0&0&1&0&0&0&0&0\end{bmatrix}\begin{bmatrix}\Phi_o(t)\\ \Phi_1(t)\\ \Phi_2(t)\\ \Phi_3(t)\\ \Phi_4(t)\\ \Phi_5(t)\\ \Phi_6(t)\\ \Phi_7(t)\end{bmatrix}$$

Hence

$$\Phi_5(t)\underline{\Phi}(t) = \Phi^5\underline{\Phi}(t)$$

where

$$\Phi^5 = \begin{bmatrix}0&0&0&0&0&1&0&0\\ 0&0&0&0&1&0&0&0\\ 0&0&0&0&0&0&0&1\\ 0&0&0&0&0&0&1&0\\ 0&1&0&0&0&0&0&0\\ 1&0&0&0&0&0&0&0\\ 0&0&0&1&0&0&0&0\\ 0&0&1&0&0&0&0&0\end{bmatrix}$$

Using this property of the Walsh functions, equation (26) becomes

$$\underline{u}^*(t) = -\sum_{\ell=0}^{M-1}\sum_{p=0}^{N-1}\hat{M}_{\ell p}Y_\ell\Phi^p\underline{\Phi}(t) = -U\underline{\Phi}(t) \tag{26a}$$

$$\text{where } U = \sum_{\ell=0}^{M-1}\sum_{p=0}^{N-1}\hat{M}_{\ell p}Y_\ell\,\Phi^p \tag{26b}$$

Similarly using the definition for $\underline{y}_s(t)$ and $\underline{Z}_s(t)$ we obtain:

$$\underline{y}_s(t) = Y_s\underline{\Phi}(t) \tag{27}$$
$$Z_s(t) = Z_s\underline{\Phi}(t)$$

where

$$Y_s = \left[\begin{array}{c|c|c|c}\Psi_o(x^1)Y_o & \Psi_1(x^1)Y_1 & \cdots & \Psi_{M-1}(x^1)Y_{M-1}\\ \vdots & \vdots & & \vdots\\ \Psi_o(x^s)Y_o & \Psi_1(x^s)Y_1 & \cdots & \Psi_{M-1}(x^s)Y_{M-1}\end{array}\right] \tag{28}$$

$$Z_s = \left[\begin{array}{c|c|c|c}\Psi_o(x^1)Z_o & \Psi_1(x^1)Z_1 & \cdots & \Psi_{M-1}(x^1)Z_{M-1}\\ \vdots & \vdots & & \vdots\\ \Psi_o(x^s)Z_o & \Psi_1(x^s)Z_1 & \cdots & \Psi_{M-1}(x^s)Z_{M-1}\end{array}\right] \tag{29}$$

$$\text{Setting } M_s(t) \simeq \sum_{i=0}^{N-1}M_s^i\Phi_i(t) \tag{30}$$

where

$$M_s^i = \int_0^1 M_s(t)\Phi_i(t)dt$$

we can write:

$$\hat{\underline{y}}(t) = M_s(t)\underline{y}_s(t) = \sum_{i=0}^{N-1}M_s^i\Phi_i(t)Y_s\underline{\Phi}(t) \tag{31}$$

or

$$\hat{\underline{y}}(t) = \sum_{i=0}^{N-1}M_s^iY_s\Phi_i(t)\underline{\Phi}(t) \tag{32}$$

Arguments similar to those leading to (26) yield

$$\hat{\underline{y}}(t) = \sum_{i=0}^{N-1}M_s^iY_s\Phi^i\underline{\Phi}(t) \tag{33}$$

Similarly:

$$\hat{\underline{Z}}(t) = M_s(t)\underline{Z}_s(t) = \sum_{i=0}^{N-1}M_s^iZ_s\Phi^i\underline{\Phi}(t) \tag{34}$$

It is shown in the appendix that using these definitions the coefficient vectors $\underline{y}_{ij}$ and $\underline{Z}_{ij}$ of the expressions (16) and (17) can be arranged in a vector form $\tilde{\underline{y}}$ and $\tilde{\underline{Z}}$, respectively, and their determination involves the solution of a linear equation {see Equations (A.25) and (A.28)}.

The computational procedure presented in [10] can also be used for this problem to determine $\tilde{\underline{y}}$ and $\tilde{\underline{Z}}$ recursively starting from the initial conditions on $\underline{y}(x,t)$.

Since all the details for developing this procedure are given in [10] they are not repeated here. The reader can find more detailed in [10] and [11]. Having determined the vectors $\tilde{\underline{y}}$ and $\tilde{\underline{Z}}$ we can return to form the matrices Y(t) and Z(t) and subsequently construct the state $\underline{Z}(x,t)$. Using the identity observer in place of the unknown state $\underline{y}(x,t)$ we can implement the control u(t) via equation (12).

A block diagram for this implementation is shown in the following figure.

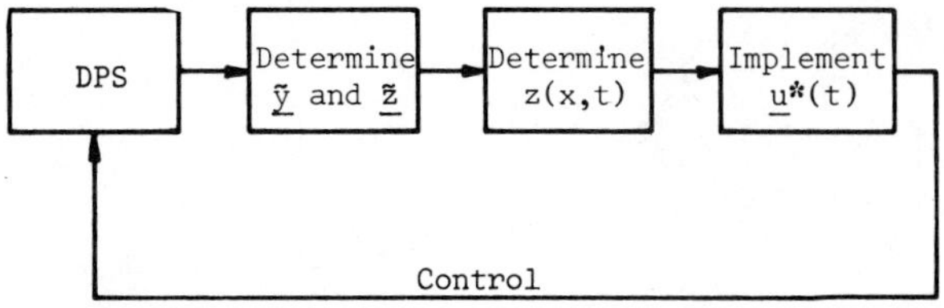

4. CONCLUSIONS

We have shown that a standard observer structure can be used in conjuction with the Double Walsh series to determine the state and thus the control of a class of distributed parameter systems, when the measurements are taken at a specified set of points. The computational procedure presented in [10] can also be applied to this class of DPS. This procedure is very useful because it starts with the initial conditions and then recursively determines the entire unknown vector. Using a computer, this algorithm can be implemented without much difficulty.

APPENDIX

Equation (1) and (8) become

$$\frac{\partial \underline{y}(x,t)}{\partial t} = A\frac{\partial \underline{y}(x,t)}{\partial x} + B\underline{y}(x,t) + C(x)\underline{u}(t) \tag{A.1}$$

$$\frac{\partial \underline{Z}(x,t)}{\partial t} = A\frac{\partial \underline{Z}(x,t)}{\partial x} + B\underline{Z}(x,t) - DM_s(t)\underline{Z}_s(t) + DM_s(t)\underline{y}_s(t) + C(x)\underline{u}(t) \tag{A.2}$$

Integration of (A.1) and (A.2) with respect to t and x gives:

$$\int_0^t\int_0^x \frac{\partial \underline{y}(x',t')}{\partial t}dx'dt' = \int_0^t\int_0^x A\frac{\partial \underline{y}(x',t')}{\partial x}dx'dt' + B\int_0^t\int_0^x \underline{y}(x',t')dx'dt' + \int_0^t\int_0^x C(x')\underline{u}(t')dx'dt' \tag{A.3}$$

and

$$\int_0^t\int_0^x \frac{\partial \underline{Z}(x',t')}{\partial t}dx'dt' = \int_0^t\int_0^x A\frac{\partial \underline{Z}(x',t')}{\partial x}dx'dt' + \int_0^t\int_0^x B\underline{Z}(x',t')dx'dt' - \int_0^t\int_0^x DM_s(t')\underline{Z}_s(t')dx'dt' + \int_0^t\int_0^x DM_s(t')\underline{y}_s(t')dx'dt' + \int_0^t\int_0^x C(x')\underline{u}(t')dx'dt' \tag{A.4}$$

Performing these integrations, we obtain

$$\int_0^x (\underline{y}(x',t)-y(x',0))dx' = \int_0^t A(\underline{y}(x,t')-\underline{y}(0,t'))dt' + B\int_0^t\int_0^x \underline{y}(x',t')dx'dt' + \int_0^t\int_0^x C(x')\underline{u}(t')dx'dt' \tag{A.5}$$

and

$$\int_0^x (\underline{Z}(x',t)-\underline{Z}(x',0))dx' = A\int_0^t (\underline{Z}(x',t')-Z(0,t'))dt' + \int_0^t\int_0^x B\underline{Z}(x',t')dx'dt' - \int_0^t\int_0^x DM_s(t')\underline{Z}_s(t')dx'dt' + \int_0^t\int_0^x DM_s(t')\underline{y}(t')dt'dx' + \int_0^t\int_0^x C(x')\underline{u}(t')dx'dt' \tag{A.6}$$

Substituting expressions (19), (26), (32) and (33) into (A.5) and (A.6) we obtain:

$$Y(t)P\underline{\Psi}(x) - \int_0^t \underline{y}(x',0)dx' = A\hat{Y}(t)\underline{\Psi}(x) - \int_0^t \underline{y}(0,t')dt' + B\hat{Y}(t)P\underline{\Psi}(x) - \hat{C}(t)\underline{\Psi}(x) \tag{A.7}$$

where

$$\hat{C}(t) = [C_0UP\underline{\Phi}(t) \vdots C_1UP\underline{\Phi}(t) \vdots \dots \vdots C_{M-1}UP\underline{\Phi}(t)]P \tag{A.8}$$

and

$$\hat{Y}(t) = [Y_0P\underline{\Phi}(t)\ Y_1P\underline{\Phi}(t) \dots\dots Y_{M-1}P\underline{\Phi}(t)] \tag{A.9}$$

The initial condition given by (2) can be written as follows:

$$\underline{y}(x,0) = \underline{g}(x) = \sum_{i=0}^{M-1} \underline{g}_i \Psi_i(x) = \Psi_0(x)[\underline{g}_0 \vdots \underline{0} \vdots \dots \vdots \underline{0}]\underline{\Phi}(t) + \Psi_1(x)[\underline{g}_1 \vdots \underline{0} \vdots \dots \vdots \underline{0}]\underline{\Phi}(t) + \dots + \Psi_{M-1}(x)[\underline{g}_{M-1} \vdots \underline{0} \vdots \dots \vdots \underline{0}]\underline{\Phi}(t)$$

Hence

$$\underline{y}(x,0) = G\underline{\Psi}(x) \tag{A.10}$$

where

$$G = \{[\underline{g}_0 \vdots \underline{0} \vdots \dots \vdots \underline{0}]\underline{\Phi}(t) \vdots [\underline{g}_1 \vdots \underline{0} \vdots \dots \vdots \underline{0}]\underline{\Phi}(t) \vdots \dots \vdots [\underline{g}_{M-1} \vdots \underline{0} \vdots \dots \vdots \underline{0}]\underline{\Phi}(t)\}$$

Substituting (A.10) into (A.7) and making use of the orthonormality property of $\underline{\Psi}(x)$ we obtain

$$Y(t)P - GP = A\hat{Y}(t) + B\hat{Y}(t)P + \hat{C}(t) \tag{A.11}$$

Equation (A.6) similarly becomes:

$$Z(t)P\underline{\Psi}(x) = A\hat{Z}(t)\underline{\Psi}(x) + B\hat{Z}(t)P\underline{\Psi}(x) - xD\sum_{i=0}^{N-1} M_s^i \underline{Z}_s \Phi^i P\underline{\Phi}(t) + xD\sum_{i=0}^{N-1} M_s^i \underline{Y}_s \Phi^i P\underline{\Phi}(t) - \hat{C}(t)\underline{\Psi}(x) \tag{A.12}$$

We have assumed that $\underline{Z}(x,0) = \underline{Z}(0,t) = \underline{Z}(1,t) = 0$. If we set

$$\sum_{i=0}^{N-1} M_s^i Z_s \Phi^i P = \hat{Z}_s, \quad \sum_{i=0}^{N-1} M_s^i Y_s \Phi^i = \hat{Y}_s \tag{A.13}$$

then equation (A.12) takes the form

$$Z(t)P\underline{\Psi}(x) = A\hat{Z}(t)\underline{\Psi}(x) + B\hat{Z}(t)P\underline{\Psi}(x) - x\hat{Z}_s\underline{\Phi}(t) + x\hat{Y}_s\underline{\Phi}(t) - \hat{C}(t)\underline{\Psi}(t) \tag{A.14}$$

Now, substituting into (A.12) an approximation for x, i.e.

$$x = \sum_{j=0}^{M-1} x_j \Psi_j(x)$$

we obtain

$$Z(t)P\underline{\Psi}(x) = A\hat{Z}(t)\underline{\Psi}(x) + B\hat{Z}(t)P\underline{\Psi}(x) - \sum_{j=0}^{M-1} x_j \Psi_j(x) DZ_s\underline{\Phi}(t) +$$

$$+\sum_{j=0}^{M-1} x_j \Psi_j(x) D\hat{Y}_s \underline{\Phi}(t) - \hat{C}(t)\underline{\Psi}(x) \qquad (A.15)$$

$$Z(t)P\underline{\Psi}(x) = A\hat{Z}(t)\underline{\Psi}(x) + B\hat{Z}(t)P\underline{\Psi}(x) - \hat{\hat{Z}}_s(t)\underline{\Psi}(x) + \hat{\hat{Y}}_s(t)\underline{\Psi}(x) - \hat{C}(t)\underline{\Psi}(x) \qquad (A.16)$$

where

$$\hat{\hat{Z}}_s(t) = [x_1 D\hat{Z}_s\underline{\Phi}(t) \vdots x_2 D\hat{Z}_s\underline{\Phi}(t) \vdots \cdots \vdots x_{M-1} D\hat{Z}_s\underline{\Phi}(t)] \qquad (A.17)$$

$$\hat{\hat{Y}}_s(t) = [x_1 D\hat{Y}_s\underline{\Phi}(t) \vdots x_2 D\hat{Y}_1\underline{\Phi}(t) \vdots \cdots \vdots x_{M-1} D\hat{Y}_s\underline{\Phi}(t)] \qquad (A.18)$$

Using the orthonormality property of $\underline{\Psi}(x)$, we obtain

$$Z(t)P = A\hat{Z}(t) + B\hat{Z}(t)P - \hat{\hat{Z}}_s(t) + \hat{\hat{Y}}_s(t) - \hat{C}(t) \qquad (A.19)$$

Putting the column vectors of $Y(t)$ in (A.11) in a vector form, we obtain [10].

$$(I\oplus P')\begin{bmatrix} Y_o \\ Y_1 \\ \vdots \\ Y_{M-1}\end{bmatrix}\underline{\Phi}(t) - (I\oplus P')\underline{g} = (A\oplus I)\begin{bmatrix} Y_o \\ Y_1 \\ \vdots \\ Y_{M-1}\end{bmatrix} P\underline{\Phi}(t)$$

$$+(B\oplus P')\begin{bmatrix} Y_o \\ Y_1 \\ \vdots \\ Y_{M-1}\end{bmatrix} P\underline{\Phi}(t) + (I\oplus P')\begin{bmatrix} C_o UP \\ C_1 UP \\ \vdots \\ C_{M-1} UP\end{bmatrix}\underline{\Phi}(t) \qquad (A.20)$$

where $\oplus$ indicates Kronecker product.

Using the orthogonality property of $\underline{\Phi}(t)$, we obtain:

$$(I\oplus P')\begin{bmatrix} Y_o \\ Y_1 \\ \vdots \\ Y_{M-1}\end{bmatrix} - [(I\oplus P')\underline{g} \vdots \underline{0} \vdots \underline{0} \cdots \vdots \underline{0}]$$

$$= (A\oplus I)\begin{bmatrix} Y_o \\ Y_1 \\ \vdots \\ Y_{M-1}\end{bmatrix} P + (B\oplus P')\begin{bmatrix} Y_o \\ Y_1 \\ \vdots \\ Y_{M-1}\end{bmatrix} + (I\oplus P')\begin{bmatrix} C_o UP \\ C_1 UP \\ \vdots \\ C_{M-1} UP\end{bmatrix} \qquad (A.21)$$

Partitioning the coefficient matrices $(A\oplus I)$, $(B\oplus P')$ and $(I\oplus P')$ of the previous expression into M submatrices, we obtain for example

$$(I\oplus P') = [(I\oplus P')_o \vdots (I\oplus P')_1 \vdots (I\oplus P')_2 \vdots \cdots \vdots (I\oplus P')_{M-1}]$$

Utilizing this partitioning, expression (A.21) becomes

$$(I\oplus P')_o Y_o + (I\oplus P)_1 Y_1 + \cdots + (I\oplus P')_{M-1} Y_{M-1} - [(I\oplus P')\underline{g} \vdots \underline{0} \vdots \underline{0} \vdots \cdots \vdots \underline{0}] =$$

$$(A\oplus I)_o Y_o P + (A\oplus I)_1 Y_1 P + \cdots + (A\oplus I)_{M-1} Y_{M-1} P$$

$$+(B\oplus P')_o Y_o P + (B\oplus P')_1 Y_1 P + \cdots + (B\oplus P')_{M-1} Y_{M-1} P$$

$$+(I\oplus P')_o C_o UP + (I\oplus P')_1 C_1 UP + \cdots +$$

$$+(I\oplus P')_{M-1} C_{M-1} UP \qquad (A.22)$$

Substituting expression (26a) into (A.22) we obtain:

$$(I\oplus P')_o Y_o + (I\oplus P')_1 Y_1 + \cdots + (I\oplus P')_{M-1} Y_{M-1}$$

$$-[(I\oplus P')\underline{g} \vdots \underline{0} \vdots \underline{0} \vdots \cdots \vdots \underline{0}] = (A\oplus I)_o Y_o P + (A\oplus I)_1 Y_1 P$$

$$+\cdots + (A\oplus I)_{M-1} Y_{M-1} P + (B\oplus P')_o Y_o P + (B\oplus P')_1 Y_1 P$$

$$+\cdots + (B\oplus P')_{M-1} Y_{M-1} P + [(I\oplus P')_o C_o + (I\oplus P')_1 C_1$$

$$+\cdots + (I\oplus P')_{M-1} C_{M-1}] \sum_{p=0}^{N-1} \hat{M}_{op} Y_o \Phi^p P + \cdots +$$

$$+[(I\oplus P')_o C_o + (I\oplus P')_1 C_1 + \cdots + (I\oplus P')_{M-1} C_{M-1}] \sum_{p=0}^{N-1} \hat{M}_{(M-1)p} Y_{(M-1)} \Phi^p P \qquad (A.23)$$

Rearranging the columns of the matrices $Y_o, Y_1, \ldots, Y_{M-1}$ into a vector form as follows:

$$\underline{y} = \begin{bmatrix} \underline{y}_{o1} \\ \underline{y}_{o2} \\ \vdots \\ \underline{y}_{oM-1} \\ \underline{y}_{11} \\ \underline{y}_{12} \\ \vdots \\ \underline{y}_{M-1,M-1} \end{bmatrix}$$

then expression (A.23) becomes:

$$[((I\oplus P')_o\oplus I) \vdots ((I\oplus P')_1\oplus I) \vdots \cdots \vdots ((I\oplus P')\oplus I)_{M-1}]\tilde{\underline{y}}$$

$$-\begin{bmatrix} (I\oplus P')\underline{g} \\ \underline{0} \\ \underline{0} \\ \vdots \\ \underline{0} \end{bmatrix} = [((A\oplus I)_o\oplus P') \vdots ((A\oplus I)_1\oplus P') \vdots + \cdots + \vdots ((A\oplus I)_{M-1}\oplus P')]\tilde{\underline{y}}$$

$$+[((B\oplus P')_o\oplus P) \vdots ((B\oplus P')_1\oplus P \vdots \cdots \vdots ((B\oplus P')_{M-1}\oplus P)]\tilde{\underline{y}}$$

$$+\{[\sum_{p=0}^{N-1} (I\oplus P')_o C_o + (I\oplus P')_1 C_1 + \cdots + (I\oplus P')_{M-1} C_{M-1}] \cdot \hat{M}_{op} \oplus (\Phi^p P)' \vdots \cdots$$

$$\cdots \vdots \sum_{p=0}^{N-1} [(I\oplus P')_o C_o + \cdots + (I\oplus P')_{M-1} C_{M-1}] \cdot \hat{M}_{(M-1)p} \oplus (\Phi^p P)'\}\tilde{\underline{y}} \qquad (A.24)$$

In general, if we collect all the coefficient matrices of $\hat{\underline{y}}$ into a common matrix, equation (A.24) can be put into the form

$$\mathbf{A}\tilde{\underline{y}} = \underline{b} \qquad (A.25)$$

Similarly equation (A.19) gives

$$(I\oplus P')\begin{bmatrix} Z_o \\ Z_1 \\ \vdots \\ Z_{M-1} \end{bmatrix}\underline{\Phi}(t)=(A\oplus I)\begin{bmatrix} Z_o \\ Z_1 \\ \vdots \\ Z_{M-1} \end{bmatrix}\underline{\Phi}(t)$$

$$+(B\oplus P')\begin{bmatrix} Z_o \\ Z_1 \\ \vdots \\ Z_{M-1} \end{bmatrix}P\underline{\Phi}(t)$$

$$-\begin{bmatrix} x_1 D\hat{Z}_s \\ x_2 DZ_s \\ \vdots \\ x_{M-1} D\hat{Z}_s \end{bmatrix}\underline{\Phi}(t)+\begin{bmatrix} x_1 D\hat{Y}_s \\ x_2 D\hat{Y}_s \\ \vdots \\ x_{M-1} D\hat{Y}_s \end{bmatrix}\underline{\Phi}(t)$$

$$-(I\oplus P')\begin{bmatrix} C_o UP \\ C_1 UP \\ \vdots \\ C_{M-1} UP \end{bmatrix}\underline{\Phi}(t) \qquad (A.26)$$

The orthogonality of $\underline{\Phi}(t)$ yields:

$$(I\oplus P')\begin{bmatrix} Z_o \\ Z_1 \\ \vdots \\ Z_{M-1} \end{bmatrix}=(A\oplus I)\begin{bmatrix} Z_o \\ Z_1 \\ \vdots \\ Z_{M-1} \end{bmatrix}+(B\oplus P')\begin{bmatrix} Z_o \\ Z_1 \\ \vdots \\ Z_{M-1} \end{bmatrix}P$$

$$-\begin{bmatrix} x_1 D\hat{Z}_s \\ x_2 DZ_s \\ \vdots \\ x_{M-1} D\hat{Z}_s \end{bmatrix}+\begin{bmatrix} x_1 D\hat{Y}_s \\ x_2 DY_s \\ \vdots \\ x_{M-1} D\hat{Y}_s \end{bmatrix}-(I\oplus P')\begin{bmatrix} C_o UP \\ C_1 UP \\ \vdots \\ C_{M-1} UP \end{bmatrix} \qquad (A.27)$$

Since the structure of (A.27) and (A.20) is the same, an expression similar to the one derived for $\underline{\tilde{y}}$ in (A.25) can be obtained, i.e.

$$\mathbf{A}_1\tilde{Z}=\underline{b}_1 \qquad (A.28)$$

REFERENCES

[1] Y.A. Liu and L Lapidus, Observer theory for distributed parameter systems, Int.J. systems Sci. 1 (1) (1976) 731-742.

[2] R. Gressang and G. Lamont, Observers for systems characterized by semigroups, IEEE Trans. Auto. Contr., AC-20 (1975) 523.

[3] P. Stavroulakis and P.E. Sarachik, Design of Optimal Controllers for Distributed Systems using finite Dimensional Observers Proc. IEEE Decision and Control, (Dec. 1973), No.W91-1, 103-109.

[4] J. Furusho, H. Kanoh and M. Masubuchi, Observers for bilinear distributed-parameter processes, Proc. 2nd IFAC Symp. on "Control of Distributed-Parameter Systems" (S.Banks and A. Prichard, Eds), Pergamon Press (1978) 219-229.

[5] S.G. Tzafestas, Optimal distributed-parameter control using classical variational theory, Int.J. Control, 16 (6) (1970) 593-608.

[6] D.L. Russell, Optimal regulation of linear symmetric hyperbolic systems with finite dimensional controls, J.SIAM Control, 4 (2), (1966) 276-295.

[7] J. Davis, The sensitivity and control of distributed parameter systems, Report R-464, University of Illinois, (April 1970)

[8] R. Goodson and R. Klein, A definition and some results of distributed systems observability, IEEE Trans. Auto Contr. systems AC-15 (1970) 165-174.

[9] K.G. Beanchamp, Walsh Functions and their Applications (Academic Press, New York, 1975).

[10] P. Stavroulakis and S.Tzafestas, Walsh series approach to observer and filter design in optimal control systems, Int.J. Control, 26, (5) (1977) 721-736.

[11] P. Stavroulakis and S. Tzafestas, A Walsh series approach to time-delay control system observer design, Int.J. Systems Sci., 9, (3) (1978) 287-299.

[12] S.G. Tzafestas, Design of distributed-parameter optimal controllers and filters via Walsh-Galerkin expansions, Proc. 2nd IFAC Symp. on "Control of Distributed-Parameter Systems" (S.Banks and A.Pritchard, Eds.) Pergamon Press (1978) 201-218.

[13] S.G. Tzafestas, Walsh series approach to lumped and distributed system identification, J.Franklin Inst., 305(4) (1978) 199-220.

21

Reprinted from *Nucl. Sci. Eng.* **62**:763–770 (1977)

NUCLEAR REACTOR CONTROL USING WALSH FUNCTION VARIATIONAL SYNTHESIS

S. Tzafestas and N. Chrysochoides

ABSTRACT

The optimal control problem of nuclear reactor systems is solved by using the Walsh function variational synthesis technique. In this technique, the control input rate is expanded in a Walsh series with coefficients to be selected such that a general criterion functional covering all practical situations is optimized. The nuclear reactor system is assumed to be in its integral equation form, and the nonlinearity due to the product of the reactivity input and the power level output is overcome by using in the criterion functional power-times-reactivity penalty terms rather than pure reactivity ones. A numerical example is presented to illustrate the feasibility of the approach. The results are first derived for point reactors and are then extended to space-dependent ones.

INTRODUCTION

The application of modern computer control theory to nuclear reactor systems is currently receiving increasing attention.[1-3] The usual approach is to formulate an optimal control problem with the nuclear reactor system being described in state-space form, either in continuous time[1] or in discrete time.[2] The approach considered in the present Note is based on the integral equation model of the reactor system. A convenient optimal (more precisely, suboptimal) control method for this model is the so-called "variational synthesis method,"[4,5] which consists in expanding the control input under selection in a truncated series of certain convenient functions, $\Phi_k(t)$ (orthogonal or not), and so reducing the problem to that of selecting the coefficients of this series. It is this method that is adopted here based on the Walsh function family, which has suitable properties.[6-8]

The difficulty due to the appearance of the product of the power level $n(t)$ and the reactivity input $\rho(t)$ in the system equations is overcome by considering a criterion function involving power-times-reactivity penalty terms rather than pure reactivity terms, and by taking as control input the variable $u(t) = n(t)\rho(t)$. Due to the fact that the Walsh functions are step-functions and possess finite jumps, it is more convenient to start by expanding in a Walsh series not the control input $u(t)$ itself, but rather its time derivative $\dot{u}(t)$. The details of the technique are developed, and a nontrivial example is computed to show the powerfulness of the approach. Then the results are applied to the more general case of space-dependent nuclear reactors equipped with N control rods inside the occupied spatial domain. Actually, this Note constitutes part of a project aiming to systematically apply Walsh function techniques to nuclear reactor estimation, computation, and control.[9-11]

WALSH FUNCTION EXPANSION

The Walsh functions constitute a family of two-valued orthonormal functions $[\Phi_k(t),\ k = 0, 1, \ldots, m - 1,\ m = 2^n]$ in the interval $t \epsilon [0, 1)$ (Ref. 7). The functions $\Phi_k(t)$ in the Paley ordering are defined in terms of the Rademacher functions $R_0(t) = 1$, $R_s(t)$, $s = 1, 2, \ldots, n$, which are a set of orthonormal square waves for $t \epsilon [0, 1)$ of unit height with repetition rate (sequency) equal to 2^{s-1}.

Specifically, $\Phi_k(t)$ is given by the formula

$$\Phi_k(t) = R_s^{\lambda_s}(t)\, R_{s-1}^{\lambda_{s-1}}(t) \ldots R_1^{\lambda_1}(t) \ ,$$

$$\Phi_0(t) = R_0(t), \ k = 0, 1, \ldots, 2^s - 1 \ ,$$

where "$\lambda_s \lambda_{s-1} \ldots \lambda_1$" is the binary expression of the decimal number k. The functions $R_s(t)$, $s = 1, 2, 3$ and $\Phi_k(t)$, $k = 0, 1, \ldots, 7$ have the form

+ + + + + + + +	R_0
+ + + + - - - -	R_1
+ + - - + + - -	R_2
+ - + - + - + -	R_3

+ + + + + + + +	$\Phi_0 = R_0$
+ + + + - - - -	$\Phi_1 = R_1$
+ + - - + + - -	$\Phi_2 = R_2 R_1^0$
+ + - - - - + +	$\Phi_3 = R_2 R_1$
+ - + - + - + -	$\Phi_4 = R_3 R_2^0 R_1^0$
+ - + - - + - +	$\Phi_5 = R_3 R_2^0 R_1$
+ - - + + - - +	$\Phi_6 = R_3 R_2 R_1^0$
+ - - + - + + -	$\Phi_7 = R_3 R_2 R_1$

with the notation "+" = "+1" and "-" = "-1".

Since all Rademacher functions are "odd" functions about $t = \frac{1}{2}$, they do not form a complete set. On the contrary, one can see that the Walsh functions constitute a complete orthonormal set of functions, and therefore one can expand a given function $F(t)$, which is absolutely integrable in $[0, 1)$, in a Walsh series as[6,7,11]

[1]E. DUNCOMBE and D. E. RATHBONE, *IEEE Trans.*, **AC-14**, 277 (1969).

[2]W. C. LIPINSKI and A. G. VACROUX, *IEEE Trans.*, **NS-17**, 510 (1970).

[3]W. CIECHANOWICS and K. SOLBERG, *Proc. ENEA-OECD Symp. Application of On-Line Computers to Nuclear Reactors*, Sandfiord, Norway, Sep. 1968, p. 239 (1968).

[4]W. M. STACEY, Jr., *Nucl. Sci. Eng.*, **34**, 257 (1968).

[5]S. H. KYONG and E. P. GYFTOPOULOS, *Trans. Am. Nucl. Soc.*, **8**, 479 (1965).

[6]C. F. CHEN and C. H. HSIAO, *Int. J. Control*, **21**, 881 (1975).

[7]H. HARMUTH, *Transmission of Information by Orthogonal Functions*, Springer-Verlag, New York-Heidelberg-Berlin (1972).

[8]S. TZAFESTAS et al., *Electron. Eng.*, 45 (Nov. 1976).

[9]S. TZAFESTAS, *Nucl. Instrum. Methods*, **84**, 207 (1970).

[10]Y. SHINOHARA and R. OGUMA, *Nucl. Sci. Eng.*, **52**, 76 (1973).

[11]S. TZAFESTAS and N. CHRYSOCHOIDES, "Time Varying Reactivity Computation Via Walsh Functions," to appear in *Nucl. Instrum. Methods.*

$$F(t) = \sum_{k=0}^{\infty} f_k \, \Phi_k(t) \quad ,$$

where the coefficients f_k, $k = 0, 1, 2, \ldots$ are given by the formula

$$f_k = \int_0^1 \Phi_k(t) \, F(t) dt \quad .$$

For example, if $F(t) = t$, $t\epsilon[0, 1)$ (ramp function), the coefficients f_k are found to be $f_0 = \frac{1}{2}$, $f_1 = -\frac{1}{4}$, $f_2 = -\frac{1}{8}$, $f_3 = 0$, $f_4 = -\frac{1}{16}$, etc., i.e.,

$$t = \Phi_0/2 - \Phi_1/4 - \Phi_2/8 - \Phi_4/16 + \ldots \quad .$$

The discrete form of the Walsh series expansion is

$$\left.\begin{aligned} F_n &= \sum_{k=0}^{m-1} f_k \Phi_{kn}, \qquad n = 0, 1, \ldots, (m-1) \\ f_k &= \frac{1}{m} \sum_{n=0}^{m-1} \Phi_{kn} F_n, \quad k = 0, 1, \ldots, (m-1) \end{aligned}\right\} \quad , \tag{1}$$

where Φ_{kn} is the value of the k'th Walsh function in the n'th time subinterval, and F_n is the mean value of $F(t)$ in the same subinterval.

In matrix notation, Eq. (1) takes the form

$$\mathbf{F} = \mathbf{\Phi}^T \mathbf{f} \text{ and } \mathbf{f} = \frac{1}{m} \mathbf{\Phi} \mathbf{F}$$

where

$$\mathbf{\Phi} = [\Phi_{kn}]$$
$$\mathbf{F} = [F_0, F_1, \ldots, F_{m-1}]^T$$
$$\mathbf{f} = [f_0, f_1, \ldots, f_{m-1}]^T.$$

Clearly, this implies that the Walsh matrix $\mathbf{\Phi}$ has the property $\mathbf{\Phi}^T = m\,\mathbf{\Phi}^{-1}$.

Using the expansion for t given above and retaining only four terms, one obtains

$$t = \int_0^t \Phi_0(\tau) d\tau = [\tfrac{1}{2}, -\tfrac{1}{4}, -\tfrac{1}{8}, 0] \boldsymbol{\phi}(t) \quad .$$

Similarly

$$\begin{aligned} &\int_0^t \Phi_1(\tau) d\tau = [\tfrac{1}{4}, 0, 0 - \tfrac{1}{8}] \boldsymbol{\phi}(t) \\ &\int_0^t \Phi_2(\tau) d\tau = [\tfrac{1}{8}, 0, 0, 0] \boldsymbol{\phi}(t) \quad , \quad \boldsymbol{\phi}(t) = \begin{bmatrix} \Phi_0(t) \\ \Phi_1(t) \\ \Phi_2(t) \\ \Phi_3(t) \end{bmatrix} \quad , \\ &\int_0^t \Phi_3(\tau) d\tau = [0, \tfrac{1}{8}, 0, 0] \boldsymbol{\phi}(t) \end{aligned} \tag{2}$$

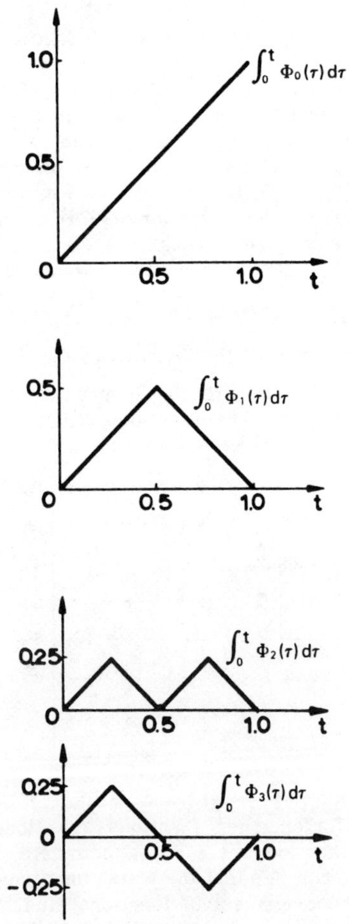

(a)

Fig. 1a. Plot of $\int_0^t \Phi_k(\tau) d\tau$ for $k = 0, 1, 2, 3$ ($m = 4$).

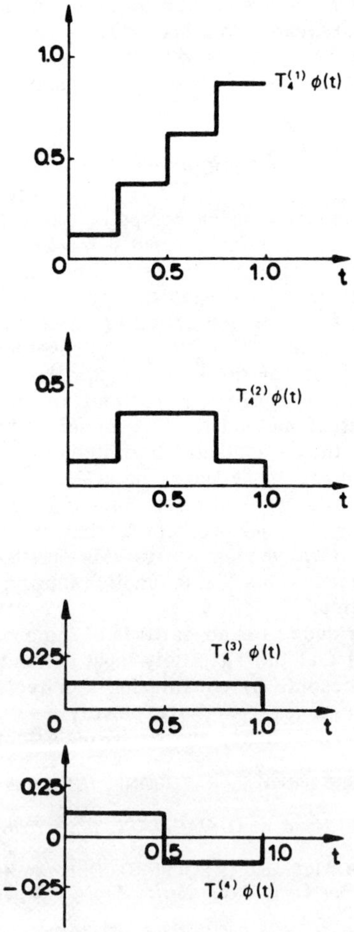

(b)

Fig. 1b. Plot of the Walsh approximation $T_4^{(k+1)} \boldsymbol{\phi}(t)$ of $\int_0^t \Phi_k(\tau) d\tau$ as given by Eq. (3). [Here $T_4^{(i)}$ represents the i'th row of $\boldsymbol{T}_4$.]

which can be written in matrix form as

$$\int_0^t \boldsymbol{\phi}(\tau)d\tau = \mathbf{T}_m\boldsymbol{\phi}(t),\ \mathbf{T}_m = \left[\begin{array}{cc|cc} \frac{1}{2} & -\frac{1}{4} & -\frac{1}{8} & 0 \\ \frac{1}{4} & 0 & 0 & -\frac{1}{8} \\ \hline \frac{1}{8} & 0 & 0 & 0 \\ 0 & \frac{1}{8} & 0 & 0 \end{array}\right] \quad (m = 4) \quad . \tag{3}$$

The plots of the four functions involved in $\mathbf{T}_4\boldsymbol{\phi}(t)$, in comparison with their exact form determined by directly evaluating $\int_0^t \boldsymbol{\phi}(\tau)d\tau$, are shown in Fig. 1.

By induction, one can find that $\mathbf{T}_m$, $m = 2^n$ has the form

$$\mathbf{T}_m = \left[\begin{array}{c|c} \mathbf{T}_{m/2} & -\frac{1}{2m} I_{m/2} \\ \hline \frac{1}{2m} I_{m/2} & 0_{m/2} \end{array}\right] \quad (m = 2^n,\ n = 1, \ldots) \quad .$$

The operational formula, Eq. (3), will be extensively used in the solution of the optimal control problem by the Walsh function variational synthesis procedure.

WALSH FUNCTION VARIATIONAL SYNTHESIS

The system model under consideration has the integral form[4,5]

$$y(t) = y_0(t) + \int_0^t h(t - t')u(t')dt' \quad , \tag{4}$$

where $u(t)$ is the r-dimensional control input vector, $h(t)$ is a known kernel $p \times r$ matrix differentiable function, $y(t)$ is the p-dimensional output vector, and $y_0(t)$ is the input-free component of $y(t)$, which is due to the initial conditions of the system.

The optimal control problem is to choose $u(t)$ in the interval $t\epsilon[0, 1)$ such that to minimize a cost functional of the Bolza type,[12,13]

$$J(y, u) = l(y, \dot{y}, \dot{u}, u)]_0^{t_f} + \int_0^{t_f} L(y, \dot{y}; \dot{u}, u, t)dt \tag{5}$$

subject to the constraint $y(t_f) = y_f$ (a known value).

The usual variational synthesis approach to the solution of this problem consists in expanding the control function $u(t)$ in a series of an orthonormal family of functions $\Phi_k(t)$, $k = 0, 1, 2, \ldots$ (although the orthonormality is not essential) as $u(t) = \Sigma_k A_k \Phi_k(t)$ and then trying to select the coefficients A_k so as to minimize the functional, Eq. (5).

In Walsh function variational synthesis, we start by a Walsh series expansion of the time derivative of $u(t)$, i.e., we set[14]

$$\dot{u}(t) = \sum_{k=0}^{m-1} U_k \Phi_k(t) = \mathbf{U}\boldsymbol{\phi}(t) \quad , \tag{6}$$

where $\mathbf{U} = [U_0, U_1, \ldots, U_{m-1}]$ is an $r \times m$ coefficient matrix and $\boldsymbol{\phi}(t)$ is as in Eq. (2). Then, by using the operational Eq. (3), we find that

$$u(t) = \int_0^t \dot{u}(\tau)d\tau + u(0) = \mathbf{U}\,\mathbf{T}_m\boldsymbol{\phi}(t) + u(0) \quad . \tag{7}$$

Having found Walsh series expansions for $\dot{u}(t)$ and $u(t)$, we need the expansions of $\dot{y}(t)$, $y(t)$, and all explicit functions of t in the cost functional J. Concerning $y(t)$, we have by the Leibnitz rule,

$$\dot{y}(t) = \dot{y}_0(t) + \int_0^t h_d(t - t')u(t')dt' + \int_0^t h(t - t')\dot{u}(t')dt' + h(0)u(t) - h(t)u(0) \quad ,$$

where $h_d(t - t') = dh(t - t')/dt'$. Hence, using Eqs. (6) and (7), we find

$$\begin{aligned} \dot{y}(t) &= \dot{y}_0(t) + \int_0^t h_d(t - t')[\mathbf{U}\,\mathbf{T}_m\boldsymbol{\phi}(t') + \boldsymbol{u}(0)]dt' \\ &\quad + \int_0^t h(t - t')\mathbf{U}\boldsymbol{\phi}(t')dt' + h(0)\,\mathbf{U}\,\mathbf{T}_m\boldsymbol{\phi}(t) \\ &\quad + h(0)u(0) - h(t)u(0) \\ &= Y_1(t) + H_1^T(t)\boldsymbol{v} \quad , \end{aligned} \tag{8a}$$

where $\boldsymbol{v}$ is the column vector composed by the columns of the matrix $\mathbf{U}$ in their natural order, and

$$\begin{aligned} Y_1(t) &= \dot{y}_0(t) + \left[\int_0^t h_d(t - t')dt' + h(0) - h(t)\right]u(0) \\ H_1^T(t) &= \int_0^t h_d(t - t') \otimes [\boldsymbol{\phi}^T(t')\mathbf{T}_m^T]dt' \\ &\quad + \int_0^t h(t - t') \otimes \boldsymbol{\phi}^T(t')dt' + h(0) \otimes [\boldsymbol{\phi}^T(t)\mathbf{T}_m^T] \quad , \end{aligned} \tag{8b}$$

where the symbol $\otimes$ denotes the Kronecker product of two matrices $\mathbf{A} = [a_{ij}]$ and $\mathbf{B} = [b_{ij}]$, being defined as

$$\mathbf{A} \otimes \mathbf{B} = \begin{bmatrix} Ab_{11} & \ldots & Ab_{1q} \\ \vdots & & \vdots \\ Ab_{p1} & \ldots & Ab_{pq} \end{bmatrix} \quad .$$

Similarly, we can write

$$\begin{aligned} y(t) &= y_0(t) + \int_0^t h(t - t')[\mathbf{U}\,\mathbf{T}_m\boldsymbol{\phi}(t') + u(0)]dt' \\ &= Y_0(t) + H_0^T(t)\boldsymbol{v} \quad , \end{aligned} \tag{9a}$$

where

$$\begin{aligned} H_0^T(t) &= \int_0^t h(t - t') \otimes [\boldsymbol{\phi}^T(t')\mathbf{T}_m^T]dt' \\ Y_0(t) &= y_0(t) + \left[\int_0^t h(t - t')dt'\right]u(0) \quad . \end{aligned} \tag{9b}$$

Equations (8) and (9) give $\dot{y}(t)$ and $y(t)$ in terms of the known matrix functions $Y_0(t)$, $Y_1(t)$, $H_0(t)$, $H_1(t)$, and the unknown vector $\boldsymbol{v}$. Due to the piecewise constant form of $\Phi_k(t)$, the integrals in Eqs. (8b) and (9b) can be evaluated in discrete time corresponding to the Walsh function subintervals.

Introducing Eqs. (6) through (9) into the cost functional, Eq. (5), yields a conventional function $J(\boldsymbol{v})$ that can be computed by performing the integration in discrete steps as

$$J(\boldsymbol{v}) = l]_0^1 + \sum_{s=0}^{m-1} J_s(\boldsymbol{v}),\ J_s(\boldsymbol{v}) = \int_{s/m}^{(s+1)/m} L dt \quad . \tag{10}$$

Note that before evaluating $J(\boldsymbol{v})$, a normalization of the time to unity is required, i.e., a change of the time variable t to $\sigma = t/t_f$, in which case the integral in Eq. (5) becomes

$$t_f \int_0^1 L d\sigma \quad .$$

The constraint $\boldsymbol{y}(t_f) = \boldsymbol{y}_f$ [i.e., $\boldsymbol{y}(\sigma)_{\sigma=1} = y_f$] yields, by virtue of Eq. (9), an algebraic constraint of the type

$$y_f = Y_0(t_f) + H_0^T(t_f)\boldsymbol{v} \quad . \tag{11}$$

Hence, the problem has been reduced to that of minimizing the function $J(\boldsymbol{v})$ in Eq. (10) under the equality constraints in Eq. (11). This can be done in the usual way either by eliminating p elements of $\boldsymbol{v}$ in Eq. (10) by solving with

---

[12] O. BOLZA, *Lectures on the Calculus of Variations*, Dover Publications, New York (1961).

[13] S. TZAFESTAS, *Int. J. Control*, **12**, 593 (1970).

[14] This expansion is made when the cost J involves $\dot{u}$ and $\dot{y}$ as in Eq. (5); otherwise an expansion of $u(t)$ is sufficient.

respect to them Eq. (11), or by using the Lagrange multiplier method. If the final value $y(t_f)$ is also free, then we simply omit the constraint Eq. (11).

POINT NUCLEAR REACTOR CONTROL

For simplicity and clarity of the presentation, consider first a point reactor model with one group of delayed neutrons, which can be written in the integral form, Eq. (4), with[4,5]:

$$y_0(t) = [\epsilon_1 + \epsilon_2 \exp(-\gamma t)]/\gamma,\ \gamma = \lambda + \beta/\Lambda$$
$$h(t - t') = \{\lambda + (\beta/\Lambda)\exp[-\gamma(t - t')]\}/\Lambda\gamma$$
$$\epsilon_1 = (n_0 + C_0)\lambda,\ \epsilon_2 = \beta n_0/\Lambda - \lambda C_0 \quad .$$

Here $y(t) = n(t)$ is the neutron power level, $u(t) = \rho(t)n(t)$, where

$\rho(t)$ = reactivity

β = delayed neutron fraction

λ = decay constant

Λ = prompt-neutron generation time

C_0 = delayed neutron precursor concentration at $t = 0$

n_0 = initial value of $n(t)$.

The problem is to select the reactivity function $\rho(t)$ so as to minimize a cost functional of the type of Eq. (5) with

$$L = P_0[y(t) - N(t)]^2 + P_1[\tau\dot{y}(t) - y(t)]^2 + P_2[\dot{y}(t)]^2 + Q_0[u(t)]^2 + Q_1[\dot{u}(t)]^2$$
$$l]_0^{t_f} = P_f\dot{y}^2(t_f) \quad , \tag{12}$$

without any constraint, where the weighting constants P_0, P_1, P_2, Q_0, Q_1, and P_f have appropriate values according to the importance of the corresponding terms, and $N(t)$ is a desired nominal trajectory. The first term in L expresses the deviation of the neutron level trajectory $y(t)$ from the nominal $N(t)$, the second term expresses the deviation of the instantaneous period $\dot{y}/y$ from a desired period τ and is used in startup control, the third term expresses the speed of change of the neutron level, and the Q terms represent the cost of control and of the speed of control.

Clearly, due to the relation $u(t) = \rho(t)n(t)$, the nuclear reactor system is nonlinear. This nonlinearity is overcome here by the assumed form, Eq. (12), of the cost functional in which the Q terms involve the control product $u(t) = n(t)\rho(t)$ of the power level and the reactivity input. The optimal reactivity is found by first computing the optimal control product $u^0(t)$ using our Walsh function variational synthesis, by then computing $y^0(t) = n^0(t)$ using Eq. (4) with $u(t') = u^0(t')$, and finally by solving with respect to $\rho^0(t)$, i.e., $\rho^0(t) = u^0(t)/n^0(t)$. Of course the optimal solution corresponding to the cost functional, Eq. (12), is slightly different from that obtained by minimizing the usual cost functional of the type of Eq. (12), with $u(t)$ being replaced by $\rho(t)$ (Ref. 1).

To compute the optimal control $u^0(t)$, we first express $u(t)$ and $\dot{u}(t)$ of Eqs. (6) and (7) in terms of the Walsh coefficient vector $\boldsymbol{v}$, i.e.,

$$u(t) = [U_0, \ldots, U_{m-1}]\begin{bmatrix} T_m^{(0)}\boldsymbol{\phi}(t) \\ \vdots \\ T_m^{(m-1)}\boldsymbol{\phi}(t)\end{bmatrix} + u(0) = A_0^T(t)\boldsymbol{v} + u(0) \tag{13a}$$

$$\dot{u}(t) = [U_0, \ldots, U_{m-1}]\begin{bmatrix} \Phi_0(t) \\ \vdots \\ \Phi_{m-1}(t)\end{bmatrix} = A_1^T(t)\boldsymbol{v} \quad . \tag{13b}$$

Here, $T_m^{(i)}$ is the $(i + 1)$'th row of $\mathbf{T}_m$, and

$$A_0^T(t) = [T_m^{(0)}\boldsymbol{\phi}(t)I, \ldots, T_m^{(m-1)}\boldsymbol{\phi}(t)I]$$
$$A_1^T(t) = [\Phi_0(t)I, \Phi_1(t)I, \ldots, \Phi_{m-1}(t)I] \quad , \tag{13c}$$

where I is the unit matrix of the same dimensionality as the vector $u(t)$.

Then, we introduce Eqs. (8a), (9), (13a), and (13b) in the cost function, Eq. (12), i.e.,

$$L = P_0[Y_0(t) - N(t) + H_0^T(t)\boldsymbol{v}]^2 + P_1\{\tau Y_1(t) - Y_0(t) + [\tau H_1(t) - H_0(t)]^T\boldsymbol{v}\}^2 + P_2[Y_1(t) + H_1^T(t)\boldsymbol{v}]^2 + Q_0[A_0^T(t)\boldsymbol{v} + u(0)]^2 + Q_1[A_1^T(t)\boldsymbol{v}]^2$$
$$l]_0^{t_f} = P_f[Y_1(t_f) + H_1^T(t_f)\boldsymbol{v}]^2$$

and solve for $\boldsymbol{v}$ the equation $\partial J/\partial\boldsymbol{v} = 0$ to obtain the optimal Walsh coefficient $\boldsymbol{v}^0$, i.e.,

$$\boldsymbol{v}^0 = -\boldsymbol{S}^{-1}\boldsymbol{q} \quad , \tag{14}$$

where

$$\boldsymbol{q} = \int_0^{t_f} \{P_0H_0(t)[Y_0(t) - N(t)] + P_1[\tau H_1(t) - H_0(t)] \times [\tau Y_1(t) - Y_0(t)] + P_2H_1(t)Y_1(t) + Q_0A_0(t)u(0)\}dt + P_fH_1(t_f)Y_1(t_f)$$

$$\boldsymbol{S} = \int_0^{t_f} \{P_0H_0(t)H_0^T(t) + P_1[\tau H_1(t) - H_0(t)][\tau H_1(t) - H_0(t)]^T + P_2H_1(t)H_1^T(t) + Q_0A_0(t)A_0^T(t) + Q_1A_1(t)A_1^T(t)\}dt + P_fH_1(t_f)H_1^T(t_f) \quad .$$

Using Eqs. (13a) and (14), we finally find

$$u^0(t) = A_0^T(t)\boldsymbol{v}^0 + u(0)$$
$$n^0(t) = n_0(t) + \left[\int_0^t h(t - t')A_0^T(t')dt'\right]\boldsymbol{v}^0 + \left[\int_0^t h(t - t')dt'\right]u(0)$$
$$\rho^0(t) = A_0^T(t)\boldsymbol{v}^0/n^0(t) \quad . \tag{15}$$

The method is applicable to more complicated nuclear reactor models. For example, if we take into account the negative temperature feedback, the control rod reactivity $\rho_c(t)$ is equal to

$$\rho_c(t) = \rho(t) - \alpha_\theta\Delta\theta(t) \quad ,$$

where α_θ is the temperature coefficient of reactivity and $\Delta\theta(t)$ is the temperature deviation. The function $\Delta\theta(t)$ satisfies the equation

$$\Delta\dot{\theta}(t) = A(y - n_0) + B\Delta\theta(t) \quad ,$$

where A and B are known constants.

NUMERICAL NUCLEAR REACTOR EXAMPLE

Consider the case where the nuclear reactor parameters have the values

$$n_0 = 10^5 \text{ n/cm}^3,\ \Lambda = 10^{-4}\text{ s}$$
$$\lambda = 0.0768\text{ s}^{-1},\ \beta = 0.0074 \quad ,$$

and the parameters of the cost functional in Eq. (12) are $P_0 = 1$, $P_1 = P_2 = Q_0 = Q_1 = P_f = 0$. This means that the cost functional to be minimized is

$$J = \int_0^{t_f} [y(t) - N(t)]^2 dt \quad .$$

The desired power output (neutron) trajectory $N(t)$ has the equation

$$N(t) = \begin{cases} 10^5 + t(10^8 - 10^5)/300, & 0 \leq t < 300 \\ 10^8, & 300 \leq t < t_f \end{cases} \quad .$$

The problem is to find the reactivity function $\rho(t)$ that would control the neutron level so as to be as close as possible to the desired trajectory $N(t)$ with minimum J and the constraint that $y(t_f) = y_f = 10^8$ n/cm³.

To solve the problem, we first normalize the time variable as $\sigma = t/t_f$, in which case we have[15]

$$J = t_f \int_0^1 [y(\sigma) - N(\sigma)]^2 d\sigma$$

$$N(\sigma) = \begin{cases} 10^5 + 2(10^8 - 10^5)\sigma, & 0 \leq \sigma < 0.5 \\ 10^8, & 0.5 \leq \sigma < 1 \end{cases}$$

$$y(\sigma) = y_0(\sigma) + t_f \int_0^\sigma h(\sigma - \sigma')u(\sigma')d\sigma'$$

$$= Y_0(\sigma) + H_0^T(\sigma)\boldsymbol{v}, \quad 0 \leq \sigma \leq 1 \quad ,$$

where

$$Y_0(\sigma) = y_0(\sigma) = [\epsilon_1 + \epsilon_2 \exp(-\gamma t_f \sigma)]/\gamma$$

$$H_0^T(\sigma) = t_f \int_0^\sigma h(\sigma - \sigma') \otimes \boldsymbol{\phi}^T(\sigma')\, d\sigma'$$

$$= [h_0(\sigma), h_1(\sigma), \ldots, h_{m-1}(\sigma)]$$

$$h_k(\sigma) = t_f \int_0^\sigma h(\sigma - \sigma')\, \Phi_k(\sigma')\, d\sigma'$$

$$h(\sigma - \sigma') = \frac{1}{\Lambda\gamma}\left\{\lambda + \frac{\beta}{\Lambda}\exp[-t_f\gamma(\sigma - \sigma')]\right\} \quad .$$

Since, in the present case we have to satisfy the final value constraint $y(t_f) = y_f$, we use the Lagrange multiplier $2p$ and minimize the augmented cost functional

$$J^* = J + 2p[y_f - Y_0(1) - H_0^T(1)\boldsymbol{v}]$$

with respect to $\boldsymbol{v}$ and p without any constraint.

Equating to zero the derivatives of J^* with respect to $\boldsymbol{v}$ and p, one obtains the linear algebraic system

$$\left[\begin{array}{c|c} \boldsymbol{S} & -H_0(1) \\ \hline -H_0^T(1) & 0 \end{array}\right] \left[\begin{array}{c} \boldsymbol{v} \\ \hline p \end{array}\right] = -\left[\begin{array}{c} \boldsymbol{q} \\ \hline y_f - Y_0(1) \end{array}\right] \quad , \tag{16}$$

where

$$\boldsymbol{S} = \int_0^1 H_0(\sigma)\, H_0^T(\sigma)\, d\sigma \simeq \frac{1}{m}\sum_{s=1}^m H_0\left(\frac{s}{m}\right) H_0^T\left(\frac{s}{m}\right)$$

$$\boldsymbol{q} = \int_0^1 H_0(\sigma)[Y_0(\sigma) - N(\sigma)]d\sigma$$

$$\simeq \frac{1}{m}\sum_{s=1}^m H_0\left(\frac{s}{m}\right)\left[Y_0\left(\frac{s}{m}\right) - N\left(\frac{s}{m}\right)\right] \quad ,$$

which can be solved for $[\boldsymbol{v}^T, p]$.

To evaluate $h_k(\sigma)$, we use the piecewise constancy of $\Phi_k(\sigma)$ and obtain the discrete time approximation,

$$h_k(s - 1) = \sum_{l=0}^{s-1} \alpha_{l,s-1}\, \Phi_{kl}$$

$$\alpha_{l,s-1} = t_f \int_{l/m}^{(l+1)/m} h\left(\frac{s-1}{m} - \sigma'\right) d\sigma' \simeq \frac{t_f}{m} h\left(\frac{s-1}{m} - \frac{l}{m}\right) \quad .$$

[15]Here the cost function J does not involve $\dot{u}(t)$, and so we expand $u(t)$ in Walsh series $u(t) = \boldsymbol{\phi}^T(t)\boldsymbol{v}$ instead of expanding $\dot{u}(t)$.

Using the present values of the various parameters, we find that

$$\alpha_{l,s-1} \simeq \begin{cases} t_f\lambda/m\Lambda\gamma, & l = 0, 1, \ldots, s-2 \\ t_f/m\Lambda, & l = s-1 \end{cases} \quad .$$

The algebraic system, Eq. (16), was solved with respect to $[\boldsymbol{v}^T, p]^0$ for m = 4, 8, 16, 32 and t_f = 600 s. Then, the optimal control $u^0(\sigma)$ was computed using $u^0(\sigma) = \Phi^T(\sigma)\boldsymbol{v}^0$, and the optimal output $y^0(\sigma) = n^0(\sigma)$ was computed using

$$y^0(\sigma) = y_0 + t_f \int_0^\sigma h(\sigma - \sigma')\, u^0(\sigma')\, d\sigma' \quad ,$$

and carrying out the integration analytically; finally, the optimal reactivity was computed using the equation $\rho^0(\sigma) = u^0(\sigma)/y^0(\sigma)$. In all cases, the optimal neutron level was found to be in perfect agreement with the desired neutron-level trajectory $N(\sigma)$, $0 \leq \sigma \leq 1$ (see Table I). The resulting optimal reactivity has the stepwise (Walsh) form shown in Fig. 2 where, for clarity, reactivity values smaller than 0.1×10^{-4} are not plotted. Figure 3 provides the plot of $y^0(\sigma)$ found with m = 8. (Only 16 points from the 80 points calculated are shown.)

SPACE-DEPENDENT NUCLEAR REACTOR CONTROL

Our purpose here is to extend the results thus far derived for space-dependent nuclear reactors.[16-19] To this end, consider a space-dependent nuclear reactor system in the integral form

$$X(x,t) = X_0(x,t) + \mathcal{L}(x,t)\, U_d(x,t), \quad x \in D, \; t \geq t_0$$

$$X_0(x,t) = \int_D K(x,t;x',0)\, X_0(x')dx'$$

$$\mathcal{L}(x,t) = \int_0^t \int_D K(x,t;x',\tau)\{\cdot\}dx'd\tau \quad , \tag{17}$$

where

D = three-dimensional spatial domain occupied by the reactor

$X(x,t)$ = p-dimensional output (or state) function

$U_d(x,t)$ = input (source) function

$K(x,t;x',\tau)$ = Green matrix function.

The input function is assumed to be of the form

$$U_d(x,t) = \sum_{k=1}^N u_k(t)\, \delta(x - x_k) \quad , \tag{18}$$

which corresponds to a situation of N control rods located at the points $x_k \in D$ (k = 1, 2, . . . , N). Introducing Eq. (18) into Eq. (17) yields the model

$$X(x,t) = X_0(x,t) + \int_0^t \boldsymbol{K}(x,t;\tau)\, \boldsymbol{u}(\tau)d\tau$$

$$\boldsymbol{K}(x,t;\tau) = [K(x,t;x_1,\tau), \ldots, K(x,t;x_N,\tau)]$$

$$\boldsymbol{u}(\tau) = [u_1(\tau), u_2(\tau), \ldots, u_N(\tau)]^T \quad .$$

The optimal control problem is to choose $\boldsymbol{u}(t)$ in the time interval $t \in [0, t_f]$ such that to minimize a functional of the type

$$J = l(X, X_t, \boldsymbol{u}, \dot{\boldsymbol{u}})\big|_0^{t_f} + \int_0^{t_f}\int_D L(X, X_t; \boldsymbol{u}, \dot{\boldsymbol{u}}; x, t)dxdt \quad , \tag{19}$$

[16]D. WIBERG, *Trans. Am. Nucl. Soc.*, **7**, 219 (1964).
[17]I. KLIGER, *Trans. Am. Nucl. Soc.*, **8**, 233 (1965).
[18]C. HSU and R. BAILEY, *Trans. Am. Nucl. Soc.*, **10**, 253 (1967).
[19]S. H. KYONG, *Nucl. Sci. Eng.*, **32**, 146 (1968).

TABLE I

Optimal Reactivity and Neutron Level Trajectory over the Normalized Interval $0 \leq \sigma \leq 1$

$m = 4$		$m = 8$		$m = 16$		$m = 32$	
$\rho^0 \times 10^4$	$y^0 \times 10^{-8}$ (n/cm^3)	$\rho^0 \times 10^4$	$y^0 \times 10^{-8}$ (n/cm^3)	$\rho^0 \times 10^4$	$y^0 \times 10^{-8}$ (n/cm^3	$\rho^0 \times 10^4$	$y^0 \times 10^{-8}$ (n/cm^3)
5.9053	0.5005	10.9160	0.2507	18.9453	0.1258	29.8934	0.0634
						21.2360	0.1258
				11.9597	0.2507	15.8843	0.1883
						12.4487	0.2507
		6.2772	0.5005	8.4048	0.3756	10.1373	0.3131
						8.5105	0.3756
				6.3892	0.5005	7.3179	0.4380
						6.4123	0.5005
3.1914	0.9999	4.2673	0.7502	5.1301	0.6253	5.7037	0.5629
						5.1353	0.6253
				4.2798	0.7502	4.6695	0.6878
						4.2810	0.7502
		3.2103	0.9999	3.6699	0.8751	3.9521	0.8126
						3.6702	0.8751
				3.2118	0.9999	3.4258	0.9375
						3.2119	0.9999
0.2546	1.0000	0.4744	1.0000	0.8271	1.0000	1.3155	1.0000
						0.5388	1.0000
				0.2130	1.0000	0.2206	1.0000
						0.0903	1.0000
		0.0701	1.0000	0.0548	1.0000	0.0370	1.0000
						0.0151	1.0000
				0.0141	1.0000	0.0062	1.0000
						0.0025	1.0000
0.0203	1.0000	0.0103	1.0000	0.0036	1.0000	0.0010	1.0000
						0.0004	1.0000
				0.0009	1.0000	0.0001	1.0000
						0.00007	1.0000
		0.0015	1.0000	0.0002	1.0000	0.00003	1.0000
						0.00001	1.0000
				0.00006	1.0000	0.000005	1.0000
						0.000002	1.0000

subject to the constraint $X(x, t_f) = X_f(x)$, $x \in D$, where $X_t = \partial X/\partial t$. Expanding $\dot{\boldsymbol{u}}(t)$ in a Walsh series, we obtain again Eqs. (6) and (7). Differentiating $X(x, t)$ yields

$$X_t(x,t) = X_{0,t}(x,t) + \boldsymbol{K}(x,t;t)\,\boldsymbol{u}(t) - \boldsymbol{K}(x,t;0)\,\boldsymbol{u}(0) + \int_0^t \boldsymbol{K}_t(x,t;\tau)\boldsymbol{u}(\tau)d\tau + \int_0^t \boldsymbol{K}(x,t;\tau)\,\dot{u}(\tau)d\tau \quad .$$

Expressing $X(x, t)$, $X_t(x, t)$, $\boldsymbol{u}(t)$, and $\dot{\boldsymbol{u}}(t)$ in terms of $\boldsymbol{v}$, and introducing the results in the cost functional, Eq. (19), we obtain the conventional function $J(\boldsymbol{v})$, which can be minimized by the procedure already described.

As an example, consider a finite-cylinder one-group reactor in the vicinity of an equilibrium point,[19] in which case $D = (0 \leqslant r \leqslant R,\ 0 \leqslant \theta \leqslant 2\pi,\ 0 \leqslant z \leqslant 1)$, $x = (r, \theta, z)$, and

$$K(x,t;x',\tau) = \frac{2}{\pi}\sum_{i,j,k} \epsilon(i)\exp\left[\gamma_{ijk}(t-\tau)\right]\cos(i\theta) \times \sin(j\pi z)R_i(\beta_{ik}r)\cos(i\theta')\sin(j\pi z')R_i(\beta_{ik}r') \quad ,$$

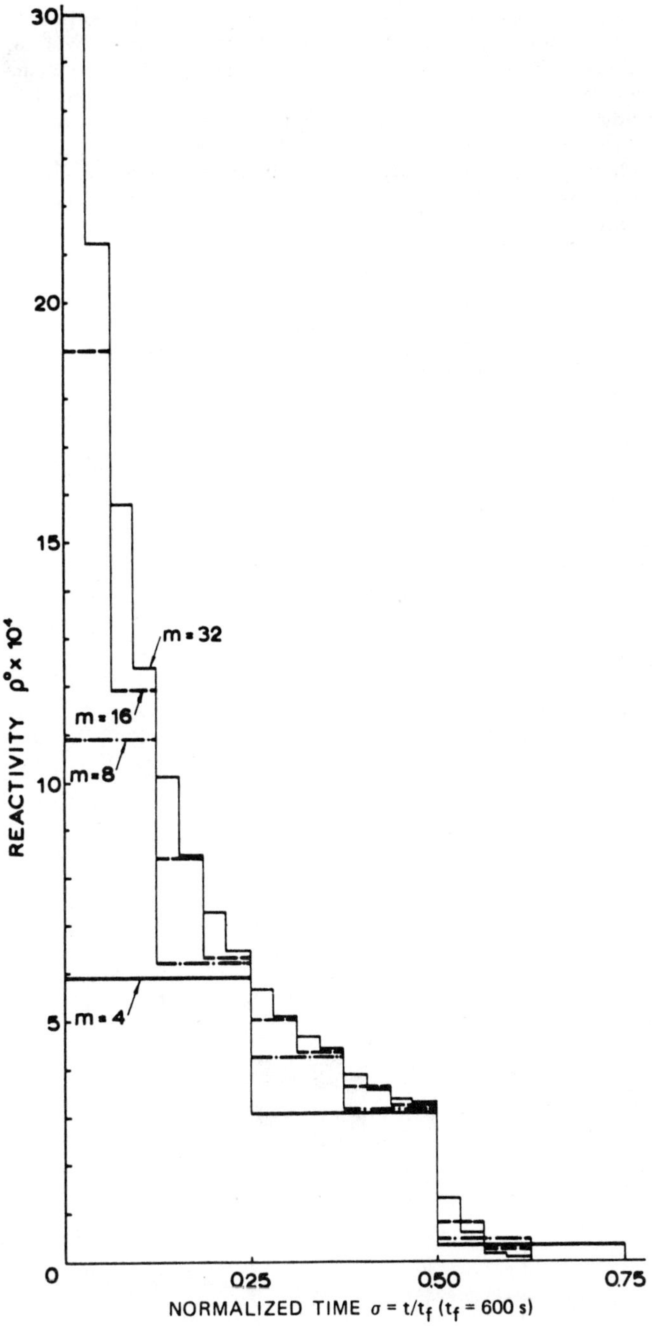

Fig. 2. Optimal step-like reactivity input for $m = 4, 8, 16$, and 32, which minimizes J under the final value constraint given.

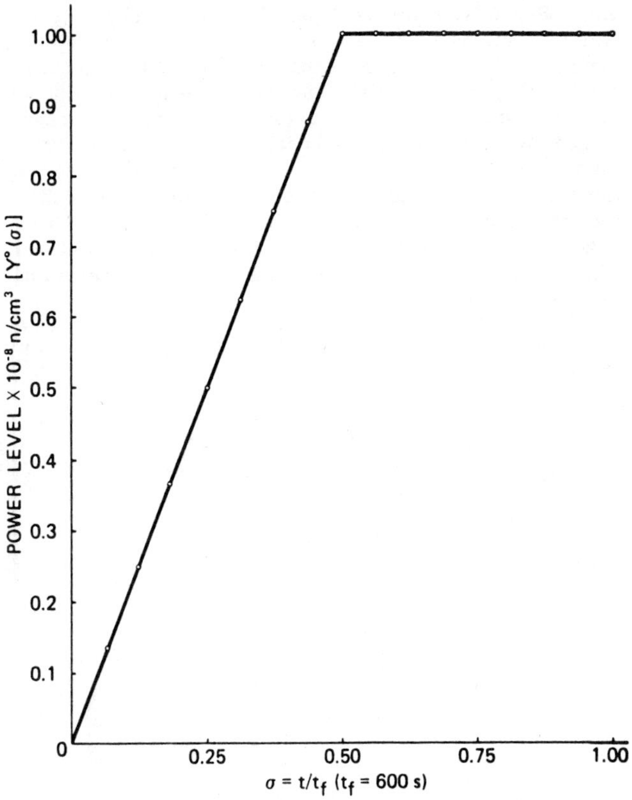

Fig. 3. Optimal neutron level trajectory computed by using the step-like reactivity input of Fig. 2 for $m = 8$. (The same response was also found for $m = 4$, 16, and 32.)

where

$$R_i(\beta_{ik}r) = J_i(\beta_{ik}r)\Big/\left[\int_0^R (\beta_{ik}r)r\,dr\right]^{1/2}$$

$$\gamma_{ijk} = \upsilon(\nu\Sigma_f - \Sigma_a) - \upsilon D_0(j\pi^2 + \beta_{ik}^2)$$

$$\epsilon(i) = \begin{cases} \frac{1}{2}, & i = 0 \\ 1, & i = 1, 2, \ldots \end{cases} ,$$

and β_{ik} is a real number such that $\zeta = \beta_{ik}R$ is the k'th root of the Bessel equations $J_i(\zeta) = 0$.

We wish to minimize the functional

$$J = \int_D [X_f(x) - X(x, t_f)]^2 dx + \mu^2 \int_0^{t_f} \boldsymbol{u}^T(t)\,\boldsymbol{u}(t)dt \tag{20}$$

without any constraint.

Since the functional J does not contain $\dot{\boldsymbol{u}}(t)$, it is sufficient to expand $\boldsymbol{u}(t)$ and not $\dot{\boldsymbol{u}}(t)$, i.e.,

$$\boldsymbol{u}(t) = \mathbf{U}\,\boldsymbol{\phi}(t) = A_1^T(t)\boldsymbol{v} \quad , \tag{21}$$

where $A_1(t)$ is given by Eq. (13c).

Using Eq. (21), $X(x, t)$ can be written in normalized time as

$$X(x, \sigma) = X_0(x, \sigma) + H_0^T(x, \sigma)\boldsymbol{v} \quad , \tag{22}$$

where

$$H_0^T(x, \sigma) = t_f \int_0^{\sigma} \boldsymbol{K}(x, \sigma, \tau)\,A_1^T(\tau)d\tau \quad .$$

Introducing Eqs. (21) and (22) into the cost functional, Eq. (20), and equating $\partial J/\partial \boldsymbol{v}$ to zero, we obtain

$$\boldsymbol{v}^0 = -\boldsymbol{S}^{-1}\boldsymbol{q} \quad ,$$

where

$$\boldsymbol{q} = \int_D [X_0(x, 1) - X_f(x)]H_0(x, 1)dx$$

$$\boldsymbol{S} = \mu^2 t_f \int_0^1 A_1(\sigma)\,A_1^T(\sigma)d\sigma + \int_D H_0(x, 1)\,H_0^T(x, 1)dx \quad .$$

CONCLUSIONS

The Walsh function family has interesting properties and has found useful applications in many areas, such as communications in the sequency domain, image processing, etc.[20] Recently, some work has been initiated on applying Walsh functions to state-space problems and to optimal control.[6] The present Note has established the successful applicability of the Walsh function expansion in point and space-dependent nuclear reactor control systems represented by their integral model. Work is in progress toward the goal of designing easily implemented reactivity meters[9,10] by the Walsh function approach.[11] In addition, the technique will be applied for solving the optimal regulation problem of complete power reactor systems.[1] The essential feature of the Walsh function variational synthesis method is the reduction of the original optimization problem to that of solving a linear algebraic system, the coefficients of which are easily computable due to the piecewise constancy of the Walsh functions. Another advantage of the approach is due to the utilization of the weighting or the Green function of the system, which can be obtained experimentally. A drawback is that the resulting controller is of the open-loop type, which, however, can be valuable in many design cases.

[20]R. LACKEY, *Proc. Symp. Walsh Functions*, p. 2, The Catholic University of America, Washington, D.C., National Technical Information Services (1972).

ACKNOWLEDGMENT

The authors wish to thank the reviewers for their valuable comments, which have been incorporated in this Note.

22

Reprinted from *IEE Proc.* **130:**300–312 (1983)

OPTIMAL CONTROL OF LINEAR SYSTEMS WITH DELAYS IN STATE AND CONTROL VIA WALSH FUNCTIONS

K. R. Palanisamy and G. Prasada Rao

Indexing terms: *Optimal control, Delay systems, Walsh functions*

Abstract: The paper presents a parameter-embedding approach to the optimal control of linear time-delay systems, and develops a simple computational algorithm via Walsh functions. The algorithm employs the concept of Walsh operational matrices for delay and advance. These operational matrices provide us with a means of numerically integrating differential equations with delayed and advanced arguments; a major task in the process of computing optimal controls for time-delay systems. An attractive feature of the present method is its ultimate simplicity and the resulting piecewise constant controls are convenient in practical implementation.

1 Introduction

Optimal control of time-delay systems has been of considerable interest to control engineers in recent years [1–20]. The early works of Kharatishvili [1] and Krasovskii [2, 3] provided the basic foundations. Eller *et al.* [4] use the basic formulation to develop a set of partial differential equations (PDE) to be solved for optimal feedback control of a linear TVP system with delay in its state alone. A parameter-embedding method of design has been suggested by Chan and Perkins [6], Gracovetsky and Vidyasagar [7] and Jamshidi and Malek-Zavarei [8] for such problems. Inoue *et al.* [20] gave a sensitivity technique and Sannuti [9] suggested a singular perturbation approach to the case of systems with small delay. Ray and Soliman [13] and Budelis and Bryson [15] gave a PDE method considering delays both in the state and control. Major problems in the process of design of optimal controls for time-delay systems are computational in nature. Often we require either the solution or the fundamental matrix of the system described by delay differential equations. Several important aspects of delay differential equations have been recently discussed by Tsoi [35–38] summarising the works of Olbrot [32–34], Popov [31], Choudhury [39], Manitius [47, 48] and others. Banks and his colleagues [40–45] and Banks [46] have done extensive work in hereditary systems using averaging approximation technique with function space analysis. Delfour [49, 52] and Delfour and Mitter [53–55] give rigorous convergence arguments for the state, co-state and Riccati operator variables, as well as for the optimal controls for the approximating discrete-system control problems. Extensive surveys are available [40, 47–50]. As we shall see later, the embedding method in this case leads simultaneously to differential equations with delayed and advanced arguments, and there seems to be little work regarding the solution of the latter kind of equations.

This paper follows the embedding approach for systems with delays in state and control in general, and presents a computational algorithm via Walsh functions (WF). The WF approach to optimal control, as suggested by Chen and Hsiao [21, 22] and Burkhardt [55], is limited to non-delay systems for want of suitable tools to deal with the delay and advance terms. The necessary tools have now been developed by the authors in the form of Walsh operational matrices for delay and advance (WDM & WAM) [27] to enlarge the scope of applications of the basic Walsh-function technique.

2 Optimal control of time-delay systems

Consider a linear system with delays in both the state and control described by

$$\dot{x}(t) = Ax(t) + Lx(t-\tau_s) + Bu(t) + Su(t-\tau_c) \quad (1)$$

with initial data

$$x(t_0) = x_0 \quad (2a)$$

$$x(t) = \gamma_x(t) \quad t \in [t_0 - \tau_s, t_0] \quad (2b)$$

$$u(t) = \gamma_u(t) \quad t \in [t_0 - \tau_c, t_0] \quad (2c)$$

x is an n-vector of state; u is an r-vector of inputs; A, L, B and S are matrices of appropriate dimensions. τ_s and τ_c are delays in state and control, respectively, and the initial functions $\gamma_x(t)$ and $\gamma_u(t)$ are continuous in their respective intervals. The problem is to minimise

$$J = 0.5x^T(t_f)Gx(t_f) + 0.5\int_0^{t_f} (x^TQx + u^TRu)\, dt \quad (3)$$

where superscript T indicates transpose, Q is a constant symmetric positive semi-definite matrix and R is a constant symmetric positive definite matrix. We wish to find a control $u(t)$, $0 \leqslant t \leqslant t_f$, which, for the initial conditions as specified in eqn. 2 for the fixed initial state x_0, free final state x_f and the fixed terminal time t_f, minimises the performance index J subject to eqn. 1. The state and co-state equations in the canonical form [16] may be shown to be

$$\begin{aligned}\dot{x}(t) &= Ax(t) + Lx(t-\tau_s) - BR^{-1}B^Tp(t) \\ &\quad - BR^{-1}S^Tp(t+\tau_c) + Su(t-\tau_c) \\ &\qquad\qquad t \in [t_0, t_0+\tau_c] \quad (4a)\\ &= Ax(t) + Lx(t-\tau_s) - [BR^{-1}B^T + SR^{-1}S^T]p(t) \\ &\quad - BR^{-1}S^Tp(t+\tau_c) - SR^{-1}B^Tp(t-\tau_c) \\ &\qquad\qquad t \in [t_0+\tau_c, t_f-\tau_c] \quad (4b)\\ &= Ax(t) + Lx(t-\tau_s) - [BR^{-1}B^T + SR^{-1}S^T]p(t) \\ &\quad - SR^{-1}B^Tp(t-\tau_c) \qquad t \in [t_f-\tau_c, t_f] \quad (4c)\end{aligned}$$

$$\dot{p}(t) = -A^Tp(t) - L^Tp(t+\tau_s) - Qx(t) \quad t \in [t_0, t_f] \quad (4d)$$

Paper 2743D, first received 20th January and in revised form 2nd June 1983

Dr. Palanisamy is with the Department of ECE, Government College of Technology, Coimbatore 641 013, India.

Prof. Rao is with the Lehrstuhl für Elektrische Steuerung und Regelung, Ruhr-Universität Bochum, 4630 Bochum 1, W. Germany, on sabbatical leave from the Department of Electrical Engineering, Indian Institute of Technology, Kharagpur 721302, India

Eqns. 4*a*–*d* satisfy the conditions in eqn. 2. In addition, $p(t) = 0$ for $t \geqslant t_f$. The optimal response $x^*(t)$ and $p^*(t)$ may be employed to get the optimal control $u^*(t)$:

$$u^*(t) = -R^{-1}B^{\mathrm{T}}p^*(t) - R^{-1}S^{\mathrm{T}}p^*(t+\tau_c) \qquad t \in [t_0, t_f - \tau_c] \quad (5a)$$

$$= -R^{-1}B^{\mathrm{T}}p^*(t) \qquad t \in [t_f - \tau_c, t_f] \quad (5b)$$

Eqns. 4 and 5 may be converted into initial-value problems by an embedding method [7, 8]. We also have

$$p(t) = k(t)x(t) + h(t) \quad (6)$$

Inserting eqn. 6 into eqns. 4 and 5 and solving, we obtain

$$\dot{k}(t) = k(t)c_1 k(t) - k(t)A - A^{\mathrm{T}}k(t) - Q \qquad t \in [t_0, t_0 + \tau_c]$$

$$k(t) = G \text{ for } t > t_f \quad (7a)$$

$$= k(t)c_5 k(t) - k(t)A - A^{\mathrm{T}}k(t) - Q \qquad t \in [t_0 + \tau_c, t_f] \quad (7b)$$

$$\begin{aligned}\dot{h}_N(t) = {} & (k(t)c_1 - A^{\mathrm{T}})h_N(t) - k(t)\{Lx_{N-1}(t-\tau_s) \\ & + c_3[k(t+\tau_c)x_{N-1}(t+\tau_c) + h_{N-1}(t+\tau_c)] \\ & + c_4[k(t-\tau_c)x_{N-1}(t-\tau_c) + h_{N-1}(t-\tau_c)]\} \\ & - L^{\mathrm{T}}h_{N-1}(t+\tau_s) - L^{\mathrm{T}}k(t+\tau_s)x_{N-1}(t+\tau_s) \\ & \qquad t \in [t_0, t_0 + \tau_c] \quad (8a)\end{aligned}$$

$$\begin{aligned}= {} & (k(t)c_5 - A^{\mathrm{T}})h_N(t) - k(t)\{Lx_{N-1}(t-\tau_s) \\ & + c_3[k(t+\tau_c)x_{N-1}(t+\tau_c) + h_{N-1}(t+\tau_c)] \\ & + c_4[k(t-\tau_c)x_{N-1}(t-\tau_c) + h_{N-1}(t-\tau_c)]\} \\ & - L^{\mathrm{T}}k(t+\tau_s)x_{N-1}(t+\tau_s) - L^{\mathrm{T}}h_{N-1}(t+\tau_s) \\ & \qquad t \in [t_0 + \tau_c, t_f - \tau_c] \quad (8b)\end{aligned}$$

$$\begin{aligned}= {} & (k(t)c_5 - A^{\mathrm{T}})h_N(t) - k(t)\{Lx_{N-1}(t-\tau_s) \\ & + c_4[k(t-\tau_c)x_{N-1}(t-\tau_c) + h_{N-1}(t-\tau_c)]\} \\ & - L^{\mathrm{T}}k(t+\tau_s)x_{N-1}(t+\tau_s) - L^{\mathrm{T}}h_{N-1}(t+\tau_s)\end{aligned}$$

$$h(t) = 0 \quad \text{for } t \geqslant t_f \quad \text{and} \quad t \in [t - \tau_c, t_f] \quad (8c)$$

$$\begin{aligned}\dot{x}_N(t) = {} & (A - c_1 k(t))x_N(t) + Lx_{N-1}(t-\tau_s) - c_1 h_N(t) \\ & - c_3[k(t+\tau_c)x_{N-1}(t+\tau_c) + h_N(t+\tau_c)] \\ & + Su(t-\tau_c) \qquad t \in [t_0, t_0+\tau_c] \quad (9a)\end{aligned}$$

$$\begin{aligned}= {} & (A - c_5 k(t))x_N(t) + Lx_{N-1}(t-\tau_s) - c_5 h_N(t) \\ & - c_3[k(t+\tau_c)x_{N-1}(t+\tau_c) + h_N(t+\tau_c)] \\ & - c_4[h_N(t-\tau_c) + k(t-\tau_c)x_{N-1}(t-\tau_c)] \\ & \qquad t \in [t_0+\tau_c, t_f-\tau_c] \quad (9b)\end{aligned}$$

$$\begin{aligned}= {} & (A - c_5 k(t))x_N(t) + Lx_{N-1}(t-\tau_s) - c_5 h_N(t) \\ & - c_4[k(t-\tau_c)x_{N-1}(t-\tau_c) + h_N(t-\tau_c)] \\ & \qquad t \in [t_f-\tau_c, t_f] \quad (9c)\end{aligned}$$

where

$$c_1 = BR^{-1}B^{\mathrm{T}}$$

$$c_2 = SR^{-1}S^{\mathrm{T}}$$

$$c_3 = BR^{-1}S^{\mathrm{T}}$$

$$c_4 = SR^{-1}B^{\mathrm{T}}$$

$$c_5 = c_1 + c_2$$

The Riccati equations (eqns. 7*a* and *b*) are solved first. Then eqn. 8 is solved by backward integration to obtain adjoint vector $h(t)$, and eqn. 9 is integrated forward in time to obtain $x(t)$. These equations are solved iteratively and subscript N denotes the iteration number.

3 Solution of Riccati differential equation

To solve eqn. 7*b*, eqns. 4*b* and 4*d* are first written in state-space form as

$$\begin{bmatrix}\dot{x}(t)\\ \dot{p}(t)\end{bmatrix} = \begin{bmatrix}A & -c_5\\ -Q & -A^{\mathrm{T}}\end{bmatrix}\begin{bmatrix}x(t)\\ p(t)\end{bmatrix} + \text{delay terms} \quad (10)$$

Let $\lambda(t_f, t)$ be the transition matrix of eqn. 10 (omitting delay terms) and $\lambda(t_f, t_f) = I$.

$\lambda(t_f, t)$ may be decomposed into the following form:

$$\lambda(t_f, t) = \begin{bmatrix}\lambda_{11}(t_f, t) & \lambda_{12}(t_f, t)\\ \lambda_{21}(t_f, t) & \lambda_{22}(t_f, t)\end{bmatrix} \quad (11)$$

We may thus write

$$\begin{bmatrix}x(t_f)\\ p(t_f)\end{bmatrix} = \begin{bmatrix}\lambda_{11}(t_f, t) & \lambda_{12}(t_f, t)\\ \lambda_{21}(t_f, t) & \lambda_{22}(t_f, t)\end{bmatrix}\begin{bmatrix}x(t)\\ p(t)\end{bmatrix} \quad (12)$$

Following Chen and Hsiao [21] from eqn. 12

$$k(t) = \lambda_{22}^{-1}(t_f, t)\lambda_{21}(t_f, t) \qquad t \in [t_0 + t_c, t_f] \quad (13)$$

Similarly $k(t)$ in eqn. 7*a* may be obtained from the transition matrix of

$$\begin{bmatrix}\dot{x}(t)\\ \dot{p}(t)\end{bmatrix} = \begin{bmatrix}A & -c_1\\ -Q & -A^{\mathrm{T}}\end{bmatrix}\begin{bmatrix}x(t)\\ p(t)\end{bmatrix} + \text{delay terms} \quad (14)$$

which is obtained from eqns. 4*a* and 4*d* in the interval $t \in [t_0, t_0 + \tau_c]$, on the lines of eqns. 11–13.

Eqns. 8 and 9 containing delayed and advanced arguments should now be solved in the respective intervals.

The additional operational matrices necessary to handle the delay and advance terms, together with the technique of extension [29] of solution over continuous intervals of time recently developed by the authors, make the WF approach computationally convenient and practically useful. The original version of the technique is relieved of its computational burden by the extension of the single-term Walsh-series technique [29].

In order to facilitate understanding of the proposed method we will briefly review the essentials of WF technique.

4 Review of Walsh operational matrices

A function $f(t)$, absolutely integrable in [0, 1), is expanded in Walsh series over the interval as

$$f(t) = \sum_{i=0}^{\infty} f_i \psi_i(t) \quad (15)$$

where $\psi_i(t)$ is the ith Walsh function and f_i the corresponding coefficient. In practice, while approximating a function, only the first m components are considered, where m is an integral power of 2. If the coefficients of the Walsh functions are concisely written as m-vectors

$$F = [f_0, f_1 \cdots f_{m-1}]^{\mathrm{T}} \quad (16)$$

and

$$\Psi(t) = [\psi_0(t), \psi_1(t) \cdots \psi_{m-1}(t)]^{\mathrm{T}} \quad (17)$$

$$f(t) \approx F^{\mathrm{T}}\Psi(t) \quad (18)$$

The coefficients f_i are chosen to minimise

$$\int_0^1 [f(t) - F^T\Psi(t)]^2\, dt \tag{19}$$

Under minimal mean-square-error condition

$$f_i = \int_0^1 f(t)\psi_i(t)\, dt \tag{20}$$

leading to an m-segment piecewise constant approximation $f(t)$ over unit interval with minimal mean-square error.

4.1 Operational matrix for integration

It has been originally shown [21, 25] that

$$\int_0^t f(t)\, dt = F^T E \Psi(t) \tag{21}$$

where the constant square matrix E, known as the operational matrix for integration, is given by

$$E_{(m\times m)} = \left[\begin{array}{c|c} E_{(m/2\times m/2)} & -\dfrac{1}{2m} I_{(m/2\times m/2)} \\ \hline \dfrac{1}{2m} I_{(m/2\times m/2)} & 0_{(m/2\times m/2)} \end{array}\right] \tag{22}$$

where $m = 2^{k_w}$, k_w being an integer.

Beginning with $E_{(1\times 1)} = 0.5$, $E_{(m\times m)}$ can be recursively built from eqn. 22. Integration of a function over [0, 1) may thus be achieved approximately (in the sense of least squares) by premultiplying the vector of Walsh spectral components of the function with a constant, nonsingular matrix. The elements of E actually stem from Walsh-series representations of the integrals of WF themselves. We will discuss operational matrices in the same sense to achieve a given delay in the following.

4.2 Walsh delay and advance matrices

4.2.1 Walsh operational matrices for delay (WDM): A function $f(t)$, absolutely integrable in [0, 1), is expanded in Walsh series as in eqn. 18. For the purpose of present development the set of Walsh functions $\Psi(t)$ is forced to be zero outside the interval. If f_i are the Walsh components of $f(t)$, then $f(t-\tau)$, where τ is a constant less than unity, may be expressed in the form

$$f(t-\tau) \approx \sum_{i=0}^{m-1} f_i \psi_i(t-\tau) \tag{23}$$

if the initial function $f(t) = 0$, $t \in [-\tau, 0)$. In particular, if $\tau = \alpha/m$, $0 \leqslant \alpha \leqslant 1$, $\psi_i(t-\tau)$ may be expressed in terms of $\psi_i(t)$ approximately in the sense of least squares as

$$\Psi(t-\tau) = D_{(m)}\Psi(t), \tag{24}$$

where $D_{(m)}$ is the Walsh delay matrix (WDM).

The effect of delaying a function may be achieved approximately (in the sense of least squares) by premultiplying the vector of Walsh spectral components of the function with a constant and sometimes singular square matrix. The elements of D stem from the Walsh-series expansion of $\Psi(t-\tau)$.

If, say $m = 2$,

$$D_{(2)} = \frac{1}{2}\begin{bmatrix} 2-\alpha & -\alpha \\ \alpha & 2-3\alpha \end{bmatrix}$$

$$= I_{(2)} + \frac{\alpha}{2}\begin{bmatrix} -1 & -1 \\ 1 & -3 \end{bmatrix}$$

$$= I_{(2)} + \frac{\alpha}{2}\mathscr{D}_{(2)} \tag{25a}$$

If $m = 4$

$$D_{(4)} = \frac{1}{4}\begin{bmatrix} 4-\alpha & -\alpha & -\alpha & -\alpha \\ \alpha & 4-3\alpha & \alpha & -3\alpha \\ \alpha & \alpha & 4-7\alpha & \alpha \\ -\alpha & 3\alpha & -\alpha & 4-5\alpha \end{bmatrix}$$

$$= I_{(4)} + \frac{\alpha}{4}\mathscr{D}_{(4)} \tag{25b}$$

and if $m = 8$,

$$D_{(8)} = \frac{1}{8}\begin{bmatrix} 8-\alpha & -\alpha & -\alpha & -\alpha & -\alpha & -\alpha & -\alpha & -\alpha \\ \alpha & 8-3\alpha & \alpha & -3\alpha & \alpha & -3\alpha & \alpha & -3\alpha \\ \alpha & \alpha & 8-7\alpha & \alpha & \alpha & \alpha & -7\alpha & \alpha \\ -\alpha & 3\alpha & -\alpha & 8-5\alpha & -\alpha & 3\alpha & -\alpha & -5\alpha \\ \alpha & \alpha & \alpha & \alpha & 8-15\alpha & \alpha & \alpha & \alpha \\ -\alpha & 3\alpha & -\alpha & 3\alpha & -\alpha & 8-13\alpha & -\alpha & 3\alpha \\ -\alpha & -\alpha & 7\alpha & -\alpha & -\alpha & -\alpha & 8-9\alpha & -\alpha \\ \alpha & -3\alpha & \alpha & 5\alpha & \alpha & -3\alpha & \alpha & 8-11\alpha \end{bmatrix}$$

$$= I_{(8)} + \frac{\alpha}{8}\mathscr{D}_{(8)} \tag{25c}$$

The $\mathscr{D}_{(2)}$s associated with $-1, -3, -5, -7, \ldots$ are defined as

$$\mathscr{D}_{(2)}^{(-1)} = \begin{bmatrix} -1 & -1 \\ 1 & -3 \end{bmatrix}$$

$$\mathscr{D}_{(2)}^{(-3)} = \begin{bmatrix} -7 & 1 \\ -1 & -5 \end{bmatrix}$$

$$\mathscr{D}_{(2)}^{(-5)} = \begin{bmatrix} -9 & -1 \\ 1 & -11 \end{bmatrix} \tag{25d}$$

$$\mathscr{D}_{(2)}^{(-7)} = \begin{bmatrix} -15 & 1 \\ -1 & -13 \end{bmatrix}$$

i.e. for any negative odd integer j, $\mathscr{D}_{(2)}^{(j-2)}$ is defined by its elements

$$\delta_{(2)12}^{(+j-2)} = (-1)\,\delta_{(2)12}^{(+j)}$$

$$\delta_{(2)21}^{(+j-2)} = (-1)\,\delta_{(2)21}^{(+j)}$$

and the diagonal elements take the values from the array

j	1		3		5		7
$\delta_{(2)11}$	−1		−7	→	−9		−15
	↓		↑		↓		↑
$\delta_{(2)22}$	−3	→	−5		−11	→	−13

It may be interesting to note that

$$\mathscr{D}_{(4)} = \left[\begin{array}{c|c} \mathscr{D}_{(2)}^{(-1)} & \mathscr{D}_{(2)}^{(-1)} \\ \hline \mathscr{D}_{(2)}^{(+1)} & \mathscr{D}_{(2)}^{(-3)} \end{array}\right]$$

$$= \left[\begin{array}{cc|cc} -1 & -1 & -1 & -1 \\ 1 & -3 & 1 & -3 \\ \hline 1 & 1 & -7 & 1 \\ -1 & 3 & -1 & -5 \end{array}\right] \tag{25e}$$

$$\mathscr{D}_{(8)} = \left[\begin{array}{c|c:c} \mathscr{D}_{(4)} & \multicolumn{2}{c}{\mathscr{D}_{(4)}} \\ \hdashline & \mathscr{D}^{(-7)}_{(2)} & -\mathscr{D}^{(+1)}_{(2)} \\ \cdashline{2-3} -\mathscr{D}_{(4)} & \mathscr{D}^{(-1)}_{(2)} & \mathscr{D}^{(-5)}_{(2)} \end{array}\right] \tag{25f}$$

and so on.

Thus $\mathscr{D}_{(m)}$, for any dimension $m = 2^k$, k an integer, may be obtained from the corresponding $\mathscr{D}_{(m)}$ built in this way. If the delay τ is greater than the segmentation interval $1/m$, i.e. if $\tau = N + \alpha/m$, where N is an integer, $0 \leqslant N \leqslant m$ and $0 \leqslant \alpha \leqslant 1$, the corresponding delay matrices may be generated in the manner similar to the above.

4.2.2 WDM for $\tau = N/m$, N being an integer: An interesting class of matrices may be generated for practical application by adjusting the time scale in the problem and choosing suitable m such that $\tau = N/m$. For $N = 1$ we obtain, by letting $\alpha = 1$ in eqns. 25*a–c*, the WDM $D^1_{(m)}$. For instance

$$D^1_{(2)} = \frac{1}{2}\begin{bmatrix} 1 & -1 \\ 1 & -1 \end{bmatrix} \tag{25g}$$

$$D^1_{(4)} = \frac{1}{4}\begin{bmatrix} 3 & -1 & -1 & -1 \\ 1 & 1 & 1 & -3 \\ 1 & 1 & -3 & 1 \\ -1 & 3 & -1 & -1 \end{bmatrix} \tag{25h}$$

and so on.

A general structural relationship exists between $D^2_{(2^{k+1})}$ and $D^1_{(2^k)}$ as

$$D^1_{(2^{k+1})} = \tfrac{1}{2}\left[\begin{array}{c:c} D^1_{(2^k)} + D^0_{(2^k)} & D^1_{(2^k)} - D^0_{(2^k)} \\ \hdashline -(D^1_{(2^k)} - D^0_{(2^k)}) & -(D^1_{(2^k)} + D^0_{(2^k)}) \end{array}\right] \tag{26}$$

where

$$D^0_{(2^k)} = I_{(2^k)} \tag{27a}$$

Given $D^1_{(2)}$ we make use of eqn. 27*a* and generate $D^1_{(m)}$ for a required m. For higher positive values of N, $D^N_{(m)}$ behaves as the Nth power of $D^1_{(m)}$, i.e.

$$D^N_{(m)} = [D^1_{(m)}]^N \tag{27b}$$

In the WDM $D^N_{(m)}$, the elements d_{ij}, have the properties

$$\left.\begin{aligned} &\sum_{j=1}^{m} d_{ij} = 0 \quad \text{for all } i \\ &\sum_{i=1}^{m} d_{ij} = W_{N+1,j} \end{aligned}\right\} \tag{28}$$

where $W_{N+1,j}$ for $j = 1, 2, 3, \ldots, m$ generates the $(N + 1)$th row of the well known Walsh matrix. For instance

$$W_{(4)} = \begin{bmatrix} 1 & 1 & 1 & 1 \\ 1 & 1 & -1 & -1 \\ 1 & -1 & 1 & -1 \\ 1 & -1 & -1 & 1 \end{bmatrix}$$

$D^1_{(m)}$ is nilpotent with index m, since

$$D^N_{(m)} = 0, \quad N \geqslant m \tag{29}$$

4.2.3 Generation of $D^N_{(m)}$ through Kronecker product formula: The Kronecker product of two matrices A and P is defined as

$$A \otimes P \triangleq \begin{bmatrix} p_{11}A & p_{12}A & \cdots & p_{1m}A \\ p_{21}A & p_{22}A & \cdots & p_{2m}A \\ \hdashline p_{m1}A & p_{m2}A & \cdots & p_{mm}A \end{bmatrix}. \tag{30a}$$

$D^N_{(m)}$ may be generated from WDMs of lower dimension and lower power through simple Kronecker product formula, without involving actual matrix powers

$$D^{2^k+M}_{(2^{k+1})} = D^1_{(2)} \otimes D^M_{(2^k)} \tag{30b}$$

For instance, $D^4_{(8)}$ may be obtained by letting $k = 2$ and $M = 0$ in eqn. 30*b* and we obtain

$$D^4_{(8)} = D^1_{(2)} \otimes D^0_{(4)}$$

$$= \frac{1}{8}\begin{bmatrix} 4 & -4 & & & & & & \\ 4 & -4 & & & & & 0 & \\ & & 4 & -4 & & & & \\ & & 4 & -4 & & & & \\ & & & & 4 & -4 & & \\ & & & & 4 & -4 & & \\ & 0 & & & & & 4 & -4 \\ & & & & & & 4 & -4 \end{bmatrix} \tag{30c}$$

Similarly we obtain $D^7_{(8)}$ as

$$D^7_{(8)} = D^1_{(2)} \otimes D^3_{(4)}$$

$$= \frac{1}{8}\begin{bmatrix} 1 & -1 & -1 & 1 & -1 & 1 & 1 & -1 \\ 1 & -1 & -1 & 1 & -1 & 1 & 1 & -1 \\ 1 & -1 & -1 & 1 & -1 & 1 & 1 & -1 \\ 1 & -1 & -1 & 1 & -1 & 1 & 1 & -1 \\ 1 & -1 & -1 & 1 & -1 & 1 & 1 & -1 \\ 1 & -1 & -1 & 1 & -1 & 1 & 1 & -1 \\ 1 & -1 & -1 & 1 & -1 & 1 & 1 & -1 \\ 1 & -1 & -1 & 1 & -1 & 1 & 1 & -1 \end{bmatrix}. \tag{30d}$$

4.2.4 Generation of WDM for $\tau = 2N/m$, N being integer, by Kronecker product formula: Higher even powers of $D^1_{(m)}$ may be generated without algebraically raising the matrix power by the Kronecker product formula

$$D^{2N}_{(2^{k-1})} = D^N_{(2^k)} \otimes D^0_{(2)} \tag{30e}$$

suggesting that even powers of $D^1_{(m)}$ are in block-diagonal form since $D^0_{(2)} = I_{(2)}$.

4.2.5 Walsh advance matrices (WAM): When τ is negative, it would mean advancing a signal by a known constant τ, as, for instance, in the case of computation of correlation functions. The related matrix for 'advance', the Walsh advance matrix (WAM), is closely related to (WDM) for the same τ. In fact, the WAM for given τ is

$$D^-_{(m)} = [D_{(m)}]^{\mathrm{T}} \tag{31}$$

i.e.

$$D^{-N}_{(m)} = [D^N_{(m)}]^{\mathrm{T}} \tag{32}$$

It is, however, very important to note that these matrices obey the power law only in one direction that is either $+1$, $2, \ldots$ or $-1, -2, \ldots$ and

$$D^{-N}_{(m)} \times D^{+N}_{(m)} \neq I_{(m)}$$

This is because certain segments of the signal overflow the chosen interval when we operate with WDM or WAM.

4.2.6 Choice of m and N in WDM and WAM in a given situation: When we wish to obtain the Walsh spectrum of $f(t - \tau)$ or $f(t + \tau)$ from that of $f(t)$ for a given constant τ, we would adopt a suitable time scaling, and choose a unit interval for computation such that on the normalised scale $\tau < 1$. Accuracy considerations suggest the maximum allowable length of the segmentation interval.

If $\delta T_{\max}$ is the maximum allowable segmentation interval for required accuracy

$$N \geqslant \tau/\delta T_{\max}$$

We choose an actual segmentation interval $\delta T \leqslant \delta T_{\max}$ so that $\tau/\delta T = N$ is an exact integer. Now, if we choose to compute the Walsh spectrum over an absolute base of time $T > \tau$, ($T = m\,\delta T$) we change time scale such that $[0 - T)$ becomes the normalised interval $[0, 1)$. We now have a normalised base $[0, 1)$ with m segments, each of length $1/m$ on the normalised scale. Having fixed N and m we employ the techniques described above to obtain suitable WDM or WAM, along with the approximate operational matrix $E_{(m)}$ for the analysis of delay systems.

5 Walsh spectrum of a delayed function with nonzero initial function

The analysis of delay systems via Walsh functions on the single interval $[0, 1)$ alone will not naturally be able to include the effects of nonzero initial functions. The Walsh spectrum of the initial function in $[-\tau, 0)$ should be suitably carried forward into $[0, 1)$. In order to understand the method of carry over, let us consider, for the sake of illustration, a function $g(t)$ in $[-1, 1)$ subjected to a delay $\tau < 1$, as shown in Fig. 1*a*. We wish to determine the Walsh spectrum of $g(t - \tau)$ on $[0, 1)$ including the effects of carry over from the previous interval $[-1, 0)$. For the sake of convenience of our discussion, let us separately designate $g(t)$ in the two unit intervals as

$$g(t) = \begin{cases} f(t), & t \in [0, 1) \\ \overleftarrow{f}(t), & t \in [-1, 0) \end{cases} \tag{33a}$$

We now define two sets of Walsh functions in these intervals

$$\psi_i(t) \qquad t \in [0, 1) \qquad i = 0, 1, \ldots, m-1$$
$$\overleftarrow{\psi}_i(t) \qquad t \in [-1, 0) \quad i = 0, 1, \ldots, m-1$$

and write them concisely as m-vectors

$$\Psi(t) = [\psi_0(t) \quad \psi_1(t) \quad \cdots \quad \psi_{m-1}(t)]^{\mathrm{T}}$$
$$\overleftarrow{\Psi}(t) = [\overleftarrow{\psi}_0(t) \quad \overleftarrow{\psi}_1(t) \quad \cdots \quad \overleftarrow{\psi}_{m-1}(t)]^{\mathrm{T}}$$

$f(t)$ and $\overleftarrow{f}(t)$ are expanded in the respective intervals in Walsh series as shown in Figs. 1*b* and 1*d*, respectively. Here $\overleftarrow{f}(t)$ plays the role of initial function. If

$$\left.\begin{aligned} f(t) &= F^{\mathrm{T}}\Psi(t) \qquad t \in [0, 1) \\ \text{and} \\ \overleftarrow{f}(t) &= \overleftarrow{F}^{\mathrm{T}}\overleftarrow{\Psi}(t) \qquad t \in [-1, 0) \end{aligned}\right\} \tag{33b}$$

in the light of the technique described in Section 4.2

$$f(t - \tau) = F^{\mathrm{T}} D_m^N \Psi(t) \qquad t \in [0, 1) \tag{34}$$

and illustrated in Fig. 1*c*.

The spectrum of $\overleftarrow{f}(t)$ for carry into $[0, 1)$ is obtained by advancing the spectrum $\overleftarrow{F}$ by $(1 - \tau)$, i.e. by $(m - N)$ segments as shown in Fig. 1*e*. Applying the operational

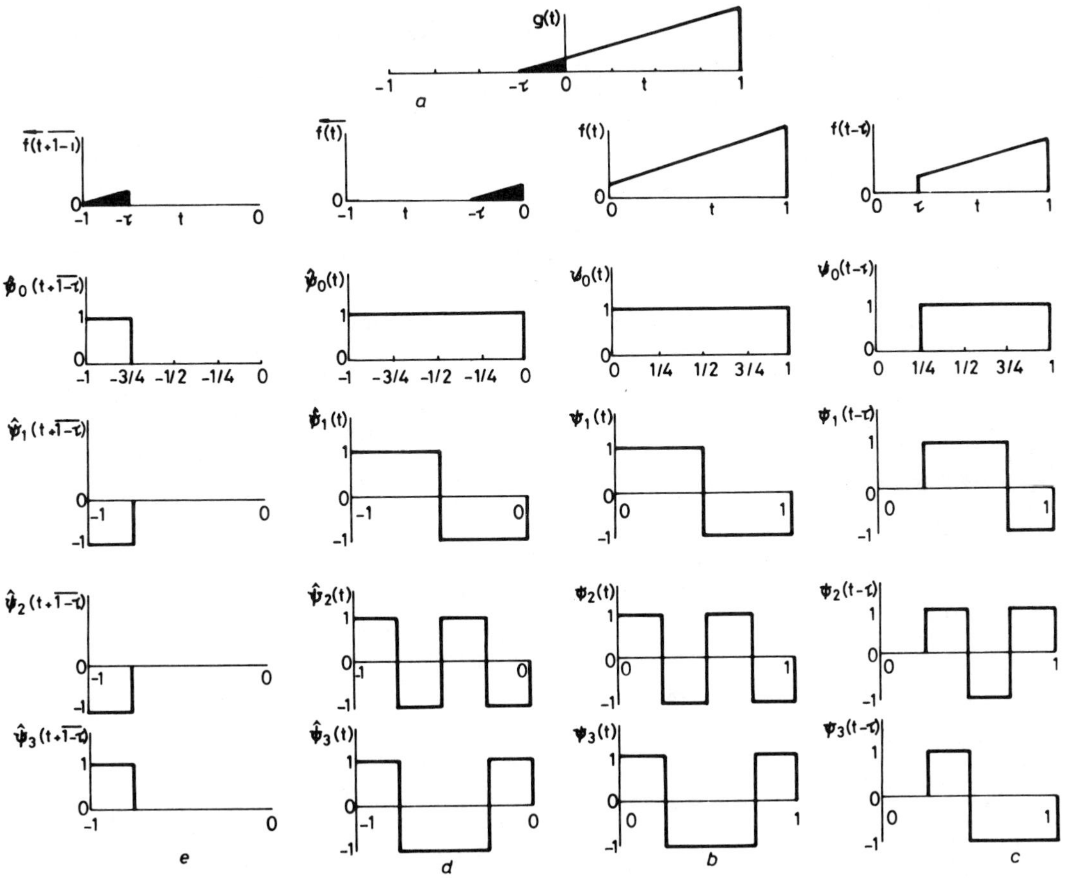

Fig. 1 *Walsh spectrum of a delayed signal and carry-over considerations*

a $g(t)$ subject to a delay $\tau < 1$
b $f(t)$
c $f(t - \tau)$
d $\overleftarrow{f}(t)$
e $\overleftarrow{f}(\overline{t + 1 - \tau})$

matrix for advance (WAM) the carry-over spectrum is given by

$$\tilde{F}_c^{\mathrm{T}} = \tilde{F}^{\mathrm{T}}[D_m^{(m-N)}]^{\mathrm{T}} \tag{35}$$

Thus the Walsh spectrum of $g(t-\tau)$ in $[0, 1)$ including the carry over due to initial function is given by

$$\tilde{G}^{\mathrm{T}} = F^{\mathrm{T}}D_m^N + \tilde{F}^{\mathrm{T}}[D_m^{m-N}]^{\mathrm{T}} \tag{36}$$

6 Solution of delay differential equations

6.1 Systems of equations with a single delay

Consider the equations

$$\dot{x}(t) = Ax(t) + Lx(t-\tau) + Bu(t)$$

$$x(t) = \gamma_x(t) \qquad t \in [-\tau, 0) \tag{37}$$

where x is an n-vector of the state, u is an input vector of l-components, A, L and B are $n \times n$, $n \times n$ and $n \times l$ matrices, respectively. We should first establish a procedure to solve eqn. 37 employing matrices D and E. First we obtain a solution over $[0, 1)$ by expanding $\dot{x}$, x and Bu in Walsh series of size m:

$$\left.\begin{aligned} \dot{x}(t) &= V\Psi(t) \\ x(t) &= X\Psi(t) \\ Bu(t) &= B_u\Psi(t) \\ x(t-\tau) &= \tilde{X}\Psi(t) \end{aligned}\right\} \tag{38}$$

where

$$V = \begin{bmatrix} v_{1,0} & v_{1,1} & \cdots & v_{1,m-1} \\ v_{2,0} & v_{2,1} & \cdots & v_{2,m-1} \\ \vdots & \vdots & & \vdots \\ v_{n,0} & v_{n,1} & \cdots & v_{n,m-1} \end{bmatrix}$$

$$\triangleq [v_0 \,\vdots\, v_1 \,\vdots \cdots \vdots\, v_{m-1}]$$

Similarly,

$$X = [x_0 \,\vdots\, x_1 \,\vdots \cdots \vdots\, x_{m-1}]$$

$$B_u = [B_{u0} \,\vdots\, B_{u1} \,\vdots \cdots \vdots\, B_{um-1}]$$

The V, X, B_u and $\tilde{X}$ are $n \times m$ matrices. The initial function $\gamma_x(t)$ will be expanded in Walsh series $\hat{\Psi}(t)$, $t \in [-1, 0)$

$$\gamma_x(t) = \Gamma_x\hat{\Psi}(t) \tag{39}$$

where

$$\Gamma_x = [\Gamma_{x0} \,\vdots\, \Gamma_{x1} \,\vdots\, \Gamma_{x2} \,\vdots \cdots \vdots\, \Gamma_{xm-1}]$$

In the above V_i, X_i, Bu_i, and Γ_{x_i} are the $(i+1)$th columns of V, X, B_u and Γ_x, respectively.

Inserting eqn. 38 in eqn. 37 we obtain

$$V\Psi(t) = AX\Psi(t) + L\tilde{X}\Psi(t) + B_u\Psi(t) \tag{40}$$

In view of eqn. 36

$$\tilde{X}\Psi(t) = [XD^N + \Gamma_x(D^{m-N})^{\mathrm{T}}]\Psi(t) \tag{41}$$

Expressing X in terms of V using the operational matrix E in eqn. 40 takes the form

$$V\Psi(t) = AVE\Psi(t) + (Ax_0 \;\underbrace{0 \;\; 0 \;\; \cdots \;\; 0}_{(m-1)\text{ columns}})\Psi(t)$$

$$+ LVED^N\Psi(t) + (Lx_0 \;\underbrace{0 \;\; 0 \;\; \cdots \;\; 0}_{(m-1)\text{ columns}})\Psi(t)$$

$$+ L\Gamma_x(D^{m-N})^{\mathrm{T}}\Psi(t) + B_u\Psi(t) \tag{42}$$

leading to

$$V = AVE + LVED^N + C \tag{43}$$

where

$$C = (Ax_0 \;\underbrace{0 \;\; 0 \;\; \cdots \;\; 0}_{(m-1)\text{ columns}}) + (Lx_0 \;\underbrace{0 \;\; 0 \;\; \cdots \;\; 0}_{(m-1)\text{ columns}})$$

$$+ L\Gamma_x(D^{m-N})^{\mathrm{T}} + B_u \tag{44}$$

and $x_0 = x(0)$.

Eqn. 43 may be solved for the elements of V as a single long column $\mathscr{V}$ using the Kronecker product formula

$$\mathscr{V} = [I - A \otimes E^{\mathrm{T}} - L \otimes (ED^N)^{\mathrm{T}}]^{-1}\mathscr{C} \tag{45}$$

where

$$\mathscr{V} = [V_0^{\mathrm{T}} \,\vdots\, V_1^{\mathrm{T}} \,\vdots\, V_2^{\mathrm{T}} \,\vdots \cdots \vdots\, V_{m-1}^{\mathrm{T}}]^{\mathrm{T}}$$

and

$$\mathscr{C} = [C_0^{\mathrm{T}} \,\vdots\, C_1^{\mathrm{T}} \,\vdots\, C_2^{\mathrm{T}} \,\vdots \cdots \vdots\, C_{m-1}^{\mathrm{T}}]^{\mathrm{T}} \tag{46}$$

Having obtained the Walsh spectrum of the rate variable we get the spectrum of the state vector

$$X = VE + [x_0 \;\underbrace{0 \;\; 0 \;\; \cdots \;\; 0}_{(m-1)\text{ columns}}] \tag{47}$$

6.2 Continuation of Walsh-series solution over an extended period

The above method provides us with an algorithm to compute the solution over $[0, 1)$ taking into account the initial-value terms and initial-function terms over $[-\tau, 0)$. It will be of much practical importance if the solution is continued over an extended period of time beyond the limit of the normal interval. This problem will be discussed here.

For eqn. 37, we have discussed a technique of computing the rate variable V and solution X in Walsh series in steps from eqn. 38 to eqn. 47.

To continue solution over the next unit interval $[1, 2)$ we consider Walsh-series expansions over $[1, 2)$ similar to those in $[0, 1)$ and require

(*a*) the Walsh spectrum X of the state in $[0, 1)$ from which the spectrum corresponding to $[1-\tau, 1)$ will be carried into $[1, 2)$

(*b*) the state $x(1)$

$x(t)$ may be obtained from the rate spectrum V in $[0, 1)$ by exact integration (not by using the matrix E) of the piecewise constant rate variable $V\Psi(t)$. If $\bar{v}_1, \bar{v}_2, \ldots, \bar{v}_m$ are the average values of the resultant rate variable V in the respective segments, then

$$x(1) = \frac{1}{m}\sum_{i=1}^{m} \bar{v}_i + x_0 \tag{48}$$

and the carry-over spectrum

$$= Lx_{[0,1)}(D^{m-N})^{\mathrm{T}} \tag{49}$$

We are now in a position to write a recursive formula for continuation from a general unit interval to the next, i.e. $[K-1, K)$ to $[K, K+1)$

$$C^{(K)} = [Ax(K-1) \;\underbrace{0 \;\; 0 \;\; \cdots \;\; 0}_{(m-1)\text{ columns}}]$$

$$+ [Lx(K-1) \;\underbrace{0 \;\; \cdots \;\; 0}_{(m-1)\text{ columns}}]D^N$$

$$+ LX^{(K-1)}(D^{m-N})^{\mathrm{T}} + B_u^{(K)}$$

$$\mathscr{V}^{(K)} = [I - A \otimes E^{\mathrm{T}} - L \otimes (ED^N)^{\mathrm{T}}]^{-1}\mathscr{C}^{(K)} \tag{50}$$

$$x(K) = \frac{1}{m}\sum_{i=1}^{m} \bar{v}_i^{(K)} + x(K-1)$$

($x(K)$ represents discrete time value of $x(t)$) $K = 1, 2, 3, \ldots$.

The superscript (K) means that the quantity belongs to the Kth subinterval. This continuation technique helps in obtaining Walsh-series solutions to any length of time. The following is the limiting case of this approach which happens to be computationally the simplest and most efficient.

6.3 Single-term Walsh-series method

Let us consider a single-segment approximation (neglecting all the Walsh components except $\psi_0(t)$). In this particular case since

$$\tau = N/m$$

$$D^1_{(1)} = 0$$

$$D^0_{(1)} = I_{(1)} = 1$$

$$E_{(1\times 1)} = 0.5$$

and

$$\text{carry-over spectrum} = X^{(K-N)}$$

Eqn. 45 reduces to a single-column form with $\hat{t} = mt$:

$$\dot{x}(\hat{t}) = \frac{A}{m}x(\hat{t}) + \frac{L}{m}x(\hat{t}-\alpha) + Bu(\hat{t})$$

$$x(\hat{t}) = \gamma_x(\hat{t})t \in [-\alpha, 0) \quad (50a)$$

$$V^{(K)} = [I - A/2m]^{-1}\left[\frac{A}{m}x(K-1) + \frac{L}{m}x^{(K-N)} + B_u^{(K)}\right] \quad (50b)$$

Also,

$$X^{(K)} = 0.5V^{(K)} + x(K-1) \quad (50c)$$

and

$$x(K) = V^{(K)} + x(K-1) \quad (50d)$$

This is equivalent to considering m segments over unit interval.

6.4 Single-term Walsh-series approach for systems with multiple delays

Let us consider a system with multiple delays

$$\dot{x} = Ax + \sum_{s=1}^{Q} L_s x(t-\tau_s) + Bu(t-\tau_u) \quad (50e)$$

where

$$\tau_1 < \tau_2 < \tau_3 < \cdots < \tau_s$$

$$x(t) = \Gamma_s(t), \quad -\tau_s \leqslant t < 0$$

$$u(t) = \Gamma_u(t), \quad -\tau_u \leqslant t < 0 \quad (50f)$$

$$x(t_0) = x_0$$

Taking a time scale $\lambda = mt$ in eqn. $50e$

$$\dot{x}(\lambda) = A/mx(\lambda) + \sum_{s=1}^{Q} L_s/mx[\lambda - (N_s + \alpha_s)]$$

$$+ 1/mBu[\lambda - (N_u + \alpha_u)] \quad (50g)$$

where

$$\tau_s = (N_s + \alpha_s)/m$$

and

$$\tau_u = (N_u + \alpha_u)/m$$

N represents the integer part of the delay and α is a fractional part of the delay. Eqn. $50g$ may be solved by single-term Walsh series as

$$V^{(k)} = [I - A/2m]^{-1}\left\{(A/m)x(K-1) + \sum_{s=1}^{Q}(L_s/m) \times [(1-\alpha_s)X^{(K-N_s)} + \alpha_s X^{(K-N_s-1)}] + \frac{1}{m}[(1-\alpha_u)B_u^{(K-N_u)} + \alpha_u B_u^{(K-N_u-1)}]\right\} \quad (50h)$$

$$X^{(K)} = 0.5V^{(K)} + x(K-1)$$
$$x(K) = V^{(K)} + x(K-1) \quad (50i)$$

$$K = 1, 2, 3, \ldots$$

Also,

$$X^{(j)} = 0.5V^{(j)} + x(j-1) \quad (51a)$$

$$x(j) = V^{(j)} + x(j-1) \quad (51b)$$

$$x(0) = \gamma_x(t_0) \qquad j = 1, 2, 3, \ldots \quad (51c)$$

and $B_u^{(j)}$ is the average value of $Bu(t)$ in $[j-1, j]$

Recursive equations of this form constitute the essential computational algorithm, hereinafter termed as the single-term Walsh-series method or the single-step method, for the analysis of delay systems. Owing to its simplicity and minimal computational effort the single-step method will be extensively employed in the following optimal control algorithms. The careful reader will notice the applications of Bellman's principle of optimality in the successive optimisation of solutions over contiguous segments in the multistage optimal decision process.

7 Convergence of the iterative method

In general, consider the state equation with a delay τ in the state with $h(t)$

$$\dot{x}(t) = Ax(t) + Lx(t-\tau) + h(t) \quad (52)$$

$$x(t) = \gamma_x(t) \qquad -\tau \leqslant t \leqslant 0$$

Now express $x(t)$, $\dot{x}(t)$, $x(t-\tau)$ and $h(t)$ in Walsh functions $\Psi(t)$ in the interval $[0, 1)$. Express $\gamma_x(t)$ also in $\hat{\Psi}(t)$, in the interval $t \in [-\tau, 0)$, since $0 \leqslant \tau \leqslant 1$ as

$$x(t) = X\Psi(t)$$

$$\dot{x}(t) = V\Psi(t)$$

$$x(t-\tau) = XD^{\alpha}\Psi(t)$$

$$\gamma_x(t) = X_I\hat{\Psi}(t) \quad \text{where} \quad \hat{\Psi}(t) \text{ is the WF in } [-1, 0)$$

$$h(t) = R_h\Psi(t) \quad \text{and} \quad \alpha = m\tau$$

We may write eqn. 52 in Walsh series as

$$V\Psi(t) = AVE\Psi(t) + LXD^{\alpha}\Psi(t) + LX_I(D_{(m)}^{m-\alpha})^{\mathrm{T}}\Psi(t) + A[x_0 \quad \underbrace{0 \quad \cdots \quad 0}_{m-1 \text{ columns}}]\Psi(t) + R_h\Psi(t) \quad (53)$$

$$X\Psi(t) = VE\Psi(t) + [x_0 \quad \underbrace{0 \quad \cdots \quad 0}_{m-1 \text{ columns}}]\Psi(t) \quad (54)$$

A Walsh spectrum X_N of $x_N(t)$ in $[0, 1)$ is the solution of the following equation:

$$V_N\Psi(t) = AV_N E\Psi(t) + LX_{N-1}D^{\alpha}_{(m)}\Psi(t) + LX_I(D_{(m)}^{m-\alpha})^{\mathrm{T}}\Psi(t) + A[x_0 \quad \underbrace{0 \quad \cdots \quad 0}_{m-1 \text{ columns}}]\Psi(t) + R_{h_{N-1}}\Psi(t)$$

$$X_N\Psi(t) = V_N E\Psi(t) + [x_0 \quad \underbrace{0 \quad \cdots \quad 0}_{m-1 \text{ columns}}]\Psi(t)$$

Gracovetsky and Vidyasagar [7] showed that the sequence of solutions in the above iterative method converges to the true solution of eqn. 52. They use functional-analysis arguments on the spectral radius of the compound operator comprising Volterra-type integral operator, multiplication and delay. Their arguments still apply in the present case with some minor changes. The general operators assume the form of operational matrices here in the method of Walsh functions.

Eqn. 53 may be expressed as

$$\mathcal{V}_N = [I - A \otimes E^{\mathrm{T}}]^{-1}[L \otimes (ED^{\alpha}_{(m)})^{\mathrm{T}}]\mathcal{V}_{N-1} + \kappa \qquad (55)$$

or

$$\mathcal{V}_N = \Xi \mathcal{V}_{N-1} + \kappa$$

where

$$\kappa = \Xi F_d$$

$$\Xi = [I - A \otimes E^{\mathrm{T}}]^{-1}[L \otimes (ED^{\alpha}_{(m)})^{\mathrm{T}}]$$

$$\mathcal{V}_N = [v_0^{\mathrm{T}} \mid v_1^{\mathrm{T}} \mid \cdots \mid v_{m-1}^{\mathrm{T}}]_N^{\mathrm{T}}$$

$$F_d = [f_0^{\mathrm{T}} \mid f_1^{\mathrm{T}} \mid \cdots \mid f_{m-1}^{\mathrm{T}}]^{\mathrm{T}}$$

$$f = \{LX_I(D^{m-\alpha}_{(m)})^{\mathrm{T}} + R_{h_{N-1}} + (Ax_0 \quad \underbrace{0 \quad \cdots \quad 0}_{m-1 \text{ columns}}] + [Lx_0 \quad \underbrace{0 \quad \cdots \quad 0}_{m-1 \text{ columns}}]D^{\alpha}_{(m)}\}$$

$$\triangleq [f_0 \mid f_1 \mid \cdots \mid f_{m-1}]$$

$$X_{N-1} = V_{N-1}E + x_0$$

$$LX_{N-1} = LV_{N-1}E + x_0$$

and

$$LX_{N-1}D^{\alpha}_{(m)} = LV_{N-1}ED^{\alpha}_{(m)} + [Lx_0 \quad \underbrace{0 \quad \cdots \quad 0}_{m-1 \text{ columns}}]D^{\alpha}_{(m)}$$

Convergence of our iteration is ensured if the spectral radius of the transition matrix Ξ is limited to unity. This can be accomplished by working with sufficiently large m, ensuring at the same time the existence of the inverse in Ξ.

Therefore as $N \to \infty$ $\mathcal{V}_N \to \mathcal{V}^*$ the solution of eqn. 55 and X^* from $\mathcal{V}^*$ becomes the solution of eqn. 52.

Since X_N and R_{h_N} correspond to Walsh spectra of $x_N(t)$ and $h_N(t)$ when $X_N \to Y^*$ and $R_{h_N} \to R_h^*$ as $N \to \infty$; $x_N(t) \to x^*(t)$ and $h_N(t) \to h^*(t)$ are the solutions of state and adjoint equations. Therefore by the Riccati transformation

$$p^*(t) = k(t)x^*(t) + h^*(t)$$

$p^*(t)$ is the solution of the co state equation and $x^*(t)$ becomes the solution of the state equation.

We have

$$u^*(t) = -R^{-1}B^{\mathrm{T}}p^*(t)$$

As $p^*(t)$ approaches the solution of the co-state equation $u(t)$ tends to $u^*(t)$, the optimal control. The same holds good for the STWS approach and also with problems of control and state delay.

8 Solution of adjoint and state equations

Let us consider eqns. 7–9. Eqns. 8 and 9 contain terms with delayed and advanced arguments. With the concept of Walsh delay and advance matrix operators we reduce the dynamics of these equations to matrix algebra in the sense of least squares, as discussed in the preceding Sections. In the backward solution of the Riccati equation (eqn. 7*b*), the procedure described in eqns. 10–14 will now take the algebraic form after a change of time scale towards normalisation of the time segment $1/m$ for a positive integer m:

$$\begin{bmatrix} \dot{x} \\ \dot{p} \end{bmatrix} = -\frac{1}{m}\begin{bmatrix} A & -c_5 \\ -Q & -A^T \end{bmatrix}\begin{bmatrix} x \\ p \end{bmatrix} \qquad \begin{matrix} x(0) = I \\ p(0) = 0.0 \end{matrix} \qquad (56)$$

The recursive equations take the form

$$\left.\begin{aligned} V^{(j)}_{xp} &= \left[I + \frac{H_b}{2m}\right]^{-1}\frac{H_b}{m}x_{xp}(j-1) \\ Y^{(j)}_{xp} &= 0.5V^{(j)}_{xp} + x_{xp}(j-1) \\ x_{xp}(j) &= V^{(j)}_{xp} + x_{xp}(j-1) \\ j &= mt_f, \ldots, m(t_0 + \tau_c) \end{aligned}\right\} \qquad (57)$$

where

$$\begin{bmatrix} \dot{x} \\ \dot{p} \end{bmatrix} = V_{xp}\psi_0(\hat{t})$$

$$\begin{bmatrix} x \\ p \end{bmatrix} = Y_{xp}\psi_0(\hat{t})$$

$$x_{xp} = \begin{bmatrix} x \\ p \end{bmatrix} = \lambda(t_f, t)$$

and

$$H_b = \begin{bmatrix} A & -C_5 \\ -Q & -A^{\mathrm{T}} \end{bmatrix}$$

The solution is in the interval $[t_0 + \tau_c, t_f]$ with the condition $k(t_f) = 0$ $t \geq t_f$. Next we consider eqn. 7*a* with the results obtained from eqn. 57 in the interval $[t_0, t_0 + \tau_0]$. $\lambda(t_f, t)$ can be found by taking

$$H_a = \begin{bmatrix} A & -C_1 \\ -Q & -A^{\mathrm{T}} \end{bmatrix} \qquad (58)$$

and $k(t_0 + \tau_c)$ can be obtained from the solution of eqn. 57.

The solution of the Riccati equations (eqns. 7*a* and *b*) can be calculated from $\lambda(t_f, t)$ by using eqn. 13. The solution of the Riccati equation is used in eqns. 8 and 9 to solve them iteratively. At first, the adjoint equation (eqn. 8) is solved by backward integration with $h(t_f) = 0$ for $t \geq t_f$. We express $k(\hat{t})$, $k(\hat{t} \pm \theta)$, $x(\hat{t} \pm \theta)$, $\dot{h}(\hat{t})$, $h(\hat{t})$, $\gamma_x(\hat{t})$, and $h(\hat{t} \pm \theta)$ in single-term Walsh components in the ith interval with θ as any known delay or advance as:

$$\left.\begin{aligned} k(\hat{t}) &= K_i\psi_0(\hat{t}) & k(\hat{t} \pm \theta) &= K_{i\pm\alpha}\psi_0(\hat{t}) \\ x(\hat{t}) &= Y_i\psi_0(\hat{t}) & x(\hat{t} \pm \theta) &= Y_{i\pm\alpha}\psi_0(\hat{t}) \\ \dot{x}(\hat{t}) &= V_i\psi_0(\hat{t}) & h(\hat{t}) &= Yh_i\psi_0(\hat{t}) \\ h(\hat{t} \pm \theta) &= Yh_{i\pm\alpha}\psi_0(\hat{t}) & \dot{h}(\hat{t}) &= Vh_i\psi_0(\hat{t}) \\ \gamma_x(t) &= Y_i\psi_0(t) & i &= -m(t_0 - \tau_s) \cdots mt_0 \end{aligned}\right\} \qquad (59)$$

$$\alpha = m\theta$$

Eqn. 8*c* can be written in each of the $1/m$ long normalised intervals for backward integration in Walsh series (single term approximation) as

$$Vh_{i,N}\psi_0(\hat{t}) = -\frac{1}{m}\{\xi_i[Vh_{i,N}E + h(i+1)] + G_{i,N-1}\}\psi_0(\hat{t}) \qquad (60)$$

where

$$\xi_i = (K_i c_5 - A^{\mathrm{T}})$$

$h(i)$ = discrete time value of $h(t)$

and

$$G_{i,N-1} = \{K_i(LY_{i-\alpha_1} + c_4K_{i-\alpha_2}Y_{i-\alpha_2} + Yh_{i-\alpha_2}) - L^{\mathrm{T}}K_{i+\alpha_1}Y_{i+\alpha_1} - L^{\mathrm{T}}Yh_{i+\alpha_1}\}_{N-1}$$

N being the iteration number $\alpha_1 = m\tau_s$, $\alpha_2 = m\tau_c$, $K_i = 0$ $(i > mt_f)$ and $Yh_i = 0$ $(i > mt_f)$

Solving eqn. 60 we obtain

$$\left.\begin{aligned} Vh_{i,N} &= \left[1 + \frac{1}{2m}\xi_i\right]^{-1} G_i \\ Yh_{i,N} &= \frac{1}{2} Vh_{i,N} + h(i+1) \\ h(i)_N &= Vh_{i,N} + h(i+1) \end{aligned}\right\} \tag{61}$$

where

$$G_i = G_{i,N-1} - \frac{1}{m}\xi_i h(i+1)$$

$$i = mt_f, \ldots, m(t_0 - \tau_c)$$

and

$$h(mt_f + 1) = 0$$

In a similar fashion, eqns. 8b and a are solved in their respective intervals. Initial starting values can be assumed as unity satisfying the given initial conditions.

After obtaining the adjoint vectors, eqn. 9 may be solved for the state. Eqn. 9a can be written in each of the $1/m$ long normalised intervals for forward integration, i.e. $\hat{t} = mt$, as:

$$V_{i,N} = \frac{1}{m}[Z_i\{V_{i,N}E + x(i-1)\} + R_{i,N-1}]$$

where

$$Z_i = (A - c_5 k_i) \qquad i = 1, 2, 3, \ldots, m\tau_s$$

and

$$\begin{aligned} R_{i,N-1} = {} & [LY_{i-\alpha_1,N-1} - c_5 Yh_{i,N} - c_4] \\ & \times \{k_{i-\alpha_2}, Y_{i-\alpha_2,N-1} + Yh_{i-\alpha_2,N}\} \end{aligned} \tag{62}$$

Solving eqn. 62 we obtain

$$\left.\begin{aligned} Y_{i,N} &= 0.5V_{i,N} + x(i-1) \\ x(i) &= V_{i,N} + x(i-1) \end{aligned}\right\} \tag{63}$$

$$i = 1, 2, \ldots, m\tau_s \quad \text{and} \quad x(0) = x_0$$

Similarly, eqns. 8b and c can be solved for other intervals.

Eqns. 8 and 9 may be solved iteratively, and finally the solution $x(t)$ converges to $y^*(t)$ which is employed to obtain the optimal value $u^*(t)$. The optimal control is calculated using eqns. 5 and 6 as follows:

$$\begin{aligned} \bar{U}_i = {} & -R^{-1}[B^T\{(K_i Y_i + Yh_i\} \\ & + S^T\{K_{i+\alpha_2} Y_{i+\alpha_2} + Yh_{i+\alpha_2}\} \\ & \times t \in [t_0, t_f - \tau_c],\ i = mt_0, \ldots, m(t_f - \tau_c) \\ = {} & -R^{-1}B^T[K_i Y_i + Yh_i] \\ & \times t \in [t_f - \tau_c, t_f] \\ & i = m(t_f - \tau_c) \cdots mt_f \\ u^*(t_i) = {} & -R^{-1}[B^T\{k(t_i)x(t_i) + h(t_i)\} \\ & + S^T\{k(t_i + \tau_c)x(t_i + \tau_c) \\ & + h(t_i + \tau_c)\}] \quad t \in [t_0, t_f - \tau_c] \\ & i = mt_0, \ldots, m(t_f - \tau_c) \\ = {} & -R^{-1}B^T[k(t_i)x(t_i) + h(t_i)] \\ & \times t \in [t_f - \tau_c, t_f] \\ & i = m(t_f - \tau_c) \cdots mt_f \end{aligned} \tag{64}$$

$u^*(t_i)$ and $\bar{U}_i$ represent the optimal discrete time value and block pulse value of control function.

9 Calculation of performance index

Differentiating eqn. 3 we obtain

$$\dot{J}(t) = 0.5(x^T Q x + u^T R u) \tag{65}$$

Now, eqn. 61 is represented by single-term Walsh series as

$$(V_j)_i = \frac{1}{2m}[Y_i^T Q Y_i + \bar{U}_i R \bar{U}_i^T] \tag{66}$$

Solving eqn. 62,

$$\left.\begin{aligned} (Y_j)_i &= 0.5(V_j)_i + J(i-1) \\ J(i) &= (V_j)_i + J(i-1) \end{aligned}\right\} \tag{67}$$

$$i = 1, 2, \ldots, mt_f$$

where

$$J(\hat{t}) = Y_{j,i}\psi_0(\hat{t}) \quad \dot{J}(\hat{t}) = V_{j,i}\psi_0(\hat{t})$$

$$x(\hat{t}) = Y_i\psi_0(\hat{t})$$

$$u(\hat{t}) = \bar{U}_i\psi_0(\hat{t}) \quad J(\hat{t}) = J(i)$$

and

$$J(0) = 0$$

10 Numerical examples

Example 1

$$\left.\begin{aligned} &\text{For a system described by References 11, 15} \\ &\dot{x}(t) = -x(t) + u(t) - Su(t - \tau_c) \\ &x(0) = 1.0 \\ &u(t) = 0 \quad t \in [-\tau_c, 0] \end{aligned}\right\} \tag{68}$$

minimise

$$J = \int_0^1 (x^2(t) + u^2(t))\, dt$$

Eqn. 68 is considered with following values of S and τ_c:

$$S = 0$$

$$S = \sqrt{0.5},\ \tau_c = 2/3$$

$$S = 0.5,\ \tau_c = 5/12$$

$$S = 0.5,\ \tau_c = 2/3$$

Comparing eqns. 68 with eqns. 1 and 3 we obtain $A = -1.0$, $L = 0.0$, $B = 1.0$, $R = 0.5$, $Q = 1.0$, $t_0 = 0$, $G = 0$ and $t_f = 1$.

With $S = 0.5$ the values of c_1, c_2, c_3, c_4, and c_5 become 2.0, 0.5, -1.0, -1.0, and 2.5, respectively.

We get the following Riccatti adjoint and state equations:

$$\dot{k}(t) = 2.5k^2(t) + 2k(t) - 1$$

$$k(t) = 0,\ t \geqslant t_f \qquad t \in [t_0 + \tau_c, t_f) \tag{69a}$$

$$\dot{k}(t) = 2k^2(t) + 2k(t) - 1 \qquad t \in [t_0, t_0 + \tau_c] \tag{69b}$$

$$\begin{aligned} \dot{h}_N(t) = {} & (2k(t) + 1)h_N(t) - k(t)\{-k(t + \tau_c)x_{N-1}(t + \tau_c) \\ & - h_{N-1}(t + \tau_c) - k(t - \tau_c)x_{N-1}(t - \tau_c) \\ & - h_{N-1}(t - \tau_c)\} \qquad t \in [t_0, t_0 + \tau_c] \end{aligned} \tag{70a}$$

$$= (2.5k(t) + 1)h_N(t) - k(t)\{-[k(t) + \tau_c)x_{N-1}(t + \tau_c) + h_{N-1}(t + \tau_c)] - [k(t - \tau_c)x_{N-1}(t - \tau_c) + h_{N-1}(t - \tau_c)]\} \quad t \in [t_0 + \tau_c, t_f - \tau_c) \quad (70b)$$

$$= (2.5k(t) + 1)h_{N-1}(t) - k(t)\{-k(t - \tau_c)x_{N-1}(t - \tau_c) + h_{N-1}(t - \tau_c)]\} \quad t \in [t_f - \tau_c, t_f] \quad (70c)$$

$$h(t) = 0, \quad \text{for } t \geqslant t_f.$$

$$\dot{x}_N(t) = (-1 - 2k(t))x_N(t) - 2h_N(t) + [k(t + \tau_c)x_{N-1}(t + \tau_c) + h_N(t + \tau_c)] \quad t \in [t_0, t_0 + \tau_c] \quad (71a)$$

$$= (-1 - 2.5k(t))x_N(t) - 2.5h_N(t) + [k(t + \tau_c)x_{N-1}(t + \tau_c) + h_N(t + \tau_c) + k(t - \tau_c)x_{N-1}(t - \tau_c) + h_N(t - \tau_c)] \quad t \in [t_0 + \tau_c, t_f - \tau_c) \quad (71b)$$

$$= (-1 - 2.5k(t))x_N(t) - 2.5h_N(t) + k(t - \tau_c)x(t - \tau_c) + h(t - \tau_c) \quad t \in [t_f - \tau_c, t_f] \quad (71c)$$

Eqns. 69 may be solved directly by single-term Walsh-series approach (STWS). Normalising eqn. 69 with $\hat{t} = mt$, and changing the sign of left-hand side for backward integration,

$$\left.\begin{aligned} \dot{k}(\hat{t}) &= \frac{-2.5}{m} k^2(\hat{t}) - \frac{2}{m} k(\hat{t}) + \frac{1}{m} \\ \dot{k}(\hat{t}) &+ \frac{-2}{m} k^2(\hat{t}) - \frac{2}{m} k(\hat{t}) + \frac{1}{m} \end{aligned}\right\} \quad (72)$$

with the expansions in STWS, namely

$$\dot{k}(\hat{t}) = V_k \psi_0(\hat{t}) \quad \text{and} \quad k(\hat{t}) = Y_k \psi_0(\hat{t})$$

we get the following quadratic equations for eqn. 72:

$$\frac{2.5}{4}[(V_k)_i]^2 + (V_k)_i[m + 2.5k(i) + 1] + [2.5k^2(i) + 2k(i) - 1] = 0 \quad (73a)$$

$$i = mt_f, \ldots, m(t_0 + \tau_c)$$

and

$$\frac{2}{4}[(V_k)_i]^2 + (V_k)_i[m + 2k(i) + 1] + [2.5k^2(i) + 2k(i) - 1] = 0 \quad (73b)$$

$$i = m(t_0 + \tau_c) \cdots m(t_0 + \tau_c)$$

$k(i)$ represents discrete time values $k(t)$ and $k(mt_f) = 0$ since $G = 0$.

At each step,

$$(Y_k)_i = 0.5(Y_k)_i + k(i)$$

$$k(i) = (V_k)_i + k(i)$$

Then eqns. 70 and 71 may be solved iteratively by STWS approach. Eqns. 64 and 67 may be used to get the control and the performance index. Figs. 2 and 3 show the control, state and performance index calculated with $m = 10$ and $m = 100$. They compare well with the results of Budelis *et al.* [15].

Example 2

Consider the example of Soliman and Ray [11]

$$\dot{x}(t) = -x(t) + x(t - 1/3) + u(t) - 0.5u(t - 2/3) \quad (74)$$

$$x(t) = 1.0 \qquad t \in [-1/3, 0]$$

$$u(t) = 0 \qquad t \in [-2/3, 0]$$

with the performance index

$$J = 0.5 \int_0^1 (x^2(t) + u^2(t))\, dt$$

The results by the present method with $m = 100$ are shown

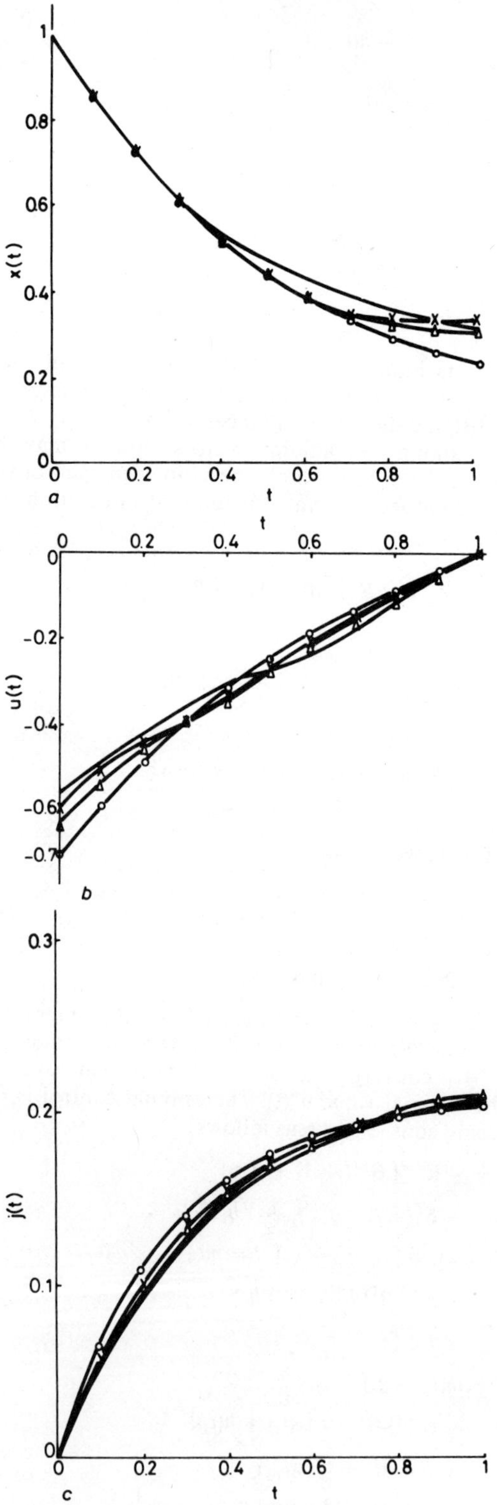

Fig. 2 *State trajectory and control in example 1*
$m = 10$, $\alpha_2 \approx 0, 6, 4, 6$
○—○—○ $S = 0$
×—×—× $S = \sqrt{0.5}$, $\tau_c = 2/3$
———— $S = 0.5$, $\tau_c = 5/12$
△—△—△ $S = 0.5$, $\tau_c = 2/3$
a $x(t)$ as a function of time
b $u(t)$ as a function of time
c Performance index in example 1

in Fig. 4 along with those by the ODE approximation method [11] ($m = 6$). The large step size in the latter case explains the disparity between the results. To compare the results at the same level of segmentation ($m = 100$) one encounters a set of 5050 Riccati equations in the ODE approximation method. In the Walsh-functions approach, however, there is only a marginal increase in the required computational effort with increased m.

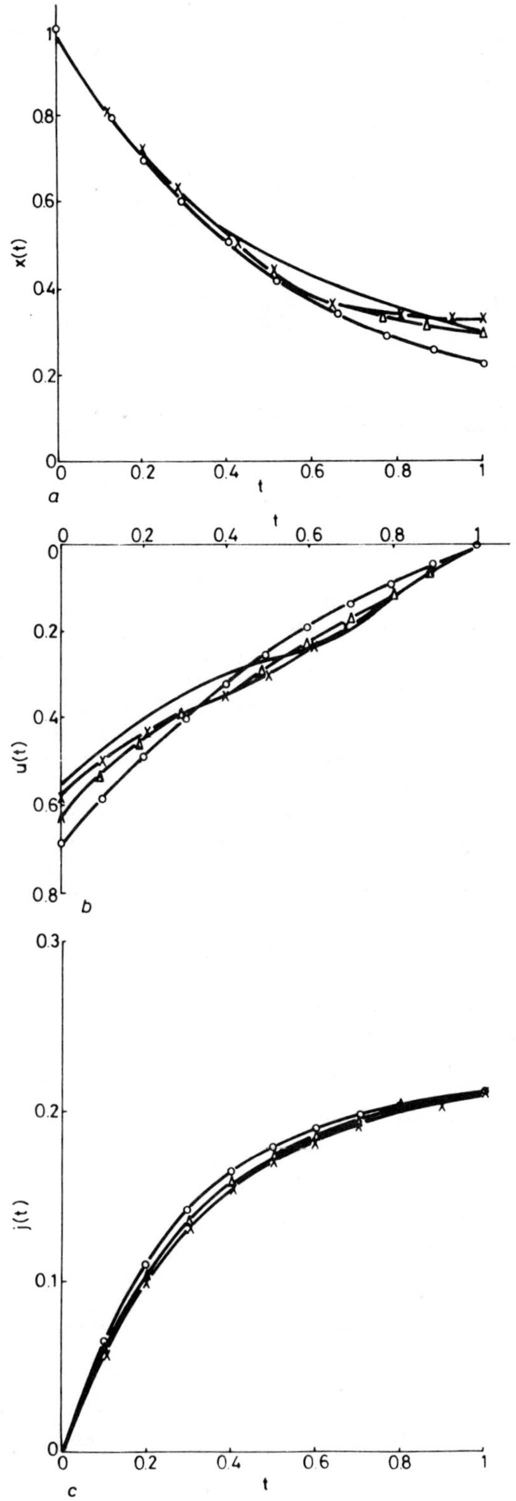

Fig. 3 *State trajectory and control in example 1*

$m = 100$, $\alpha_2 \approx 0, 66, 42, 66$

○ ○ ○ $S = 0$

× × × $S = 0.5$, $\tau_c = 2{:}3$

.... $S = 0.5$, $\tau_c = 5{:}12$

△ △ △ $S = 0.5$, $\tau_c = 2{:}3$

a $x(t)$ as a function of time

b $u(t)$ as a function of time

c Performance index in example 1

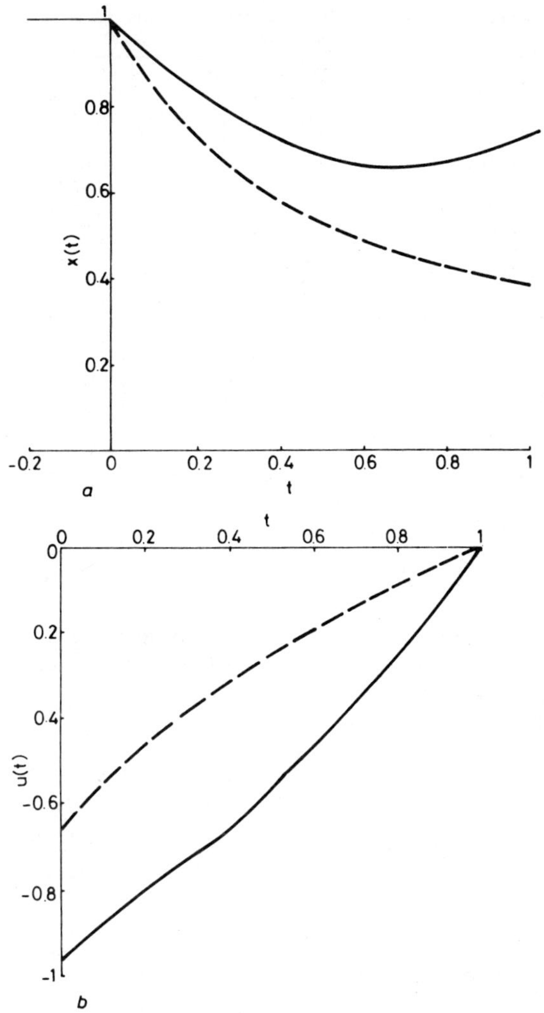

Fig. 4 *State trajectory and control in example 2*

$m = 100$, $\alpha_1 \approx 33$, $\alpha_2 \approx 66$

STWS

WH ray

a $x(t)$ as a function of time

b $u(t)$ as a function of time

Example 3

Consider the example of Banks and Burns [40] and Delfour [49]

$$\dot{x}(t) = x(t-1) + u(t) \tag{75}$$

$$x(t) = 1 \qquad -1 \leqslant t \leqslant 0$$

with the performance index

$$J = 0.5 \int_0^2 (x^2 + u^2)\, dt$$

The results obtained are given in Table 1. The results compare well with the solutions obtained by Banks *et al.* [40].

The performance index

$$J = 1.6497 \quad \text{with} \quad m = 100$$

$$= 1.6504 \quad \text{with} \quad m = 10$$

Banks and Burns [40] obtain J as 1.6419 with index $N_B = 20$. The solution by WF method converged within four iterations.

Table 1: Results of example 3 with $m = 100$

Time	Control $u(t)$ $N_B = 20$ Banks and Burns [40]	Control $u(t)$ STWS	State $x(t)$ Banks and Burns [40]	State $x(t)$ STWS
0.00	−1.9908	−1.9870	1.0000	1.0000
0.20	−1.6559	−1.6566	0.8339	0.8364
0.41	−1.3798	−1.3691	0.7296	0.7299
0.61	−1.1511	−1.1143	0.6763	0.6794
0.81	−0.9604	−0.9547	0.6641	0.6703
1.02	−0.7974	−0.7947	0.6804	0.6971
1.22	−0.6481	−0.6525	0.7124	0.7321
1.43	−0.4981	−0.5031	0.7562	0.7716
1.63	−0.3386	−0.3362	0.8174	0.8310
1.83	−0.1641	−0.1631	0.9057	0.9163
2.00	0.0000	0.0000	1.0091	1.0189

Example 4

Consider the example of Banks and Burns [40]

$$\dot{x}(t) = x(t) + x(t-1) + u(t) \tag{76}$$

$$x(t) = 1 \qquad -1 \leqslant t \leqslant 0$$

with performance index

$$J = 1.5[x(2)]^2 + 0.5\int_0^2 u(t)^2\,dt$$

The results obtained for the above problem are given in Table 2. The results compare well with the exact solution [40] and solution by Banks *et al.* [40].

The performance index

$$J = 3.0876 \quad \text{with} \quad m = 100$$

$$= 3.0889 \quad \text{with} \quad m = 10$$

Banks and Burns [40] obtain an average J as 3.0833, with the averaging approximation technique, with the index $N_B = 20$. The solution via WF converged within six iterations.

Table 2: Results of example 4 with $m = 100$

Time	Control $u(t)$ Exact [40]	Control $u(t)$ $N_B = 20$ Banks *et al* [40]	Control $u(t)$ STWS	State $x(t)$ STWS
0.00	−3.9787	−3.9734	−3.9730	1.0000
0.19	−3.1182	−3.1201	−3.1178	0.6772
0.39	−2.4083	−2.3973	−2.4079	0.4374
0.59	−1.8533	−1.8376	−1.8528	0.2845
0.80	−1.4014	−1.4082	−1.4008	0.2050
1.00	−1.0687	−1.0884	−1.0680	0.1988
1.20	−0.8750	−0.8604	−0.8747	0.2110
1.41	−0.7093	−0.6961	−0.7093	0.2004
1.61	−0.5807	−0.5678	−0.5810	0.1773
1.81	−0.4755	−0.4635	−0.4759	0.1506
2.00	−0.3932	−0.3842	−0.3937	0.1312

Example 5

Consider the example of Banks and Burns [40]

$$\dot{x}(t) = x(t-1) + u(t) \tag{77}$$

$$x(t) = 1 \qquad -1 \leqslant t \leqslant 0$$

With the performance index

$$J = 1.5[x(3)]^2 + 0.5\int_0^5 [u(t)]^2\,dt$$

The results obtained are given in Table 3 with the results obtained by Banks and Burns [40].

The value of performance index

$$J = 1.7088 \quad \text{with} \quad m = 100$$

$$= 1.7098 \quad \text{with} \quad m = 10$$

Banks and Burns [40] obtain J as 1.7338 (average value).

Table 3: Results of example 5 with $m = 100$

Time	Control $u(t)$ Banks and Burns [40] $N_B = 20$	Control $u(t)$ Exact [40]	Control $u(t)$ STWS	State $x(t)$ STWS
0.00	−1.9663	−1.9681	−1.9765	1.0000
0.30	−1.6518	−1.6656	−1.6630	0.7553
0.61	−1.3898	−1.3867	−1.3916	0.5933
0.91	−1.1731	−1.7775	−1.1799	0.5088
1.20	−1.0016	−1.0122	−1.0120	0.4633
1.50	−0.8309	−0.8435	−0.8403	0.4043
1.80	−0.6813	−0.6748	−0.6684	0.3530
2.10	−0.5761	−0.5623	−0.5538	0.3269
2.41	−0.5534	−0.5623	−0.5544	0.2950
2.72	−0.5529	−0.5623	−0.5550	0.2440
3.00	−0.5528	−0.5623	−0.5554	0.1851

11 Conclusion

An algorithm using Walsh functions in an embedding approach to optimal control of system with state and control delays is presented. The computations in the optimal-control algorithm are through optimal approximations in the sense of least squares. The resulting optimal control becomes naturally piecewise constant (owing to application of Walsh functions) and is convenient in practical implementation. The solutions at the transition points are also available in addition to the respective Walsh spectra. The figures in all the illustrative examples show only the solution as obtained at the transition points shown as continuous curves for the purpose of comparison with those obtained by others. The available piecewise constant solutions therefore are not presented here. The computational algorithms are simple and present little convergence problems.

12 Acknowledgment

Prof. G.P. Rao thanks the Alexander von Humboldt Foundation for supporting his visit to the Ruhr University Bochum.

13 References

1 KHARATISHVILI, G.L.: 'The maximum principle in the theory of optimal processes with time lags', *Dokl. Akad. Nauk SSSR*, 1961, **136,** (1)

2 KRASOVSKII, N.N.: 'Analytic construction of an optimal regulator in a system with time lags', *Prickl Mat. & Mekh. (USSR)*, 1962, **26,** (1)

3 KRASOVSKII, N.N.: 'Optimal processes in systems with time lags'. Proc. 2nd IFAC Conf., Switzerland, 1963

4 ELLER, D.H., AGGARWAL, J.K., and BANKS, H.T.: 'Optimal control of time delay systems', *IEEE Trans.*, 1969, **AC-14,** pp. 678–687

5 AGGARWAL, J.K.: 'Computation of optimal control for time delay systems', *ibid.*, 1970, **AC-15,** pp. 683–685

6 CHAN, H.C., and PERKINS, W.R.: 'Optimization of time delay systems using parameter imbedding', *Automatica*, 1973, **9,** pp. 257–261

7 GRACOVETSKY, S.A., and VIDYASAGAR, M.: 'A simple iterative method for sub-optimal control of linear time delay systems with quadratic cost', *Int. J. Control*, 1972, **16,** pp. 997–1002

8 JAMSHIDI, M., and MALEK-ZAVAREI, M.: 'Suboptimal design of linear control systems with time delay', *Proc. IEE*, 1972, **119,** (12), pp. 1743–1746

9 SANNUTI, P.: 'Near-optimal design of time delay systems by singular perturbation method'. Proc. Joint Automatic Computer Conference, Atlanta, USA, 1970, pp. 489–496

10 SANNUTI, P., and KOKOTOVIC, P.V.: 'Near optimum design of linear systems by a singular perturbation method', *IEEE Trans.*, 1969, **AC-14,** pp. 15–22

11 SOLIMAN, M.A., and RAY, W.H.: 'Optimal control for linear quadratic systems having time delays', *Int. J. Control*, 1972, **15,** pp. 609–627

12 SOLIMAN, M.A., and RAY, W.H.: 'Optimal control of multivariable systems with pure time delays', *Automatica*, 1976, **17,** pp. 681–689

13 RAY, W.H., and SOLIMAN, M.A.: 'The optimal control processes containing pure time delays—I. Necessary conditions for an optimum', *Chem. Eng. Sci.*, 1970, **25,** pp. 1911–1925

14 RAY, W.H.: 'The optimal control processes modelled by transfer functions containing pure time delays', *ibid.*, 1969, **24,** pp. 209–216

15 BUDELIS, J.J., and BRYSON, A.E.: 'Some optimal control results for differential difference systems', *IEEE Trans.*, 1970, **AC-15,** pp. 237–241

16 KOIVO, H.N., and LEE, E.B.: 'Controller synthesis for linear systems with retarded state and control variables and quadratic cost', *Automatica*, 1972, **18,** pp. 203–208

17 ROSS, D.W.: 'Controller design for time lag systems via a quadratic criterion', *IEEE Trans.*, 1971, **AC-16,** pp. 664–672

18 ALEKAL, Y., BRUNOVSKY, P., CHUNG, D.H., and LEE, E.B.: 'The quadratic problem for system with time delays', *ibid.*, 1971, **AC-16,** pp. 673–687

19 AGGARWAL, J.K.: 'Feedback control of linear systems with distributed delay', *Automatica*, 1973, **9,** pp. 367–379

20 INOUE, K., AKASHI, H., OGINO, K., and SAWARAGI, Y.: 'Sensitivity approaches to optimization of linear systems with time delay', *ibid.*, 1971, **17,** pp. 671–679

21 CHEN, C.F., and HAIAO, C.H.: 'Design of piecewise constant gains for optimal control via Walsh functions', *IEEE Trans.*, 1975, **AC-20,** pp. 596–603

22 CHEN, C.F., and HSIAO, C.H.: 'Walsh series analysis in optimal control', *Int. J. Control*, 1975, **21,** pp. 881–897

23 SANNUTI, P.: 'Analysis and synthesis of dynamic systems via block-pulse functions', *Proc. IEE*, 1972, **124,** (6), pp. 569–571

24 RAO, G.P., and SRINIVASAN, T.: 'Analysis and synthesis of dynamic systems containing time delays via block-pulse functions', *ibid.*, *IEE*, 1978, **125,** (9), pp. 1064–1068

25 RAO, G.P., and SIVAKUMAR, L.: 'System identification via Walsh functions', *ibid.*, 1975, **122,** pp. 1160–1161

26 PALANISAMY, K.R., and RAO, G.P.: 'Minimum energy control of time-delay systems via Walsh functions', *Optimal control appl. & methods*, 1983, **4,** pp. 213–226

27 RAO, G.P., and PALANISAMY, K.R.: 'A new operational matrix for delay via Walsh functions and some aspects of its algebra and applications'. Proc. National Systems Conference, 1978, pp. 1, 60–61

28 RAO, G.P., and PALANISAMY, K.R.: 'Optimal control of time delay systems via Walsh functions'. Accepted for presentation at 9th IFIP Conf. on optimization, Warsaw, Poland, 4–8 Sept. 1979

29 RAO, G.P., PALANISAMY, K.R., and SRINIVASAN, T.: 'Extension of computation beyond the limit of initial normal interval in Walsh series analysis of dynamical systems', *IEEE Trans.*, 1980, **AC-25,** pp. 317–319

30 SHIEH, L.S., YEUNG, C.K., and McINNIS, B.C.: 'Solution of state space equations via block pulse functions', *Int. J. Control*, 1978, **28,** pp. 383–392

31 POPOV, V.M.: 'Pointwise degeneracy of linear time invariant delay differential equations', *J. Differ. Equations*, 1972, **11,** pp. 541–561

32 OLBROT, A.W.: 'On degeneracy and related problems for linear constant time-lag systems', *Ric. Autom.*, 1972, **3,** pp. 203–220

33 OLDBROT, A.W.: 'Algebraic criteria of controllability to zero function for linear constant time lag systems', *Control & Cybern.*, 1973, **2,** pp. 59–77

34 OLBROT, A.W.: 'Observability tests for constant time lag systems', *ibid.*, **4,** pp. 71–84

35 TSOI, A.C.: 'Recent advances in the algebraic system theory of delay differential equations', *in* GREGIEN, M. (Ed.): 'Recent theoretical advances in control theory'. (Academic Press, 1978) pp. 68–127

36 TSOI, A.C.: 'An explicit solution to a class of delay-differential equations', *Int. J. Control*, 1975, **21,** pp. 39–48

37 TSOI, A.C.: 'An explicit solution to a class of delay-differential equations', *ibid.*, 1975, **22,** pp. 787–799

38 TSOI, A.C.: 'An explicit solution to a class of functional differential equations', *ibid.*, 1975, **22,** pp. 869–875

39 CHOUDHURY, A.K.: 'A contribution to the controllability of time lag systems', *ibid.*, 1973, **17,** pp. 365–373

40 BANKS, H.T., and BURNS, J.A.: 'Hereditary control problems: Numerical methods based on averaging approximations', *SIAM J. Control & Optimiz.*, 1978, **16,** pp. 169–208

41 BANKS, H.T., and BURNS, J.A.: 'An abstract frame work for approximate solutions to optimal control problems governed by hereditary systems'. *in* ANTOSIEWICZ, H.A. (Ed.): Proc. of International Conference on differential equations. (Academic Press, New York, 1975) pp. 10–25

42 BANKS, H.T., BURNS, J.A., CLITT, E.M., and THRIFT, P.R.: 'Numerical solutions of hereditary control problems via approximation technique'. Brown. University, LCDS Technical Report 75–6, 1975, RI, USA

43 BANKS, H.T., and KENT, G.A.: 'Control of functional differential equations of retarded and neutral type to target sets in function space', *SIAM J. Control & Optimiz.*, 1972, **10,** pp. 567–593

44 BANKS, H.T., and MANITIUS, A.: 'Application of abstract variational theory to hereditary systems—A survey', *IEEE Trans.* 1974, **AC-19,** pp. 524–533

45 BANKS, H.T., and MANITIUS, A.: 'Projection series for retarded functional differential equations with applications to optimal control problems', *J. Differ. Equations*, 1975, **18,** pp. 296–332

46 BANKS, H.T.: 'Approximation of non-linear functional differential equation control systems', *J. Optimiz. Theory & Appl.*, 1979, **29,** pp. 383–408

47 MANITIUS, A.: 'Optimal control of time lag systems with quadratic performance indexes'. Proc. Fourth IFAC Congress, Warsaw, 1969

48 MANITIUS, A.: 'Sufficient conditions for function space controllability and feedback stabilizability of linear retarded systems'. Proc. 1976, IEEE Conf. on Decision and Control

49 DELFOUR, M.C.: 'The linear quadratic optimal control problem for hereditary differential systems: Theory and numerical solution', *Appl. Math. & Optimiz.*, 1977, **3,** pp. 101–162

50 DELFOUR, M.C.: 'Linear hereditary systems and their control'. *in* KIRBY, B.J. (Ed.): 'Optimal control and its applications'. (Springer-Verlag, New York, 1974) Part 2, pp. 92–154

51 DELFOUR, M.C.: 'Numerical solution of the optimal control problem for linear hereditary differential systems with a linear quadratic cost function and approximation of Riccati differential equation', CRM. Report 408, Universite de Montreal, 1974

52 DELFOUR, M.C.: 'Numerical solution of the operational Riccati differential equation in the optimal control theory of linear hereditary differential systems with a linear-quadratic cost function'. Proc. 1974 IEEE Conference on Decision and Control, Arizona, USA, pp. 784–790

53 DELFOUR, M.C., and MITTER, S.K.: 'Hereditary differential systems with constant delays. I. General case', *J. Differ. Equations*, 1972, **12,** pp. 213–235

54 DELFOUR, M.C., and MITTER, S.K.: 'Hereditary differential equations with constant delays. II. A class of affine systems and the adjoint problem', *ibid.*, 1975, **18,** pp. 18–28

55 BURKHARDT, H.: 'Ein Beitrag zur Lösung optimaler Steuerungs und Regelungsprobleme mit Hilfe der Walsh-Transformation'. Dr.-Ing. Thesis, University Karlsruhe, June 1974

56 RAO, G.P.: 'Piecewise constant orthogonal functions and their application to systems and control'. *in* BALAKRISHNAN, A.V., and THOMA, M. (Ed.): LNCIS series, Vol. 55 (Springer-Verlag, 1983)

Part VI

BLOCK-PULSE FUNCTIONS

Editor's Comments on Papers 23 Through 27

23 SANNUTI
Analysis and Synthesis of Dynamic Systems Via Block-Pulse Functions

24 PRASADA RAO and SRINIVASAN
Analysis and Synthesis of Dynamic Systems Containing Time Delays Via Block-Pulse Functions

25 SHIH, HWANG, and CHIA
Parameter Estimation of Delay Systems Via Block Pulse Functions

26 KWONG and CHEN
Linear Feedback System Identification Via Block-Pulse Functions

27 RAO and RAO
Optimal Feedback Control Via Block-Pulse Functions

Block pulse functions constitute another complete set of orthogonal basis functions that have a one-to-one relationship with Walsh functions, and naturally find applications in analysis, synthesis, and control of systems. The type of approximation is the same as with Walsh functions, the only difference being in the simplicity of computations.

The analysis and synthesis of linear and nonlinear dynamic systems in state space via block pulse functions is made in Paper 23 by Sannuti. It is shown that the final results coincide with those obtained through Walsh functions, but here we have a simpler integration operational matrix. The connection between block pulse and trapezoidal approximation is also revealed. Alternative results concerning the solution of state space models using block pulse functions are provided in a paper by Shieh, Yeung, and McInnis (1978). They develop an algorithm that involves the inversion of only one matrix having the dimensionality of the system matrix.

Prasada Rao and Srinivasan study systems involving time delays in Paper 24 by using Sannuti's results together with an appropriate "delay matrix" Δ specially constructed in the paper. The effect of this matrix is shown by the equation $\phi(t - 1/m) = \Delta\phi(t)$ where $\phi(t)$ is the vector of block pulse functions, m is the number of terms in the expansion, and $\tau = 1/m$ is the basic time delay. An error analysis is also included, which can be used as a basis for selecting the number of segments m to obtain specified *r.m.s.* error bounds. Systems containing a piece-wise constant delay are considered in the work of

Chen and Jeng (1981), where a delay operational matrix $S_{(m)}$ is introduced to deal with the time delay, which is not a multiple of $1/m$. The solution is then found by using this delay operational matrix, the properties of block pulse functions, and some integral operations.

In Paper 25 the identification of delay systems is considered and solved through block pulse functions. The key tool is the delay operational matrix, called shift identity matrix, by which delayed block pulse functions can be obtained. Using the block-pulse functions integration matrix together with the shift identity matrix, the state-space delay model at hand is converted into an algebraic model that is then treated, as usual, by the least squares method.

Paper 26 deals with the problem of identifying the impulse response $g(t)$ of an unknown plant involved in a feedback control system with known feedback path (impulse responce $h(t)$) on the basis of an input-output record. The solution has an algorithmic form and is derived by first expressing the convolution and deconvolution operations in terms of the block pulse functions.

Paper 27 by V.P. Rao and K.R. Rao treats the optimal feedback control problem using block-pulse function expansions. The results are derived for a simple third-order model of a power system consisting of a generator connected to an infinite bus through a transmission line, but the methodology can be generally applied to any linear or linearized system. It is useful to relate the results of this paper to the results of Chen and Hsiao (1975) and Gopalsami and Deekshatulu (1976) (see Part V). Some further results on the utilization of block pulse functions may be found in the works of Chen (1981), Wang and Shih (1982), Tzafestas (1984), Hsu and Cheng (1981), and Prasada Rao and Srinivasan (1980).

REFERENCES

Chen, W. L., 1981, Block Pulse Series Analysis of Scaled Systems, *Int. J. Systems Sci.* **12:**885–891.

Chen, C. F., and C. H. Hsiao, 1975, Design of Piecewise Constant Gains for Optimal Control Via Walsh Functions, *IEEE Trans. Autom. Control* **AC-20:**596–602.

Chen, W. L., and B. S. Jeng, 1981, Analysis of Piecewise Constant Delay Systems Via Block Pulse Functions, *Int. J. Systems Sci.* **12:**625–633.

Gopalsami, N., and B. L. Deekshatulu, 1976, Comments on "Design of Piecewise Constant Gains for Optimal Control Via Walsh Functions," *IEEE Trans. Autom. Control* **AC-21:**634–636.

Hsu, N. S., and B. Cheng, 1981, Analysis and Optimal Control of Time-Varying Linear Systems via Block-Pulse Functions, *Int. J. Control* **33:**1107–1122.

Prasada Rao, G., and T. Srinivasan, 1980, Multidimensional Block-Pulse Functions and their Use in the Study of Distributed Parameter Systems, *Int. J. Syst. Sci.* **11:**689–708.

Shieh, L. S., C. K. Yeung, and B. C. McInnis, 1978, Solution of State-Space Equations Via Block-Pulse Functions, *Int. J. Control* **28:**383–392.

Tzafestas, S. G., C. N. Papastergiou, and J. N. Anoussis, 1984, Dynamic Reactivity Computation in Nuclear Reactors Using Block-Pulse Function Expansion, *Int. J. Modelling Simulation* **4:**73–76.

Wang, C. H., and Y. P. Shih, 1982, Explicit Solutions of Integral Equations via Block Pulse Functions, *Int. J. Syst. Sci.* **13:**773–782.

23

Reprinted from *IEE Proc.* **124:**569–571 (1977)

ANALYSIS AND SYNTHESIS OF DYNAMIC SYSTEMS VIA BLOCK-PULSE FUNCTIONS

P. Sannuti

Indexing terms: Differential equations, Function approximation, Integration, Numerical methods

Abstract

The paper presents a method of numerically integrating a system of differential equations based on an idea of orthogonal approximation of functions. Here, block-pulse functions are chosen as the orthogonal set. The method gives piecewise constant solutions with minimal mean-square error and is computationally similar to the familiar trapezoidal rule of integration. Design of piecewise constant controls or feedback gains for dynamic systems can be simplified following this approach.

1 Introduction

Approximating a function as a linear combination of a set of orthogonal basis functions is a standard tool in a numerical analysis. Recently, Corrington[1] proposed a method of solving nonlinear differential and integral equations using a set of Walsh functions as the basis. His method is aimed at obtaining piecewise constant (approximate) solutions of dynamic equations, and requires previously prepared tables of coefficients for integrating Walsh functions. To alleviate the need for such tables, Chen and Hsiao[2-4] introduced an operational matrix to perform integration of Walsh functions. This operational matrix approach has been applied to various problems such as time-domain analysis[2] and synthesis[3] of linear systems, piecewise constant-feedback-gain determination for optimal control of linear systems[4] and for inverting irrational Laplace transforms.[5]

This paper simplifies, enormously, the method of Chen and Hsiao by using block-pulse functions instead of Walsh functions as the basis. Since there is a one-to-one relationship between Walsh and block-pulse functions, the final results and type of approximations are identical to those in References 2–4; i.e. the difference between the methods is not in the final results, but rather in their computation. However, there is a strong connection between the present method and the well known trapezoidal method of integrating a system of differential equations. Thus, the method can be extended easily to the analysis and synthesis of time-varying and nonlinear systems, whereas Chen and Hsiao's approach is limited to linear time-invariant systems.

2 Block-pulse functions and an operational matrix of integration

To incorporate a piecewise constant constraint in constructing an approximate solution of a system of differential equations, we utilise here a set of block-pulse functions defined on a unit interval (0, 1] as follows:

For each integer i, $1 \leqslant i \leqslant m$, the function $\phi_i(t)$ is given by

$$\phi_i(t) = \begin{cases} 1 & \text{for } \dfrac{i-1}{m} < t \leqslant \dfrac{i}{m} \\ 0 & \text{otherwise} \end{cases} \tag{1}$$

This set of functions can be concisely described by an m-vector $\phi(t)$ with $\phi_i(t)$ as its ith component. It is well known that a function which is integrable in (0, 1) can be approximated as

$$f(t) \simeq \sum_{i=1}^{m} a_i \phi_i(t) \tag{2}$$

where the coefficients a_i are determined such that the following integral square error is minimised:

$$\epsilon = \int_0^1 (f(t) - \sum_{i=1}^{m} a_i\phi_i(t))^2 dt$$

In fact, a_i is given by

$$a_i = m\int_{\frac{i-1}{m}}^{\frac{i}{m}} f(t)dt = \text{average value of } f(t) \text{ over the interval } \frac{i-1}{m} < t \leqslant \frac{i}{m} \tag{3}$$

Note that the approximation in (eqn. 2) gives the best m-segment piecewise constant approximation of $f(t)$ and is unique whatever may be the set of basis functions. In particular, use of Walsh functions for the basis will give the same approximation.

Now we would like to expand the intregal, $\int_0^t \phi(t)\, dt$, into the set of basic functions $\phi(t)$. The coefficients of expansion can be calculated using eqn. 3. Arranging the coefficients in a matrix form, we have

$$\int_0^t \phi(t)dt \simeq P\,\phi(t) \tag{4}$$

where P is an $m \times m$ matrix given by

$$P = \frac{1}{m}\begin{bmatrix} \frac{1}{2} & 1 & 1 & 1 & . & . & . & 1 \\ 0 & \frac{1}{2} & 1 & 1 & . & . & . & 1 \\ 0 & 0 & \frac{1}{2} & 1 & . & . & . & 1 \\ . & . & . & . & & & & . \\ . & . & . & . & & & & . \\ . & . & . & . & & & & . \\ 0 & 0 & 0 & 0 & . & . & . & \frac{1}{2} \end{bmatrix} \tag{5}$$

The significance of the matrix P is that the integration of $\phi(t)$ can be approximately (in the least-square error sense) achieved by premultiplying $\phi(t)$ by the constant matrix P. For this reason, P is called the operational matrix for block-pulse functions.

3 Solution of dynamic equations

In this Section, we utilise the notions introduced in the previous Section to construct a piecewise constant solution to a system of differential equations. To fix our ideas, at first we consider a time-invariant linear system and then discuss a straightforward extension of the method for nonlinear and timevarying systems. Let

$$\dot{X} = AX + Bu, X(0) = X_0 \tag{6}$$

where X is an n-component state vector, u is an r-component input vector, the dot denotes d/dt, and A and B are matrices of appropriate dimension. For solving this problem by the block-pulse function

Paper 7878 C, first received 23rd November 1976 and in revised form 1st March 1977

Prof. Sannuti is with the Department of Electrical Engineering, Rutgers University, PO Box 909, University Heights Campus, Piscata Way, New Jersey 08854, USA

approach, we expand the rate variable $\dot{X}$, the state variable X and the overall control vector Bu into block-pulse functions

$$\dot{X}(t) = \sum_{i=1}^{m} C_i\phi_i(t) = C\phi(t) \qquad X(t) = \sum_{i=1}^{m} D_i\phi_i(t) = D\phi(t)$$
$$Bu(t) = \sum_{i=1}^{m} E_i\phi_i(t) = E\phi(t) \tag{7}$$

Here C_i, D_i and E_i are n-vectors and form the ith columns of the $n \times m$ matrices C, D and E, respectively. The coefficient matrices C and D are yet unknown, whereas E can be obtained for a given input $u(t)$ by applying eqn. 3. Determining a piecewise constant approximation to the solution of eqn. 6 is tantamount to determining the matrices C and D. To get an interrelationship between the matrices, we use the following identity:

$$\int_0^t \dot{X}\,dt = X(t) - X(0) \tag{8}$$

Inserting the expansion (eqn. 7) and replacing the integral by the operational matrix P, we have

$$CP\phi(t) = D\phi(t) - X(0) = D\phi(t) - \tilde{X}_0\phi(t) \tag{9}$$

where $\tilde{X}_0$ is an $n \times m$ matrix each column of which is X_0. Equating like coefficients of $\phi_i(t)$ in eqn. 9, we get

$$[CP]_i = D_i - X_0. \tag{10}$$

Here $[CP]_i$ is the ith column of CP and in view of eqn 5 is given by

$$[CP]_i = \frac{1}{m}\sum_{j=1}^{i-1} C_j + \frac{1}{2m} C_i. \tag{11}$$

It is convenient to rewrite eqn. 11 as a recurrence equation

$$[CP]_1 = \frac{1}{2m} C_1$$

$$[CP]_{i+1} = [CP]_i + \frac{1}{2m}(C_{i+1} + C_i) \tag{12}$$

for all $i \geqslant 1$. Combining eqns 10 and 12 yields

$$D_1 = \frac{1}{2m} C_1 + X_0$$

$$D_{i+1} = D_i + \frac{1}{2m}(C_{i+1} + C_i) \tag{13}$$

for all $i \geqslant 1$.

Eqn. 13 is a basic identity, and is the consequence of eqn. 8, in which the approximation of eqn 4 is embedded. It does not depend on the dynamic eqn. 6. However, eqn. 6 can be used to get another interrelationship between the coefficient matrices. Again, inserting the expansion given by eqn. 7 into eqn. 6 yields

$$C\phi(t) = AD\phi(t) + E\phi(t) \tag{14}$$

Equating like coefficients of $\phi_i(t)$, we have

$$C_i = AD_i + E_i \tag{15}$$

for all $i \geqslant 1$. Now it is relatively easy to obtain C and D. Substituting for C_i from eqn. 15 in eqn. 13, we get

$$D_1 = \alpha X_0 + \frac{\alpha}{2m} E_1$$

$$D_{i+1} = \beta D_i + \frac{\alpha}{2m}(E_{i+1} + E_i) \tag{16}$$

for all $i \geqslant 1$, where $\alpha = \left(I - \frac{A}{2m}\right)^{-1}$ and $\beta = \alpha\left(I + \frac{A}{2m}\right)$.

Since the eigenvalues of $I - \frac{A}{2m}$ are given by $1 - \frac{\lambda_K}{2m}$ where $\lambda_K, K = 1$ to n are the eigenvalues of A, the matrix inverse, $\left(I - \frac{A}{2m}\right)^{-1}$, always exists, except possibly for some finite values of m. Solution of the difference eqn. 16 yields an m-segment piecewise constant approximation of $X(t)$.

If $X(t)$ is a known function of t, and if a piecewise constant approximation of it is constructed using eqn. 3, the result will differ slightly from that obtained via eqn. 16. This is because eqn. 16 incorporates the approximation of eqn. 4. However, since there is a one-to-one relationship between Walsh and block-pulse functions[5], the solution via eqn. 16 will be identical to the result obtained by Chen and Hsiao[2-4] who introduced the concept of the Walsh operational matrix. In References 2–4, the solution of eqn. 6 was reduced through the use of a Kronecker product formula to an

Table 1

COMPARISON OF BLOCK-PULSE AND WALSH-FUNCTION METHODS

Type of operation	Number of operations in	
	Block-pulse-function method	Walsh-function method
Matrix inverse of order $n \times n$	1	$\geqslant l$
Multiplication of two $n \times n$ matrices	1	$\geqslant 3l$
Multiplication of an $n \times n$ matrix and an n-vector	$2m$	$\geqslant \frac{l(l-1)}{2} + 2m$

algebraic problem involving the inverse of a $nm \times nm$ matrix which was subsequently solved by inverting $n \times n$ matrices. It is appropriate, here, to compare numerically the present method with the method of References 2–4. Table 1, where $l = \log_2 m$, enumerates the number of numerical operations (other than simple additions and subtractions) involved in both the methods. Obviously, the number of critical operations involved in the block-pulse-function method is much smaller than the Walsh-series method. Therefore, we have saved computing time, storage and, consequently, have reduced the round-off errors significantly.

3.1 Comparison of the block-pulse method with conventional trapezoidal method

The solution of eqn. 6 by the conventional trapezoidal method[6-8], with $1/m$ as step size, is given by

$$X_{i+1} = \beta X_i + \frac{\alpha}{2m}((Bu)_{i+1} + (Bu)_i) \quad 0 \leqslant i \leqslant m$$

where X_i and $(Bu)_i$ are $X(t)$ and $Bu(t)$ evaluated at $t = i/m$, and α and β are as defined previously. If $E_i = \frac{1}{2}((Bu)_i + (Bu)_{i-1})$, it can be easily verified that $D_1 = \frac{1}{2}(X_0 + X_1)$ and then the method of induction can be used to show that $D_i = \frac{1}{2}(X_{i-1} + X_i)$ for all $1 \leqslant i \leqslant m$. In other words, the approximation introduced by replacing the integral in eqn. 4 by the operational matrix P is tantamount to the approximation involved in solving the system of differential eqns. 6 by the trapezoidal rule with step size equal to $1/m$ and then taking the average of the co-ordinates at the two end points of any internal to represent $X(t)$ during that interval. Since the block-pulse function method is closely related to the trapezoidal method, the analysis regarding convergence, stability and discretisation error and round-off error behaviour of the algorithm of eqn. 16 can be done by paralleling the established procedures.[8]

3.2 Extension to nonlinear systems

If the dynamic equation 6 is replaced by

$$\dot{X} = f(x, u, t) \quad X(0) = X_0 \tag{17}$$

then eqn. 14 should be replaced by

$$\sum_{i=1}^{m} C_i\phi_i(t) = f\left(\sum_{i=1}^{m} D_i\phi_i(t) \quad \sum_{i=1}^{m} u_i\phi_i(t) \quad \sum_{i=1}^{m} t_i\phi_i(t)\right) \tag{18}$$

where u_i and t_i are, respectively, the coefficients of expansion of $u(t)$ and t into block-pulse functions

$$u(t) = \sum_{i=1}^{m} u_i\phi_i(t) \qquad t = \sum_{i=1}^{m} t_i\phi_i(t)$$

Since the functions $\phi_i(t)$ are mutually disjointed, (i.e., if $\phi_K(t)$ is nonzero for a given t, then $\phi_i(t) = 0$ for all i except $i = K$) we can rewrite eqn. 18 as

$$C_i = f(D_i, u_i, t_i) \quad 1 \leqslant i \leqslant m \tag{19}$$

Eqn. 13 is still valid and can be used, together with eqn. 19, to solve for D_i:

$$D_1 = \frac{1}{2m} f(D_1, u_1, t_1) + X_0$$

$$D_{i+1} = D_i + \frac{1}{2m}(f(D_{i+1}, u_{i+1}, t_{i+1}) + f(D_i, u_i, t_i)), i \geqslant 1 \tag{20}$$

Again, eqn. 20 is related to the trapezoidal method which, in general, is an implicit method of integration, and, hence, requires iterative or predictor-corrector type of algorithms for its solution. An extensive discussion of such algorithms can be found in References 6 and 7.

3.3 Determination of piecewise constant gains for linear systems

It is known that the optimal control of a linear system with respect to a quadratic performance criteria over a finite interval often involves the determination of time-varying feedback gains. As it is hard to implement such time-varying gains, methods[4,9] have been proposed to approximate them by piecewise constant functions. In Reference 4, the fundamental matrix of the state, and the costate equations that result from the optimal control theory, is first calculated using the Walsh-function method and then the gains are calculated as they are directly related to the fundamental matrix. Approximation of the fundamental matrix by piecewise constant functions will also yield the same type of approximation for the gains. Since the calculation of a fundamental matrix involves the solution of an initial value problem, the block-pulse-function method eqn. 16 can be used effectively. This would simplify the calculation of piecewise constant gains enormously, and yet will allow the same accuracy as in Reference 4.

4 Conclusions

A method of numerical integration of a system of differential equations is given. The method is based on block-pulse functions, and yields a recursive algorithm analogous to the known trapezoidal method of integration. Compared with the Walsh-function approach, the method is computationally much simpler, and yet allows the same numerical accuracy. Design of piecewise constant gains for linear systems can be much simplified following this approach.

5 References

1 CORRINGTON, M.S.: 'Solution of differential and integral equations with Walsh functions', *IEEE Trans.*, 1973, **CT-20**, pp. 470–476

2 CHEN, C.F., and HSIAO, C.H.: 'A state-space approach to Walsh series solution of linear systems', *Int. J. Syst. Sci.*, 1975, **6**, 9, pp. 833–858

3 CHEN, C.F., and HSIAO, C.H.: 'Time domain synthesis via Walsh functions', *Proc. IEE*, 1975, **122**, (5), pp. 565–570

4 CHEN, C.F., and HSIAO, C.H.: 'Design of piecewise constant gains for optimal control via Walsh functions', *IEEE Trans.*, 1975, **AC-20**, 5, pp. 596–603

5 WU, T.T., CHEN, C.F., and TSAY, Y.T.: 'Walsh operational matrices for fractional calculus and their application to distributed systems'. IEEE International Symposium on Circuits and Systems, Munich, Germany, April 1976

6 HILDEBRAND, F.B.: 'Introduction to numerical analysis' (McGraw-Hill, 1974)

7 CHUA, L.O.: and LIN, P.M.: 'Computer-aided analysis of electronic circuits: algorithms and computational techniques' (Prentice-Hall, 1975)

8 HENRICI, P.: 'Error propagation for difference methods' (John Wiley, 1963)

9 KLEINMAN, D.L., FORTMANN, T.F., and ATHANS, M.: 'On the design of linear systems with piecewise constant feedback gains', *IEEE Trans.*, 1968, **AC-13**, pp. 354–361

Reprinted from *IEE Proc.* **125**:1064–1068 (1978)

ANALYSIS AND SYNTHESIS OF DYNAMIC SYSTEMS CONTAINING TIME DELAYS VIA BLOCK-PULSE FUNCTIONS

G. Prasada Rao and T. Srinivasan

Indexing terms: *Delay-differential systems, Function approximation, Integration, Numerical methods*

Abstract

This paper presents a method of numerically integrating differential equations containing time delays via block-pulse functions. The resulting solutions are piecewise constant with minimal mean-square error. The method is based on the use of a newly developed 'delay matrix' and enlarges the scope of useful applications of block-pulse functions to problems of dynamic systems. Examples of linear and nonlinear delay systems are included.

1 Introduction

Block-pulse functions (b.p.f.) and Walsh functions (w.f.) are closely related. As basis functions in an approximation the two sets of functions lead to the same results. The results are piecewise constant with minimal mean-square-error. B.P.F. have received less attention than w.f. in applications to control problems. These functions have only recently attracted the attention of a few researchers who sought and found computational convenience and simplification in the related algorithms. Gopalsami and Deekshatulu[1] discussed the application of b.p.f. in the computation of piecewise-constant gains for optimal control of lumped linear systems. Sannuti[2] has shown an enormous reduction of the computational effort in these problems by b.p.f. and has indicated an extension of the b.p.f. approach to nonlinear and time-varying systems. This paper presents a method of treating problems containing time-delays with b.p.f. The method as simple as that of Sannuti, is based on the use of a newly developed 'delay matrix' and extends the capability of the b.p.f. approach to tackle a large class of systems.

2 Block-pulse functions (b.p.f.)

A set of b.p.f. on a unit interval is defined as

$$\phi_i(t) = \begin{cases} 1 & \text{for } \dfrac{i-1}{m} < t \leq \dfrac{i}{m} \\ 0 & \text{otherwise} \end{cases}, \quad i = 1, 2, \ldots, m. \tag{1}$$

Represented concisely as an m-vector $\Phi(t)$, the set may be used to approximate an integrable function $f(t)$ in the interval $(0, 1)$ as

$$f(t) \approx F'\Phi(t) \tag{2}$$

where the elements f_i of the coefficient vector F are chosen such that

$$\epsilon = \int_0^1 [f(t) - F'\Phi(t)]^2 dt,$$

is minimised. For such a least-squares fit

$$f_i = m \int_{\frac{i-1}{m}}^{\frac{i}{m}} f(t)dt, \frac{i-1}{m} < t \leq \frac{i}{m}, \quad i = 1, 2, \ldots, m \tag{3}$$

This optimal m-segment piecewise-constant approximation is unique whatever may be the set of piecewise constant basis functions chosen for approximation. B.P.F. have the following useful properties:

(a) $\phi_i(t)\phi_j(t) = 0$, for all $i \neq j$,

$= \phi_i(t)$, for $j = i$,

(disjoint property), and (4)

(b) $\int_0^1 \phi_i(t)\phi_j(t)dt = 0$, for all $i \neq j$

(orthogonality) (5)

Paper 8167 C, received 24th February 1978

Prof. Ganti Prasada Rao and Mr. Srinivasan are with the Department of Electrical Engineering, Indian Institute of Technology, Kharagpur-2 Pin 721302, W. Bengal, India. Mr. Srinivasan is on leave from the Department of Electrical Engineering, Institute of Technology, Banaras Hindu University, Varanasi-5, 221005, India

It is known[1,2] that

$$\int_0^t \Phi(t)dt = P\Phi(t) \tag{6}$$

where the operational matrix for integration P is given by

$$P = \frac{1}{m}\begin{bmatrix} \frac{1}{2} & 1 & 1 & 1 & \ldots & 1 \\ & \frac{1}{2} & 1 & 1 & \ldots & 1 \\ & & \ldots & \ldots & \ldots & \ldots \\ & & & & \frac{1}{2} & 1 \\ & & & & & \frac{1}{2} \end{bmatrix} \tag{7}$$

The work on the application of b.p.f. to control problems so far is based mainly on the use of this matrix which reduces the calculus of dynamic systems to the interestingly simple and elegant algebra of triangular transformation matrices. However, at present, these techniques are limited to only lag-free systems. A new 'delay matrix' to remove this limitation will now be presented.

3 'Delay matrix'

If a set of b.p.f. $\Phi(t)$ is delayed by $\tau = \frac{1}{m}$, it can be shown that

$$\Phi\left(t - \frac{1}{m}\right) = \Delta\Phi(t) \tag{8}$$

where

$$\Delta = \begin{bmatrix} 0 & 1 & 0 & 0 & \ldots & 0 \\ 0 & 0 & 1 & 0 & \ldots & 0 \\ \ldots & \ldots & \ldots & \ldots & \ldots & \ldots \\ 0 & \ldots & \ldots & \ldots & 0 & 1 \\ 0 & \ldots & \ldots & \ldots & \ldots & 0 \end{bmatrix} \tag{9}$$

The significance of the matrix Δ is such that the delay operation on $\Phi(t)$ can be approximately (in the least-squared-error sense) achieved by premultiplying $\Phi(t)$ by the constant matrix Δ. We would therefore call such a matrix a 'delay matrix' for b.p.f. In general, when the delay $\tau = (1/m)(N + \delta), N = 0, 1, 2, \ldots, m-1, 0 < \delta < 1$,

$$\Phi(t - \tau) = D\,\Phi(t) \tag{10}$$

where

$$D = (1 - \delta)\Delta^N + \delta\Delta^{N+1} \tag{11}$$

The delay matrix represents a linear and sometimes singular transformation of the vector $\Phi(t)$ leading to an appropriate $\Phi(t - \tau)$. The matrix P, on the other hand, represents a linear nonsingular transformation of $\Phi(t)$ leading to $\int_0^t \Phi(t)dt$.

4 Solution of system of equations containing time delay

Consider a class of systems described by the equation

$$\dot{X} = AX + LX(t - \tau) + Bu(t) + Cu(t - \tau), X(0) = X_0, \tag{12}$$

where X is an n-vector of the state, u is an input vector of l components, A, L, B and C are $n \times n$, $n \times n$, $n \times l$, and $n \times l$ matrices, respectively. We would like to establish a procedure to solve eqn. 12 via b.p.f. employing the matrix D in addition to P. At the outset, we expand $\dot{X}$, X, Bu, $Cu(t-\tau)$ etc. in block-pulse-function series of size m.

$$\dot{X}(t) = \sum_{i=1}^{m} V_i\phi_i(t) = V\Phi(t)$$

$$X(t) = \sum_{i=1}^{m} Y_i\phi_i(t) = Y\Phi(t)$$

$$Bu(t) = \sum_{i=1}^{m} R_i\phi_i(t) = R\Phi(t)$$

$$Cu(t-\tau) = \sum_{i=1}^{m} \tilde{R}_i\phi_i(t) = \tilde{R}\Phi(t)$$

and

$$X(t-\tau) = \sum_{i=1}^{m} \tilde{Y}_i\phi_i(t) = \tilde{Y}\Phi(t) \tag{13}$$

Substituting eqn. 13 in eqn. 12 we obtain

$$V\Phi = AY\Phi + L\tilde{Y}\Phi + R\Phi + \tilde{R}\Phi \tag{14}$$

V, Y, $\tilde{Y}$, R and $\tilde{R}$ are $n \times m$ matrices composed of the respective b.p.f. components. For a given value of τ, R and $\tilde{R}$ may be directly computed. Denoting $R + \tilde{R}$ as G for convenience and since $\tilde{Y}\Phi(t) = Y\Phi(t-\tau)$, we may write eqn. 14 as

$$V\Phi(t) = AY\Phi(t) + LY\Phi(t-\tau) + G\Phi(t) \tag{15}$$

Making use of the known relation $\int_0^t \Phi(t)dt = P\Phi(t)$, we write eqn. 15 in the form

$$V\Phi(t) = A[VP\Phi(t) + X_0] + LY\Phi(t-\tau) + G\Phi(t) \tag{16}$$

We use eqn. 10 in eqn. 16 and employ the P matrix in the second term on the right-hand side of eqn. 16 to get

$$V = AVP + LVPD + K, \tag{17}$$

where

$$K = H + G, H = \underbrace{[AX_0 AX_0 \ldots AX_0]}_{m \text{ columns}}$$

Eqn. 17 may be solved for V from the form

$$\begin{bmatrix} v_1 \\ v_2 \\ \cdot \\ \cdot \\ \cdot \\ v_n \end{bmatrix} = [P' \otimes A] \begin{bmatrix} v_1 \\ v_2 \\ \cdot \\ \cdot \\ \cdot \\ v_n \end{bmatrix} + [(PD)' \otimes L] \begin{bmatrix} v_1 \\ v_2 \\ \cdot \\ \cdot \\ \cdot \\ v_n \end{bmatrix} + \begin{bmatrix} k_1 \\ k_2 \\ \cdot \\ \cdot \\ \cdot \\ k_n \end{bmatrix} \tag{18}$$

as

$$\begin{bmatrix} v_1 \\ v_2 \\ \cdot \\ \cdot \\ \cdot \\ v_n \end{bmatrix} = [I - P' \otimes A - (PD)' \otimes L]^{-1} \begin{bmatrix} k_1 \\ k_2 \\ \cdot \\ \cdot \\ \cdot \\ k_n \end{bmatrix} \tag{19}$$

where the Kronecker products are defined as

$$P' \otimes A = \begin{bmatrix} a_{11}P' & a_{12}P' & \ldots & a_{1n}P' \\ a_{21}P' & a_{22}P' & \ldots & a_{2n}P' \\ \vdots & & & \\ a_{22}P' & a_{n2}P' & \ldots & a_{nn}P' \end{bmatrix} \quad \text{and}$$

$$(PD)' \otimes L = \begin{bmatrix} l_{11}(PD)' & l_{12}(PD)' & \ldots & l_{1n}(PD)' \\ l_{21}(PD)' & l_{22}(PD)' & \ldots & l_{2n}(PD)' \\ \vdots & & & \\ l_{n1}(PD)' & l_{n2}(PD)' & \ldots & l_{nn}(PD)' \end{bmatrix} \tag{20}$$

while v_i and k_i are the n state components each of size m. Having obtained the rate variable V, we obtain the state variable as

$$Y\Phi(t) = VP\Phi(t) + X_0 \tag{21}$$

Solution of eqn. 18 via eqn. 19 involving the inversion of $(n \times m) \times (n \times m)$ matrices in this way is computationally not attractive. The problem may be reduced to a certain number inversions of $n \times n$ matrices. However, since the disjoint property (eqn. 4) of b.p.f. lends triangular structure to the matrices P and D, it is possible to solve for the b.p.f. components of the state vector by developing a simple recursive formula as illustrated in the following Section.

5 Solution by recursive formula

To solve eqn. 12, we begin with eqn. 15 and take a different course. We find the connections of the matrices V and $\tilde{Y}$ with Y. The matrices Y and V are shown by Sannuti[2] to be connected by the following relation:

$$Y_1 = \frac{1}{2m}V_1 + X_0, \quad \text{and}$$

$$Y_{i+1} = Y_i + \frac{1}{2m}(V_{i+1} + V_i), \quad \text{for all} \quad i \geq 1 \tag{22}$$

In obtaining eqn. 22 use is made of the triangular P matrix. In eqn. 22 Y_i is the ith column of Y and the disjoint property (which implies orthogonality as well) of b.p.f. facilitates this simple reduction. We equate the coefficients of the like components $\phi_i(t)$ in eqn. 15 to obtain

$$V_i = (AY)_i + (L\tilde{Y})_i + G_i, \quad \text{for all} \quad i \geq 1 \tag{23}$$

Substituting eqns. 10 and 22 in eqn. 15 we have

$$Y_{i+1} = Y_i + \frac{1}{2m}[AY_{i+1} + L(YD)_{i+1} + G_{i+1} + AY_i + L(YD)_i + G_i] \tag{24}$$

The delay matrix suggests the following:

$$(YD)_i = (1-\delta)Y_{i-N} + \delta Y_{i-N-1} \tag{25}$$

Using eqn. 25 eqn. 24 may be finally written as

$$Y_1 = \alpha X_0 + \frac{1}{2m}G_1, \quad \text{and}$$

$$Y_{i+1} = \beta Y_i + \frac{\alpha}{2m}L[(1-\delta)Y_{i+1-N} + Y_{i-N} + \delta Y_{i-N-1}] + \frac{\alpha}{2m}(G_{i+1} + G_i), \quad \text{for all} \quad i \geq 1,$$

where

$$\alpha = \left(I - \frac{A}{2m}\right)^{-1} \text{ and } \beta = \alpha\left(I + \frac{A}{2m}\right). \tag{26}$$

If $\delta = 0$ by suitable choice of m, eqn. 26 takes the form

$$Y_1 = \alpha X_0 + \frac{1}{2m}G_1, \quad \text{and}$$

$$Y_{i+1} = \beta Y_i + \frac{\alpha}{2m}L(Y_{i-N} + Y_{i+1-N}) + \frac{\alpha}{2m}(G_{i+1} + G_i), \quad \text{for all} \quad i \geq 1. \tag{27}$$

In eqns. 26 and 27, α and β are $n \times n$ matrices. α exists except for some finite values of m. The solution for Y via eqn. 26 or 27, as the case may be, offers enormous reduction in the computational effort required. Sannuti[2] has compared the computational effort in the two methods of solution, namely the Kronecker product method and the recursive formula, in his treatment of lag-free problems by b.p.f.

6 Illustrative examples

6.1 Example 1: A linear delay system

To illustrate the present method, let us consider a unity negative feedback system[3]. The closed-loop transfer function of the system

$$\frac{X}{R}(s) = \frac{(s+0{\cdot}3)\exp(-s)}{s^2+(s+0{\cdot}3)\exp(-s)} \tag{28}$$

The corresponding differential equation is

$$\ddot{x}(t)+\dot{x}(t-1)+0{\cdot}3x(t-1) = u(t-1)+0{\cdot}3r(t-1),$$
$$x(0)=0, \dot{x}(0)=0 \tag{29}$$

where $u(t-1)$ and $r(t-1)$ are unit step and ramp functions, respectively, each delayed by 1 s. Let $x(t)=x_1, \dot{x}(t)=\dot{x}_1=x_2, \ddot{x}(t)=\dot{x}_2$. Then eqn. 29 may be written as

$$\begin{bmatrix}\dot{x}_1\\ \dot{x}_2\end{bmatrix} = \begin{bmatrix}0 & 1\\ 0 & 0\end{bmatrix}\begin{bmatrix}x_1\\ x_2\end{bmatrix} + \begin{bmatrix}0 & 0\\ -0{\cdot}3 & -1\end{bmatrix}\begin{bmatrix}x_1(t-1)\\ x_2(t-1)\end{bmatrix} + \begin{bmatrix}0\\ u(t-1)+0{\cdot}3r(t-1)\end{bmatrix} \tag{30}$$

For the sake of simplicity in illustration, we will obtain the solution of eqn. 30 over 6 seconds. To begin with, we change the time scale letting $\lambda = \frac{1}{6}t$ to facilitate the use of b.p.f. over the unit interval and obtain

$$\dot{X}(\lambda) = AX(\lambda)+LX(\lambda-\tfrac{1}{6})+Bu(\lambda), \tag{31}$$

where

$$A = \begin{bmatrix}0 & 1\\ 0 & 0\end{bmatrix} \quad L = \begin{bmatrix}0 & 0\\ -10{\cdot}8 & -6{\cdot}0\end{bmatrix}$$

$$Bu(\lambda) = \begin{bmatrix}0\\ 36u(\lambda-\tfrac{1}{6})+64{\cdot}8r(\lambda-\tfrac{1}{6})\end{bmatrix}$$

Note that $G=\tilde{R}$ in this case. Choosing $m=12$, we have $\tau = 2/m$, i.e. $N=2$. Since the input functions are known, we compute G as

$$G = \begin{bmatrix}0 & 0 & 0 & 0 & 0 & 0 & 0 & 0 & 0 & 0 & 0 & 0\\ 0 & 0 & 38{\cdot}7 & 44{\cdot}1 & 49{\cdot}5 & 54{\cdot}9 & 60{\cdot}3 & 65{\cdot}7 & 71{\cdot}1 & 76{\cdot}5 & 81{\cdot}9 & 87{\cdot}3\end{bmatrix} \tag{32}$$

The solution using eqn. 27 with $N=2$ is obtained as

$$Y = \begin{bmatrix}0{\cdot}0 & 0{\cdot}0 & 0{\cdot}0671875 & 0{\cdot}3453125 & 0{\cdot}911631 & 1{\cdot}7263774 & 2{\cdot}6799608 & 3{\cdot}633975 & 4{\cdot}4789222 & 5{\cdot}1484883 & 5{\cdot}6290601 & 5{\cdot}9671279\\ 0{\cdot}0 & 0{\cdot}0 & 1{\cdot}6125 & 5{\cdot}0625 & 8{\cdot}529144 & 11{\cdot}024769 & 11{\cdot}861233 & 11{\cdot}035651 & 9{\cdot}242544 & 6{\cdot}8270424 & 4{\cdot}7066795 & 3{\cdot}4069482\end{bmatrix} \tag{33}$$

Table 1 compares this solution, valid almost everywhere in the interval except at the boundaries of the blocks, with those obtained by some other numerical methods for the same step size. The solutions by the other numerical methods on the other hand are valid only at the sampling instants.

Table 1
SOLUTION OF EXAMPLE 1

Time	Fourth-order Runge-Kutta method	Trapezoidal method	Z-transform method	b.p.f. method
0·0	0·0	0·0	0·0	0·0
0·5	0·0	0·0	0·0	0·0
1·0	0·0	0·0625	0·0	0·0671875
1·5	0·1729166	0·321875	0·131	0·3453125
2·0	0·6298610	0·8519531	0·550	0·911631
2·5	1·3698869	1·6181835	1·260	1·7263774
3·0	2·2734285	2·5203343	2·18	2·6799608
3·5	3·1367584	3·4300836	3·17	3·6339975
4·0	4·0131436	4·2363253	4·10	4·4789222
4·5	4·7515479	4·8754244	4·86	5·1484883
5·0	5·2870788	5·3427102	5·41	5·6290601
5·5	5·6752991	5·6840703	5·78	5·9671279
6·0	5·9844159	5·9738988	6·05	

6.2 Example 2: nonlinear delay system

Consider the well known Lienard equation

$$\ddot{x}(t)+\{x^2(t-\tau)-1\}\dot{x}(t-\tau)+x(t) = 0 \tag{34}$$

Since in general

$$\left.\begin{aligned} h[x(t)] &= h\left[\sum_{i=1}^{m} y_i\Phi_i(t)\right] = h[Y\phi(t)]\\ x^2(t) &= \sum_{i=1}^{m} y_i^2\,\phi_i(t) \quad \text{and}\\ x^2(t-\tau) &= \sum_{i=1}^{m} y_{i-N}^2\,\phi_i(t)\end{aligned}\right\} \tag{35}$$

Rewriting eqn. 34 as

$$\left.\begin{aligned}\dot{x}_1(t) &= x_2(t)\\ \dot{x}_2(t) &= -[x_1^2(t-\tau)-1]\,x_2(t-\tau)-x_1(t),\end{aligned}\right\} \tag{36}$$

where $x_1(t)=x(t)$, the solution in terms of b.p.f. will be written as

$$\begin{bmatrix}x_1\\ x_2\end{bmatrix} = \begin{bmatrix}Y1_1 & Y1_2 & \ldots & Y1_i & \ldots & Y1_m\\ Y2_1 & Y2_2 & \ldots & Y2_i & \ldots & Y2_m\end{bmatrix} \tag{37}$$

where

$$Y1_1 = \frac{1}{2m}Y2_1 + x_1(0)$$

$$Y2_1 = \frac{1}{2m}\{-(Y1_{1-N}^2-1)\,Y2_{1-N}-Y1_1\}+x_2(0) \tag{38}$$

and

$$\left.\begin{aligned} Y1_{i+1} &= Y1_i+\frac{1}{2m}[Y2_{i+1}+Y2_i]\\ Y2_{i+1} &= Y2_i+\frac{1}{2m}[-(Y1_{i+1-N}^2-1)\,Y2_{i+1-N}-Y1_{i+1}\\ &\qquad -(Y1_{i-N}^2-1)\,Y2_{i-N}-Y1_i]\\ &\text{for}\quad i=1,2,\ldots,m-1\end{aligned}\right\} \tag{39}$$

We consider a specific case of eqn. 34 with $\tau = 0{\cdot}75$ s, $x_1(0)=0$, $x_2(0)=2{\cdot}0$, $x_1(t)=x_2(t)=0$ for $-\tau<t<0$. To compute the solution over a period of 10 s, we let $t = 10\lambda$ in eqn. 36 and obtain

$$\left.\begin{aligned}\dot{x}_1(\lambda) &= x_2(\lambda)\\ \dot{x}_2(\lambda) &= -10[x_1^2(\lambda-0{\cdot}075)-1]\,x_2(\lambda-0{\cdot}075)-100x_1(\lambda)\\ &\text{with}\ x_1(0)=0, x_2(0)=20 \quad\text{and}\quad x_1(\lambda)=x_2(\lambda)=0,\\ &\text{for}\ -0{\cdot}075<\lambda<0.\end{aligned}\right\} \tag{40}$$

We choose $m=40$, so that $N=3$. Eqns. 38 and 39 then take the particular form

$$\left.\begin{aligned} Y1_1 &= \frac{1}{80}Y2_1\\ Y2_1 &= \frac{1}{80}[-(Y1_{-2}^2-1)\,Y2_{-2}-100\,Y1_1]+20\end{aligned}\right\} \tag{41}$$

and

$$\left.\begin{aligned} Y1_{i+1} &= Y1_i + \frac{1}{80}[Y2_{i+1} + Y2_i] \\ Y2_{i+1} &= Y2_i + \frac{1}{80}[-10(Y1^2_{i-2} - 1)Y2_{i-2} - 100Y1_{i+1} \\ &\quad - 10(Y1^2_{i-3} - 1)Y2_{i-3} - 100Y1_i] \end{aligned}\right\} \quad (42)$$

The solution is shown in phase plane in Fig. 1. It shows a periodic solution with a period of approximately 5·75 s, and agrees well with that obtained by Grafton.[4] The b.p.f. solution in phase plane is a set of points. At each of these points a period of time equal to $1/m$ on the normalised scale is lumped. The dotted line through these points is shown to indicate the phase trajectory.

7 Error analysis

7.1 Representation error in block-pulse approximation

If a function $f(t)$ is represented by a b.p.f. $\phi_i(t)$ of magnitude f_i in the interval $(i-1)/m < t \leqslant i/m$, the representation error is defined as

$$e(t) = f_i - f(t)$$

over the ith interval. The integral of squared error (i.s.e.)

$$\int_{\frac{i-1}{m}}^{\frac{i}{m}} [e(t)]^2\, dt$$

is minimised by choosing f_i such that

$$f_i = m \int_{\frac{i-1}{m}}^{\frac{i}{m}} f(t)\, dt$$

The minimum integral squared error (m.i.s.e.), ϵ_i^2 say, is a measure of the error of approximation in the process of representation. In fact,

$$\epsilon_i^2 = \int_{\frac{i-1}{m}}^{\frac{i}{m}} [f(t)]^2\, dt - m\left[\int_{\frac{i-1}{m}}^{\frac{i}{m}} f(t)\, dt\right]^2$$

It is possible to show via a Taylor-series expansion of $f(t)$ that

$$\epsilon_i^2 = \frac{1}{12m^3}[f'(\xi_i)]^2,\ \xi_i \in \left[\frac{i-1}{m}, \frac{i}{m}\right]$$

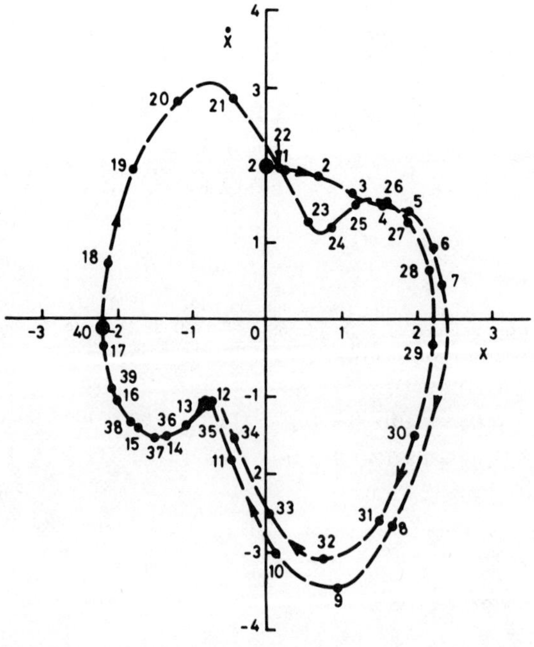

Fig. 1
Solution of example 2

The mean-squared-error over the m blocks of the unit interval

$$\sum_{i=1}^{m} \epsilon_i^2 = \frac{1}{12m^3}\sum_{i=1}^{m}[f'(\xi_i)]^2,\ \xi_i \in \left[\frac{i-1}{m}, \frac{i}{m}\right]$$

If f'_{max} is the largest among all $f'(\xi_i)$, the upper bound of

$$\sum_{i=1}^{m} \epsilon_i^2 \triangleq E^2 = \frac{1}{12m^2} f'^2_{max}$$

7.2 Choice of m in segmenting a signal for specified r.m.s. error bound

If a signal $f(t) = A \sin \omega t$ has to be segmented into a set of m b.p.f. and if the allowable error of representation should be such that

$$\frac{E}{A} \leqslant \sigma_r, \text{ a specified constant,}$$

then

$$m \geqslant \frac{\omega}{2\sqrt{3}\sigma_r}$$

In general, if a composite signal $f(t)$ has A_{max} and ω_{max} as the highest amplitude and highest frequency in its spectrum and if the maximum allowable error is such that

$$\frac{E}{A_{max}} \leqslant \sigma_r, \text{ a specified constant.}$$

then

$$m \geqslant \frac{\omega_{max}}{2\sqrt{3}\,\sigma_r}$$

The number of segments may be determined on the basis of an upper bound $A_{max}\,\omega_{max}$ for f' in the signal.

It may be observed that if a signal is approximated by m blocks, the signal is averaged over each subinterval of length $1/m$, for b.p.f. representation. This is similar to a process of lowpass filtering in which all signal components with period less than or equal to $1/m$ are stopped.

7.3 Error due to re-representation in b.p.f. of the integral of a set of b.p.f.

The integral $\int_0^t f_i(t)\,\phi_i(t)\,dt$, $t \in [(i-1)/m, i/m]$ is a ramp function with slope f_i and the corresponding re-representation has a b.p.f. value of $f_i/2m$. Then the m.i.s.e. over the ith block in the process of re-representation, ν_i^2, is given by

$$\nu_i^2 = \int_{\frac{i-1}{m}}^{\frac{i}{m}} \left[\frac{f_i}{2m} - f_i\left(t - \frac{i-1}{m}\right)\right]^2 dt = \frac{f_i^2}{12m^3}$$

The mean-square-error over the m blocks of the unit interval

$$= \sum_{i=1}^{m} \nu_i^2 = \frac{1}{12m^3}\sum_{i=1}^{m} f_i^2$$

If f_{max} is the largest coefficient in the b.p.f. expansion, the upper bound of

$$\sum_{i=1}^{m} \nu_i^2,\ \Gamma^2 \text{ say, is given by } \Gamma^2 = (1/12m^2) f^2_{max}.$$

7.4 Behaviour of ∫ e (t) dt

In the ith segment the error

$$e(t) = m\int_{\frac{i-1}{m}}^{\frac{i}{m}} f(t)\,dt - f(t),$$

and

$$\int_{\frac{i-1}{m}}^{\frac{i}{m}} e(t)\,dt = \frac{1}{2}\left(t - \frac{i-1}{m}\right)\left[\frac{1}{m} - \left(t - \frac{i-1}{m}\right)\right] f'(\xi_i),$$

$$\xi_i \in \left[\frac{i-1}{m}, \frac{i}{m}\right],$$

which vanishes at $t = (i-1)/m$ and i/m. Thus the integral of the representation error vanishes at the boundaries of the segments. That is, the error in computing the integral of a function from its approximate b.p.f. version does not propagate.

7.5 Error implications of nonzero δ in the delay operation

The re-representation error in re-representing a set of b.p.f. delayed by $\tau = (N+\delta)/m$, $0 < \delta < 1$, $N = 0, 1, \ldots, m-1$, in terms of the original b.p.f. has a m.i.s.e. value

$$\epsilon_{\delta i}^2 = \int_{\frac{i-1}{m}}^{\frac{i}{m}} e^2\,dt$$

$$= \frac{\delta}{m}\{[\delta f_{i-N-1} + (1-\delta) f_{i-N}] - f_{i-N-1}\}^2$$

$$+ \frac{(1-\delta)}{m}\{[\delta f_{i-N-1} + (1-\delta) f_{i-N}] - f_{i-N}\}^2$$

The mean-squared-error over the unit interval.

$$\sum_{i=1}^{m} \epsilon_{\delta i}^2 = \frac{\delta(1-\delta)}{m} \sum_{i=N+1}^{m} (f_{i-N} - f_{i-N-1})^2$$

This error vanishes when $\delta = 0$ or 1, and has a maximum, at $\delta = 1/2$, equal to $(1/4m) \sum_{i=N+1}^{m} (f_{i-N} - f_{i-N-1})^2$, suggesting that m, while satisfying all other conditions, should be chosen so as to make $\delta = 0$.

8 Conclusion

A method of solving delay differential equations via block pulse functions is presented. The new 'delay matrix' introduced in this paper becomes an additional tool useful in the process of transforming given differential equations into computationally convenient matrix-algebraic form. The suggested error analysis provides a basis for choosing the number of segments m, for specified bounds of r.m.s. error. Increase of m demands very little additional computational effort in the recursive algorithm. Under certain conditions[2] the recursive algorithm resembles the conventional trapezoidal formula. A similar treatment of delay systems via Walsh functions is possible but it requires a comparatively heavier computational effort to yield results with the same accuracy.

9 References

1 GOPALASAMI, N., and DEEKSHATULU, B.L.: 'Comments on design of piecewise constant gains for optimal control via Walsh Func'ions,' *IEEE Trans.*, 1976, AC-21, pp. 634–635

2 SANNUTI, P.: 'Analysis and synthesis of dynamic systems via block-pulse functions', *Proc. IEE*, 1977, **124**, (6), pp. 569–571

3 JURY, E.I.: 'Theory and application of the z-transform mehtod' (John Wiley & Sons, 1964), p. 53 & 229

4 GRAFTON, R.B.: 'Periodic solutions of Lienard equations with dealy: some theoretical and numerical results', *in* Schmitt, K. (Ed.): 'Delay and functional differential equations and their applications' (Academic Press, 1972) pp. 321–334

Reprinted from *J. Dyn. Syst., Meas., Control* **102:**159–162 (1980)

Yen-Ping Shih
Chyi Hwang
Department of Chemical Engineering.

Wei-Kong Chia
Department of Electrical Engineering.

National Cheng Kung University, Tainan, Taiwan, China

Parameter Estimation of Delay Systems Via Block Pulse Functions

Linear time-invariant delay-differential equation systems are approximately represented by a set of linear algebraic equations with the block pulse functions. A least squares estimate is then used to determine the unknown parameters. Examples with satisfactory results are given.

Significance

A new method for the parameter estimation of differential equation system is proposed with the choice of time-invariant delay-differential equation systems as an example for a detailed study. First, the differential equations are transformed into a set of simultaneous linear algebraic equations. Then, a least squares estimate is used to find the unknown parameters.

Introduction

Systems with time-delay occur frequently in mechanical systems, transmission lines, and industrial processes [1,2]. There are many systems such as population growth, epidemic growth, neural networks, etc. modelled by delay-differential equations [3,4]. The analysis, state estimation, and identification of delay systems have been studied frequently [5–15].

Much attention has been given to the stability of delay systems [5]. N-Nagy and Al-Tikriti [5] used a time-partition method to analyze delay systems using the Laplace transformation. Shih [6] extended the time-partition method to time-domain. In another approach, Tsoi [7–9] obtained the transition matrices for certain classes of delay-differential equations. State estimation and identification of delay systems are given by Kwakernaak [10], Lindquist [11], Koivo [12], Koivo and Stoller [13], Gabay and Merkav [14], and Bellman, et al. [15].

In this paper, parameter estimation of delay systems is studied with the block pulse functions. Notice that Chen and Shih [16] used the Walsh functions to obtain piecewise-constant solutions of delay-differential equations. The analysis of delay systems can be greatly simplified in computation with the block pulse functions. First, a very brief review of the block pulse functions and their important properties is given.

Block Pulse Functions

The block pulse vector $\psi(t)$, $t\epsilon[0,1)$ is defined as

$$\psi(t) = [\psi_0(t), \psi_1(t), \ldots, \psi_{m-1}(t)]' \tag{1}$$

where " ′ " means transpose and $\psi_i(t)$'s are the orthogonal block pulse functions,

$$\psi_i(t) = \begin{cases} 1, & i/m \le t < (i+1)/m \\ 0, & \text{otherwise} \end{cases} \tag{2}$$

$$\int_0^1 \psi_i(t)\psi_j(t)dt = \begin{cases} 1/m, & i = j \\ 0, & i \neq j \end{cases} \tag{3}$$

A time function $f(t)$ which is absolutely integrable in $t\epsilon[0,1)$ can be approximately represented by a block pulse series,

$$f(t) = \sum_{i=0}^{m-1} f_i \psi_i(t) = \mathbf{f}'\psi(t) \tag{4}$$

where $\mathbf{f}$ is a block pulse coefficient vector

$$\mathbf{f} = [f_0, f_1, \ldots, f_{m-1}]' \tag{5}$$

f_i's which are obtained by the minimization of the integral square error

$$\epsilon = \int_0^1 [f(t) - \mathbf{f}'\psi(t)]^2 dt \tag{6}$$

are the mean values of $f(t)$ in the interval $t\epsilon[i/m,(i+1)/m]$

$$f_i = m\int_0^1 f(t)\psi_i(t)dt$$

$$= m\int_{i/m}^{(i+1)/m} f(t)dt \tag{7}$$

The integration of the block pulse vector is [17],

$$\int_0^t \psi(s)ds = \mathbf{H}\psi(t) \tag{8}$$

where $\mathbf{H}$ is the operational matrix of integration

Contributed by the Dynamic Systems and Control Division for publication in the JOURNAL OF DYNAMIC SYSTEMS, MEASUREMENT, AND CONTROL. Manuscript received at ASME Headquarters, August 1, 1980.

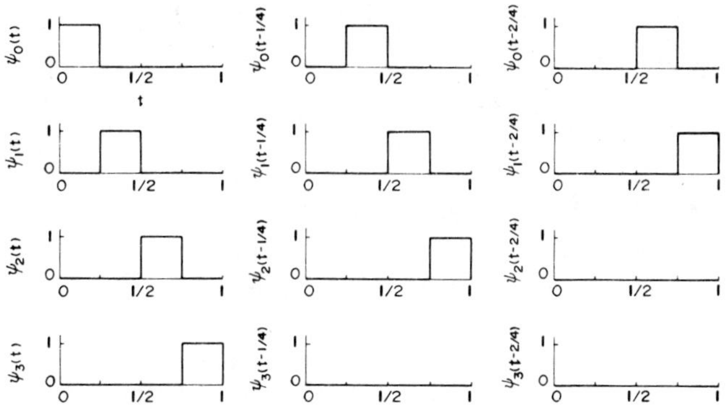

Fig. 1 Block pulse and delay block pulse functions

$$\mathbf{H}=\frac{1}{m}\begin{bmatrix} \frac{1}{2} & 1 & 1 & \cdot\cdot & 1 \\ 0 & \frac{1}{2} & 1 & \cdot\cdot & 1 \\ \cdot\cdot & \cdot\cdot & \cdot\cdot & \cdot\cdot & \cdot\cdot \\ 0 & 0 & 0 & \cdot\cdot & \frac{1}{2} \end{bmatrix} \tag{9}$$

Delay Block Pulse Functions

The delay block pulse function $\psi_i\,(t-k/m)$ is the shift of the block pulse function $\psi_i(t)$ along the time axis by k subintervals. That is

$$\psi_i(t-k/m) \;=\; \begin{cases} 1, & (i+k)/m \le t \le (i+k+1)/m \\ 0, & \text{otherwise} \end{cases} \tag{10}$$

as shown in Fig. 1 for $m = 4$, $k = 1,2,3$. Hence, for $m = 4$ and $k = 1$,

$$\begin{bmatrix} \psi_0(t-1/4) \\ \psi_1(t-1/4) \\ \psi_2(t-1/4) \\ \psi_3(t-1/4) \end{bmatrix} = \begin{bmatrix} 0 & 1 & 0 & 0 \\ 0 & 0 & 1 & 0 \\ 0 & 0 & 0 & 1 \\ 0 & 0 & 0 & 0 \end{bmatrix} \begin{bmatrix} \psi_0(t) \\ \psi_1(t) \\ \psi_2(t) \\ \psi_3(t) \end{bmatrix} \tag{11}$$

The general expression of the delay block pulse functions in terms of the block pulse functions is

$$\psi(t-k/m)=\mathbf{I}[m;-k]\psi(t) \tag{12}$$

Here $\mathbf{I}[m;\,-k]$ is called shift identity matrix

$$\mathbf{I}[m;\,-k] \;=\; \begin{bmatrix} \mathbf{0} & \mathbf{I}_{(m-k)} \\ \mathbf{0} & \mathbf{0} \end{bmatrix} \tag{13}$$

$\mathbf{I}_{(m-k)}$ an $(m-k)\times(m-k)$ identity matrix. (13) is a key for the analysis of delay-differential equations.

Block Pulse Approximations of Delay Systems

Consider a time-invariant linear delay differential equation,

$$\dot{\mathbf{x}}(t)=\mathbf{A}\mathbf{x}(t)+\mathbf{B}\mathbf{x}(t-k/m)+\mathbf{C}\mathbf{u}(t)+\mathbf{D}\mathbf{u}(t-k/m) \tag{14}$$

$$\mathbf{x}(0)\ \text{ specified} \tag{14a}$$

$$\mathbf{x}(t)=\mathbf{h}(t),\quad -k/m\le t<0 \tag{14b}$$

$$\mathbf{u}(t)=\mathbf{g}(t),\quad -k/m\le t<0 \tag{14c}$$

where $\mathbf{x}(t)$ is an $n\times l$ vector, $\mathbf{u}(t)$ is a $q\times l$ vector, $\mathbf{A}$, $\mathbf{B}$, $\mathbf{C}$, and $\mathbf{D}$ are matrices of appropriate dimension. Let the block pulse approximations of $\mathbf{x}(t)$ be

$$\mathbf{x}(t)=[\mathbf{x}_0,\mathbf{x}_1,\ldots,\mathbf{x}_{m-1}]\psi(t) \tag{15}$$

with the block pulse coefficients $\mathbf{x}_i$,

$$\mathbf{x}_i=m\int_{i/m}^{(i+1)/m}\mathbf{x}(t)\,dt \tag{16}$$

Likewise

$$\mathbf{u}(t)=[\mathbf{u}_0,\mathbf{u}_1,\ldots,\mathbf{u}_{m-1}]\psi(t) \tag{17}$$

$$\mathbf{h}(t-k/m)=[\mathbf{h}_0,\mathbf{h}_1,\ldots,\mathbf{h}_{m-1}]\psi(t) \tag{18}$$

$$\mathbf{g}(t-k/m)=[\mathbf{g}_0,\mathbf{g}_1,\ldots,\mathbf{g}_{m-1}]\psi(t) \tag{19}$$

Integration of (14) gives

$$\mathbf{x}(t)-\mathbf{x}(0)=\int_0^t\mathbf{A}\mathbf{X}(s)\,ds+\int_0^t\mathbf{B}\mathbf{x}(s-k/m)\,ds + \int_0^t\mathbf{C}\mathbf{u}(s)\,ds+\int_0^t\mathbf{D}\mathbf{u}(s-k/m)\,ds \tag{20}$$

For $0\le t<k/m$, substituting (14*b,c*), (15), (17)-(19) into (20) and applying the property of (8), one obtains

$$[\mathbf{x}_0,\mathbf{x}_1,\ldots,\mathbf{x}_{m-1}]\psi(t)-\underbrace{[\mathbf{x}(0),\mathbf{x}(0),\ldots,\mathbf{x}(0)]}_{m\text{ columns}}\psi(t)$$

$$=\mathbf{A}[\mathbf{x}_0,\mathbf{x}_1,\ldots,\mathbf{x}_{m-1}]\mathbf{H}\psi(t)+\mathbf{B}[\mathbf{h}_0,\mathbf{h}_1,\ldots,\mathbf{h}_{m-1}]\mathbf{H}\psi(t) +\mathbf{C}[\mathbf{u}_0,\mathbf{u}_1,\ldots,\mathbf{u}_{m-1}]\mathbf{H}\psi(t)+\mathbf{D}[\mathbf{g}_0,\mathbf{g}_1,\ldots,\mathbf{g}_{m-1}]\mathbf{H}\psi(t) \tag{21}$$

Since $\psi_i(t)$'s are linearly independent, equating the coefficients of (21) and rearranging give

$$\mathbf{x}_0=\mathbf{x}(0)+\frac{1}{2m}(\mathbf{A}\mathbf{x}_0+\mathbf{B}\mathbf{h}_0+\mathbf{C}\mathbf{u}_0+\mathbf{D}\ \mathbf{g})$$

$$\mathbf{x}_1=\mathbf{x}_0+\frac{1}{2m}[\mathbf{A}(\mathbf{x}_1+\mathbf{x}_0)+\mathbf{B}(\mathbf{h}_1+\mathbf{h}_0)+\mathbf{C}(\mathbf{u}_1+\mathbf{u}_0)+\mathbf{D}(\mathbf{g}_1+\mathbf{g}_0)]$$

$$\mathbf{x}_{k-1}=\mathbf{x}_{k-2}+\frac{1}{2m}[\mathbf{A}(\mathbf{x}_{k-1}+\mathbf{x}_{k-2})+\mathbf{B}(\mathbf{h}_{k-1}+\mathbf{h}_{k-2}) +\mathbf{C}(\mathbf{u}_{k-1}+\mathbf{u}_{k-2})+\mathbf{D}(\mathbf{g}_{k-1}+\mathbf{g}_{k-2})] \tag{22}$$

Notice that (21) is valid only for $0\le t<k/m$, therefore only $\mathbf{x}_0,\mathbf{x}_1,\ldots,\mathbf{x}_{k-1}$ are evaluated by (22) which is a set of linear algebraic equations, $\mathbf{x}_i$ can be calculated successively from (22). Whereas, in the Walsh function approach, the Walsh coefficients have to be calculated by solving a large number of simultaneous linear algebraic equations [16].

For $k/m\le t<1$, using the unit step functions, $S(t)$;

$$S(t) \;=\; \begin{cases} 1, & t\ge 0 \\ 0, & t<0 \end{cases} \tag{23}$$

(20) becomes

$$\mathbf{x}(t)-\mathbf{x}(0)=\int_0^t\mathbf{A}\mathbf{x}(s)\,ds+\int_0^t\mathbf{B}\mathbf{x}(s-k/m)S(s-k/m)\,ds + \int_0^t\mathbf{C}\mathbf{u}(s)\,ds+\int_0^t\mathbf{D}\mathbf{u}(s-k/m)S(s-k/m)\,ds+\mathbf{v} \tag{24}$$

where

$$\mathbf{v}=\int_0^{k/m}\mathbf{B}\mathbf{h}(s-k/m)\,ds+\int_0^{k/m}\mathbf{D}\mathbf{g}(s-k/m)\,ds \tag{25}$$

Substituting (15), (17)-(19) into (24), using the property of (8), and applying (12) to the second and third terms on the righthand side of (24), one obtains

$$[\mathbf{x}_0, \mathbf{x}_1 \ldots, \mathbf{x}_{m-1}]\psi(t) + \underbrace{[\mathbf{x}(0), \mathbf{x}(0), \ldots, \mathbf{x}(0)]}_{\text{m columns}}\psi(t)$$

$$= \mathbf{A}[\mathbf{x}_0, \mathbf{x}_1, \ldots, \mathbf{x}_{m-1}]\mathbf{H}\psi(t)$$
$$+ \mathbf{B}[\mathbf{0}, \ldots, \mathbf{0}, \mathbf{x}_0, \mathbf{x}_1, \ldots, \mathbf{x}_{m-k-1}]\mathbf{H}\psi(t)$$
$$+ \mathbf{C}[\mathbf{u}_0, \mathbf{u}_1, \ldots, \mathbf{u}_{m-1}]\mathbf{H}\psi(t)$$
$$+ \mathbf{D}[\mathbf{0}, \ldots, \mathbf{0}, \mathbf{u}_0, \mathbf{u}_1, \ldots, \mathbf{u}_{m-k-1}]\mathbf{H}\psi(t)$$
$$+ \underbrace{[\mathbf{v}, \ldots, \mathbf{v}]}_{\text{m columns}}(t) \tag{28}$$

Since (26) is valid for $k/m \le t < 1$, equating the coefficients of (26) and using the expressions of $\mathbf{x}_0, \mathbf{x}_1, \ldots, \mathbf{x}_{k-1}$ of (22), and rearranging give a set of algebraic equations for the calculation of $\mathbf{x}_k, \mathbf{x}_{k-1}, \ldots, \mathbf{x}_{m-1}$. That is

$$\mathbf{x}_k = \mathbf{x}_{k-1} + \frac{1}{2m}[\mathbf{A}\mathbf{x}_k + \mathbf{x}_{k-1}) + \mathbf{B}(\mathbf{x}_0 + \mathbf{h}_{k-1}) + \mathbf{C}(\mathbf{u}_k + \mathbf{u}_{k-1}) + \mathbf{D}(\mathbf{u}_0 + \mathbf{g}_{k-1})]$$

$$\mathbf{x}_{k+1} = \mathbf{x}_k + \frac{1}{2m}[\mathbf{A}(\mathbf{x}_{k+1} + \mathbf{x}_k) + \mathbf{B}(\mathbf{x}_1 + \mathbf{x}_0) + \mathbf{C}(\mathbf{u}_{k+1} + \mathbf{u}_k) + \mathbf{D}(\mathbf{u}_1 + \mathbf{u}_0)]$$

$$\mathbf{x}_{m-1} = \mathbf{x}_{m-2} + \frac{1}{2m}[\mathbf{A}(\mathbf{x}_{m-1} + \mathbf{x}_{m-2}) + \mathbf{B}(\mathbf{x}_{m-k-1} + \mathbf{x}_{m-k-2}) + \mathbf{C}(\mathbf{u}_{m-1} + \mathbf{u}_{m-2}) + \mathbf{D}(\mathbf{u}_{m-k-1} + \mathbf{u}_{m-k-2})] \tag{27}$$

(22) and (27) form the block pulse approximation of the delay systems (14), (14*a–c*). Once, **A**, **B**, **C**, **D**, $\mathbf{h}_i$, $\mathbf{g}_i$, and $\mathbf{u}_i$ are specified, $\mathbf{x}_i$'s can be calculated successively. Consider an example,

$$\begin{bmatrix} \dot{x}_1(t) \\ \dot{x}_2(t) \end{bmatrix} = \begin{bmatrix} 0 & 1 \\ -25 & -5 \end{bmatrix} \begin{bmatrix} x_1(t-1/4) \\ x_2(t-1/4) \end{bmatrix} + \begin{bmatrix} 0 \\ 1 \end{bmatrix} \tag{28}$$

with $x_1(t) = x_2(t) = 0$ for $t \epsilon [-1/4, 0]$. The block pulse approximate solution which is piecewise-constant is shown in Fig. 2 for $m = 8$. The exact solution obtained by time-partition method [6] is also shown in the figure for comparison.

5 Parameter Estimation

The parameter estimation problem is the determination of the $2n^2 + 2nq$ unknown elements in **A**, **B**, **C**, and **D** from (22) and (27) provided $\mathbf{x}(0)$, $\mathbf{x}_i$, $\mathbf{u}_i$, and $\mathbf{g}_i$ are known. (22) and (27) form a set of mn equations. Since m can be chosen arbitrarily, one can have

$$m \ge 2n + 2q \tag{29}$$

Let **y** be a $(2n^2 + 2nq)$ vector formed from the matrices **A**, **B**, **C**, **D**, α be an $mn \times (2n^2 + 2nq)$ matrix formed by the known elements on the right sides of (22) and (27), β be an mn-vector formed by the known elements on the left sides of (22) and (27). Neglecting the mathematical detail, (22) and (27) become [18]

$$\alpha \mathbf{y} = \beta \tag{30}$$

The least squares estimate [19] of **y** is

$$\mathbf{y} = (\alpha' \alpha)^{-1} \alpha' \beta \tag{31}$$

6 Example

The parameters a and b are to be determined for a delay system

$$\dot{x}(t) = ax(t) - bx(t-1/4) + bu(t-1/4) \tag{32}$$

with

$$x(t) = u(t) = 0, \quad -1/4 \le t < 0 \tag{32a}$$
$$x(0) = 0 \tag{32b}$$

The response of $x(t)$ with $u(t) = 1$, $t \ge 0$ is

t	0	1/8	1/4	3/8	1/2	5/8	3/4	7/8	1
$x(t)$	0	0	0	0.2350	0.4423	0.5967	0.6809	0.7119	0.7117

(33)

The values of x_i, the block pulse coefficients of $x(t)$, are obtained by taking the average values at the sampling points. That is

$$x_i = \frac{1}{2}[(x(iT) + x(i+1)T)], \quad i = 0, 1, \ldots, 7 \tag{34}$$

The block pulse approximation of (32) and (32*a*,*b*) is

$$x_0 = \frac{1}{16} a x_0$$

$$x_1 = x_0 + \frac{1}{16}(ax_1 + ax_0)$$

$$x_2 = x_1 + \frac{1}{16}(ax_2 + ax_1 - bx_0 + bu_0) \tag{35}$$

$$x_3 = x_2 + \frac{1}{16}(ax_3 - bx_1 + bu_1 + ax_2 - bx_0 + bu_0)$$

$$\ldots$$

$$x_7 = x_6 + \frac{1}{16}(ax_7 - bx_5 + bu_5 + ax_6 - bx_4 + bu_4)$$

Since $u(t) = 1$ for $t \ge 0$,

$$u_0 = u_1 = \ldots = u_7 = 1 \tag{36}$$

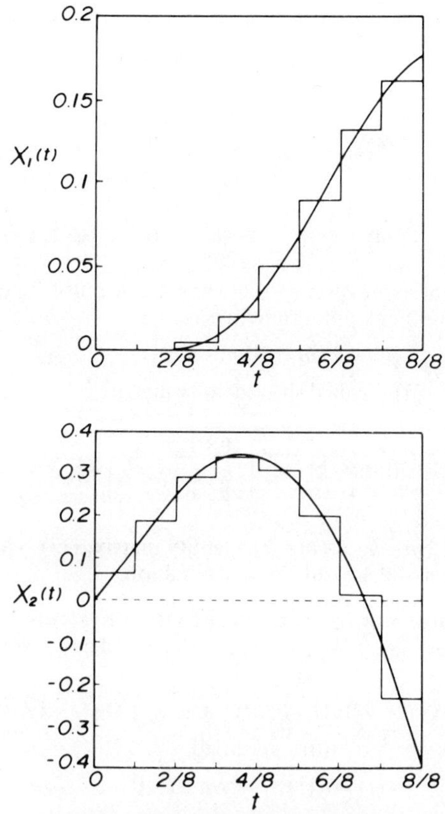

Fig. 2 Response of a delay system

The least squares estimate of a and b from (35)–(36) is

$$a = -0.9708$$
$$b = 1.974 \tag{37}$$

The exact values are $a = -1$, $b = 2$ from which (33) is calculated.

Discussion

From the definition of the block pulse functions, it is found that

$$\psi_i(t)\psi_j(t) = \begin{cases} \psi_i(t), & i=j \\ 0, & i \neq j \end{cases} \tag{38}$$

Hence the product of two time functions, $f(t)$ and $g(t)$, becomes

$$f(t)g(t) = \mathbf{f}'\psi(t)\psi'(t)\mathbf{g} = [f_0g_0, f_1g_1, \ldots, f_{m-1}g_{m-1}]\psi(t) \tag{39}$$

where $\mathbf{g}$ is the block pulse coefficient vector of $g(t)$. Based on (39), the analysis and parameter estimation method of the previous sections can be easily extended to time-varying systems.

Conclusions

Linear delay differential equation systems are approximately represented by a set of linear algebraic equations via block pulse functions. For given input, the linear algebraic equations are used to obtain the piecewise-constant approximate solutions of the systems. On the other hand, for given input and state variables, the unknown parameters can be determined by least squares estimate. The method of parameter estimation via block pulse functions is simple and straightforward and would be extended to other systems.

References

1 Takahashi, Y., Rabins, M. J., and Auslander, D. M., *Control*, Addison-Wesley, Reading, 1970.

2 Koppel, L. B., *Introduction to Control Theory*, Prentice-Hall, Englewood Cliffs, N. J., 1968.

3 Hale, J. K., *Functional Differential Equations*, Springer-Verlag, Berlin, 1971.

4 Halanay, A., *Differential Equations, Stability, Oscillation, Time Lags*, Academic Press, New York, 1966.

5 N-Nagy, F. L., and Al-Tikriti, M. N., "Time-Partition Method for Dynamic Simulation of Control Systems With Transport Delays," *Proc. IEE*, Vol. 120, Jan. 1973, pp. 131–137.

6 Shih, Y. P., "Time-Partition Solutions for Delay Differential Equations," *Trans. of Aeronautical and Astronautical Society*, Republic of China, Nov. 1977, pp. 80–94.

7 Tsoi, A. C., "Explicit Solution of a Class of Delay-Differential Equations," *Int. J. Control*, Vol. 21, Apr. 1975, pp. 39–49.

8 Tsoi, A. C., "Explicit Solution to a Class of Neutral Delay Differential Equations," *Int. J. Control*, Vol. 22, Dec. 1975, pp. 787–799.

9 Tsoi, A. C.,, "Explicit Solution to a Class of Functional Differential Equations," *Int. J. Control*, Vol. 22, Dec. 1975, pp. 869–875.

10 Kwakernaak, H., "Optimal Filtering in Linear Systems With Time Delays," *IEEE Trans. Automatic Control*, Vol. AC-12, Apr. 1967, pp. 169–173.

11 Lindquist, A., "A Theorem on Duality Between Estimation and Control for Linear Stochastic Systems With Time Delay," *J. Math. Analysis and Applications*, Vol. 37, 1972, pp. 56–536.

12 Koivo, A. J., "Determination of a Dynamical Model for Time-Lag Systems Using a Second-Order Method," *Mathematical Bioscience*, Vol. 7, 1970, pp. 15–25.

13 Koivo, A. J., and Stoller, R. L., "Least-Squares Estimator for Nonlinear Systems With Transport Delay," ASME JOURNAL OF DYNAMIC SYSTEMS, MEASUREMENT, AND CONTROL, Vol. 96, Sept. 1974, pp. 301–306.

14 Gabay, E., and Merhav, S. J., "Identification of Linear Systems With Time Delay Operating in a Closed Loop in the Presence of Noise," *IEEE Trans. on Automatic Control, Vol. AC-21*, Oct. 1976, pp. 711–716.

15 Bellman, R., Kagiwada, H. H., and Kalaba, R. E., "Quasilinearization and the Estimation of Time Lags," *Mathematical Biosciences*, Vol. 1, 1967, pp. 39–44.

16 Chen, W. L., and Shih, Y. P., "Shift Walsh Matrix and Delay-Differential Equations," *IEEE Trans. on Automatic Control*, Vol. AC-23, Dec. 1978, pp. 1023–1028.

17 Chen, C. F., Tsay, T,. Y., and Wu, T. T., "Walsh Operational Matrices for Fractional Calculus and Their Application to Distributed Systems," *J. Franklin Institute*, Vol. 303, Mar. 1977, pp. 267–284.

18 Lancaster, P., *Theory of Matrices*, Academic Press, New York, 1969.

19 Box, G. E. P., and Jenkins, G. M., *Time Series Analysis, Forecasting and Control*, Revised Edition, Holden-Day, San Francisco, 1976.

26

Reprinted from *Int. J. Syst. Sci.* **12:**635-642 (1981)

Linear feedback system identification via block-pulse functions

C. P. KWONG† and C. F. CHEN†

Given a continuous, linear, feedback control system, the impulse response of the unknown plant is identified from the known measurement function in the feedback path and the given input and output records. The problem is approached via block-pulse approximation of the continuous-time functions and the algorithm-type solution for integral equations. The method is easily adapted to digital computations.

1. Introduction

System identification is a common problem encountered in the design of control systems. The unknown components, usually the plant under control, are assumed to be described satisfactorily by their respective models. Then the problem of identification is the characterization of the assumed model based on some observations or measurements. It is well known that one may set up more than one model for a dynamic system and in control system design the choice of the most suitable one depends heavily on the design method being used. For example, in linear control problems, a parametric model such as a state model is usually adopted in ' modern ' design while in ' classical ' design a non-parametric model such as an impulse-response function is more appropriate (Eykhoff 1974). In this paper, we present a method to identify an unknown plant modelled by an impulse-response function. The situation is depicted in Fig. 1. A feedback control system is formed with $g(t)$ as feed-forward impulse-response function and $h(t)$ as feedback impulse-response function ; given $h(t)$, input $x(t)$ and output $y(t)$, it is required to find $g(t)$ without breaking the feedback loop. This is a typical identification problem in control. The system under consideration is continuous. However, in our proposed method, we opproximate the continuous-time functions with a set of piecewise constant functions known as the ' block-pulse functions ' (Harmuth 1969). The approximation enables the use of digital computation in the identification process, which is the main advantage of this approach.

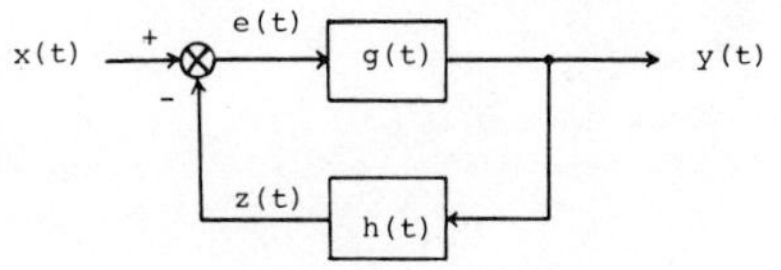

Figure 1. The main problem. Unknown plant modelled by impulse-response function.

Received 24 October 1980.
† Department of Electronics, The Chinese University of Hong Kong, Shatin, N.T., Hong Kong.

In § 2, we review the theory of block-pulse function approximation. In § 3, we derive the convolution integral and its reverse process of 'deconvolution' in terms of our block-pulse formulation. Finally in § 4 we tackle our main problem of Fig. 1 using the basic results obtained in the previous sections.

2. Review of block-pulse function approximation

The block-pulse functions $\{\phi_i(t)\}$, $i=1, 2, \dots, m$ form a set of orthogonal functions defined over a time interval $t \in [0, T]$:

$$\left.\begin{aligned} &\phi_1(t)=1 \quad \text{for } t\in[0, T/m], \\ &\phi_i(t)=\begin{cases} 1 & \text{for } t\in((i-1)T/m,\ iT/m] \\ 0 & \text{elsewhere} \end{cases} \\ &\qquad \text{for } i=2, 3, \dots, m \end{aligned}\right\} \tag{1}$$

Thus, as shown in Fig. 2, the m-order block-pulse functions are time functions having a unit height and T/m width. By using the orthogonal property that

$$\int_0^T \phi_i(t)\phi_j(t)\, dt = \begin{cases} T/m & \text{if } i=j \\ 0 & \text{if } i\neq j \end{cases} \tag{2}$$

we can approximate a time function $f(t)$ by a finite Fourier series

$$f(t) \simeq \sum_{i=1}^{m} a_i \phi_i(t) \tag{3}$$

provided that $f(t)$ is absolutely integrable. The ith Fourier coefficient a_i is determined by

$$a_i = \frac{m}{T} \int_{(i-1)T/m}^{iT/m} f(t)\, dt \quad \text{for } i=1, 2, \dots, m \tag{4}$$

such that the mean square error of approximation is minimized. In practice, the integral in (4) can be approximated by

$$a_i \simeq \tfrac{1}{2}[f((i-1)T/m)+f(iT/m)] \quad \text{for } i=1, 2, \dots, m \tag{5}$$

if $f(t)$ is continuous and sufficiently smooth over the interval $t\in[(i-1)T/m, iT/m]$. This can usually be ensured if m is taken to be sufficiently large.

Harmuth (1969) was the first to introduce the block-pulse functions to electrical engineers. However, not until very recently have the functions found application in the areas of systems and control. In obtaining the Walsh operational matrices for fractional calculus, Chen *et al.* (1977) derived an integration matrix for the block-pulse functions and applied the result to the solution of distributed systems. At about the same time Gopalsami and Deekshatulu (1976) and Sannuti (1977) applied the block-pulse functions to the synthesis and the solution of linear and non-linear problems. Since then the block-pulse functions have received increasing attention for other possible applications (Rao and Srinivasan 1978, Rao and Rao 1979).

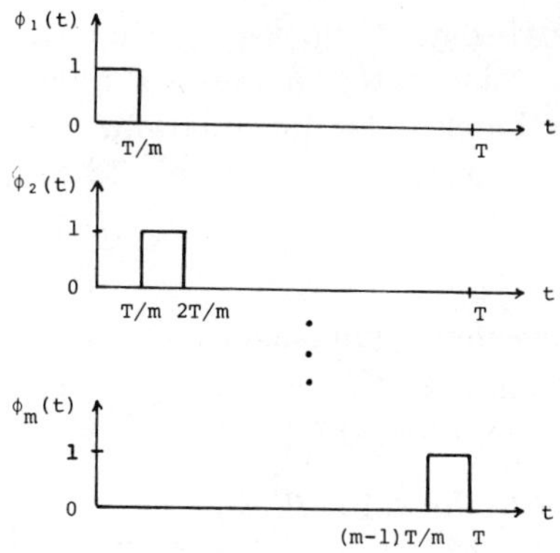

Figure 2. Block-pulse functions.

3. Convolution and deconvolution

First of all, we state clearly our basic assumptions for the problem of Fig. 1:

(1) The single-input single-output plant and measurement can be modelled by their respective impulse-response functions. This implies that the systems are linear and causal.

(2) The initial conditions are zeros. The input is applied at $t = 0$.

(3) The systems are time-invariant.

(4) The time interval considered is $t \in [0, T]$ where T is finite. All the time functions, including $g(t)$ and $h(t)$, are absolutely integrable over $t \in [0, T]$. It follows that each of the time-functions can be approximated by a block-pulse series as given by (3).

Referring to Fig. 1, the output $y(t)$ is given by the convolution integral

$$y(t) = \int_0^t e(\lambda) g(t - \lambda)\, d\lambda \tag{6}$$

We approximate $e(\lambda)$ and $g(\lambda)$ by the block-pulse series:

$$e(\lambda) \simeq \sum_{i=1}^{m} c_i \phi_i(\lambda) \tag{7}$$

$$g(\lambda) \simeq \sum_{i=1}^{m} d_i \phi_i(\lambda) \tag{8}$$

Pictorially, the block-pulse approximated waveforms can be viewed as being composed of consecutive weighted rectangular blocks as shown in Fig. 3 (*a*) and (*b*). We shall derive the solution of the convolution integral using this interpretation. It is easy to see that Fig. 3 (*c*) and (*d*) are respectively the approximated functions $g(-\lambda)$ and $g(t - \lambda)$.

The idea of developing the solution is as follows: We first consider the result of convolving two single blocks. Secondly, we derive the block-pulse coefficients for the block-pulse approximation of the result of convolving a single block with m blocks. Finally, the result of convolving two m-block

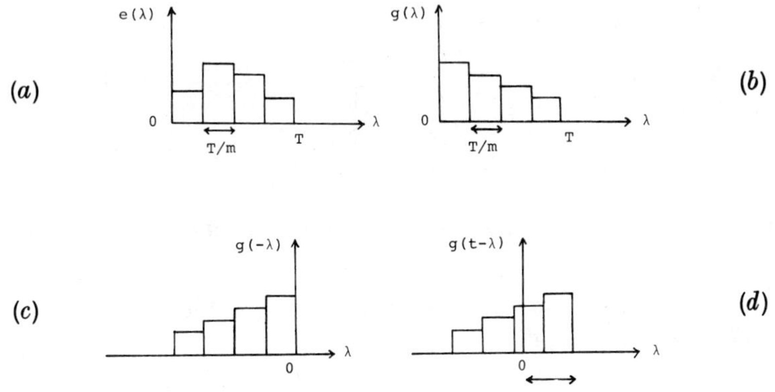

Figure 3. Block-pulse approximated waveforms.

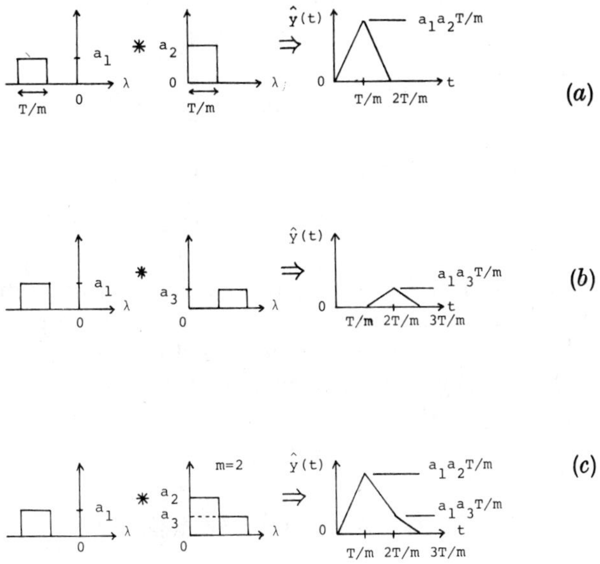

Figure 4. Convolution of block-pulse functions. Asterisk denotes convolution.

functions is obtained by the principle of superposition and by noting the difference in delay. The following are the steps of the development :

Step 1. Fig. 4 (*a*) and (*b*) depict the result of convolving two single blocks, which is a triangular waveform. Note that the convolution obtained is exact.

Step 2. By the use of the superposition theorem and the result of Step 1, the convolution of a single block with m consecutive blocks is shown in Fig. 4 (*c*). Then, the block-pulse approximation of the resulting piecewise linear waveform can be easily derived as shown in Fig. 5.

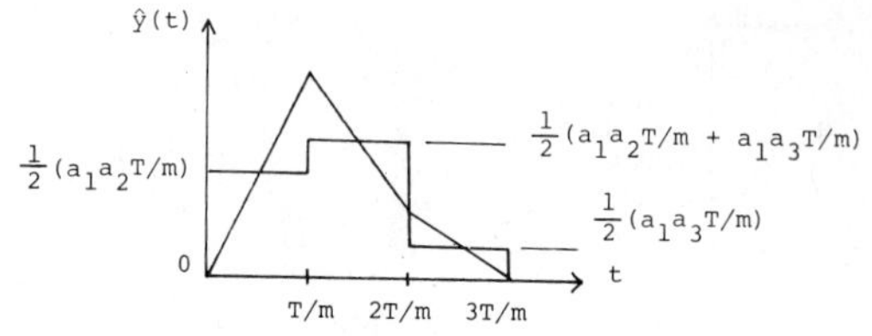

Figure 5. Block-pulse approximation of the piecewise linear waveform of Fig. 4 (*c*).

Step 3. Consider the convolution of the two m-block functions given in (7) and (8), we define

$$C=[c_1\ c_2\ \dots\ c_m] \tag{9}$$

$$D=[d_1\ d_2\ \dots\ d_m] \tag{10}$$

We also define the block-pulse coefficients for the piecewise linear waveform $\hat{y}(t)$ in Fig. 5 by

$$\hat{R}=[\hat{r}_1\ \hat{r}_2\ \dots\ \hat{r}_m] \tag{11}$$

i.e., we let

$$\hat{y}(t)\simeq \sum_{i=1}^{m} \hat{r}_i\phi_i(t) \tag{12}$$

Note that we write (12) rather than

$$y(t)\simeq \sum_{i=1}^{m} r_i\phi_i(t) \tag{12'}$$

because $y(t)$ and $\hat{y}(t)$ are different and we are representing $\hat{y}(t)$ as a block-pulse series. Of course $\hat{y}(t)$ is a close approximation of $y(t)$ for large m. From the results obtained in Steps 1 and 2, and taking the different delay for individual blocks into account, we have, by the superposition theorem,

$$\left.\begin{aligned} \hat{r}_1&=d_1c_1(\tfrac{1}{2}T/m)\\ \hat{r}_2&=[d_1(c_1+c_2)+d_2c_1](\tfrac{1}{2}T/m)\\ &\vdots\\ \hat{r}_m&=[d_1(c_{m-1}+c_m)+\ \dots\ +d_mc_1](\tfrac{1}{2}T/m) \end{aligned}\right\} \tag{13}$$

Define

$$Q_c=\begin{bmatrix} c_1 & c_1+c_2 & \cdots & c_{m-1}+c_m \\ & c_1 & \cdots & c_{m-2}+c_{m-1} \\ & & \ddots & \vdots \\ & \bigcirc & & c_1 \end{bmatrix} \tag{14}$$

to be the ' convolution matrix ', we have

$$\hat{R}=\tfrac{1}{2}\frac{T}{m}DQ_c \tag{15}$$

which is the solution of the block-pulse coefficients for $\hat{y}(t)$.

Suppose in (6), $e(t)$ and $y(t)$ are given and approximated by their respective block-pulse series. Then, from (15), the block-pulse coefficients for $g(t)$ are simply

$$D = \frac{2m}{T} RQ_c^{-1} \tag{16}$$

where $R = [r_1 \ r_2 \dots r_m]$ is the block-pulse coefficient vector for $y(t)$, D and Q_c are defined as before. Since Q_c is an upper triangular matrix with diagonal elements c_1, Q_c^{-1} always exists as long as $c_1 \neq 0$. As (15) is a process of convolution, (16) is the corresponding ' deconvolution ' process.

4. Identification of $g(t)$ for the closed-loop system

In § 3, we have derived the identification algorithm for $g(t)$ in the open-loop situation. In this section, we perform the identification with the feedback loop closed, which is the main problem posed in this paper.

Referring to Fig. 1, we have the following integral equation which describes the feedback system :

$$y(t) = \int_0^t g(t-\lambda)[x(\lambda) - \int_0^\lambda h(\lambda-\tau)y(\tau)\,d\tau]\,d\lambda \tag{17}$$

It is required to solve for $g(t)$ given $x(t)$, $y(t)$ and $h(t)$.

The four time functions $g(\lambda)$, $x(\lambda)$, $y(\tau)$, $h(\tau)$ are approximated by their respective block-pulse functions :

$$\left.\begin{aligned} g(\lambda) &\simeq \sum_{i=1}^m d_i\phi_i(\lambda) \\ x(\lambda) &\simeq \sum_{i=1}^m k_i\phi_i(\lambda) \\ y(\tau) &\simeq \sum_{i=1}^m r_i\phi_i(\tau) \\ h(\tau) &\simeq \sum_{i=1}^m l_i\phi_i(\tau) \end{aligned}\right\} \tag{18}$$

let

$$\begin{aligned} z(\lambda) &= \int_0^\lambda h(\lambda-\tau)y(\tau)\,d\tau \\ &\simeq \sum_{i=1}^m s_i\phi_i(\lambda) \end{aligned} \tag{19}$$

Then, from the result given in § 3, we have

$$S = [s_1 \ s_2 \dots s_m]$$

given by

$$S = \tfrac{1}{2}\frac{T}{m} LQ_r \tag{20}$$

where $L=[l_1\ l_2 \dots l_m]$ and

$$Q_r = \begin{bmatrix} r_1 & r_1+r_2 & \cdots & r_{m-1}+r_m \\ & r_1 & \cdots & r_{m-2}+r_{m-1} \\ & & \ddots & \vdots \\ \bigcirc & & & r_1 \end{bmatrix} \tag{21}$$

From Fig. 1, we have

$$\begin{aligned} e(\lambda) &= x(\lambda) - z(\lambda) \\ &\simeq \sum_{i=1}^{m} k_i \phi_i(\lambda) - \sum_{i=1}^{m} s_i \phi_i(\lambda) \\ &= \sum_{i=1}^{m} (k_i - s_i) \phi_i(\lambda) \end{aligned} \tag{22}$$

From (7), we have

$$c_i = k_i - s_i \quad \text{for } i = 1, 2, \dots, m \tag{23}$$

Finally, from the result given in § 3, we have

$$D = \frac{2m}{T} R Q_c^{-1} \tag{24}$$

which is (16). In summary, the algorithm to compute the block-pulse coefficient vector for $g(t)$ is of three steps, i.e. (20)→(23)→(24).

Example

We have a feedback control system given in Fig. 1 with $4/s$ as the measurement transfer function. The input $x(t)$ is a unit-step function, the output $y(t)$ can be measured and is approximated by a block-pulse series. Using $m=16$, $T=1$, we obtain the following vectors :

$$K = [k_1\ k_2 \dots k_m] = [\underbrace{1\ 1 \dots 1}_{16}]$$

$$L = [\underbrace{4\ 4 \dots 4}_{16}]$$

$$\begin{aligned} R = [&0{\cdot}05501\ 0{\cdot}15135\ 0{\cdot}22221\ 0{\cdot}27126 \\ &0{\cdot}30198\ 0{\cdot}31758\ 0{\cdot}32097\ 0{\cdot}31476 \\ &0{\cdot}30124\ 0{\cdot}28241\ 0{\cdot}25995\ 0{\cdot}23529 \\ &0{\cdot}20960\ 0{\cdot}18381\ 0{\cdot}15865\ 0{\cdot}13469] \end{aligned}$$

Applying the algorithms given by (20), (23) and (24), we have the solution

$$D = [1{\cdot}77251\ 1{\cdot}37776\ 1{\cdot}07063\ 0{\cdot}83164$$
$$0{\cdot}64628\ 0{\cdot}50179\ 0{\cdot}38968\ 0{\cdot}30236$$
$$0{\cdot}23446\ 0{\cdot}18200\ 0{\cdot}14044\ 0{\cdot}10915$$
$$0{\cdot}08389\ 0{\cdot}06498\ 0{\cdot}04945\ 0{\cdot}03869]$$

In fact, $g(t) = 2\exp(-4t)$.

If we define a vector W whose components $w_1, w_2, \ldots, w_m$ are the exact solution at $t = \frac{1}{2}(2n-1)T/m$, $n = 1, 2, \ldots, m$, we can define a measurement of accuracy of approximation as :

Average percentage error

$$\text{Average percentage error} = \left\{\frac{1}{m}\sum_{i=1}^{m}\frac{|w_i - d_i|}{|w_i|}\right\} \times 100\%$$

We found that, for this example, the average percentage error was 2·41%.

5. Conclusion

In this paper, we have introduced the block-pulse functions as a mean to approximate time functions. The convolution and deconvolution algorithms are derived and the results are applied to an identification problem commonly encountered in control systems. The solution is of algorithm type which is easily adapted to digital computations.

References

Chen, C. F., Tsay, Y. T., and Wu, T. T., 1977, *Franklin Inst.*, **303,** 267.
Eykhoff, P., 1974, *System Identification* (New York : Wiley).
Gopalsami, N., and Deekshatulu, B. L., 1976, *Proc. Instn elect. Engrs*, **123,** 461.
Harmuth, H. F., 1969, *I.E.E.E. Spectrum*, **6,** 82.
Sannuti, P., 1977, *Proc. Instn. elect. Engrs*, **124,** 569.
Rao, V. P., and Rao, K. R., 1979, *I.E.E.E. Trans. autom. Control*, **24,** 372.
Rao, G. P., and Srinivasan, T., 1978, *Proc. Instn. elect. Engrs*, **125,** 1064.

Reprinted with permission from *IEEE Trans. Autom. Control* **AC-24:**372–374 (1979)

Optimal Feedback Control Via Block-Pulse Functions

V. PURNACHANDRA RAO AND K. RANGANATHA RAO

***Abstract*—Using block-pulse functions, a method is presented to determine the piecewise constant feedback controls for a finite linear optimal control problem of a power system. The method is simple and computationally advantageous.**

Introduction

It is known that the optimal control of a linear system with respect to a quadratic performance criteria over a finite interval of time often involves the determination of time-varying feedback gains. Walsh functions have been used for determining piecewise constant feedback gains [1], [2]. Recently, a method based on block-pulse functions for numerical integration of a system of differential equations were presented [3]. The block-pulse functions can be used for determining the piecewise constant gains for dynamic systems with certain computational advantages.

This paper presents the results of piecewise constant feedback gains for a simple power system using the method of block-pulse functions. Taking the average for each pair of consecutive values of the piecewise constant gains obtained for m-segments for a finite time and continuing this process of taking averages till we obtain for two subintervals, the set of gains corresponding to the first subinterval were found to be same as the gains obtained for an infinite-time linear optimal control problem.

Problem Formulation

A simple model of a power system, consisting of a single generator connected to an infinite bus through a transmission line is considered. A simplified third-order model, for studying dynamic performance with optimal feedback controls for excitation, can be obtained as in [4], [5].

$$\begin{aligned}\dot{x}_1&=x_2\\ \dot{x}_2&=a_1-a_2\sin x_1\cdot x_3-a_3\sin 2x_1\\ \dot{x}_3&=u-a_4x_3+a_5\cos x_1\end{aligned}\tag{1}$$

where $x_1=\delta$, $x_2=p\delta$, $x_3=E_q'$ and

$$a_1=\frac{P_i}{M_i},\qquad a_2=\frac{V}{M_i(x_d'+x_e)},\qquad a_3=\frac{(x_d'-x_q)V^2}{2M_i(x_d'+x_e)(x_q+x_e)},$$

$$a_4=\frac{(x_d+x_e)}{T_{d0}'(x_d'+x_e)},\quad a_5=\frac{(x_d-x_d')V}{T_{d0}'(x_d'+x_e)},\quad u=\frac{E_{ex}}{T_{d0}'}.$$

x_e is the external reactance to the synchronous machine and p_i is the prime-mover input. u is the control variable corresponding to excitation.

For the study of transients involving small disturbances, the original nonlinear system (1) can be linearized about the operating point, giving rise to a linear vector matrix equation

$$\Delta\dot{x}=A\Delta x+B\Delta u\tag{2}$$

where $\Delta x=[\Delta\delta,\Delta\omega,\Delta E_q']'$ and A and B are matrices of appropriate dimensions.

It is assumed that the transients in the system states Δx's are to be minimized within a finite-time interval t_f s. The quadratic cost function

$$J=\int_0^{t_f}(\Delta x'Q\Delta x+\Delta u'R\Delta u)\,dt\tag{3}$$

is minimized using the optimal control given by

$$\Delta u^*=R^{-1}B'P(t)\tag{4}$$

where $P(t)$ satisfies the following canonical equation:

$$\begin{bmatrix} \Delta\dot{x} \\ \dot{P} \end{bmatrix} = \begin{bmatrix} A & BR^{-1}B' \\ Q & -A' \end{bmatrix} \begin{bmatrix} \Delta x \\ P \end{bmatrix} \tag{5}$$

and the boundary conditions are specified as

$$\Delta x(0) = \Delta x_0 \tag{6}$$

$$P(t_f) = 0. \tag{7}$$

Equation (7) is the transversality condition. It is convenient to change the independent variable by defining

$$T = t_f - t. \tag{8}$$

Then (5) becomes

$$\begin{bmatrix} \Delta\dot{x}(T) \\ \dot{P}(T) \end{bmatrix} = \begin{bmatrix} -A & -BR^{-1}B' \\ -Q & A' \end{bmatrix} \begin{bmatrix} \Delta x(T) \\ P(T) \end{bmatrix} = -M \begin{bmatrix} \Delta x(T) \\ P(T) \end{bmatrix}. \tag{9}$$

The transition matrix of (9) is

$$e^{-Mt} = \begin{bmatrix} \eta_{11}(T) & \eta_{12}(T) \\ \eta_{21}(T) & \eta_{22}(T) \end{bmatrix} \tag{10}$$

since $P(T=0)=0$, the solution of (9) can be written as

$$\Delta x(T) = \eta_{11}(T)\Delta x(T=0) \tag{11}$$

$$P(T) = \eta_{21}(T)\Delta x(T=0). \tag{12}$$

From (11) we have

$$\Delta x(T=0) = \eta_{11}^{-1}(T)\Delta x(T) \tag{11a}$$

substituting (11a) into (12)

$$P(T) = \eta_{21}(T)\eta_{11}^{-1}(T)\Delta x(T). \tag{12a}$$

Then the optimal control is reduced to

$$\begin{aligned} \Delta u^*(t) &= R^{-1}B'\eta_{21}(t_f - t)\eta_{11}^{-1}(t_f - t)\Delta x(t_f - t) \\ &= K(t_f - t)\Delta x(t_f - t) \end{aligned} \tag{13}$$

where $K(t_f - t)$ is the optimal feedback gain matrix.

Block-Pulse Functions Solution to the Problem

For a piecewise constant approximate solution of a system of differential equations, a set of block-pulse functions $\phi_i(t)$ defined on a unit interval $(0,1)$ as [3]

$$\phi_i(t) = \begin{cases} 1, & \text{for } \dfrac{i-1}{m} < t < \dfrac{i}{m} \\ 0, & \text{otherwise.} \end{cases} \tag{14}$$

This set of functions is concisely described by an m-vector $\phi(t)$ with $\phi_i(t)$ as its ith component. A function which is integrable in $(0,1)$ can be approximated as

$$f(t) = \sum_{i=1}^{m} a_i\phi_i(t) \tag{15}$$

where

$$\begin{aligned} a_i &= m\int_{(i-1)/m}^{i/m} f(t)\,dt \\ &= \text{average value of } f(t) \text{ over the interval } \frac{i-1}{m} < t < \frac{i}{m}. \end{aligned} \tag{16}$$

The approximation in (15) gives the best m-segment piecewise constant approximation of $f(t)$ and is unique.

For solving the optimal control problem in (2)–(4) by the block-pulse function approach, the variables $\Delta\dot{x}$, $\dot{P}$, P, Δx in (9) will be expanded into block-pulse functions as

$$\begin{bmatrix} \Delta\dot{x}(t) \\ \dot{P}(t) \end{bmatrix} = \sum_{i=1}^{m} C_i\phi_i(t) = C\phi(t) \tag{17a}$$

$$\begin{bmatrix} \Delta x(t) \\ P(t) \end{bmatrix} = \sum_{i=1}^{m} D_i\phi_i(t) = D\phi(t). \tag{17b}$$

C_i and D_i are n-vectors and form the ith columns of the $n \times m$ matrices C, D, respectively. The integral $\int_0^t \phi(t)\,dt$ can be expanded into a set of basic functions $\phi(t)$ as

$$\int_0^t \phi(t)\,dt = P_b\phi(t) \tag{18}$$

where P_b is an $m \times m$ operational matrix [3].

Determining a piecewise constant approximation to the solution of (9) is tantamount to determining the matrices C and D. The following identity is used to get an interrelationship between the matrices:

$$\int_0^t \begin{bmatrix} \Delta\dot{x} \\ \dot{P} \end{bmatrix} dt = \begin{bmatrix} \Delta x(t) \\ P(t) \end{bmatrix} - \begin{bmatrix} \Delta x(0) \\ P(0) \end{bmatrix}. \tag{19}$$

Substituting the relations (17) and (18) in (19) we have

$$CP_b\phi(t) = D\phi(t) - \Delta\underline{x}(0). \tag{20}$$

$\Delta\underline{x}(0)$ is an $n \times m$ matrix each column of which is $\Delta x(0)$. Equating like coefficients of $\phi_i(t)$ in (20), we get

$$[CP_b]_i = D_i - \begin{bmatrix} \Delta x_0 \\ P_0 \end{bmatrix}. \tag{21}$$

Here $[CP_b]_i$ is the ith column of CP_b and in view of P_b value is given by [3]

$$[CP_b]_i = \frac{1}{m}\sum_{j=1}^{i-1} C_j + \frac{1}{2m}C_i. \tag{22}$$

The above equation is written as a recurrence equation

$$\begin{aligned} [CP_b]_i &= \frac{1}{2m}C_1 \\ [CP_b]_{i+1} &= [CP_b]_i + \frac{1}{2m}(C_{i+1} + C_i) \end{aligned} \tag{23}$$

for all $i > 1$. Combining (21) and (22) gives

$$\begin{aligned} D_1 &= \frac{1}{2m}C_1 + \begin{bmatrix} \Delta x_0 \\ P_0 \end{bmatrix} \\ D_{i+1} &= D_i + \frac{1}{2m}(C_{i+1} + C_i). \end{aligned} \tag{24}$$

Equation (24) is a basic identity and is the result of (19), in which the approximation of (18) is used. It does not depend on the dynamic equation (2).

The feedback gains for the optimal control problem equations (2)–(4) can be obtained from (13) when the η_{11} and η_{21} components of the transition matrix for (9) are determined.

The solution of (9) can be written as

$$\begin{bmatrix} \Delta x(T) \\ P(T) \end{bmatrix} = \begin{bmatrix} \eta_{11}(T) & \eta_{12}(T) \\ \eta_{21}(T) & \eta_{22}(T) \end{bmatrix} \begin{bmatrix} \Delta x(T=0) \\ P(T=0) \end{bmatrix}. \tag{25}$$

Applying block-pulse function relations in (17) to the system in (9), we have

$$\begin{bmatrix} \Delta\dot{x}(T) \\ \dot{P}(T) \end{bmatrix} = -M \begin{bmatrix} \Delta x(T) \\ P(T) \end{bmatrix} \tag{26}$$

$$C\phi(T) = -MD\phi(T). \tag{27}$$

Equating like coefficients of $\phi(T)$, we get

$$C_i = -MD_i, \qquad i > 1. \tag{28}$$

Substituting (25) in (24), we have

$$D_1 = \frac{1}{2m}(-MD_1) + \begin{bmatrix} \Delta x(T=0) \\ P(T=0) \end{bmatrix}.$$

Rearranging, we get

$$D_1 = \left[I - \frac{-M}{2m}\right]^{-1} \begin{bmatrix} \Delta x(T=0) \\ P(T=0) \end{bmatrix} = \alpha \begin{bmatrix} \Delta x(T=0) \\ P(T=0) \end{bmatrix} \tag{29}$$

and

$$D_{i+1} = \beta D_i \tag{29b}$$

for all $i > 1$. Where $\beta = \alpha[I + (-M/2m)]$, $\alpha = [I - (-M/2m)]^{-1}$ from (29),

$$D_1 = \alpha \begin{bmatrix} \Delta x(T=0) \\ P(T=0) \end{bmatrix}$$

$$D_2 = \beta D_1 = (\beta\alpha) \begin{bmatrix} \Delta x(T=0) \\ P(T=0) \end{bmatrix}$$

$$\vdots \qquad \vdots$$

$$D_m = \beta D_{m-1} = (\beta^{m-1}\alpha) \begin{bmatrix} \Delta x(T=0) \\ P(T=0) \end{bmatrix}.$$

Comparing (25) and (17b), we can determine η_{11} and η_{21} for m intervals from $D_1, D_2, \cdots, D_m$.

Results

A simple power system problem with the following numerical data has been chosen for the study. $x_d = 2.0$, $x_d' = 0.232$, $x_q = 2.0$, $x_s = 0.18$, $T_{do}' = 1.466$, $H = 5.3$, and $P_e = 0.8 p{\cdot}u$ at $0.8 p{\cdot}f$ (lag). The A, B, Q, and R matrices in (2)–(5) are

$$A = \begin{bmatrix} 0 & 1 & 0 \\ -52.458 & 0 & -43.367 \\ -0.409 & 0 & -0.837 \end{bmatrix}; \quad B = \begin{bmatrix} 0 \\ 0 \\ 1 \end{bmatrix};$$

$$Q = \text{diag}(500, 50, 500) \text{ and } R = 1.$$

The time $t_f = 1$ s and the number of segments $m = 32$. The feedback gain constants k_1, k_2, and k_3 in (13) corresponding to the system state variables Δx in (2) are given in Table I.

Taking the average for each pair of consecutive values of the above obtained optimal gains with 32 subintervals, we obtain a new set of piecewise feedback gains for 16 subintervals. Continuing this procedure, we obtain constant feedback gains for 6, 4, 2, and 1 subintervals.

It is found that the gains corresponding to the first subinterval in the above procedure for 2 subintervals, are same as the gains obtained for an infinite-time linear optimal control problem from the Riccati matrix equation solution [5].

Conclusions

Block-pulse functions have been applied to determine piecewise constant feedback gains for a simple power system problem. Taking the averages for each pair of consecutive values of piecewise constant gains, the optimal feedback gains for the infinite control problem can be obtained.

TABLE I

	Feedback gains			Average values of gains		
m	k_1	k_2	k_3	k_1	k_2	k_3
1	—	—	—			
2	−1.94	1.95	−19.03			
3	−7.10	3.55	−25.01			
4	−13.06	4.61	−26.0			
5	−17.36	5.14	−29.41			
6	−19.38	5.36	−30.0			
7	−19.45	5.52	−30.3			
8	−18.17	5.71	−30.56	−11.14	5.427	−28.0
9	−16.16	5.97	−30.88			
10	−13.93	6.27	−31.27			
11	−11.82	6.58	−31.7			
12	−10.04	6.87	−32.09			
13	−8.66	7.10	−32.43			
14	−7.64	7.28	−32.7			
15	−6.95	7.42	−32.91			
16	−6.51	7.51	−33.05			
17	−6.24	7.57	−33.14			
18	−6.06	7.61	−33.21			
19	−5.94	7.63	−33.25			
20	−5.91	7.64	−33.27			
21	−5.87	7.65	−33.29			
22	−5.66	7.65	−33.29			
23	−5.80	7.67	−33.32			
24	−5.83	7.66	−33.30	−5.88	7.655	−33.29
25	−5.83	7.66	−33.30			
26	−5.76	7.69	−33.34			
27	−5.77	7.69	−33.36			
28	−5.81	7.67	−33.33			
29	−5.84	7.66	−33.32			
30	−5.87	7.65	−33.26			
31	−5.87	7.70	−33.39			
32	−5.87	7.69	−33.36			

Optimal gains for infinite control problem: $K = [-5.825, 7.666, -33.83]$.

For the finite-time optimal control problem solution, the method should prove to be computationally economical compared to that obtained from the Riccati matrix equation.

Nomenclature

x_d, x_q = direct and quadrature axis reactances
M_i = moment of inertia
T_{do}' = transient open circuit time constant
x_e = transmission line reactance
E_{ex} = excitation voltage
E_q' = voltage proportional to field flux linkages
δ = load angle
V = infinite bus voltage
P_i = primemover input

References

[1] C. F. Chen and C. H. Hsias, "Design of piecewise constant gains for optimal control via Walsh functions," *IEEE Trans. Automat. Contr.*, vol. AC-20, pp. 596–603, Oct. 1975.

[2] ——, "Time domain synthesis via Walsh functions," *Proc. Inst. Elec. Eng.*, vol. 122, pp. 565–570, May 1975.

[3] P. Sannuti, "Analysis and synthesis of dynamic systems via block-pulse functions," *Proc. Inst. Elec. Eng.*, vol. 124, pp. 569–571, June 1977.

[4] V. P. Rao, "Optimal excitation controls to improve power system stability," Ph.D. dissertation, Dep. Elec. Eng., Banaras Hindu Univ., India, Apr. 1977.

[5] V. P. Rao and V. S. S. Rao, "Improvement of power system stability using optimal excitation control via the new approach," *Proc. IEEE* (Lett.), vol. 65, pp. 280–282, Feb. 1977.

Part VII

MISCELLANEOUS PROPERTIES: WALSH-TO-FOURIER CONVERSION

Editor's Comments on Papers 28 Through 31

28 BLACHMAN
Sinusoids versus Walsh Functions

29 SIEMENS and KITAI
A Nonrecursive Equation for the Fourier Transform of a Walsh Function

30 KITAI
Walsh-to-Fourier Spectral Conversion for Periodic Waves

31 KITAI
Synthesis of Periodic Sinusoids from Walsh Waves

The study of the properties of Walsh functions and Walsh transforms and their interrelationships with the conventional sinusoidal functions and transforms has helped to obtain a deep understanding of the advantages and disadvantages achieved when using them in various engineering applications. In this part a representative set of papers dealing with several aspects of Walsh function expansions is reviewed. Yuen (1972) considers a variety of cases in which the Walsh transform of a function can be expressed by means of its derivatives. For both the continuous and discrete data cases, upper bounds are determined, which indicate that a rapidly converging Walsh series with a small number of significant terms can be found. However, if the function to be expanded is discontinuous, or has discontinuous derivatives, these bounds are larger and the speed of convergence of the Walsh series is smaller. For the case of random functions it is concluded that in general the Walsh power spectra are slowly convergent and therefore many Walsh transform components contain approximately equal signal power.

Ahmed, Rao, and Abdussattar (1973) devote their study to the relationship between the cyclic autocorrelation and Walsh-Hadamard transform (WHT). It is shown that the complete set of circular shift invariants of WHT, called the Q-Spectrum, is invariant to the cyclic shift of a periodic sequence $\{X(n)\}$, and that the modified WHT of the cyclic autocorrelation provides the Q-spectrum in a manner analogous to the discrete Fourier transform. Yuen (1975) develops an algorithm by which one can compute the correlations of Walsh functions. Walsh-Paley functions are used instead of the sequency-ordered Walsh functions because of their mathematical simplicity. The correlations of sequency-ordered Walsh functions are found through their Paley orderings, which are determined using the Gray code. A FORTRAN program is also included.

Robinson (1972) derives the recursive relationships between arithmetic and logical autocorrelation functions of a wide sense stationary process. The Fourier (or Walsh) power spectra are computed from the weighted arithmetic (or logical) autocorrelation function with the aid of the fast Fourier transform (FFT) or the fast Walsh transform (FWT). Examples from the image and speech processing fields are provided. A derivation of the natural moments of a stochastic process through their corresponding Walsh characteristic function and the dyadic derivative operator is provided by Maqusi (1973).

In his paper on function approximation Yuen (1975) considers the problem of selecting the coefficients in a finite Walsh series so as to obtain a best approximation of a given function. His paper provides two methods, namely the process of *threshold sampling* and the method of *degrees and subdegrees,* which also includes an error analysis. It is shown that data compression is obtained if terms with sufficiently large subdegrees are omitted. The maximum error obtained by a selected Walsh series is further reduced by modifying the coefficients according to Lawson's algorithm. This is actually a Walsh expansion approximation with minimax errors. The theoretical results are sufficiently verified in a set of computed examples. In the paper by Eris (1978) the relationships between the Rademacher-Walsh spectra of Boolean functions and the spectrum of the Boolean product (AND) and sum (OR) of these functions are studied. These relationships enable one to execute the logical operations AND and OR in the Walsh spectra domain, using the spectra of the functions being logically connected rather than transforming back to the Boolean domain whenever such an operation is required.

Paper 28 reviews the properties of Walsh functions, presents a number of relations between sinusoids and Walsh functions, and investigates the truncation and round-off errors occurring when using Fourier and Walsh expansions. It is shown that for both smooth and discontinuous signals the Walsh series needs more terms than the Fourier series for a given total r.m.s. error. This fact somewhat counterbalances the computational advantages of FWT over FFT.

Papers 29, 30, and 31 are concerned with the Fourier transforms of Walsh functions and with the conversion of the Walsh series expansion of a signal to its Fourier series representation. Paper 29 derives the nonrecursive expression for the two-sided Fourier transform of Walsh functions in Gray code ordering, and develops a useful algorithm for its computation. The Fourier transforms of Walsh functions are needed in the conversion of Walsh series to the corresponding Fourier series. The expression of this paper is derived by regarding each continuous Walsh function as the convolution of the sequence of unit impulses representing the discrete Walsh function over $\theta \epsilon (-\frac{1}{2}, \frac{1}{2})$ with a rectangular pulse of height 1 and width equal to the spacing between the unit impulses.

Paper 30 derives matrix relation for converting truncated Walsh spectra of frequency-limited periodic signals to their corresponding Fourier spectra. According to the sequency-based sampling theorem, if the highest harmonic in the signal to be expanded is F, then one needs $2F$ sequency values plus the direct current value for representing the signal. The matrix expressions of the paper are simplified when the highest-order component in the Walsh spectrum to be converted is 2^n where n is an integer.

In Paper 31 Kitai proposes a hardware set-up for the synthesis of a periodic sinusoid using a truncated Walsh series expansion. This set-up involves a synchronous Walsh array generator, a level-converting device, and a summing circuit.

Robinson (1974) derives an alternative closed-form expression for the Fourier transform of a Walsh function defined on a finite interval. The derivation is made by first determining the closed-form expression for the pulse transfer function of the system, which produces a Walsh function from an elementary rectangular pulse.

Kitai, Renyi, and Vajda (1976) propose the realization of the conversion process from Walsh spectrum to Fourier spectrum using a microprocessor. In this realization the matrix conversion equations derived by Kitai are used. The microprocessor selection is primarily based on the availability of benchmark programs in order to regard the required matrix multiplication as benchmark. A comparison is made of the Fairchild PPS-25, Rockwell PPS-4, Intel MCS-4, and Intel 8080 microprocessors able to perform BCD operations.

Siemens and Kitai provide a digital method for obtaining the Walsh series of a periodic signal and then derive the conversion formulas by which the corresponding Fourier series coefficients are obtained from the Walsh series ones. This indirect way of determining the Fourier series of a signal via its Walsh series has the advantage of making the measurement process of the signal simpler, while the Walsh to Fourier conversion is independent of the measurement process.

Harmuth (1971) explains that a sequency-limited signal represented by a superposition of Walsh functions can be converted to a frequency-limited signal consisting of a superposition of sine-cosine functions (and vice versa) without wasting either frequency or sequency bandwidth. This result is useful for determining at what rate sequency-multiplexed signals can be transmitted through standard frequency band-limited communication channels.

REFERENCES

Ahmed, N., K. R. Rao, and A. L. Abdussatar, 1973, On Cyclic Autocorrelation and the Walsh-Hadamard Transform, *IEEE Trans. Electromagn. Compat.* **EMC-15:**141–146.

Eris, E., 1978, Relationships Between Rademacher-Walsh Spectra of Boolean Functions, *Computers Digital Tech.* **1:**45–48.

Harmuth, H. F., 1971, Conversion of Sequency-Limited Signals into Frequency-Limited Signals and Vice Versa, *IEEE Trans. Inf. Theory* **IT-17:**343–344.

Kitai, R., I. Renyi, and F. Vajda, 1976, Using a Microprocessor in a Walsh-Fourier Spectral Analyzer, *Computer* **9:**27–31.

Maqusi, M., 1973, On Moments and Walsh Characteristic Function, *IEEE Trans. Commun.* **COM-21:**768–770.

Robinson, G. S., 1972, Logical Convolution and Discrete Walsh and Fourier Power Spectra, *IEEE Trans. Audio Electroacoust.* **AU-20:**271–280.

Robinson, G. S., 1974, Fourier Transforms of Walsh Functions, *IEEE Trans. Electromagn. Compat.* **EMC-16:**183–185.

Siemens, K., and R. Kitai, 1969, Digital Walsh-Fourier Analysis of Periodic Waveforms, *IEEE Trans. Instrum. Meas.* **IM-18:**316–321.

Yuen, C. K., 1972, Upper Bounds on Walsh Transforms, *IEEE Trans. Computers* **C-21:** 1273–1280.

Yuen, C. K., 1975, An Algorithm for Computing the Correlation Functions of Walsh Functions, *IEEE Trans. Electromagn. Compat.* **EMC-17:**177–180.

Yuen, C. K., 1975, Function Approximation by Walsh Series, *IEEE Trans. Computers* **C-24:**590–598.

Reprinted with permission from *IEEE Proc.* **62**:346–354 (1974)

Sinusoids versus Walsh Functions

NELSON M. BLACHMAN, FELLOW, IEEE

***Abstract*—In most of the applications contemplated for Walsh functions these binary waveforms would replace the more usual sinusoids, as the fast-Walsh-transform algorithm appears to make them very attractive for many kinds of signal processing. This paper begins with a brief review of the characteristics of Walsh functions and of their applications. Some old and some new interrelations are presented between sinusoids and Walsh functions, but the principal aim of the paper is to investigate the truncation and roundoff errors associated with the use of Fourier and of Walsh series.**

By employing simplifying approximations it is found that, for long samples of smooth signals, far more terms are required in the Walsh-series representation and greater accuracy is required of their coefficients for a given rms total error. Even for discontinuous signals the Walsh series may require substantially more terms, thus counterbalancing the computational advantage of the fast Walsh transform. This relative inefficiency of the Walsh-series representation of long waveforms may explain why it has not proven particularly effective in applications.

I. Introduction and Summary

IT IS A FAMILIAR FACT that any waveform $g(t)$ having a finite energy in the interval $(0, T)$ can be expressed in this interval as a Fourier series

$$g(t) = F_0 + \sqrt{2}\,F_1 \sin\frac{2\pi t}{T} + \sqrt{2}\,F_2 \cos\frac{2\pi t}{T} + \sqrt{2}\,F_3 \sin\frac{4\pi t}{T} + \cdots \tag{1}$$

with coefficients

$$F_0 = \frac{1}{T}\int_0^T g(t)\,dt$$
$$F_1 = \frac{\sqrt{2}}{T}\int_0^T g(t)\sin\frac{2\pi t}{T}\,dt$$
$$F_2 = \frac{\sqrt{2}}{T}\int_0^T g(t)\cos\frac{2\pi t}{T}\,dt$$
$$\cdots \tag{2}$$

In 1923 Joseph L. Walsh devised a new complete set of orthonormal functions [1] in order to demonstrate the convergence properties of expansions like (1). His functions, which we denote by wal (k, t), bear a number of resemblances to the foregoing sines and cosines, appearing to be squared-up versions of them, as shown in Fig. 1.

They take only the values ± 1 and they change sign only when t is a multiple of a power of $\frac{1}{2}$. The index k is the number of sign changes of wal (k, t) inside the unit interval. In terms of these functions, which are explicitly defined in the next section, we have the expansion over the interval $(0, T)$

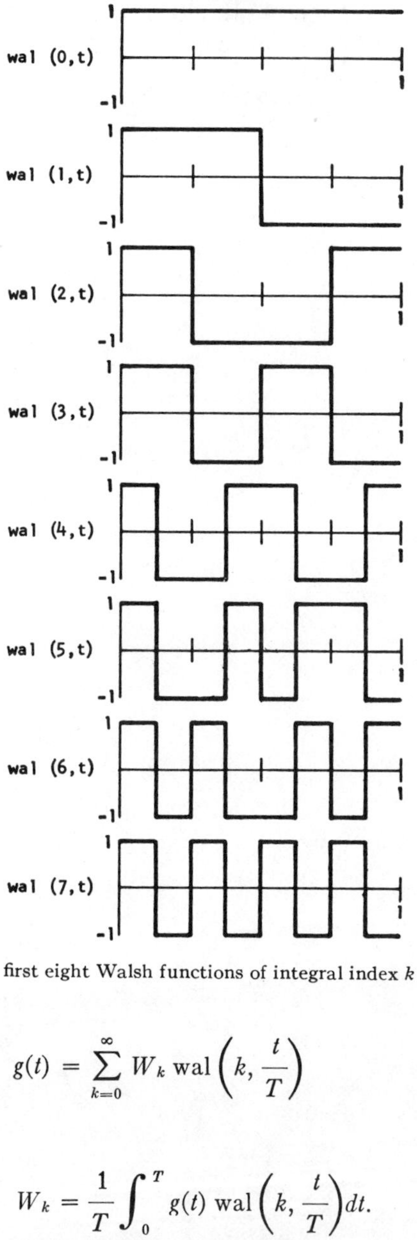

Fig. 1. The first eight Walsh functions of integral index k for $0 \leq t \leq 1$.

$$g(t) = \sum_{k=0}^{\infty} W_k \,\mathrm{wal}\left(k, \frac{t}{T}\right) \tag{3}$$

with

$$W_k = \frac{1}{T}\int_0^T g(t)\,\mathrm{wal}\left(k, \frac{t}{T}\right)dt. \tag{4}$$

The number of sign changes per unit time, $s = k/T$, of any term of (3) will be called its sequency, although some authors use this word for $\frac{1}{2}s$. The Fourier and Walsh coefficients (2) and (4), of course, depend on T.

In the past few years Walsh functions have been suggested for a wide variety of applications involving expansions like (3). As a perusal of the proceedings of Walsh-function symposia [2]–[4] reveals, the coefficients $\{W_k\}$ can serve for the analysis, filtering, and bandwidth reduction of speech, electrocardio- and electroencephalograms, and other signals as well as for the multiplexing of communication channels and

Manuscript received October 5, 1972; revised November 10, 1973.
The author is with GTE Sylvania Electronic Systems, Mountain View, Calif. 94042.

the processing of images for pattern recognition, redundancy reduction, and image enhancement.

These operations are carried out in the sequency domain in much the same way as in the frequency domain. For pattern recognition or waveform analysis, instead of looking in the time domain at $g(t)$, one can search for recognizable configurations of Walsh coefficients W_k. To reduce redundancy and bandwidth one may transmit the Walsh coefficients instead of $g(t)$ itself, omitting those hopefully many coefficients that are nearly always small. For filtering and image enhancement each W_k obtained from (4) is multiplied by an appropriate function of k (and, possibly, added to some linear combination of other coefficients) before being substituted into (3) to yield an improved $g(t)$. In Walsh-function multiplexing the coefficients $\{W_k\}$ of the transmitted signal $g(t)$ represent the information associated with the respective subcarriers $\{\text{wal}(k, t/T)\}$ during the interval $(0, T)$; during successive intervals of length T new numbers $\{W_k\}$ are transmitted by forming new composite signals $g(t)$.

Walsh functions are particularly attractive for such applications because, taking only the values ± 1, they obviate the need for multiplication in (3) and (4). Moreover, as a result of the product identity (13), there exists a fast-Walsh-transform algorithm [5] for computing the Walsh coefficients $\{W_k\}$, and it is accordingly far faster than the fast Fourier transform. Walsh functions have thus been found useful for some of the applications mentioned above, although the advantage obtained from them is significantly inferior to that afforded by Fourier analysis, which requires much more computational effort.

The kth term of (3) has much in common with the kth term of (1). Both have k sign changes inside the interval $(0, T)$, both have the same even or odd symmetry about $t=\frac{1}{2}T$, and there is the same quarter-of-a-period cyclic-shift relation between wal $(k-1, t/T)$ and wal $(k, t/T)$ when k is even that exists between the corresponding sine and cosine (of frequency $k/2T$). The resemblances are fully described in the next section, and the relationship between Fourier and Walsh coefficients is presented in Section III. Section IV deals with the error involved in retaining only a finite number of terms in (1) or (3), and Section V treats the error resulting from a finite accuracy in the representation of the coefficients.

If the rounding off of the coefficients is consistent with the truncation of the series, the total of the two errors is found to have rms value

$$\delta_F = \frac{\sqrt{6}}{\pi\sqrt{N}} g_{\text{rms}} \tag{5}$$

for the N-term Fourier-series representation of a sample of a zero-mean stationary random process $g(t)$ containing only insignificant power at frequencies beyond $N/2T$. The total rms error of the N-term Walsh-series representation with N large enough to satisfy the foregoing condition is found to be

$$\delta_W = \frac{\pi BT}{N} g_{\text{rms}} \tag{6}$$

where B is the rms frequency (in hertz) in the spectrum of $g(t)$.

If, for example, the spectrum of $g(t)$ is flat up to 2 kHz, we have $B=1.15$ kHz. If we want a 1-percent rms total error, we set $\delta=0.01\ g_{\text{rms}}$ and, if $T=1$ s, we find from (5) that $N=6080$ terms are needed in the Fourier series while from (6) $N=362\ 000$ terms are required in the Walsh series. The ratio of the numbers of terms needed in the two series is

$$\frac{N_W}{N_F} = \frac{\pi^3 BT\delta}{6 g_{\text{rms}}} \tag{7}$$

for $\delta_F=\delta_W=\delta$. It would be less unfavorable to the Walsh series if BT were smaller or a smaller rms error were required, but for large time-bandwidth products the greater number of terms needed in the Walsh series will outweigh the speed of the fast Walsh transform.

The reason for the superiority of the Fourier series in this case is considered in Section VI, and in Section VII the case of a discontinuous $g(t)$ is studied. It is found that

$$\delta_F = \sqrt{\frac{4ST}{\pi^2 N}}\, g_{\text{rms}} \tag{8}$$

for a $g(t)$ whose Walsh spectrum is flat up to a sequency S for which ST is a power of 2 and that

$$\delta_W = \sqrt{\frac{2ST}{3N}}\, g_{\text{rms}} \tag{9}$$

is the rms total error for the Walsh series if the mean (in **rms**) includes averaging over a uniform distribution of synchronization errors. If the jumps of $g(t)$ coincide with those of the N Walsh functions in (3), of course, there is no truncation error, but $g(t)$ is then most conveniently represented by its ST successive, uncorrelated, constant values. Comparing (8) with (9), we see that in the absence of synchronization the Walsh series requires $\pi^2/6$ times as many terms as the Fourier series for the same rms error.

This difficulty may be attributed to the fact that after time shifting, a Walsh function generally becomes the sum of an infinite number of Walsh functions while a sinusoid simply turns into the sum of a sine and a cosine of the same frequency. Thus a change in the time scale or a shift of the time origin usually will grossly alter a Walsh spectrum but has only a minor effect on the Fourier spectrum. It also follows that Walsh functions can be used for the multiplexing of communication channels only when all of the channels are in synchronism, as the Walsh functions are otherwise not orthogonal. This form of multiplexing is then equivalent to time division except for the need to mix and separate channels with the aid of Walsh functions or transforms.

Although the Walsh functions in (3) have much in common with the sinusoids in (1), we thus see that their coefficients (4) represent a much less efficient coding of the information in $g(t)$ than do the Fourier coefficients (2) for large time-bandwidth products and even for some discontinuous $g(t)$'s. The discontinuous nature of the Walsh functions may make them difficult to generate and necessitates precise synchronization. These are probably some of the reasons why Walsh functions, although very interesting from a mathematical viewpoint, have not yet met with much success in applications despite the computational convenience of the fast Walsh transform.

II. Walsh Functions

To define the Walsh function wal (k, t) of index $k \geq 0$ and argument t it is first necessary to express the integer k in the form

$$k = \sum_{r=0}^{m-1} 2^r k_r \tag{10}$$

where $\{k_r\}$ are the digits, 0 or 1, of the binary numeral for k, and 2^m is any power of 2 that exceeds k. Then [6] for all t we define

$$\text{wal}(k, t) = \prod_{r=0}^{m-1} \text{sgn}(\cos^{k_r} 2^r \pi t) \tag{11}$$

the signum function sgn x being ± 1 accordingly as $x>0$ or $x<0$. There is no need to worry about its value when $x=0$, as that will happen on a set of t's of measure zero. (For other variants of the Walsh functions see [7].)

The signum function can be described as an ideal limiter, and in (11) it turns the cosine (when $k_r=1$) into a regular symmetric squarewave like the wal $(1, t)$, wal $(2, t)$, and wal $(4, t)$ shown in Fig. 1 for $r=0$, 1, and 2, respectively. When $k_r=0$, the corresponding factor in (11) becomes the constant 1. Thus wal (k, t) is the product of as many such squarewaves as there are 1's in the binary numeral for k. For example, wal $(6, t)=$ wal $(4, t)$ wal $(2, t)$ since $6=4+2$ has the binary numeral 110.

Because $\cos 2^r\pi t$ has its sign changes where t is an odd multiple of 2^{-r-1}, no two factors in (11) change sign at the same time, and so the sign changes of wal (k, t) are the total of those of its squarewave factors. The factor sgn $(\cos 2^r\pi t)$ has 2^r sign changes in the unit interval, and it thus follows from (10) that wal (k, t) has k sign changes per unit time.

From (10) and (11) it can be seen that

$$\text{wal}(2k, t) = \text{wal}(k, 2t) \tag{12}$$

i.e., doubling the index k compresses the time scale in half, as we see in Fig. 1. Note that the argument t of the Walsh function (11) is not confined to the unit interval. Multiplying two Walsh functions together, we similarly find that their product is a Waslh function

$$\text{wal}(h, t)\, \text{wal}(k, t) = \text{wal}(h \oplus k, t) \tag{13}$$

since the square of any squarewave factor in (11) is identically 1. Here $h \oplus k$ denotes the number whose binary numeral has 0's in those positions where the binary numerals for h and k are alike and has 1's where they are different. For example, we have wal $(3, t)$ wal $(6, t)=$ wal $(5, t)$ because $011 \oplus 110 = 101$.

From (4) and (13) it follows that the Walsh transform of the product of two functions is the "dyadic convolution" of their Walsh transforms [8]. Hence, there is an analog of Parseval's theorem for Walsh transforms, and the Walsh spectrum of a stationary random process is the Walsh transform of its "dyadic correlation function" [9], but we shall not need these results below.

We turn now to the symmetries of the Walsh functions. Replacing t by $1-t$ in (11), we see that all of the factors are unchanged except sgn $(\cos \pi t)$, which changes sign. This factor enters only if k is odd, and so we have

$$\text{wal}(k, 1-t) = (-1)^k\, \text{wal}(k, t). \tag{14}$$

Thus wal (k, t) has even or odd symmetry about $t=\frac{1}{2}$ accordingly as k is even or odd.

The other symmetries of the Walsh functions involve their periodicities. In one sense the period of any Walsh function is just the period of its slowest squarewave factor—2^{1-q} if k_q is the least significant 1 in the binary numeral for k. In connection with the expansion (3), however, we want to confine our attention to wal (k, t) with $0 \leq t \leq 1$, and so we map this interval into a circle and consider cyclic periodicity, addition to t being modulo 1.

More precisely, we define cyclic shifting by an amount u as turning wal (k, t) (with $0 \leq t < 1$) into wal $(k, \{t+u\})$, where $\{x\}$ is the fractional part of x, i.e., x minus the largest integer not exceeding x. The cyclic period of wal (k, t) is defined as the smallest $u>0$ for which the result of this cyclic shift is wal (k, t) whenever $0 \leq t < 1$. This cyclic period cannot exceed 1, as $u=1$ shifts the Walsh function exactly once around the circle, leaving it where it started.

Since the ordinary period of any Walsh function of even index is 1 or a submultiple of 1, its cyclic period is the same, but for odd k it is smaller than the ordinary period. In Appendix A it is shown that the cyclic period of wal (k, t) for odd k is the same as that of wal $(k+1, t)$, and a cyclic shift of any Walsh function by half its cyclic period P always changes its sign:

$$\text{wal}(k, t \pm \tfrac{1}{2}P) = -\,\text{wal}(k, t) \tag{15}$$

the sign on the left being chosen so as to leave the argument in the unit interval.

In Appendix A it is also shown that, if k is odd (i.e., if $k_0=1$) and if k_q is the least significant 0 in the binary numeral for k [the cyclic period of wal (k, t) thus being $P=2^{1-q}$], we have

$$\text{wal}(k, t \pm \tfrac{1}{4}P) = \pm(-1)^{k_{q+1}}\, \text{wal}(k+1, t). \tag{16}$$

Thus each Walsh function of odd index is related to its successor in (3) by a quarter-period cyclic shift in almost the same way that the corresponding sine and cosine in (1) are related to each other. (See Fig. 1.) These sinusoids also share the symmetry (14) and the half-period cyclic-shift property (15) with the Walsh functions, and a relation like (12) is also applicable to them, since doubling the frequency is equivalent to doubling the time.

Except for the constant wal $(0, t)=1$, which has no period, every Walsh function has an integral number of cyclic periods in the unit interval and so, by (15), its integral over this interval is zero. From (13), then, it follows that every Walsh function is orthogonal to each of the other Walsh functions, as $h \oplus k$ can be zero only if $h=k$. The Walsh functions are, therefore, an orthonormal set.

To see that they constitute a complete set and thus suffice for representing any finite-energy waveform $g(t)$ in the form (3), we notice that the 2^n Walsh functions of indices $k=0, 1, \cdots, 2^n-1$ (shown in Fig. 1 for $n=3$) are not only orthogonal and, hence, linearly independent but also have all of their sign changes where t is a multiple of 2^{-n}. In the unit interval, therefore, any stepfunction whose jumps occur at these times can be expressed as a linear combination of these

Walsh functions, since there are as many terms as steps. As $n \to \infty$, the width of the steps goes to zero, and so (3) is able to represent any finite-energy $g(t)$ with a vanishing integral-squared error.

It is worthwhile to note that, when n is finite, i.e., when (3) is summed only from $k=0$ to $k=2^n-1$, the truncated series will approximate any $g(t)$ as a stepfunction each of whose steps has a height equal to the average value of $g(t)$ over the width of that step. The integration in (4) can be seen to carry out this averaging.

III. Interrelation of the Two Series Expansions

The Fourier coefficients (2) are the components of a Hilbert-space vector $\boldsymbol{F}$ representing $g(t)$ over the interval $(0, T)$. The square of the length of this vector, $\boldsymbol{F}^\dagger \boldsymbol{F} = \Sigma F_k{}^2$, is found from (1) to be the mean-squared value of $g(t)$ over this interval. Likewise, the Walsh coefficients (4) are the components of another vector $\boldsymbol{W}$ that has the same length and represents $g(t)$ in terms of a different basis.

By substituting (3) into (2) or (1) into (4), we find that the components of either vector are linear combinations of the components of the other. This relationship can be expressed in matrix form as

$$\boldsymbol{W} = \boldsymbol{A}\boldsymbol{F} \tag{17}$$

where any element A_{kK} of the matrix $\boldsymbol{A}$ is the result of substituting the Kth sinusoid (along with its factor $\sqrt{2}$, if any) that appears in (1) for $g(t)$ in (4). Thus, $A_{00}=1$, and $A_{k0}=0$ for $k \neq 0$. When K is even and positive, we have

$$A_{k,K-1} = \sqrt{2} \int_0^1 \text{wal}\,(k,\, t) \sin K\pi t \, dt$$

$$A_{kK} = \sqrt{2} \int_0^1 \text{wal}\,(k,\, t) \cos K\pi t \, dt. \tag{18}$$

With $k < 2^m$, the integrations here are easily carried out over the 2^m intervals of width 2^{-m} over which the Walsh functions are constant. The left end of each such interval is represented by an m-digit binary numeral, and the summation of the integrals over these intervals is readily carried out by summing over the two values taken by each of the m digits. The results are most conveniently expressed by stating that [10] A_{kK} and $A_{k,K-1}$ are, respectively, the real and imaginary parts of

$$\sqrt{2}\, i^K \frac{\sin 2^{-m-1}\pi K}{2^{-m-1}\pi K} \left(-i \begin{matrix}\sin\\ \cos\end{matrix}\, 2^{-m-1}\pi K \right) \cdot \left(-i \begin{matrix}\sin\\ \cos\end{matrix}\, 2^{-m}\pi K \right) \cdots \left(-i \begin{matrix}\sin\\ \cos\end{matrix}\, \tfrac{1}{4}\pi K \right) \tag{19}$$

where $-i$ sin or cos is to be used in the factor with argument $2^{-r-1}\pi K$ accordingly as $\cos\,(2^{-r}\pi k + 0)$ is negative or positive, i.e., accordingly as $k_{r-1}+k_r$ is odd or even.

The work of computing these matrix elements is considerably reduced by utilizing some of the properties that are discussed below. Thus we find that the Fourier-to-Walsh transform matrix $\boldsymbol{A}$ is [18]

$$\begin{bmatrix}
1 & 0 & 0 & 0 & 0 & 0 & 0 & 0 & 0 & 0 & 0 & 0 & 0 & 0 & 0 & 0 & 0 & \cdots \\
0 & 0.900 & 0 & 0 & 0 & 0.300 & 0 & 0 & 0 & 0.180 & 0 & 0 & 0 & 0.129 & 0 & 0 & 0 & \cdots \\
0 & 0 & 0.900 & 0 & 0 & 0 & -0.300 & 0 & 0 & 0 & 0.180 & 0 & 0 & 0 & -0.129 & 0 & 0 & \cdots \\
0 & 0 & 0 & 0.900 & 0 & 0 & 0 & 0 & 0 & 0 & 0 & 0.300 & 0 & 0 & 0 & 0 & 0 & \cdots \\
0 & 0 & 0 & 0 & 0.900 & 0 & 0 & 0 & 0 & 0 & 0 & 0 & -0.300 & 0 & 0 & 0 & 0 & \cdots \\
0 & -0.373 & 0 & 0 & 0 & 0.724 & 0 & 0 & 0 & 0.435 & 0 & 0 & 0 & -0.053 & 0 & 0 & 0 & \cdots \\
0 & 0 & 0.373 & 0 & 0 & 0 & 0.724 & 0 & 0 & 0 & -0.435 & 0 & 0 & 0 & -0.053 & 0 & 0 & \cdots \\
0 & 0 & 0 & 0 & 0 & 0 & 0 & 0.900 & 0 & 0 & 0 & 0 & 0 & 0 & 0 & 0 & 0 & \cdots \\
0 & 0 & 0 & 0 & 0 & 0 & 0 & 0 & 0.900 & 0 & 0 & 0 & 0 & 0 & 0 & 0 & 0 & \cdots \\
0 & -0.074 & 0 & 0 & 0 & -0.484 & 0 & 0 & 0 & 0.650 & 0 & 0 & 0 & 0.268 & 0 & 0 & 0 & \cdots \\
0 & 0 & -0.074 & 0 & 0 & 0 & 0.484 & 0 & 0 & 0 & 0.650 & 0 & 0 & 0 & -0.268 & 0 & 0 & \cdots \\
0 & 0 & 0 & -0.373 & 0 & 0 & 0 & 0 & 0 & 0 & 0 & 0.724 & 0 & 0 & 0 & 0 & 0 & \cdots \\
0 & 0 & 0 & 0 & 0.373 & 0 & 0 & 0 & 0 & 0 & 0 & 0 & 0.724 & 0 & 0 & 0 & 0 & \cdots \\
\cdots & \cdots & \cdots & \cdots & \cdots & \cdots & \cdots & \cdots & \cdots & \cdots & \cdots & \cdots & \cdots & \cdots & \cdots & \cdots & \cdots & \cdots
\end{bmatrix} \tag{20}$$

Because $\boldsymbol{A}$ represents an orthogonal transformation, i.e., a rotation of the Hilbert-space coordinate system, its inverse $\boldsymbol{A}^{-1}$ is its transpose $\boldsymbol{A}^\dagger$, and we have $\boldsymbol{F} = \boldsymbol{A}^\dagger \boldsymbol{W}$.

The kth row of (20) consists of the Fourier coefficients (2) of $g(t) = \text{wal}\,(k,\, t/T)$, the counting starting from $k=0$. The second row, for example, can be recognized as representing the squarewave wal $(2,\, t/T)$. From (12) it follows that, whenever k is even [and the cyclic period of wal $(k,\, t)$ is the same as its ordinary period]

$$A_{2k,2K} = A_{kK} \tag{21}$$

and so the pattern of the second row of (20), for example, repeats itself in stretched-out form, in the 4th, 8th, 16th, $\cdots$ rows. Moreover, each odd-numbered row of $\boldsymbol{A}$, by virtue of the quarter-period cyclic-shift relation (16), is identical with the next row except for a change in every other sign and a shift from sines to cosines of the same frequency.

On account of the change of sign (15) when shifted by half their period, the Walsh functions contain only odd harmonics of their fundamental frequencies in their Fourier spectra; hence the large number of zeros in (20). For example, when k is twice an odd number, the Fourier spectrum of wal $(k,\, t)$ consists entirely of cosines whose frequencies are odd integers. The symmetry relation (14) implies that wal $(k,\, t)$ consists entirely of cosines when k is even and sines when k is odd.

The columns of $\boldsymbol{A}$ are the Walsh coefficients of the sinusoids, and by (16) each odd-numbered column is the same as the one to its right except for a change in every other sign and a shift from odd- to even-indexed Walsh functions. In Appendix B it is shown that A_{KK}, which by (19) is always positive, is the largest element (regardless of sign) in the Kth column of $\boldsymbol{A}$, and it also appears that A_{kk} is always the largest element in the kth row. Hence the wal $(K, t/T)$ component of the kth sinusoid in (1) is its strongest Walsh component, and the strongest Fourier component of wal $(k, t/T)$ when expanded in the form (1) is the kth term.

These two relationships between corresponding terms of (1) and (3) augment the list of their already mentioned common properties. It is shown in Appendix B, however, that the strongest Walsh component of a sinusoid (and vice versa) can be arbitrarily weak and usually contains only a small fraction of the total power, while there are always other components of comparable (and, often, almost equal) strength. The relation between corresponding terms of the two series is, therefore, not so close nor so unique as one might hope, and the use of the Walsh transform to discover Fourier components [11] can be misleading.

IV. Truncation Error in Fourier and Walsh Series

When a sample function $g(t)$ of a zero-mean continuous stationary random process is expanded as an infinite series (1) or (3), only a finite number of terms N can actually be used, and their coefficients (2) or (4) can be determined with only a finite number of significant digits. Two kinds of errors, therefore, result—one due to the truncation of the series and the other due to the rounding off of the coefficients.

If we retain the terms of the Walsh series (3) with $k<N$ with N a large power of 2, the truncated series will give a stepfunction approximation to $g(t)$ with steps of width T/N, and the error will be close to the slope $g'(t)$ times the time difference from the center of the step, as $g(t)$ will depart very little from a straight line over the width of the step. Integrating the square of this error with respect to the time difference from $-T/2N$ to $T/2N$, averaging over the distribution of $g'(t)$, and dividing by the width of the step T/N, we find that the mean-squared truncation error is approximately, say, $Q_W = g'_{\text{rms}}{}^2 T^2/12N^2$.

Since the differentiation of (1) interchanges sines and cosines and multiplies their coefficients by $\pm 2\pi$ times their frequencies, $g'_{\text{rms}}{}^2$ is $4\pi^2 g_{\text{rms}}{}^2$ times B^2, the "mean-squared frequency" in the spectrum of $g(t)$ [12]. If $D(f)$ is the power spectral density of $g(t)$, we have

$$B^2 = \int_0^\infty f^2 D(f)df/g_{\text{rms}}{}^2 = \int_0^\infty f^2 D(f)df \Big/ \int_0^\infty D(f)df.$$

We can thus express the mean-squared Walsh-series truncation error as

$$Q_W = \pi^2 g_{\text{rms}}{}^2 B^2 T^2/3N^2. \tag{22}$$

Just as the truncation error for (3) is due principally to the discontinuities of the Walsh functions, the truncation error for (1) is due mainly to the jump from $g(T)$ to $g(0)$ when the interval $(0, T)$ is mapped into a circle, as the sines and cosines are smooth functions on this circle. If T is big enough that $g(0)$ and $g(T)$ can be regarded as statistically independent, the mean-squared magnitude $g(0)-g(T)$ of this jump is $2g_{\text{rms}}{}^2$.

The high-frequency spectrum of such a discontinuity is easily approximated by noticing that its derivative is an impulse of rms area $\sqrt{2}\, g_{\text{rms}}$ minus the latter's dc component. For such an impulse we have from (2) $F_{K\,\text{rms}} = 2g_{\text{rms}}/T$ for even K and $F_{K\,\text{rms}} = 0$ for odd K. Dividing by $K\pi/T$ to account for the effect of integration, we see that the mean-squared component of (1) at any frequency well above the highest significant frequencies in the spectrum of $g(t)$ is $(2g_{\text{rms}}/\pi K)^2$.

To approximate the summation of this quantity over all even values of K from N, the number of terms retained in (1), to ∞, we take half of its integral with respect to K, thus finding that the mean-squared truncation error for the Fourier series (1) is approximately

$$Q_F = 2g_{\text{rms}}{}^2/\pi^2 N. \tag{23}$$

Here the truncation error shows up mainly near $t=0$ and $t=T$ because Fourier components outside the spectrum of $g(t)$ arise only on account of end effects, which extend over times of the order of T/N. In addition to (23) there may be a truncation error arising from the tail of the spectrum of $g(t)$, but with $g(t)$ continuous this tail falls faster than inversely as the square of the frequency, and its contribution will be of a smaller order than (23) for large N.

Comparing (22) with (23), we see that the truncation errors for Walsh and Fourier series behave quite differently as functions of the time-bandwidth product BT and the number N of retained terms. As N grows, Q_W shrinks faster than Q_F, but for a given N (22) will exceed (23) if BT is sufficiently large. In Section VI we shall see that the superiority of the Fourier series when BT is large stems from its closer relationship to the optimum series, the Karhunen–Loève expansion [13].

V. Roundoff Error

We turn now to the error due to using only a finite number of digits in representing the coefficients of the terms in (1) and (3) that we retain. Since all of the functions that these coefficients multiply have unit time-averaged mean-squared values, the mean-squared error due to rounding off any retained coefficient depends only upon the weight of the least significant digit that is used in expressing it. The roundoff error is presumably uniformly distributed between $-\frac{1}{2}$ and $\frac{1}{2}$ of the weight of this roundedoff digit. Hence, the available digits should be apportioned among the retained coefficients so that they occupy the most significant positions they can, thus leaving all of the roundedoff digits with approximately the same weight. All retained coefficients will then contribute more or less equally to the mean-squared roundoff error.

The truncation error due to the omitted term whose coefficient has the largest mean-squared value should be comparable with this single-term roundoff error, for, if it were larger, it would merit retention and a digit to represent it, while, if it were substantially smaller, there would presumably be retained coefficients too small to make significant use of the digits allotted to them. Thus when the digits to be used in representing the coefficients are optimally allocated, there is a close relationship between the roundoff error and the truncation error. Although the least significant digits for the retained coefficients should all have about the same weight, the number of digits allocated to any coefficient will depend on its mean-squared value, the dependence being logarithmic, so that they can, with high probability, cover the range of the coefficient.

In the course of obtaining the Fourier-series truncation

error (23) we found that the mean-squared contribution from the omitted terms of frequency $K/2T$ is approximately $4g_{\rm rms}^2/\pi^2K^2$. Setting $K=N$, we thus get an approximation for the mean-squared roundoff error contributed by each of the N retained terms and, multiplying by N, we obtain the mean-squared Fourier-series roundoff error

$$R_F = 4g_{\rm rms}^2/\pi^2N. \qquad (24)$$

Adding (23) to (24) and taking the square root, we get (5) as the rms total error for Fourier series truncated in a manner consistent with the precision used for the coefficients of the retained terms. Although several rough approximations are involved in the derivation of this result, it should be of the right order of magnitude, and its simple form facilitates understanding and comparison with the Walsh-series result.

In order to get a simple approximation for the largest mean-squared truncation error contributed by a term omitted from (3), we suppose that (22) can be regarded as the summation or integral of such contributions in the same way that (23) was in its derivation. Hence we differentiate (22) with respect to N, getting $2\pi^2g_{\rm rms}^2B^2T^2/3N^3$. Multiplying as in the Fourier case by N, we get the approximation

$$R_W = 2g_{\rm rms}^2B^2T^2/3N^2 \qquad (25)$$

for the mean-squared Walsh-series roundoff error. Adding (22) and taking the square root, we get (6) for the rms total error for Walsh series truncated consistently with the precision of their coefficients.

Although the mean-squared roundoff errors (24) and (25) turned out in both cases to be just twice the mean-squared truncation errors (23) and (22), respectively, these results should be taken to mean that the roundoff and truncation errors ought to be of the same order of magnitude.

VI. Optimum Series Expansion

In order to understand why the Fourier series needs fewer terms than the Walsh series when the time-bandwidth product BT is sufficiently large, as indicated by (7), we direct our attention now to the covariance matrices of the series coefficients $\boldsymbol{F}=\{F_k\}$ and $\boldsymbol{W}=\{W_k\}$. The Fourier covariance matrix $\boldsymbol{M}_F$ is the expected value of $\boldsymbol{F}\boldsymbol{F}^\dagger$, with $(\boldsymbol{M}_F)_{hk} = E\{F_hF_k\}$, and the Walsh covariance matrix is $\boldsymbol{M}_W = E\{\boldsymbol{W}\boldsymbol{W}^\dagger\}$. Since by (17) $\boldsymbol{W}^\dagger=\boldsymbol{F}^\dagger\boldsymbol{A}^\dagger$, we have

$$\boldsymbol{M}_W = \boldsymbol{A}\boldsymbol{M}_F\boldsymbol{A}^\dagger. \qquad (26)$$

More generally, $g(t)$ can be expanded over $(0, T)$ as

$$g(t) = \sum_{k=0}^{\infty} C_k\phi_k(t, T), \quad \text{with } C_k = \frac{1}{T}\int_0^T g(t)\phi_k(t, T)dt \qquad (27)$$

where $\{\phi_k(t, T)\}$ is any complete orthonormal set of functions on this interval, such as the sinusoids or Walsh functions in (1)–(4). The covariance matrix of these coefficients $\boldsymbol{C}=\{C_k\}$ is

$$\boldsymbol{M} = E\{\boldsymbol{C}\boldsymbol{C}^\dagger\} \qquad (28)$$

which is related to $\boldsymbol{M}_F$ and to $\boldsymbol{M}_W$ by orthogonal transformations like (26).

When the series (27) is truncated after N terms, the resulting mean-squared error is $E\{\sum_{k=N}^{\infty}C_k^2\}$. Since $E\{\sum_{k=0}^{\infty}C_k^2\} = g_{\rm rms}^2$, this is $g_{\rm rms}^2 - E\{\sum_{k=0}^{N-1}C_k^2\}$, which is $g_{\rm rms}^2$ minus the sum of the first N diagonal elements of $\boldsymbol{M}$. Hence the mean-squared truncation error is minimized by using a set of functions $\{\phi_k(t, T)\}$ that maximizes the sum of the first N diagonal elements of $\boldsymbol{M}$.

We now make use of a matrix inequality (see Appendix C) proved independently around 1949 by G. Polya, Ky Fan, and H. Weyl [14]: The sum of N diagonal elements of a self-adjoint matrix $\boldsymbol{M}=\boldsymbol{M}^\dagger$ cannot exceed the sum of its N largest eigenvalues, with equality only if the matrix is of block-diagonal form with a block corresponding to these diagonal elements and these eigenvalues. It follows that, if the truncation error is to be a minimum for every N, the matrix must be diagonal.

Orthogonal transformations, such as (26), do not affect the eigenvalues of the covariance matrix, but the eigenvectors of, say, $\boldsymbol{M}_F$ indicate the linear combinations of the trigonometric functions which, serving as the $\{\phi_k(t, T)\}$, will diagonalize $\boldsymbol{M}$. The diagonality of $\boldsymbol{M}$ means that the expansion coefficients $\{C_k\}$ are uncorrelated with each other; this is the defining property of the Karhunen–Loève expansion [13]. We thus see that the mean-squared truncation error is minimized for all N by the use of such an expansion [15] with the $\{\phi_k(t, T)\}$ taken in order of decreasing eigenvalues. (In case one is interested only in a specific N, the first N of them can be replaced by any orthonormal combination of themselves and the remaining ones likewise.)

The Fourier coefficients of an ergodic random process are not uncorrelated but, as $T\to\infty$, they become uncorrelated [13]. For large T, however, the Walsh coefficients do not become uncorrelated (although those of bounded index will do so if the spectral density of the process at zero frequency does not vanish, since for them the process is effectively white). To see this we make use of (17), (20), and the iterated form of (21), which together tell us, for example, that

$$W_{2^n} = 0.900F_{2^n} - 0.300F_{3\cdot 2^n} + 0.180F_{5\cdot 2^n} - \cdots$$
$$W_{3\cdot 2^n} = 0.373F_{2^n} + 0.724F_{3\cdot 2^n} - 0.435F_{5\cdot 2^n} + \cdots$$

for any integer $n>0$. The first equation here shows the Walsh coefficient of sequency $2^n/T$ to be a linear combination of the Fourier cosine coefficients of frequency $2^{n-1}/T$ and its odd harmonics, while the second equation shows the Walsh coefficient of sequency $3\cdot 2^n/T$ to be a function of the same Fourier coefficients. Letting n and T grow in such a way as to keep these frequencies constant, we see that, if the spectral density of $g(t)$ is nonzero at frequency $2^{n-1}/T$ but vanishes at its odd harmonics, the two Walsh coefficients will tend to be proportional to F_{2^n} as $T\to\infty$, and their correlation coefficient will tend toward unity.

This asymptotic correlatedness of the Walsh coefficients makes them inefficient at representing the information in $g(t)$ and accounts for the greater truncation error for the Walsh series when T is large. Being asymptotically uncorrelated, the Fourier coefficients resemble [16] the Karhunen–Loève coefficients, which are optimum for all T but suffer from the considerable difficulty that the Karhunen–Loève functions are the eigenfunctions of an integral equation and depend in a nontrivial way upon T.

Another matrix inequality, the Hadamard determinant inequality [14], explains why for large T the Walsh-series roundoff error exceeds that for the Fourier series. This inequality, which is proved in Appendix C, states that the determinant of a self-adjoint nonnegative-definite matrix never exceeds the product of its diagonal elements, with equality only if the matrix is diagonal. We shall apply it to the $N \times N$ covariance matrix of the N retained coefficients.

Since these coefficients are all rounded off to about the same number of digits after the decimal (or binary) point, the number of digits used for each is a linear function of the logarithm of its variance, which is the corresponding diagonal element of the covariance matrix. Hence the total number of digits is a linear function of the logarithm of the product of the diagonal elements of the matrix, and we see that it is least when we use basis functions $\{\phi_k(t, T)\}$ that result in a diagonal covariance matrix, since the determinant of this matrix is invariant under changes of the orthonormal basis spanning the same N space.

The Karhunen–Loève expansion is thus again optimum, apart from its difficulty of implementation, and the Fourier coefficients asymptotically have the property that is required in order to minimize the mean-squared roundoff error for a given total number of digits to be used in representing the coefficients. The Walsh coefficients, however, lack this property because they are asymptotically correlated and their covariance matrix does not become diagonal. They therefore again are a less efficient representation of $g(t)$ over $(0, T)$ for large T than are the Fourier coefficients.

VII. Series Representation of Discontinuous Waveforms

In order to determine whether this inferiority of the Walsh series is simply the result of our assumption above that $g(t)$ is continuous, we now turn to the case of a discontinuous waveform. Thus instead of supposing it to have a Fourier spectrum with a rapidly falling high-frequency tail, we suppose that its Walsh spectrum is flat up to sequency $S=2^n/T$, where it cuts off sharply. Hence $g(t)$ is now a stepfunction that takes a new, uncorrelated, zero-mean value every $1/S$ seconds.

If the Walsh analysis of $g(t)$ is synchronized with its jumps, i.e., if t is taken to be 0 at one of these jumps, $g(t)$ will be represented without any truncation error by the sum of the first ST terms of its Walsh series. Here the coefficients are uncorrelated, and they represent $g(t)$ very efficiently. Of course, the values of $g(t)$ itself on its ST steps would be an equally efficient representation in this case and would avoid the necessity of Walsh transformation. To avoid this triviality, we therefore assume that the analysis is not synchronized with the jumps of $g(t)$ and that the first jump in $(0, T)$ occurs at a time distributed uniformly between 0 and $1/S$.

With the number of terms N retained in the Walsh series a power of 2 greater than or equal to ST, their sum equals the average value of $g(t)$ across each interval of width T/N starting at $t=0$, and there is no truncation error in any such interval unless it contains a jump of $g(t)$.

In the latter case, if $g(t)$ has the value g_1 for the first τ seconds of the interval and the value g_2 for the remaining $T/N-\tau$, the truncated series takes the value $(N\tau/T)g_1 + (1-N\tau/T)g_2$ throughout this interval, giving integral-squared error $(1-N\tau/T)\tau(g_1-g_2)^2$. This becomes $2\tau(1-N\tau/T)g_{\rm rms}^2$ when averaged over the two uncorrelated values of g. Multiplying by ST, the number of intervals in which $g(t)$ jumps, and dividing by T, we find the mean-squared truncation error $2S\tau(1-N\tau/T)g_{\rm rms}^2$ which, when averaged over the uniform distribution of τ, becomes

$$Q_W = \frac{ST}{3N} g_{\rm rms}^2. \tag{29}$$

Following the same approach as in Section V to find the roundoff error when the rounding off of the coefficients is consistent with the truncation of the series, we find that $R_W=Q_W$. Finally, adding these two mean-squared errors and taking the square root, we obtain (9).

To find the Fourier-series truncation error for our discontinuous $g(t)$, we observe that a square pulse of height g and width $1/S$ centered at time t has, by (2), Fourier coefficients

$$F_k = \frac{\sqrt{8}}{\pi k} g \sin \frac{\pi k}{2ST} \cos \frac{\pi k t}{T}$$

$$F_{k-1} = \frac{\sqrt{8}}{\pi k} g \sin \frac{\pi k}{2ST} \sin \frac{\pi k t}{T}$$

for all positive even values of k. The mean-squared error [averaged over the whole interval $(0, T)$] due to truncating the Fourier series for this pulse at N terms is thus

$$\sum_{k=N}^{\infty} F_k^2 = \sum_{h=N/2}^{\infty} \frac{2g^2}{\pi^2 h^2} \sin^2 \frac{\pi h}{ST}$$

$$\approx \int_{N/2}^{\infty} \frac{g^2}{\pi^2 h^2}\, dh = \frac{2g^2}{\pi^2 N} \cdot \tag{30}$$

Averaging over the pulse height g and multiplying by ST, the number of such pulses that together form $g(t)$ over $(0, T)$, we find that the mean-squared Fourier-series truncation error is

$$Q_F = \frac{2ST}{\pi^2 N} g_{\rm rms}^2. \tag{31}$$

From (30) we can find the mean-squared contribution of the first omitted term to the truncation error, and from it we obtain the corresponding mean-squared roundoff error just as in Section V. Again it turns out to equal the mean-squared truncation error: $R_F=Q_F$. From this result and (31) we find (8) as the total rms error when the rounding off of Fourier coefficients is consistent with the truncation of the series.

Comparing (31) with (29), we see that, in the absence of synchronization between the waveform and its Walsh analysis, the Fourier series needs only about $6/\pi^2=0.61$ times as many terms as the Walsh series in order to represent our discontinuous sequency-limited waveform to within a given rms error.

Appendix A

Cyclic Periodicity of Walsh Functions

The periodicity of wal (k, t) was determined very simply in Section II for even k. When k is odd, however, a more involved analysis is required. Here $k_0=1$, and we suppose that $k_0=k_1=\cdots=k_{q-1}=1$ but $k_q=0$ is the least significant 0 in the binary numeral for k. Accordingly, (11) includes the factors

$$\text{sgn}(\cos \pi t)\, \text{sgn}(\cos 2\pi t) \cdots \text{sgn}(\cos 2^{q-1}\pi t).$$

To simplify this product when t is confined to the unit interval, it is convenient to multiply it by sgn $(\sin \pi t)$, which is 1 when $0<t<1$. Making repeated use of the identity sgn $(\sin x)$ sgn $(\cos x)=$ sgn $(\sin 2x)$ (which follows from $\sin x \cos x=\frac{1}{2}\sin 2x$), we find that the foregoing product is sgn

($\sin 2^q\pi t$) in the unit interval, and so for $0<t<1$

$$\operatorname{wal}(k, t) = \operatorname{sgn}(\sin 2^q\pi t) \prod_{r=q+1}^{m-1} \operatorname{sgn}(\cos^{k_r} 2^r\pi t).$$

Note that the right-hand side is unchanged if t is replaced by $t\pm 1$.

Replacing t by $t\pm 2^{-q-1}$, we see that, if this is in (0, 1)

$$\begin{aligned}\operatorname{wal}(k, t \pm 2^{-q-1}) &= \pm(-1)^{k_{q+1}} \operatorname{sgn}(\cos 2^q\pi t) \prod_{r=q+1}^{m-1} \operatorname{sgn}(\cos^{k_r} 2^r\pi t)\\ &= \pm(-1)^{k_{q+1}} \operatorname{wal}(k+1, t).\end{aligned}$$

Since the difference between the two arguments on the left is 2^{-q}, we see that a cyclic shift by this amount changes the sign of wal (k, t). A shift by twice as much will leave wal (k, t) unaffected, and hence $P=2^{1-q}$ is its cyclic period. With this substitution, we thus obtain (15) and (16). (Alternatively, we can establish the cyclic period for odd k by noticing that the foregoing equation shows it to be the same as that of the Walsh function of index $k+1$.)

Appendix B

Principal Walsh Components of Sinusoids and Vice Versa

Because of (14) and (16), it suffices to consider only cosines and Walsh functions of even index in studying the principal Walsh components of a sinusoid and vice versa. With k and K thus even, (19) is real, giving us A_{kK}, and we shall consider its square, $|A_{kK}|^2$, which is the fraction of the unit total power of $\sqrt{2}\cos K\pi t$ contained in its wal (k, t) component and vice versa.

For any K, it is easy to see what k maximizes $|A_{kK}|^2$, viz., the k that gives a sine or a cosine in each factor of (19) accordingly as a sine or a cosine would be larger in absolute value. Since $\cos 2\theta=\cos^2\theta-\sin^2\theta$, the choice between sine and cosine is made according to the sign of the cosine of twice the argument. Referring to the second line after (19), we conclude that the k in question is simply K. The strongest Walsh component of any sine or cosine appearing in (1) is, therefore, the one in (3) having the same number of sign changes inside the interval (0, T) and, of course, the same odd or even symmetry about $t=\frac{1}{2}T$.

While this result shows a new relationship between corresponding terms of (1) and (3), it has greater import for some values of K than for others. $|A_{KK}|^2$ can be as large as $8/\pi^2=0.81057$ for $K=1$ or, by (21), whenever K is a power of 2, but $|A_{KK}|^2$ can also be arbitrarily small. For example, when K is close to one-third of a large power of 2, nearly all of the trigonometric functions in (19) are close to $\frac{1}{2}\sqrt{3}$ in absolute value and, hence, $|A_{KK}|^2$ is of the order of

$$(3/4)^{\log_2 K} = K^{-0.4158}. \tag{B-1}$$

For large K, small values of $|A_{KK}|^2$ are more typical than large ones, as the following argument shows. The binary numeral for a large random K contains a long sequence of statistically independent digits that are equally likely to be 0 or 1. As the argument $2^{-m-1}\pi K$ of the first trigonometric function in (19) is repeatedly doubled to obtain the arguments of the successive factors, this sequence of digits shifts left, and the digits with weight π or more are "lost" in that they do not affect the value of the square of the sine or cosine.

Because K is large, many such left shifts are involved in (19), and the angles that constitute the arguments of the trigonometric functions can, therefore, be considered uniformly distributed in direction. For $k=K$, the factor whose argument is θ will have absolute value $|\sin\theta|$ or $|\cos\theta|$, whichever is larger, and so we can treat θ as uniformly distributed between 0 and $\pi/4$.

To log $|A_{KK}|^2$, each trigonometric factor in (19) will hence contribute, on the average, [17]

$$\frac{4}{\pi}\int_0^{\pi/4} \log\cos\theta\, d\theta = \frac{2}{\pi}G - \log 2 = -0.11003$$

where $G=1-1/3^2+1/5^2-\cdots=0.91597$ is Catalán's constant. $|A_{KK}|^2$ will, therefore, be of the order of

$$\exp(-0.22005\log_2 K) = K^{-0.3173}. \tag{B-2}$$

This expression is only a rough approximation, however, because of the large variance of the sum of the logarithms of the sines and cosines in (19), and relatively large values of $|A_{KK}|^2$ will occur whenever the binary numeral for K consists mainly of long strings of 1's or of 0's.

To identify the second strongest Walsh component of $\cos K\pi t$ on the unit interval, we look at (19) with $k=K$, in which all of the trigonometric factors (beyond the first sine) are at least $1/\sqrt{2}$ in absolute value. If the one closest to this value but exceeding it is changed from a sine to a cosine or vice versa, $|A_{kK}|^2$ will be reduced minimally from its maximum value, and we find the second largest Walsh component. One of the trigonometric factors in (19) always has absolute value $1/\sqrt{2}$; it can equally well be a sine or a cosine, thus permitting the realization of the k that otherwise makes only the foregoing change in the trigonometric functions.

The absolute value closest to but exceeding $1/\sqrt{2}$ can be at most $\cos\pi/8$ as the repeated halving and doubling of other sorts of angles will yield at least one value closer to $1/\sqrt{2}$ while still exceeding it. (This can be established by using the binary numeral for K.) Hence the second largest value of $|A_{kK}|^2$ for a given K is always at least $\tan^2\pi/8=0.1716$ times its greatest value $|A_{KK}|^2$. This bound is attained whenever K is a power of 2.

For most large values of K, however, the ratio of the second largest $|A_{kK}|^2$ to the largest is close to 1, as the binary numerals for most K's include a string of 0's or of 1's long enough to yield a sine or cosine in (19) with absolute value close to $1/\sqrt{2}$. In fact, there are usually many values of k for which $|A_{kK}|^2$ comes close to $|A_{KK}|^2$.

While the k that maximizes $|A_{kK}|^2$ for a given K was easy to determine, the K that maximizes it for a given k is not. In attacking this problem it suffices to confine our attention to even values of k on account of (16) and, more particularly, to values of k that are twice an odd number on account of (21), as doubling k will simply double the maximizing K. Since the Fourier spectrum of wal (k, t) contains only odd harmonics of its fundamental frequency, which is then unity, the maximizing K must also be twice an odd number.

To shorten the search for this K, we observe that it must be less than 2^m for $k<2^m$, as the trigonometric factors in (19) are all symmetric about this value and repeat themselves periodically while the denominator increases monotonically with K. A search was therefore programmed to compare $|A_{kK}|^2$ for $K=2, 6, 10, \cdots, 2^m-2$ with $k=2, 6, 10, \cdots, 4094$. The computation was facilitated by omitting the last factor in (19), canceling the $\sqrt{2}$ against the second last factor, expressing each remaining trigonometric factor in terms of the cosine of twice the argument, and generating a first quadrant table of the required cosines by repeated use of the half-angle formula.

The compilation from Fortran and execution took 28.5 min on a CDC 3200 computer and printed out $|A_{kk}|^2$ for each k along with the largest other value observed for $|A_{kK}|^2$ and the corresponding K. In every case, $|A_{kk}|^2$ was the larger of the two, showing that cos $k\pi t$ is the strongest Fourier component of wal (k, t) and that sin $k\pi t$ is the strongest Fourier component of wal $(k-1, t)$ for every even k up to at least 4096.

Again this strongest component can carry as much as $8/\pi^2 = 0.81057$ of the energy, viz., when k is a power of 2, and it can also be very weak, as described by $(B-1)$, for example, when k is close to one-third of a larger power of 2. $|A_{kk}|^2$ tends to be large when the binary numeral for k consists of a small number of strings of 1's or of 0's, but for large k small values predominate, as described by (B-2).

The second strongest Fourier component of wal (k, t) can have as little as one-ninth of the energy of the strongest component, as in the case of the regular symmetric squarewaves, where k is a power of 2 and the second strongest component is the third harmonic of its fundamental frequency. The second strongest component can also be nearly as strong as the strongest; for example, $|A_{2050,2050}|^2 = 0.328878$ and $|A_{2050,2046}|^2 = 0.328145$. When k is 2 more than a large power of 2, the ratio of the two largest values of $|A_{kK}|^2$ differs from 1 by approximately $(4\pi - 8)/k = 4.566/k$, both values being close to $32/\pi^4 = 0.32851$. This is the result of modulation of the fast squarewave factor in (11) by the slow one.

Appendix C

Matrix Inequalities

To prove the Polya–Fan–Weyl inequality quoted in Section VI, we suppose that $\boldsymbol{M}$ is diagonalized by the unitary matrix $\boldsymbol{U}^\dagger = \boldsymbol{U}^{-1}$, so that $\boldsymbol{M} = \boldsymbol{UDU}^\dagger$ with $\boldsymbol{D}$ diagonal. Then the hth diagonal element of $\boldsymbol{M}$ is $M_{hh} = \Sigma_k |U_{hk}|^2 D_k$ and, hence, $\Sigma_{h=0}^{N-1} M_{hh} = \Sigma_k (\Sigma_{h=0}^{N-1} |U_{hk}|^2) D_k$. Since $\Sigma_{h=0}^{N-1} |U_{hk}|^2 \leq 1$ for a unitary matrix, the N largest eigenvalues D_k cannot have more than unit weight each in this sum over k, and the smaller eigenvalues get the remainder of the total weight $N = \Sigma_k \Sigma_{h=0}^{N-1} |U_{hk}|^2$. If the N largest eigenvalues are to get the entire weight N, it is necessary that $U_{hk} = 0$ whenever $h \geq N$ and D_k is among the N largest eigenvalues or $h < N$ and D_k is not among them, and $\boldsymbol{M}$ must thus be block diagonal. This completes the proof.

To prove the Hadamard determinant inequality for the self-adjoint positive-definite matrix $\boldsymbol{M} = (M_{hk})$ we observe first that all of the diagonal elements M_{hh} must be positive, since there would otherwise be a nonzero vector $\boldsymbol{X}$ violating the assumption that $\boldsymbol{X}^\dagger \boldsymbol{MX} > 0$. Hence we can let

$$\boldsymbol{V} = \begin{bmatrix} 1 & 0 & 0 & \cdots \\ -M_{10}/M_{00} & 1 & 0 & \cdots \\ -M_{20}/M_{00} & 0 & 1 & \cdots \\ \cdots & \cdots & \cdots & \cdots \end{bmatrix}.$$

Then $\boldsymbol{VMV}^\dagger$ has zeros throughout its $k=0$ column and $h=0$ row except at their intersection, where it has the element M_{00}, and the hth diagonal element apart from this one is $M_{hh} - |M_{h0}|^2/M_{00}$. This transformation evidently does not increase the diagonal elements of $\boldsymbol{M}$, but they remain positive because $\boldsymbol{VMV}^\dagger$ is positive-definite, $\boldsymbol{Y}^\dagger \boldsymbol{VMV}^\dagger \boldsymbol{Y}$ being positive for any nonzero vector $\boldsymbol{Y}$ $(= \boldsymbol{V}^{\dagger-1}\boldsymbol{X})$. The determinant of $\boldsymbol{V}$ is 1, and so the determinant of $\boldsymbol{VMV}^\dagger$ is the same as that of $\boldsymbol{M}$.

The same procedure can now be applied to the next row and column and so forth until the matrix finally becomes diagonal. Its determinant is unchanged, and the matrix remains positive-definite, with positive diagonal elements that have not been increased. In fact, they can all remain constant through this whole process only if $\boldsymbol{M}$ is diagonal, thus proving that the determinant of $\boldsymbol{M}$ never exceeds the product of its diagonal elements, equaling it only when $\boldsymbol{M}$ is diagonal.

Acknowledgment

I am indebted to David Slepian for many a careful thorough reading of ealier versions of this paper and for his very helpful comments.

References

[1] J. L. Walsh, "A closed set of normal orthogonal functions," *Amer. J. Math.*, vol. 45, pp. 5–24, 1923.
[2] C. A. Bass, Ed., *Proc. Symp. Applications of Walsh Functions* (Mar.–Apr. 1970), ASTIA Doc. AD-707 431.
[3] R. W. Zeek and A. E. Showalter, Eds., *Proc. Symp. Applications of Walsh Functions* (*IEEE Trans. Electromagn. Compat.*, vol. EMC-13, pp. 1–218, Aug. 1971), ASTIA Doc. AD-727 000.
[4] ——, *Proc. Symp. Applications of Walsh Functions* (Mar. 1972), ASTIA Doc. AD-744 650.
[5] J. E. Whelchel and D. F. Guinn, "Fast Fourier-Hadamard transform and its use in signal representation and classification," in *EASCON 1968 Rec.*, pp. 561–573.
[6] C. Cardot, "Définition analytique simple des fonctions de Walsh et application à la détermination exacte de leurs propriétés spectrales," *Ann. Télécomm.*, vol. 72, pp. 31–47, Jan.–Feb. 1972.
[7] N. M. Blachman, "Comments concerning Walsh functions," *IEEE Trans. Inform. Theory* (Corresp.), vol. IT-18, pp. 427–428, May 1972.
[8] B. L. N. Kennett, "A note on the finite Walsh transform," *IEEE Trans. Inform. Theory* (Corresp.), vol. IT-16, pp. 489–491, July 1970.
[9] J. E. Gibbs, "Walsh spectrometry, a form of spectral analysis well suited to binary digital computation," Nat. Phys. Lab., Teddington, Middlesex, England, 1967, unpublished.
[10] N. M. Blachman, "Spectral analysis with sinusoids and Walsh functions," *IEEE Trans. Aerosp. Electron. Syst.*, vol. AES-7, pp. 900–905, Sept. 1971.
[11] J. E. Gibbs and H. A. Gebbie, "Application of Walsh functions to transform spectroscopy," *Nature*, vol. 224, pp. 1012–1013, Dec. 6, 1969.
[12] P. A. Bello, "On the rms bandwidth of nonlinearly envelope detected narrow-band Gaussian noise," *IEEE Trans. Inform. Theory*, vol. IT-11, pp. 236–239, Apr. 1965.
[13] W. B. Davenport, Jr., and W. L. Root, *An Introduction to the Theory of Random Signals and Noise.* New York: McGraw-Hill, 1958, pp. 96–99.
[14] M. Marcus and H. Minc, *A Survey of Matrix Theory and Matrix Inequalities.* Boston, Mass.: Allyn and Bacon, 1964.
[15] L. E. Franks, *Signal Theory.* Englewood Cliffs, N. J.: Prentice-Hall, 1969, sec. 7.5, pp. 189–194.
[16] J. Pearl, "On coding and filtering stationary signals by discrete Fourier transforms," *IEEE Trans. Inform. Theory* (Corresp.), vol. IT-19, pp. 229–232, Mar. 1973.
[17] I. M. Ryshik and I. S. Gradstein, *Tables of Series, Products, and Integrals.* Berlin, Germany: Deutscher Verlag der Wissenschaften, 1957, p. 184, eq. 3.6157.
[18] J. R. Cohen, "A pragmatic view of Walsh functions through the Fourier series," in *Proc. Symp. Applications of Walsh Functions*, R. W. Zeek and A. E. Showalter, Eds. (Apr. 1973), pp. 257–258, ASTIA Doc. AD-763 000. Fig. 2 of this reference is a CRT representation of (20).

29

Reprinted with permission from *IEEE Trans. Electromagn. Compat.* **EMC-15:**81–83 (1973)

A Nonrecursive Equation for the Fourier Transform of a Walsh Function

KARL H. SIEMENS AND REUVEN KITAI

Abstract—Convolution of a discrete Walsh function with a rectangular pulse simplifies the derivation of an expression for the Fourier transform of a Walsh function. The nonrecursive transform equation that is developed is a function of the bits of the Gray code number for the order of the Walsh function.

INTRODUCTION

Considerable attention has recently been accorded to the orthogonal set of Walsh functions [1]–[3]. Much of the theory [4] and hardware application [5], [6] has involved signal analysis using the Walsh spectrum domain. If one wishes to convert the Walsh series of a signal to the more usual Fourier series representation, then the Fourier transforms of the Walsh functions are needed in the conversion equations [5]. Partial listings of the transforms are available [7], [8]. Beyond this range, unlisted transforms may be obtained using a recursive equation [8], which can be tedious, or by using an expression due to Blachman [9] for the Walsh transform of sinusoids, which can be modified to yield the Fourier transform of a Walsh function. In the latter process, sine or cosine terms in the transform are selected accordingly as the sum of adjacent bits in the binary representation of the order is odd or even.

This paper develops an alternative expression for the Fourier transform of a Walsh function which differs from previous expressions in that it incorporates the Gray code representation of the order of the function. The expression is nonrecursive, and it is also unified in the sense that no sine- or cosine-factor selection process is involved in conjunction with it. This expression leads to an algorithm whereby the transform may be obtained simply from inspection of the bits of the Gray code representation.

DERIVATION OF TRANSFORM EXPRESSION

Let the Fourier transform of a Walsh function, wal (m, θ), be defined by

$$\mathfrak{F}[\text{wal}(m,\theta)] = \int_{-1/2}^{1/2} \text{wal}(m,\theta)\exp[j2\pi f\theta]\, d\theta \qquad (1)$$

where m is the order of the function and θ is the normalized time. The even and odd Walsh functions, cal (s, θ) and sal (s, θ), respectively, have the Fourier transforms

$$\mathfrak{F}[\text{cal}(s,\theta)] = C(f,s) = \int_{-1/2}^{1/2} \text{cal}(s,\theta)\cos 2\pi f\theta\, d\theta \qquad (2)$$

$$\mathfrak{F}[\text{sal}(s,\theta)] = jS(f,s) = j\int_{-1/2}^{1/2} \text{sal}(s,\theta)\sin 2\pi f\theta\, d\theta \qquad (3)$$

where f and s are normalized frequency and sequency [4], respectively. Since the Walsh functions are discontinuous, evaluation of (1) to (3) would normally involve a summation of integrals.

It is convenient to view a continuous Walsh function as the convolution of the sequence of unit impulses that form the discrete Walsh function [10] over the interval $-\frac{1}{2} \leqslant \theta < \frac{1}{2}$ with a rectangular pulse of unit magnitude and of width $1/2^M$ equal to the spacing of the unit impulses, where M is the number of bits in the binary representation of m. The Fourier transform of the Walsh function is then the product of the transforms of the discrete Walsh function and the rectangular pulse. This will be shown to be

$$F[\text{wal}(m,\theta)] = (-1)^{g_0}(-j)^{\alpha} \cdot \left[\prod_{x=0}^{M-1} \cos\left(\frac{\pi f}{2^{x+1}} - g_x\frac{\pi}{2}\right)\right] \text{sinc}(f/2^M) \qquad (4)$$

where $j = \sqrt{-1}$, g_x is the xth bit in the Gray code representation of m, and α is the number of Gray code bits of value ONE, and

$$\text{sinc}(f/2^M) = \frac{\sin(\pi f/2^M)}{\pi f/2^M}. \qquad (5)$$

If the cal and sal functions are given, the order m is found from

$$\text{wal}(m,\theta) = \text{cal}(s,\theta), \qquad m = 2s \qquad (6)$$

or

$$\text{wal}(m,\theta) = \text{sal}(s,\theta), \qquad m = 2s - 1. \qquad (7)$$

To derive (4), we begin by defining the discrete Walsh function pulse sequence over one period. Referring to Fig. 1(a) a discrete wal (m, θ) in the range $0 \leqslant \theta < 1$ consists of a sequence of 2^M positive or negative unit impulses, where $\delta(\theta - \theta_y)$ denotes an impulse at $\theta_y = y/2^M$, $y = 0, 1, 2, \cdots, 2^M - 1$. Each impulse can also be described in terms of a delay operator d^y, where

$$d^y\,\delta(\theta) = \delta(\theta - \theta_y). \qquad (8)$$

Manuscript received March 28, 1972. This work was supported by the National Research Council of Canada.
K. H. Siemens is with General Motors of Canada, St. Catharines, Ont., Canada.
R. Kitai is with McMaster University, Hamilton, Ont., Canada.

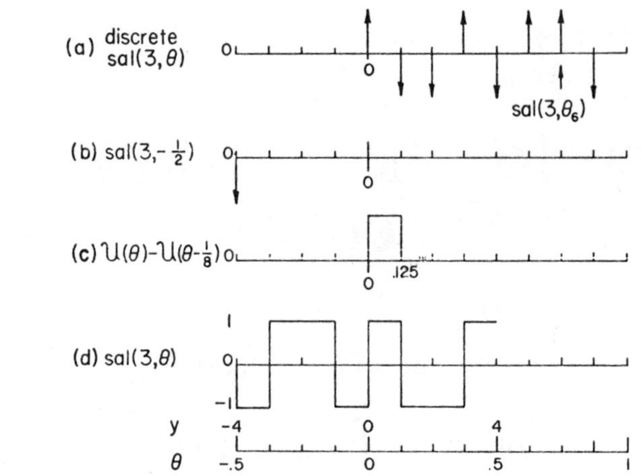

Fig. 1. Convolution quantities to form sal (3, θ) over range $-\frac{1}{2} \leqslant \theta < \frac{1}{2}$.

The delay operator has the property

$$d^y d^z = d^{y+z}. \tag{9}$$

The sign of an impulse at $\theta_y = y/2^M$ can be obtained [11] from a parity check for an even number of ONE's in the Gray code for m that are enabled by the bits y_{M-1-x} of binary y. That is

$$\mathrm{wal}\,(m, \theta_y) = \prod_{x=0}^{M-1} (-1)^{g_x y_{M-1-x}}. \tag{10}$$

If m is written in binary code

$$m = \sum_{x=0}^{M-1} m_x 2^x, \qquad m_x = 0 \text{ or } 1, \qquad m_M = 0 \tag{11}$$

then the corresponding Gray code bits g_x in (10) are found in the usual manner from

$$g_x = m_{x+1} \oplus m_x \tag{12}$$

where ⊕ denotes EXCLUSIVE OR.

In (10) it suffices to consider only those terms for which $y_{M-1-x} = 1$. For example, if $\theta_y = 6/8$, $y = 6 = 110_2$, $M = 3$, and

$$(-1)^{g_0 y_2} (-1)^{g_1 y_1} (-1)^{g_2 y_0} = (-1)^{g_0} (-1)^{g_1}. \tag{13}$$

The value of the discrete Walsh function at $\theta_y = 6/8$ can then be written

$$\mathrm{wal}\,(m, \theta_6) = (-1)^{g_0} (-1)^{g_1} d^6\, \delta(\theta). \tag{14}$$

If now $m = 5$, wal $(5, \theta_6)$ = sal $(3, \theta_6)$; then $m = 101_2 = 111$ Gray. So $g_0 = g_1 = 1$ and sal $(3, \theta_6) = +1\, d^6\, \delta(\theta)$. (See Fig. 1(a).) Similarly

$$\mathrm{wal}\,(m, \theta_0) = d^0\, \delta(\theta). \tag{15}$$

The summation of impulses, each of the form of (14), that define the discrete wal (m, θ) over the interval $0 \leqslant \theta < 1$ may be factored to yield

$$\text{discrete wal}\,(m, \theta) = \prod_{x=0}^{M-1} \left[d^0 + (-1)^{g_x} d^{2^{M-1-x}} \right] \delta(\theta). \tag{16}$$

Convolution of a delta function at $\theta = -\frac{1}{2}$ with (16) achieves a shift in the range from $0 \leqslant \theta < 1$ to $-\frac{1}{2} \leqslant \theta < \frac{1}{2}$. However, the sign of this δ function is negative for functions of odd sequency and positive for functions of even sequency [1]. Since the Gray code bit $g_0 = 1$ and the sequency is odd, the required delta function at $\theta = -\frac{1}{2}$ is

$$\text{discrete wal}\,(m, -\tfrac{1}{2}) = (-1)^{g_0} d^{-2^{M-1}}\, \delta(\theta). \tag{17}$$

The final convolving quantity is the unit amplitude pulse of width $1/2^M$. If $\mathcal{U}(\theta - \theta_y)$ defines a unit step at $\theta_y = y/2^M$, then the required pulse is

$$P(\theta_1) = \mathcal{U}(\theta) - \mathcal{U}(\theta - \theta_1). \tag{18}$$

Hence

$$\mathrm{wal}\,(m, \theta) = \prod_{x=0}^{M-1} \left[d^0 + (-1)^{g_x} d^{2^{M-1-x}} \right] \delta(\theta) \tag{19}$$
$$* (-1)^{g_0} d^{-2^{M-1}}\, \delta(\theta)$$
$$* [\mathcal{U}(\theta) - \mathcal{U}(\theta - \theta_1)]$$

where $-\frac{1}{2} \leqslant \theta < \frac{1}{2}$.

The three convolving quantities are shown in Figs. 1(a)-1(c), resulting in the continuous Walsh function sal $(3, \theta)$ shown in Fig. 1(d).

The Fourier transform of wal (m, θ) is the product of the transforms of (16)-(18). Each impulse that comprises the discrete Walsh function has the Fourier transform

$$\mathcal{F}\left[\prod_{x=0}^{M-1} (-1)^{g_x y_{M-1-x}} d^y\, \delta(\theta) \right]$$
$$= \left[\prod_{x=0}^{M-1} (-1)^{g_x y_{M-1-x}} \right] \int_0^1 \delta(\theta - \theta_y) \exp(j2\pi f\theta)\, d\theta$$
$$= \left[\prod_{x=0}^{M-1} (-1)^{g_x y_{M-1-x}} \right] \exp\left(j\pi f y/2^{M-1} \right). \tag{20}$$

The only change, then, in the summation of impulses that form discrete wal (m, θ) and in its transform is that $\exp(j\pi f y/2^{M-1})$ replaces $d^y\, \delta(\theta)$. Hence, the sum of the Fourier transforms of the impulses of discrete wal (m, θ), can be factored in a manner similar to (16). Replacing $d^0\, \delta(\theta)$ and $d^{2^{M-1-x}}\, \delta(\theta)$ by their Fourier transforms, 1 and $\exp(j\pi f/2^x)$, respectively

$$\mathcal{F}[\text{discrete wal}\,(m, \theta)] = \prod_{x=0}^{M-1} \left[1 + (-1)^{g_x} \exp(j\pi f/2^x) \right]. \tag{21}$$

Each factor in (21) can be rewritten as a sine or cosine, depending on the value of the Gray code bit. If $g_x = 1$

$$-[1 - \exp(j\pi f/2^x)] = 2j \exp(j\pi f/2^{x+1}) \sin(\pi f/2^{x+1}). \tag{22}$$

If $g_x = 0$

$$[1 + \exp(j\pi f/2^x)] = 2 \exp(j\pi f/2^{x+1}) \cos(\pi f/2^{x+1}). \tag{23}$$

The ONE bits in the Gray code number $G = g_0g_1g_2 \cdots g_x \cdots$ for the order m correspond to the sine factors of the transform (22). If there are α ONE's in G, then the sinusoidal form of (21) is

$\mathfrak{F}$ [discrete wal (m, θ)]

$$= 2^M (-j)^\alpha \exp j\pi f(\tfrac{1}{2} + \tfrac{1}{4} + \cdots + 1/2^M) \prod_{x=0}^{M-1} \cos\left(\frac{\pi f}{2^{x+1}} - g_x \frac{\pi}{2}\right)$$

$$= 2^M (-j)^\alpha \exp j\pi f(1 - 1/2^M) \prod_{x=0}^{M-1} \cos\left(\frac{\pi f}{2^{x+1}} - g_x \frac{\pi}{2}\right) \quad (24)$$

where

$$\cos\left(\frac{\pi f}{2^{x+1}} - g_x \frac{\pi}{2}\right) = \sin \frac{\pi f}{2^{x+1}} \text{ if } g_x = 1.$$

The Fourier transforms of (17) and (18), respectively, are

$$\mathfrak{F}\left[(-1)^{g_0} d^{-2^{M-1}} \delta(\theta)\right] = (-1)^{g_0} \exp(-j\pi f) \quad (25)$$

$$\mathfrak{F}[\mathfrak{U}(\theta) - \mathfrak{U}(\theta - \theta_1)] = \frac{1}{2^M} \exp(j\pi f/2^M) \operatorname{sinc}(f/2^M). \quad (26)$$

The product of (24)-(26) is the nonrecursive equation (4).

TRANSFORM ALGORITHM

Using (4), a simple algorithm is now developed for the Fourier transform of wal (m, θ). It uses only the bits in the Gray code number G that represent the order m. In (4) it is seen that each factor depends on G; thus g_0 and g_x are bits in G, α is the number of bits that equal ONE, and M is the number of bits in G.

The first step of the algorithm is to write G as $g_0g_1g_2 \cdots g_x \cdots g_{M-1}$. Wherever $g_x = 0$, substitute $\cos(\pi f/2^{x+1})$ and where $g_x = 1$, substitute $\sin(\pi f/2^{x+1})$ to yield an expression in the form of a product of cosine and sine terms. There are as many terms (i.e., M) in the product expression as there are bits in G. The next step is to multiply the expression obtained thus far by $\operatorname{sinc}(f/2^M) = [\sin(\pi f/2^M)]/(\pi f/2^M)$.

The sign of the transform is given by $(-1)^{g_0}(-j)^\alpha$. A simple counting procedure is convenient. Form

$$\beta = \alpha + \langle \alpha/2 \rangle + g_0 \quad (27)$$

where $\langle \alpha/2 \rangle$ denotes the integer part of $\alpha/2$. If β is odd, the sign of the transform is negative. If $\alpha/2$ is not an integer, multiply the transform by $j = \sqrt{-1}$. An example to illustrate the algorithm follows.

EXAMPLE

Let us find, say, the Fourier transform of sal $(5, \theta)$. From (7), sal $(5, \theta)$ = wal $(9, \theta)$, so the order $m = 9 = 1001_2$. Applying (12), the corresponding Gray code number is $G = 1101 = g_3g_2g_1g_0$. Substituting sines and cosines for $g_0 \to g_3$ yields

$$\sin \frac{\pi f}{2} \cos \frac{\pi f}{4} \sin \frac{\pi f}{8} \sin \frac{\pi f}{16}.$$

Multiply the above by $[\sin(\pi f/16)]/(\pi f/16)$.

To obtain the sign, there are three ONE's in G so $\alpha = 3$. Also, $g_0 = 1$. Hence

$$\beta = 3 + \langle 3/2 \rangle + 1 = 5 \quad (28)$$

which is odd, so the sign is negative. Since α is odd, multiply by j. Consequently, the Fourier transform of wal $(9, \theta)$ is

$$jS(f, 5) = -j \sin \frac{\pi f}{2} \cos \frac{\pi f}{4} \sin \frac{\pi f}{8} \frac{\sin^2(\pi f/16)}{\pi f/16}. \quad (29)$$

CONCLUSION

A nonrecursive expression for the continuous two-sided Fourier transform of a Walsh function of any given order m has been derived, and a convenient algorithm for its evaluation has been developed. To obtain the Fourier series coefficients of a periodic Walsh function, the transform expression is evaluated for integral values of the normalized frequency f. Since the order m is the only required information, the algorithm should find application in the computer evaluation of either the Fourier transform or the Fourier series coefficients of a Walsh function.

ACKNOWLEDGMENT

The authors wish to thank R. de Buda for suggesting the convolution technique and D. Taylor for some useful comments.

REFERENCES

[1] H. F. Harmuth, "Transmission of Information by Orthogonal Functions," Springer-Verlag, New York, 1969.

[2] C. A. Bass (Editor), "Proc. 1970 Symposium and Workshop on Applications of Walsh Functions," Washington, April 1970.

[3] R. W. Zeek and A. E. Showalter (Editors), "Proc. 1971 Symposium on Applications of Walsh Functions," Special issue, *IEEE Trans. Electromagnetic Compatibility*, vol. EMC-13, Aug. 1971.

[4] Henning F. Harmuth, "A Generalized Concept of Frequency and some Applications," *IEEE Trans. on Information Theory*, vol. 14, May 1968, pp. 375–382.

[5] K. H. Siemens and R. Kitai, "Digital Walsh-Fourier Analysis of Periodic Waveforms," *IEEE Trans. on Instrumentation and Measurement*, vol. IM-18, Dec. 1969, pp. 316–321.

[6] C. Boesswetter, 'Analog Sequency Analysis and Synthesis of Voice Signals," *Proc. Symp. Appl. of Walsh Functions*, Washington, 1970, pp. 220–229.

[7] K. H. Siemens, "Digital Walsh-Fourier Analyser for Periodic Waveforms," M.Eng. Thesis, McMaster University, Hamilton, Ontario, Canada, 1969.

[8] Heinz H. Schreiber, "Bandwidth Requirements for Walsh Functions," *Proc. Symp. Appl. of Walsh Functions*, Washington, 1970, pp. 46–51.

[9] N. M. Blachman, "Spectral Analysis with Sinusoids and Walsh Functions," *IEEE Trans. Aerospace and Electronic Systems*, vol. AES-7, Sept. 1971, pp. 900–905.

[10] John L. Shanks, "Computation of the Fast Walsh-Fourier Series," *IEEE Trans. on Computers*, vol. C-18, May 1969, pp. 457–459.

[11] R. Kitai and K. H. Siemens, "Comment on a Simplified Definition of Walsh Functions," *IEEE Trans. on Computers*, vol. C-21, May 1972, p. 512.

30

Reprinted with permission from *IEEE Trans. Electromagn. Compat.* **EMC-17:**266–269 (1975)

Walsh-to-Fourier Spectral Conversion for Periodic Waves

REUVEN KITAI

***Abstract*—Relations are developed for the determination of the Fourier spectra of frequency-limited periodic waves from truncated Walsh spectra. The matrix conversion process is simplest if the highest-order Walsh coefficient in the spectrum to be converted is 2^n, where n is an integer. For such cases, compensation for truncation consists of a diagonal matrix that premultiplies the Walsh to Fourier conversion matrix and the elements of which are $[(\sin x)/x]^{-2}$ terms. Element values range between unity and less than $\pi^2/4$. The same compensation matrix is used for determining the Walsh spectra of sequency-limited waves from 2^n Fourier expansion terms. Examples are included which demonstrate the spectral conversion processes, Walsh to Fourier and Fourier to Walsh.**

INTRODUCTION

The Walsh expansion of a periodic wave $g(t)$ warrants attention as a practical alternative to the Fourier expansion because of the hardware simplicity and speed of Walsh spectral analyzers. The multiplication by sines and cosines in Fourier processing is obviated in Walsh processing; in the latter, signal values are simply accumulated with or without sign change, depending on the signs of signal, sal, and cal values at any instant. Also, arrays of Walsh waves with low orthogonality errors are easily generated, so that rapid parallel processing of a signal is practicable in the Walsh domain.

If the highest harmonic in $g(t)$ is F, then $2F$ sequency values and the dc value suffice to describe the function [1]. Generally, conversion from a Walsh spectrum to the equivalent Fourier spectrum is achieved by using the matrix relations

$$[a] = [\Lambda_a][A]$$
$$[b] = [\Lambda_b][B] \tag{1}$$

where $[A]$, $[B]$, $[a]$, and $[b]$ are vectors whose elements are cal, sal, cosine, and sine expansion coefficients, respectively. $[\Lambda_a]$ and $[\Lambda_b]$ are infinite conversion matrices whose properties have been examined in some detail by Blachman [2].

Frequency-limited signals have infinite Walsh spectra, but Walsh analysers have finite range. For such signals $[A]$ and $[B]$ are necessarily truncated, with the consequent possibility of substantial error in conversion when using (1). The theory of spectral conversion developed here arises from a computer evaluation of such errors [3]. It establishes that the finite vectors $[a]$ and $[b]$ may be obtained without error from the truncated vectors $[A]$ and $[B]$, provided that the dimension of the latter vector pair equals or exceeds that of the former. Compensation coefficients K_{af} and K_{bf} defined by (2), are derived for each Fourier coefficient of normalised frequency f. The case of particular interest is shown to be that for which the dimension of $[A]$ and $[B]$ is $S = 2^{n-1}$, where n is an integer. The compensation coefficient for a term of frequency f is a function of f and n only, and is given by

$$K_{af} = K_{bf} = \operatorname{sinc}^{-2}(f/2^n), \qquad \text{for } f < 2^{n-1}$$
$$K_{bf} = \tfrac{1}{2}\operatorname{sinc}^{-2}(1/2) \approx 1.23, \qquad \text{for } f = 2^{n-1} \tag{2}$$

Manuscript received March 25, 1975. This work was supported by a grant from the National Research Council of Canada.
The author is with the Department of Electrical Engineering, McMaster University, Hamilton, Ont., Canada.

where $\operatorname{sinc} z = (\sin \pi z)/\pi z$. Values of K_{af} and K_{bf} range between unity and less than $\pi^2/4 \approx 2.47$. For frequency-limited periodic signals (1) is modified to be:

$$[a] = [K_a][\Lambda_a][A]$$
$$[b] = [K_b][\Lambda_b][B] \tag{3}$$

where $[\Lambda_a]$ and $[\Lambda_b]$ are square and of order 2^{n-1}, and where $[K_a]$ and $[K_b]$ are diagonal matrices of the same order, whose elements are given by (2). As the number of Walsh coefficients n approaches infinity, the compensation matrices approach identity matrices.

The dual of the previous case is that in which a sequency-limited periodic signal is expanded in the form of a truncated Fourier series. Again it is advantageous to have the dimension of the truncated vectors $[a]$ and $[b]$ to be 2^{n-1}, no less than that of $[A]$ and $[B]$; then the same conversion and compensation matrices as in (3) are used in the conversion relations

$$[A] = \tfrac{1}{2}[\Lambda_a]^T[K_a][a]$$
$$[B] = \tfrac{1}{2}[\Lambda_b]^T[K_b][b] \tag{4}$$

The next step is to summarize some established general relationships between Walsh and Fourier expansions. These relations are applied to the conversion process, and examples are given to illustrate the applications of (3) and (4).

WALSH AND FOURIER INTERRELATIONSHIPS

A natural Walsh set comprises 2^n functions $\text{wal}(k,\theta), 0 \leq k \leq 2^n$. n is the number of digits in the binary representation of the order k. The normalised time is $\theta = t/T$. For example, the Walsh set of Fig. 1 has $n = 3$. If $\text{wal}(0,\theta)$ is taken as a cal function, then a set comprises 2^{n-1} sal functions and an equal number of cal functions.

From the point of view of analyser and converter implementation, it is convenient to adopt Boesswetter's expansions [4] of $g(t)$ in terms of frequency and sequency components:

$$g(\theta) = a_0 + \sum_f a_f \cos(f2\pi\theta) + \sum_s \sin(f2\pi\theta), \qquad \text{where } f = 1,2,3,\cdots$$
$$= A_0 + \sum_s A_s \operatorname{cal}(s,\theta) + \sum_s B_s \operatorname{sal}(s,\theta), \qquad \text{where } s = 1,2,3,\cdots. \tag{5}$$

In (5) the sum of the cosine in terms may be equated to the sum of the cal terms, and similarly for the sine and sal terms. If, now, the Fourier expansions of Walsh functions are written

$$\operatorname{cal}(s,\theta) = \sum_f a_{fs} \cos(f2\pi\theta)$$
$$\operatorname{sal}(s,\theta) = \sum_f b_{fs} \sin(f2\pi\theta) \tag{6}$$

where

$$a_{fs} = 2\int_0^1 \operatorname{cal}(s,\theta)\cos(f2\pi\theta)\, d\theta$$
$$b_{fs} = 2\int_0^1 \operatorname{sal}(s,\theta)\sin(f2\pi\theta)\, d\theta \tag{7}$$

then

$$a_f = \sum_s a_{fs}A_s$$
$$b_f = \sum_s b_{fs}B_s. \tag{8}$$

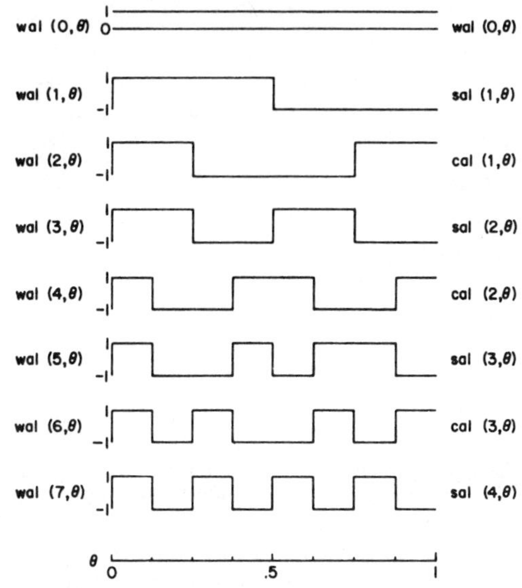

Fig. 1. The natural set of Walsh functions for $n = 3$.

Equations (1) are the matrix form of (8), and we see that the elements of $[\Lambda_a]$ and $[\Lambda_b]$ are Fourier coefficients of Walsh functions. Conversely, we may replace cosines and sines in (5) by their corresponding cal and sal expansions, whence

$$A_s = \sum_f A_{sf} a_f$$

$$B_s = \sum_f B_{sf} b_f, \tag{9}$$

where A_{sf}, B_{sf} are Walsh coefficients of trigonometric functions. They are also evaluated from (7), but with the factor 2 omitted since the average value of a squared trigonometric function over a cycle is 1/2 whereas it is unity for a squared Walsh function. (The d.c. component is, of course, the exception.) Hence

$$A_{sf} = \tfrac{1}{2} a_{fs}$$

$$B_{sf} = \tfrac{1}{2} b_{fs} \tag{10}$$

and conversion to a Walsh spectrum from a Fourier spectrum is achieved by using

$$[A] = \tfrac{1}{2}[\Lambda_a]^T[a],$$

$$[B] = \tfrac{1}{2}[\Lambda_b]^T[b]. \tag{11}$$

For a given f and s, a_{fs}, and b_{fs} have the same magnitudes, but their signs may differ. They have been evaluated in different ways, [5]; [6] which relate to the expression used for representing Walsh functions. The nonrecursive Gray code representation [6], [7] is convenient here. For $0 \le \theta < 1$, the one-sided Fourier transform of a Walsh function of order k is

$$\mathfrak{F}\{\text{Wal}\,(k,\theta)\} = 2(-j)^{\alpha} \exp(-j\pi f) \left(\text{sinc}\,\frac{f}{2^n}\right) \prod_{x=0}^{n-1} \cdot \cos\left(\frac{\pi f}{2^{x+1}} - g_x \frac{\pi}{2}\right) \tag{12}$$

where $k = 2s$ for cal functions, $k = 2s - 1$ for sal functions, $k_{\text{Gray}} = g_{n-1} \cdots g_x \cdots g_1 g_0$, α is the number of one's in k_{Gray}. α is even for cal functions and odd for sal functions, so the transform of a sal function is imaginary. A term within the product in (12) is a sine or a cosine, depending on whether g_x is 1 or 0. Substitution of integer values of f in (12) yields a_{fs} and b_{fs}. For example, the Fourier expansion of the periodic wave cal $(3,\theta)$ is obtained by putting $k = 6 = 101_{\text{Gray}} = g_2 g_1 g_0$ in (12), whence

$$\mathfrak{F}\{\text{cal}\,(3,\theta)\} = -2 \exp(-j\pi f) \left(\frac{8}{\pi f} \sin\frac{\pi f}{8}\right)\left(\sin\frac{\pi f}{8}\right) \cdot \left(\cos\frac{\pi f}{4}\right)\left(\sin,\frac{\pi f}{2}\right).$$

On substituting $f = 1,2,3,\cdots$ we get $a_{13} = 0.52739$, $a_{33} = 1.0246$, $a_{53} = -0.61477$ etc., all even-frequency coefficients being zero. The first 4×4 elements a_{fs} and b_{fs} are listed in (13) and (14). In (13) the first element is $a_{00} = 1$

$$[\Lambda_a] = \begin{bmatrix} 1 & 0 & 0 & 0 & \cdots \\ 0 & 4/\pi & 0 & 0.52739 & \cdots \\ 0 & 0 & 4/\pi & 0 & \cdots \\ 0 & -4/3\pi & 0 & 1.0246 & \cdots \\ \cdot & \cdot & \cdot & \cdot & \\ \cdot & \cdot & \cdot & \cdot & \\ \cdot & \cdot & \cdot & \cdot & \end{bmatrix} \tag{13}$$

$$[\Lambda_b] = \begin{bmatrix} 4/\pi & 0 & -0.52739 & 0 & \cdots \\ 0 & 4/\pi & 0 & 0 & \cdots \\ 4/3\pi & 0 & 1.0246 & 0 & \cdots \\ 0 & 0 & 0 & 4/\pi & \cdots \\ \cdot & \cdot & \cdot & \cdot & \\ \cdot & \cdot & \cdot & \cdot & \\ \cdot & \cdot & \cdot & \cdot & \end{bmatrix} \tag{14}$$

The Fourier coefficients of a given Walsh function are obtained from the corresponding column s, and the Walsh coefficients of a given trigonometric function from the corresponding row f. The familiar Fourier expansion of sal $(1,\theta)$ is seen in the first column of (14), and that of cal $(1,\theta)$ in the second column of (13).

CONVERSION USING TRUNCATED SPECTRA

Assume that $g(\theta)$ is a frequency-limited periodic wave whose Fourier expansion coefficients do not extend beyond a_{S-1} and b_S, where S is arbitrary, and let its Walsh expansion be truncated at coefficients A_{S-1} and B_S. Cal and sal expansions are now considered separately. Let $\hat{a}_f$ be the approximation to a_f when only S coefficients are summed in (8). Then from (9) and (10)

$$\hat{a}_f = \sum_{s=0}^{S-1} a_{fs} A_s = \sum_{s=0}^{S-1} \tfrac{1}{2} a_{fs}(a_{1s}a_1 + a_{2s}a_2 + \cdots + a_{fs}a_f + \cdots + a_{S-1,s}a_{S-1}). \tag{15}$$

Each term in (15) is of the form

$$\sum_{s=0}^{S-1} \tfrac{1}{2} a_{fs} a_{hs} a_h. \tag{16}$$

We now employ (12), using integer values for f and h, and write

$$a_{fs} a_{hs} = 4\{\cos \pi(f+h)\}(\text{sinc}\, f/2^n)(\text{sinc}\, h/2^n)P \tag{17}$$

where

$$P = \prod_{x=0}^{n-1} \frac{1}{2}\left[\cos\frac{\pi}{2^{x+1}}(f-h) + \cos\left\{\frac{\pi}{2^{x+1}}(f+h) - g_x \pi\right\}\right].$$

The second cosine term in P may be written

$$(-1)^{g_x} \cos\frac{\pi}{2^{x+1}}(f+h).$$

Let G_x denote the sign of $(-1)^{g_x}$, i.e., G_x is + for $g_x = 0$ and − for $g_x = 1$. Then

$$P = \frac{1}{2^n} \prod_{x=0}^{n-1} \left\{ \cos \frac{\pi}{2^{x+1}} (f - h) + G_x \cos \frac{\pi}{2^{x+1}} (f + h) \right\}. \quad (18)$$

The expansion of (18) produces a sum of 2^n terms, all but one of which include a sign coefficient G_x or a product of sign coefficients. Assume now that (16) is summed in sequency over a natural Walsh set, i.e., $S = 2^{n-1}$. Since a Gray-coded number set is complete for a given natural Walsh set, all but two terms cancel in the expanded form of (18), leaving the reduced P given by (19) to be summed:

$$P = \frac{1}{2^n} \left\{ \prod_{x=0}^{n-1} \cos \frac{\pi}{2^{x+1}} (f - h) + \prod_{x=0}^{n-1} G_x \cos \frac{\pi}{2^{x+1}} (f + h) \right\}. \quad (19)$$

To illustrate, assume that $n = 3$ so that $k = g_2 g_1 g_0$. The corresponding seven sign coefficients in the expansion of (18) are G_0, G_1, G_2, G_0G_1, G_0G_2, G_1G_2, and $G_0G_1G_2$. These are listed in the upper half of Table 1 for the cal sequencies within the range. We see from the columns that summing results in the cancelling of all terms in (18) that contain G_x and G_x products, with the exception of those that contain $G_0G_1G_2$, i.e., the product of all of the sign coefficients.

The upper bound for f and h in the cosine set is $2^{n-1} - 1$. Substitution of all allowed values of f and h in (19) yields $P = 1/2^n$ for $f = h$ and $P = 0$ for all other values since for all $f \neq h$ there is one term of the form $\cos (2r + 1)\pi/2$ in each product in (19), r being an integer. For any natural Walsh set, (15) therefore reduces to

$$\hat{a}_f = \sum_{s=0}^{S-1} \frac{1}{2} a_{fs}{}^2 a_f = a_f \sum_{s=0}^{2^{n-1}-1} \left(\frac{1}{2^{n-1}} \operatorname{sinc}^2 \frac{f}{2^n} \right)$$

$$= a_f \operatorname{sinc}^2 \frac{f}{2^n}.$$

The same procedure is followed in evaluating sine coefficients from

$$\hat{b}_f = \sum_{s=1}^{S} b_{fs} B_s \quad (20)$$

the maximum values of f and h now being $S = 2^{n-1}$. The same cancelling of terms in (18) occurs, as illustrated by the sal sequencies in the lower half of Table 1, while (19) is again zero for all $f \neq h$, and $1/2^n$ for all $f = h < 2^{n-1}$. However, for $f = h = 2^{n-1}$, $P = 1/2^{n-1}$, so (20) reduces to

$$\hat{b}_f = \sum_{s=1}^{S} \frac{1}{2} b_{fs}{}^2 b_f = b_f \operatorname{sinc}^2 f/2^n, \qquad \text{for } f < S$$

$$\hat{b}_f = 2b_f \operatorname{sinc}^2 1/2, \qquad \text{for } f = S.$$

Since the compensation for a Fourier coefficient of frequency f is a function of f and n only, spectral conversion is achieved without error by the compensated process (3).

The dual case, which is more of theoretical than practical interest, is the evaluation of the Walsh spectrum of a sequency-limited signal from its truncated Fourier spectrum. If the highest sequency component in $g(\theta)$ is S, then $2S$ Fourier spectral values and the dc value suffice to define it. Nevertheless, it is desirable to acquire $2^n \geq 2S$ Fourier coefficients, for then the conversion relations are given by (4). In transposing $[\Lambda_a]$ for use in (4), element a_{00} is replaced by 2.

Two examples illustrate the conversion processes, one wave being frequency-limited and the other sequency-limited. In each example 4×4 matrices $[\Lambda_a]$ and $[\Lambda_b]$ suffice; element values used are those in (13) and (14) to three significant figures.

EXAMPLE 1

We assume that $g(\theta)$ consists of three harmonically related sinusoids

$$g(\theta) = \sin (2\pi\theta) + 5 \sin 2(2\pi\theta) + 4 \sin 3(2\pi\theta).$$

TABLE I

	s	k	$g_2g_1g_0$	G_0	G_1	G_2	G_0G_1	G_0G_2	G_1G_2	$G_0G_1G_2$
	0	0	0 0 0	+	+	+	+	+	+	+
CAL	1	2	0 1 1	-	-	+	+	-	-	+
	2	4	1 1 0	+	-	-	-	-	+	+
	3	6	1 0 1	-	+	-	-	+	-	+
	1	1	0 0 1	-	+	+	-	-	+	-
SAL	2	3	0 1 0	+	-	+	-	+	-	-
	3	5	1 1 1	-	-	-	+	+	+	-
	4	7	1 0 0	+	+	-	+	-	-	-

The corresponding Walsh spectrum is obtained from (11); for $[b]^T = [1\ 5\ 4\ 0]$ the first four elements of the infinite vector $[B]$ are 1.49, 3.18, 1.79, and 0. These would be the measured values of $g(\theta)$, using a synchronised Walsh spectrum analyzer that truncates at $n = 3$. From (1), the uncompensated Fourier spectrum $[\hat{b}]$ is

$$\begin{bmatrix} 1.27 & 0 & -0.527 & 0 \\ 0 & 1.27 & 0 & 0 \\ 0.424 & 0 & 1.02 & 0 \\ 0 & 0 & 0 & 1.27 \end{bmatrix} \begin{bmatrix} 1.49 \\ 3.18 \\ 1.79 \\ 0 \end{bmatrix} = \begin{bmatrix} 0.95 \\ 4.05 \\ 2.46 \\ 0 \end{bmatrix}$$

which is substantially in error. In compensating for truncation, the elements of the 4×4 matrix $[K_b]$ to be used are $\operatorname{sinc}^{-2} (1/8)$, $\operatorname{sinc}^{-2} (1/4)$, $\operatorname{sinc}^{-2} (3/8)$, and $1/2 \operatorname{sinc}^{-2} (1/2)$, so that $[b] = [K_b][\hat{b}]$ is

$$\begin{bmatrix} 1.05 & 0 & 0 & 0 \\ 0 & 1.23 & 0 & 0 \\ 0 & 0 & 1.63 & 0 \\ 0 & 0 & 0 & 1.23 \end{bmatrix} \begin{bmatrix} 0.95 \\ 4.05 \\ 2.46 \\ 0 \end{bmatrix} = \begin{bmatrix} 1 \\ 5 \\ 4 \\ 0 \end{bmatrix}.$$

The Fourier spectrum of $g(\theta)$ would be recovered for any other initial value of θ from the corresponding eight cal and sal coefficients in its truncated Walsh spectrum.

EXAMPLE 2

We assume that $g(\theta)$ is a rectangular periodic pulse

$$g(\theta) = \begin{cases} 1 & \text{for } 0 \leq \theta \leq 1/8 \\ 0 & \text{for } 1/8 \leq \theta \leq 1 \end{cases}.$$

This wave is sequency-limited, with Walsh spectrum

$$[A]^T = [B]^T = (1/8)[1\ 1\ 1\ 1].$$

Its Fourier spectrum may be obtained by using these vectors in (1), or in the more familiar way, from

$$a_f = \frac{1}{\pi f} \sin f \frac{\pi}{4}$$

$$b_f = \frac{1}{\pi f} \left(1 - \cos f \frac{\pi}{4} \right).$$

When truncated at $f = 3$ for a_f and at $f = 4$ for b_f, this spectrum is

$$[a] = \begin{bmatrix} 0.125 \\ 0.225 \\ 0.159 \\ 0.075 \end{bmatrix}$$

$$[b] = \begin{bmatrix} 0.093 \\ 0.159 \\ 0.181 \\ 0.159 \end{bmatrix}$$

In (4), then, $[K_a][a]$ is given by

$$\begin{bmatrix} 1 & 0 & 0 & 0 \\ 0 & 1.05 & 0 & 0 \\ 0 & 0 & 1.23 & 0 \\ 0 & 0 & 0 & 1.63 \end{bmatrix} \begin{bmatrix} 0.125 \\ 0.225 \\ 0.159 \\ 0.075 \end{bmatrix} = \begin{bmatrix} 0.125 \\ 0.237 \\ 0.196 \\ 0.122 \end{bmatrix}$$

so that $[A]$ is

$$\frac{1}{2} \begin{bmatrix} 2 & 0 & 0 & 0 \\ 0 & 1.27 & 0 & -0.424 \\ 0 & 0 & 1.27 & 0 \\ 0 & 0.527 & 0 & 1.02 \end{bmatrix} \begin{bmatrix} 0.125 \\ 0.237 \\ 0.196 \\ 0.122 \end{bmatrix} = \begin{bmatrix} 0.125 \\ 0.125 \\ 0.125 \\ 0.125 \end{bmatrix}$$

The same result is obtained for $[B]$ by using (14) and the matrix $[K_b]$ of Example 1. If compensation is not applied, the calculated spectrum of $[A]^T$ is [0.125, 0.127, 0.101, 0.098].

ACKNOWLEDGMENT

The author is grateful to R. deBuda, S. Cohn-Sfetcu, and C. Kuczer for helpful discussions.

REFERENCES

[1] H. F. Harmuth and R. deBuda "Conversion of Sequency-Limited Signals into Frequency-Limited Signals and Vice Versa". IEEE Trans. IT-17, 1971, pp. 343–4.

[2] N. M. Blachman "Sinusoids versus Walsh Functions", Proc. IEEE, 62, 1974, pp. 346–354.

[3] K. H. Siemens and R. Kitai "Walsh Series to Fourier Series Conversion". Symp. Appl. Walsh Fns. Washington 1972, pp. 295–297.

[4] C. A. Boesswetter "The Mutual Spectral Representation of Trigonometric Functions and Walsh Functions" Symp. Appl. Walsh Fns Washington 1971, pp. 43–46.

[5] N. M. Blachman "Spectral Analysis with Sinusoids and Walsh Functions". IEEE Trans. AES-7, 1971, pp. 900–905.

[6] K. H. Siemens and R. Kitai "A Non-recursive Equation for the Fourier Transform of a Walsh Function" IEEE Trans EMC-15, 1973, pp. 81–83.

[7] C. K. Yuen "Walsh Functions and Gray Code" IEEE Trans. EMC-13, 1971, pp. 68–73.

31

Reprinted with permission from *IEEE Trans. Instrum. Meas.* **IM-24:**313–317 (1975)

Synthesis of Periodic Sinusoids from Walsh Waves

REUVEN KITAI, SENIOR MEMBER, IEEE

***Abstract*—The Walsh expansion of a periodic sinusoid is an infinite series; however, if one synthesizes a sinusoid from a truncated Walsh expansion, the result is a stepped approximation to the sinusoid that has minimum mean-square error. The first 2^m nonzero terms in the Walsh expansion yield a wave that has 2^m steps per quarter-cycle. A notable feature of this wave is the wide separation between its harmonic pairs. For example, eight nonzero Walsh terms yield a wave whose only nonzero harmonics are the 31st and 33rd, the 63rd and 65th, etc., harmonic magnitudes being inversely proportional to frequency. Waves of this type are readily synthesized using a synchronous Walsh array generator with true and complement outputs, a level converter system, and a summing circuit.**

INTRODUCTION

THE WALSH EXPANSION of a periodic wave $g(\theta) = \cos 2\pi\theta$ is the infinite series $\Sigma_s a_s$ cal (s,θ) where s is the normalized sequency and $\theta = t/T$ is the normalized time. If this series is truncated at the 2^{n-1}th cal term in a set of 2^n Walsh functions, then the resultant approximation to a cosine wave, $\hat{g}(\theta)$, has 2^{n-2} equally spaced steps per quarter-cycle, and its Fourier expansion is shown in the Appendix to be

$$\hat{g}(\theta) = \left(\operatorname{sinc}^2 \frac{1}{2^n}\right)\left\{\cos 2\pi\theta + \sum_q \frac{(-1)^q}{q2^n + 1} \cos\left[(q2^n + 1)\, 2\pi\theta\right] - \sum_q \frac{(-1)^q}{q2^n - 1} \cos\left[(q2^n - 1)\, 2\pi\theta\right]\right\} \quad (1)$$

where $q = 1,2,3, \cdots$, and sinc $z = (\sin \pi z)/\pi z$. Thus the spectrum consists of a fundamental of magnitude $\operatorname{sinc}^2 (\frac{1}{2^n})$, and harmonic pairs of normalized frequency $f = q2^n \pm 1$ whose amplitudes are inversely proportional to frequency. For example, if $n = 5$, the wave has eight steps per quarter-cycle, the fundamental is of magnitude 0.9968, the first harmonic pair is the 31st and 33rd, of magnitude 0.9968/31 and 0.9968/33, respectively; the next pair is the 63rd and 65th, of magnitude 0.9968/63 and 0.9968/65, respectively, and so on.

Being orthogonal, the Walsh functions as basis functions meet the requirement of finality of coefficients, the property whereby one may determine any given coefficient in an expansion without needing to know any of the other coefficients in the expansion. Also, with regard to accuracy of representation when not all of the terms for an exact representation can be used, a truncated Walsh expansion minimizes the "integral-squared error," the integral of the square of the difference be-

Manuscript received May 15; revised July 12, 1975. This work was supported by the National Research Council of Canada.

The author is with the Department of Electrical Engineering, McMaster University, Hamilton, Ont. L8S 4L7, Canada.

TABLE I

Sequency	1	3	5	7	9	11	13	15
Magnitude	$2/\pi$	0.2637	0.0525	0.1266	0.0125	0.0052	0.0260	0.0627
Sign								
cal	+	+	−	+	−	−	−	+
sal	+	−	−	−	−	+	−	−

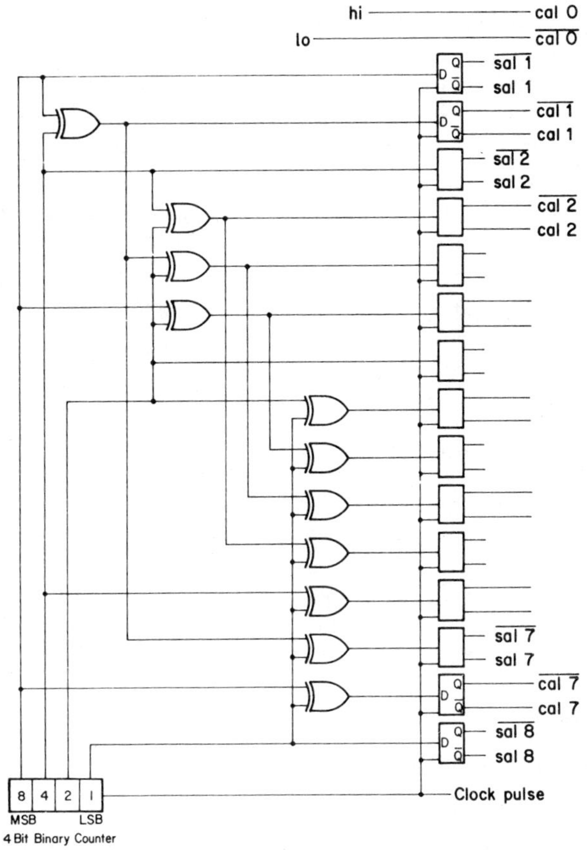

Fig. 1. Harmuth-type generator for a synchronous array of 16 Walsh functions, with clocked D-type flip-flop outputs.

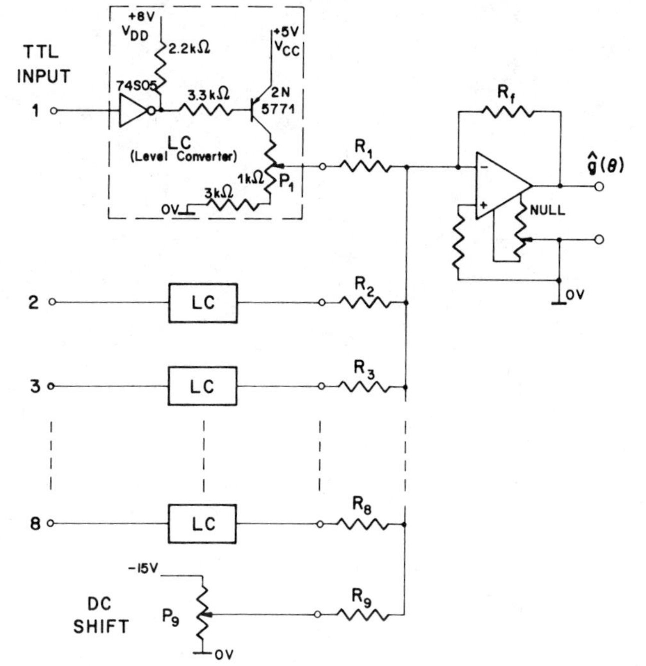

Fig. 2. Circuit for the synthesis of a sinusoid from eight TTL level Walsh inputs.

tween the ideal signal and its approximation [1]. Of course this minimization property is not restricted to Walsh functions; it can be shown fairly readily that the approximation to a sinusoid that has 2^{n-2} equally spaced steps per quarter-cycle and which minimizes the integral-squared error has the spectrum given by (1). The steps could be formed by using other basis functions or by a READ-ONLY memory in conjunction with a digital-to-analog (D/A) converter, and so on. However, Walsh functions as basis functions are of interest because highly orthogonal and precisely timed Walsh wave arrays can be generated very simply [2]. Also, since all even-sequency coefficients of the expansion are zero, only 2^{n-2} terms are required. For example, if $n = 5$, there are eight nonzero Walsh coefficients in the truncated series. Synchronous cal waves combine to form a stepped cosine wave, or alternatively, a sine wave is formed from sal waves; coefficient magnitudes for a particular sequency are the same in each case, but their signs may differ, according to Table I. The synthesis process is readily implemented and the spectrum (1) confirmed by measurement. The coefficient magnitudes and signs in Table I are derived in the Appendix.

IMPLEMENTATION

The synthesis of a stepped sinusoid is readily implemented by using a synchronous Walsh array generator, a level converter, and a summing amplifier. A Harmuth generator [2, p. 298] with an output stage consisting of synchronously clocked D-type flip-flops type 74175 was constructed to provide an array of 64 Walsh functions, each with a true and a complement output. The complement outputs were used for terms of negative sign. The simplicity of the generator is evident from Fig. 1, which shows the circuit for an array of 16 orthogonal functions. The output voltage levels of the generator were converted from TTL to 0 V and +5 V using level converter (LC) circuits, one of which is shown in detail in Fig. 2. The TTL-level inputs 1 to 8 are applied to open-collector hex inverters type 74S05, with 2.2-kΩ pullup resistors connected to a supply V_{DD} whose voltage may lie anywhere within the range +6 V to +9.5 V. The p-n-p transistors have an unusually low saturation emitter–collector voltage drop of near 30 mV; also the temperature coefficient of variation is low because both junctions are forward biased when the transistor is in saturation. A common regulated power supply V_{CC}, slightly in excess of +5 V, is used for all eight LC's. Po-

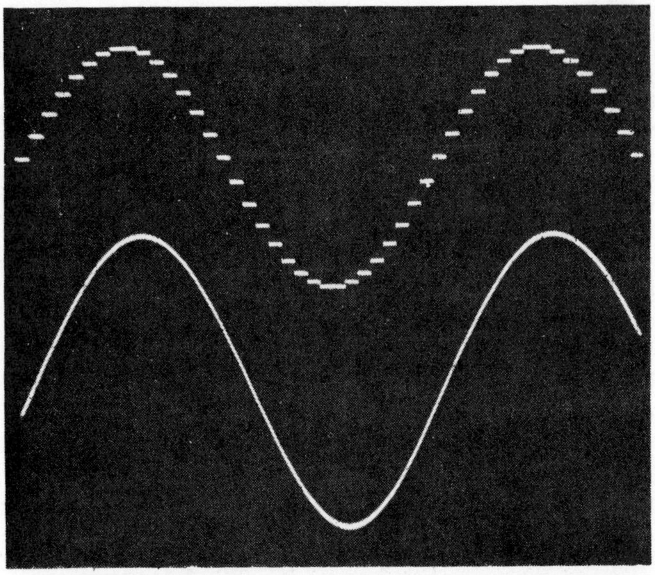

Fig. 3. Upper trace: Stepped sinusoid at the output of the summing circuit of Fig. 1, f = 100 Hz. Lower trace: Waveform at the output of a 0–3-kHz low-pass filter.

tentiometers P are 1-kΩ 15-turn trim potentiometers, which can be readily adjusted to within 0.2 percent of full-range resistance value. Up to eight appropriately scaled waves may be summed, plus a dc voltage to remove the dc component from the amplifier output.

Walsh coefficient magnitudes in Table I are scaled up by a factor 1.25π so that sal 1 (or cal 1) is 5 V peak to peak, i.e., potentiometer P_1 is near its full-range setting. Correspondingly good utilization of the remaining trim potentiometer ranges is achieved by appropriate choice of resistors R_2 to R_9, according to Table I.

Coefficient magnitudes are set individually by connecting a dc digital voltmeter at the operational amplifier output with the dc shift input grounded. Since the meter is of the averaging type, its reading is half the peak to peak value of an offset Walsh wave. First, Walsh input terminals 1 to 8 and the dc shift input are grounded, and the operational amplifier null potentiometer is adjusted to give zero output voltage. Then with only the sal 1 (or cal 1) input applied to its LC, and with $R_f = R_1$, potentiometer P_1 is adjusted so that the voltmeter reads 2.5 V. The remaining seven Walsh coefficients are set individually in the same way. Finally, all eight Walsh inputs are applied together, and P_9 is adjusted to remove the dc component from the output. For eight Walsh inputs, the output $\hat{g}(\theta)$ has a peak value of 3.9 V. With individual Walsh coefficients set to at least three decimal digits on the voltmeter, the spectrum of the output was measured on a General Radio Company wave analyzer, type 1800A. Measured spectral components were 3.2 percent and 3 percent for the 31st and 33rd harmonics, respectively, 1.6 percent and 1.5 percent for the 63rd and 65th harmonics, respectively, and so on, in accordance with (1). Theoretically, there should be no harmonics in the range 2 to 30; however, imperfections in settings gave rise to 15th and 17th harmonic spectral lines, each of magnitude 0.007 percent of the fundamental. All other harmonics in this range had measured magnitudes below 0.005 percent. The separation of the harmonics from the fundamental is readily demonstrated by connecting a low-pass filter to the amplifier output. Fig. 3 shows an oscilloscope picture of $\hat{g}(\theta)$ at 100 Hz, applied to the input of a 3-kHz two-stage m-derived low-pass filter. The sine wave at the filter output is shown in the lower trace. For an ideal 0–3-kHz low-pass filter, distortion in the output would arise when the input frequency is reduced to 3000/31 = 96.8 Hz. The demonstration shows that, in practice, a single low-pass filter may be used to provide a sinusoidal output over a 30 to 1 frequency range of a stepped wave that is synthesized from eight Walsh waves. Orthogonality of cosine and sine waves may be demonstrated by using cal $(1,\theta)$ and sal $(1,\theta)$ alternatively as the external trigger signal for the oscilloscope time base.

Other waves that are readily synthesized are the following.

a) Triangular wave:

$$g(\theta) = \text{cal}\,(1,\theta) + \tfrac{1}{2}\,\text{cal}\,(3,\theta) + \tfrac{1}{4}\,\text{cal}\,(7,\theta) + \tfrac{1}{8}\,\text{cal}\,(15,\theta) + \cdots .$$

b) Ramp:

$$g(\theta) = \text{sal}\,(1,\theta) + \tfrac{1}{2}\,\text{sal}\,(2,\theta) + \tfrac{1}{4}\,\text{sal}\,(4,\theta) + \tfrac{1}{8}\,\text{sal}\,(8,\theta) + \cdots .$$

c) Pulse of duty cycle 1 in 8:

$$g(\theta) = \text{cal}\,(0,\theta) + \text{cal}\,(1,\theta) + \text{cal}\,(2,\theta) + \text{cal}\,(3,\theta) + \text{sal}\,(1,\theta) + \text{sal}\,(2,\theta) + \text{sal}\,(3,\theta) + \text{sal}\,(4,\theta).$$

Waves a) and b) are obtained by simply connecting the appropriate TTL level Walsh array generator outputs directly to a conventional D/A converter. The sequency-limited pulse c) requires voltage level conversion followed by summing resistors of equal value. The spectra a), b), and c) are easily derived.

APPENDIX

DERIVATION OF THE FOURIER SPECTRUM OF $\hat{g}(\theta)$

Any periodic wave $g(\theta)$ of even symmetry may be expressed in terms of Fourier or Walsh expansions:

$$g(\theta) = \sum_f a_f \cos(f2\pi\theta) = \sum_s A_s \,\text{cal}\,(s,\theta) \qquad (2)$$

the normalized frequency being $f = 0,1,2, \cdots$ and the normalized sequency being $s = 0,1,2, \cdots$. Wal $(0,\theta)$ is taken to be a cal function; then a complete set of 2^n Walsh functions has 2^{n-1} cal functions and an equal number of sal functions. The Fourier expansion of cal (s,θ) is now written

$$\text{cal}\,(s,\theta) = \sum_f a_{fs} \cos(f2\pi\theta) \qquad (3)$$

where

$$a_{fs} = 2\int_0^1 \text{cal }(s,\theta) \cos (f2\pi\theta)\, d\theta. \tag{4}$$

When (3) is substituted in (2), we obtain

$$a_f = \sum_s a_{fs}A_s. \tag{5}$$

In the case of a sinusoidal signal, we put $g(\theta) = \cos (2\pi\theta)$. Then

$$A_s = \int_0^1 \cos (2\pi\theta) \text{ cal }(s,\theta)\, d\theta = \frac{1}{2}a_{1s}$$

because of (4). Consequently, for a single sinusoid, (5) becomes

$$a_f = \sum_{s=0}^{\infty} \frac{1}{2}a_{1s}a_{fs} = \begin{cases} 1, & \text{for } f = 1 \\ 0, & \text{for } f \neq 1. \end{cases} \tag{6}$$

Assume now that the sequency in the expansion (6) is truncated at an arbitrary value S, and let $\hat{a}_f$ and $\hat{g}(\theta)$ be the corresponding approximations to a_f and $g(\theta)$, respectively. Then

$$\hat{g}(\theta) = \sum_f \hat{a}_f \cos (f2\pi\theta)$$

where

$$\hat{a}_f = \sum_{s=0}^{S} \frac{1}{2}a_{1s}a_{fs} \tag{7}$$

in which a_{fs} is the general coefficient of frequency f in the Fourier expansion of cal (s,θ). It is evaluated by substituting integer values f in the Fourier transform of wal (k,θ) where $k = 2s$. The nonrecursive Fourier transform expression which uses the Gray-coded ordering of Walsh functions [3] is convenient for our purposes. For $0 \le \theta < 1$, the one-sided transform of wal (k,θ), where $0 \le k \le 2^n$, is

$$F\{\text{wal }(k,\theta)\} = 2(-j)^{\alpha} e^{-j\pi f} \text{ sinc } \frac{f}{2^n} \cdot \left\{\prod_{x=0}^{n-1} \cos \left(\frac{\pi f}{2^{x+1}} - g_x\frac{\pi}{2}\right)\right\} \tag{8}$$

where $k = 2s$ for cal functions, $k = 2s - 1$ for sal functions, $k_{\text{Gray}} = g_{n-1} \cdots g_x \cdots g_2g_1g_0$ (n bits), and α is the number of ones in k_{Gray}. α is even for cal functions and odd for sal functions. Substituting (8) in (7), we get

$$\frac{1}{2}a_{1s}a_{fs} = 2(\text{sinc } 1/2^n)(\text{sinc } f/2^n)P$$

where

$$P = \prod_{x=0}^{n-1} \frac{1}{2}\left[\cos \frac{\pi}{2^{x+1}} (f-1) + \cos \left\{\frac{\pi}{2^{x+1}}(f+1) - g_x\pi\right\}\right].$$

The second cosine term in P may be rewritten

$$(-1)^{g_x} \cos \frac{\pi}{2^{x+1}}(f+1).$$

TABLE II

s	k	$g_2g_1g_0$	G_0	G_1	G_2	G_0G_1	G_0G_2	G_1G_2	G_0-G_1G_2
0	0	000	+	+	+	+	+	+	+
1	2	011	−	−	+	+	−	−	+
2	4	110	+	−	−	−	−	+	+
3	6	101	−	+	−	−	+	−	+

Let G_x denote the sign of $(-1)^{g_x}$, i.e., G_x is + for $g_x = 0$ and − for $g_x = 1$. Then

$$\frac{1}{2}a_{1s}a_{fs} = \frac{(\text{sinc } 1/2^n)(\text{sinc } f/2^n)}{2^{n-1}} \cdot \prod_{x=0}^{n-1}\left\{\cos \frac{\pi(f-1)}{2^{x+1}} + G_x \cos \frac{\pi(f+1)}{2^{x+1}}\right\}. \tag{9}$$

Expansion of the product term in (9) yields a sum of 2^n terms, all but one of which include a coefficient G_x or a product of G_x coefficients. For example, if $n = 3$, seven of the terms in the expansion of the product in (9) contain one of the coefficients G_0, G_1, G_2, G_0G_1, G_0G_2, G_1G_2, $G_0G_1G_2$, while the eighth term is

$$\left\{\cos \frac{\pi(f-1)}{2}\right\}\left\{\cos \frac{\pi(f-1)}{4}\right\}\left\{\cos \frac{\pi(f-1)}{8}\right\}.$$

Assume, now, that $S = 2^{n-1}$, signifying that the summation in (7) ranges over the complete subset of cal sequencies. Then all but two terms in the expanded form of (9) cancel; one remaining term is that which does not include a coefficient G_x or G_x products, while the other contains the product of all coefficients G_x. To illustrate, if $n = 3$, the highest Walsh order in the set is 7. Table II lists the sequencies s in this set, the corresponding cal Walsh orders $k = 2s = g_2g_1g_0$ in Gray code, and the associated coefficients G_x and their products. Canceling of terms is evident from inspection of the columns in Table II, the exception being the column for $G_0G_1G_2$. Accordingly, for a complete Walsh set, substitution of (9) in (7) yields

$$\hat{g}(\theta) = \sum_f \left\{(\text{sinc } 1/2^n)(\text{sinc } f/2^n) \cdot \left[\prod_{x=0}^{n-1} \cos \frac{\pi(f-1)}{2^{x+1}} + \prod_{x=0}^{n-1} \cos \frac{\pi(f+1)}{2^{x+1}}\right]\right\} \cos (f2\pi\theta). \tag{10}$$

The first product term in (10) is zero for all integrals f other than those given by $\pi(f-1)/2^n = q\pi$ where $q = 0,1,2,3, \cdots$, i.e., $f = q2^n + 1$. For all other integers f, one term in the expansion of the product is zero. Likewise, the second product in (10) is nonzero only if $f = q2^n - 1$ where $q = 1,2,3, \cdots$. Finally, for $f + q2^n \pm 1$,

$$(\text{sinc } 1/2^n)(\text{sinc } f/2^n) = \pm \frac{(-1)^q}{q2^n \pm 1} \text{sinc}^2 (1/2)^n.$$

so that (10) reduces to the spectrum given by (1).

WALSH EXPANSION OF A SINUSOID

The Walsh expansion of cos $(f2\pi\theta)$ is written as the following:

$$\sum_s A_{sf} \text{ cal } (s,\theta) \tag{11}$$

where A_{sf} is the Walsh coefficient of sequency s, i.e.,

$$A_{sf} = \int_0^1 \cos(f2\pi\theta) \text{ cal } (s,\theta)\, d\theta = \frac{1}{2}a_{fs} \tag{12}$$

from (4). Substitution of $f = 1$, $s = 1,2,3,\cdots$, in (8) yields the required coefficients. To illustrate, for $s = 7$ (cal) in a set of 16 Walsh functions, $k = 14 = 1001$ in Gray code, so $\alpha = 2$. A term within the product in (8) is a cosine or a sine, depending on whether g_x is 0 or 1. Hence

$$A_{71} = \frac{1}{2}a_{17} = (\text{sinc } 1/16)(\sin \pi/2) \cdot (\cos \pi/4)(\cos \pi/8)(\sin \pi/16) \approx 0.1266.$$

The sal coefficients may be evaluated in the same way.

ACKNOWLEDGMENT

The author wishes to thank A. S. Gladwin for his helpful comments.

REFERENCES

[1] G. R. Cooper and C. D. McGillem, *Methods of Signal and System Analysis.* New York: Holt, Rinehart, 1967, pp. 33–38.

[2] H. Schreiber and G. F. Sandy, Eds., *Applications of Walsh Functions and Sequency Theory.* New York: IEEE, 1974, Cat. no. 74CH0861-5EMC, ch. 15.

[3] K. H. Siemens and R. Kitai, "A nonrecursive equation for the Fourier transform of a Walsh function," *IEEE Trans. Electromagn. Compat.*, vol. EMC-15, pp. 81–83, May 1973.

Part VIII

WALSH TRANSFORM APPLICATIONS

Editor's Comments on Papers 32 Through 38

Walsh transform theory encompasses a wide spectrum of application areas. Paper 32 presents an application of Walsh spectral analysis to electroencephalogram (EEG) signals. To show the feasibility of using Walsh spectral features for classification of sleep EEG data, a comparison is made of Walsh and Fourier features extracted from a given data set. The conclusion is that Walsh spectral features offer about the same level of success on classification as the Fourier spectral features. Hence the advantage is the mathematical simplicity and computational superiority of fast Walsh transform (FWT) over fast Fourier transform (FFT).

Paper 33 uses the fast Walsh transform for implementing a real-time microprocessor-based electrocardiogram data compression system, and investigates the possibility of enhancing the quality of the signals reconstructed via linear filtering. In Paper 34 an application of the Walsh-Hadamard transform (WHT) to two-dimensional image coding is reported. Some of the advantages of transmitting the WHT of an image are increase of speed (by an order of magnitude), Walsh-Hadamard samples being quantized to as few as 64 levels, increased tolerance of channel errors, and possibility of bandwidth reduction.

In Paper 35 and the work of Shum, Elliott, and Brown (1973) the WHT technique is applied to speech coding. It is reported that intelligible speech has been reconstructed with a few dominant (4, 6, or 8) Walsh-Hadamard coefficients out of a field of 64 coefficients. In both works the audio signal was sampled at a rate in the range 6.5 to 13 kbits/sec. In the work of Frangoulis and Turner (Paper 35) the results of a preliminary series of listening tests are presented that are in agreement with the theoretical conclusions drawn from the statistics of the speech transformed. It is also explained that approximately half of the transmission rate is required for the sending of labelling information. A combined application of the Hadamard and Haar transforms as a technique of speech coding is reported in Frangoulis and Turner (1978), where in a series of listening tests it was found that intelligible but slightly noisy speech can be achieved at bit rates of approximately 3.5 kbits/sec. A comparison of the effectiveness of discrete Fourier, Hadamard-Walsh, and Karhunen-Loève transforms, with conventional companded PCM processing is provided by Campanella and Robinson (1971). Using the quantization error (or noise) as a measure of effectiveness, it is found that data rate reductions of 13.5, 10, and 7.5 kbits/sec are possible employing discrete Karhunen-Loève, Fourier, and Walsh representations, correspondingly, as compared with 56 kbits/sec PCM. These reduction levels are found to be almost independent of the bit rate.

In Paper 36 a fast-digital hardware system for both signal analysis and synthesis is presented, which can be used in speech processing and other applications. This system makes use of the Walsh function generator and analyzer of Tzafestas, Frangakis, and Pimenidis (1976) and Frangakis and Tzafestas (1979) (see Parts I and II) and can be easily realized using standard Large Scale Integration (LSI) components. The dominant term concept of Walsh series expansions is described by Chen and Sun (1980) and a FWT/Inverse FWT subroutine, as well as an approximate inverse FWT subroutine in FORTRAN code, are developed and applied to a variety of functions. Based on the dominant term concept a digital, fully programmable signal synthesizer is also designed and implemented, giving results coinciding with the computer simulation ones. Paper 37 provides the theory for using sequency filters to process two-dimensional signals and images.

In Paper 38 Hurst provides a unified survey of concepts and techniques existing in the fields of the threshold logic synthesis and general binary synthesis using Rademacher coefficients. Some other applications of Walsh functions and transforms are described by Pearl (1971), Edwards (1975), Besslich (1978), Seed and Albrecht (1976), Albrecht and Wright (1974), and Kartalopoulos (1976). Pearl's (1971) paper on the application of Walsh transform to stochastic analysis explores the properties of the WHT of probability functions of discrete random variables. It is shown that in stochastic systems with dyadic symmetry the convolution property of the Walsh transform is meaningful, and the job of treating statistical data can be best carried out in the transform domain. For systems without any symmetry the Walsh transform still has many properties of the Fourier transform with a considerable improvement in the speed of computation.

The paper by Edwards (1975) defines five operations in the Rademacher-Walsh domain, and uses them for a concise classification of Boolean functions. A method is also developed that can synthesize any Boolean function using an optimized, universal-threshold logic gate as a basic component together with exclusive OR gates. In the paper by Besslich (1978) the role of the WHT in extracting prime implicants covering a true minterm of a Boolean function is studied. It is shown that to this end, it suffices to inspect the elements of a single inversion transform. A procedure for this purpose is given, which can be implemented through an algorithm enabling computer-aided prime implicants extraction from incomplete Boolean functions.

In the two-part paper by Seed and Albrecht (1976) the Walsh function expansion technique is applied to the analysis of neutron transport problems. In the first part entitled *Theory* the Walsh expansion is applied to the one-speed, isotropic-scattering, rectangular geometry form of the neutron transport equation, and the partial differential equations for the expansion coefficients together with the relevant boundary conditions are derived. These equations are of the hyperbolic type, and can be transformed to S_n-like (quadrature) equations, which are coupled only via the scattering term. The experimental computational work and the results obtained for both one-dimensional and two-dimensional problems are given in the second part entitled *Analysis*. Some important results of the two-dimensional analysis are as follows.

The polar-angle cosine expansion does not considerably affect the ray-effect performance.

The utilization of unequal azimuthal expansion on the various μ-level expansions is detrimental to the flux shape.

The optimal flux shape is obtained from equal azimuthal-angle expansions on all μ-levels.

It is possible to use special quadrature sets for providing effective mitigation of the ray effects while maintaining acceptable integral quantity accuracies.

An application of Walsh function sequences to the construction of binary test signals for reactor dynamics measurement (identification) may be found in the work of Albrecht and Wright (1974). The discussion on Walsh function applications is closed with the paper of Kartalopoulos (1976) where Walsh functions are employed to design a simple device that measures "response time," that is, the time that elapses between an objective stimulus to one of the senses and the reflex action. This response-time measuring device is useful for manmachine interaction systems, and with some minor changes can be used to measure the response time in medical, physiological, psychological, and biological applications. In these applications special electrodes and amplifiers are needed for providing the stimulation signals and monitoring the desired response.

REFERENCES

Albrecht, R. W., and S. A. Wright, 1974, Binary Signals Constructed from Walsh Sequences for Reactor Dynamics Measurements, *Nucl. Sci. Eng.* **55:**296–306.

Besslich, P. W., 1978, On the Walsh-Hadamard Transform and Prime Implicant Extraction, *IEEE Trans. Electromagn. Compat.* **EMC-20:**516–519.

Campanella, S. J., and G. S. Robinson, 1971, A Comparison of Orthogonal Transformations for Digital Speech Processing, *IEEE Trans. Commun. Theory* **COM-19:** 1045–1050.

Chen, B. D., and Y. Y. Sun, 1980, Waveform Synthesis Via Walsh Transform, *Int. J. Electron.* **48:**243–256.

Edwards, C. R., 1975, The Application of the Rademacher-Walsh Transform to Boolean Function Classification and Threshold Logic Synthesis, *IEEE Trans. Computers* **C-24:**48–62.

Frangakis, G., and S. Tzafestas, 1979, A Digital Walsh Function Analyzer, *Electron. Eng.* **151:**625.

Frangoulis, E., and L. F. Turner, 1978, Speech Coding by the Hadamard-Haar Transform, *Computers Digital Tech.* **1:**106–112.

Kartalopoulos, S. U., 1976, Response Time Measurement Using Walsh Functions, *IEEE Trans. Instrum. Meas.* **IM-25:**116–119.

Pearl, J., 1971, Application of Walsh Transform to Statistical Analysis, *IEEE Trans. Syst. Man, Cybern.* **SMC-1:**111-119.

Seed, T. J., and R. W. Albrecht, 1976, Application of Walsh Functions to Neutron Transport Problems—I. Theory and II. Analysis, *Nucl. Sci. Eng.* **60:**337–356.

Shum, Y. Y., A. R. Elliott, and W. O. Brown, 1973, Speech Processing with Walsh-Hadamard Transforms, *IEEE Trans. Audio Electroacoust.* **AU-21:**174–179.

Tzafestas, S., G. Frangakis, and T. Pimenidis, 1976, Global Walsh Function Generators, *Electron. Eng.* **48:**585.

Reprinted with permission from *IEEE Trans. Biomed. Eng.* **BME-27:**485-492 (1980)

Walsh Spectral Estimates with Applications to the Classification of EEG Signals

HUGH LARSEN AND DAVID C. LAI

***Abstract*–A basis for the processing of EEG signals using the discrete, orthogonal set of Walsh functions is presented. The Walsh power spectrum is examined from the point of view of its statistical properties, especially as it relates to spectral resolution. Features, selected from the spectrum of sleep EEG data are compared to corresponding Fourier features. Each feature set is used to classify the data using a minimum-distance clustering algorithm. The results show that the Walsh spectral features classify the data in much the same way as the Fourier spectral features. This provides sufficient justification for usage of Walsh spectral features in place of Fourier spectral features, enabling one to take advantage of the vast computational superiority of the fast Walsh transform over the fast Fourier transform.**

INTRODUCTION

ELECTRICAL activity in the cortical and subcortical layers of the brain gives rise to a potential at the surface of the scalp known as the electroencephalogram, or more simply, the EEG. The EEG is of fundamental importance in diagnosing certain brain defects and mental disorders, and in characterizing sleep. When the desired signal properties are very subtle, or when the amount of data to be processed is large, computer processing is desirable. It is often useful to examine the frequency properties of the EEG using the fast Fourier transform [FFT]. There are, however, computational limitations to the amount of processing which is possible using Fourier analysis due to the computational efficiency of the FFT, particularly if real-time processing is required. In this work we consider a viable alternative to Fourier analysis, namely the analysis of the EEG in the Walsh domain.

Very efficient algorithms exist for the Walsh transform in its various forms: Hadamard, or natural order [1], Paley, or dyadic order [2], and Walsh, or sequency order [3]-[5]. A well-written fast Walsh transform [FWT] algorithm [Appendix I] is about three times faster than an efficient radix-four FFT. This is because the FFT and FWT algorithms are structurally similar, but the FWT contains no multiplies, and only $N\log_2 N$ real additions, compared to $N\log_2 N$ complex multiplies and another $N\log_2 N$ complex additions for the FFT. These efficient FWT algorithms enhance the incentive to find useful applications. For example, a fast algorithm has been developed for the autocorrelation of a wide-sense stationary process from the transform of the Walsh power spectrum [6]-[10]. This algorithm, however, has not so far demonstrated great success in applications. The most widely used application, however, is the use of the Walsh power spectrum to characterize the distribution of spectral energy. It is widely believed that the discriminatory power of the Walsh spectrum for classifying sleep EEG data is poor relative to that of the Fourier power spectrum [11]. However, this hypothesis has not been rigorously tested, and, in fact, our results show that this pessimism is unwarranted. By computing Walsh spectral estimates which are appropriate for EEG signals, favorable results are possible. Here, we present some results reported in [12] which support this fact.

The Walsh transform of a process x, consisting of the values $x(k)$, where $k = 0, 1, \cdots, N-1$, is given by

$$X(m) = \frac{1}{\sqrt{N}} \sum_{k=0}^{N-1} x(k)\,\mathrm{wal}\,(m,k), \quad m = 0, 1, \cdots, N-1 \tag{1}$$

with inverse

$$x(k) = \frac{1}{\sqrt{N}} \sum_{m=0}^{N-1} X(m)\,\mathrm{wal}\,(k,m), \qquad k = 0, 1, \cdots, N-1. \tag{2}$$

The forward and inverse transforms are mathematically equivalent, since the discrete Walsh functions are symmetric: $\mathrm{wal}\,(m,k) = \mathrm{wal}\,(k,m)$. An excellent review of some basic properties of the discrete Walsh functions and the transform are given in [13], and some properties of the Walsh power spectrum for a random process are given in [7], among others. For a random process consisting of an ensemble of N-point sample members x_i, the logical autocorrelation is defined as

$$L(k) = E[L_i(k)] = E\left[\frac{1}{N}\sum_{j=0}^{N-1} x_i(j)\,x_i(j \oplus k)\right]$$

where $E[\ ]$ is the expectation operator, and $\oplus$ denotes modulo 2 (binary) addition with no carry. The Walsh power spectrum is defined as

$$P(m) = E[P_i(m)] = E[X_i^2(m)]$$

where $P(m)$ and $L(k)$ are Walsh transform pairs. This relationship is known as the logical Weiner-Khintchine theorem.

Walsh Spectral Estimation: Let us consider a sequence of

Manuscript received November 26, 1979; revised May 1, 1980. This work was supported in part by NASA under Grant 46-001-041-1.
H. Larsen is with the Hewlett-Packard Company, Andover, MA 01810.
D. C. Lai is with the Department of Electrical Engineering, University of Vermont, Burlington, VT 05401.

real-valued data points, sectioned into a series of consecutive intervals, i.e.,

$$x_i(k) = x(k + (i-1)L), \quad k = 0, 1, \cdots, L-1$$
$$i = 1, 2, \cdots, M.$$

The "local" Walsh spectral estimate for the ith segment is $P_i(m) = X_i^2(m)$, and the overall estimate is

$$\hat{P}(m) = \frac{1}{M}\sum_{i=1}^{M} P_i(m). \tag{3}$$

The covariance for this estimate is

$$\begin{aligned}\text{cov}\,[\hat{P}(m),\hat{P}(n)] &= E\,[(\hat{P}(m) - E\,[\hat{P}(m)])\,(\hat{P}(n) - E\,[\hat{P}(n)])] \\ &= E\,[\hat{P}(m)\,\hat{P}(n)] - E\,[\hat{P}(m)]\;E\,[\hat{P}(n)] \\ &= \frac{1}{M^2}\sum_{k=1}^{M}\sum_{i=1}^{M} E\,[P_k(m)\,P_i(n)] - P(m)\,P(n).\end{aligned}$$

If the segments are long enough so that data separated by L points are assumed statistically independent, then we can write

$$E\,[P_k(m)\,P_i(n)] = \begin{cases} P(m)P(n), & k \neq i \\ E[P_i(m)P_i(n)], & k = i \end{cases}$$

and, therefore,

$$\text{cov}\,[\hat{P}(m),\hat{P}(n)] = \frac{1}{M^2}\sum_{i=1}^{M} E\,[P_i(m)\,P_i(n)] - \frac{1}{M}P(m)P(n).$$

A general expression for $E[P_i(m)\,P_i(n)]$ is not possible to obtain analytically for most cases. However, for a zero-mean Gaussian white noise process, with variance σ^2, for which

$$E\,[x_i(n)\,x_i(m)] = \sigma^2\,\delta(m-n),$$

which implies that

$$\begin{aligned}E\,[x_i(n)x_i(m)x_i(p)x_i(q)] = \sigma^4\{&\delta(m-n)\delta(p-q) \\ &+ \delta(n-p)\delta(m-q) \\ &+ \delta(n-q)\delta(m-p)\},\end{aligned}$$

the following results are readily shown:

$$E\,[P_k(m)\,P_k(n)] = \sigma^4 + 2\sigma^4\,\delta(m-n)$$

$$\text{cov}\,[\hat{P}(m),\hat{P}(n)] = \frac{2}{M}\sigma^4\,\delta(m-n)$$

$$\text{var}\,[\hat{P}(m)] = \frac{2}{M}\sigma^4.$$

The last expression shows that to improve the estimate by reducing spectral variance, the number of intervals M must be large. However, if $N = ML$ is a constant, where N is the number of points for each estimate of $\hat{P}(m)$, the segment length L must be kept small. If L is too small, however, spectral estimate will degrade, since the spacing between adjacent spectral components will be too large to properly characterize the activity. Thus, we have conflicting requirements; so a compromise choice between M and L must be selected for the best overall estimate.

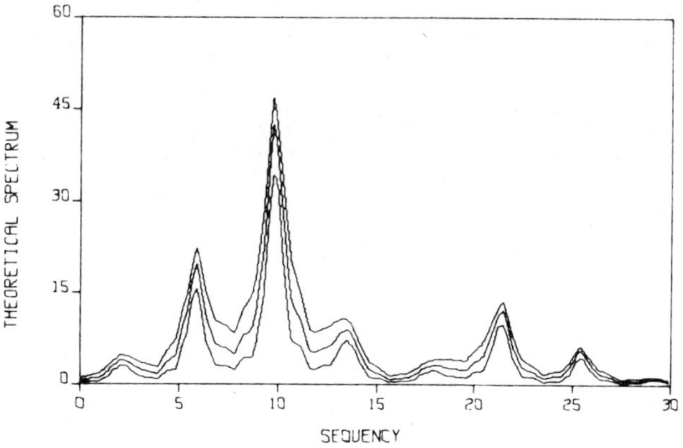

Fig. 1. Theoretical Walsh power spectra from (4) and (5) with $f_0 = 10$ Hz and $\alpha = 2$–4. (Increasing α corresponds to increasing width of the spectral lobes.)

Walsh Spectral Estimates for EEG Signals: To determine an appropriate spectral spacing for EEG signals, we consider a model with known first- and second-order statistics. We then determine the theoretical power spectrum for this process, which is algebraically related to the autocorrelation by

$$\begin{aligned}P(m) &= E\,[X^2(m)] \\ &= \frac{1}{N}\sum_{n=0}^{N-1}\sum_{k=0}^{N-1} E\,[x(n)x(k)]\;\text{wal}\,(m,k)\,\text{wal}\,(m,n).\end{aligned}$$

Rearranging terms, and using $E\,[x(n)x(m)] = R(m-n) = R(n-m)$ for a wide-sense stationary process, we obtain

$$P(m) = \frac{2}{N}\sum_{n=0}^{N-1} R(n)\sum_{k=0}^{N-n-1} \text{wal}\,(m,k)\,\text{wal}\,(m,k+n) - R(0). \tag{4}$$

The second summation in (4) is the linear autocorrelation of the mth Walsh function, which can be evaluated for all n by the fast algorithm by Yuen [15]. It is noted, however, that the indices of the Walsh functions must be converted from sequency to the corresponding Paley indices before direct application of the algorithm. The expression (4) represents a template to compare estimates taken from the simulated process. Fig. 1 shows a family of theoretical spectra computed from (4) when

$$R(n) = e^{-\alpha n \Delta T}\cos\,(2\pi f_0 n \Delta T) \tag{5}$$

with $\Delta T = 4$ ms, $f_0 = 10$ Hz, and $\alpha = 1, 2, 3$, respectively. Notice that the spectral lobe width increases with increasing α, so that α is a bandwidth parameter in the Walsh spectrum, as it is in the Fourier spectrum. [Here, and elsewhere in this paper, the spectral components are represented as a function of sequency by adding adjacent components $P(m)$ in a pairwise fashion; that is, the power at the kth sequency, denoted as s_k, is given by $P(s_k) = P(2k) + P(2k-1)$, $k = 1, 2, \cdots, (N-1)/2$, where $P(s_0) = P(0)$, and $P(s_N/2) = P(N-1)$.] Although the

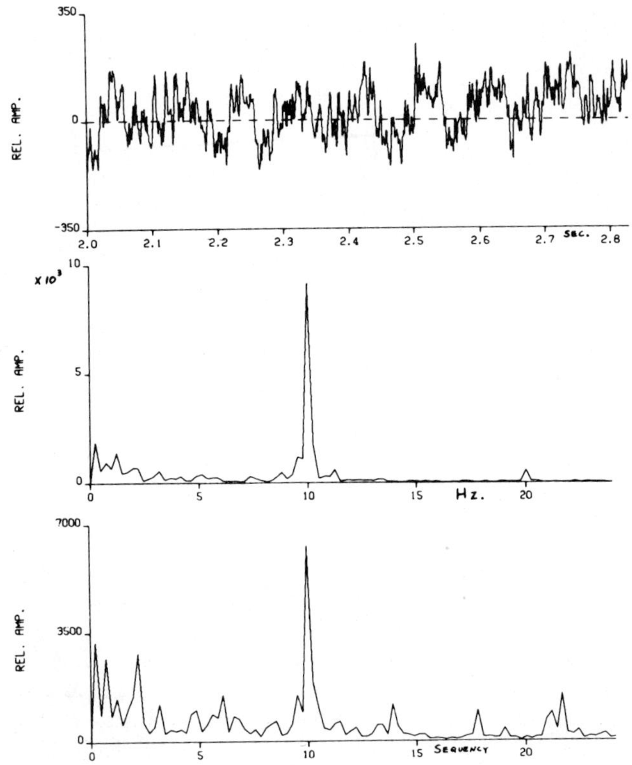

Fig. 2. Top: high alpha (8-12 Hz) content EEG data. Middle: the Fourier power spectral estimate. Bottom: the Walsh power spectral estimate.

EEG is the result of a highly complex neurological process, researchers have found that autocorrelation functions of the type of (5) are often appropriate [16]. As a selected example, a segment of real EEG data was recorded from the left occipital region of the scalp of a subject during entrainment to a 10 Hz strobe light. A portion of the data, its Fourier and Walsh power spectra, is shown in Fig. 2. The spectral spacing for both estimates is (approximately) 0.25 Hz, and both estimates were obtained by ensemble averaging four local spectra.

A simulated process, generated to study the effect of spectral resolution is given by

$$x(k) = s(k) \cos (2\pi f_0 k\Delta T + \phi) \tag{6}$$

where $f_0 = 10$ Hz, $\Delta T = 4$ ms, ϕ is a random phase angle with uniform distribution over $[-\pi, \pi]$, and $s(k)$ is the output of Gaussian white noise through a simple low-pass filter. That is

$$s(k) = \sum_{j=-\infty}^{k} h(k-j)\, n(j) \tag{7}$$

where $n(j)$ has zero mean and variance σ^2. The impulse response is

$$h(k) = Be^{-bk\Delta T}, \quad k \geqslant 0$$
$$= 0, \quad k < 0. \tag{7a}$$

The corresponding autocorrelation function of $s(k)$ is $R_s(k) = (B^2\sigma^2/4b)e^{-b\Delta T|k|}$. Therefore, the autocorrelation function for the entire process $x(k)$ is given by $R(k) = Ae^{-b\Delta T|k|} \cos (2\pi f_0 k\Delta T)$, where $A = B^2\sigma^2/8b$, which is consistent with (5).

The process $x(k)$ was obtained by using the FFT to evaluate (7) by estimating $h(k)$ with a finite length impulse response. The filter length must be long enough so that the impulse response is effectively zero at the truncation point. In addition, sufficient zeros must be added to the impulse response prior to Fourier transformation to avoid circular convolution. Samples of the resulting simulated process $x(k)$ are shown in Fig. 3 and their autocorrelation estimates in Fig. 4.

From the simulated data, Walsh spectral estimates were computed for different spectral spacings, and the results compared to the theoretical. When $N = 4096$ and $\Delta T = 4$ ms, estimates corresponding to a spectral spacing of 0.25-0.50 Hz were found to give the best results, as shown in Fig. 5. Spectral spacing closer than this (larger L) causes spurious peaks due to the larger spectral variance, but coarser spectral spacing (smaller L) causes a loss in spectral detail.

Walsh and Fourier Spectral Features: Walsh power spectral estimates were used to compute features for the purpose of automatic classification of real-sleep EEG data. Sleep data were used because of their well-known property of undergoing several distinct stages, known as sleep stages [17]. The various stages are known to be reasonably well characterized (but not completely) [18] by spectral features. The task here is to examine traditional Fourier spectral features and corresponding Walsh spectral features in two different ways: 1) visually and 2) automatically, by measuring their relative performance on classification of the data, using a clustering algorithm.

Approximately 1 h after the onset of sleep, occipital data (analog filtered from 0.5-25 Hz) were digitized at a 250 sample/s rate and stored on digital tape in 4096 point segments. About 4 h of digital data were recorded. The records were edited in a semiautomatic way, by computing, for each record, the number of Fourier spectral components required to obtain 80 percent of the signal energy for each epoch. If a particular record takes too few terms, then this means that there is an excessive amount of low-frequency artifact, whereas, too many terms imply excessive high frequency noise. Questionable records were examined visually with a CRT display, and eliminated where necessary. A single Walsh or Fourier spectral estimate was obtained for each 4096 point record by breaking it into four subintervals, computing a local spectrum for each subinterval, and ensemble averaging the four local spectra. The final estimate was obtained by further ensemble averaging three consecutive such spectra. Each spectral estimate had a spectral spacing of 0.25 Hz. The next 4096 point interval was then arbitrarily ignored, to make successive spectral estimates somewhat independent from each other. Certain features were extracted from these spectra estimates. A set of six features was obtained from each spectral estimate by summing the components over a specified frequency (or sequency) range. These features, and their respective ranges are: delta 1 (0.5-2.0 Hz), delta 2 (2.0-4.0 Hz), theta (4.0-8.0 Hz), alpha (8.0-12.0 Hz), sigma (12.0-14.0 Hz), and beta (14.0-20.0 Hz). The summation for each feature includes the bottom limit, up to, but not including, the upper limit. The first four features, which had the most important effect on classification, are shown in Figs. 6-9.

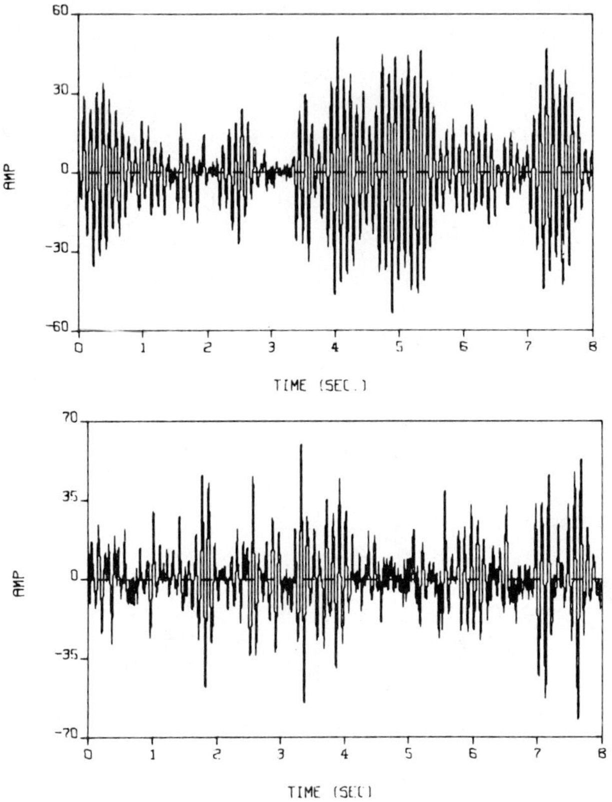

Fig. 3. Simulated EEG process $x(k)$ obtained from (6), (7), and (7a). Top: $b = 1$, $f_0 = 10$ Hz. Bottom: $b = 5$, $f_0 = 10$ Hz. The constant $B = (8b/\sigma^2)^{1/2}$.

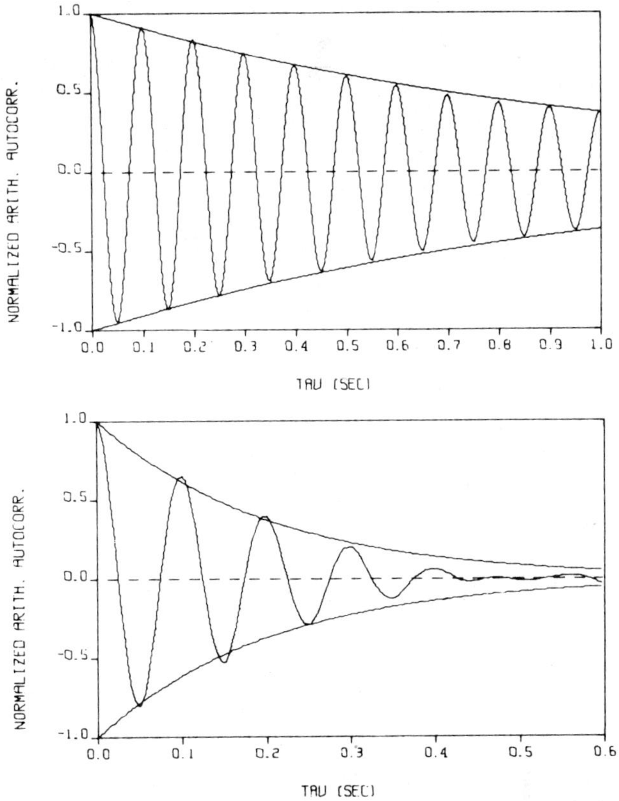

Fig. 4. Normalized arithmetic autocorrelation estimates from simulated EEG process $x(k)$ showing theoretical envelopes. Top: $b = 1, f_0 = 10$ Hz. Bottom: $b = 5, f_0 = 10$ Hz.

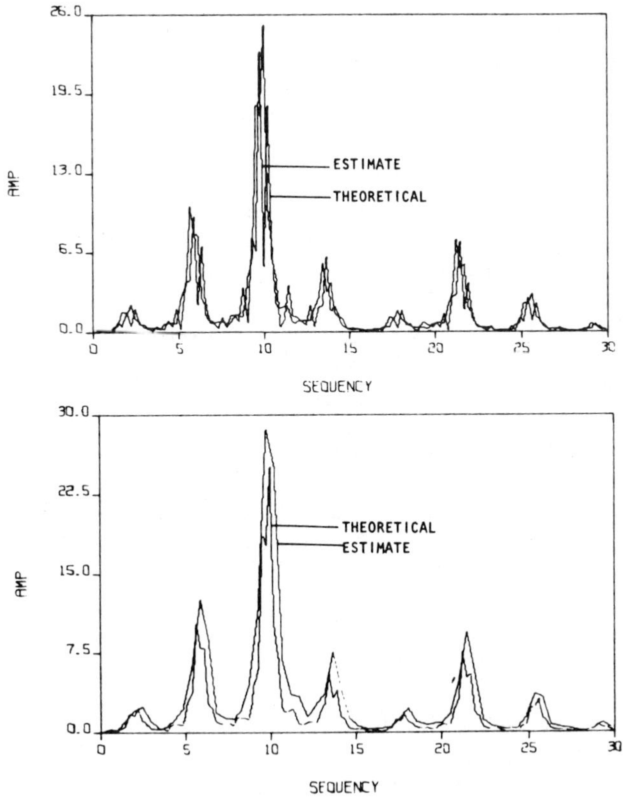

Fig. 5. Theoretical Walsh power spectrum compared to estimates. Top: spectral resolution of estimate is 0.25 Hz. Bottom: spectral resolution of estimate is 0.50 Hz.

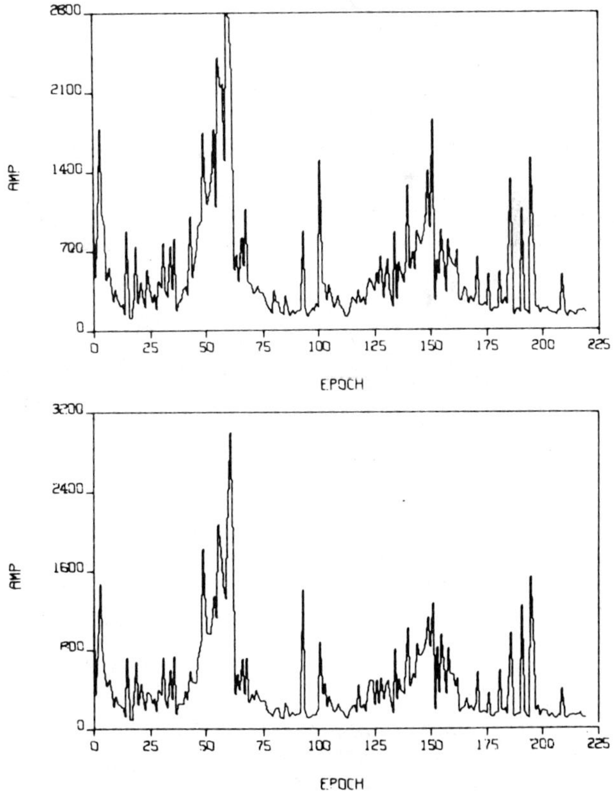

Fig. 6. The delta 1 spectral feature (0.5 Hz–2.0 Hz) extracted from sleep EEG data. Top: the Walsh feature. Bottom: the corresponding Fourier feature.

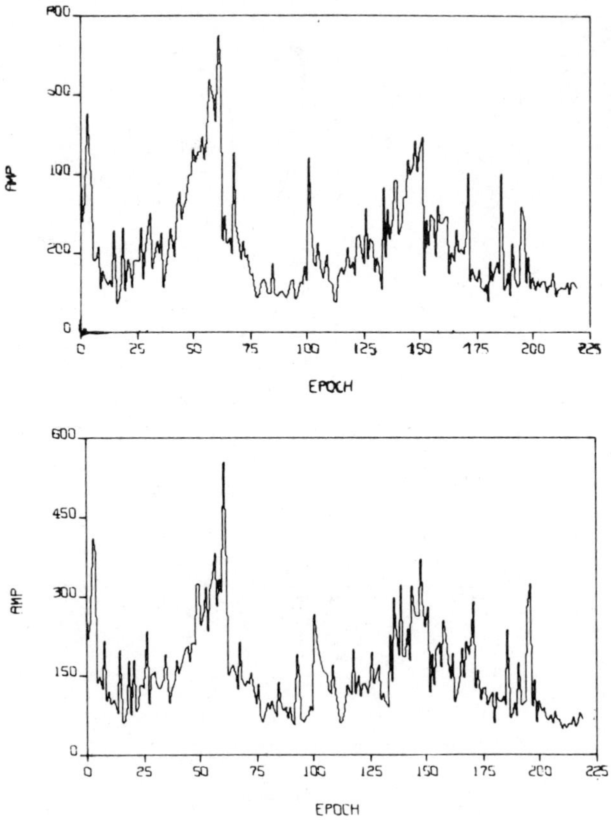

Fig. 7. The delta 2 spectral feature (2.0–4.0 Hz) extracted from sleep EEG data. Top: the Walsh feature. Bottom: the Fourier feature.

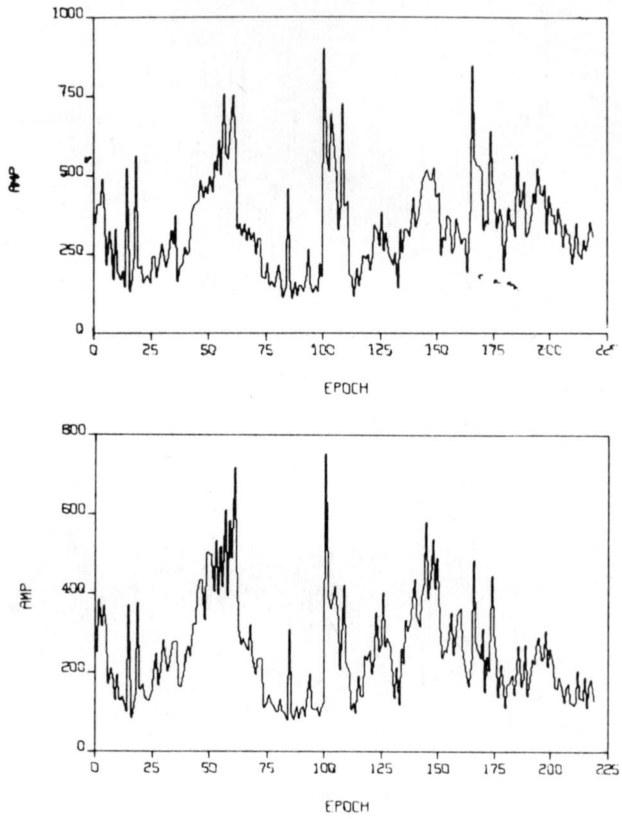

Fig. 8. The theta spectral feature (4.0–8.0 Hz) extracted from sleep EEG data. Top: the Walsh feature. Bottom: the Fourier feature.

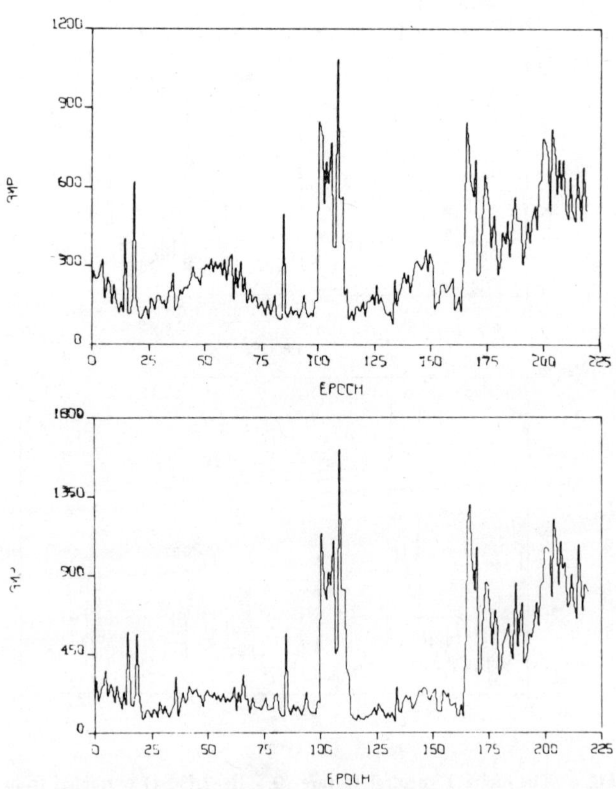

Fig. 9. The alpha spectral feature (8.0–12.0 Hz) extracted from sleep EEG data. Top: the Walsh feature. Bottom: the Fourier feature.

Classification

In order to classify the data, each epoch is represented by a single feature vector, whose components are the six spectral features defined earlier. Each of these components has been normalized by dividing it by the total power in the epoch, and then multiplying the result by 10. This brings the components into a convenient range and guarantees the magnitude of the vector $\leqslant 10$. This scaling must be remembered when interpreting the results of Figs. 10, 11, 13, and 14. We will comment further on these results after a brief description of the clustering algorithm.

Each set of feature vectors, the Walsh and Fourier, are separately classified using a clustering algorithm where the number of clusters must be specified. Figs. 10–12 summarize the results when four clusters are specified, and Figs. 13–15 summarize the results for five clusters. The clustering technique is basically the ISODATA [19] algorithm, which iteratively maximizes the intercluster distance while minimizing the average within-cluster radius. Basically, initial cluster centers are selected, one for each cluster, and each feature vector in the set is tested against each cluster center and assigned the "label" of the closest (minimum Euclidean distance) cluster center. When all feature vectors have been labeled, or classified, the cluster centers are recomputed to be the average of all feature vectors in that class. On the next iteration, these new cluster centers are used to reclassify all feature vectors. The iterative cycle stops at convergence, when no feature vectors change classification.

	DELTA 1	DELTA 2	THETA	ALPHA	SIGMA	BETA
Class 0	1.037	0.740	2.151	3.024	0.390	0.514
Class 1	2.303	1.369	2.267	1.886	0.270	0.320
Class 2	3.068	1.678	2.119	1.428	0.235	0.269
Class 3	4.394	1.563	1.691	1.004	0.183	0.171

	DELTA 1	DELTA 2	THETA	ALPHA	SIGMA	BETA
Class 0	1.079	0.701	1.479	5.590	0.231	0.411
Class 1	2.586	1.313	1.953	2.922	0.183	0.279
Class 2	3.535	1.571	2.498	1.533	0.158	0.190
Class 3	5.041	1.249	1.792	0.995	0.098	0.125

Fig. 10. The Walsh (top) and Fourier (bottom) cluster centers when four clusters are found.

Intercluster Distance	Average Intracluster Radius	Separability Measure
R_{01} = 1.83	r_0 = .595	S_{01} = 1.68
R_{02} = 2.76	r_1 = .489	S_{02} = 2.56
R_{03} = 4.04	r_2 = .479	S_{03} = 3.56
R_{12} = 0.95	r_3 = .538	S_{12} = 0.98
R_{13} = 2.35		S_{13} = 2.28
R_{23} = 1.46		S_{23} = 1.43

Intercluster Distance	Average Intracluster Radius	Separability Measure
R_{01} = 3.16	r_0 = .786	S_{01} = 1.93
R_{02} = 4.93	r_1 = .847	S_{02} = 3.13
R_{03} = 6.10	r_2 = .788	S_{03} = 3.98
R_{12} = 1.78	r_3 = .746	S_{12} = 1.08
R_{13} = 3.13		S_{13} = 1.96
R_{23} = 1.77		S_{23} = 1.15

Fig. 11. The Walsh (top) and Fourier (bottom) cluster statistics when four clusters are found.

Selection of the initial cluster centers is done automatically. The first initial cluster center is chosen (arbitrarily) equal to the first feature vector in the set. The second is found by scanning all through the set and selecting the center equal to the feature vector which maximizes the Euclidean distance from the first center. The third center is found by again scanning all features vectors and finding the one which maximizes the sum of the distances to the other two. These three initial cluster centers are then used in the algorithm to determine three major groupings of the data. At convergence, that cluster whose center has the largest "alpha" feature is split by selecting two vectors from within it and reclassifying all vectors into now four groups, or clusters. At convergence of the algorithm, the cluster center with the largest "delta 2" feature is found, and that cluster split as before to obtain five clusters in all.

Walsh

Fourier	0	1	2	3
0	50	0	0	0
1	2	29	9	0
2	0	5	64	7
3	0	0	4	34

$$A = \begin{bmatrix} 50 & 0 & 0 & 0 \\ 2 & 29 & 9 & 0 \\ 0 & 5 & 64 & 7 \\ 0 & 0 & 4 & 34 \end{bmatrix}$$

Fig. 12. The comparison matrix between Walsh and Fourier classifications when four clusters are found. Element A_{ij} equals the number of occurrences of data classified as class i when using the Fourier features and as class j when using the Walsh features.

	DELTA 1	DELTA 2	THETA	ALPHA	SIGMA	BETA
Class 0	1.003	0.728	2.133	3.053	0.393	0.525
Class 1	1.876	1.017	2.298	2.241	0.301	0.337
Class 2	2.674	1.603	2.183	1.600	0.253	0.298
Class 3	3.554	1.647	1.994	1.275	0.210	0.232
Class 4	4.665	1.527	1.579	0.924	0.176	0.155

	DELTA 1	DELTA 2	THETA	ALPHA	SIGMA	BETA
Class 0	1.055	0.689	1.453	5.641	0.232	0.414
Class 1	2.084	1.183	2.017	3.633	0.184	0.303
Class 2	3.178	1.408	1.912	2.333	0.197	0.277
Class 3	3.477	1.641	2.720	1.361	0.144	0.162
Class 4	4.933	1.253	1.832	1.027	0.099	0.124

Fig. 13. The Walsh (top) and Fourier (bottom) cluster centers when five cluster centers are found.

Figs. 10-12 pertain to four clusters. Fig. 10 shows all four cluster centers in both Walsh and Fourier feature space. Notice that the magnitude of each six-dimensional vector is larger in Fourier space than in Walsh. This implies that Fourier "packs" more energy in the 0.5-20 Hz range than Walsh. This accounts for the difference in scale seen in Fig. 11. In Fig. 11, R_{ij} is the Euclidean distance between cluster centers i and j, and r_i is the average radius of cluster i, essentially a measure of standard deviation. The "separability" is a performance measure: $S_{ij} = R_{ij}/r_i + r_j$. Since S_{ij} is a ratio, the scaling difference between Walsh and Fourier apparent in R_{ij} and r_i tends to disappear. The separability measure, therefore, represents a means to compare how well Walsh tends to separate the clusters relative to Fourier. Note that (Fig. 11) the separability is about the same, comparing Walsh to Fourier, but there is a trend to indicate that "alpha dominant" (class ϕ) data are separated better in Fourier than Walsh and "delta dominant" (class 3) are better separated in Walsh. The confusion matrix in Fig. 12 indicates very good agreement between Walsh and Fourier on the final classification.

Equivalent information for five clusters is shown in Figs. 13-15. In Fig. 13, class ϕ corresponds to alpha dominant data, and class 4 corresponds to delta 1 dominant activity. Fig. 14 shows the same trend as Fig. 11, namely that alpha activity

Intercluster Distance	Average Intracluster Radius	Separability Measure
$R_{01} = 1.25$	$r_0 = .577$	1.12
$R_{02} = 2.39$	$r_1 = .533$	2.36
$R_{03} = 3.26$	$r_2 = .434$	3.26
$R_{04} = 4.36$	$r_3 = .423$	4.21
$R_{12} = 1.18$	$r_4 = .457$	1.22
$R_{13} = 2.06$		2.15
$R_{14} = 3.21$		3.24
$R_{23} = 0.96$		1.12
$R_{24} = 2.19$		2.45
$R_{34} = 1.24$		1.40

Intercluster Distance	Average Intracluster Radius	Separability Measure
$R_{01} = 2.38$	$r_0 = .750$	1.54
$R_{02} = 4.02$	$r_1 = .794$	2.86
$R_{03} = 5.17$	$r_2 = .652$	3.67
$R_{04} = 6.11$	$r_3 = .657$	4.03
$R_{12} = 1.71$	$r_4 = .763$	1.18
$R_{13} = 2.79$		1.92
$R_{14} = 3.91$		2.51
$R_{23} = 1.32$		1.00
$R_{24} = 2.25$		1.59
$R_{34} = 1.83$		1.28

Fig. 14. The Walsh (top) and Fourier (bottom) cluster statistics when five cluster centers are used.

Walsh

Fourier	0	1	2	3	4
0	47	1	0	0	0
1	3	11	5	1	0
A = 2	0	2	34	7	0
3	0	0	27	27	1
4	0	1	1	15	25

Fig. 15. The Walsh versus Fourier comparison matrix when five clusters are used. Element A_{ij} equals the number of occurrences of data classified as class i when the Fourier features are used, and classified as class j when the Walsh features are used.

is better discriminated in Fourier, and delta activity is better discriminated in Walsh. The confusion matrix of Fig. 15 has a rather large A_{32} term, meaning a significant number of epochs were labeled "class 2" in Walsh feature space, but labeled "class 3" in Fourier feature space. We might hypothesize a mapping into feature space such that either the Walsh or the Fourier feature vectors provide better discrimination between classes 2 and 3, but this is not actually the case. The separability measure S_{23} is close to 1 for both feature sets, which indicates neither set unambiguously discriminates between classes 2 nor 3. Therefore, additional features are evidently required to enhance the discrimination between these two classes. One possibility is to augment the feature vectors with other non-spectral features. Alternatively, one might combine the six spectral features from each of two simultaneously recorded EEG signals from different electrode sites and form one twelve-dimensional feature vector.

The main point that we have attempted to establish is that the traditional role played by Fourier spectral features in sleep discrimination may be successfully played by Walsh spectral features instead.

Conclusions

We have characterized the theoretical Walsh power spectrum as an algebraic function of the autocorrelation for a stationary process, and demonstrated its utility in establishing the resolution appropriate for Walsh power spectral estimates of EEG signals. By this method, spectral resolution in the range 0.25-0.50 Hz was found to be appropriate. To demonstrate the feasibility of using Walsh spectral features for classification of sleep EEG data, both Walsh and Fourier features were extracted from a commom data set and compared. The method of comparison was to test the performance of each feature set using a minimum distance clustering algorithm, to classify the data. On this basis, we conclude that Walsh spectral features can be used with about the same level of success on classification as Fourier spectral features. The advantage is the simplicity and computational superiority of the Walsh transform relative to the fast Fourier transform.

Appendix

```
PROGRAM TO COMPUTE THE FAST WALSH TRANSFORM.
SEE REFERENCE 3 FOR THE ALGORITHM STRUCTURE.
INPUT IS IN NORMAL ORDER, OUTPUT IS IN SEQUENCY ORDER.
N = NUMBER OF POINTS IN TRANSFORM, WHERE N=2**M.
THE SYMBOL $ IMPLIES A NEW LINE IN THE SUBROUTINE.

      SUBROUTINE FWT(X,M)
      DIMENSION X(1)
      N=2**M
      DO 5 J=1,N,2
      B=X(J+1)
      X(J+1)=X(J)-B
    5 X(J)=X(J)+B
      DO 4 I=2,M
      IA=2**I   $IB=IA/2   $IB2=IB/2
      DO 4 J=1,N,IA
      IE=J+IB2-1
      DO 4 K=J,IE
      IC=K+IB   $K2=K+IB2   $IC2=K2+IB
      B=X(IC)   $B1=X(IC2)   $X(IC)=X(K)-B
      X(K)=X(K)+B   $X(IC2)=X(K2)+B1
    4 X(K2)=X(K2)-B1
C  NOW SIMPLY BIT REVERSE THE DATA.
      N1=N-1
      N2=N/2
      I=1
      DO 1 J=1,N1
      IF(I.LE.J)GO TO 2
      TEMP=X(I)   $X(I)=X(J)   $X(J)=TEMP
    2 K=N2
    3 IF(I.LE.K)GO TO 1
      I=I-K   $K=K/2   $GO TO 3
    1 I=I+K
      RETURN
      END
```

Acknowledgment

The authors would like to thank C. Grant and N. Nelson for their careful typing of this manuscript.

References

[1] J. E. Whelchel and D. F. Guinn, "The fast Fourier-Hadamard transform and its use in signal representation and classification," in *1968 EASCON Rec.*, pp. 561-573.

[2] J. L. Shanks, "Computation of the fast Walsh-Fourier transform," *IEEE Trans. Comput.*, vol. C-18, pp. 457-459, May 1969.

[3] H. G. Larsen, "An algorithm to compute the sequency-ordered Walsh transform," *IEEE Trans. Acoust., Speech, Signal Processing*, vol. ASSP-24, pp. 334-335, Aug. 1976.

[4] J. W. Manz, "A sequency-ordered fast Walsh transform," *IEEE Trans. Audio Electroacoust.*, vol. AU-20, pp. 204-205, Aug. 1972.

[5] R. D. Brown, "A recursive algorithm for sequency-ordered fast Walsh Transformers," *IEEE Trans. Comput.*, vol. C-26, pp. 819-822, Aug. 1977.

[6] F. Pichler, "Walsh functions and linear systems theory," in *Proc. Symp. Walsh Functions Appl.*, Apr. 1970, pp. 175-182.

[7] G. S. Robinson, "Logical convolution and discrete Walsh and Fourier power spectra," *IEEE Trans. Audio Electroacoust.*, vol. AU-20, pp. 271-280, Oct. 1972.

[8] P. V. Lopresti and H. L. Suri, "A fast algorithm for the estimation of autocorrelation functions," *IEEE Trans. Acoust. Speech, Signal Processing*, vol. ASSP-22, pp. 449-453, Dec. 1974.

[9] N. Ahmed and T. Natarajan, "On logical and arithmetic autocorrelation," *IEEE Trans. Electromagn. Compat.*, vol. EMC-16, pp. 177-183, Aug. 1974.

[10] H. Larsen, "Comments on: A fast algorithm for the estimation of autocorrelation functions," *IEEE Trans. Acoust. Speech, Signal Processing*, vol. ASSP-24, pp. 432-434, Oct. 1976.

[11] J. R. Smith and W. C. Yeo, "Walsh power spectra of human encephalograms," in *Proc. Symp. Walsh Functions*, 1972, pp. 159-162.

[12] H. Larsen, "Walsh analysis and the EEG," Ph.D. dissertation, Univ. Vermont, Burlington, Oct. 1976.

[13] B. L. N. Kennett, "A note on the finite Walsh transform," *IEEE Trans. Inform. Theory*, vol. IT-16, pp. 489-491, July 1970.

[14] H. G. Larsen and D. C. Lai, "Estimating Walsh power spectra for EEG signals," in *Proc. 29th Annu. ACEMB Conf.*, 1976, p. 29.5.

[15] C. K. Yuen, "An algorithm for computing the correlation functions of Walsh functions," *IEEE Trans. Electromagn. Compat.*, vol. EMC-17, pp. 177-180, Aug. 1975.

[16] L. H. Zetterberg, "Estimation of parameters for a linear difference equation with application to EEG analysis," *Math. Biosci.*, vol. 5, pp. 227-275, 1969.

[17] W. C. Dement and N. Kleitman, "Cyclic variations in EEG during sleep and their relation to eye movements, body mobility, and dreaming," *Electroencephalogr. Clin. Neurophysiol.*, vol. 9, pp. 673-690, Nov. 1957.

[18] L. E. Larsen and D. O. Walter, "On automatic methods of sleep staging by EEG spectra," *Electroencephalogr. Clin. Neurophysiol.*, vol. 28, pp. 459-690, May 1970.

[19] G. H. Ball and D. J. Hall, "Isodata: A novel method of data analysis and classification," Stanford Res. Inst., Menlo Park, CA, Tech. Rep., 1966.

Reprinted from *Med. Biol. Eng. Comput.* **21:**465–472 (1983)

Fast Walsh transform data-compression algorithm: e.c.g. applications

W. S. Kuklinski

Bioengineering Programme, Industrial Engineering Department, Texas A&M University, College Station, Texas 77843, USA

Abstract—*The feasibility of using a fast Walsh transform algorithm to implement a real-time microprocessor-based e.c.g. data-compression system was studied. Using the mean square error between the original and reconstructed e.c.g. signals as a measure of the utility of the reconstructed signals, the limit to which an e.c.g. signal could be compressed and still yield an acceptable reconstruction was determined. The possibility of enhancing the quality of the reconstructed signals using linear filtering techniques was also investigated.*

Keywords—*Algorithm, Data compression, Electrocardiogram, Filtering, Microprocessor, Walsh transform*

1 Introduction

THE ADVANTAGES of digital transmission, storage and processing of electrophysiological signals have been documented in a number of studies. The use of Walsh functions for real-time analysis of e.m.g. signals was presented by GOLD (1977) and for e.e.g. signals by YEO and SMITH (1972), JANSEN *et al.* (1981) and WEIDE *et al.* (1978). The application of digital signal-processing techniques to the electrocardiogram has been discussed by MCFEE and BAULE (1972), COX *et al.* (1972) and GRAHAM (1976).

One factor which has limited a more widespread application of digital signal-processing techniques to e.c.g. signals is the number of bits typically used to represent these signals. Using American Heart Association standards (PIPBERGER *et al.*, 1967) of 500 samples per s with 9 bits per sample, real-time transmission of a 3-lead vector cardiogram would require a channel capacity of 13·5 kbits per s, which is far in excess of that provided by the voice-grade telephone lines commonly used to transmit e.c.g. data. Approximately 18 k 8 bit words would be required to store a corresponding 10 s data record.

In response to this situation a number of computer-based e.c.g. data-compression algorithms have been proposed, implemented and evaluated. Among these are data-compression algorithms which use an orthogonal transformation technique (SHRIDHAR and STEVENS, 1979; AHMED *et al.*, 1975; WOMBLE *et al.*, 1977; ZIED and WOMBLE, 1979).

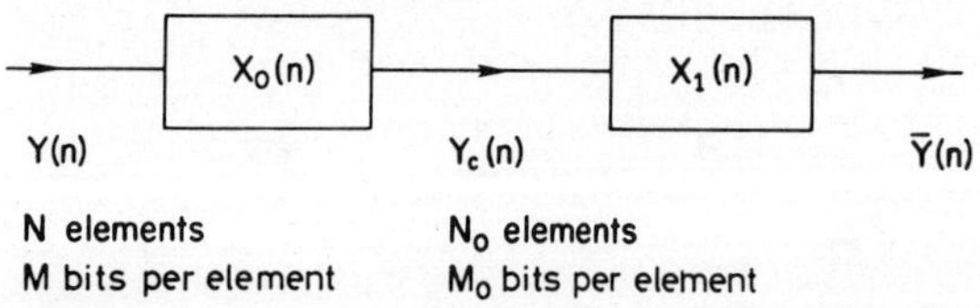

Fig. 1 *Orthogonal transformation data-compression system*

Present address: Department of Electrical Engineering, Worcester Polytechnic Institute, Worcester, MA 01609, USA.

First received 19th October 1981 and in final form 2nd April 1982

0140-0118/83/040465+08 $02·00/0

The operation of an orthogonal transformation data-compression algorithm is illustrated in Fig. 1. The digital representation of the e.c.g., $Y(n)$, an N-element long sequence of M-bit binary numbers, is operated on by the process $X_0(n)$ to produce the sequence $Y_c(n)$. The terms of $Y_c(n)$ are the magnitudes of the projection of $Y(n)$ into an N-dimensional vector space. To be useful as a data-compression algorithm the process $X_0(n)$ must produce a sequence $Y_c(n)$ that can be used to reconstruct $Y(n)$ within some acceptable error, and simultaneously yield $N_0 M_0 < NM$. The reconstructed signal $\bar{Y}(n)$, produced by the process $X_1(n)$ operating on $Y_c(n)$, can be represented as the sum of the input sequence $Y(n)$ and an error sequence $E(n)$.

$$\bar{Y}(n) = Y(n) + E(n) \quad . \quad . \quad . \quad . \quad . \quad . \quad . \quad (1)$$

The overall utility of a data-compression algorithm is a function of at least three interrelated factors:

(*a*) the degree to which $\bar{Y}(n)$ approximates $Y(n)$

(*b*) the degree to which the data are compressed, typically expressed as the ratio $(NM/N_0 M_0)$

(*c*) the complexity of the compression and reconstruction processes $X_0(n)$ and $X_1(n)$.

In developing the orthogonal transformation e.c.g. compression algorithm presented here, although an attempt was made to address all of these factors, special emphasis was placed on the issue of complexity as it relates to eventual real-time microprocessor implementation of the algorithm. For this reason Walsh functions were chosen as the basis functions in this algorithm.

An additional concern in developing a data-compression algorithm specifically for e.c.g. signals is the location of any error relative to the main features of the e.c.g. Because small errors at the beginning of the QRS complex can have greater diagnostic significance than errors of the same magnitude in other portions of the e.c.g., the most obvious error criterion would be a limit on the maximum difference between corresponding points on the original and reconstructed signals.

The most natural error criterion for analysing orthogonal transformation data-compression algorithms is, however, a normalised mean square error (AHMED *et al.*, 1975). Although certain data-compression algorithms can produce reconstructed signals with low mean square error caused by a few large differences concentrated in a small region, this is not the case when an orthogonal transformation data-compression algorithm is used. Since the mean square error of a reconstructed signal obtained from an orthogonal transformation data-compression technique is well correlated to the maximum difference that can be expected at any point, it can be used as a measure of the utility of a reconstructed e.c.g.

The primary objective of this study was to investigate the operational characteristics of a fast Walsh transform e.c.g. data-compression algorithm. Although certain aspects of this data-compression algorithm's behaviour can only be determined from an actual microprocessor implementation, its basic operational characteristics were determined from a minicomputer Fortran implementation.

2 Fast Walsh transform data-compression algorithm

The data-compression technique used in this study employed Walsh functions as the orthogonal basis set. In this method a Walsh–Fourier series is used to represent the sequence obtained by sampling an e.c.g. signal. The Walsh–Fourier series representation of an N-element long sampled e.c.g. signal $f(n)$ can be expressed as

$$f(n) = \sum_{m=0}^{N-1} a_m \phi_m(n). \qquad (2)$$

where ϕ_m is the m-th discrete Walsh function. A data-compression system using this orthogonal representation of the sampled e.c.g. signals is shown in Fig. 2. To exactly reconstruct $f(n)$ all Na_m coefficients

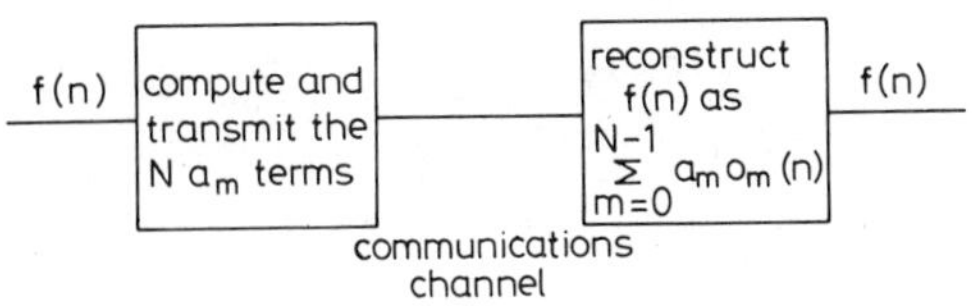

Fig. 2 *Discrete Walsh transform data-compression system*

must be computed and transmitted. The data can be compressed in either of two ways: first by computing and transmitting less than the entire set of Na_m coefficients, or second by using fewer bits to represent the a_m coefficients than were used to represent the samples of the original signal.

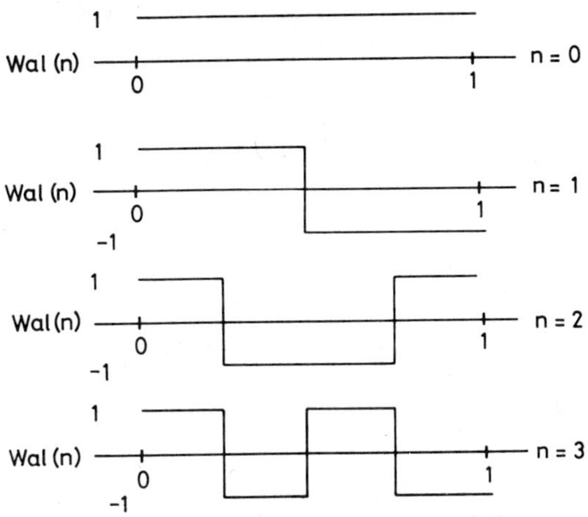

Fig. 3 *Sequence-ordered continuous Walsh functions*

Because the values of the Walsh functions are only either $+1$ or -1, (see Fig. 3), it is possible to compute the Walsh transform representation of a sampled e.c.g. signal by adding and subtracting the samples from each other in a particular sequence. The Walsh–Fourier transform $F(m)$ of an N-point discrete signal $f(n)$ can be computed as

$$F(m) = \sum_{n=0}^{N-1} f(n)\,\mathrm{Wal}(m, n) \quad m = 0, 1, 2, \ldots, N-1 \qquad (3)$$

with an associated inverse transform

$$f(n) = \frac{1}{N} \sum_{m=0}^{N-1} F(m)\,\mathrm{Wal}(n, m) \quad n = 0, 1, 2, \ldots, N-1 \qquad (4)$$

If the transform is computed using this technique, N^2 operations would be required.

A number of algorithms have been developed that can compute an N-point Walsh transform in $N \log_2 N$ rather than N^2 operations (SHANKS, 1969; LINDENBERG, 1974; BROWN, 1977). A recursive sequence-ordered fast Walsh transform algorithm presented by BROWN (1977) was used in this study. Using Brown's notation the algorithm may be summarised as follows:

$$F_0(n, 0) = f(n), \text{ for } 1 \leqslant n \leqslant N \qquad (5)$$

$$F_{i+1}(k,j) = F_i\left(2k-1, \left[\frac{j}{2}\right]\right)$$

$$+(-1)^{[(j+1)/2]}F_i\left(2k, \left[\frac{j}{2}\right]\right). \quad \quad (6)$$

for

$$i = 0, 1, 2, \ldots, p-1, p = \log_2 N$$

$$j = 0, 1, 2, \ldots, 2^{i+1}-1$$

$$K = 1, 2, 3, \ldots, \frac{N}{2^{i+1}}$$

$$F(j) = F_p(i,j) \quad \quad (7)$$

A flow diagram of the algorithm for $N = 8$ is given in Fig. 4.

3 Methods and results

The e.c.g. data used in this study were sampled at 500 Hz using 12 bits per sample. The algorithm was tested using 11 channels of data, lead I and lead II, six praecordial leads and three vector leads. The data were segmented into 512 samples, 1·024 s records and operated on by the fast Walsh transform algorithm, which produced the corresponding 512 Walsh coefficients. A typical data record and corresponding sequence ordered Walsh spectrum is seen in Fig. 5.

The first technique used to obtain a compressed representation of an e.c.g. signal involved quantising the 512 Walsh coefficients using from 3 to 11 bits per coefficient. Reconstructed e.c.g. signals were produced from each of these Walsh spectra. A second technique used to compress the data consisted of using only a fraction of a Walsh spectrum to reconstruct an e.c.g., in which reconstructed e.c.g. signals were produced using from 1/16 to the entire Walsh spectrum. In doing so the lowest sequence fraction of the spectrum was retained, while the remaining portion of the spectrum was set to zero.

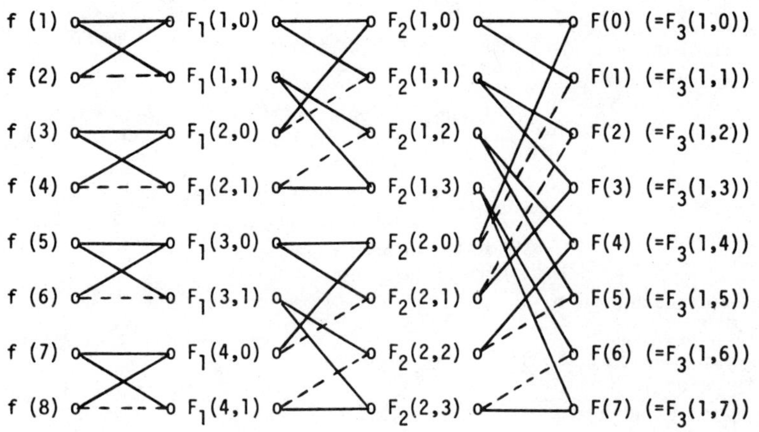

Fig. 4 *Flow diagram of fast Walsh transform algorithm for $N = 8$*
——— *add*
— — *subtract*

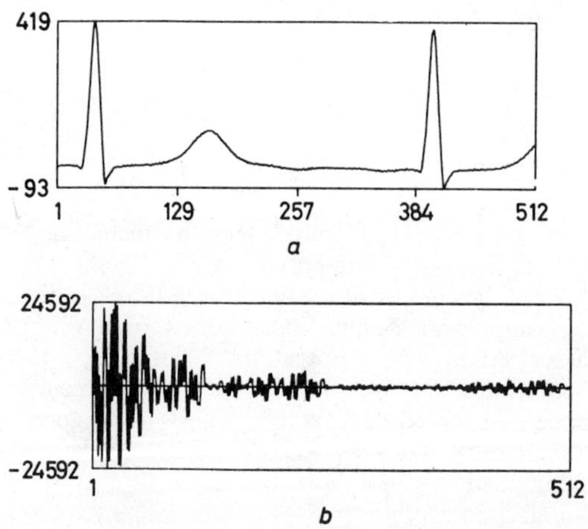

Fig. 5 *Typical e.c.g. and corresponding sequence ordered Walsh spectrum*

Plots of reconstructed e.c.g. signals are seen in Figs. 6, 7 and 8. These three sets of five reconstructed e.c.g. signals were obtained using all, one quarter and one eighth, respectively, of the Walsh spectrum with 3, 5, 7, 9 and 11 bits per coefficient. One can note a degradation of these signals as both the fraction of the Walsh spectrum used in the reconstruction process and the number of bits used to represent each Walsh coefficient are reduced.

The normalised mean square error between the original and reconstructed e.c.g. signals was computed as

$$\text{m.s.e.} = \frac{\sum_{n=1}^{N} [y(n)-\bar{y}(n)]^2}{\sum_{n=1}^{N} y(n)^2} \quad \quad (8)$$

Fig. 9 is a contour plot of the normalised mean square

error (m.s.e.) for an entire set of reconstructed e.c.g. signals. Figs. 6, 7 and 8 were generated using a portion of this set. This contour plot depicts the relationship between the m.s.e. and both the fraction of the Walsh spectrum used in the reconstruction process and the number of bits used to represent each coefficient.

The primary source of degradation in the reconstructed e.c.g. signal is the additive high-frequency noise associated with representing the signals with less than the entire Walsh spectrum and/or using a reduced number of bits to represent each coefficient. The effects of using a 9th-order finite impulse response (f.i.r.) low-pass filter to reduce the high-frequency noise components of the reconstructed e.c.g. signals are seen in Figs. 10, 11 and 12, which are filtered versions of Figs. 6, 7 and 8, respectively. The z-plane zero locations of the filter are shown in Fig. 13.

4 Discussion

The results obtained in this study demonstrate the feasibility of using a fast Walsh transform algorithm to implement a microprocessor-based real-time e.c.g.

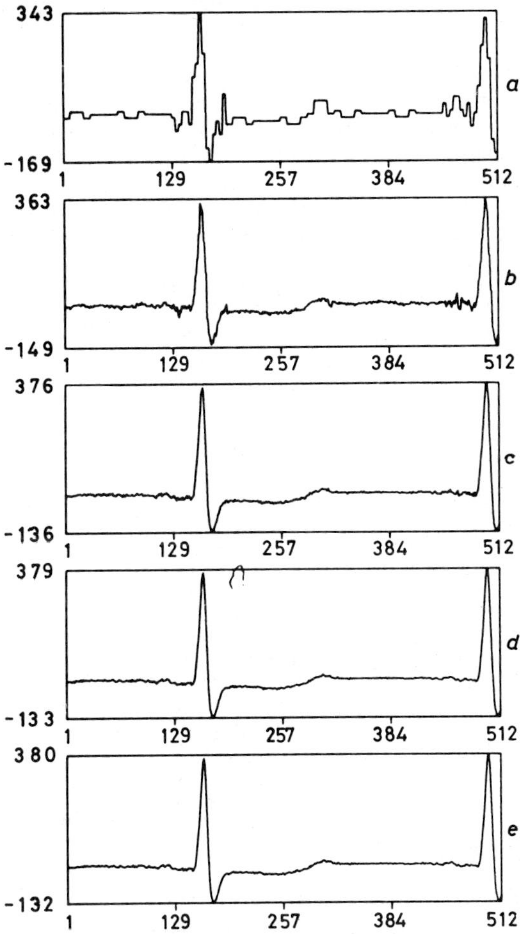

Fig. 6 *Reconstructed praecordial lead V_4 e.c.g. signals obtained using the entire Walsh spectrum*
(*a*) *normalised m.s.e.* = 0·1726, *coefficient ratio* = 1 : 1, *bits per coefficient* = 3
(*b*) *normalised m.s.e.* = 0·0229, *coefficient ratio* = 1 : 1, *bits per coefficient* = 5
(*c*) *normalised m.s.e.* = 0·0020, *coefficient ratio* = 1:1, *bits per coefficient* = 7
(*d*) *normalised m.s.e.* = 0·0002, *coefficient ratio* = 1 : 1, *bits per coefficient* = 9
(*e*) *normalised m.s.e.* = 0·0000, *coefficient ratio* = 1 : 1, *bits per coefficient* = 11

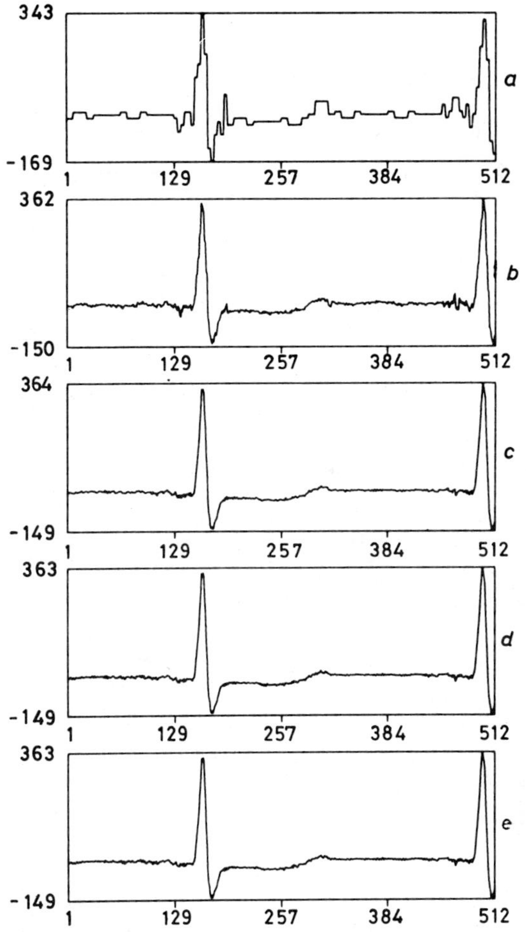

Fig. 7 *Reconstructed praecordial lead V_4 e.c.g. signals obtained using only one half of the Walsh spectrum.*
(*a*) *normalised m.s.e.* = 0·1726, *coefficient ratio* = 2:1, *bits per coefficient* = 3
(*b*) *normalised m.s.e.* = 0·0313, *coefficient ratio* = 2:1, *bits per coefficient* = 5
(*c*) *normalised m.s.e.* = 0·0253, *coefficient ratio* = 2:1, *bits per coefficient* = 7
(*d*) *normalised m.s.e.* = 0·0249, *coefficient ratio* = 2:1, *bits per coefficient* = 9
(*e*) *normalised m.s.e.* = 0·0249, *coefficient ratio* = 2:1, *bits per coefficient* = 11

data-compression system. Because the fast Walsh transform of an e.c.g. signal can be computed using $N \log_2 N$ additions and subtractions the time required to obtain this orthogonal representation is at least ten times less than the time required to compute a corresponding fast Fourier transform, where an additional $N \log_2 N$ complex multiplications are required. In a preliminary test of a microprocessor (T1990/189) implementation of this fast Walsh transform algorithm a 512-point transform, generated from a 1·024 s e.c.g. segment, was computed in approximately 75 ms.

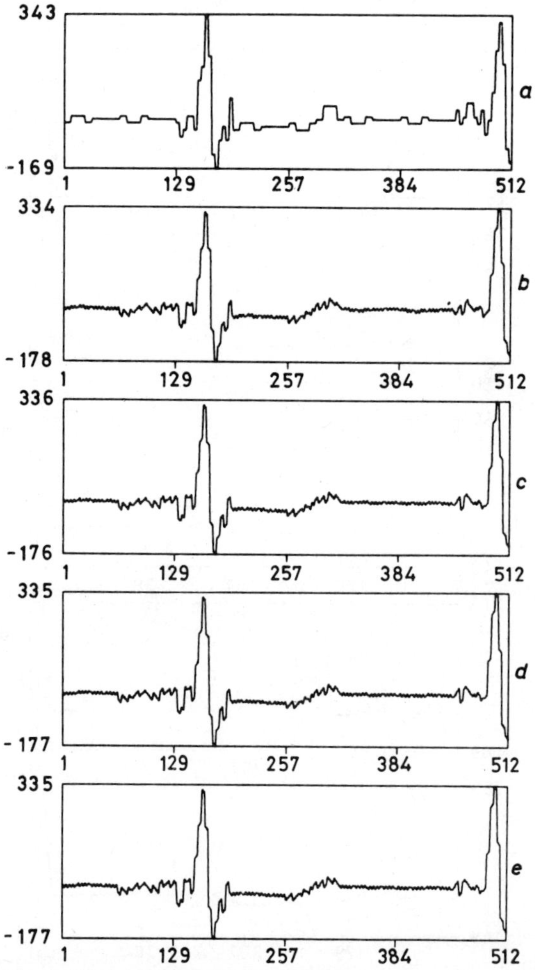

Fig. 8 *Reconstructed praecordial lead V_4 e.c.g. signals obtained using one eighth of the Walsh spectrum*
(*a*) *normalised m.s.e.* = 0·1726, *coefficient ratio* = 8 : 1, *bits per coefficient* = 3
(*b*) *normalised m.s.e.* = 0·1043, *coefficient ratio* = 8 : 1, *bits per coefficient* = 5
(*c*) *normalised m.s.e.* = 0·1028, *coefficient ratio* = 8 : 1, *bits per coefficient* = 7
(*d*) *normalised m.s.e.* = 0·1026, *coefficient ratio* = 8 : 1, *bits per coefficient* = 9
(*e*) *normalised m.s.e.* = 0·1022, *coefficient ratio* = 8 : 1, *bits per coefficient* = 11

The data obtained in this study can also be used to elucidate the relationship between the mean square error of a reconstructed e.c.g. and both the number of bits per coefficient and the fraction of the Walsh spectrum used to represent the signal in compressed form. Specifically, for any desired mean square error, there was a unique combination of bits per coefficient and fraction of Walsh spectrum retained that produced maximum data compression. The optimum operating points occur at the 'knees' of the contours seen in Fig. 9. For example, for a maximum normalised mean square error level of 0·080 the optimum operating point occurs at approximately 3 bits per coefficient and one eighth of the Walsh spectrum retained. Using a larger fraction of the Walsh spectrum will only decrease the normalised mean square error when more than 3 bits are used to represent each coefficient. Likewise, using more than 3 bits per coefficient will only decrease the normalised mean square error when one quarter or greater of the Walsh spectrum is retained.

The frequency-domain representation of the error associated with using less than the entire Walsh spectrum to reconstruct a signal can be obtained as follows. Consider the discrete Walsh transform of an N-point signal $X(n)$.

$$W(m) = \sum_{n=0}^{N-1} X(n)\,\mathrm{Wal}\,(m, n) \quad m = 0, 1, 2, \ldots, N-1 \qquad (9)$$

with the corresponding inverse transform

$$X(n) = \frac{1}{N} \sum_{m=0}^{N-1} W(m)\,\mathrm{Wal}\,(n, m) \qquad n = 0, 1, 2, \ldots, N-1 \qquad (10)$$

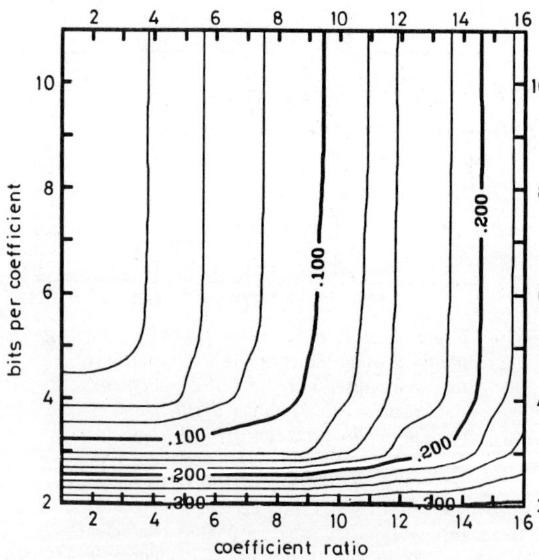

Fig. 9 *Contour plot of normalised mean square error against bits per coefficient and fraction of Walsh spectrum*

The discrete Fourier transform of the same signal can be computed as:

$$F(m) = \sum_{n=0}^{N-1} X(n)e^{-j2\pi nm/N} \quad m = 0, 1, 2, \ldots, N-1 \quad (11)$$

with the corresponding inverse transform

$$X(n) = \frac{1}{N}\sum_{m=0}^{N-1} F(m)e^{j2\pi nm/N} \quad n = 0, 1, 2, \ldots, N-1 \quad (12)$$

Using matrix notation, where $X(n)$, $W(m)$ and $F(m)$ are represented as N-dimensional vectors $\boldsymbol{X}$, $\boldsymbol{W}$, and $\boldsymbol{F}$, respectively, eqns. 9 and 11 become

$$\boldsymbol{W} = W\boldsymbol{X} \quad (13)$$

$$\boldsymbol{F} = F\boldsymbol{X} \quad (14)$$

where W and F are the $N \times N$ discrete Walsh and

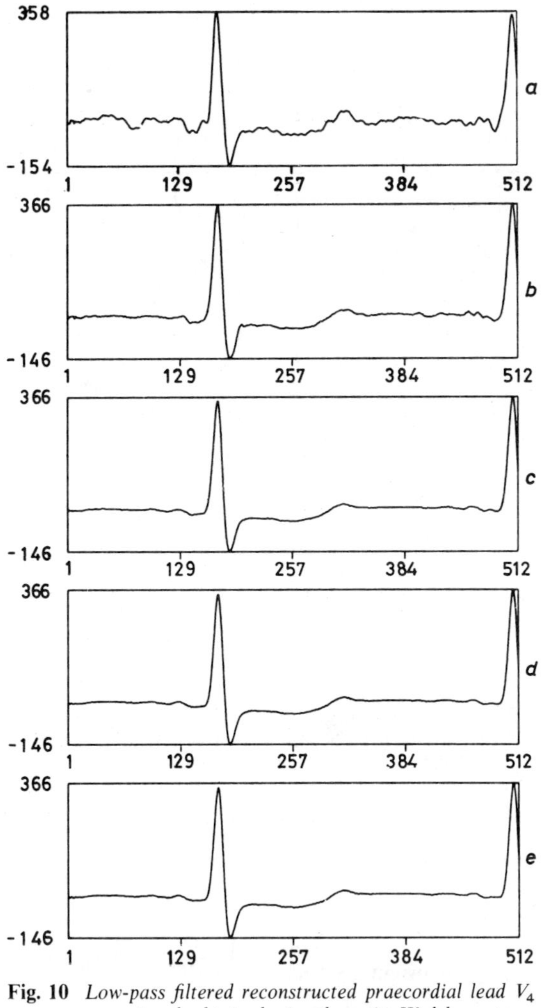

Fig. 10 *Low-pass filtered reconstructed praecordial lead* V_4 *e.c.g. signals obtained using the entire Walsh spectrum*
(*a*) *normalised m.s.e.* = 0·0410, *coefficient ratio* = 1 : 1, *bits per coefficient* = 3
(*b*) *normalised m.s.e.* = 0·0374, *coefficient ratio* = 1 : 1, bits per coefficient = 5
(*c*) *normalised m.s.e.* = 0·0057, *coefficient ratio* = 1 : 1, *bits per coefficient* = 7
(*d*) *normalised m.s.e.* = 0·0011, *coefficient ratio* = 1 : 1, *bits per coefficient* = 9
(*e*) *normalised m.s.e.* = 0·0000, *coefficient ratio* = 1 : 1, *bits per coefficient* = 11

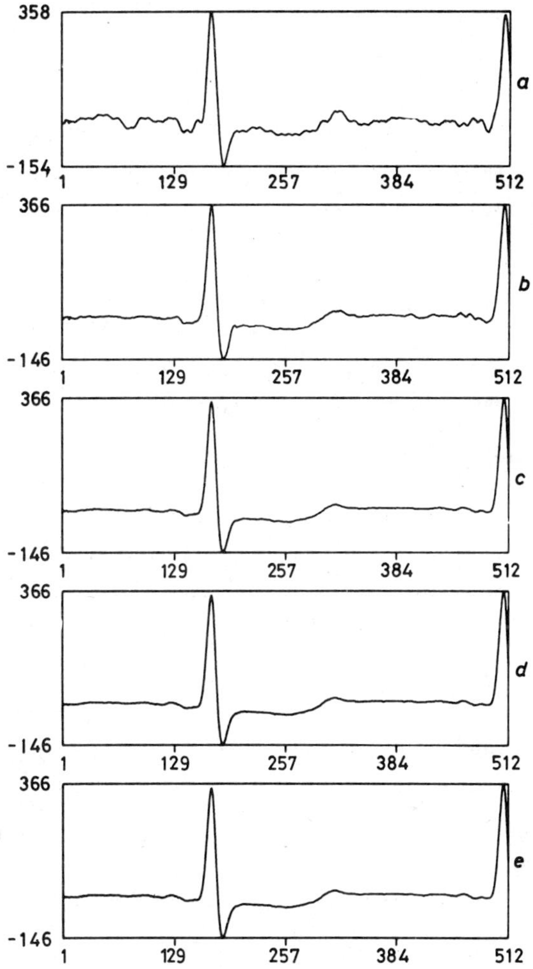

Fig. 11 *Low-pass filtered reconstructed praecordial lead* V_4 *e.c.g. signals obtained using one half of the Walsh spectrum*
(*a*) *normalised m.s.e.* = 0·0410, *coefficient ratio* = 2:1, *bits per coefficient* = 3
(*b*) *normalised m.s.e.* = 0·0309, *coefficient ratio* = 2:1, *bits per coefficient* = 5
(*c*) *normalised m.s.e.* = 0·0231, *coefficient ratio* = 2:1, *bits per coefficient* = 7
(*d*) *normalised m.s.e.* = 0·0213, *coefficient ratio* = 2:1, *bits per coefficient* = 9
(*e*) *normalised m.s.e.* = 0·0209, *coefficient ratio* = 2:1, *bits per coefficient* = 11

Fourier transformation matrices, respectively. The corresponding inverse relationships are

$$X = W^{-1}W \quad (15)$$

and

$$X = F^{-1}F. \quad (16)$$

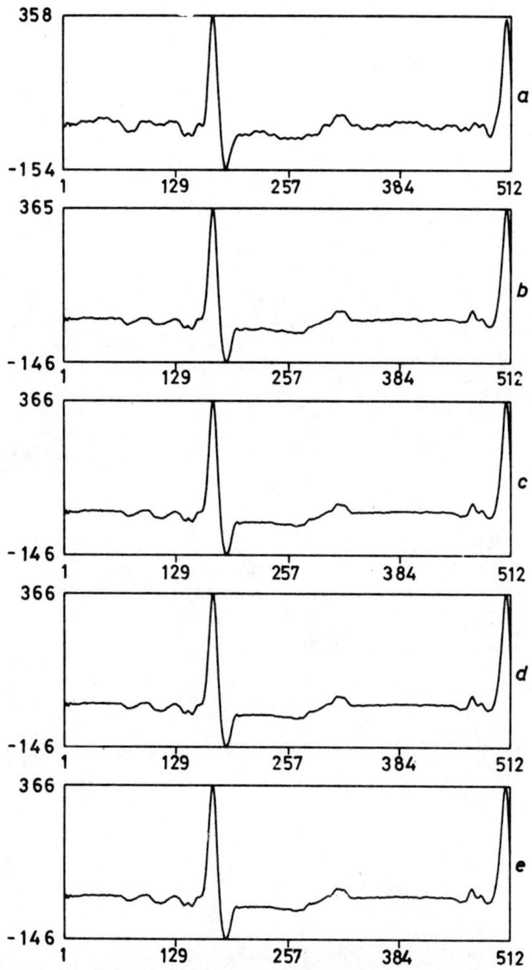

Fig. 12 *Low-pass filtered reconstructed praecordial lead V_4 e.c.g. signals obtained using one eighth of the Walsh spectrum*

(*a*) *normalised m.s.e.* = 0·0410, *coefficient ratio* = 8 : 1, *bits per coefficient* = 3
(*b*) *normalised m.s.e.* = 0·0452, *coefficient ratio* = 8 : 1, *bits per coefficient* = 5
(*c*) *normalised m.s.e.* = 0·0396, *coefficient ratio* = 8 : 1, *bits per coefficient* = 7
(*d*) *normalised m.s.e.* = 0·0396, *coefficient ratio* = 8 : 1, *bits per coefficient* = 9
(*e*) *normalised m.s.e.* = 0·0398, *coefficient ratio* = 8 : 1, *bits per coefficient* = 11

Substituting eqn. 15 into 14 we obtain the Walsh–Fourier conversion as

$$F = TW \quad (17)$$

where T is an $N \times N$ matrix given by:

$$T = FW^{-1}. \quad (18)$$

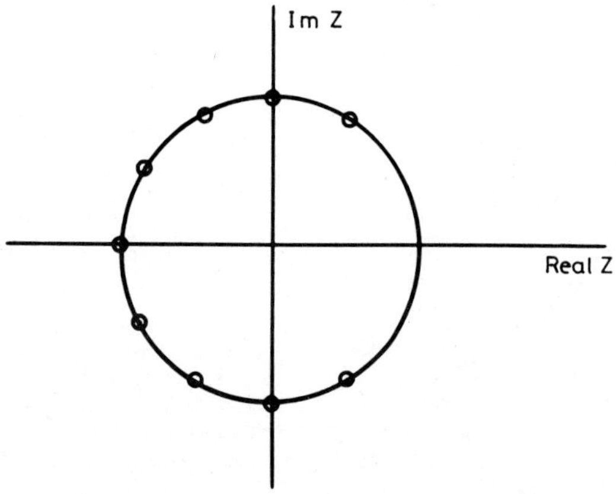

Fig. 13 *Low-pass filter z-plane zero locations*

$T = 0{\cdot}002 \quad f_c = 83{\cdot}33\,Hz$

$H(z) = (z+0{\cdot}5 \pm 0{\cdot}866j)$

$(z-0{\cdot}5 \pm 0{\cdot}866j)$

$(z+0{\cdot}866 \pm 0{\cdot}5j)$

$(z+j)(z+l)$

By defining the noise associated with the reconstruction process as the difference between the reconstructed and original signals we obtain

$$E = y_R - y_o. \quad (19)$$

where

E = noise signal
y_R = reconstructed signal
y_o = original signal

The Walsh transform of eqn. 19 yields

$$W_E = W_{y_R} - W_{y_o}. \quad (20)$$

where

W_E = Walsh transform of the noise signal
W_{y_r} = Walsh transform of the reconstructed signal
W_{y_o} = Walsh transform of the original signal

The frequency-domain representation of the noise can be obtained from eqn. 17 as

$$F_E = TW_E \quad . \quad . \quad . \quad . \quad . \quad . \quad . \quad . \quad (21)$$

For those cases where the lowest sequence portion of Walsh spectrum is retained W_E is given as

$$W_E = -\begin{bmatrix} 0 \\ \vdots \\ W_{N/2} \\ \vdots \\ W_{N-1} \end{bmatrix} \quad . \quad . \quad . \quad . \quad . \quad . \quad (22)$$

This results in an additive noise signal with the following frequency-domain representation:

$$F_E = -\begin{bmatrix} t_{1(N/2)+1}W_{N/2} + \dots t_{1N}W_{N-1} \\ \vdots \\ t_{N(N/2)+1}W_{N/2} + \dots t_{NN}W_{N-1} \end{bmatrix}$$

Although the data-compression algorithm corresponds with bandlimiting the original signal in the sequence domain, both the reconstructed signal and associated additive noise have finite energy at all frequencies. As the diagonal elements of the Walsh–Fourier transformation matrix are larger in magnitude than the nondiagonal elements, (MAQUSI, 1981), the additive noise will be primarily high-frequency in nature, hence the use of a low-pass filter to reduce the effects of this noise.

Acknowledgment—This work was supported by the USAF Summer Faculty Fellowship Programme.

References

AHMED, N., MILNE, P. J. and HARRIS, S. G. (1975) Electrocardiographic data compression via orthogonal transforms. *IEEE Trans.*, **BME-22,** 484–487.

BROWN, R. D. (1977) A recursive algorithm for sequency-ordered fast Walsh transforms. *IEEE Trans.*, **C-26,** 819–823.

COX, J. R., NOLLE, F. M. and ARTHUR, R. M. (1972) Digital analysis of the electroencephalogram, the blood pressure wave and the electrocardiogram. *Proc. IEEE*, **60,** 1137–1154.

GOLD, C. (1977) Real-time computation of e.m.g. power density spectra by fast Walsh–Hadamard transformation. Proceedings of fifth New England Bioengineering Conference, 166–170.

GRAHAM, M. (1976) Latent components in the electrocardiogram. *IEEE Trans.*, **BME-23,** 220–224.

JANSEN, B. H., BOURNE, J. R. and WARD, J. W. (1981) Spectral decomposition of e.e.g. intervals using Walsh and Fourier transforms. *IEEE Trans.*, **BME-28,** 836–838.

LINDENBERG, K. W. (1974) A sequency ordered in-place fast Walsh transform algorithm. Proceedings of the San Diego Biomedical Symposium, **V13,** 495–498.

MAQUSI, M. (1981) *Applied Walsh analysis.* Heyden & Son.

MCFEE, R. and BAULE, G. M. (1972) Research in electrocardiography and magnetocardiography. *Proc. IEEE*, **60,** 290–321.

PIPBERGER, H. V., BAULE, G., BENSON, A. S., BRILLER, S. A., GESELOWITZ, D. B., HORAN, L. G. and SCHMITT, O. H. (1967) Recommendations for standardization of instruments in electrocardiography. Report of subcommittee on instrumentation committee on electrocardiography, American Heart Association. *IEEE Trans.*, **BME-14,** 60–68.

SHANKS, J. L. (1969) Computation of the fast Walsh–Fourier transform. *IEEE Trans.*, **C-18,** 457–459.

SHRIDHAR, M. and STEVENS, M. F. (1979) Analysis of e.c.g. data, for data compression. *Int. J. Bio-Med. Comput.*, **10,** 113–128.

WEIDE, B. W., ANDREWS, L. T. and TANNONE, A. M. (1978) Real-time analysis of e.c.g. using Walsh transforms. *Comput. Biol. Med.*, **8,** 225–263.

WOMBLE, M. E., HALLIDAY, J. S., MITTEN, S. K., LANCASTER, M. C. and TRIBWASSER, J. H. (1977) Data compression for storage and transmitting e.c.g.s/v.c.g.s. *Proc. IEEE*, **65,** 702–706.

YEO, W. C. and SMITH, J. R. (1972) Walsh power spectra of human electroencephalograms. *Proc. Appl. Walsh Functions*, pp. 159–172

ZIED, A. M. and WOMBLE, M. E. (1979) Application of a partitioned Karhunen–Loeve expansion scheme to e.c.g.–v.c.g. data compression. Proceedings of eighth New England Bioengineering Conference, 102–105.

Reprinted with permission from *IEEE Proc.* **57**:58–68 (1969)

Hadamard Transform Image Coding

WILLIAM K. PRATT, MEMBER, IEEE, JULIUS KANE, SENIOR MEMBER, IEEE, AND HARRY C. ANDREWS, MEMBER, IEEE

***Abstract*—The introduction of the fast Fourier transform algorithm has led to the development of the Fourier transform image coding technique whereby the two-dimensional Fourier transform of an image is transmitted over a channel rather than the image itself. This development has further led to a related image coding technique in which an image is transformed by a Hadamard matrix operator. The Hadamard matrix is a square array of plus and minus ones whose rows and columns are orthogonal to one another. A high-speed computational algorithm, similar to the fast Fourier transform algorithm, which performs the Hadamard transformation has been developed. Since only real number additions and subtractions are required with the Hadamard transform, an order of magnitude speed advantage is possible compared to the complex number Fourier transform. Transmitting the Hadamard transform of an image rather than the spatial representation of the image provides a potential toleration to channel errors and the possibility of reduced bandwidth transmission.**

I. Introduction

IN PREVIOUS papers a new technique of digital image coding, Fourier transform coding, has been introduced [1]–[3]. By this technique a two-dimensional Fourier transform of a digitized image is performed by a digital computer using the fast Fourier transform [4]–[8]. If $f(x, y)$ represents the amplitude of image samples over a square array of N^2 points then the two-dimensional Fourier transform $F(u, v)$ is defined as[1]

$$F(u, v) = \sum_{x=0}^{N-1} \sum_{y=0}^{N-1} f(x, y) \exp\left\{-\frac{2\pi i}{N}(ux + vy)\right\}. \quad (1)$$

The Fourier transform of the image is quantized, coded, and transmitted over the channel. Then, at the receiver, the inverse Fourier transform

$$\hat{f}(x, y) = \sum_{u=0}^{N-1} \sum_{v=0}^{N-1} \hat{F}(u, v) \exp\left\{\frac{2\pi i}{N}(xu + yv)\right\}, \quad (2)$$

is taken of the decoded transform $\hat{F}(u, v)$ to reconstruct a close approximation of the original image. Fig. 1 contains photographs taken from a cathode ray tube of an original test scene, the logarithm of the magnitude of its two-dimensional Fourier transform, and the inverse Fourier transform of the Fourier transform. The logarithm of the magnitude of the Fourier transform is displayed rather than the magnitude itself because of dynamic range limitations of the recording film.

The Fourier transform image coding technique has been shown to provide good quality image transmission with the same number of bits as required to code the spatial domain of an image by conventional pulse code modulation. Furthermore, transmission of the Fourier transform of an image rather than the image itself offers a certain immunity to channel errors and the ability to achieve a bandwidth reduction [2].

The tolerance of channel errors is due to the averaging process of the Fourier transform as indicated by (2). Each point in a reconstructed image is a weighted sum of all points in the spatial frequency domain. For subjective image viewing the overall loss of resolution that occurs when channel errors are introduced in the Fourier transform is often less objectionable than discrete defects that occur when channel errors affect the spatial representation of an image.

Bandwidth reduction is possible because the image energy, which is usually uniformly distributed in the spatial domain, tends to be concentrated near the origin of the Fourier domain, as evidenced by Fig. 1(b). Many of the higher spatial frequency components are of very low magnitude, and need not be transmitted.

The Fourier transform was a natural operator to be applied to the image coding problem because of its widespread use in other fields and the fact that a very efficient computational algorithm exists. A question that naturally arises is: are there other applicable transforms beside the Fourier transform? The answer is affirmative. For the image coding application, strictly speaking, all that is required is a two-dimensional operator which has an inverse, possesses the averaging property, and redistributes the image energy properly. It is advantageous from the standpoint of implementation that a fast computational algorithm exists and that the operator be its own inverse. The symmetric Hadamard matrix transform fulfills all of these requirements, and as will be discussed later, the Hadamard transform is better in many respects than the Fourier transform for image coding. Both transforms, in fact, are special cases of a matrix multiplication of the function to be transformed by a general matrix multiplier. Which matrix multipliers may be factored to reduce computation remains an open question [9]. Because of this, the broader questions as to what other transforms exist and what is the best transform for image coding will not be considered here.

Manuscript received July 22, 1968; revised October 14, 1968. This work was supported by NASA Grant NGR-05-018-044, Jet Propulsion Laboratory Grant 952312, and U. S. Air Force Grant AFCRL-68-C-0342.

W. K. Pratt and H. C. Andrews are with the Department of Electrical Engineering, University of Southern California, Los Angeles, Calif. 90007

J. Kane was with the Department of Electrical Engineering, University of Southern California. He is currently with the Department of Mathematics and Mathematical Ecology, University of British Columbia, Vancouver, B. C., Canada.

[1] Some authors insert the multiplying factors $1/N$ or $1/N^2$ before the double summation for purposes of normalization. The multiplying factor is not inherent to the computational algorithm and will be omitted here for both the Fourier and Hadamard transform formulations in series form.

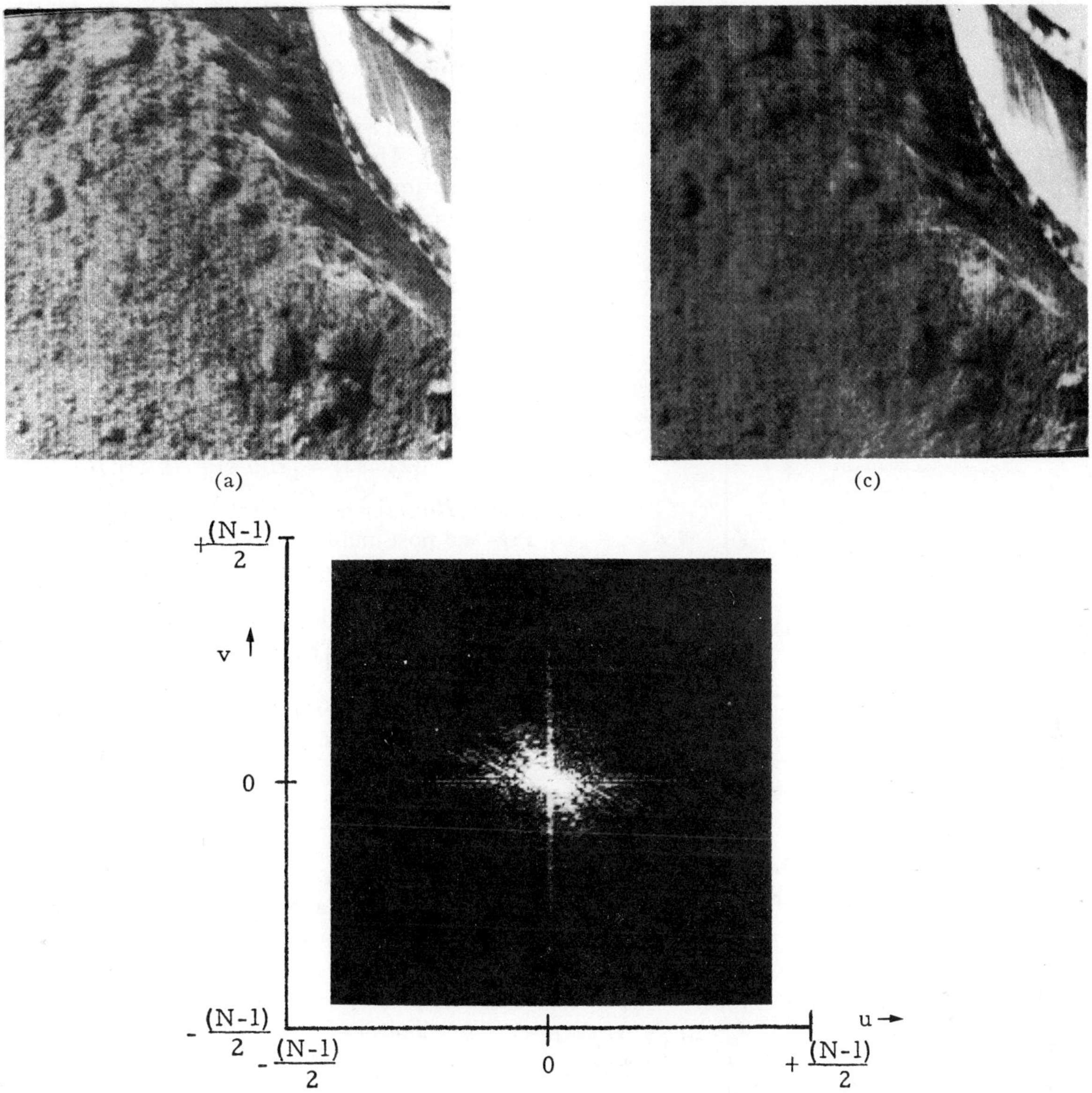

(b)

Fig. 1. Fourier transforms of Surveyor footpad: (a) footpad, (b) the logarithm of the magnitude of the Fourier transform of (a), and (c) the inverse Fourier transform of the Fourier transform of (a).

The Hadamard transform has been applied to the coding of vocoder speech signals by Crowther and Rader [10]. They have built coder and decoder networks whose designs are based upon a Hadamard matrix. Because of the correlation properties of the vocoder speech signals, it was found possible to reduce the number of quantization levels of the Hadamard transformed speech signals.

II. Hadamard Matrices [11], [13]

The Hadamard matrix[2] is a square array of plus and minus ones whose rows (and columns) are orthogonal to one another. If $\boldsymbol{H}$ is an $N \times N$ Hadamard matrix then the product of $\boldsymbol{H}$ and its transpose is the identity matrix. Thus,

$$\boldsymbol{H}\boldsymbol{H}^{+} = N\boldsymbol{I} \tag{3}$$

where $\boldsymbol{I}$ is the identity matrix. If $\boldsymbol{H}$ is a symmetric Hadamard matrix then (3) reduces to

$$\boldsymbol{H}\boldsymbol{H} = N\boldsymbol{I}. \tag{4}$$

The rows (and columns) of a Hadamard matrix may be exchanged with one another without affecting the orthogonality properties of the matrix.

The lowest-order Hadamard matrix is of order two.

$$\boldsymbol{H} = \begin{bmatrix} 1 & 1 \\ 1 & -1 \end{bmatrix}. \tag{5}$$

It is known that if a Hadamard matrix of order N exists ($N > 2$), then $N \equiv 0 \pmod 4$. The existence of a Hadamard matrix for every value of N satisfying this requirement has not been shown, but constructions are available for nearly all permissible values of N up to 200. The simplest con-

[2] Boldface symbols represent matrices.

$$N=2 \quad \begin{array}{c} \text{matrix} \\ \begin{bmatrix} + & + \\ + & - \end{bmatrix} \end{array} \quad \begin{array}{c} \text{Sequency} \\ 0 \\ 1 \end{array}$$

$$N=4 \quad \begin{array}{c} \text{matrix} \\ \begin{bmatrix} + & + & + & + \\ + & - & + & - \\ + & + & - & - \\ + & - & - & + \end{bmatrix} \end{array} \quad \begin{array}{c} \text{Sequency} \\ 0 \\ 3 \\ 1 \\ 2 \end{array}$$

$$N=8 \quad \begin{array}{c} \text{matrix} \\ \begin{bmatrix} + & + & + & + & + & + & + & + \\ + & - & + & - & + & - & + & - \\ + & + & - & - & + & + & - & - \\ + & - & - & + & + & - & - & + \\ + & + & + & + & - & - & - & - \\ + & - & + & - & - & + & - & + \\ + & + & - & - & - & - & + & + \\ + & - & - & + & - & + & + & - \end{bmatrix} \end{array} \quad \begin{array}{c} \text{Sequency} \\ 0 \\ 7 \\ 3 \\ 4 \\ 1 \\ 6 \\ 2 \\ 5 \end{array}$$

Fig. 2. Hadamard matrices of order $N = 2^n$.

struction is for a Hadamard matrix of order $N = 2^n$ where n is an integer. In this case if $\boldsymbol{H}$ is a Hadamard matrix of order N, the matrix

$$\boldsymbol{G} = \begin{bmatrix} \boldsymbol{H} & \boldsymbol{H} \\ \boldsymbol{H} & -\boldsymbol{H} \end{bmatrix} \tag{6}$$

is a Hadamard matrix of order $2N$. Fig. 2 contains several Hadamard matrices of order $N = 2^n$. Another simple construction is possible if $\boldsymbol{A}$ and $\boldsymbol{B}$ are Hadamard matrices of orders M and N, respectively. Then there exists a Hadamard matrix of order $M \cdot N$ given by

$$\boldsymbol{H}_{M \cdot N} = \begin{bmatrix} a_{11}\boldsymbol{B} & a_{12}\boldsymbol{B} \cdots a_{1M}\boldsymbol{B} \\ a_{21}\boldsymbol{B} & \vdots \\ \vdots & \vdots \\ a_{M1}\boldsymbol{B} \cdots\cdots & a_{MM}\boldsymbol{B} \end{bmatrix} \tag{7}$$

Other constructions are given in [13]–[15]. The set of known Hadamard matrices is sufficiently numerous to satisfy almost all size requirements for image coding.

A frequency interpretation can be given to the Hadamard matrix. Along each row of the Hadamard matrix the frequency is called the number of changes in sign. Harmuth has coined the word "sequency" to designate the number of sign changes [16]. Fig. 2 gives the sequency interpretation for several Hadamard matrices of binary order. It is possible to construct a Hadamard matrix of order $N = 2^n$ that has frequency components at every integer from 0 to $N-1$.

This frequency interpretation of the rows of a Hadamard matrix leads one to consider the rows to be equivalent to rectangular waves ranging between ± 1 with a subperiod of $1/N$ units. Such functions are called Walsh functions [17]–[21] and are further related to the Rademacher functions [22]. Thus, in this context the Hadamard matrix merely performs the decomposition of a function by a set of rectangular waveforms rather than the sine-cosine waveforms associated with the Fourier transform.

III. Hadamard Transformation of Images

Let the array $f(x, y)$ represent the intensity samples of an original image over an array of N^2 points. Then the two-dimensional Hadamard transform, $F(u, v)$, of $f(x, y)$ is given by the matrix product

$$[F(u, v)] = [H(u, v)][f(x, y)][H(u, v)] \tag{8}$$

where $[H(u, v)]$ is a symmetric Hadamard matrix of order N. Pre- and post-multiplication of $[F(u, v)]$ by the Hadamard matrix gives

$$\begin{aligned} &[H(u, v)][F(u, v)][H(u, v)] \\ &\quad = [H(u, v)][H(u, v)][f(x, y)][H(u, v)][H(u, v)]. \end{aligned} \tag{9}$$

But, from (4), for a symmetric Hadamard matrix, $\boldsymbol{HH} = N\boldsymbol{I}$. Hence,

$$[f(x, y)] = \frac{1}{N^2}[H(u, v)][F(u, v)][H(u, v)] \tag{10}$$

and, aside from the constant scaling factor N^2, the arrays $f(x, y)$ and $F(u, v)$ are two-dimensional transform pairs.

For symmetric Hadamard matrices of order $N = 2^n$, the two-dimensional Hadamard transform may be written in series form as

$$F(u, v) = \sum_{x=0}^{N-1} \sum_{y=0}^{N-1} f(x, y)(-1)^{p(x,y,u,v)} \tag{11}$$

where

$$p(x, y, u, v) \equiv \sum_{i=0}^{n-1} (u_i x_i + v_i y_i).$$

The terms u_i, v_i, x_i and y_i are the binary representations of u, v, x, and y respectively. For example,

$$(u)_{\text{decimal}} = (u_{n-1}u_{n-2} \cdots u_1 u_0)_{\text{binary}} \tag{12}$$

where $u_i \in \{0, 1\}$. In (11) the summation in the exponent is performed modulo two. This representation of the Hadamard transform is for the Hadamard matrix in "natural" form as given by (6). Another series representation exists for a Hadamard matrix in "ordered" form in which the sequency of each row is larger than the preceding row. By this representation

$$F(u, v) = \sum_{x=0}^{N-1} \sum_{y=0}^{N-1} f(x, y)(-1)^{q(x,y,u,v)} \tag{13}$$

where

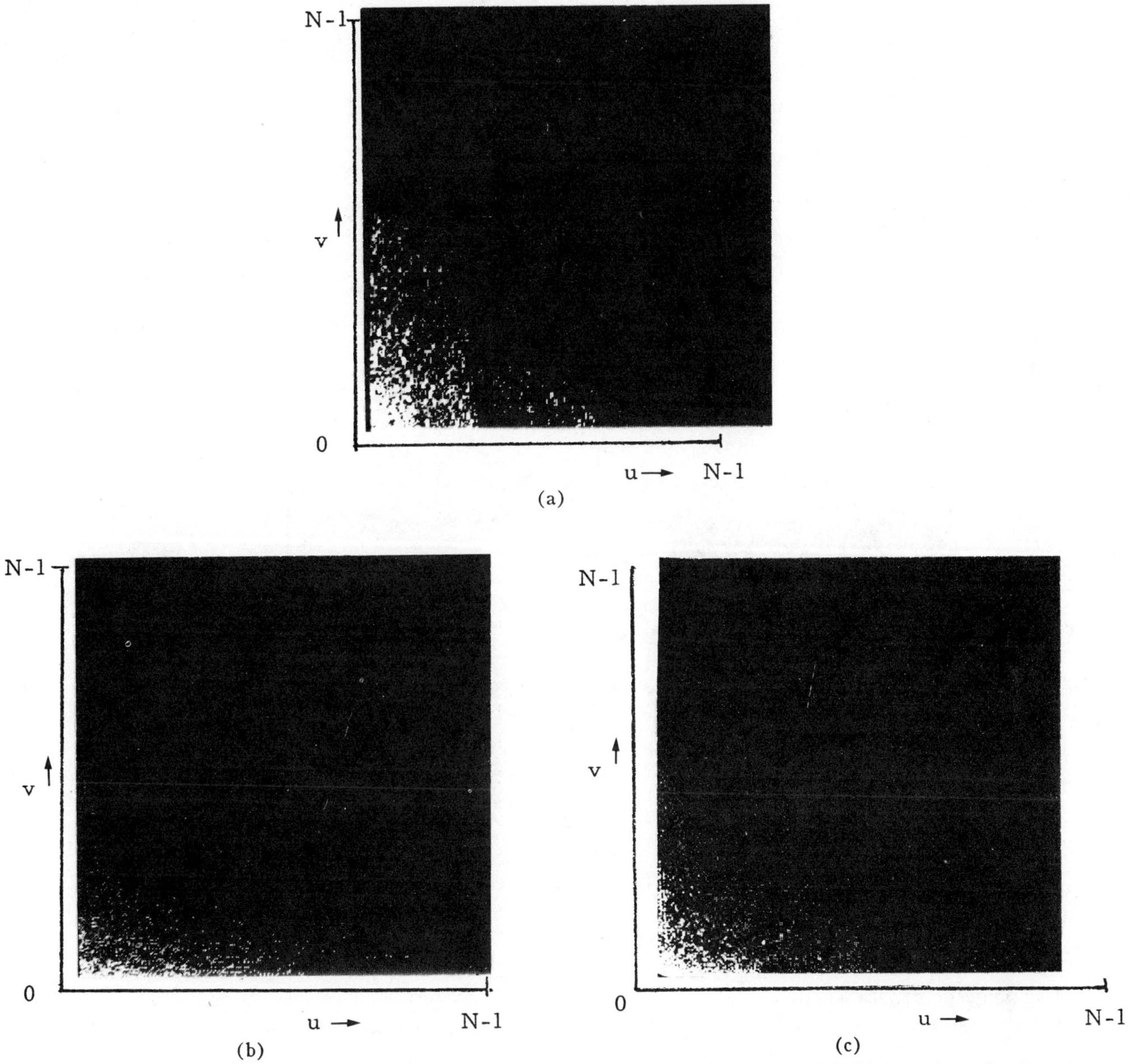

Fig. 3. Hadamard transforms of Surveyor scenes (displays of the logarithm of the magnitude of the Hadamard transform); (a) footpad, (b) boom, and (c) experimental box.

$$q(x, y, u, v) \equiv \sum_{i=0}^{n-1} \left[g_i(u)x_i + g_i(v)y_i\right]$$

and

$$\begin{aligned} g_0(u) &\equiv u_{n-1} \\ g_1(u) &\equiv u_{n-1} + u_{n-2} \\ g_2(u) &\equiv u_{n-2} + u_{n-3} \\ &\vdots \\ g_{n-1}(u) &\equiv u_1 + u_0. \end{aligned}$$

The two-dimensional Hadamard transform may be computed in either natural or ordered form with an algorithm analogous to the fast Fourier transform computer algorithm. A computational algorithm for an ordered transform is considered later. All experimental results in this paper have been obtained with the Hadamard transform given by (13).

Experiments have been performed to determine the nature of the two-dimensional Hadamard transform of an image, and to assess the effects of a double Hadamard transformation. Fig. 3 contains cathode ray tube displays of the logarithm of the magnitude of the ordered Hadamard transform of several Surveyor scenes. In these photographs the origin (zero sequency) appears in the lower left corner. As the sequency increases, the magnitudes of the samples tend to decrease. This is an indication that there are relatively few high-amplitude brightness transitions between elements in the original scenes. Fig. 4 shows the double Hadamard transforms of the three scenes. There is no

(a)

(b) (c)

Fig. 4. Double Hadamard transforms of Surveyor scenes; (a) footpad, (b) boom, and (c) experimental box.

noticeable image degradation between the originals and the double Hadamard transforms [compare, for example, Fig. 1(a) and Fig. 4(a)].

IV. Properties of Hadamard Image Transforms

The Hadamard transform has several interesting properties. The most important properties from the standpoint of image coding are dynamic range, conservation of energy, and entropy.

The zero sequency term

$$F(0, 0) = \sum_{x=0}^{N-1} \sum_{y=0}^{N-1} f(x, y) \tag{14}$$

is a measure of the average brightness of a scene. If $f(x, y)$ is a positive real function, then the maximum possible value for the zero sequency term is N^2A where A is the maximum value of $f(x, y)$. All Hadamard domain samples other than the zero sequency sample range between $\pm N^2A/2$. The magnitude of the zero sequency term is a bound for the magnitude of all other Hadamard domain samples.

A conservation of energy property exists between the spatial domain and the Hadamard domain. Specifically,

$$\sum_{x=0}^{N-1} \sum_{y=0}^{N-1} |f(x, y)|^2 = \frac{1}{N^2} \sum_{u=0}^{N-1} \sum_{v=0}^{N-1} |F(u, v)|^2. \tag{15}$$

This equation is analogous to Parseval's relationship for the Fourier transform [23]. The implication of this equation for image coding is that if a few of the Hadamard domain samples are of large magnitude, then many of the remaining samples must necessarily be of very low magnitude. Conceivably, the low-magnitude samples may be discarded, as is possible with the Fourier transform coding technique, to obtain a bandwidth reduction.

If $f(x, y)$ is considered to be a random two-dimensional function with a specified entropy, then the entropy of $F(u, v)$ is the same as the entropy of $f(x, y)$ since the Jacobian of the transformation matrix is unity [24]. Hence, with proper coding it is possible to transmit either a scene or its Hadamard transform over a channel with the same channel capacity.

V. Computational Algorithm

The computation of (13) is performed in two steps. First a one-dimensional Hadamard transform is taken along each row of the array $f(x, y)$ yielding

$$F(u, y) = \sum_{x=0}^{N-1} f(x, y)(-1)^{\sum_{i=0}^{n-1} g_i(u)x_i} \tag{16}$$

Then a second one-dimensional Hadamard transform is taken along each column of $F(u, y)$ giving the desired result

$$F(u, v) = \sum_{y=0}^{N-1} F(u, y)(-1)^{\sum_{i=0}^{n-1} g_i(v)y_i} \tag{17}$$

Computation of the one-dimensional Hadamard transform by brute force methods requires N^2 operations where an operation is either an addition or subtraction. An algorithm for obtaining the one-dimensional Hadamard transform in $N \log_2 N$ operations has been developed. The algorithm is quite similar to the fast Fourier transform algorithm; computational savings are realized by storage of intermediate results [8]. A "fast" Hadamard transform algorithm was outlined in 1937 by Yates [25]. In 1958, Good described a matrix decomposition technique which can be implemented to perform the Hadamard transform with $N \log_2 N$ operations [26].

Fig. 5 illustrates the computations performed for our one-dimensional Hadamard transformation with eight data points. The data points are arranged in a column at level 3 and then summed by pairs to produce intermediate results for level 2. A dotted line linking two modes indicates that the data point at the higher level is multiplied by minus one before addition, or equivalently, the data point forms the subtrahend of a subtraction operation. Operations follow the tree graph to level 0 which is the ordered Hadamard transform of $f(x)$. There are two operations performed at each node of levels 1, 2, and 3 yielding a total of $8 \log_2 8 = 24$ operations.

The fast Hadamard transform algorithm performs all of the operations indicated in Fig. 5, but in a certain selected order. All operations of level k are not completed before proceeding to level $k-1$, but rather operations are performed according to a sieving sequence. Fig. 6 describes the basic sequence computations. The first sequence is the "S" sequence in which the sum of all data points is formed to produce $F(0)$. A "1" sequence subtracts the lower node from the upper node of a pair of nodes at level $k-1$ to produce a result at level k. In the "2" sequence, operations begin at level $k-2$ where pairs are subtracted from one another to produce the results of level $k-1$ which in turn are added together. The "3" sequence and higher sequences to the "n" sequence follow directly. Fig. 6 also indicates the storage requirements for the computational procedure.

Original data are stored in a block of N words corresponding to level n. Intermediate results are stored $n-1$ different blocks of sizes 2^{n-1} to 2^2 words. These storage locations correspond to levels $n-1$ to 1 in the computation procedure.

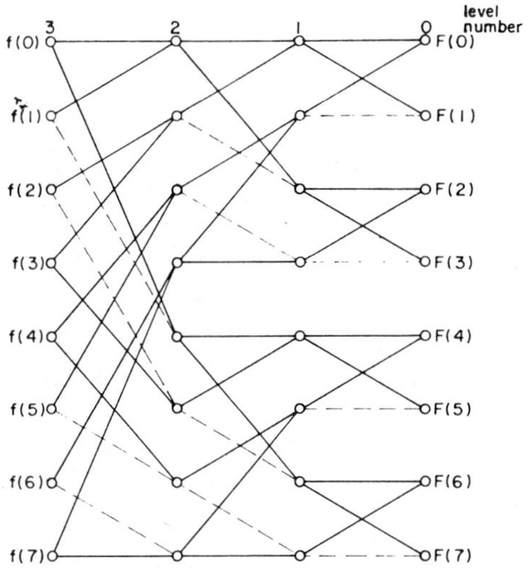

Fig. 5. Computations for one-dimensional third-order Hadamard matrix.

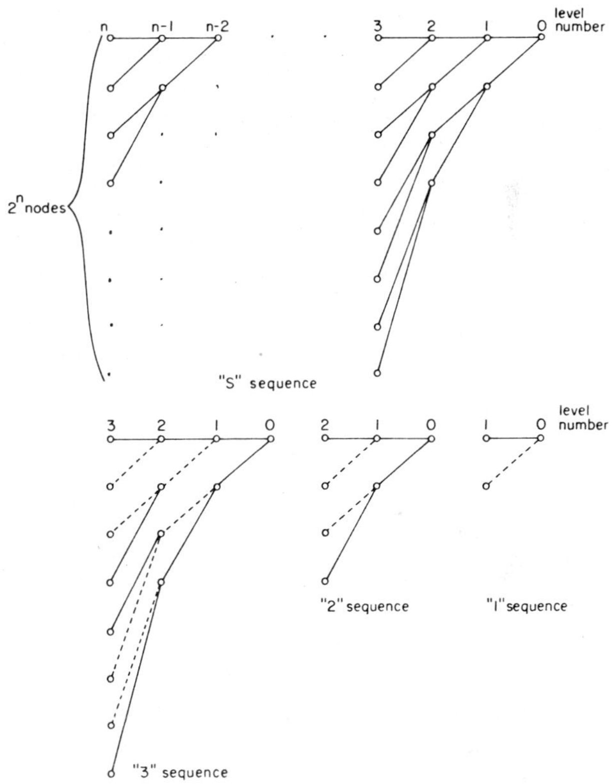

Fig. 6. Hadamard transform computational sequences.

Computation of the Hadamard transform begins with the "S" sequence which computes $F(0)$. Subsequent calculations are controlled by the following sieving sequence of integers building up to integer n:

$$\{1, 2, 1, 3, 1, 2, 1, 4, 1, 2, 1, 3, 1, 2, 1, \cdots, 1, n, 1, 2, 1, \cdots, 1, 2, 1, 3, 1, 2, 1\}.$$

For example, after the "S" sequence, the "1" sequence computes $F(1)$ using the intermediate results stored in level 1. Then, the "2" sequence computes $F(2)$ using the intermediate results in level 2.

	Level 3	2	1	0
f(0)	A	_I	_M	F(0) P
f(1)	B	_J	_N	F(1) Q
f(2)	C	_K		F(2) R
f(3)	D	_L		F(3) S
f(4)	E			F(4) T
f(5)	F			F(5) U
f(6)	G			F(6) V
f(7)	H			F(7) W

Fig. 7. Storage locations for computation of one-dimensional third-order Hadamard transform.

Sequence number	Level number	Operation	Contents of storage locations	Result
—	3	—	f(0) A f(1) B f(2) C f(3) D f(4) E f(5) F f(6) G f(7) H	
S	2	A+B → I	f(0) + f(1) I	
		C+D → J	f(2) + f(3) J	
		E+F → K	f(4) + f(5) K	
		G+H → L	f(6) + f(7) L	
	1	I+J → M	f(0) + f(1) +f(2) +f(3) M	
		K+L → N	f(4) + f(5) +f(6) +f(7) N	
	0	M+N → P	f(0) +f(1) +f(2) +f(3) +f(4) +f(5) +f(6) +f(7) P	F(0)
1	0	M-N → Q	f(0) +f(1) +f(2) +f(3) -f(4) -f(5) -f(6) -f(7) Q	F(1)
2	1	I -J → M	f(0) +f(1) -f(2) -f(3) M	
		L-K → N	-f(4) -f(5) +f(6) + f(7) N	
	0	M+N → R	f(0) +f(1) -f(2) -f(3) -f(4) -f(5) +f(6) +f(7) R	F(2)
1	0	M-N → S	f(0) +f(1) -f(2) -f(3) +f(4) +f(5) -f(6) -f(7) S	F(3)
3	2	A-B → I	f(0) -f(1) I	
		D-C → J	-f(2) +f(3) J	
		E-F → K	f(4) -f(5) K	
		H-G → L	-f(6) +f(7) L	
	1	I + J → M	f(0) -f(1) -f(2) +f(3) M	
		K+L → N	f(4) -f(5) -f(6) +f(7) N	
	0	M+N → T	f(0) -f(1) -f(2) +f(3) +f(4) -f(5) -f(6) +f(7) T	F(4)
1	0	M-N → U	f(0) -f(1) -f(2) +f(3) -f(4) +f(5) +f(6) -f(7) U	F(5)
2	1	I -J → M	f(0) -f(1) +f(2) -f(3) M	
		L-K → N	-f(4) +f(5) -f(6) +f(7) N	
	0	M+N → V	f(0) -f(1) +f(2) -f(3) -f(4) +f(5) -f(6) +f(7) V	F(6)
1	0	M-N → W	f(0) -f(1) +f(2) -f(3) +f(4) -f(5) +f(6) -f(7) W	F(7)

Fig. 8. Computer operations for computation of one-dimensional third-order Hadamard matrix.

Fig. 7 gives the storage locations for computation of a one-dimensional third-order Hadamard transform. The computer operations for this example are listed in Fig. 8.

Since computation of the Hadamard transform requires only real additions and subtractions, whereas the Fourier transform computations are composed of complex multiplications, additions, and subtractions, considerable computational savings are afforded with the Hadamard transform. Both the Fourier and Hadamard transforms have been programmed on a TRW-530 digital computer. For a 256 by 256 point scene, the Fourier transform can be computed in 20 minutes and the Hadamard transform in 3 minutes.

VI. Hadamard Domain Quantization

The Hadamard transform of an image must be quantized for subsequent digital coding and transmission over a channel. In order to quantize Hadamard domain samples it is necessary to determine the number and placement of quantization levels. Both of these factors affect the degree of image degradation caused by quantization errors and the entropy of the quantized symbols. The approach taken here has been to select quantum levels to maximize the source entropy, and then to subjectively evaluate image quality.

Two models for the probability density of Hadamard domain samples that have been considered are shown in Fig. 9. The first is the linear quantization model, commonly used for spatial domain quantization, in which the quantum levels are equispaced over some maximum range. The second model is the Gaussian quantization model in which the quantization levels are chosen to partition the Gaussian curve into equal areas.

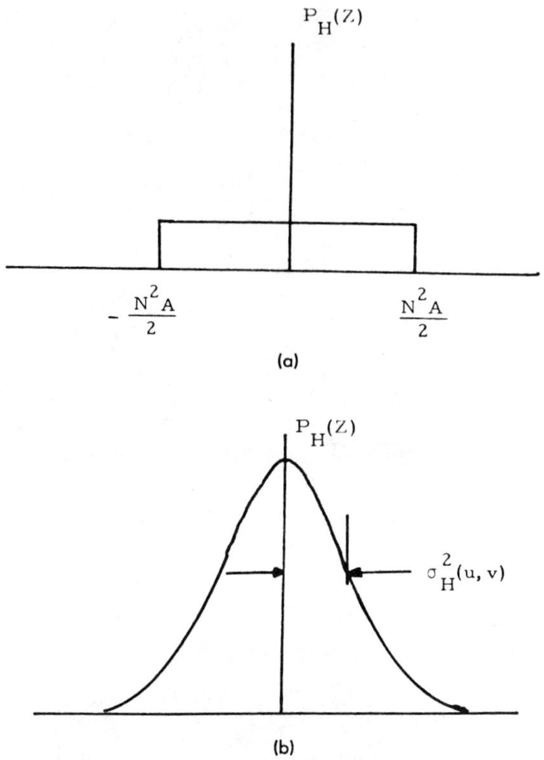

Fig. 9. Hadamard domain quantization rules: (a) linear, and (b) Gaussian.

In the Gaussian quantization model the probability density of Hadamard domain samples is assumed to be of the form

$$P_H(Z) = \left[2\pi\sigma_H^2(u, v)\right]^{-\frac{1}{2}} \exp\left\{-\frac{Z^2}{2\sigma_H^2(u, v)}\right\} \tag{18}$$

where the variance function $\sigma_H^2(u, v)$ controls the placement of quantization levels as a function of the Hadamard spatial frequency. Referring to Fig. 3, it is evident that the variance function should be highest at the origin in the Hadamard domain, be circularly symmetric over one quadrant, and decrease monotonically toward the higher Hadamard dimensions. A convenient two-dimensional function possessing these properties is the Gaussian curve described by

$$\sigma_H^2(u, v) = S \exp\left\{-\frac{u^2 + v^2}{p}\right\} \tag{19}$$

where S is an amplitude scaling constant and p is a spread control constant.

Reconstructions of the Hadamard transform of the footpad scene with the linear and Gaussian quantizer are shown in Fig. 10. In both cases, 64 quantization levels have been employed. Results with the linear quantizer are poor because of the large quantization errors at high sequencies. The reconstruction using the Gaussian quantizer with the variance parameter $p = 1500$, on the other hand, shows negligible image degradation. Fig. 11 contains additional examples of reconstructions with the 64 level Gaussian quantizer with $p = 1500$. Tests have been conducted to determine the effect of fewer quantization levels. Fig. 12 shows reconstructions of the footpad scene with the Gaussian quantizer with $p = 1500$ for 32 and 16 quantization levels. The loss of resolution apparent in these pictures is due to the quantization errors at high sequencies.

If the Gaussian quantization rule is to be practical it is imperative that a variance function can be chosen for a wide class of scenes without detailed knowledge of the content of these scenes. Figs. 10 and 11 show that the Gaussian variance function, with the same variance parameter ($p = 1500$), provides satisfactory reconstructions for three different scenes. It is also of interest to determine the effect of changes in the variance parameter. Fig. 13 contains reconstructions using the Gaussian quantizer with $p = 500$, 1000, 2000, and 5000 and with 64 quantization levels. For $p = 1000$ and $p = 2000$ the reconstructions are quite satisfactory. At $p = 500$ a distinct checking effect appears. This checking is due to quantization errors of large-magnitude low-sequency Hadamard samples. The quantizer

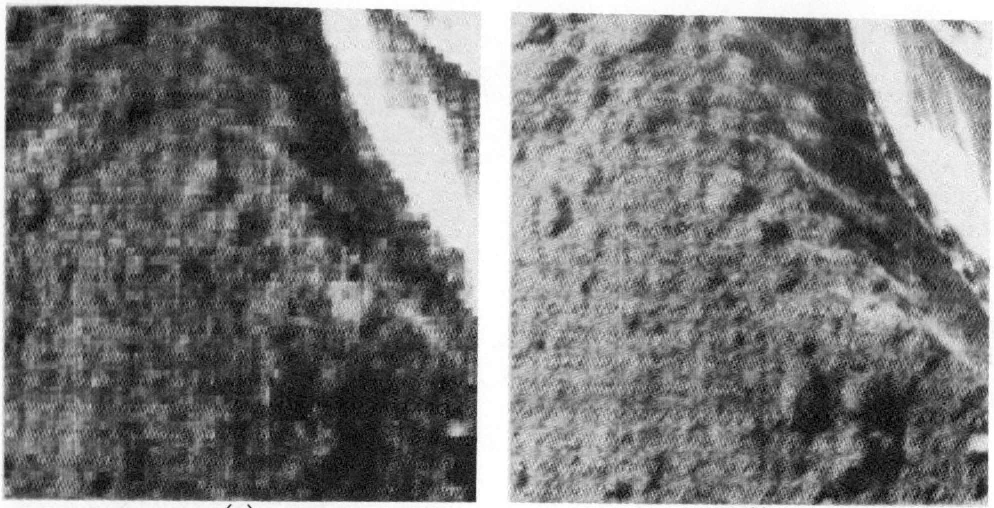

(a) (b)

Fig. 10. 64-level quantization of footpad; (a) linear quantization, and (b) Gaussian quantization, $p=1500$.

(a) (b)

Fig. 11. 64-level Gaussian quantization of boom and experimental box, $p=1500$; (a) boom, and (b) experimental box.

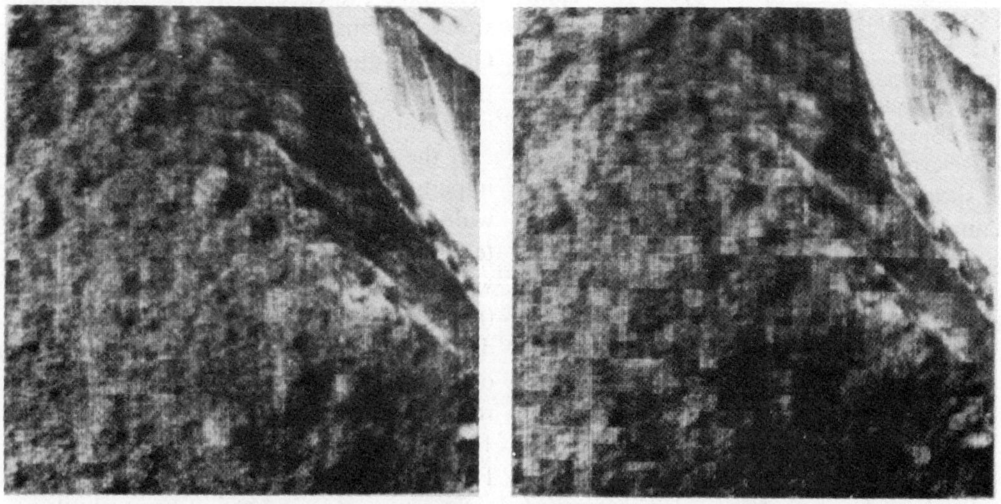

(a) (b)

Fig. 12. 32- and 16-level Gaussian quantization of footpad, $p=1500$; (a) 32-level quantization, and (b) 16-level quantization.

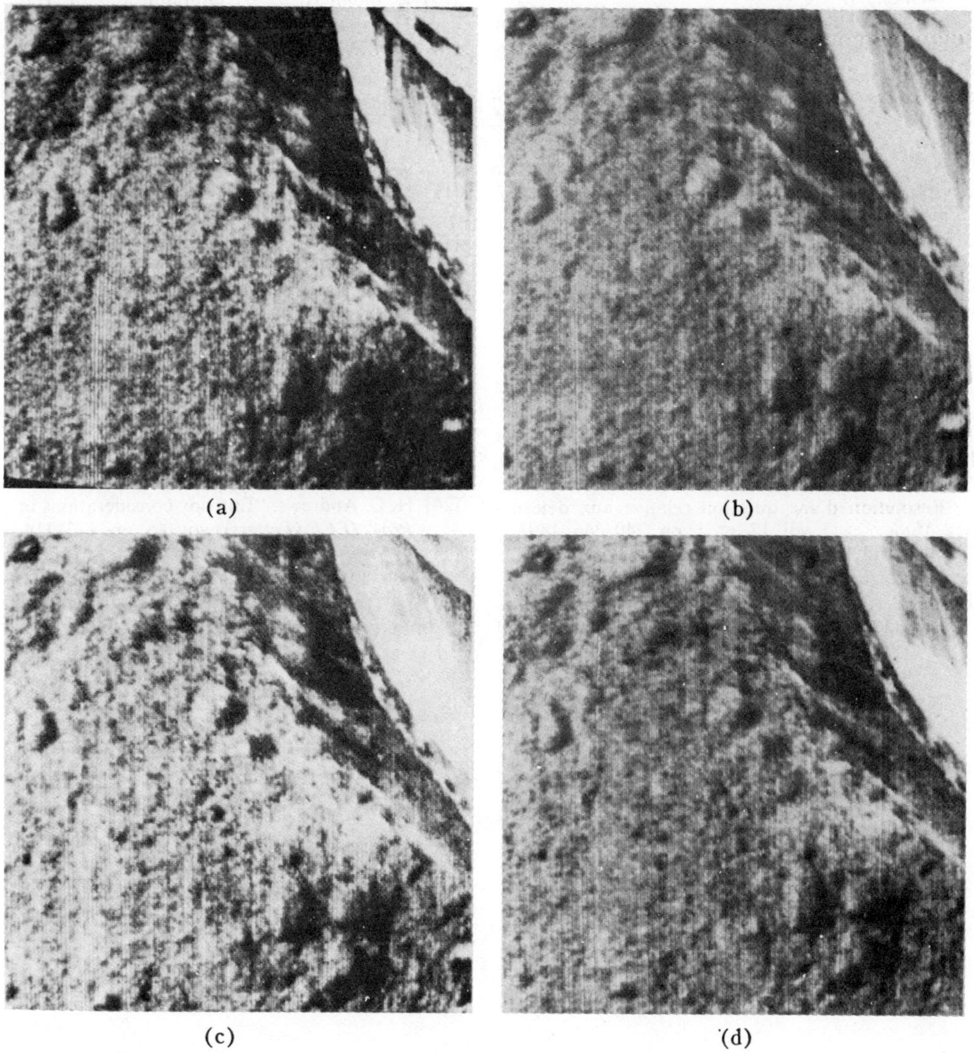

Fig. 13. 64-level Gaussian quantization of footpad with different variance functions; (a) $p=500$, (b) $p=1000$, (c) $p=2000$, (d) $p=5000$.

has clipped the magnitudes of these samples to a relatively low level. At the other extreme, for $p=5000$, the high-sequency samples, which are of low magnitude, are all given the same magnitude of the first quantum level. The result is a loss of resolution.

Conclusions

It has been shown that it is possible to transform an image by a Hadamard matrix operator which is similar to the two-dimensional Fourier transform. Use of the Hadamard rather than the Fourier transform results in an order of magnitude speed increase. The Hadamard samples can be quantized to as few as 64 levels, using a Gaussian quantization rule. Reconstruction of the quantized transform by a second two-dimensional Hadamard transform introduces negligible image degradation. With conventional PCM coding it is possible to transmit the Hadamard transform of an image at the same bit rate as the image itself. The potential advantages of the Hadamard image coding technique are a tolerance of channel errors and the possibility of bandwidth reduction.

Acknowledgment

Professor L. R. Welch of the University of Southern California suggested series representations of the Hadamard transform given by (11) and (13). The assistance of D. Ketcham of USC, who developed many of the experimental computer programs, and J. Pepin also of USC, who carried out the experiments, is greatly appreciated. Original digitized images were provided by F. Billingsley of the Jet Propulsion Laboratory.

References

[1] H. C. Andrews and W. K. Pratt, "Fourier transform coding of images," *Hawaii Internat'l Conf. on System Sciences*, pp. 677–679, January 1968.

[2] H. C. Andrews and W. K. Pratt, "Television bandwidth reduction by Fourier image coding," presented at the Soc. of Motion Picture and Television Engrs., 103rd Tech. Conf., May 1968.

[3] H. C. Andrews and W. K. Pratt, "Television bandwidth reduction by

encoding spatial frequencies," to be published in *J. Soc. Motion Picture and Television Engrs.*

[4] J. W. Cooley and J. W. Tukey, "An algorithm for the machine calculation of complex Fourier series," *Math. of Comp.*, vol. 19, pp. 297–301, 1956.

[5] Subcommittee on Measurement Concepts, Audio and Electroacoustic Group, "What is the fast Fourier transform?" *Proc. IEEE*, vol. 55, pp. 1664–1673, October 1967.

[6] J. W. Cooley, P. A. W. Lewis, and P. D. Welch, "Historical notes on the fast Fourier transform," *Proc. IEEE*, vol. 55, pp. 1675–1677, October 1967.

[7] E. O. Brigham and R. E. Morrow, "The fast Fourier transform" *IEEE Spectrum*, vol. 4, pp. 63–70, December 1967.

[8] H. C. Andrews, "A high speed algorithm for the computer generation of Fourier transforms," *IEEE Trans. Computers (Short Notes)*, vol. C-17, pp. 373–375, April 1968.

[9] W. M. Gentlemen, "Matrix multiplication and fast Fourier transformations," *Bell Sys. Tech. J.*, vol. 47, pp. 1099–1103, July–August 1968.

[10] W. R. Crowther and C. M. Rader, "Efficient coding of vocoder channel signals using linear transformation," *Proc. IEEE (Letters)*, vol. 54, pp. 1594–1595, November 1966.

[11] J. Hadamard, "Resolution d'une question relative aux determinants," *Bull. Sci. Math.*, ser. 2, vol. 17, pt. 1, pp. 240–246, 1893.

[12] H. J. Ryser, *Combinatorial Mathematics*. New York: Wiley, 1963.

[13] R. E. A. C. Paley, "On orthogonal matrices," *J. Math. Phys.*, vol. 12, pp. 311–320, 1933.

[14] J. Williamson, "Hadamard's determinant theorem and the sum of four squares," *Duke Math. J.*, vol. 11, pp. 65–81, 1944.

[15] L. Baumert, S. W. Golomb, and M. Hall, Jr., "Discovery of an Hadamard matrix of order 92," *Bull. Am. Math. Soc.*, vol. 68, pp. 237–238, 1962.

[16] H. F. Harmuth, "A generalized concept of frequency and some applications," *IEEE Trans. Information Theory*, vol. IT-14, pp. 375–382, May 1968.

[17] J. L. Walsh, "A closed set of orthogonal functions," *Am. J. Math.*, vol. 55, pp. 5–24, 1923.

[18] N. J. Fine, "On the Walsh functions," *Trans. Am. Math. Soc.*, vol. 65, pp. 372–414, 1949.

[19] N. J. Fine, "The generalized Walsh functions," *Trans. Am. Math. Soc.*, vol. 69, pp. 66–77, 1950.

[20] G. W. Morgenthaler, "On Walsh-Fourier series," *Trans. Am. Math. Soc.*, vol. 84, pp. 472–507, 1957.

[21] K. W. Henderson, "Some notes on the Walsh functions," *IEEE Trans. Electronic Computers (Correspondence)*, vol. EC-13, pp. 50–52, February 1964.

[22] H. Rademacher, "Einige Sätze von allgemeinen Orthogonal-Funktionen," *Math. Ann.*, vol. 87, pp. 122–138, 1922.

[23] A. Papoulis, *The Fourier Integral and Its Applications*. New York: McGraw-Hill, 1962, p. 27.

[24] H. C. Andrews, "Entropy considerations in the frequency domain," *Proc. IEEE (Letters)*, vol. 46, pp. 113–114, January 1968.

[25] F. Yates, *The Design and Analysis of Factorial Experiments*. Harpenden: Imperial Bureau of Soil Analysis, 1937.

[26] I. J. Good, "The interaction algorithm and practical Fourier analysis," *J. Royal Stat. Soc.*, ser. B, vol. 20, pp. 361–372, 1958.

Reprinted from *IEE Proc.* **124:**845-852 (1977)

HADAMARD-TRANSFORMATION TECHNIQUE OF SPEECH CODING: SOME FURTHER RESULTS

E. Frangoulis and L. F. Turner

Indexing terms: Digital communication systems, Encoding, Transforms, Voice communication

Abstract

The results of an extensive investigation of the properties of 64-point Hadamard transformed speech are presented. Detailed information is given about the probability density functions of the Hadamard coefficients, the average power-density spectrum in the Hadamard domain and the logical-autocorrelation function. The results indicate that good-quality speech can be reconstructed from 6 to 8 dominant Hadamard coefficients, but that the use of fewer coefficients is unlikely to lead to the reconstruction of speech of acceptable quality. The results of a preliminary series of listening tests are presented and these confirm conclusions drawn from the statistical properties of the transformed speech. It is shown that the number of bits needed for coefficient labelling constitutes a significant proportion of the total number of bits needed to represent Hadamard transformed speech. A technique is presented for reducing by more than 50% the number of labelling bits needed, and it is explained how, by using this technique, it should be possible to obtain good quality speech when using a transmission bit rate of 8 k bits/s.

List of principal symbols

C_i = Hadamard coefficient
x_i = data sample
H = Hadamard matrix
$L_S(k)$ = local, or short term, logical-autocorrelation function
$\hat{L}(k)$ = average value of logical-autocorrelation function
W = transmission bit rate in bits/s
L = number of dominant coefficients used to reconstruct speech signal
M = number of bits used to encode each coefficient
l = number of bits needed to label a coefficient
R = rate at which speech signal is sampled
Q = number of components in Walsh-Fourier spectrum
G_i = component of Walsh-Fourier spectrum
γ_i^2 = sequency spectral component
$a_s(n)$ = sal (n) coefficient
$a_c(n)$ = cal (n) coefficient

1 Introduction

In recent years, interest has grown in the possibility of using orthogonal transformations as a means of reducing the bit rate necessary for the transmission of speech signals. This interest stems mainly from the ease with which techniques of this kind can be implemented in practice.

One of the earliest works in which orthogonal transformations were applied to the processing of speech signals was that of Boesswetter,[1] who examined a 1·4 s section of German speech, and showed that if 2 ms samples of the speech were transformed using a 16-point Hadamard transformation, then reasonably intelligible speech could be reconstructed using only the two largest coefficients from the set of 16 Hadamard coefficients. This was then followed by the work of Campanella and Robinson.[2,3] In an early paper,[2] Campanella and Robinson performed a sequency analysis on the vowels a, e, i, o and u and then went on in a later paper[3] to show mathematically that, by using a 16-point Hadamard transformation, it should be possible, for a given signal/quantisation noise ratio, to transmit speech using 48·5 k bits/s, rather than 56 k bits/s as in conventional p.c.m. In a recent important paper, Shum *et al.*[4] have reported on the application of a 64-point Walsh-Hadamard transformation to speech compression, and have stated that their tentative results indicate that fair to good reproduction of speech can be obtained when using from 4 to 8 of the 64 Hamamard coefficients. In a practical system the complete set of coefficients would be determined and the dominant (largest) coefficients would be obtained and transmitted. The 4 to 8 coefficient representation corresponds to bit rates ranging from 6·5 k bits/s to 13 k bits/s.

The results of Shum *et al.*,[4] although only of a provisional nature, are indicative and sufficiently important to warant a more detailed examination of the question of speech compression by orthogonal transformations. This paper is a report on the results of an extensive and systematic investigation that has been conducted into the application of the Hadamard transformation to speech compression. In Section 2 of the paper the question of compression by use of a limited number of coefficients is examined, with particular attention being paid to the existence and time varying nature of dominant coefficients, and to the sensitivity of these coefficients to factors such as the sampling rate and changes in speaker. The quantitative results contained in this Section are new and are of considerable importance, since, if speech compression is to be achieved by using a limited number of coefficients, then it is essential that the variation and sensitivity of these coeffcients to various parameters be clearly understood. In Section 3 the question of the subjective quality of speech reconstructed from dominant coefficients is considered briefly, and some comments are made as to how the speech quality varies with variations in the number of coefficients used, the sampling rate and the number of bits used to represent each coefficient. The comments made in Section 3 concerning speech quality should only be considered as indicative, since an extensive series of subjective tests is necessary to obtain a quantitative assessment of speech quality. In the fourth Section of the paper, the nature and occurrence of the dominant coefficients is considered, and it is shown that the dominant coefficients are always members of a fixed subset of the 64 Hadamard coefficients, and that by making use of this fact it is possible to reduce the necessary transmission bit rate without affecting speech quality. In Section 5 of the paper, some initial results concerning window widths are presented, and it is argued that, if the results that have been obtained so far can be shown to be generally true, then it should be possible to obtain acceptable speech quality when using transmission bit rates of approximately 6 k bits/s.

2 Some properties of Hadamard-transformed speech signals

2.1 Introduction

In this Section, some of the results of an extensive study of the properties of Hadamard transformed speech are presented. The results, although only a selection from a very large number of results, are totally representative of the general pattern of behaviour that has emerged from the tests, and conclusions drawn from them are of general validity.

The specific properties that are considered in this Section are as follows:

(a) the probability density function (p.d.f.) of the Hadamard coefficients for the 64-point transformation
(b) the average power-density function in the Hadamard domain
(c) the variation of the logical autocorrelation function in terms of the number of coefficients.

These properties are studied with respect to various parameters, such as the length of the speech section transformed, the sampling rate, speaker variation and sentence variation.

Paper 7971 E, first received 9th December 1976 and in revised form 12th May 1977

Mr. Frangoulis and Dr. Turner are with the Department of Electrical Engineering, Imperial College of Science Technology, South Kensington, London SW7 2BT, England

2.2 Probability density function of the Hadamard coefficients for the 64-point transformation

In all the tests reported in this paper the speech signals were passed through low-pass filter of cutoff frequency 3·4 kHz and then sampled. The sampled values were then analogue-digital converted using a 10-bit/sample convertor and the digitised samples were then processed in a PDP-15 computer. After being processed, the digitised sampled speech was then reconverted to analogue form using a 10-bit digital-analogue convertor and an associated low-pass filter.

In applying the Hadamard transformation technique to speech signals, the n Hadamard coefficients $C_1, C_2, \ldots, C_n$ are obtained by taking blocks of $n(=r^2)$ data samples $x_1, \ldots, x_n$ and then performing the following matrix multiplication:

$$\begin{bmatrix} C_1 & \cdots & C_r \\ \vdots & & \vdots \\ \vdots & & \vdots \\ C_{r^2-r+1} & \cdots & C_n \end{bmatrix} = [H] \begin{bmatrix} x_1 & \cdots & x_r \\ \vdots & & \vdots \\ \vdots & & \vdots \\ x_{r^2-r+1} & \cdots & x_n \end{bmatrix} [H]$$

where $[H]$ is an $r \times r$ Hadamard matrix.

In the first series of tests long sections* of speech, with the lengths ranging from parts of a sentence to a number of sentences, were Hadamard transformed and the p.d.f.s associated with the 64 ($= n$) Hadamard coefficients were obtained. A selection from the results are shown in Figs. 1 to 9. It is not possible in these Figures to show separately each of the 64 coefficients. What is done, where appropriate, is to display the coefficient p.d.f.s in bounded sets. Figs. 1 and 2 show the coefficient p.d.f.s associated with a complete sentence spoken by both male and female speakers. These two Figures are perhaps the most important of the first set of nine Figures, since they appear to indicate the existence of a small set of dominant (large valued) coefficients. The Figures indicate that the dominant coefficients exist irrespective of whether the speaker is male or female. Figs. 3 and 4, which relate to parts of a sentence, also indicate the existence of dominant coefficients. Although it is not clear from Figs. 1 to 4, the dominant set of coefficients may change, and in fact do change, from speaker to speaker and from one part of a sentence to another. This is a point that is taken up in more detail later in the paper. Figs. 5 to 8 are essentially a repeat of Figs. 1 to 4, but with the difference that attention is focused on isolated dominant coefficients. Figs. 5 and 6 show how the p.d.f. of a dominant coefficient varies as the length of speech considered is changed from a whole sentence to part of a sentence. Also, within each of the two Figures, a comparison is given as to how the p.d.f. of the coefficient varies from the male to the female speaker. Figs. 7 and 8 show how the p.d.f. of a dominant coefficient varies from sentence to sentence. Figs. 1 to 8 indicate that dominant coefficients tend to have higher average values for female speakers than for male speakers, which suggests that it may be necessary to use a larger average number of binary digits to represent female speech. Fig. 9 shows how the p.d.f. of a dominant coefficient varies as a function of the rate at which the speech signal is sampled. It should be noted from Fig. 9 that the dominant coefficient tends to have a higher average value for lower sampling rates, and, thus, this again indicates that an increased number of bits/dominant-coefficient may be necessary at lower sampling rates.

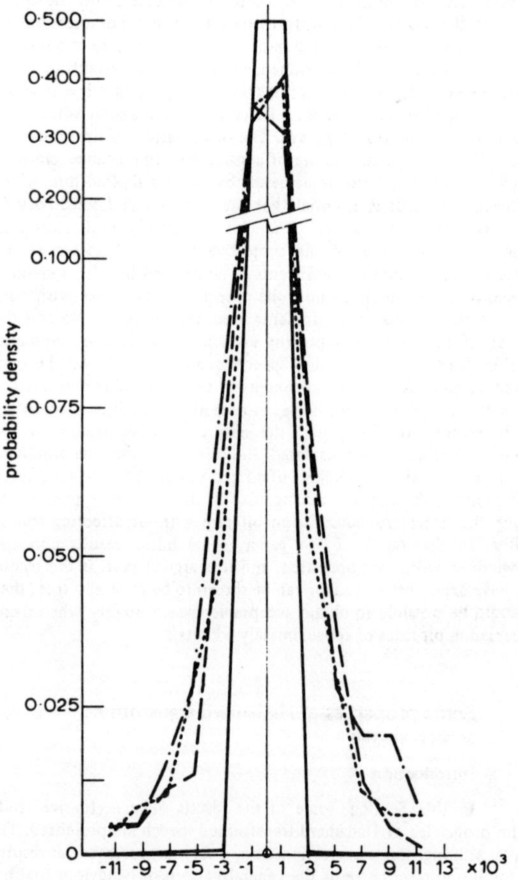

Fig. 1
Probability density functions of Hadamard coefficients

Complete sentence spoken by male, sampling rate 8 k samples/s

---- bounds on set of 56 nondominant coefficients
-·-·- bounds on set of 8 dominant coefficients

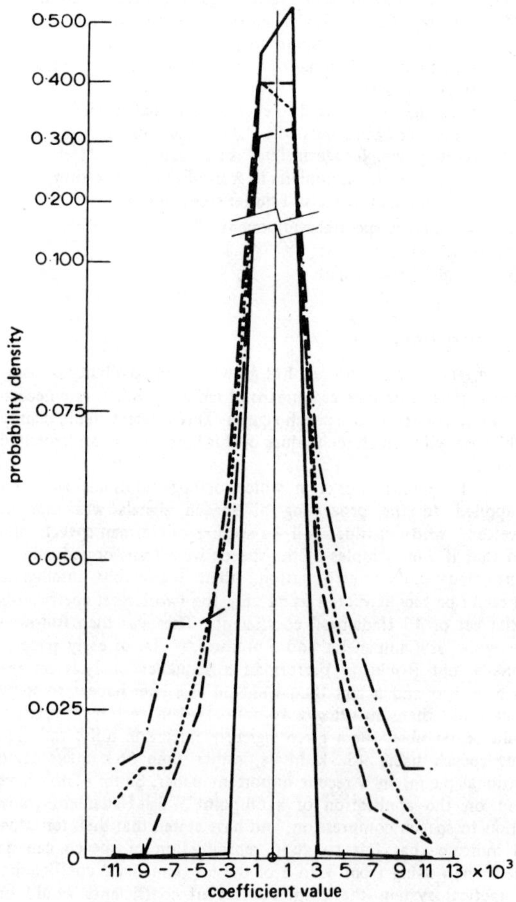

Fig. 2
Probability density functions of Hadamard coefficients

Complete sentence spoken by female, sampling rate 8 k samples/s

---- bounds of set on 56 nondominant coefficients
-·-·- bounds on set of 8 dominant coefficients

*Long in comparison with the duration of the 64-sample set used to compute each set of Hadamard coefficients

2.3 Power-density functions for 64-point Hadamard transformed speech

Although the p.d.f.s of the Hadamard coefficients indicate the existence of certain large valued dominant Hadamard coefficients, this information provides little real evidence as to how accurately speech waveforms can be represented when using a few dominant coefficients. In order to obtain some quantitative information relating to this matter, speech signals were examined in terms of their average power-density spectrum in the Hadamard domain. The results of this examination are summarised in Figs. 10 to 12 and in Table 1. Fig. 10, which was obtained using a complete sentence spoken by a male speaker, shows how, as a function of sampling rate, the power-density spectrum in the Hadamard domain is distributed among the Hadamard coefficients. Fig. 11 shows the way in which the Hadamard power-density spectrum changes from male to female speakers, and Fig. 12 shows how, for a sentence spoken by a male, the percentage of the total power contained in 5 and 8 dominant coefficients varies with increasing sampling rate. Table 1 provides important general information about the percentage of the total waveform power that is contained in 5 and 8 dominant coefficients for various sentences spoken by both male and female speakers.

The results of this Section indicate that, for both male and female speakers, between 60% and 90% of the total average power is contained in the 6 to 8 dominant coefficients when using a sampling rate of 8 k samples/s. The results also indicate that there is approximately a 10% reduction in the power contained in the dominant coefficients as the sampling rate is reduced from 10·8 k samples/s to 6·1 k samples/s. In addition, it is clear that for female speakers the dominant coefficients contain a smaller percentage of the total power than for male speakers. The results contained in Table 1 indicate that the average power contained in 5 dominant coefficients is between 20% and 25% less than that contained in 8 dominant coefficients, and this thus casts considerable doubt on the feasibility of obtaining good-quality speech reconstruction from 5, or fewer, coefficients.

2.4 Logical autocorrelation function for 64-point Hadamard transformed speech

Further insight into the acceptability of speech reconstructed from a limited number of Hadamard coefficients can be gained by considering the normal (arithmetic) autocorrelation function of the reconstructed speech, or, alternatively, by considering the so-called[5] logical autocorrelation function.

The local, or short term, logical autocorrelation function of a sample of speech is defined[5, 6] to be

$$L_s(k) = \frac{1}{N}\sum_{j=0}^{N-1} x(j \oplus k)x(j)$$

where $k = 0, 1, 2, \ldots, N-1$ and $j \oplus k$ denotes the bit by bit modulo-2 addition of the integers j and k expressed in binary notation, and $N = 2^n$ is the length of the sequence. The logical autocorrelation function $L(k)$ is defined to be the expected (or average) value of the local logical autocorrelation function, and it can be estimated by its sample

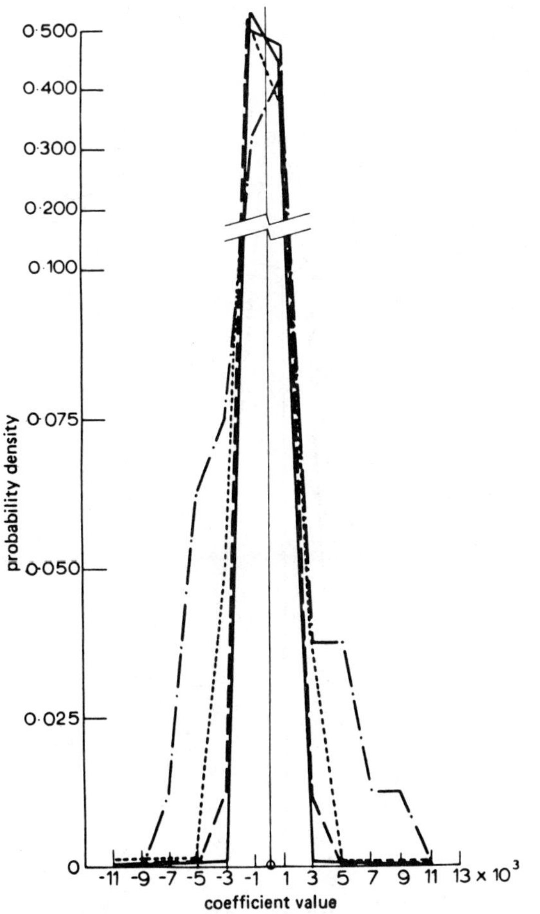

Fig. 3
Probability density functions of Hadamard coefficients

Part sentence spoken by male, sampling rate 8 k samples/s

---- bounds on set of 56 nondominant coefficients
-·-·- bounds on set of 8 dominant coefficients

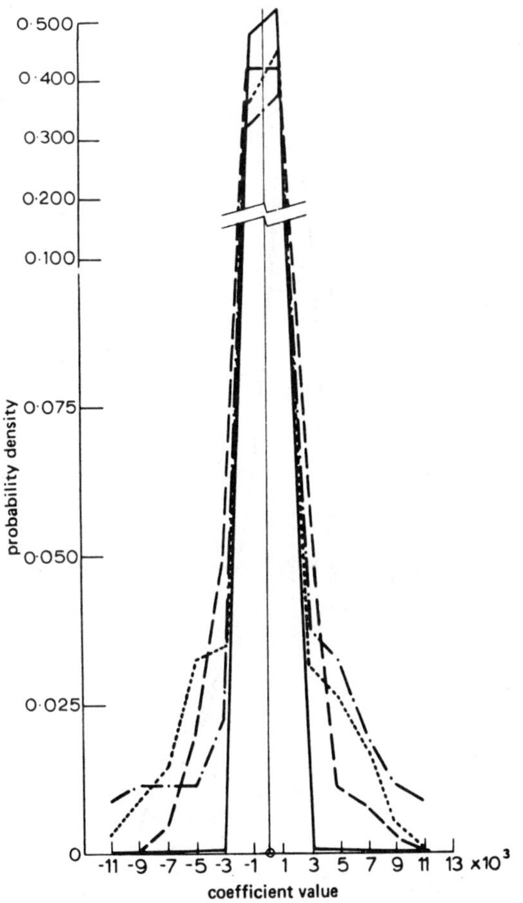

Fig. 4
Probability density functions of Hadamard coefficients

Part sentence spoken by female, sampling rate 8 k samples/s

---- bounds on set of 56 nondominant coefficients
-·-·- bounds on set of 8 dominant coefficients

Table 1
PERCENTAGE OF TOTAL POWER CONTAINED IN 5 AND 8 DOMINANT COEFFICIENTS

Sentence Number	Percentage of total power contained in 8 dominant coefficients		Percentage of total power contained in 5 dominant coefficients	
	male	female	male	female
1	62	59	41	26
2	67	63	44	43
3	73	67	48	46
4	75	68	50	46
5	88	84	69	64

mean $\hat{L}(k)$, where

$$\hat{L}(k) = \frac{1}{M} \sum_{s=1}^{M} L_s(k)$$

where $k = 0, 1, \ldots, N-1$. The logical autocorrelation function is considerably easier to compute than the normal arithmetic autocorrelation function, and it has the property[5,6] that the arithmetic autocorrelation function can be derived from it by a relatively simple non-singular linear transformation.

Figs. 13 and 14 show some of the results of a study and comparison of the logical autocorrelation function of speech and logical autocorrelation function of the corresponding speech signals reconstructed from dominant Hadamard coefficients. Figs. 13 to 15 show the logical autocorrelation functions for complete sentences spoken by a male speaker when the speech is reconstructed from 8, 7 and 5 dominant

Table 2
SPEECH QUALITY AGAINST SAMPLING RATE, BITS/COEFFICIENT AND NUMBER OF COEFFICIENTS

Sampling rate	Number of coefficients used	Number of bits/coefficient	Resultant transmission bit rate	Speech quality
k sample/s			k bits/s	
10·8	8	7	17·55	very good
	6	7	13·16	very good
	6	6	13·14	very good
10·0	8	7	16·25	very good
	6	7	12·18	very good
	6	6	11·24	very good
9·0	8	7	14·62	very good
	6	7	10·96	very good
	6	6	10·12	very good
8·0	8	7	13·00	very good
	6	7	9·74	very good
	6	6	8·99	good
7·0	8	7	11·37	good
	6	7	8·53	good
	6	6	7·87	average
6·1	8	7	9·91	poor
	6	7	7·43	poor
	6	6	6·85	poor

Sentences used in all tests were taken from the Harvard list number 1. A tape recording of the results is available from the authors on request.
Very good = very little different from original speech and of low noise
Good = highly intelligible but noticeably noisy
Average = less acceptable than 'good', intelligible but noisy
Poor = of low intelligibility and very noisy

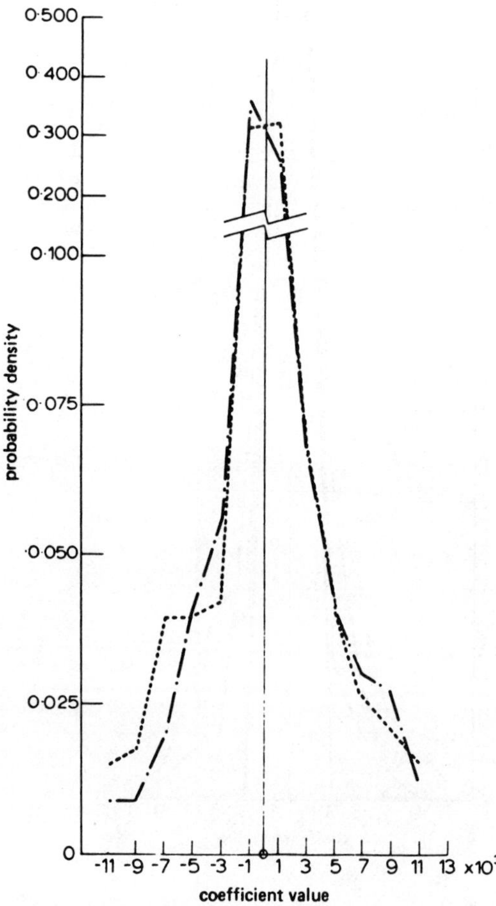

Fig. 5
Probability density function of dominant Hadamard coefficient
Complete sentence; sampling rate 8 k samples/s
------ female speaker — — — male speaker

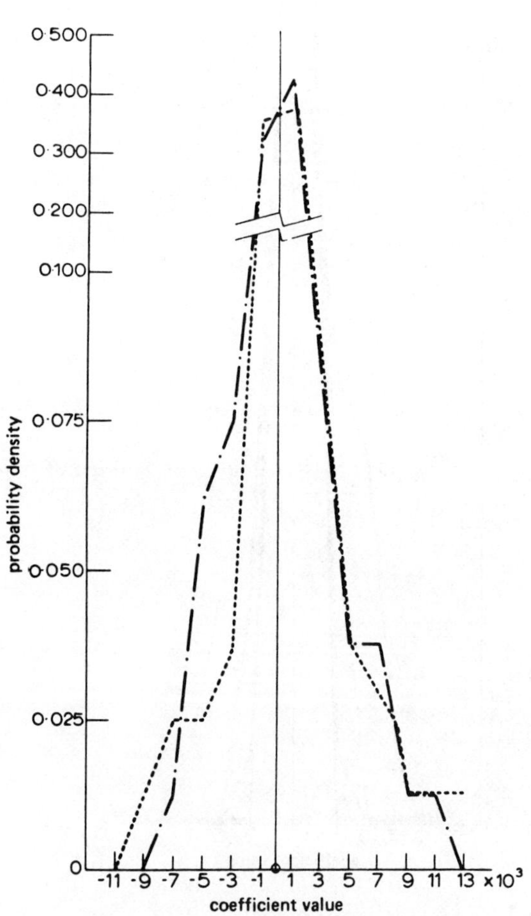

Fig. 6
Probability density function of dominant Hadamard coefficient
Part sentence; sampling 8 k samples/s
------ female speaker —·— male speaker

coefficients, respectively, and Figs. 16 to 18 show the corresponding functions for the same sentences spoken by a female speaker. Examination of these Figures shows that the extent of the agreement between the true logical autocorrelation function and that associated with speech reconstructed from 7 or 8 dominant Hadamard coefficients is quite remarkable. This is a feature that was found to be true for all sentences examined in the current investigation. The extent and generality of the agreement is such that it adds much weight to the idea of using 7 or 8 dominant Hadamard coefficients to represent speech that can be reconstructed from a limited number of the 64 dominant coefficients is not likely to be so satisfactory.

3 Quality of speech reconstructed from limited number of Hadamard coefficients

In addition to performing a detailed investigation of the statistical properties of Hadamard transformed speech signals, an initial series of tests were performed to gain some 'feel' for the quality of speech that can be reconstructed from a limited number of the 64 Hadamard coefficients. The tests were of a provisional nature and a detailed and extensive series of subjective tests are to be performed to obtain a quantitative assessment of speech quality and to obtain a comparison of the effectiveness of the technique as compared with other methods of speech coding. In the tests,† speech quality was considered, though somewhat superficially, as a function of various factors such as sampling rate, number of coefficients used, and the number of bits used to encode each coefficient.

The results of the tests are shown in Table 2, with the transmission rates W quoted in the table being derived using the expression

$$W = L\frac{(m+l)}{64}R \qquad (1)$$

where L is the number of coefficients used in reconstructing the signal, m is the number of bits used to encode each coefficient, l is the number of bits needed to indicate to the receiver which of the 64 coefficients a particular received coefficient is, and R is the sampling rate.

From Table 2 it can be seen that when using 8 coefficients and transmission bit rates in excess of 13 k bits/s the speech quality is very little different from the original speech and is virtually noise free. Speech of this quality is designated as 'very good'. When using 8 coefficients and a sampling rate of 7 k samples/s the speech is highly intelligible, but slightly more noisy than at higher bit rates. When the sampling rate is lowered to 6·1 k samples/s, with a corresponding transmission bit rate of 9·91 k bits/s, the speech is very noisy and is of low intelligibility. In the light of the information contained in Fig. 12, this fall-off in performance when using a sampling rate of 6·1 k samples/s is to be expected. When using 6 coefficients, and 7 bits/coefficient, the quality of the reconstructed speech was found to be very little different from that obtained when using 8 coefficients. When using 6 coefficients, and 6 bits/coefficients, the speech quality was found to be slightly more noisy than when using 7 bits to encode each coefficient. However, despite this slight increase in noise level, the speech remains highly intelligible for transmission bit rates of 8·99 k bit/s, and greater.

4 Possibility of further bit-rate reductions

The results presented in the previous Sections indicate that, when using a 64-point Hadamard transformation, it is unlikely that acceptable quality speech can be obtained when using fewer than 6 coefficients, and fewer than 6 bits to encode each of the coefficients. In view of these facts the question arises as to whether it is at all possible to reduce the transmission bit rates below those values quoted in Table 2, and still maintain acceptable speech quality.

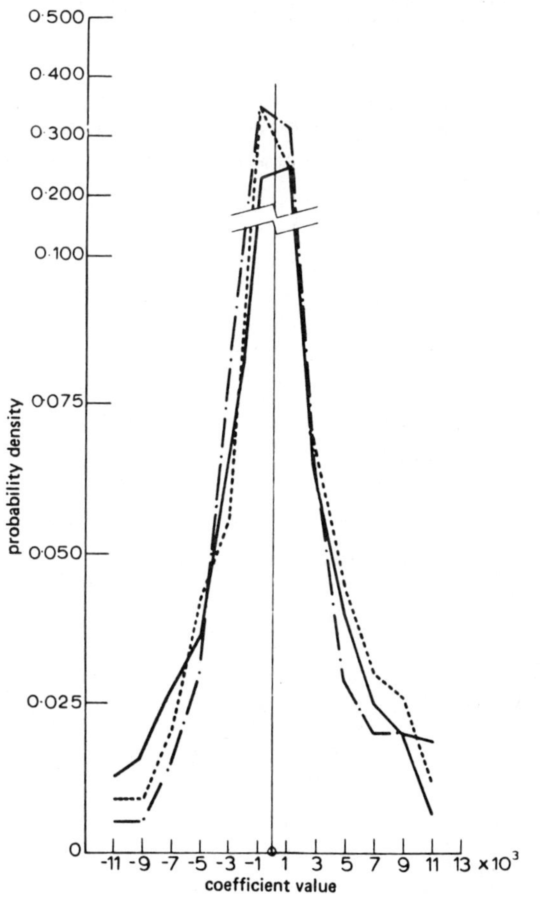

Fig. 7
Probability density function of dominant coefficient for two separate sentences and the combination of the sentences

Sentences spoken by male; sampling rate 8 k samples/s
— · — sentence 1
——— sentence 2
- - - - - - sentence 1 and 2

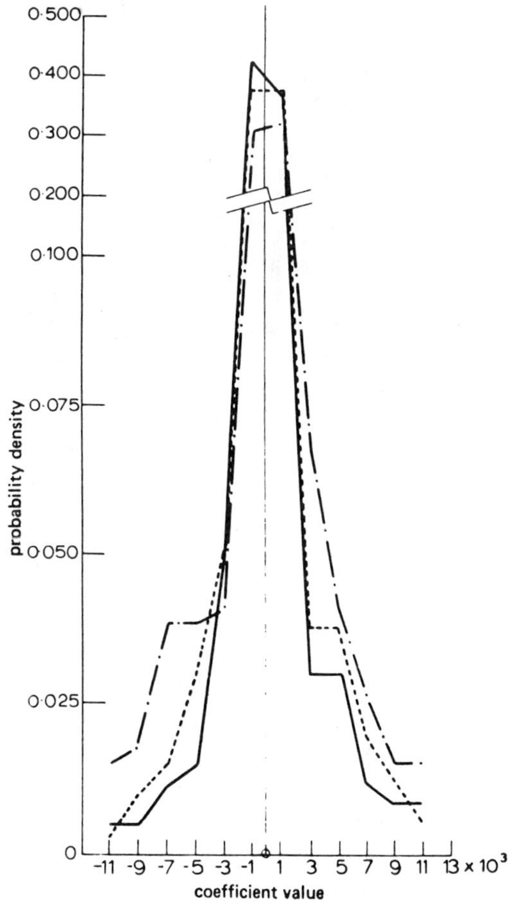

Fig. 8
Probability density function of dominant coefficient for two separate sentences and the combination of the sentences

Sentences spoken by female; sampling rate 8 k samples/s
— · — sentence 1
——— sentence 2
- - - - - - sentences 1 and 2

†The sentences used in all tests were taken from the Harvard list number 1.

For the 64-point Hadamard transformation, the transmission bit rate is given by eqn. 1, from which it is clear that, if L, m and R are fixed, then the only way in which the transmission bit rate can be reduced is by reducing the number of bits l used to indicate to the receiver which of the possible 64 coefficients a particular transmitted coefficient is. It is necessary to provide this information since the coefficients which dominate change as the speech varies. In the work of Shum *et al.*,[4] and in obtaining the bit rates quoted in Table 2, it was assumed that to provide this information it was necessary to use 6 labelling bits per coefficient. This means that approximately half of the transmission bit rate is being used to provide the necessary labelling information.

In an attempt to determine whether it is possible to reduce the number of labelling bits, an investigation was carried out, with attention being focused on the question of the occurrence of the dominant coefficients.

Examination of the occurrence of the dominant coefficients reveals that the 6 or 8 dominant coefficients are always members of the following group of 32 coefficients:

1, 4, 9, 11, 13, 15, 17, 20, 21, 25, 27, 29, 31, 33, 35, 37, 39,
40, 41, 43, 45, 47, 49, 51, 53, 55, 57, 59, 60, 61, 63, 64

This means that the number of labelling bits can be reduced by 1 bit/coefficient without the speech quality being affected in any way. Although this saving is of value, it is not a major saving. A second aspect of the investigation, involving a consideration of the Walsh-Fourier spectrum of speech, has revealed, however, that it is possible to reduce the average number of labelling bits by more than 50%.

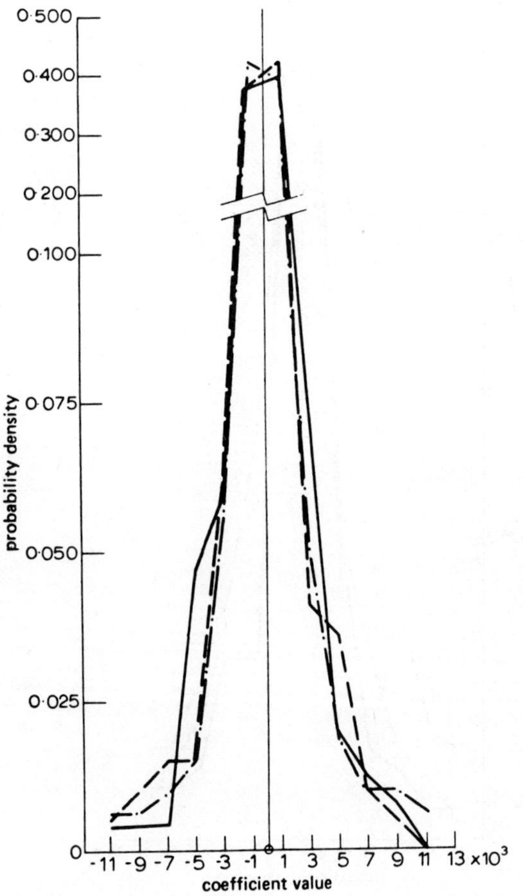

Fig. 9
Variation of probability density function of Hadamard coefficients with sampling rate

Complete sentence spoken by male
—·— 6 k samples/s
– – – 8 k samples/s
——— 10 k samples/s

The Walsh-Fourier spectrum of a signal has been considered previously[2] and it has been shown that, for an N-point transformation, the Walsh-Fourier spectrum consists of $Q = 1 + \log_2 N$ components $G_0, G_1, \ldots, G_{Q-1}$, and that the groups are comprised sequency spectral components γ_n^2 as follows:

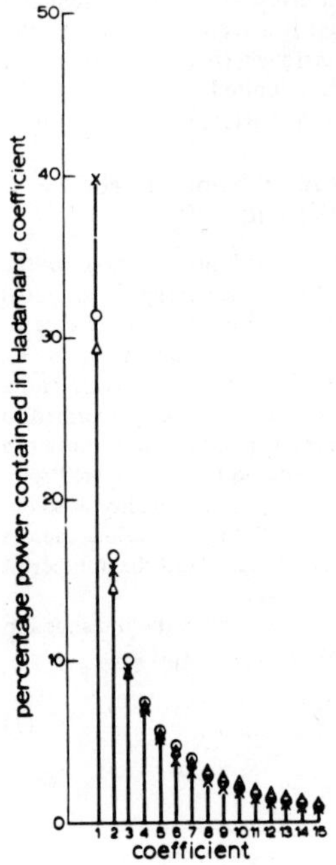

Fig. 10
Power density against Hadamard coefficient

Complete sentence spoken by male; sampling rate 8 k samples/s
X 10 k samples/s
o 8 k samples/s
Δ 6 k samples/s

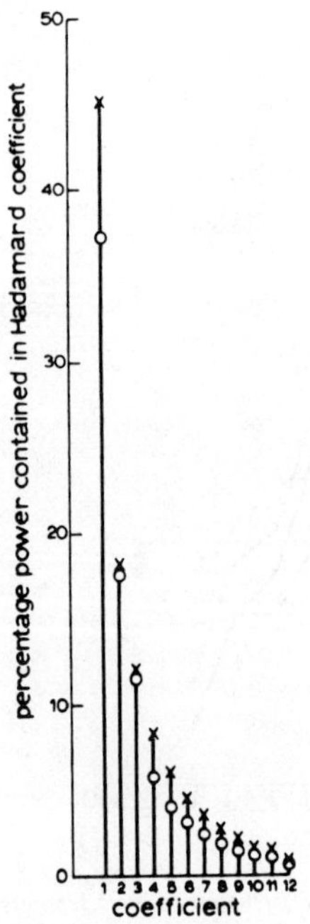

Fig. 11
Variation of power density from male to female speaker

Complete sentence; sampling rate 8 k samples/s
X male speaker
o female speaker

$$\left.\begin{aligned}
G_0 &= a_0^2 \\
G_1 &= \gamma_1^2 + \gamma_3^2 + \gamma_5^2 + \ldots \\
G_2 &= \gamma_2^2 + \gamma_6^2 + \gamma_{10}^2 + \ldots \\
G_3 &= \gamma_4^2 + \gamma_{12}^2 + \gamma_{20}^2 + \ldots \\
G_4 &= \gamma_8^2 + \gamma_{24}^2 + \gamma_{40}^2 + \ldots \\
&\vdots \\
G_{\log_2 N} &= a_s^2(N/2)
\end{aligned}\right\} \qquad (2)$$

where $\gamma_n^2 = a_c^2(n) + a_s^2(n)$ and where $a_s(n)$ and $a_c(n)$ are the sal(n) and cal(n) coefficients, respectively, obtained by Walsh transforming the signal. In the case of the 64-point transformation, the set of eqns. 2 becomes

$$\left.\begin{aligned}
G_0 &= a_0^2 \\
G_1 &= \gamma_1^2 + \gamma_3^2 + \gamma_5^2 + \ldots + \gamma_{31}^2 = \sum_{i=1}^{i=N/4} \gamma_{2i-1}^2 \\
G_2 &= \gamma_2^2 + \gamma_6^2 + \gamma_{10}^2 + \ldots + \gamma_{30}^2 \\
G_3 &= \gamma_4^2 + \gamma_{12}^2 + \gamma_{20}^2 + \gamma_{28}^2 \\
G_4 &= \gamma_8^2 + \gamma_{24}^2 \\
G_5 &= \gamma_{16}^2 \\
G_{\log_2 N} &= G_{\log_2 64} = G_6 = a_s^2(32)
\end{aligned}\right\} \qquad (3)$$

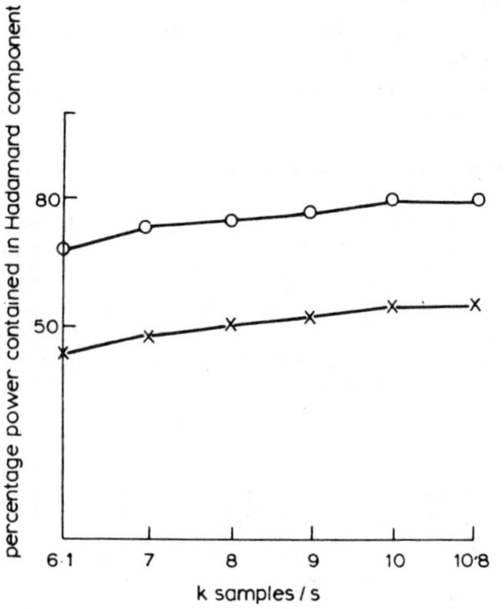

Fig. 12
Variation of percentage power contained in limited number of coefficients against sampling rate

Complete sentence spoken by male

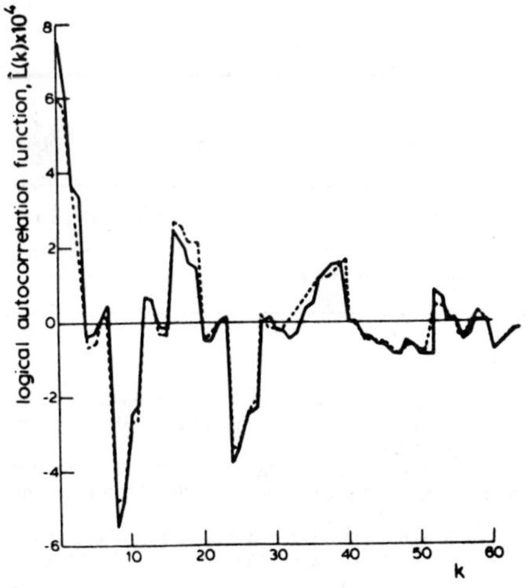

Fig. 13
Logical autocorrelation function

Complete sentence spoken by male; sampling rate 8 k samples/s
------ logical autocorrelation function reconstructed from 8 dominant coefficients
——— logical autocorrelation function of original sentence

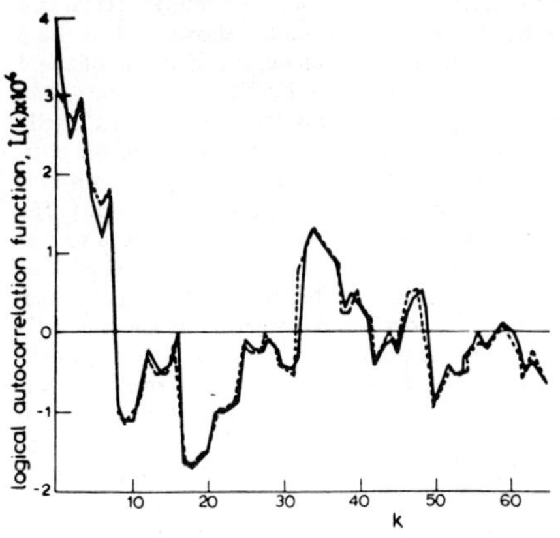

Fig. 14
Logical autocorrelation function

Complete spoken by male; sampling rate 8 k samples/s
------ logical autocorrelation function reconstructed from 7 dominant coefficients
——— logical autocorrelation function of original sentence

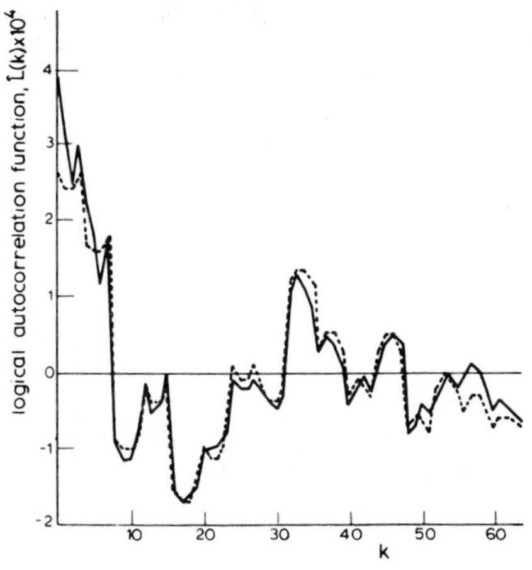

Fig. 15
Logical autocorrelation function

Complete sentence spoken by male; sampling rate 8 k samples/s
------ logical autocorrelation function reconstructed from 5 dominant coefficients
——— logical autocorrelation function of original sentence

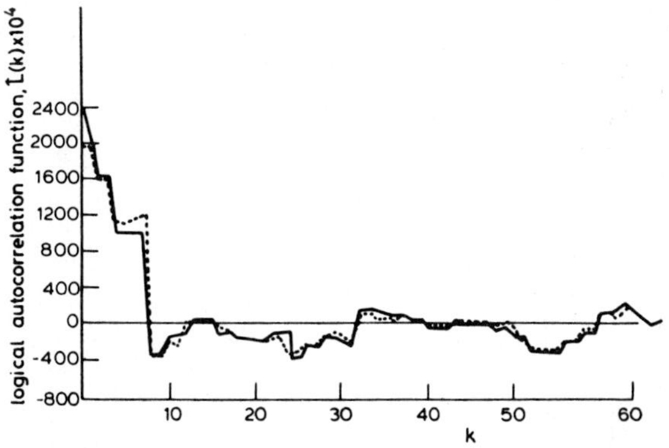

Fig. 16
Logical autocorrelation function

Complete sentence spoken by female; sampling rate 8 k samples/s
------ logical autocorrelation function reconstructed from 8 dominant coefficients
——— logical autocorrelation function of original sentence

An examination of the relationship between the dominant Hadamard coefficients and the Walsh-Fourier spectrum has shown that of the 8 dominant coefficients, 7 tend to predominate, and that one of the 7 coefficients occurs in each of the Walsh-Fourier components, and further that the Walsh-Fourier component is determined almost totally by the dominant coefficient. The occurrence of one dominant coefficient in each of the Walsh-Fourier components is highly significant from the point of view of reducing the number of labelling bits. The number of labelling bits can be reduced, since, if it is arranged so that the first dominant coefficient to be transmitted is that associated with G_0, the second dominant coefficient to be transmitted is that associated with G_1 etc., then, in order for the receiver to know exactly which Hadamard coefficient is dominating each G_i term, it is only necessary to supply a number of bits as indicated in Table 3. This is highly significant, since it means that only 15 bits have to be used to label 7 dominant coefficients, rather than the 42 (=6 x 7) bits that would be necessary when using the labelling scheme adopted in connection with Table 2. Expressed in terms of transmission bit rates this means that, if with a 64-point transformation a sampling rate of 8 k samples/s is used, and 7 dominant coefficients are used to represent the speech signal, then, with the simplified labelling scheme, the transmission bit rate is 8 k bit/s. Clearly, this constitutes a significant bit-rate reduction when compared with the 11·375 k bit/s transmission rate that would be necessary when using the labelling scheme associated with Table 2.

5 16-point and 512-point Hadamard transformed speech signals

In addition to considering the 64-point Hadamard transformed speech, some initial consideration has also been given to 16-point and 512-point Hadamard transformed speech. An analysis of a number of sentences has shown that, in the case of the 16-point transformation, excellent quality speech can be obtained if the 16 Hadamard coefficients are encoded in the manner indicated in Table 4. This encoding corresponds to a transmission rate of 16 k bits/s. Furthermore, an initial examination of the coefficient p.d.f.s and the average power spectrum in the Hadamard domain, indicates that it may be possible to obtain good-quality speech reconstruction when using only 3 dominant coefficients from the set of 16 Hadamard coefficients. The results suggest that 2 coefficients may not be sufficient for the reconstruction of good quality speech, and this thus suggests that the question of intelligibility touched upon by Boesswetter[1] should be examined carefully.

In the case of the 512-point transformation, a study of the average power spectrum in the Hadamard domain suggest that the use of 30 dominant coefficients from the set of 512 coefficients may be sufficient to reconstruct good quality speech. Some initial calculations, based on a 7 k sample/s sampling rate, the use of 6 bits/coefficient and the use of 9 labelling bits/coefficient, indicate that if 30 coefficients are sufficient, then it should be possible to obtain good quality speech with a transmision bit rate of 6·15 k bits/s.

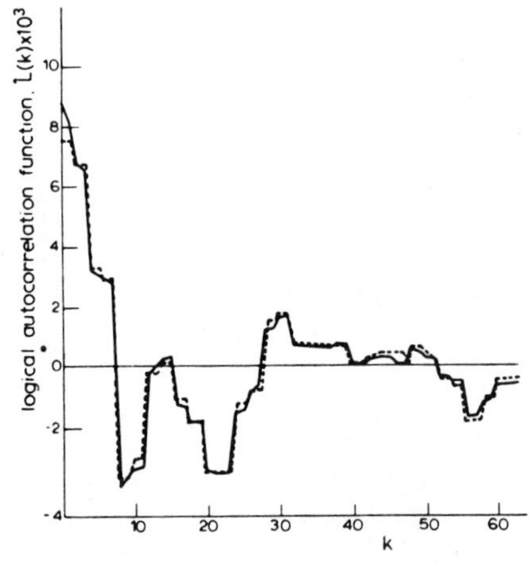

Fig. 17
Logical autocorrelation function

Complete sentence spoken by female; sampling rate 8 k samples/s
------ logical autocorrelation function reconstructed from 7 dominant coefficients
——— logical autocorrelation function of original sentence

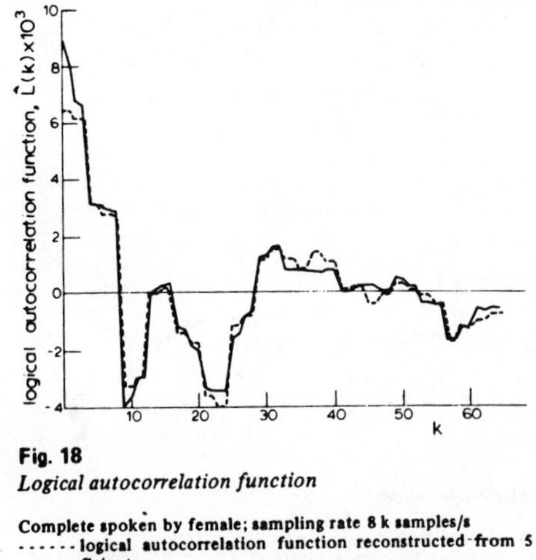

Fig. 18
Logical autocorrelation function

Complete spoken by female; sampling rate 8 k samples/s
------ logical autocorrelation function reconstructed from 5 dominant coefficients
——— logical autocorrelation function of original sentence

6 Conclusions

In the paper the results of an extensive analysis of the properties of 64-point Hadamard transformed speech are presented. Detailed information is given in Section 2 about the p.d.f.s of the Hadamard coefficients, the average power-density spectrum in the Hadamard domain, and the logical autocorrelation function of speech reconstructed from a limited number of Hadamard coefficients. The results contained in Section 2 indicate that good quality speech can be reconstructed from 6 to 8 dominant (large valued) Hadamard coefficients, and the initial listening tests reported in Section 3 confirm this. The indications are that the quality of the speech that can be reconstructed from 5, or fewer, Hadamard coefficients will not be acceptable. In Section 4 of the paper it is explained that, at first sight, it appears that approximately 50% of the transmission bit rate is needed for the sending of labelling information. However, the results of an investigation are presented, and from these results it is argued that the necessary number of labelling bits can be reduced by more than 50%. When using the modified-labelling technique it should be possible to obtain good quality speech when using a bit rate of 8 k bits/s.

Table 3
NUMBER OF BITS NEEDED TO IDENTIFY DOMINANT HADAMARD COEFFICIENTS ASSOCIATED WITH WALSH-FOURIER COMPONENT

Walsh-Fourier component	Number of bits needed to identify dominant Hadamard coefficients associated with Walsh-Fourier component
G_0	0
G_1	5
G_2	4
G_3	3
G_4	2
G_5	1
G_6	0

Table 4
SCHEME FOR ENCODING THE COEFFICIENTS OF THE 16-POINT HADAMARD TRANSFORMATION

Coefficient Number	1	2	3	4	5	6	7	8	9	10	11	12	13	14	15	16
Number of bits/coefficient	5	1	2	1	2	1	2	1	5	1	2	1	4	1	2	1

7 References

1 BOESSWETTER, C.: 'Analog sequency analysis and synthesis of voice signals'. Proceedings of the symposium on applied Walsh functions. Washington, D.C., 1970, pp. 220–229

2 CAMPANELLA, S.J., and ROBINSON, G.S.: 'Digital sequency decomposition of voice signals'. Proceedings of the symposium on applied Walsh functions, Washington, DC, 1970, pp. 230–237

3 CAMPANELLA, S.J., and ROBINSON, G.S.: 'A comparison of orthogonal transformations for digital speech processing', *IEEE Trans.*, 1971, **COM-19**, pp. 1045–49

4 SHUM. Y.Y., ELLIOTT, A.R., and BROWN, O.W.: 'Speech processing with Walsh-Hadamard transforms', *ibid.*, 1973, **AU-21**, pp. 174–179

5 LOPRESTI, P.V., and SURI, H.L.: 'A fast algorithm for the estimation of autocorrelation functions', *ibid.*, 1974, **ASSP-22**, pp. 449–453

6 ROBINSON, G.S.: 'Logical convolution and discrete Walsh and Fourier power spectra', *ibid.*, 1972, **AU-20**, pp. 271–280

Reprinted from *Math. Computers Simulation* **23:**163-169 (1981)

A FAST DIGITAL WALSH–HADAMARD SIGNAL PROCESSING SYSTEM

S. TZAFESTAS
Control Systems Laboratory, University of Patras, Patras, Greece

G. FRANGAKIS
N.R.C. Demokritos, Aghia Paraskevi, Attiki, Greece

A fast digital hardware system for signal analysis and synthesis is presented. This system, combined with data compression or information rate reduction techniques, which are also discussed in the paper, provides a powerful tool for a variety of control, communication and signal processing applications. An illustrative example is given.

1. INTRODUCTION

Walsh functions constitute a family of binary orthogonal functions and find important applications in both the control and communication fields [1]-[8]. Among the various processes that can be accomplished using Walsh functions we mention sequency filtering, spectroscopy, two-dimensional image transformation, optimal controller implementation, system identification, system state reconstruction (observer design), and speech processing. The salient features of Walsh function expansions (transforms) which make them attractive in the above applications are:

(i) The Walsh coefficients are computed by simply adding and subtracting sampled values of the signal waveform at hand.

(ii) The signals can be reconstructed with good accuracy using only a few Walsh transform components (coefficients).

(iii) Due to property (ii), fast algorithms can be designed in system and control applications, and a considerable "data compression" or "bit rate reduction" can be obtained in signal transmission applications.

Our purpose in this paper is to present a new hardware system for processing (analysing and synthesizing) analogue signals using Walsh functions, which can be easily implemented using LSI components. Actually signal analysis and synthesis with Walsh function transforms can be accomplished using software means, but the hardware method, which is cheap and fast, is preferable in on-line processing applications where dedicated special purpose equipment is required. The results reported in [6]-[8] were obtained using general purpose computing facilities (CDC-1700, PDP-15). Although the performance of Walsh-Hadamard transform in terms of bit rate reduction is not as good as that of the Fourier transform or the Karhunen-Loève transform [6], it gives good results, in speech signal coding, if 4 to 8 dominant Walsh-Hadamard coefficients are used [7]. Actually, in trying to reconstruct a sequency (Walsh frequency) band-limited signal using the Walsh function transform one must be compatible with the following sampling theorem which states [9]:

"A signal $x(t)$, sequency-band-limited to Z zeros/sec can be completely reconstructed from samples at every $T=1/2^{k+1}$ seconds, where k is an integer such that $2^k \geq Z$. In particular: $x(t)=x(i/2^{k+1})$, $i/2^{k+1} \leq t<(i+1)/2^{k+1}$."

The present Walsh function processing system makes use of the Walsh function generator and analyzer presented in [10],[11]. Some work making use of microprocessors for converting the Walsh coefficients to Fourier coefficients or for simulating speech signals may be found in [12], [13]. Before presenting our signal processing system a short account of the basic digital processing theory using orthogonal transforms, followed by a review of some important experimental results, is given.

2. DIGITAL SIGNAL PROCESSING USING ORTHOGONAL TRANSFORMS

Consider a discrete data vector $X^T=[x_o, x_1,\ldots,x_{N-1}]$ and an orthogonal transform $P=[\phi_o,\phi_1,\ldots,\phi_{N-1}]^T$, where $\phi_o,\phi_1,\ldots,\phi_{N-1}$ are orthonormal real-valued vectors, i.e. vectors having the property $\phi_i^T\phi_j=\delta_{ij}$ (Kronecker delta) or $P^TP=I$ (unit NxN matrix). Applying P upon X gives the transformed vector components $y_i=\phi_i^TX$ or in matrix form

$$Y=PX,\quad Y^T=[y_o,y_1,\ldots,y_{N-1}] \qquad (1)$$

Premultiplying (1) by P^T and using the orthogonality of P gives the inverse transform

$$X=P^TX=\sum_{i=0}^{N-1} y_i\psi_i \qquad (P^T=P^{-1}) \qquad (2)$$

Clearly, P and P^{-1} are linear inverse transforms and are used in both the communication and control fields. In both fields the basic problem is to select a family of vector functions $\psi_o,\psi_1,\ldots,\psi_{N-1}$ such that the signal X can be reconstructed or estimated using a smaller number of terms $y_i\psi_i$, i=0,1,...,M-1. This implies that only M<N data values y_i have to be used or transmitted for reconstructing X as

$$\hat{X}=\sum_{i=0}^{M-1} y_i\psi_i+\sum_{i=M}^{N-1} c_i\psi_i \qquad (3)$$

where c_i are preselected constants. Using fewer data points or transform coefficients y_i we have what is known as "data compression" which allows to transmit the information signal X approximately with reduced information rate, a result which is mostly important in practice.

The optimum transform (or expansion) with respect to the mean square criterion is that in which ψ_i, i=0,1,...,N-1 are the eigenvectors of the data covariance matrix $\Sigma_x=E\{(X-\bar{X})(X-\bar{X})^T\}$, where $\bar{X}=E[X]$ is the mean value of X. This is the well known Karhunen-Loéve expansion. Indead, the mean square error V is given by

$$V=E\{(X-\hat{X})^T(X-\hat{X})\}=\sum_{i=M}^{N-1} E\{(y_i-c_i)^2\} \qquad (4)$$

and the optimality conditions $\partial V/\partial c_i=0, \partial V/\partial\psi_i=0$ subject to $\psi_i^T\psi_i=1$ give, respectively:

$$c_i=E[y_i]=\psi_i^T\bar{X},\quad \Sigma_x\psi_i=\lambda_i\psi_i \qquad (5)$$

where λ_i are the Lagrange coefficients for the constraints $\psi_i^T\psi_i=1$. This shows that ψ_i is an eigenvector of Σ_x and λ_i is the correspon- ding eigenvalue. Now from Y=PX we obtain

$\Sigma_y=P\Sigma_xP^{-1}=P\Sigma_xP^T=\text{diag.}[\lambda_o,\lambda_1,\ldots,\lambda_{N-1}]$, i.e. Σ_y is diagonal which implies that the components of y_i are uncorrelated.

The minimum value V^o is obtained by introducing (5) into (4), i.e.

$$V^o=\sum_{i=M}^{N-1}\psi_i^T\Sigma_x\psi_i=\sum_{i=M}^{N-1}\psi_i^T\lambda_i\psi_i=\sum_{i=M}^{N-1}\lambda_i$$

This shows that V^o is equal to the sum of the eigenvalues corresponding to the data components y_i which are discarded. Thus in practice one must keep the M data components having the largest eigenvalues. The discarded data components are replaced by the constants $c_i=\psi^T\bar{X}$ which can be reduced to zero by preprocessing the given data such that $\bar{X}=0$. This data compression criterion for selecting the set of y_i components with the largest variances is known as "variance data compression criterion". Note that for all other orthogonal transforms the covariance matrix Σ_y is not diagonal.

In the case of discrete-time signals we have $x_k=x(kT)$, where k is the sample index and T is the time between samples, Eq. (2) gives

$$x(kT)=\sum_{i=0}^{N-1} y_i\psi_i(kT) \qquad (6)$$

where $\psi_i(kT)$ is the kth component of the vector ψ_i, and actually is the value of $\psi_i(t)$ at the kth sampling instant t=kT. Here NT is the window size. Some common types of $\psi_i(kT)$, besides the Karhunen-Loéve vectors, are the discrete Fourier, Slant, and Walsh-Hadamard vectors.

In [6] a comparison is made of the Fourier, Walsh-Hadamard and Karhunen-Loéve transforms regarding their ability to reduce the bit rate in speech signal transmission. To this end a general expression is derived for the information rate needed for the transmission of the coefficients y_i such that the reconstructed signal x(kT) has the same signal-to-quantizing noise ratio as that of the standard PCM transmission. If N samples are required to determine the expansion coefficients (or, equivalently, if N coefficients are used to represent a set of time series samples of size N), then the quantized noise (error) due to the quantized transmission (using n_i bits) of the coefficients y_i, i=0,1,...,N-1 is given by

$$Q=\frac{1}{N^2}\sum_{i=0}^{N-1} C_i\sigma_i^2/2^{2n_i} \qquad (7)$$

where σ_i^2 is the variance associated with y_i, and C_i is a constant expressing the consequence of the scaling and companding associated with the coefficient quantization. The quantization noise of the PCM transmission of the same information is $Q_{PCM}=C_{PCM}\sigma^2/2^{2M}$, where M is the total number of bits used to quantize each sample, σ^2 is the signal power, or variance in the zero bias case, and C_{PCM} is the scaling-companding constant. The quantization noise Q in (7) is minimized when all terms of the summation are equal. Hence, to ensure that $Q=Q_{PCM}$ one must select the number of bits n_i such that $C_i\sigma_i^2/N^2 2^{2n_i}=C_{PCM}\sigma^2/N2^{2M}$, i.e.

$$n_i=M+\log_2(\sigma_i^2/N\sigma)+\frac{1}{2}\log_2(NC_i/C_{PCM})$$

The total bit rate for the transform (coefficient) transmission is $R_o=(1/NT)\sum_{i=0}^{N-1} n_i$, and for the PCM transmission is $R_{PCM}=M/T$. Therefore, to reduce the bit rate by the orthogonal transform (i.e. to ensure that $R_o<R_{PCM}$) one must have $(1/N)\sum_{i=0}^{N-1} n_i<M$.

For a window size NT the frequency resolution in the Fourier transform case is 1/NT. Thus, for N=16 and T=125μsec, the frequency resolution is 500Hz. This resolution is sufficient to resolve long-terms spectral distribution of speech in terms of the variances of the Fourier coefficients (components). Higher values of N give higher resolution by which short-term spectral distribution of speech can be resolved. Due to the similarity of the spectral distributions of the Fourier, Hadamard and Karhunen-Loéve transforms the above procedure of determining the sample window holds approximately for the Hadamard and Karhunen-Loéve transforms as well.

Experimental work on both man and woman speeches of duration 65sec [6], sampled every T=125 μsec (8000 samples per second) has showed that to obtain a S/Q_n ratio (Q_n denotes quantized noise) equal to that of a standard 56 kbits/sec PCM, the bit rates required in kbits/sec are as follows.

Processing Transform	Fourier	Hadamard	Karhunen-Loéve
Bit rate	46	48.5	42.5
Bit rate reduction	10	7.5	13.5

As expected, the best performance is obtained using the Karhunen-Loéve processing transform. The behaviour of the above transforms was also studied by comparing their S/Q_n ratios with that obtained using logarithmically compånded PCM at bit rates 56,28 and 14 kbits/sec. The S/Q_n improvements are 5db for Fourier, 3db for Hadamard, and 9 db for Karhunen-Loéve processing, averaged over the three bit rates used. A somewhat independency of the S/Q_n improvement of each transform on bit rate was also observed.

In [7] the Hadamard-Haar processing scheme was applied on speech signals sampled again at a rate of 8000 samples/sec, and broken down into blocks of N=64 samples. Each block of samples was Walsh-Hadamard transformed yielding the corresponding N=64 Hadamard coefficients. The Hadamard coefficients were then Haar transformed yielding N=64 Haar coefficients which were to be transmitted in digitally coded form.

To realize the above processing scheme they passed the input speech signals through a lowpass filter (cutoff frequency 3.4 kHz), and used 10 bit A/D and D/A converters for signal conversion, and a PDP 15 computer for processing. By examining the p.d.f. of the Haar coefficients for the 64-point transformation, they found that there is a small group of M=6 coefficients that tend to predominate for both male and female speakers. This shows that a strong data compression can be used, and that it may be possible to transmit intelligible speech using fewer than N=64 coefficients. Actually, a good quality speech was reconstructed (tested by natural listening) using M=6 coefficients which cooresponds to a transmission bit rate of 10kbits/sec.

Using 3 coefficients the quality of reconstructed speech was again intelligible but much more noisy.

Finally, using 2 coefficients the resulting speech was again intelligible with "metallic" sound.

Similar results were obtained in [8] where it was shown that an intelligible speech transmission can be made using Hadamard processing at bit rates of about 13kbits/sec, and in [15] where by an appropriate coefficient labelling technique the bit rate of transmission was reduced to 8kbits/sec (using 6-8 dominant Hadamard coefficients) without any loss of speech intelligibility.

3. THE PRESENT WALSH-HADAMARD SIGNAL PROCESSING SYSTEM

The Walsh-Hadamard Signal Analyzer

The present Walsh-Hadamard analyzer (Fig. 1) is a faster version of the analyzer proposed in [11]. Here a parallel A/D converter in place of the V/F converter, and an accumulator

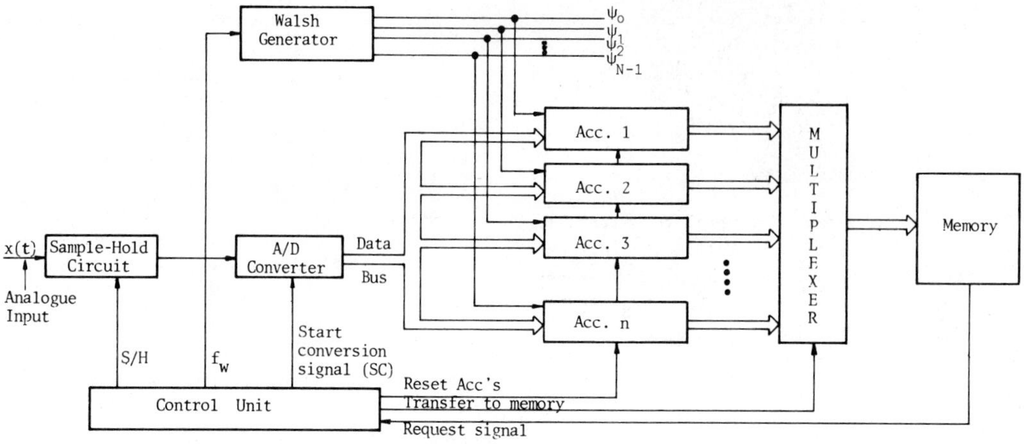

Fig. 1. Block diagram of the analyzer.

(ALU+Register) in place of each counter of the direct fast Walsh transform circuit of [11] are employed. The operation of the present analyzer is as follows. The analogue signal x(t) is sampled at the middle of each time subinterval (of duration 1/N) of the Walsh signal (see pulse train S/H in Fig. 2), and converted to a digital value by the A/D converter. The resulting digital "word" is passed, via the databus, at the inputs of the accumulators. The accumulators perform additions or subtractions, according to the value 0 or 1 of the Walsh signals ψ_i, i=0, 1,2,...,N-1 provided by a

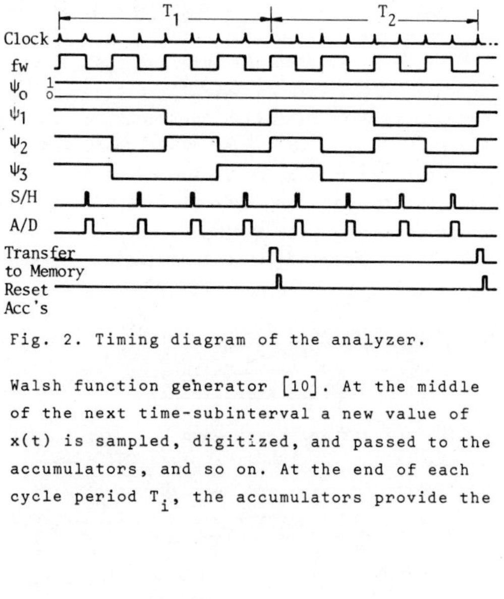

Fig. 2. Timing diagram of the analyzer.

Walsh function generator [10]. At the middle of the next time-subinterval a new value of x(t) is sampled, digitized, and passed to the accumulators, and so on. At the end of each cycle period T_i, the accumulators provide the Walsh coefficients y_{ij} where i indicates the cycle T_i in which the computations are made, and j indicates the coefficient order into this cycle. These coefficients y_{ij} are transferred and stored in the available memory through a multiplexer. When this transfer is completed, the accumulators are cleared before the beginning of the first sampling in the next cycle period T_{i+1}. The A/D converter which is of a successive approximation type (SAR) is composed by a proper connection of the MLM301A voltage comparator, the MC1408 D/A and the MC14559 SAR, and converts each sampled value to an 8-bit word. Each 8-bit accumulator consists of two MC14581 ALUs and two MC14035 registers. The signals ψ_i=1 (for addition) and ψ_i=0 (for subtraction) are applied at the "function select" inputs of the ALUs. The multiplexer is formed by the 8-channel MC14152 data selector units properly connected. For the data storage a RAM memory can be used.

Some examples of analysis of known signals in Walsh series were given in [11]. Consider the case of the unit ramp signal, and suppose that N=4. Then at the end of the period T_1 the accumulators of Fig. 1 will have the following contents, in 2's complement form:

$ACC_1:(00010000)_2=16_{10}=y_{10}$
$ACC_2:(11111000)_2=-8_{10}=y_{11}$
$ACC_3:(11111100)_2=-4_{10}=y_{12}$
$ACC_4:(00000000)_2=0 \ =y_{13}$

The Walsh-Hadamard Synthesizer

To synthesize an analogue signal x(t) from its Walsh-Hadamard coefficients, Eq.(6) must be implemented. The present signal-synthesizer, which is based on the fact that the Walsh functions can take the values 0 or 1, is shown in Fig.3 and works as follows.

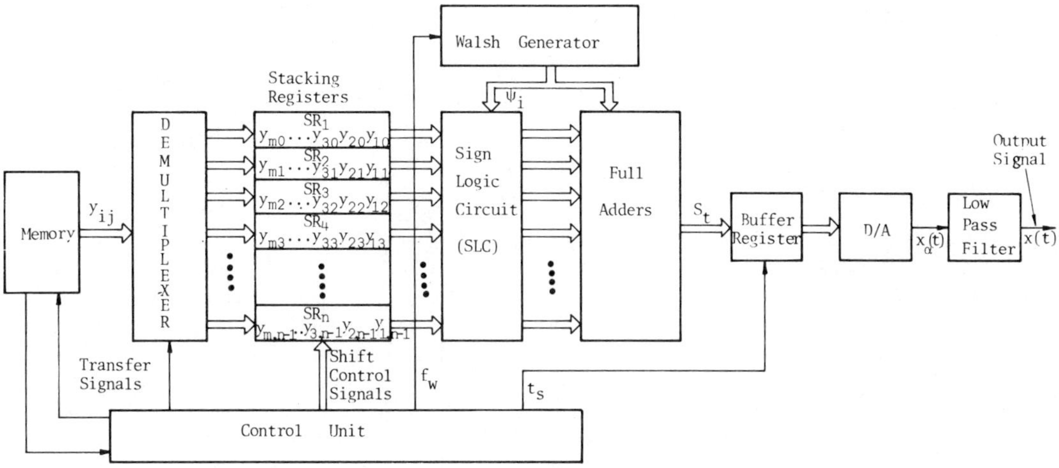

Fig.3. Block diagram of the synthesizer.

The Walsh coefficients y_{ij}, stored in 2's complement form in the memory, are called and stacked sequentially in the stacking register block [14] through the demultiplexer. The coefficients y_{ij} lying in the last column of the stacking register block are passed at the inputs of the "full-adder" block (FA) through the "Sign Logic Circuits" block (SLC), and provide the signs of the numbers entered into the FA according to the value of the Walsh function ψ_i. The output S_t of the FA block gives the value of the synthesized signal x(t) at each subinterval in digital form. This value is written in a buffer register and is then converted to analogue form $x_a(t)$ and passed through an appropriate filter to give the desired signal x(t). The stacking register is a two-dimensional FIFO buffer memory and works as follows [14]. Each column of flip-flops is connected with the flip-flops of the next column and so each row of flip-flops forms a shift register. A "word" entered into the first column is shifted to the right until it reaches the last column. The next word is shifted up to the second from the end column, and so on. Thus in the present application (Fig.3), where the synthesis is started in the period T_1, the coefficient y_{10} is called first and stored in the first stacking register (SR1), then y_{11} is called and stored in SR2, next y_{12} is stored in SR3, and finally y_{13} is stored in SR4. Similarly the coefficients y_{20}, y_{21}, y_{22} and y_{23}, corresponding to the next computation period T_2, are called and stored in SR1, SR2, SR3 and SR4 respectively, and so on. We note that when the synthesis operation is completed for the period T_i, by resetting the last flip-flop of the control unit of the stacking register block, all vertical words of the stacking register are shifted one position to the right, before the operations for the period T_{i+1} start.

Since the coefficients y_{ij} may be positive or negative and the same holds for the Walsh functions ψ_i, the term $y_{ij}\psi_i$ can be positive or negative. The sign of the product $y_{ij}\psi_i$ is obtained by the SLC block which is controlled by ψ_i as follows. When ψ_i=1, the coefficient y_{ij} of the last column of the stacking register appears at the input of the FA unchanged, whereas when ψ_i=0, its 1's complement appears, and the CARRY-IN (C_{in}) input of the FA is set to "1". The detailed diagram of the last column of the stacking register block for N=4, along with the SLC and the FA blocks, are shown in

Fig. 4.

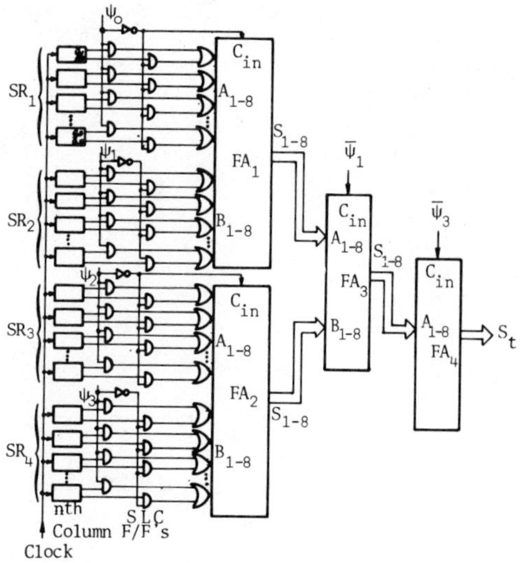

Fig.4. Sign logic and full adders.

4. AN ILLUSTRATIVE EXAMPLE

The problem is to synthesize the ramp signal by using the circuit of Fig.3. The Walsh coefficients y_{ij}, j=0,1,2,3,of this signal, determined in the analyzer section, are assumed to be available in a memory. These coefficients are transferredinto the stacking registers as explained above. In the first subinterval of T_1, the generator outputs are $\psi_o=\psi_1=\psi_2=\psi_3=1$, and so at the inputs of FA1 (MC14008) the coefficients y_{10} and y_{11} are fed unchanged. The output of FA1 is then $(00001000)_2$, since the overflow C_o is ignored. Similarly y_{12} and y_{13} are fed unchanged to the inputs of FA2 and give the output $(11111100)_2$. The output of FA4 which is the same with the output of FA3 is $S_t=(00000100)_2$.

This value is written in the buffer register by the t_s signal (Fig.5) and is converted to the analogue form $x_a(t)$ via the MC1508 D/A converter. The signal $x_a(t)$ is then filtered (integrated) to obtain the desired ramp signal x(t).

In the second subinterval, where $\psi_o=\psi_1=1$ and $\psi_2=\psi_3=0$,the output of FA1 is $(00001000)_2$ and the output of FA2 is $(00000011)_2$, since at the FA2 input the 1's complements of y_{12} and y_{13} are applied via SLC, and $C_{in}=1$. The FA3 output is $(00001011)_2$ and the FA4 output is $S_t=(0001100)_2$ since $C_{in}=1$..

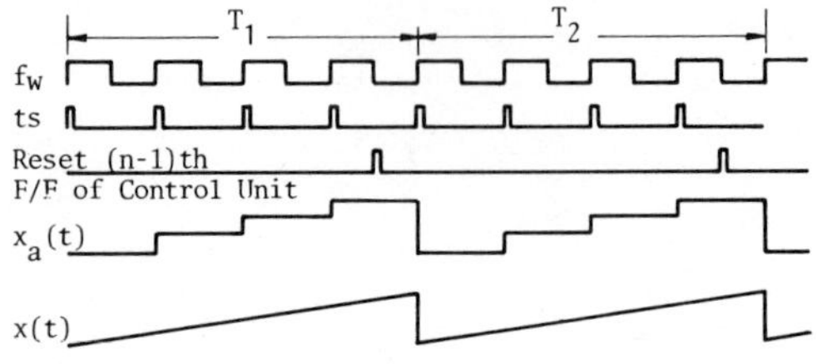

Fig.5. Timing diagram of the synthesizer and output signals $x_a(t)$ and x(t).

In the third subinterval, where $\psi_o=1$, $\psi_1=0$, $\psi_2=1$, and $\psi_3=0$, the output of FA1 is $(00010111)_2$ since y_{11} is complemented, the output of FA2 is $(11111011)_2$ since y_{13} is complemented, and the FA3 output is $(00010011)_2$ since its $C_{in}=1$.The output of FA4 will give $S_t=(00010100)_2$.

Finally in the fourth subinterval, where $\psi_o=1$, $\psi_1=\psi_2=0$, and $\psi_3=1$, the FA1 output is $(00010111)_2$, the FA2 output is $(00000100)_2$,and the output of FA4, which is equal to that of FA3, is $S_t=(00011100)_2$. The above values S_t are written in the buffer register by the signal t_s and are then converted to analogue form.

During the above procedure other sets of y_{ij} coefficients are stored in the stacking register block to be available for the synthesis over the next period T_{i+1}, and so on.

It is noted that only at the beginning of the synthesis, a delay time is required to tranfer the first four coefficients in the last column of the stacking registers. After this transfer is completed no other delay time is needed, since the coefficients required for the synthesis over the next computation intervals are stacked into the stacking registers from the memory.

Fig. 5 gives the timing diagram of the synthesizer along with the resulting step-wise signal $x_a(t)$ and the continuous output x(t), under the assumption that the available Walsh coefficients y_{ij} are periodic with respect to i, and have been obtained from a periodic ramp signal with period T.

5. CONCLUSION

The present analyzer/synthesizer is appropriate for fast hardware Walsh-Hadamard analysis and synthesis of continuous-time signals by fully digital means. Since by a careful application of data compression techniques one can reduce substantially the number of Walsh-Hadamard coefficients required for the signal reconstruction (synthesis), the present system can be efficiently used in on-line applications involving very rapidly varying signals. Work is in progress for the actual analysis and synthesis of speech waveforms using the present system, and for the realization of the system using current microprocessor components.

REFERENCES

[1] H. Harmuth, Transmission of information by orthogonal functions (Springer, 1969).

[2] R. Lackey, The wonderful world of Walsh functions, Proc. 1972 Symp. on Appl. of Walsh functions, The Catholic Univ. of America, Washington D.C., pp. 2-7 (1972).

[3] C. Chen and C. Hsiao, Walsh series analysis in optimal control, Int. J. Control, Vol. 21, pp.881-897 (1975).

[4] P. Stavroulakis and S. Tzafestas, Walsh series approach to observer and filter design in optimal control systems, Int. J. Control, Vol. 26, pp. 721-736, (1977).

[5] S. Tzafestas, Walsh series approach to lumped and distributed system identification, J. Franklin Inst., Vol. 305, pp. 199-220, (1978).

[6] S. Campanella and G. Robinson, A comparison of orthogonal transformations for digital speech processing, IEEE Trans. on Commun., Vol. COM-19, pp. 1045-1050 (1971).

[7] E. Frangoulis and L. Turner, Speech coding by the Hadamard-Haar transform, Computers and Digital Techniques, Vol. 1, pp.106-112, (1978).

[8] Y. Shum, A. Elliott and W. Brown, Speech processing with Walsh-Hadamard transforms, IEEE Trans. on Audio and Electroacoustics, Vol. AU-21, pp. 174-179, (1973).

[9] S. Kaak, Sampling theorem in Walsh Fourier analysis, Electronics Lett., Vol.6, pp. 447-448, (1970).

[10] S. Tzafestas, G. Frangakis and T. Pimenidis, Global Walsh function generators, Electronic Engineering, pp. 45-49 (Nov. 1976).

[11] G. Frangakis and S. Tzafestas, A digital Walsh function analyzer, Electronic Engineering, pp. 89-91, (July 1979).

[12] R. Kitai, I.Renyi and F. Vajda, Using a microprocessor in a Walsh-Fourier signal spectral analyzer, Computer, pp.27-31, (April 1976).

[13] P. Terell, Speech activity simulation using microprocessors, Electronic Engineering, pp. 77-82, (July 1979).

[14] G. Frangakis, Fast Stacking-Register Buffer has Minimum Delays, Electronic Engineering, pp. 34-35, (April 1972).

[15] E. Frangoulis and L. Turner, Hadamard-Transformation Technique of Speech Coding: Some Further Results, Proc. IEE, pp. 845-852, (Oct. 1977).

Two-Dimensional Sequency Filters

HENNING F. HARMUTH, MEMBER, IEEE, HARRY C. ANDREWS, MEMBER, IEEE, AND KOSHO SHIBATA

***Abstract*—This paper provides a tutorial exposition of the possible use of sequency filters for the processing of signals that are functions of two space variables. Historically, optical and, more recently, digital techniques have been used for two variable processing of images with the computer providing a domain for other than sine-cosine filtering. Sequency filters based on Walsh or similar functions can be implemented with about equal effort for time and space variable signals. They could actually be built with electronic components up to the ultimate case of signals with one time and three space variables. Two kinds of filters have emerged so far: instantaneous filters and sampling filters. In the following sections we briefly discuss other possible transforms for space filters before investigating sequency-oriented filtering techniques. The sequency techniques include instantaneous filters, efficient practical implementation, and some potential circuits for imagery applications.**

I. Introduction

FILTERS for time signals have been used in communications for half a century. There has long been a practical interest to extend the theory to space signals, since photographs, weather maps, etc., are signals of two spatial variables, while the ubiquitous television picture is a signal of two spatial and one time variable. The extension of filter theory from time to space signals runs into the difficulty that coils and capacitors work best with one-dimensional time waveforms.

A practical way to implement filters for signals with two spatial variables was provided by optical lenses and coherent light that produced the two-dimensional Fourier transform of an image. Digital computers provided a second means for implementing spatial filters when they became large and fast enough to perform two-dimensional Fourier transformations. A number of scientists have published computer-filtered photographs [1]–[4] and many more are working in this field [5]–[8].

Coils, capacitors, and optical lenses favor sine–cosine functions. The digital computer has other favorites, such as Walsh, Hadamard, and Haar functions, which have become increasingly significant in communications since the era of semiconductors. Computer filtering or processing of images can be used whenever time is not too critical. A typical application is to increase the visibility of a tumor in an X-ray photograph. Computers are not yet fast and inexpensive enough to filter TV signals. However, the techniques learned from computer processing can be implemented by analog circuits, which are somewhat loosely referred to as sequency filters [9]–[11].

Sequency filters are rather different from the usual "frequency" filters and a good deal of rethinking and relearning is required before one can use them. To show that it is worth going to this trouble, we start out by showing some examples and then discuss the underlying theory. Fig. 1 shows two TV pictures side by side. The original analog picture is on the right. The left was obtained by passing the original signal through a very simple sequency filter, encoding it into PCM, transmitting it in a closed loop, reconverting it from digital to analog form, and finally passing it through the inverse sequency filter. The important point is that only 2.5 bits/sample were transmitted. The quality is so good that most readers will not be able to identify the original picture. Fig. 2 shows a second example where again 2.5 bits/sample were used. In addition, color pictures have also been transmitted requiring 3.75 bits/sample for comparable quality [12].

The equipment from which Figs. 1 and 2 were obtained was developed by the Kokusai Denshin Denwa Company, Ltd., in cooperation with the Oki Electric Company [13], [14]. Similar equipment for black and white TV pictures was developed by Hatori, Taki, and others at Tokyo University [15].

II. Fourier, Walsh, Hadamard, and Other Transforms

It is well known that a time signal $F(\theta)$ that is only of interest in a finite time interval $-\frac{1}{2} < \theta < \frac{1}{2}$, where θ is the normalized time variable, can be expanded into a Fourier series:

$$F(\theta) = a(0) + \sqrt{2} \sum_{i=1}^{\infty} [a(i) \cos 2\pi i\theta + b(i) \sin 2\pi i\theta]$$

$$a(i) = \sqrt{2} \int_{-1/2}^{1/2} F(\theta) \cos 2\pi i\theta \, d\theta,$$

$$b(i) = \sqrt{2} \int_{-1/2}^{1/2} F(\theta) \sin 2\pi i\theta \, d\theta,$$

$$a(0) = \int_{-1/2}^{1/2} F(\theta) \, d\theta. \tag{1}$$

Consider now the Walsh functions shown in Fig. 3 [16]. They assume the two values +1 and −1, an obvious advantage for their use in semiconductor circuits and digital computers. Using Walsh functions, one

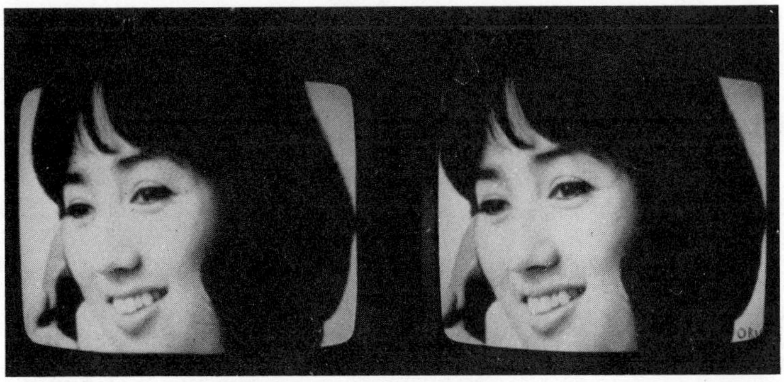

Fig. 1. Original analog TV signal (right) and processed signal transmitted by PCM using 2.5 bits/sample.

Fig. 2. Original analog TV signal (right) and processed signal transmitted by PCM using 2.5 bits/sample

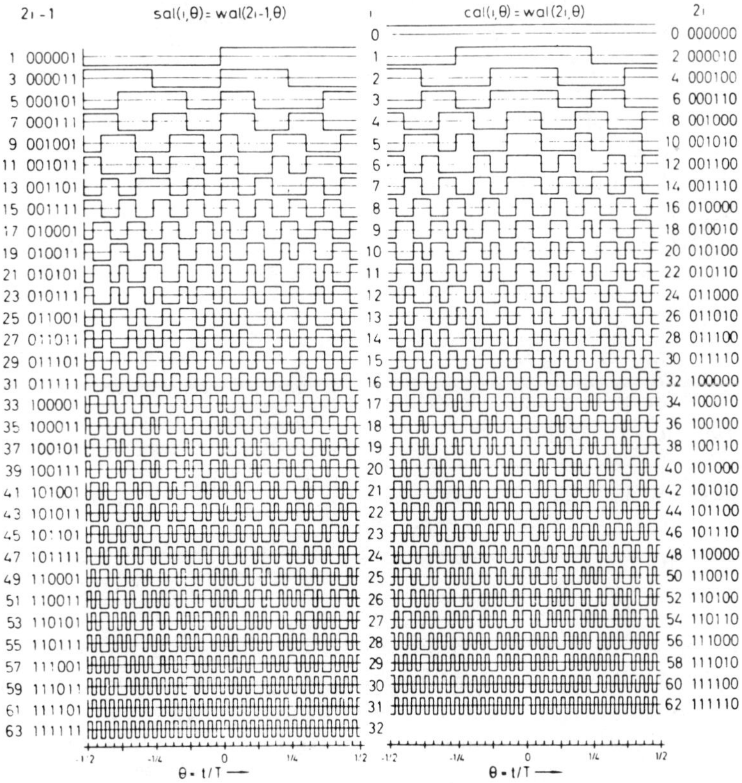

Fig. 3. Walsh functions sal (i, θ) and cal (i, θ), $\theta = t/T$ of the time variable t.

may write a Walsh series instead of a Fourier series for the signal $F(\theta)$:

$$F(\theta) = a(0) + \sum_{i=1}^{\infty} [a'(i) \operatorname{cal}(i, \theta) + b'(i) \operatorname{sal}(i, \theta)]$$

$$a'(i) = \int_{-1/2}^{1/2} F(\theta) \operatorname{cal}(i, \theta)\, d\theta,$$

$$b'(\theta) = \int_{-1/2}^{1/2} F(\theta) \operatorname{sal}(i, \theta)\, d\theta. \tag{2}$$

Since the Walsh functions $\operatorname{cal}(i, \theta)$ and $\operatorname{sal}(i, \theta)$ are either $+1$ or -1, there are no multiplications to be performed in the transformation from $F(\theta)$ to $a'(i)$ and $b'(i)$ and in the inverse transformation back to $F(\theta)$. This translates into faster operation in computers and simpler equipment in analog circuits compared with the Fourier series of (1).

Many other systems of functions besides sine–cosine and Walsh functions can be used in the transformations (1) and (2). The Walsh functions are closely related to certain Hadamard matrices, which are orthogonal matrices with the entries $+1$ and -1 [17]. One can derive similar systems of functions from all the other Hadamard matrices and one obtains a whole class of transformations called Hadamard transforms. Another interesting system of functions are the Haar functions, which in essence assume the three values $+1$, 0, and -1 only (the normalized Haar functions are multiplied by powers of $\sqrt{2}$ [18]). There exists a fast Haar transform that is faster than the fast Walsh transform, which in turn is faster than the fast Fourier transform [19], [22]. Analog circuits based on Haar functions are sometimes simpler than those based on Walsh functions.

Let us now turn from functions of time to those with two spatial variables. Fig. 4 shows the systems of block pulses $\operatorname{blo}(k, x)$ and of Walsh functions $\operatorname{wal}(k, x)$. The notation $\operatorname{sal}(i, \theta)$ and $\operatorname{cal}(i, \theta)$ is discarded in preference to the simpler notation $\operatorname{wal}(k, x)$ for more dimensions:

$$\operatorname{cal}(i, x) = \operatorname{wal}(2i, x),$$

$$\operatorname{sal}(i, x) = \operatorname{wal}(2i - 1, x).$$

A block pulse with the two variables x and y is defined by the product $\operatorname{blo}(k, x)\operatorname{blo}(m, y)$. Fig. 5 shows 64 functions with $k, m = 0 \cdots 7$ in the interval $-\frac{1}{2} < x, y < \frac{1}{2}$. The function $\operatorname{blo}(k, x)\operatorname{blo}(m, y)$ is found at the intersection of the column denoted $\operatorname{blo}(k, x)$ and the row denoted $\operatorname{blo}(m, y)$. Black areas indicate that the function has the value $+1$, white areas that it has the value 0. Walsh functions $\operatorname{wal}(k, x)\operatorname{wal}(m, y)$ with two variables x and y are shown in Fig. 6 with $k, m = 0 \cdots 7$ in the same interval. Again, the function $\operatorname{wal}(k, x)\operatorname{wal}(m, y)$ is found at the intersection of the column denoted $\operatorname{wal}(k, x)$ and the row denoted $\operatorname{wal}(m, y)$. Black areas indicate the value $+1$ but white areas indicate the value -1 rather than 0.

A space function $F(x, y)$ that is of interest in a finite interval $-\frac{1}{2} < x, y < \frac{1}{2}$ can be expanded into a two-

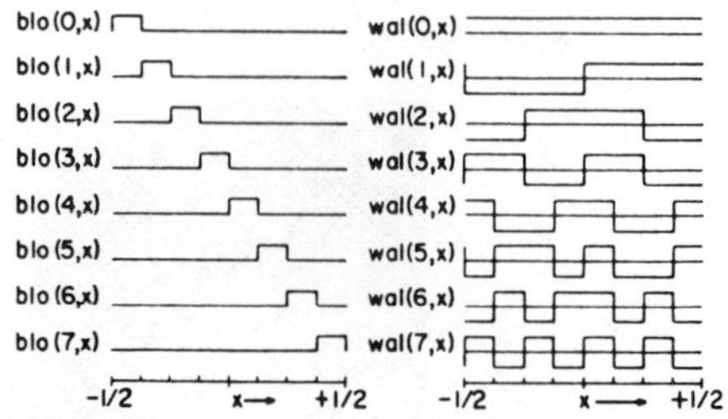

Fig. 4. Block pulses blo (k, x) and Walsh functions wal (k, x) of one space variable x in the interval $-\frac{1}{2} \leq x < \frac{1}{2}$ for $k = 0 \cdots 7$.

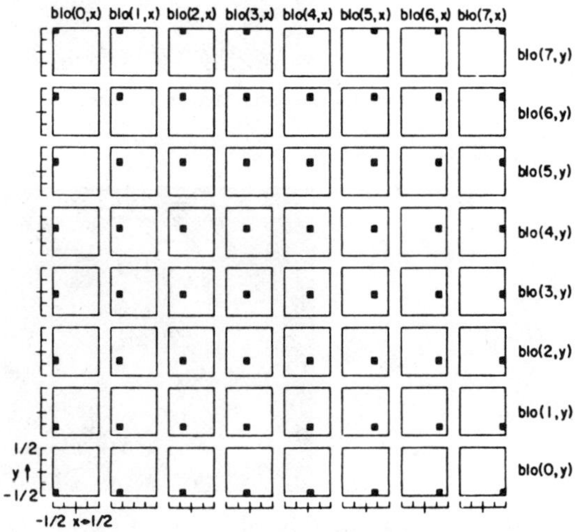

Fig. 5. Block pulses blo (k, x) blo (m, y) of two space variables x and y in the interval $-\frac{1}{2} < x < \frac{1}{2}$, $-\frac{1}{2} < y < \frac{1}{2}$ for $k, m = 0 \cdots 7$. Black areas indicate the value $+1$, white areas the value 0. The function blo (k, x) blo (m, y) is located at the intersection of the column denoted blo (k, x) and the row denoted blo (m, y).

dimensional Walsh series:

$$F(x, y) = \sum_{k=0}^{\infty} \sum_{m=0}^{\infty} a(k, m) \operatorname{wal}(k, x) \operatorname{wal}(m, y)$$

$$a(k, m) = \int_{-1/2}^{1/2} \int_{-1/2}^{1/2} F(x, y) \operatorname{wal}(k, x) \operatorname{wal}(m, y)\, dx\, dy. \tag{3}$$

Photographs or TV pictures are space limited so there is no interest in considering infinite space intervals. Hence, the double sum of the Walsh series (3) suffices and we do not need to consider its replacement by a double integral.

III. Instantaneous Filters for Space Signals

Fig. 7 shows a circuit that generates the Walsh transform $a(k, m)$ of a space signal $F(x, y)$. The signal is represented by 4×4 voltages $a, b, \cdots, p$ in the x, y plane as shown in the lower right corner.

Four printed-circuit boards denoted by $\operatorname{wal}(0, x)$ to $\operatorname{wal}(3, x)$ are intersected by 4×4 wires; only seven of them are shown by dashed lines to avoid obscuring the picture. The voltages $a \cdots p$ are applied to these 4×4 wires. Resistors R feed them to the adding (+) or subtracting (−) inputs of summing amplifiers A. At the

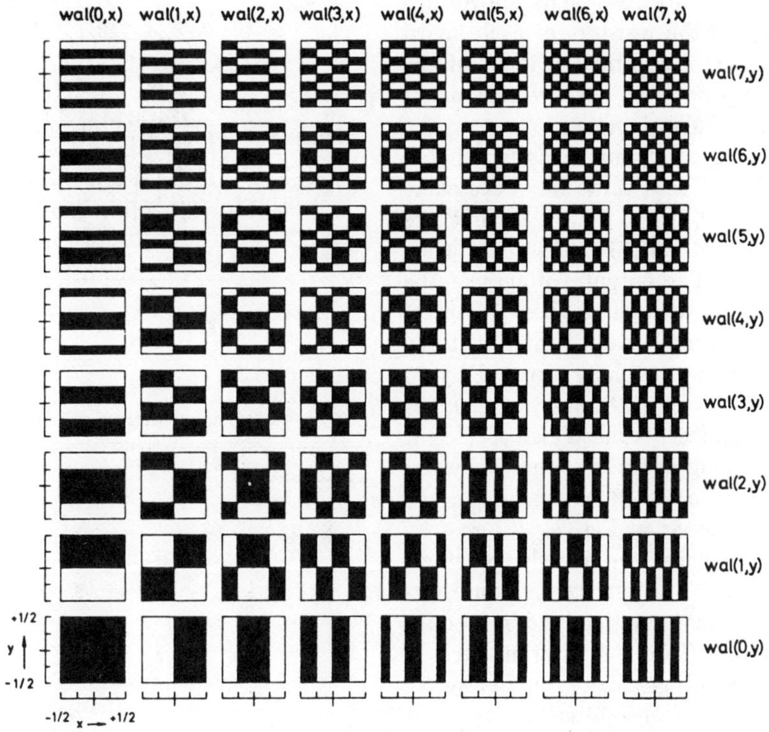

Fig. 6. Walsh functions wal (k,x) wal (m,y) of two space variables x and y in the interval $-\frac{1}{2} < x < \frac{1}{2}$, $-\frac{1}{2} < y < \frac{1}{2}$ for $k, m = 0 \cdots 7$. Black areas indicate the value $+1$, white areas the value -1. The function wal (k, x) wal (m, y) is located at the intersection of the column denoted wal (k, x) and the row denoted wal (m, y).

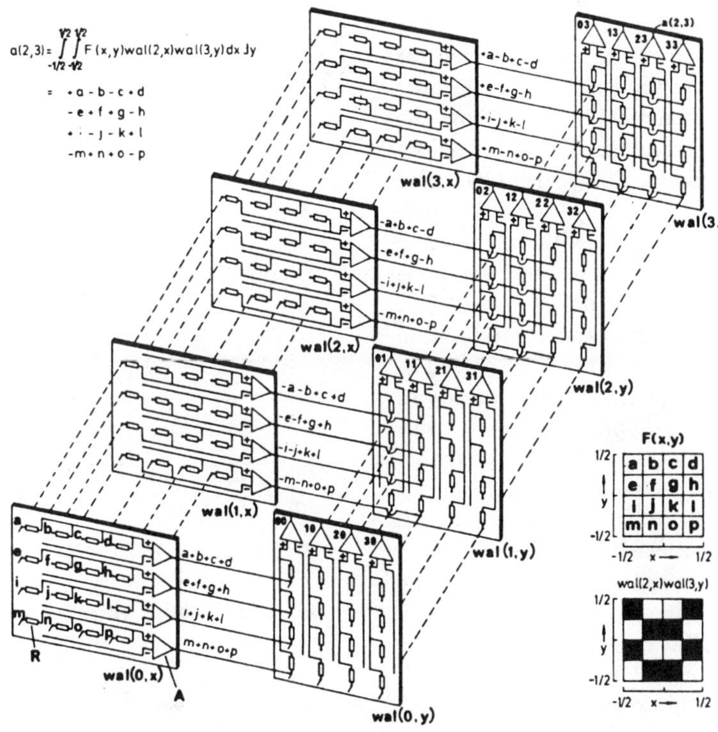

Fig. 7. Principle of a sequency filter for functions with two space variables x and y. The circuit shown transforms $F(x, y)$ into $a(k, m)$. A similar circuit is required for the inverse transformation.

outputs of the board wal(0, x) occur the four voltages $a+b+c+d$, $e+f+g+h$, $i+j+k+l$, and $m+n+o+p$; the topmost output of the board wal(1, x) yields the voltage $-a-b+c+d$, etc. Note that the resistors of each board are connected to the adding or subtracting input of the summing amplifier according to the positive and negative sign of the respective Walsh function.

A second set of four printed-circuit cards is denoted by wal(0, y) to wal(3, y). These cards have the same circuitry as the cards wal(0, x) to wal(3, x) but are rotated 90°. Again, only 7 of the 16 intersecting wires are shown by dashed lines. The output voltages of the cards wal(0, x) to wal(3, x) are fed to these 16 intersecting wires as shown. One may readily infer that the output voltage of the summing amplifier denoted by 00 is the sum $(a+b+c+d)$ + $(e+f+g+h)$ + $(i+j+k+l)$ + $(m+n+o+p)$ and thus represents the Walsh coefficient a(0, 0). Generally, a summing amplifier denoted by km yields the Walsh coefficient $a(k, m)$, as shown in the example in the upper left corner of Fig. 7.

The circuit of Fig. 7 produces the Walsh transform $a(k, m)$ of the signal $F(x, y)$. An essentially equal circuit will do the inverse transformation and reproduce $F(x, y)$ from the transform $a(k, m)$. A filtering effect is obtained by modifying the voltages $a(k, m)$. Before we can enter into a discussion of this filtering, we must develop a suitable and simple representation of the processes to be studied. When investigating the filtering of time signals one may plot the signal as a function of time, its Fourier transform as a function of frequency, the attenuation and phase shift of the filter as functions of frequency, and finally the filtered signal again as a function of time. The very same plots can be drawn for space signals $F(x, y)$ but one needs three-dimensional perspective. The sampled function $F(h, i)$ of the space signal, Fig. 8, shows an example. Table I provides the numerical values to the figure. The Walsh transform $a(k, m)$ of $F(h, i)$ is represented by Fig. 9 and Table II. Now consider a filter function $K(k, m)$ with values 0 and 1 as indicated in Fig. 10, or varieties of other "zero–one filters" displayed in Fig. 11. A filter defined as in Fig. 10 could be referred to as a two-dimensional attenuation low-pass sequency filter. The low-pass filters discussed so far let certain parts of the Walsh–Fourier transform $a(k, m)$ pass unchanged and suppressed others completely. However, the representation of Fig. 11 is capable of handling more complicated cases, i.e., the addition of variable weights at each of the $a(k, m)$ sequency points.

Let us now turn from attentuation low-pass filters to resolution filters. The output voltages $a(k, m)$ of the 16 operational amplifiers in Fig. 7 denoted by 00 to 33 shall be fed to 16 analog/digital converters. Let some of the voltages be converted into binary numbers with many digits and other voltages into binary numbers with a few digits only. The voltages $a(k, m)$ are thus represented with different "resolution" and the resolution is a function of k and m. This will be equivalent to variable-length source encoding of the sequency coefficients. Given a fidelity criterion such encoding can provide a data-compression technique. We can define a resolution function $R(k, m) = 5$ to mean that the voltage $a(k, m)$ is represented by five binary digits. Coupling this concept with the low-pass filters of Fig. 11, one such resolution

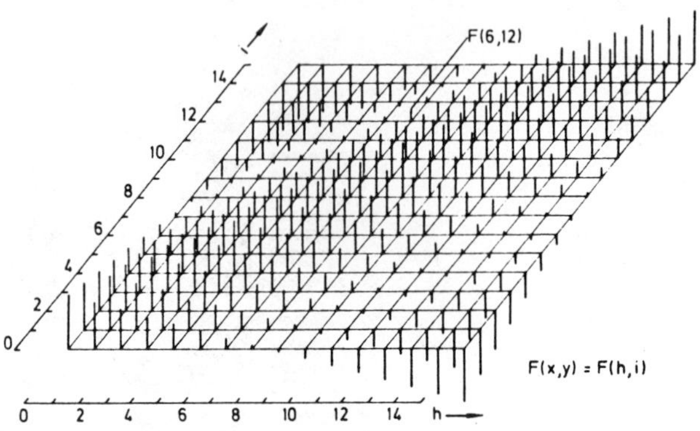

Fig. 8. Sample function $F(h,i)$.

TABLE I
REPRESENTATION OF THE FUNCTION $F(h,i)$

i \ h	0	1	2	3	4	5	6	7	8	9	10	11	12	13	14	15
15	−15	−13	−11	−9	−7	−5	−3	−1	+1	+3	+5	+7	+9	+11	+13	+15
14	−13	−11	−9	−7	−5	−3	−1	+1	+3	+5	+7	+9	+11	+13	+15	+13
13	−11	−9	−7	−5	−3	−1	+1	+3	+5	+7	+9	+11	+13	+15	+13	+11
12	−9	−7	−5	−3	−1	+1	+3	+5	+7	+9	+11	+13	+15	+13	+11	+9
11	−7	−5	−3	−1	+1	+3	+5	+7	+9	+11	+13	+15	+13	+11	+9	+7
10	−5	−3	−1	+1	+3	+5	+7	+9	+11	+13	+15	+13	+11	+9	+7	+5
9	−3	−1	+1	+3	+5	+7	+9	+11	+13	+15	+13	+11	+9	+7	+5	+3
8	−1	+1	+3	+5	+7	+9	+11	+13	+15	+13	+11	+9	+7	+5	+3	+1
7	+1	+3	+5	+7	+9	+11	+13	+15	+13	+11	+9	+7	+5	+3	+1	−1
6	+3	+5	+7	+9	+11	+13	+15	+13	+11	+9	+7	+5	+3	+1	−1	−3
5	+5	+7	+9	+11	+13	+15	+13	+11	+9	+7	+5	+3	+1	−1	−3	−5
4	+7	+9	+11	+13	+15	+13	+11	+9	+7	+5	+3	+1	−1	−3	−5	−7
3	+9	+11	+13	+15	+13	+11	+9	+7	+5	+3	+1	−1	−3	−5	−7	−9
2	+11	+13	+15	+13	+11	+9	+7	+5	+3	+1	−1	−3	−5	−7	−9	−11
1	+13	+15	+13	+11	+9	+7	+5	+3	+1	−1	−3	−5	−7	−9	−11	−13
0	+15	+13	+11	+9	+7	+5	+3	+1	−1	−3	−5	−7	−9	−11	−13	−15

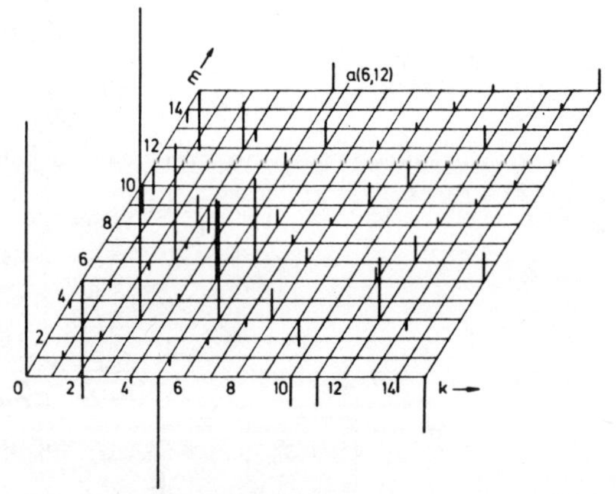

Fig. 9. Walsh–Fourier transform $a(k, m)$ of the sample function $F(h,i)$ of Fig. 8.

TABLE II

Representation of the Walsh Transform $a(k, m)$ of the Function $F(h,i)$ of Table I

m ↑ \ k →	0	1	2	3	4	5	6	7	8	9	10	11	12	13	14	15
15	−16	0	0	0	0	+8	0	0	0	0	0	+2	0	0	0	+6
14	−4	0	0	0	0	0	0	0	0	0	+2	0	0	0	+2	0
13	0	0	0	−4	0	0	0	0	0	+2	0	0	0	+2	0	0
12	0	0	0	+16	0	0	+8	0	+2	0	0	0	+6	0	0	0
11	−8	0	0	0	0	+4	0	0	0	0	0	+2	0	0	0	+2
10	−8	0	0	0	0	0	0	0	0	0	+6	0	0	0	+2	0
9	0	0	0	−8	0	0	0	0	0	+6	0	0	0	+2	0	0
8	0	0	0	+8	0	0	+4	0	+2	0	0	0	+2	0	0	0
7	0	0	0	−2	0	0	0	+2	0	0	0	0	0	0	0	0
6	0	0	−2	+32	0	0	+22	0	+4	0	0	0	+8	0	0	0
5	−32	−2	0	0	0	+22	0	0	0	0	0	+4	0	0	0	+8
4	−2	0	0	0	+2	0	0	0	0	0	0	0	0	0	0	0
3	0	0	0	+86	0	0	+32	−2	+8	−8	0	0	+16	−4	0	0
2	0	0	+2	0	0	0	−2	0	0	0	0	0	0	0	0	0
1	0	+2	0	0	0	−2	0	0	0	0	0	0	0	0	0	0
0	+70	0	0	0	−2	−32	0	0	0	0	−8	−8	0	0	−4	−16

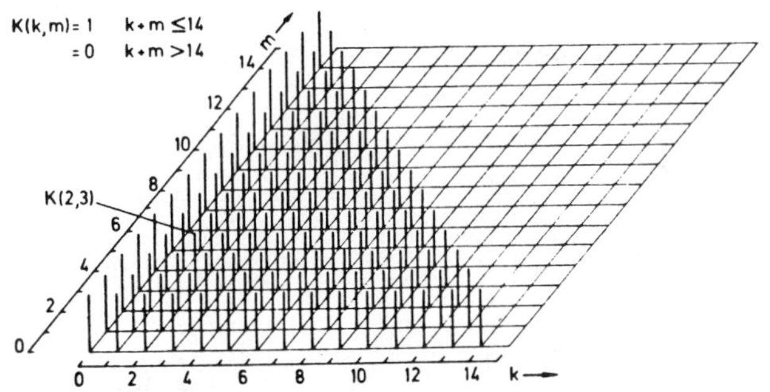

Fig. 10. Attenuation function $K(k, m)$ for a sequency low-pass filter.

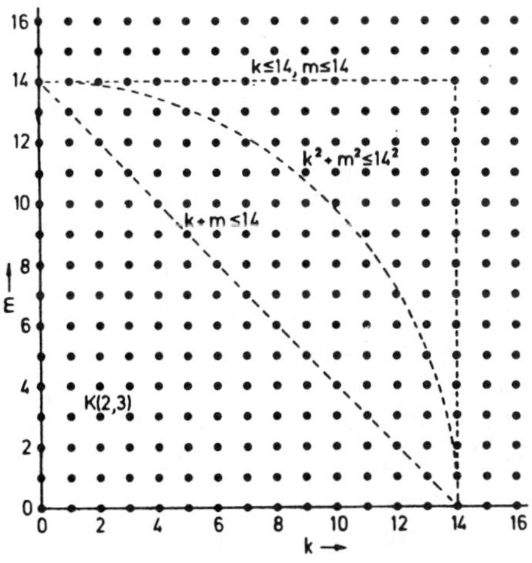

Fig. 11. The k, m plane of the two-dimensional sequency domain with three band limits for low-pass filters.

low-pass filter could be

$$R(k, m) = 5, \qquad k + m \leqq 14$$

$$R(k, m) = 4, \qquad 14 < k + m, \;\; k^2 + m^2 \leqq 14^2$$

$$R(k, m) = 3, \qquad 14^2 < k^2 + m^2, \;\; k \leqq 14, m \leqq 14$$

$$R(k, m) = 0, \qquad 14 < k, 14 < m.$$

In order to apply these results to TV pictures let the function $F(x, y)$ be sampled in a 512×512 array. The implementation of a filter with 512^2 inputs would be a formidable task if done according to Fig. 7. There are two methods to simplify such a filter. First, one can decompose the square of 512^2 picture elements into smaller squares. For instance, 32 squares with 16^2 picture elements each is a good choice. A filter with 16^2 inputs is then needed that transforms sequentially the 32 square picture sections; this simplification results in a somewhat reduced performance. Second, the circuit of Fig. 7 can be brought into a more practical form, the topic of the following section.

IV. Practical Implementation of a Filter for Space Variables

If the continuous function $F(x, y)$ is replaced by the sampled function $F(h, i)$ the Walsh expansion is replaced by a double sum,

$$a(k, m) = \sum_{h=0}^{2^n-1} \sum_{i=0}^{2^n-1} F(h, i) \cdot \operatorname{wal}\left(k, \frac{h+\frac{1}{2}}{2^n} - \frac{1}{2}\right) \operatorname{wal}\left(m, \frac{i+\frac{1}{2}}{2^n} - \frac{1}{2}\right). \tag{4}$$

The relative position of the scales for x, y, and h, i is shown in Fig. 12.

The summation can be performed in two steps:

$$a(k \mid i) = \sum_{h=0}^{2^n-1} F(h, i) \operatorname{wal}\left(k, \frac{h+\frac{1}{2}}{2^n} - \frac{1}{2}\right)$$

$$a(k, m) = \sum_{i=0}^{2^n-1} a(k \mid i) \operatorname{wal}\left(m, \frac{i+\frac{1}{2}}{2^n} - \frac{1}{2}\right). \tag{5}$$

Consider the case $2^n = 16$ and let the voltages $F(h, i)$ for a fixed value of i and for $h = 0 \cdots 15$ be fed to the input terminals $0, \cdots, 15$ of the circuit in Fig. 13. The voltages $-a(0 \mid i) \cdots -a(15 \mid i)$ are obtained at the output terminals $i = 0, \cdots, 15$.

The 16 horizontal printed-circuit cards in Fig. 12 each contain the circuit of Fig. 13. If the function $F(h, i)$ is fed to the 16^2 input terminals denoted 0, 0 to 15, 15 one obtains the voltages $a(0 \mid 15), a(1 \mid 15) \cdots a(15 \mid 15)$ at the output terminals of the topmost card and generally the voltages $a(0 \mid i) \cdots a(15 \mid i)$, $i = 0 \cdots 15$ at any one of the 16 cards indexed by i. The second summation transforms $a(k \mid i)$ into $a(k, m)$. This process is the same as

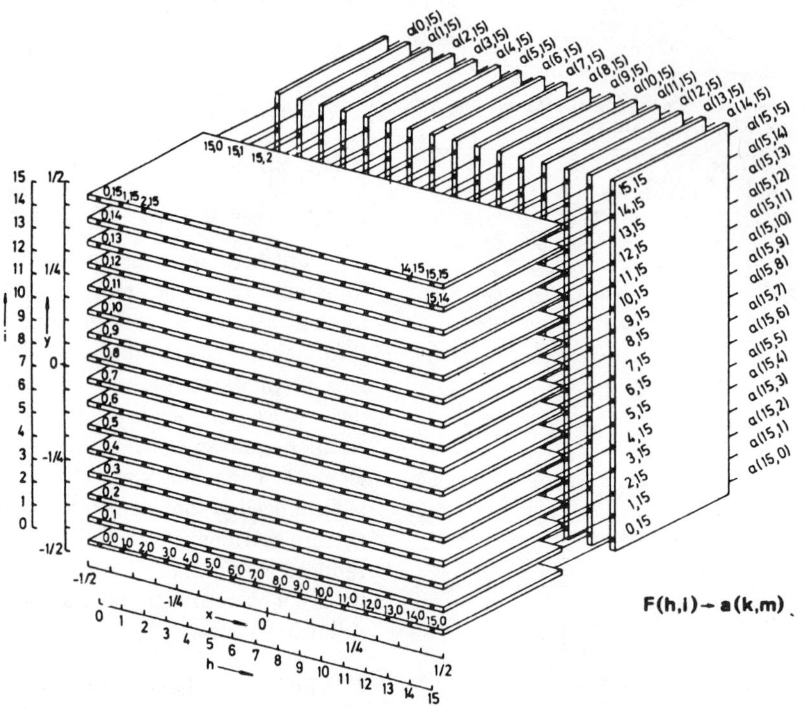

Fig. 12. Practical version of a sequency filter for functions of two space variables x and y or h and i.

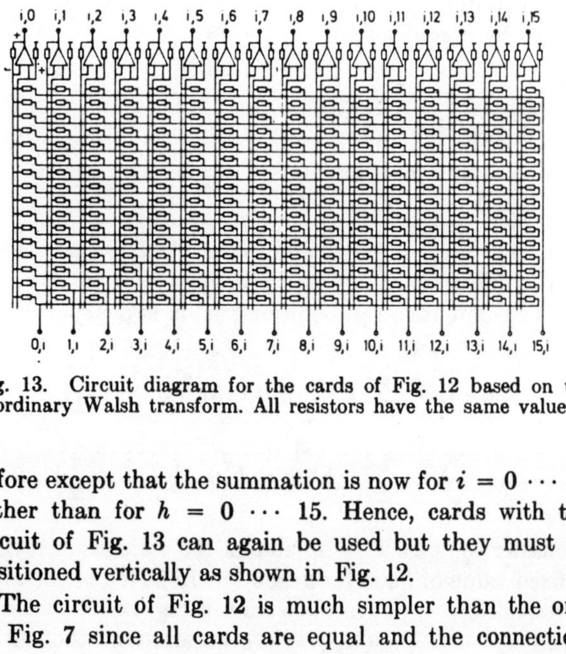

Fig. 13. Circuit diagram for the cards of Fig. 12 based on the ordinary Walsh transform. All resistors have the same value.

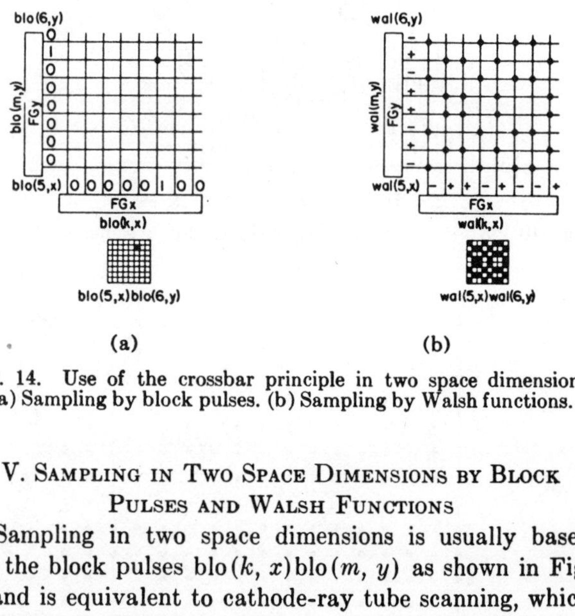

Fig. 14. Use of the crossbar principle in two space dimensions. (a) Sampling by block pulses. (b) Sampling by Walsh functions.

before except that the summation is now for $i = 0 \cdots 15$ rather than for $h = 0 \cdots 15$. Hence, cards with the circuit of Fig. 13 can again be used but they must be positioned vertically as shown in Fig. 12.

The circuit of Fig. 12 is much simpler than the one of Fig. 7 since all cards are equal and the connection between the horizontal and vertical cards can be made by means of a socket rather than by wires that have to be routed through the cards. Note that its operation does not depend on the use of Walsh functions: the circuit of Fig. 13 can be implemented, e.g., for Haar functions; it can also be implemented according to the fast Walsh transform rather than the ordinary Walsh transform used in Fig. 13. The arrangement of the boards in Fig. 12 is not affected by such changes.

V. Sampling in Two Space Dimensions by Block Pulses and Walsh Functions

Sampling in two space dimensions is usually based on the block pulses $\text{blo}(k, x)\text{blo}(m, y)$ as shown in Fig. 5 and is equivalent to cathode-ray tube scanning, which may be considered to represent the decomposition of a function $F(x, y)$ by the system of block pulses $\text{blo}(k, x) \cdot \text{blo}(m, y)$. This decomposition can also be represented by the crossbar scanner in Fig. 14(a). The function generator FGx produces the function $\text{blo}(5, x)$ and the generator FGy produces the function $\text{blo}(6, y)$. The crossing of the two bars with voltage 1 is indicated by a black dot. This dot represents the function $\text{blo}(5, x)\text{blo}(6, y)$.

Walsh function crossbar scanning is displayed in Fig. 14(b). The function $\text{wal}(5, x)$ represented by positive and negative voltages applied to the vertical bars is supplied by the function generator FGx, while the generator FGy supplies the function wal(6, y) to the horizontal bars. The crossings of bars with equal applied voltages are indicated by black dots. One may readily see that this dot pattern corresponds to the black areas of the function wal(5, x)wal(6, y). The white areas correspond to the crossings where a voltage difference exists. Crossbar sampling according to Walsh functions provides a way for implementation of filters for signals with two space variables. The filtering is accomplished in the simplest case by not sampling the signal $F(x, y)$ in the particular Walsh function pattern that one wants to suppress. An amplitude resolution filter is implemented by representing not all sampled values by an equal number of binary digits. There is no great difficulty to apply the results previously stated for the instantaneous filters to the sampling filters.

Let us look at Fig. 14 not from the standpoint of filtering and TV picture generation but from the standpoint of information storage. The crossbars may represent a magnetic core storage with a core at each crossing. The usual way is to store one bit in one core or one word in one storage location. This corresponds to storage according to block pulses. It is interesting to note that the destruction of one storage location destroys the information stored in this location completely and leaves all other information unchanged in case of the block pulse storage. Storage according to Walsh functions would distribute each bit over all cores or each word over all storage locations. In this case by the loss of one storage location all information would be degraded—resulting, e.g., in smaller output pulses—but no information would be completely destroyed. In theory, such a storage would remain operable as long as more than half the storage locations are operating. This has an equivalent interpretation in Reed–Muller codes [20].

VI. Some Circuits of Sampling Filters for Space Signals

We have shown by means of Fig. 14 that the crossbar sampler can be operated according to Walsh functions. It remains to be shown how the required generator for space variable Walsh functions wal(k, x) can be constructed and how a picture can be decomposed and recomposed by means of Walsh functions.

The generation of space-variable Walsh functions demonstrates the importance of simple systems of functions. Fig. 15 shows a circuit diagram of a tunable generator for space-variable Walsh functions and Fig. 16 shows its pulse diagram. Using binary indexing any Walsh function from wal(0001, x) to wal(1111, x) can be generated by a product of the four functions wal(0001, x), wal(0010, x), wal(0100, x), and wal(1000, x). Fig. 16 shows as example the function wal(1101, x):

$$\text{wal}(1101, x) = \text{wal}(1000, x)\text{wal}(0100, x)\text{wal}(0001, x).$$

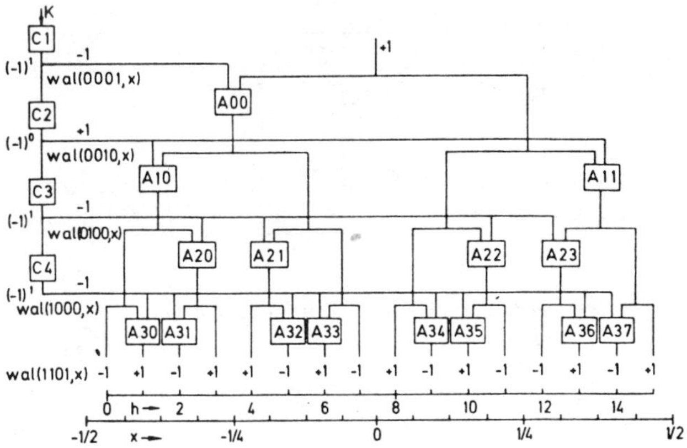

Fig. 15. Tunable generator for space variable Walsh functions. $A00 \cdots A37$ half-adders (exclusive OR gates). $C1 \cdots C4$ binary counter stages. K input terminal for trigger pulses.

This is simply the result of the dyadic property: wal($k \oplus j, x$) = wal(k, x) $\times$ wal(j, x). Multiplication by wal(0001, x) requires a sign reversal in the interval $-\frac{1}{2} \leqq x < 0$ and no reversal in the interval $0 \leqq x < \frac{1}{2}$ according to Fig. 16. Similarly, multiplication by wal(0010, x), wal(0100, x), or wal(1000, x) requires a sign reversal in the intervals where these functions are negative and no reversal where they are positive. The circuit of Fig. 15 is comprised of exclusive OR gates (denoted $A00$–$A15$ and flip-flops forming the counter C.

One may see that the arrangement of the exclusive OR gates $A20$–$A23$ corresponds to the function wal(0100, x) and that of the gates $A30$–$A37$ to the function wal(1000, x). The voltage +1 fed to the top of the circuit of Fig. 15 will thus be multiplied in the levels denoted wal(0001, x), wal(0100, x) and wal(1000, x) according to the positive and negative signs of these functions.

The 16 output voltages denoted +1 or −1 represent the function wal(1101, x) by voltage samples. Note that 1101 is the number stored in the binary counter stages $C1$ to $C4$. These counter stages may of course be replaced by switches if the generator is to be set or "tuned" manually. In the crossbar scanning application, all Walsh functions wal(k, x) with $k = 0, 1, 2 \cdots$ must be produced and this is most easily done by using binary counter stages in Fig. 15. Replacing them by a shift register makes it possible to generate a Walsh function according to the number shifted in. Furthermore, by feeding a time variable function wal(j, θ) instead of a fixed voltage +1 into this generator one obtains time–space variable functions wal(j, θ) wal(k, x).

Fig. 17 shows one of the several possible circuits for the decomposition of a picture by means of Walsh functions according to Fig. 14(b). Two Walsh function generators WGx and WGy feed functions wal(k, x) and wal(m, y) to the vertical and horizontal crossbars. A half-adder $A00$–$A77$ and a single-pole double-throw switch $S00$–$S77$ is located at each bar crossing. If the voltages at two crossing bars are equal, the input I of the switch is connected to the positive (+) input of the

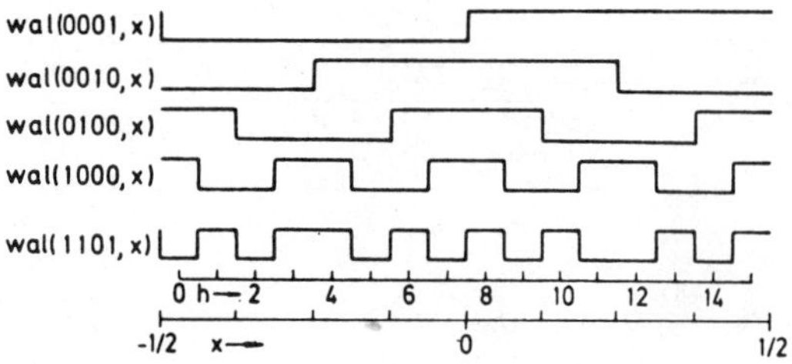

Fig. 16. Pulse diagram for the Walsh function generator of Fig. 15.

summing amplifier (SA); if they are not equal, I is connected to the negative input (−) of SA. The output voltage of the summing amplifier SA represents the Walsh–Fourier transform $a(k, m)$ of the signal $F(x, y)$ applied to the inputs $I00$–$I77$ of the circuit.

The filtering process requires that each sample $a(k, m)$ be multiplied by an attenuation coefficient $K(k, m)$ and the inverse transformation be performed. In the simplest cases $K(k, m)$ is either 1 or 0. The coefficients $a(k, m)$ for which $K(k, m)$ is zero do not need to be produced at all.

Let us consider resolution filters. The voltage $a(k, m)$ in Fig. 17 would be fed to an analog/digital converter. The generators WGx and WGy would produce the functions wal(k, x) and wal(k, y) as above. The number of bits into which $a(k, m)$ would be decomposed would be made a function of k and m.

The input signal in Fig. 17 could be provided by an array of microphones, e.g., for the production of images by sound waves in water, by photocells in a pattern reader, or by any other transducer that converts an observable quantity into a voltage. Progress in large-scale integration may eventually make it possible to produce such a sampler with 512 × 512 input terminals instead of the 8 × 8 shown. Using phototransistors as input devices one would then have an image scanner with a filter built in. Only three external connections beyond the power supply leads would be required: the terminal Kx and Ky for trigger pulses and the output of the summing amplifier SA. The Walsh function generators for 512 functions would not be much more complex than the one shown by Fig. 15. A very large number of half-adders (A) and switches (S) would be required ($512^2 = 262\,144$) but they would all be equal. It is known from computer simulation that one may expect a reduction of the number of bits by a resolution source encoding filter to about $\frac{1}{8}$ compared with sampling by block pulses. This not only decreases the bandwidth required for transmission but also the necessary switching speed of the sampling device.

Let us turn to the problem of reconverting the filtered transform $a(k, m)$ to the space signal. Fig. 18 shows the principle of this reconversion. A crossbar device like that of Fig. 17 drives at each crossing a half-adder Ahi, which in turn operates a single-pole double-throw switch Shi. The voltages $+a(k, m)$ and $-a(k, m)$ are fed to all switches. Either $+a(k, m)$ or $-a(k, m)$ passes through the switch to the output terminal Ohi. This completes the reconversion for those cases where a positive or negative output voltage is wanted.

A problem arises in the reconversion of optical images. A positive voltage is easily transformed into light by means of a glow tube or similar devices. However, a negative voltage cannot be used to subtract already emitted light. The storage device Hhi is provided in Fig. 18 to overcome this problem. A positive or negative charge proportionate to $+a(k, m)$ or $-a(k, m)$ is fed

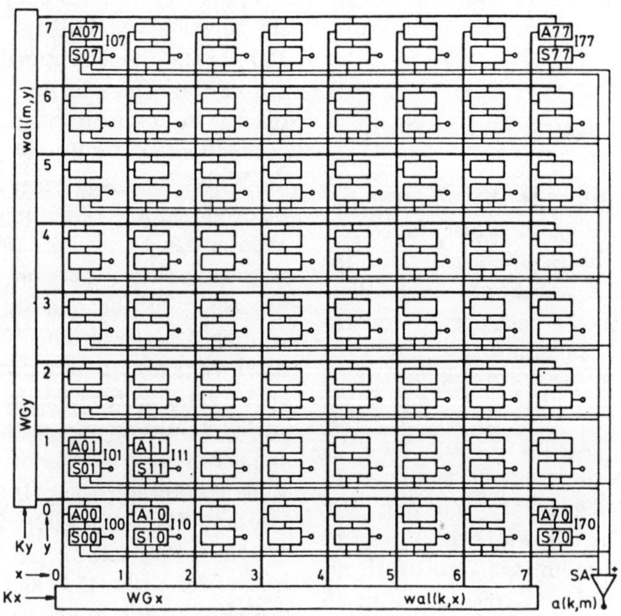

Fig. 17. Principle of a Walsh function sampler with built-in filtering capability.

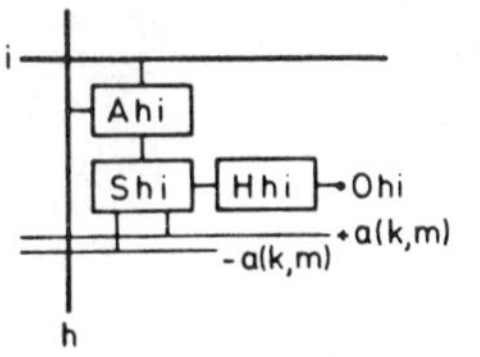

Fig. 18. Retransformation of the transform $a(k, m)$ into a spatial signal $F(h, i)$. A—half-adder; S—single-pole double-throw switch; H—storage.

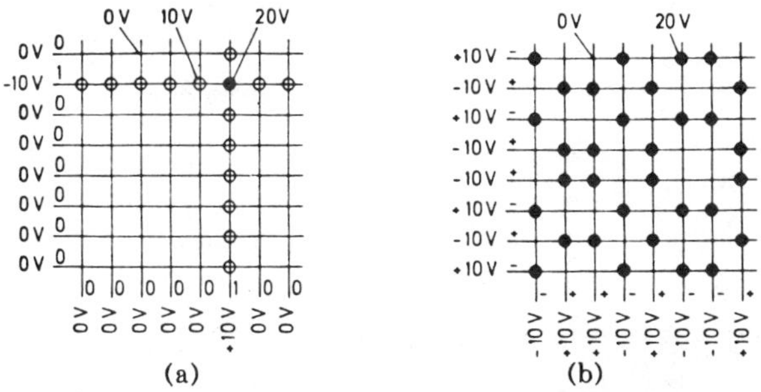

Fig. 19. Voltage differences in a crossbar sampler operated according to block pulses (a) and Walsh functions (b).

through Shi into Hhi. At the end of a scanning cycle—that is, after the voltages $+a(k, m)$ and $-a(k, m)$ for all values of k and m have been applied to Shi—the charge in Hhi is converted into light. Only charges of one polarity can be in the storage devices Hhi at this moment if all coefficients $a(k, m)$ have been transmitted unchanged. This is not necessarily so if filtering is used, but the problem is overcome by input and output transducers with logarithmic characteristic [2]. The implementation of the circuit of Fig. 18 in semiconductor technology using, e.g., metastable states for storage and the exponential variation of distribution curves to obtain a logarithmic characteristic might be a challenging task for scientists working in the semiconductor field. The practical incentive for such a development is the expected reduction of the bandwidth of TV signals, but there is also the prospect of a significant increase of brightness compared with crossbar displays using block pulses.

For a partial explanation of the last statement, consider Fig. 19(a). The block pulse blo$(5, x)$ represented by a voltage +10 V is applied to the vertical bars and a block pulse blo$(6, y)$ represented by a voltage −10 V is applied to the horizontal bars. A voltage difference of 20 V is produced where the function blo$(5, x)$blo$(6, y)$ is +1. The voltage difference at most bar crossings is zero, but there are a few with a difference of 10 V. Let a voltage-difference-to-light converter be placed at each bar crossing, e.g., a glow tube. The largest permissable voltage difference would be limited to twice the difference that just starts light emission. Furthermore, light can only be emitted from one bar crossing at a time, unless some storage mechanism is used. These two restrictions limit the brightness of a produced image.

Fig. 19(b) shows the function wal$(5, x)$ applied to the vertical and the function wal$(6, y)$ to the horizontal bars. Note that the sign of 10 V corresponds to that of the Walsh function at the vertical bars but is reversed at the horizontal bars, just like it was for the block pulses. Only voltage differences of 20 and 0 V occur. There is no intermediate voltage difference and hence no restriction on the permissable largest voltage difference. Furthermore, light is emitted from half the bar crossings at a time. The latter effect becomes even more important in a three-dimensional display to which the crossbar principle can be extended.

VII. Conclusions

This paper has attempted to present a few tutorial ideas on the practical use of sequency filtering for image transmission. The state of the art of semiconductor technology suggests that switching function decompositions might be useful and the Walsh system becomes a likely candidate. Computer simulations have also suggested that the Walsh transform might provide a useful domain for picture transmission. We have attempted to illustrate these ideas with existing system results, Figs. 1 and 2, and have endeavored to present a few circuit schematics suggesting possible large-scale implementation procedures. Indeed, it is hoped that this tutorial exposition may serve as a possible motivation for further studies by other individuals and institutions.

References

[1] H. C. Andrews, *Computer Techniques in Image Processing*. New York: Academic Press, 1970.

[2] A. V. Oppenheim, R. W. Schafer, and T. G. Stockham, "Nonlinear filtering of multiplied and convolved signals," *Proc. IEEE*, vol. 56, pp. 1264–1291, Aug. 1968.

[3] J. D. Kennedy, S. J. Clark, and W. A. Parkyn, "Digital imagery data compression techniques," McDonnell-Douglas Astronautics Co., Huntington Beach, Calif., Rep. MDC-G0402, Jan. 1970.

[4] A. Habibi and P. A. Wintz, "Image coding by linear transformation and block quantization," *IEEE Trans. Commun. Technol.*, vol. COM-19, pp. 50–62, Feb. 1971.

[5] V. K. Agarwal, "A new approach to the fast Hadamard transform algorithm," in *Proc. 1971 Hawaii Int. Conf. System Sciences*, pp. 408–410.

[6] W. A. Parkyn, "Digital image processing aspects of the Walsh transform," in *Proc. 1970 Walsh Functions Symp.*, pp. 152–156, AD 707-431, Washington, D. C.

[7] R. J. Lechner, "Invariant properties of Hadamard transform under affine group," in *Proc. 1970 Walsh Functions Symp.*, pp. 88–94, AD 707-431, Washington, D. C.

[8] M. Kabrisky and J. W. Carl, "Sequency filtered densely connected transforms for pattern recognition," in *Proc. 1971 Hawaii Int. Conf. System Sciences*, pp. 223–228.

[9] H. F. Harmuth, "Survey of analog sequency filters based on Walsh functions," in *Proc. 1970 Walsh Functions Symp.*, pp. 208–219, AD 707-431, Washington, D. C.

[10] ——, *Transmission of Information by Orthogonal Functions*. New York: Springer, 1969.

[11] ——, "A generalized concept of frequency and some applications," *IEEE Trans. Inform. Theory*, vol. IT-14, pp. 375–382, May 1968.

[12] H. Enomoto and K. Shibata, "Orthogonal transform coding system for television signals," *IEEE Trans. Electromagn. Compat.* (*Special Issue on Walsh Functions Symp. 1971*), vol. EMC-13, pp. 11–17, Aug. 1971.

[13] H. Enomoto and K. Shibata, "Features of a Hadamard transformed television signal" (in Japanese), in *Proc. 1965 Nat. Conf. IECEJ*, Paper 881.

[14] K. Shibata, T. Ohira, and S. Terauchi, "Color television

signal orthogonal transformation PCM terminal equipment" (in Japanese), *1970 IECEJ Papers Tech. Group Commun. Syst.*, CS 70-74 (1970-07).
[15] M. Hatori *et al.*, "On the band compression of television signals by the E-sequence transformation technique" (in Japanese), *1970 IECEJ Papers Tech. Group Inform. Theory*, IT70-13(1970-05).
[16] J. L. Walsh, "A closed set of orthogonal functions," *Amer. J. Math.*, vol. 45, pp. 5–24, 1923.
[17] R. E. A. C. Paley, "On orthogonal matrices," *J. Math. Phys.*, vol. 12, pp. 311–320, 1933.
[18] A. Haar, "Zur Theorie der orthogonalen Funktionensysteme," *Math. Ann.*, vol. 69, pp. 331–371, 1910; also vol. 71, pp. 38–53, 1912.
[19] H. C. Andrews and K. L. Caspari, "A generalized technique for spectral analysis," *IEEE Trans. Comput.*, vol. C-19, pp. 16–25, Jan. 1970.
[20] I. S. Reed, "A class of multiple error-correcting codes and the decoding scheme," *IRE Trans.*, vol. PGIT-4, pp. 38–49, Sept. 1954.
[21] J. D. Lee, "Review of recent work on applications of Walsh functions in communications," in *Proc. 1970 Walsh Functions Symp.*, AD 707-431, pp. 26–35, Washington, D. C.
[22] H. F. Harmuth, *Transmission of Information by Orthogonal Functions*, 2nd ed. New York: Springer, 1972.

Henning F. Harmuth (M'53) was born in Vienna, Austria, on July 27, 1928. He received the Dipl.-Ing. and Dr. Tech. degrees from the Technische Hochschule, Vienna, in 1951 and 1953, respectively.

He has worked for the Signal Corps, the General Electric Company, and Battelle Institute. Presently, he is with the Department of Electrical Engineering, Catholic University of America, Washington, D. C. Since 1965 he has been a Consulting Engineer in the field of sequency theory and Walsh functions. He has lectured at several universities. He is the author of *Transmission of Information by Orthogonal Functions* and numerous technical papers.

Harry C. Andrews (S'66–M'68) was born in Glendale, Calif., on May 7, 1943. He received the B.S. degree in electrical engineering from Stanford University, Stanford, Calif., in 1964, and the M.S. and Ph.D. degrees in electrical engineering from the University of Southern California, Los Angeles, in 1966 and 1968, respectively.

He worked part time as a Hughes Master's Fellow at Hughes Aircraft, Culver City, Calif. He is now an Assistant Professor in the Department of Electrical Engineering, University of Southern California. He has written many articles in the area of bandwidth compression and image coding. He is the author of *Computer Techniques in Image Processing* (Academic Press, 1970). He is currently a Consultant with the ITT Electrophysics Laboratories, Inc., Hyattsville, Md.

Dr. Andrews is a member of Tau Beta Pi, Eta Kappa Nu, and Sigma Xi.

Kosho Shibata was born in Tokushima, Japan, on March 15, 1933. He received the B.E. degree from the University of Electro-Communication, Chofu City, Tokyo, Japan, in 1957.

From 1961 he was engaged in research on bandwidth compression of TV signals at the Research and Development Laboratory, Kokusai Denshin Denwa Co., Ltd. (Japan's overseas radio and cable system), Tokyo. At first he was engaged in the research of TV bandwidth compression by the wobbling method in which a sinusoidal signal was superposed in encoding and dot interlace methods. Since the end of 1964 he has been engaged in research using Hadamard transform and modified Hadamard transform coding systems for TV signals.

Reprinted from *Computer J.* **16:**165–173 (1973), by permission of The British Computer Society

THE APPLICATION OF CHOW PARAMETERS AND RADEMACHER-WALSH MATRICES IN THE SYNTHESIS OF BINARY FUNCTIONS

S. L. Hurst

This paper surveys and brings together separate existing information from the field of (*a*) threshold logic synthesis, and (*b*) general binary synthesis using Rademacher-Walsh coefficients. The correlation between the Chow parameters widely used in the former and the latter Rademacher-Walsh coefficients is pursued. With a better understanding of the common features of both fields, inspiration towards further profitable developments may be hastened.

Index terms
Boolean functions, Chow parameters, linear-separability, Rademacher-Walsh matrices, threshold functions, spectra.

List of symbols used

a_i, $i = 0$ to n, = minimum-integer weighting in a threshold realisation.
b_i, $i = 0$ to n, = threshold function characterising parameter, (Chow parameter).
$f(x)$ = binary function, value 0 (false) or 1 (true)
$f(y)$ = ditto, value -1 or $+1$
$f(z)$ = ditto, value $+1$ or -1
n = number of binary input variables per system.
R_i, $i = 0$ to n, = Rademacher variables or functions, $+1$, -1 valuation.
R_i, $i = 0, 1, \cdots, 12, \cdots$, = Rademacher-Walsh functions, $+1$, -1 valuation.
$\hat{R}_i$, $i = 0, 1, \cdots, 12, \cdots$, = modified Rademacher-Walsh functions, 0, 1 valuation.
$[R] = 2^n \times 2^n$ Rademacher-Walsh matrix.
$[\hat{R}] = 2^n \times 2^n$ modified Rademacher-Walsh matrix.
p = minterm, e.g. $p = 1$ is minterm $\cdots 001$
W_i, $i = 0, 1, \cdots 2^n$, = Walsh function, $+1$, -1 valuation
W_i', $i = 0, 1, \cdots 2^n$, = ditto, but taken in different matrix row order from W_i
$[W]$ and $[W'] = 2^n \times 2^n$ Walsh matrix
w_i = real-number weight associated with R_i, W_i, or W_i', $i = 0$, 1, etc.
$\hat{w}_i$ = real-number weight associated with $\hat{R}_i$, $i = 0$, 1 etc. (note $w_i = \hat{w}_i$, except for $i = 0$)
x_i, $i = 1, 2, \cdots n$, = binary input variable, value 0 or 1
y_i, $i = 1, 2, \cdots n$, = ditto, value -1 or $+1$
z_i, $i = 1, 2, \cdots n$, = ditto, value $+1$ or -1

Introduction

A considerable volume of work has been covered in recent years in investigating the nature of linear-separability in binary functions. A binary function is said to be linearly-separable if, when represented on a n-dimensional hypercube such as that shown in **Fig. 1,** a separating plane can be drawn which precisely divides all true minterms of the function from all false minterms (Hurst, 1971). Fig. 1 illustrates the 3-variable linearly-separable function $f(x) = [x_1 + x_2x_3]$.

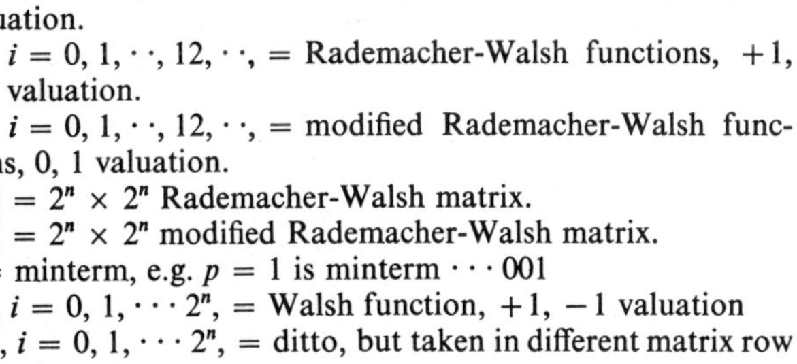

Fig. 1. Hypercube illustration of linearly-separable function $f(x) = [x_1 + x_2x_3]$, threshold realisation $\langle 2x_1 + x_2 + x_3\rangle_{2:1}$

All such linearly-separable functions are realisable by a single threshold logic gate of appropriate weights and thresholds (Hurst, 1971; Lewis, 1967; Dertouzos, 1965). The example given above is realisable by a gate which weights the x_1 input by two units and the x_2 and x_3 inputs each by one unit, a nominal input summation ≥ 2 being required to operate the gate, ≤ 1 being insufficient to operate the gate. This threshold gate realisation therefore may be expressed arithmetically as $f(x) = \langle 2x_1 + x_2 + x_3\rangle_{2:1}$.

Associated with this work has been a classification and tabulation procedure for all possible linearly-separable functions, involving the characteristic vectors of such functions and canonic positive Chow parameter classifications (Dertouzos, 1965; Winder, 1964; Winder, 1965; Muroga, Tsuboi and Baugh, 1967).

In the field of general Boolean synthesis, not restricted to the above linearly-separable class of functions, work using orthogonal matrices has been pursued. The matrices involved are normally the $2^n \times 2^n$ Rademacher-Walsh matrices, which lend themselves directly to the specification of any Boolean function of n variables (Coleman, 1961; Lockheed, 1964; Cooper, 1963; Liedl and Pichler, 1971). To date, however, direct circuit realisation using this approach has not been an economic practical proposition.

In the following sections of this paper, therefore, the existing state of the art of binary circuit synthesis in both the above fields is examined and compared. Strong common features will be found, which it is hoped may assist in furthering the necessary continuing development in this field.

1. Characteristic vectors, Chow parameters

The 'characteristic vector' b_i of any linearly-separable (threshold) function uniquely defines the given function (Dertouzos, 1965). The coefficients of a characteristic vector for an n-variable Boolean function $f(x_1, \cdots, x_n)$ are $(n + 1)$ in number, and are defined by:

$$b_0 = [(\text{number of true minterms}) - (\text{number of false minterms})]$$

$$b_i,\ i = 1 \text{ to } n, = [(\text{number of agreements between the value of } x_i \text{ and } f(x) \text{ taken over all minterms}) - (\text{number of disagreements between } x_i \text{ and } f(x))]\,.$$

Chow (1961) formally proved that $(n + 1)$ coefficients as above are sufficient to uniquely define any linearly-separable function. Also, if the magnitudes $|b_i|$ of the coefficients of all possible threshold functions are taken and written in lexicographical order, (e.g. $-2, -6, +2, +10, -6$ becomes 10, 6, 6, 2, 2) then such invariance operations produce the most compact form of classifying and listing all the possible threshold functions of n variables, giving the canonic positive CHOW PARAMETER or CHOW COEFFICIENT tabulations.

2. +1, −1 valuations

If instead of the false/true valuations 0, 1 for $f(x)$ and $x_1, x_2, \ldots x_n$, we adopt the alternative false/true valuations $-1/+1$, identified hereafter by $f(y)$ and $y_1, y_2, \cdots y_n$, we have that:

$$f(y) = [2f(x) - 1]$$

and

$$y_i, i = 1 \text{ to } n, = [2x_i - 1]$$

Thence:

$$b_0 = \sum_{p=0}^{2^n-1} \{f(y)\},$$

the summation being taken over all the 2^n minterms from $p = 0$ to $2^n - 1$,

and

$$b_i, i = 1 \text{ to } n, = \sum_{p=0}^{2^n-1} \{f(y).y_i\}.$$

(NOTE: $\sum$ indicates normal arithmetic summation, *not* modulo-two)

If now an additional 'variable' y_0 is added to the system, where y_0 by definition is always equal to $+1$, then b_0 may be redefined as:

$$b_0 = \sum_{p=0}^{2^n-1} \{f(y).y_0\}$$

Hence all the b_i may now be defined by the one relationship:

$$b_i, i = 0 \text{ to } n, = \sum_{p=0}^{2^n-1} \{f(y).y_i\}$$

Should yet another alternative terminology be adopted, with $+1/-1$ representing the 0/1 false/true values, identified hereafter by $f(z)$ and $z_1, z_2, \cdots z_n$, we have that:

$$f(z), = -f(y), = [1 - 2f(x)],$$

and

$$z_i, i = 1 \text{ to } n, = -y_i, = [1 - 2x_i]$$

Then:

$$b_0 = -\sum_{p=0}^{2^n-1} \{f(z)\},$$

$$b_i, i = 1 \text{ to } n, = \sum_{p=0}^{2^n-1} \{f(z).z_i\}.$$

In a similar manner to $f(y)$ above, if an additional parameter z_0 is added, where $z_0 \triangleq -1$, b_0 may be redefined as:

$$b_0 = \sum_{p=0}^{2^n-1} \{f(z).z_0\},$$

and hence all the b_i may be jointly defined by:

$$b_i, i = 0 \text{ to } n, = \sum_{p=0}^{2^n-1} \{f(z).z_i\}$$

Given the b_i values for any threshold function, the corresponding minimum-integer realising weights a_i required for the threshold realisation have been fully tabulated. Thus the optimum threshold-gate realisation of

$$f(x) = \langle a_1x_1 + a_2x_2 + \cdots a_nx_n \rangle_{t_1:t_2},$$

where t_1 and t_2 are the upper and lower gate thresholds, therefore is immediately available.

3. Signs of $|b_i|$

As published, each $|b_i|$ Chow parameter listing defines one of several possible threshold functions, with the same coefficient values $|b_i|$. To uniquely define any specific threshold function, reinstatement of the signs of the b_i, $i = 0$ to n, is necessary. The b_0 term with its appropriate sign is essential and cannot be omitted.

For example consider the three linearly-separable functions plotted in **Fig. 2.** If the b_i values for these functions are computed, we obtain:

	b_0	b_1	b_2	b_3
Function (*a*):	−2	−6	+2	+2
(*b*):	−2	+6	+2	+2
(*c*):	+2	+6	+2	+2

Thus only the sign of b_1 differentiates between the two quite distinct functions (*a*) and (*b*), and, similarly, only the sign of b_0 differentiates between functions (*b*) and (*c*).

This therefore illustrates the absolute necessity to take into account the signs of all the b_i when specific functions are being considered. The three functions illustrated in Fig. 2 are of course three of many with the same canonic $|b_i|$ classification of 6, 2, 2, 2.

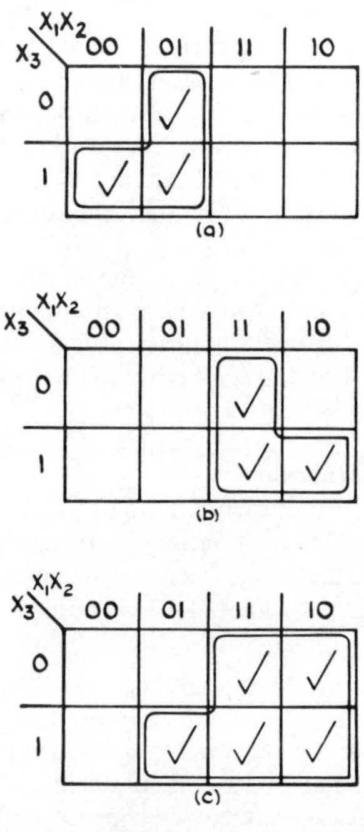

Fig. 2. Linearly-separable functions of three variables
(*a*) $f(x) = [\bar{x}_1(x_2 + x_3)], = \langle 2\bar{x}_1 + x_2 + x_3 \rangle_{3:2}$
(*b*) $f(x) = [x_1(x_2 + x_3)], = \langle 2x_1 + x_2 + x_3 \rangle_{3:2}$
(*c*) $f(x) = [x_1 + x_2x_3], = \langle 2x_1 + x_2 + x_3 \rangle_{2:1}$

4. Non-linearly-separable functions, Rademacher-Walsh coefficients

The $(n + 1)$ Chow parameters are not adequate to define all binary functions, i.e. the non-linearly-separable class.

Consider for example the two dissimilar functions plotted in **Fig. 3,** neither of which are linearly-separable functions (Hurst, 1970).

If the b_i values for each of these functions are computed, both will be found to yield:

b_0	b_1	b_2	b_3
0	0	0	0

Clearly, therefore, from this extreme example, the b_i's, $i = 0$ to n, are not sufficient to explicitly define functions which are not linearly-separable. Thus to define or classify such non-

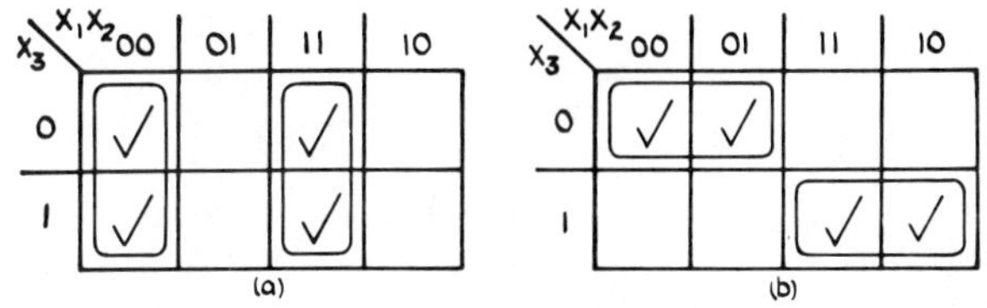

Fig. 3. Simple non-linearly-separable functions
(*a*) $f(x) = [\bar{x}_1\bar{x}_2 + x_1x_2]$
(*b*) $f(x) = [\bar{x}_1\bar{x}_3 + x_1x_3]$

linearly-separable binary functions, we must either augment the $(n + 1)$ Chow parameter coefficients with additional information, or derive an alternative classification more powerful than these basic Chow coefficients.

One algebraic method of defining and also classifying all binary functions is by means of the Rademacher-Walsh coefficients or 'spectra' (Dertouzos, 1965; Henderson, 1964; Golomb, 1959; Liedl and Pichler, 1971). As will be developed below, the Rademacher-Walsh parameters form an augmented form of the Chow parameters, being 2^n in number compared with the $(n + 1)$ total of the Chow classification.

The 2^n Rademacher-Walsh functions or variables—the terminology 'variable' is frequently used in the present context of binary synthesis, although 'function' is possibly more correct —consist of two parts, namely:

(*a*) the Rademacher functions, which constitute the first $(n + 1)$ of the 2^n total,
(*b*) the remaining $2^n - (n + 1)$ Walsh functions, which may be formed from the previous $(n + 1)$ Rademacher set.

The former may be termed the 'primary set', and the latter the 'secondary set' (Dertouzos, 1965).

When these Rademacher-Walsh functions are appropriately multiplied by the minterm values of any given binary function $f(x)$, the result is the Rademacher-Walsh coefficients or 'spectra' of the given function $f(x)$ (Dertouzos, 1965; Lockheed, 1964; Cooper, 1963; Liedl and Pichler, 1971). These 2^n coefficients uniquely define $f(x)$ in a similar way that the previous $(n + 1)$ Chow coefficients uniquely define any given threshold function $f(x)$.

(Note: although outside our present interests, it is of interest to note that these Rademacher-Walsh coefficients will uniquely define *any* function which exhibits discrete values in its output spectra; a binary function with two discrete output values 0 and 1 is therefore merely a particular case of the more general possible application of these Rademacher-Walsh coefficients.)

The Rademacher functions (or variables) may be found defined in several dissimilar but equivalent ways (Dertouzos,

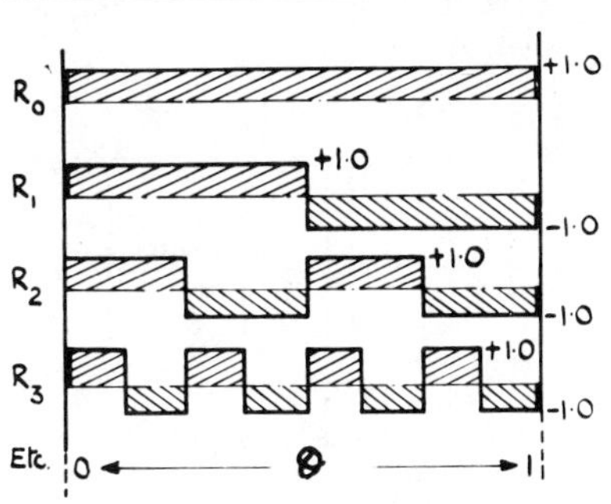

Fig. 4. Rademacher functions in the range 0 to +1.

1965; Cooper, 1963; Henderson, 1964; Barrett and Gordon, 1971). All definitions define a series of square waves with two discrete values, normally -1 and $+1$, which form a complete set over a unit 'range', or 'interval', normally taken as either from $-\frac{1}{2}$ to $+\frac{1}{2}$, or from 0 to 1.

One definition for the Rademacher functions $R_n(\theta)$, $n = 0, 1, 2, \cdots$, over the range of 0 to 1 is:

$$R_n(\theta) = \text{sign}\,\{\sin(2^n\,\pi\,\theta)\}\,.\,1{\cdot}0, \text{ where } 0 \leqslant \theta \leqslant 1\,.$$

An alternative definition, also over the range 0 to 1 is:

$$R_n(\theta) = +1 \text{ if } \frac{m}{2^n} \leqslant \theta \leqslant \frac{m+1}{2^n}\,,\ m = \text{even integer}\,,$$

$$= -1 \text{ if } \frac{m}{2^n} \leqslant \theta \leqslant \frac{m+1}{2^n}\,,\ m = \text{odd integer}\,.$$

Both these definitions are as illustrated in **Fig. 4.** Alternative definitions over the range $-\frac{1}{2}$ to $+\frac{1}{2}$ are similar, but displaced so as to be symmetrical about zero.

It will be apparent from the above that in a binary system the Rademacher functions R_1, R_2, etc. correspond to the binary variables x_1, x_2, etc. of a n-variable binary function, where the full 2^n minterms of the binary function are considered as contained in the range $\theta = 0$ to 1.

Hence a further definition for Rademacher functions (variables) in an n-variable binary system $f(x_1, \cdots x_n)$ is:

$$R_0 \triangleq +1$$
$$R_i,\ i = 1 \text{ to } n,\ = [1 - 2x_i]\,,$$
$$\left.\begin{aligned} &= [-y_i] \\ &= [z_i] \end{aligned}\right\}\ \text{see section 2}$$

Thus for example, for 3 binary variables x_1, x_2, x_3, we have the following correlation between the variously defined binary variables and the Rademacher variables:

Minterm	*Binary* x_i			*Binary* y_i			*Binary* z_i			*Rademacher* R_i			
p	x_1	x_2	x_3	y_1	y_2	y_3	z_1	z_2	z_3	R_0	R_1	R_2	R_3
0	0	0	0	−1	−1	−1	+1	+1	+1	+1	+1	+1	+1
1	0	0	1	−1	−1	+1	+1	+1	−1	+1	+1	+1	−1
2	0	1	0	−1	+1	−1	+1	−1	+1	+1	+1	−1	+1
3	0	1	1	−1	+1	+1	+1	−1	−1	+1	+1	−1	−1
4	1	0	0	+1	−1	−1	−1	+1	+1	+1	−1	+1	+1
5	1	0	1	+1	−1	+1	−1	+1	−1	+1	−1	+1	−1
6	1	1	0	+1	+1	−1	−1	−1	+1	+1	−1	−1	+1
7	1	1	1	+1	+1	+1	−1	−1	−1	+1	−1	−1	−1

The additional $2^n - (n + 1)$ Walsh functions may now be formed by obtaining *all possible different products* of these Rademacher valuations at each minterm p, taken two-at-a-time, three-at-a-time, up to n-at-a-time. (This is the simplest definition of the Walsh functions, although they may be defined in their own right without reference to the Rademacher set (Walsh, 1923; Lackey and Meltzer, 1971)). Thus, denoting the product of $R_1 \times R_2$ by R_{12}, etc. and noting that:

$$R_0 \times R_0 = R_0\,,$$
$$R_i \times R_i,\ i = 1 \text{ to } n,\ = R_0\,,$$
$$R_0 \times R_i,\ i = 1 \text{ to } n,\ = R_i\,,$$
$$R_{ij} \times R_i,\ i, j = 1 \text{ to } n, i \neq j,\ = R_j\,,$$

we have that all the possible Rademacher-Walsh valuations for a 3-variable binary system are as follows:

Rademacher-Walsh R_i: Minterms p:	0	1	2	3	4	5	6	7		
R_0	+1	+1	+1	+1	+1	+1	+1	+1	Prmry Set	Cmplt Set
R_1	+1	+1	+1	+1	−1	−1	−1	−1		
R_2	+1	+1	−1	−1	+1	+1	−1	−1		
R_3	+1	−1	+1	−1	+1	−1	+1	−1		
R_1R_2	+1	+1	−1	−1	−1	−1	+1	+1	Scndry Set	
R_1R_3	+1	−1	+1	−1	−1	+1	−1	+1		
R_2R_3	+1	−1	−1	+1	+1	−1	−1	+1		
$R_1R_2R_3$	+1	−1	−1	+1	−1	+1	+1	−1		

Note: because of the relationships $R_i \times R_i = R_0$ etc. there will always be exactly 2^n rows in the complete set for any n.

5. Alternative matrix row order

The vertical order of the Rademacher-Walsh functions as tabulated above follows from using the primary set to generate the secondary set. If Walsh's original definition and format is followed, the tabulation is made in order of *increasing number of zero crossings* of the +1, −1 values in the given interval. This is analogous to increasing frequency of the Fourier series in sinusoidal working. If frequency is defined as one half the number of zero crossings per unit time, then the corresponding 'frequency' of the Walsh functions, which is termed the 'sequency', is similarly defined as:

sequency = $\frac{1}{2}$ (average number of zero crossings per unit time).

The rows of the previous tabulation for $n = 3$ therefore may be arranged in vertical order of Walsh function sequency as follows:

$$
\begin{aligned}
W_0 &= R_0 \text{ (zero sequency)}\\
W_1 &= R_1\\
W_2 &= R_1R_2\\
W_3 &= R_2\\
W_4 &= R_2R_3\\
W_5 &= R_1R_2R_3\\
W_6 &= R_1R_3\\
W_7 &= R_3 \text{ (max. sequency)}
\end{aligned}
$$

A yet further alternative ordering of the rows of the matrix has been proposed by Lechner, Colman and possibly others (Henderson, 1964). For $n = 3$, the equivalence of the modified set W'_i is:

$$
\begin{aligned}
W'_0 &= W_0 = R_0\\
W'_1 &= W_7 = R_3\\
W'_2 &= W_3 = R_2\\
W'_3 &= W_4 = R_2R_3\\
W'_4 &= W_1 = R_1\\
W'_5 &= W_6 = R_1R_3\\
W'_6 &= W_2 = R_1R_2\\
W'_7 &= W_5 = R_1R_2R_3
\end{aligned}
$$

which gives a matrix row ordering of:

Minterms p:	0	1	2	3	4	5	6	7
Walsh W'_i: W'_0	+1	+1	+1	+1	+1	+1	+1	+1
W'_1	+1	−1	+1	−1	+1	−1	+1	−1
W'_2	+1	+1	−1	−1	+1	+1	−1	−1
W'_3	+1	−1	−1	+1	+1	−1	−1	+1
W'_4	+1	+1	+1	+1	−1	−1	−1	−1
W'_5	+1	−1	+1	−1	−1	+1	−1	+1
W'_6	+1	+1	−1	−1	−1	−1	+1	+1
W'_7	+1	−1	−1	+1	−1	+1	+1	−

The interesting feature of this re-ordering of the rows of the matrix is that is has a simple recursive structure, such that a matrix of order 2^{n+1} is formed by:

$$[W'_{2^{n+1}}] = \begin{bmatrix} [W'_{2^n}] & [W'_{2^n}] \\ [W'_{2^n}] & [-W'_{2^n}] \end{bmatrix}$$

Also the binary 0, 1 equivalent matrix (see next section) may be obtained by a simple relationship, when required.

6. Rademacher-Walsh matrix, 0, 1 valuation

If we convert the R_i Rademacher-Walsh values of +1, −1 into x_i 0, 1 values, i = 1, 2, etc., by the direct conversion of +1 to 0, −1 to 1, the first 3-variable Rademacher-Walsh matrix given above becomes as follows:

Minterms:	0	1	2	3	4	5	6	7
$\hat{R}_0$	1	1	1	1	1	1	1	1
$\hat{R}_1 (= x_1)$	0	0	0	0	1	1	1	1
$\hat{R}_2 (= x_2)$	0	0	1	1	0	0	1	1
$\hat{R}_3 (= x_3)$	0	1	0	1	0	1	0	1
$\hat{R}_1\hat{R}_2$	0	0	1	1	1	1	0	0
$\hat{R}_1\hat{R}_3$	0	1	0	1	1	0	1	0
$\hat{R}_2\hat{R}_3$	0	1	1	0	0	1	1	0
$\hat{R}_1\hat{R}_2\hat{R}_3$	0	1	1	0	1	0	0	1

Note: the designation $\hat{R}_0$, $\hat{R}_1$, etc. will be used to differentiate between an 0, 1 Rademacher-Walsh matrix, and a +1, −1 matrix. $\hat{R}_0$, the first row of the matrix, by definition remains +1.

If we now examine the secondary $\hat{R}_1\hat{R}_2$ etc. Rademacher-Walsh products, we see that in this converted 0, 1 set they will be generated by modulo-two addition of the appropriate primary Rademacher values, with any carry of the modulo-two additions discarded.

$$
\begin{aligned}
\text{e.g. } \hat{R}_1\hat{R}_2 &= \hat{R}_1 \oplus \hat{R}_2, = x_1 \oplus x_2 .\\
\hat{R}_1\hat{R}_3 &= \hat{R}_1 \oplus \hat{R}_3, = x_1 \oplus x_3 .\\
\hat{R}_2\hat{R}_3 &= \hat{R}_2 \oplus \hat{R}_3, = x_2 \oplus x_3 .\\
\hat{R}_1\hat{R}_2\hat{R}_3 &= \hat{R}_1 \oplus \hat{R}_2 \oplus \hat{R}_3, = x_1 \oplus x_2 \oplus x_3 .
\end{aligned}
$$

To summarise therefore, if these modified 0, 1 Rademacher-Walsh variables $\hat{R}_i$ are considered for binary synthesis and/or analysis, the first $(n + 1)$ *R-W* variables R_0 to $\hat{R}_n$ constitute the actual binary inputs x_i of the system, with $\hat{R}_0 \triangleq 1$, whilst the remaining *R-W* variables are modulo-two additions of these actual inputs. A binary system may therefore be considered as shown schematically in **Fig. 5.**

7. Orthogonality

The complete −1, +1 Rademacher-Walsh matrix as detailed in Sections 4 and 5 will be seen to be *orthogonal* and *normal*, i.e. 'orthonormal', as the integral of the product of any two of the 2^n function R_i, R_j taken over the full interval θ = 0 to 1 is:

$$
\int_{\theta=0}^{1} R_i\, R_j . d\theta = 0,\ i \neq j \text{ (orthogonal)}
$$
$$
= 1,\ i = j \text{ (normal).}
$$

As the *R-W* functions are discrete-valued, the above

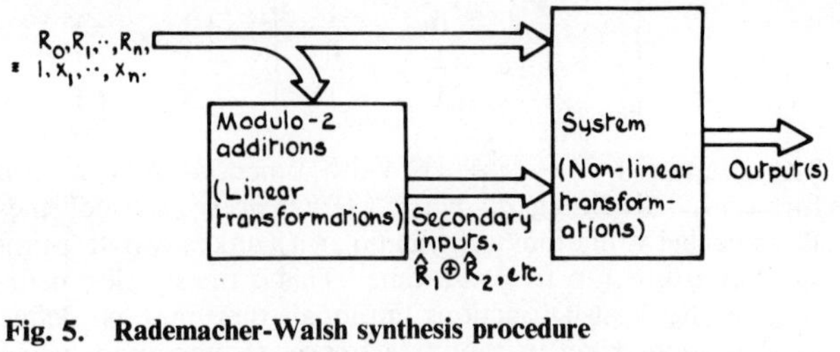

Fig. 5. Rademacher-Walsh synthesis procedure

integration may be replaced by the summation:

$$\frac{1}{2^n}\left\{\sum_{0}^{2^n-1}\{R_i \, . \, R_j\}\right\} \begin{array}{l} = 0, \, i \neq j \, , \\ = 1, \, i = j \, , \end{array}$$

the summation being taken at each discrete interval 0 to $2^n - 1$ in the full interval of $\theta = 0$ to 1.

Note:

(*a*) without the $\frac{1}{2^n}$ in the latter summation, the resultant for $\sum R_i . R_j$ when $i = j$ will sum to 2^n rather than to unity. The $\frac{1}{2^n}$ is thus a normalising factor;

(*b*) irrespective of the order of the rows of the matrix, for example the Lechner/Colman order given in Section 5, the matrix will remain orthogonal;

(*c*) when, however, the +1, −1 values are converted into binary 0, 1 values, see above, the resulting 0, 1 matrix $[\hat{R}_{2^n}]$ is no longer orthogonal for $\hat{R}_i . \hat{R}_j$. Orthogonality now holds for $\hat{R}_i \circledast \hat{R}_j$, as

$$\frac{1}{2^n}\left\{\sum_{0}^{2^n-1}\{\hat{R}_i \circledast \hat{R}_j\}\right\} \begin{array}{l} = 0, \, i \neq j \, , \\ = 1, \, i = j \, , \end{array}$$

where $\{\hat{R}_i \circledast \hat{R}_j) \triangleq [1 - 2\{\hat{R}_i \oplus \hat{R}_j\}]$.

Note: $\left\{\sum_{0}^{2^n-1}\{\hat{R}_i \circledast \hat{R}_j\}\right\}$ = [(agreements between $\hat{R}_i$ and $\hat{R}_j$) − (disagreements between $\hat{R}_i$ and $\hat{R}_j$)].

8. Rademacher-Walsh coefficients or spectra, +1, −1 configuration

Any arbitrary switching function of n variable can be described by the arithmetic sum of its 2^n Rademacher-Walsh variables R_i, each individually weighted by an appropriate real-number w_i, i.e.

$f(z)_p \triangleq +1$ or -1 at any minterm p,

$= (\sum w_i R_i)_p$, where w_i is the weight associated with each respective R_i,
$i = 0, 1, 2, \cdots, 12, \cdots$.

These weights w_i of the Rademacher-Walsh variables may now be termed the Rademacher-Walsh 'coefficients' or 'spectra' of the switching function.

The required weights may be expressed by the matrix relationship:

$$w_i] = \frac{1}{2^n}[R] . f_z] \, ,$$

where $w_i]$ = column matrix whose elements are the required w_i,

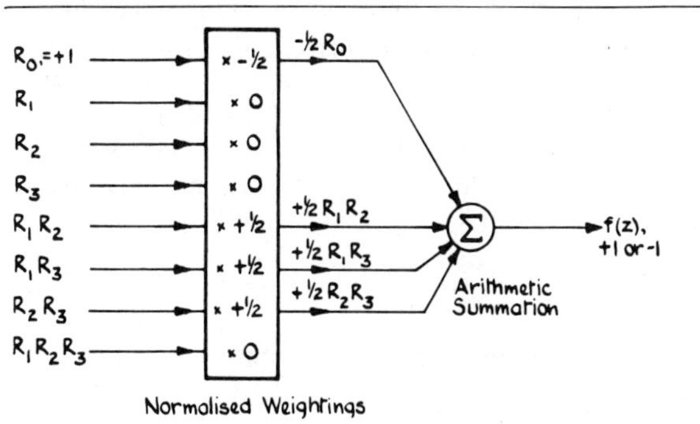

Fig. 6. Rademacher-Walsh realisation for $f(x) = [x_1\bar{x}_2 + x_2\bar{x}_3 + \bar{x}_1x_3]$

$[R]$ = the $2^n \times 2^n$ square matrix whose rows are the R_i's, and $f_z]$ = column matrix whose elements are the output values +1 or −1 of the function $f(z)$.

Proof of the above follows from the orthogonality of the matrix, see Dertouzos (1965), and others.

For example, given the 0, 1 function

$$f(x) = [x_1\bar{x}_2 + x_2\bar{x}_3 + \bar{x}_1x_3] \, .$$

Converting from 0, 1 to +1, −1, we have:

p	x_1	x_2	x_3	$f(x)$	z_1	z_2	z_3	$f(z)$
0	0	0	0	0	+1	+1	+1	+1
1	0	0	1	1	+1	+1	−1	−1
2	0	1	0	1	+1	−1	+1	−1
3	0	1	1	1	+1	−1	−1	−1
4	1	0	0	1	−1	+1	+1	−1
5	1	0	1	1	−1	+1	−1	−1
6	1	1	0	1	−1	−1	+1	−1
7	1	1	1	0	−1	−1	−1	+1

Hence:

$$\begin{array}{ccccccccc} p = & 0 & 1 & 2 & 3 & 4 & 5 & 6 & 7 \quad f(z) \end{array}$$

$$w_i] = \frac{1}{2^3}\begin{bmatrix} +1 & +1 & +1 & +1 & +1 & +1 & +1 & +1 \\ +1 & +1 & +1 & +1 & -1 & -1 & -1 & -1 \\ +1 & +1 & -1 & -1 & +1 & +1 & -1 & -1 \\ +1 & -1 & +1 & -1 & +1 & -1 & +1 & -1 \\ +1 & +1 & -1 & -1 & -1 & -1 & +1 & +1 \\ +1 & -1 & +1 & -1 & -1 & +1 & -1 & +1 \\ +1 & -1 & -1 & +1 & +1 & -1 & -1 & +1 \\ +1 & -1 & -1 & +1 & -1 & +1 & +1 & -1 \end{bmatrix}\begin{bmatrix} +1 \\ -1 \\ -1 \\ -1 \\ -1 \\ -1 \\ -1 \\ +1 \end{bmatrix}$$

$$= \frac{1}{8}\begin{bmatrix} -4 \\ 0 \\ 0 \\ 0 \\ +4 \\ +4 \\ +4 \\ 0 \end{bmatrix}$$

or:

$$\begin{aligned} w_0 &= -\tfrac{1}{2} \, , \\ w_1 &= 0 \, , \\ w_2 &= 0 \, , \\ w_3 &= 0 \, , \\ w_1w_2 &= +\tfrac{1}{2} \, , \\ w_1w_3 &= +\tfrac{1}{2} \, , \\ w_2w_3 &= +\tfrac{1}{2} \, , \\ w_1w_2w_3 &= 0 \, . \end{aligned}$$

∴ The normalised Rademacher-Walsh spectra for the given function is:

$$-\tfrac{1}{2}, 0, 0, 0, +\tfrac{1}{2}, +\tfrac{1}{2}, +\tfrac{1}{2}, 0,$$

giving the realisation for $f(z)$ of:

$$f(z)_p = (\textstyle\sum\{-\tfrac{1}{2} R_0 + 0 + 0 + 0 + \tfrac{1}{2}(R_1R_2) + \tfrac{1}{2}(R_1R_3) + \tfrac{1}{2}(R_2R_3) + 0\})_p \, ,$$

or, equally:

$$(\tfrac{1}{8}\textstyle\sum\{-4R_0 + 4(R_1R_2) + 4(R_1R_3) + 4(R_2R_3)\})_p \, ,$$

which can be quickly shown to correctly realise $f(z) = -1$ or +1 over all the 2^n minterms p.

Thus if we generate the +1, −1 Rademacher-Walsh variables, it would be possible to synthesise the given function to give the required output $f(z) = +1$ or -1, as shown in **Fig. 6.** The +1, −1 output $f(z)$ could subsequently be converted to an 0, 1 output $f(x)$, if required.

If the above matrix definition for the required weights w_i is re-examined, it will be seen that they may be re-defined as follows:

$$w_0 = -\frac{1}{2^n}[(\text{number of true minterms}) - (\text{number of false minterms})],$$

$$= \frac{1}{2^n}\left[\sum_{p=0}^{2^n-1}\{f(z)\}\right],$$

and

w_i, $i = 1, 2, \cdots, 12, \cdots, =$

$\frac{1}{2^n}$ [(number of agreements between the value of R_i and $f(z)$ taken over all minterms p) − (number of disagreements between R_i and $f(z)$)].

Both these results may be jointly expressed as:

$$w_i = \frac{1}{2^n}\left[\sum_{p=0}^{2^n-1}\{f(z).R_i\}\right],$$

$$i = 0, 1, 2, \cdots, 12, \cdots.$$

It is now interesting to compare the above, which apply to *all* Boolean functions expressed in the $+1$, -1 notation, with the threshold Chow parameters b_i, $i = 0$ to n, given in Sections 1 and 2. A basic 'sameness' will be noted, except that the threshold b_i's are not normalised by dividing by 2^n.

9. Rademacher-Walsh spectra, 0, 1 configuration

To redefine the above procedure in terms of the normal 0, 1 binary notation and the resultant modified Rademacher-Walsh variables $\hat{R}_i$, it can be shown (Cooper, 1963) that:

$f(x)_p$, $\triangleq$ 0 or 1 at any minterm p,

$$= \left(\hat{w}_0\hat{R}_0 + \sum \hat{w}_i\hat{R}_i\right)_p,$$

where $\hat{w}_0$ = weight associated with $\hat{R}_0$,

$\hat{w}_i$ = weight associated with each $\hat{R}_i$, $i = 1, 2, \cdots, 12, \cdots$.

These weights $\hat{w}_i$ of the modified Rademacher-Walsh variables may now be termed the modified or binary Rademacher-Walsh 'coefficients' or 'spectra' of the given switching function.

The value of the weights is given by:

$$\hat{w}_0 = f(x)_{p=0},$$

i.e. the value of the function $f(x)$ at minterm $p = 0$, and

$$\hat{w}_i,\ i = 1, 2, \cdots, 12, \cdots, =$$

$$\frac{1}{2^n}\left[\sum_{p=0}^{2^n-1}\{(f(x) \oplus \bar{R}_i) - (f(x) \oplus \hat{R}_i)\}\right].$$

Note:

As $\hat{R}_0$ is always 1, $\hat{w}_0\hat{R}_0$ is therefore always $f(x)_{p=0}$.

Also the above expression for $\hat{w}_i$, $i \neq 0$, will on examination be seen to define precisely the same summation as for the $+1$, -1 case, namely:

$\hat{w}_i$, $i = 1, 2, \cdots, 12, \cdots, =$

$\frac{1}{2^n}$ [(number of agreements between the value of $\hat{R}_i$ and $f(x)$ taken over all minterms p) − (number of disagreements between $\hat{R}_i$ and $f(x)$)].

Considering our previous example

$f(x) = [x_1\bar{x}_2 + x_2\bar{x}_3 + \bar{x}_1x_3]$

again, we now have the following development:

p:	0	1	2	3	4	5	6	7
$f(x)$:	0	1	1	1	1	1	1	0
$\hat{R}_0$	1	1	1	1	1	1	1	1
$\hat{R}_1$	0	0	0	0	1	1	1	1
$\hat{R}_2$	0	0	1	1	0	0	1	1
$\hat{R}_3$	0	1	0	1	0	1	0	1
$\hat{R}_1\hat{R}_2$	0	0	1	1	1	1	0	0
$\hat{R}_1\hat{R}_3$	0	1	0	1	1	0	1	0
$\hat{R}_2\hat{R}_3$	0	1	1	0	0	1	1	0
$\hat{R}_1\hat{R}_2\hat{R}_3$	0	1	1	0	1	0	0	1

whence:

$$\hat{w}_0 = 0$$
$$\hat{w}_1 = 0$$
$$\hat{w}_2 = 0$$
$$\hat{w}_3 = 0$$
$$\hat{w}_1\hat{w}_2 = +\tfrac{1}{2}$$
$$\hat{w}_1\hat{w}_3 = +\tfrac{1}{2}$$
$$\hat{w}_2\hat{w}_3 = +\tfrac{1}{2}$$
$$\hat{w}_1\hat{w}_2\hat{w}_3 = 0$$

The normalised modified binary Rademacher-Walsh spectra for the given function is therefore:

$$0, 0, 0, 0, +\tfrac{1}{2}, +\tfrac{1}{2}, +\tfrac{1}{2}, 0,$$

giving the realisation for $f(x)$ of:

$f(x)_p =$

$(\sum\{0 + 0 + 0 + 0 + \frac{1}{2}(\hat{R}_1\hat{R}_2) + \frac{1}{2}(\hat{R}_1\hat{R}_3) + \frac{1}{2}(\hat{R}_2\hat{R}_3) + 0\})_p$,

$= (\sum\{\frac{1}{2}(\hat{R}_1 \oplus \hat{R}_2) + \frac{1}{2}(\hat{R}_1 + \hat{R}_3) + \frac{1}{2}(\hat{R}_2 + \hat{R}_3)\})_p$,

$\equiv (\sum\{\frac{1}{2}(x_1 \oplus x_2) + \frac{1}{2}(x_1 \oplus x_3) + \frac{1}{2}(x_2 + x_3)\})_p$.

Thus the function may be realised as shown in **Fig. 7**.

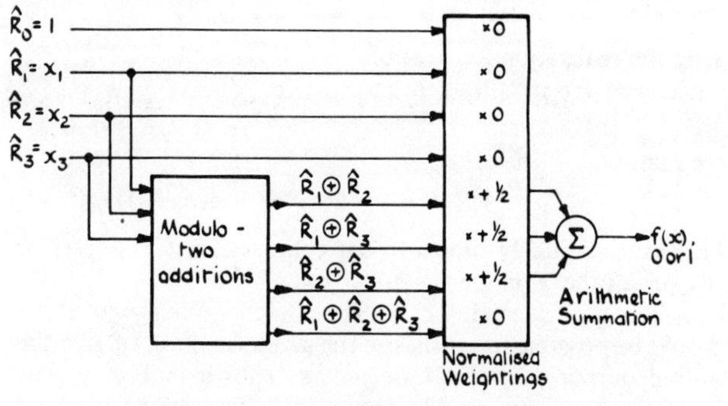

Fig. 7. Modified 0, 1 Rademacher-Walsh realisation for $f(x) = [x_1\bar{x}_2 + x_2\bar{x}_3 + \bar{x}_1x_3]$

10. Alternative definitions for w_i and $\hat{w}_i$

In both the $+1$, -1 case, see Section 8, and the 0, 1 case, see Section 9, it was shown that the respective required weights w_i and $\hat{w}_i$ for the Rademacher-Walsh variables, $i \neq 0$, can be defined as:

w_i or $\hat{w}_i$, $i = 1, 2, \cdots, 12, \cdots, =$

$\frac{1}{2^n}$ [(No. of agreements ——) − (No. of disagreements ——)].

(NOTE: the w_0, $\hat{w}_0$ weight is *not* the same in the $+1$, -1 and the 0, 1 cases).

This expression may be algebraically re-arranged as follows:

Let a = number of minterms where $f(x) = 1$ and $x_i = 1$,
b = number of minterms where $f(x) = 1$ and $x_i = 0$,
c = number of minterms where $f(x) = 0$ and $x_i = 1$,
d = number of minterms where $f(x) = 0$ and $x_i = 0$,

for example see **Fig. 8**(*a*), where a $\bar{x}_1$, x_1 map division is illustrated. Then w_i and $\hat{w}_i$, $i \neq 0$, are given by:

$$w_i, \hat{w}_i = \frac{1}{2^n}[(a + d) - (b + c)]$$

Considering just the expression within the brackets, re-arranging we have:

$$[(a - b) - (c - d)],$$
$$= [(a - b) - ((2^{n-1} - a) - (2^{n-1} - b))],$$
$$= [(a - b) - (-a + b)], = 2[(a - b)].$$

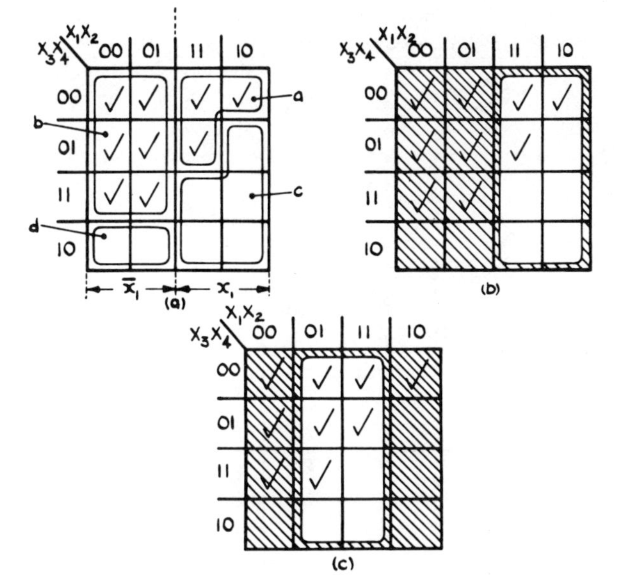

Fig. 8. Karnaugh-map plots of $f(x) = [\bar{x}_1x_4 + x_2\bar{x}_3 + \bar{x}_3x_4]$
(a) $\bar{x}_1$, x_1 division
(b) x_1 'window'
(c) x_2 'window'

Now $(a + b)$ = number of true minterms, = say m.
$\therefore$ Substituting we obtain the expression:

$$2[a - (m - a)],$$
$$= [4a - 2m].$$

$\therefore$ The normalised equation for the required w_i or $\hat{w}_i$, $i \neq 0$, may be written as

$$\frac{1}{2^n}[4a - 2m],$$

$$= \frac{1}{2^n}[4(\text{number of true minterms of the function when } x_i \text{ is true}) - 2(\text{total number of true minterms})].$$

This alternative definition lends itself to reading off the required w_i or $\hat{w}_i$, $i \neq 0$, from Karnaugh maps by merely counting the number of true minterms in the appropriate part of the map. For example the function shown in Fig. 8(*a*) is further illustrated in **Fig. 8**(*b*) and (*c*), from which the following results may be read:

$$w_1 \text{ or } \hat{w}_1 = \frac{1}{2^n}[(4 \times 3) - (2 \times 9)]$$
$$= \frac{1}{2^n}[-6],$$
$$w_2 \text{ or } \hat{w}_2 = \frac{1}{2^n}[(4 \times 5) - (2 \times 9)]$$
$$= \frac{1}{2^n}[+2],$$

and so on for w_3, etc.
NOTE: for the $+1$, -1 situation w_0 may be rewritten as:

$$w_0 = -\frac{1}{2^n}[m - (2^n - m)]$$
$$= -\frac{1}{2^n}[2m - 2^n],$$
$$= \frac{1}{2^n}[2^n - 2m].$$

For the 0, 1 situation, $\hat{w}_0$ remains as the function value $f(x)$ at $p = 0$, see Section 9.

11. Karnaugh-map identification of all 2^n Rademacher-Walsh functions

It is of interest to further consider the areas or parts of a Karnaugh-map layout that all the 2^n Rademacher-Walsh functions (variables) define. We have previously seen that the Rademacher variables, the 'primary set', are equivalent to the binary input variables of the system, R_0 always being $+1$, the 'secondary set' being all different modulo-two additions of this primary set in 0, 1 working.

Thus for $n = 3$, we have the full situation shown in **Fig. 9.** Similar maps for $n = 4$ etc. may be drawn.

It is now of great interest to note that although the primary set $R_0, \cdots, R_n$ are all linearly-separable (threshold) functions, that is each may be realised by one threshold logic gate, NONE OF THE MODULO-2 SECONDARY SET ARE LINEARLY SEPARABLE. In fact, the final Rademacher-Walsh variable $R_{12\cdots n}$, $(\hat{R}_{12\cdots n})$, is as far removed from a threshold function as is possible—it may be considered as 'opposite' in some sense to a threshold function.

This may help to non-mathematically indicate why a combination of all these variables, appropriately employed, is able to realise *any* given binary function, whereas threshold functions, which form a more restricted class of binary functions, require only the appropriately-weighted primary set $R_0, \cdots, R_n$ for their realisation, see **Fig. 10** (*a*) and (*b*).

It may also help to explain the interest that one or two researchers have shown, in attempting to realise any given binary function by the use of some minimum combination of threshold gates plus modulo-two gates. By definition a linearly-separable function requires only one threshold gate; if a given function is not linearly-separable, then the minterms which cannot be included by a threshold gate may be included by some modulo-two gate, such as to give some overall minimum network realisation as indicated in **Fig. 11.**

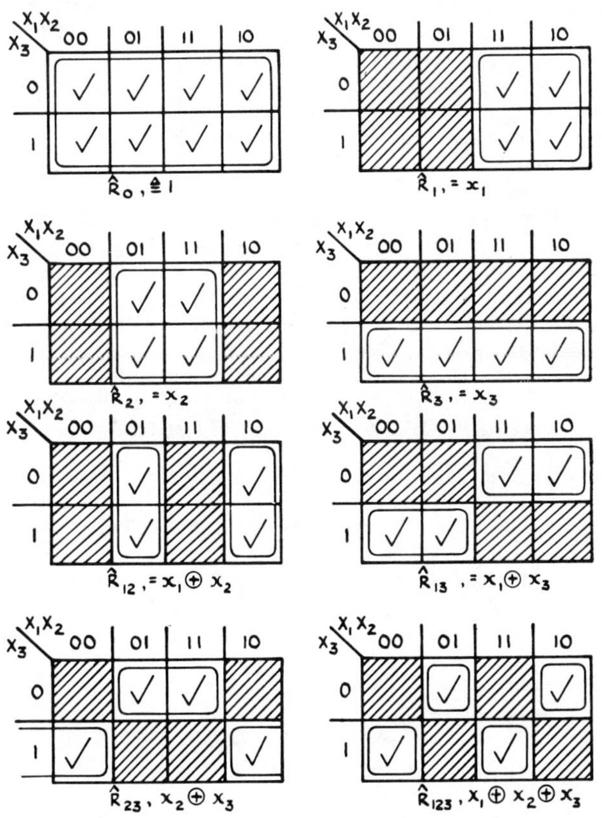

Fig. 9. Karnaugh-map identification of modified Rademacher-Walsh variables

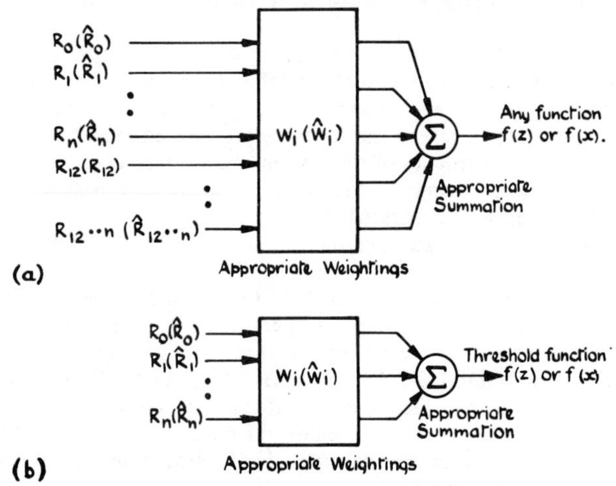

Fig. 10. Comparison of Rademacher-Walsh and threshold synthesis
(*a*) Rademacher-Walsh
(*b*) threshold

It should be noted that the weighting of all the 2^n Rademacher-Walsh variables and their subsequent appropriate summation, whilst mathematically elegant and universal, does not normally provide an economic realisation if a direct one-to-one translation into hardware is contemplated.

Considerable work in this field of minimum network synthesis remains to be investigated.

12. Classification of Boolean functions

The magnitude of the Chow parameters $|b_i|$ taken in a lexicographical order forms the most compact form of classifying and listing all the possible linearly-separable (threshold) functions (Dertouzos, 1965; Winder, 1965), see Section 1. This form of classification is a numerical form of SD ('self-dualised' or 'hyperfunction') classification.

It may therefore be of interest to consider the Rademacher-Walsh w_i to see if they form a classification for all binary functions, in a manner similar to the $|b_i|$'s for threshold functions.

As an exercise, take as a starting point the Chow parameter $|b_i|$ classification 6, 2, 2, 2. This classification covers a series of threshold functions of three variables $f(x_1, x_2, x_3)$, characterised by having either 7, 5, 3 or 1 true minterms. (The 7 and the 1, and the 5 and the 3, are complements of each other; respectively, these two groups being related by the SD classification.)

Let us therefore list a few linearly-separable and non-linearly-separable functions with the above minterm count. The Rademacher-Walsh weights will be listed un-normalised, that is not divided through by $\frac{1}{2^n}$, so as to make the w_0 to w_3 numerically identically to the b_0 to b_3, respectively. **Fig. 12** illustrates the functions chosen.

From these arbitrary examples, the following features will be seen:

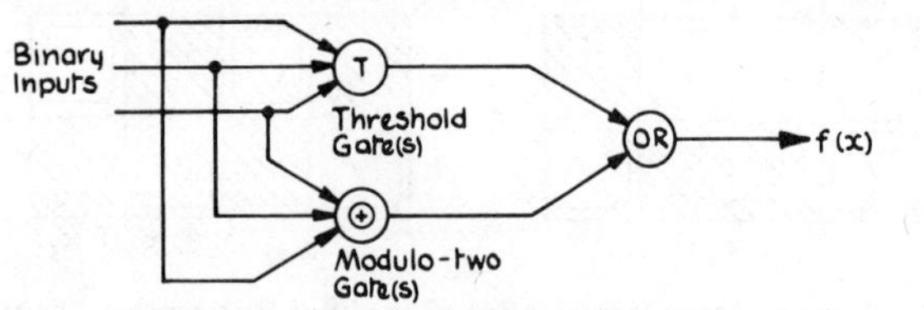

Fig. 11. Threshold and modulo-two synthesis

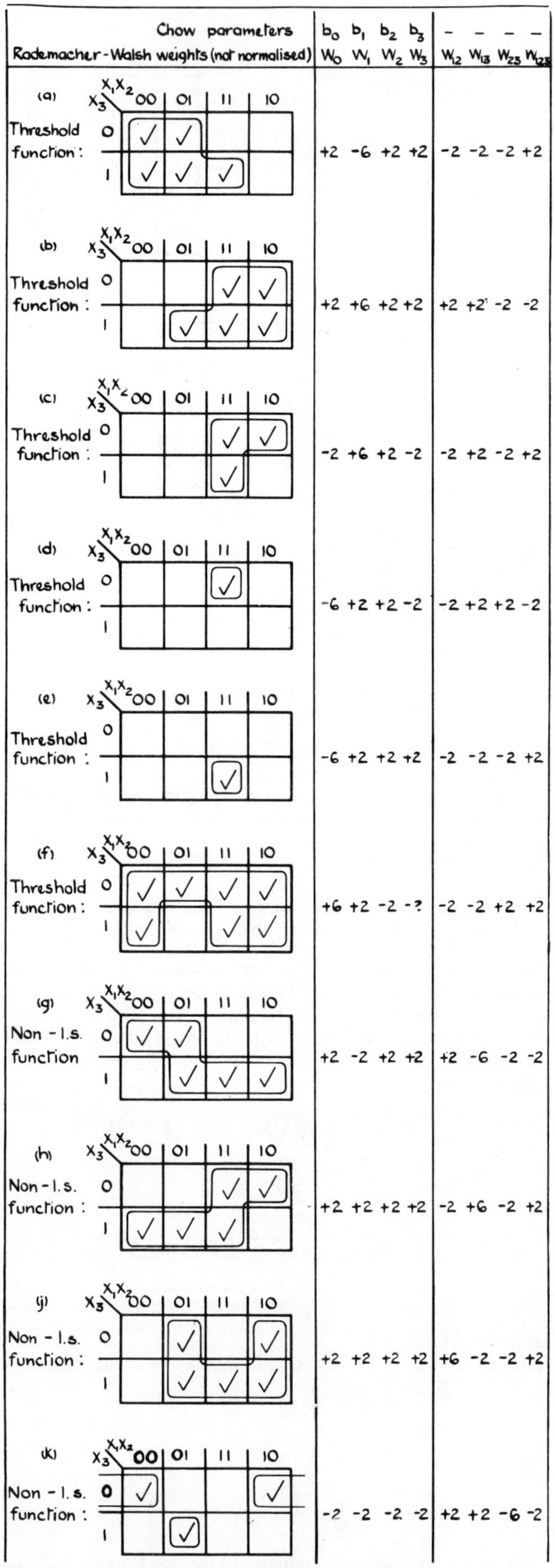

Fig. 12. Classification of example binary functions

(i) all given functions are characterised by un-normalised Rademacher-Walsh weights $|w_i|$ of

6, 2, 2, 2, 2, 2, 2, 2

when listed in descending magnitude;

(ii) if however the '6' appears in the left-hand 'primary set', then the function must be a threshold function; conversely if the '6' appears in the right-hand 'secondary set', leaving $|b_0|$ to $|b_3|$ as 2, 2, 2, 2, then such a function cannot be a threshold function;

(iii) with the threshold functions, the values of the 'primary set' w_0 to w_3 with their appropriate signs, uniquely define the function, as these primary sets are the Chow parameters, see Section 3. Hence the remaining w_i values, $i = 12$, 13 etc., do not add any further information about the function—their values must therefore be determinable from the w_i of the primary set.

The precise correlation between w_i, $i = 0$ to n, and w_i, $i = 12$, 13, etc. of these and all other threshold functions, however, is not directly obvious;

(iv) when the function is not a threshold function, the values of the primary set do not define the function, see Figs. 12(*h*) and (*j*) covering functions with the same w_0 to w_3 values. Equally, however, the secondary set w_i values do not explicitly define the function, see further examples illustrated in **Fig. 13.**

Thus for a non-linearly separable function, all 2^n Rademacher-Walsh weights with their appropriate signs are essential to define the function.

To attempt to summarise the above, it would appear that all binary functions could be classified by their $|w_i|$ values, in the same way that all threshold functions are classified by their $|b_i|$ values. At this stage it is not quite clear just what invariance operations are involved per classification. (The invariance operations involved in the threshold function $|b_i|$ classifications is well understood (Dertouzos, 1965).)

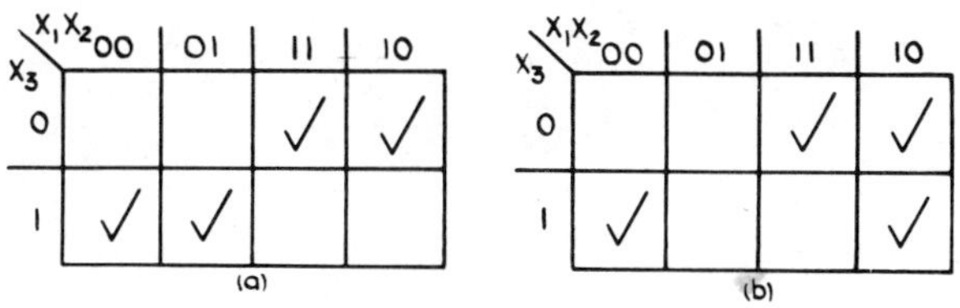

Fig. 13. Classification of dissimilar functions
(*a*) $f(x) = [x_2\bar{x}_3 + \bar{x}_1x_3]$, $w_i = 0, -4, 4, 0; 0, 4, 4, 0$
(*b*) $f(x) = [x_1x_3 + x_2\bar{x}_3]$, $w_i = 0, 4, -4, 0; 0, 4, 4, 0$

The precise correlation between the w_i values, $i = 0$ to n, and the w_i values, $i = 12$, 13, etc. for the threshold functions also requires some further thought.

Conclusions

A considerable area of research and consolidation is still open in the field of synthesis of Boolean switching functions. With the increasing practicality of economically generating complex switching functions in integrated-circuit form, such as threshold gates and modulo-two functions, etc., a new approach to the most economical practical realisation of given switching problems may arise. The present-day NAND/NOR gate realisations may well be superceded by more compact realisations, with a smaller gate population, in future generations of equipment.

Acknowledgements

Profitable discussions with colleagues of the School of Electrical Engineering of the University of Bath, particularly Mr J. D. Martin and Mr C. R. Edwards, must be gratefully acknowledged.

References

BARRETT, R. and GORDON, J. A. (1971). *Walsh functions—an introduction*, Proc. Symposium. Theory and Application of Walsh Functions, Hatfield Polytechnic, England, June, 1971.

CHOW, C. K. (1961). *On the characterization of threshold functions*, IEEE Special Publication S.134, August 1961, pp. 34-38.

COLEMAN, R. P. (1961). Orthogonal functions for the logical design of switching circuits, *Trans. I.R.E.*, EC-10, pp. 379-383.

COOPER, J. A. (1963). *Orthogonal expansion applied to the design of threshold-element networks*, Technical Report No. 6204-1, U.S. Office of Naval Research, Contract Nonr. 225(24), NR373360, December 1963.

DERTOUZOS, M. L. (1965). *Threshold logic: a synthesis approach*, Research Monograph No. 32, M.I.T. Press.

GOLOMB, S. B. (1959). On the classification of Boolean Functions, *Trans. I.R.E.*, CT6, (Special Supplement), pp. 176-186.

HENDERSON, K. W. (1964). Some notes on the Walsh functions, *Trans. I.E.E.E.*, EC13, pp. 50-52.

HENDERSON, K. W. (1970). Previous investigation of realization of an arbitrary switching function with a network of threshold and parity elements, *Trans. I.E.E.E.*, C19, pp. 941-942.

HURST, S. L. (1970). Correlation between Karnaugh-map patterns and Chow parameters for 3-variable linearly-separable functions, *Electronic Letters*, Vol. 6, No. 11, pp. 324-326.

HURST, S. L. (1971). *Threshold Logic*, Mills and Boon, London.

LACKEY, R. B. and MELTZER, D. (1971). A simplified definition of Walsh functions, *Trans. I.E.E.E.*, C20, pp. 211-213.

LEWIS, P. M. (1967). A practical guide to threshold logic, *Electronic Design*, Vol. 22, pp. 66-88.

LIEDL, R. and PICHLER, F. (1971). *On harmonic analysis of switching functions*, Proc. Symposium Theory and Application of Walsh Functions, Hatfield Polytechnic, England, June 1971.

MUROGA, S., TSUBOI, T., and BAUGH, C. R. (1967). *Enumeration of threshold functions of eight variables*, Report No. 245, University of Illinois, August 1967.

OSTAIKO, D. L. and YAU, S. S. (1970). Realization of an arbitrary switching function with a two-level network of threshold and parity elements, *Trans. I.E.E.E.*, C19, pp. 262-269.

Research on Automatic computer electronics, Vol. II, Logical Design Research, Lockheed Missile and Space Co., California, Tech. Document Report No. RTD-TDR-63-4173, February 1964.

WALSH, J. L. (1923). A closed set of normal orthogonal functions, *Am. Journal Math.*, Vol. 45, pp. 5-24.

WINDER, R. O. (1964). *Threshold functions through n = 7*, Scientific Report No. 7, AFCRL Contract AF19(604)-8423, October 1964.

WINDER, R. O. (1965). Enumeration of seven-argument threshold functions, *Trans. I.E.E.E.*, EC14, pp. 315-325.

AUTHOR CITATION INDEX

SUBJECT INDEX

About the Editor

SPYROS G. TZAFESTAS is Professor of Control and Automation at the computer engineering division of the National Technical University, Athens, Greece. He received the Ph.D. in electrical engineering, and the D.Sc. for his published work in the systems and control field from Southampton University, England in 1969 and 1978 respectively. From 1969 to 1973 Tzafestas was leader of the computer control program at the Nuclear Research Center Democritos in Athens, and from 1973 to 1984 he was Professor of Systems and Control Engineering at Patras University, Patras, Greece.

Tzafestas's research interests include distributed parameter systems, reliability and maintenance optimization, nuclear reactor safety and control, Walsh function techniques, and multidimensional systems. He has published 10 books and 200 technical papers. He is an active reviewer of 15 journals, associate editor of 3 journals, the Editor-in-Chief of the *International Journal of Modelling and Simulation*, and the Editor of the Reidel book series entitled "Microprocessor-Based Systems Engineering." Tzafestas has served as guest editor of three special issues on distributed systems and orthogonal expansions, and has organized and chaired many international conferences on systems and control. He is a fellow of the Institution of Electrical Engineers (United Kingdom), a senior member of the Institute of Electrical and Electronics Engineers (United States), a member of the American Society of Mechanical Engineers (United States), and the Chairman of the Greek section of the International Association for Mathematics and Computers in Simulation (Greece).